Advances in Morphometrics

NATO ASI Series

Advanced Science Institutes Series

A series presenting the results of activities sponsored by the NATO Science Committee, which aims at the dissemination of advanced scientific and technological knowledge, with a view to strengthening links between scientific communities.

The series is published by an international board of publishers in conjunction with the NATO Scientific Affairs Division

A	**Life Sciences**	Plenum Publishing Corporation
B	**Physics**	New York and London
C	**Mathematical and Physical Sciences**	Kluwer Academic Publishers
D	**Behavioral and Social Sciences**	Dordrecht, Boston, and London
E	**Applied Sciences**	
F	**Computer and Systems Sciences**	Springer-Verlag
G	**Ecological Sciences**	Berlin, Heidelberg, New York, London,
H	**Cell Biology**	Paris, Tokyo, Hong Kong, and Barcelona
I	**Global Environmental Change**	

PARTNERSHIP SUB-SERIES

1. Disarmament Technologies	Kluwer Academic Publishers
2. Environment	Springer-Verlag
3. High Technology	Kluwer Academic Publishers
4. Science and Technology Policy	Kluwer Academic Publishers
5. Computer Networking	Kluwer Academic Publishers

The Partnership Sub-Series incorporates activities undertaken in collaboration with NATO's Cooperation Partners, the countries of the CIS and Central and Eastern Europe, in Priority Areas of concern to those countries.

Recent Volumes in this Series:

Volume 282 — Vaccines: New Generation Immunological Adjuvants
edited by Gregory Gregoriadis, Brenda McCormack, and Anthony C. Allison

Volume 283 — Eicosanoids: From Biotechnology to Therapeutic Applications
edited by Gian Carlo Folco, Bengt Samuelsson, Jacques Maclouf, and G. P. Velo

Volume 284 — Advances in Morphometrics
edited by Leslie F. Marcus, Marco Corti, Anna Loy, Gavin J. P. Naylor, and Dennis E. Slice

Series A: Life Sciences

Advances in Morphometrics

Edited by

Leslie F. Marcus

Queens College of the City University of New York
Flushing, New York and
American Museum of Natural History
New York, New York

Marco Corti

University of Rome "La Sapienza"
Rome, Italy

Anna Loy

University of Rome "La Sapienza"
Rome, Italy

Gavin J. P. Naylor

Yale University
New Haven, Connecticut

and

Dennis E. Slice

State University of New York
Stony Brook, New York

Springer Science+Business Media, LLC

Based on proceedings of a NATO Advanced Study Institute on
Advances in Morphometrics,
held July 18 – 30, 1993,
in Il Ciocco, Tuscany, Italy

Library of Congress Cataloging-in-Publication Data

Advances in morphometrics / edited by Leslie F. Marcus ... [et al.].
p. cm. -- (NATO ASI series. Series A, Life sciences ; v. 284)
"Based on proceedings of a NATO Advanced Study Institute on Advances in Morphometrics, held July 18-30, 1993, in Il Ciocco, Tuscany, Italy"--T.p. verso.
Includes bibliographical references and index.
ISBN 978-1-4757-9085-6 ISBN 978-1-4757-9083-2 (eBook)
DOI 10.1007/978-1-4757-9083-2
1. Morphology--Mathematics--Congresses. 2. Morphology--Statistical methods--Congresses. 3. Biology--Classification--Statistical methods--Congresses. I. Marcus, Leslie Floyd, 1930- . II. NATO Advanced Study Institute on Advances in Morphometrics (1993 : Il Ciocco, Tuscany, Italy) III. Series.
QH351.A38 1996
574.4'01'5195--dc20 96-18484
CIP

ISBN 978-1-4757-9085-6

Originally published by Plenum Press, New York in 1996
Softcover reprint of the hardcover 1st edition 1996

10 9 8 7 6 5 4 3 2

To the memory of Peter J. Scott

(February 24, 1944 – September 5, 1994)

PREFACE

This volume is based on the NATO Advanced Study Institute, "Advances in Morphometrics" held in Il Ciocco, Tuscany, Italy from July 18–30, 1993, and directed by Leslie F. Marcus.

The "Advances in Morphometrics" ASI was advertised in *Nature* and a number of professional journals. Announcements were sent to relevant institutions and departments throughout the world. Because NATO required that the majority of attendees be from NATO countries, the 71 persons attending represented nine NATO countries, four eastern European countries, now recognized as equal partners for ASIs, and a few participants from non-NATO countries. Participants were all active scholars in different disciplines within biology, as well as computer science, statistics, geology and paleontology. Their experience ranged from that of graduate students to senior faculty, as well as one emeritus scholar. A complete list of the those attending and their addresses, phone and FAX numbers and, where available, e-mail addresses is given in the participants list.

All the local arrangements were made by Marco Corti and Anna Loy of the University of Rome "la Sapienza." They made the initial contact with the Il Ciocco conference center and then arranged for computer and Xerox rentals, design of logos, organization of posters, and publication of poster abstracts. Carlo Fadda, a graduate student at la Sapienza, and Angelo Loy, from the University of Rome at Tor Vegata, worked to make the meeting run smoothly—including setting up and keeping all the computer equipment working, tape-recording sessions, keeping track of other equipment and replenishing supplies as they were needed. Anna Loy, one of the original organizers and an editor of the volume, was unable to attend the meeting—she had a baby just prior to the meeting. Stanley G. Marcus played an invaluable administrative role at the meetings by keeping track of textbooks and bills and acting as liaison with the management of Il Ciocco. He also made all social arrangements and helped to make the time-off activities a success. Bruno Gianassi, our contact in the Il Ciocco management, made our chores much easier and made all of the day-to-day happenings run without a problem.

The instructors were chosen to represent a wide variety of points of view and experience. Richard Reyment, Fred Bookstein, Jim Rohlf, and Leslie Marcus had all participated in the first National Science Foundation, sponsored morphometrics workshop at Michigan in 1988. Anna Loy, Marco Corti, Dennis Slice, Gavin Naylor, and Antonio García-Valdecasas participated in the Stony Brook workshop in 1990 and became faculty at Il Ciocco. Dennis Slice, Marco Corti, and Anna Loy also participated as staff in an earlier Spanish workshop directed by García-Valdecasas. Michel Baylac attended that workshop and directed another in Paris in 1992. Michel Baylac, David Dean and Andras Demeter were welcome additions completing the faculty at Il Ciocco.

The first day of the program began with a panel discussion, "Revolution in Morphometrics: Real or Imagined," by the entire faculty. The afternoon was taken up by a demonstration of special acquisition hardware and software available during the institute. The program then settled down to a daily schedule consisting of morning lectures and afternoon demonstrations presented by the faculty—augmented by special lectures and demonstrations by some of the participants.

Well in advance of the conference all participants received copies of "Proceedings of the Michigan Morphometrics Workshop," edited by F. James Rohlf and Fred L. Bookstein (University of Michigan Museum of Zoology Special Publication 2, 1990), together with an assortment of reprints. At the ASI they were all given copies of "Contributions to Morphometrics," edited by Leslie F. Marcus, Elisa Bello, and Antonio García-Valdecasa (Monografias del Museo Nacional de Ciencias Naturales 8, Madrid, 1993), courtesy of the Madrid Museum.

Early in the program, the participants presented posters on projects already completed or contemplated prior to their attendance at the ASI. Abstracts of all the posters were distributed to the participants.

Twenty advanced IBM clones and five MacIntosh computers, together with a sufficient number of laser printers, were rented and were available for practice and data collection and analysis every afternoon and evening. Additional IBM clones were borrowed from la Sapienza. Some participants and faculty brought their own portable computers, and color monitors were available for their use as well. Some faculty were present in the "computer room" at most times.

A variety of modern data acquisition systems were demonstrated, and some were available for use by participants for the duration of the conference. Most notable was a Reflex microscope for three-dimensional coordinate acquisition for small specimens, which was made available by Peter Scott and manned by Peter's son Paddy and Andrew Peakman, and a 3Draw digitizer and Mira software for larger specimens, courtesy of the Polhemus Corporation. (See Dean, this volume, for more information on the Reflex and Polhemus.) Television image capture systems were available along with Morphosys and MTV software packages, as well as a new home-grown Budapest Video with its newly developed software package, IMAGOES (see Demeter et al., this volume). A variety of digitizing tablets were available for two-dimensional data capture. A laser scanner was demonstrated by Dr. G. Quinci of ATTA, using Surfacer software from Imageware (see Dean, this volume, for details on Imageware) during the first day of the conference.

At the conclusion of each day's formal program, time was set aside for discussion of all or any of the material presented up to that time. This time period was also used for general review. Several ad hoc evening sessions were arranged to cover topics not on the program, including conservation biology, matrix algebra, a MATLAB tutorial, and other tutorials.

On the last day of the ASI most of the software used during the two-week period was distributed free to all attendees. This software is now available by anonymous FTP from the State University of New York at Stony Brook (see Appendix II).

All faculty contributed articles to the proceedings volume. Participants who had already prepared applications or methods papers, or who began research that could be completed for this volume were invited to present shorter papers. Some faculty contributed more than one paper or essay, and 24 student participants contributed articles or served as coauthors on papers in this volume.

Dennis Slice organized a glossary of technical terms (Appendix I), which will be on the Internet by the time this volume is published (see Appendix II on how to view it on-line). This contribution is very important, and in its Internet form will become a dynamic document.

The editorial committee was chosen to represent the directors and organizers; Gavin Naylor and Dennis Slice were chosen to represent younger faculty members. F. James Rohlf and Fred Bookstein made major contributions as editorial consultants to the volume, in addition to their articles. They read each and every article and contributed valuable criticisms and suggestions at every stage of the editorial work. A meeting of the editors was held in New York in May 1994 and was attended by all editors and consultants. The organization of this volume grew out of that meeting. Also, at that meeting was born the idea of the essay contributed by Fred Bookstein, "Combining the tools of Geometric Morphometrics"—a kind of what to do and not to do guide.

The various sections of the volume were assigned to editors, along with the responsibility for writing introductory essays for their respective sections. F. James Rohlf kindly agreed to serve as editor for the section on Outlines and contributed the introductory essay for that section. He also helped to produce Appendix II on the Internet resources available at Stony Brook.

The senior editor assumes all responsibility for any shortcomings or errors in the volume.

Leslie F. Marcus

CONTENTS

Section 3. Outlines With and Without Landmarks

PART II: COMPARING SHAPES OF ORGANISMS

Section 1. Geographic Comparisons

Section 2. Low Taxonomic Level and Intrapecific Comparisons

Section 3. Higher Taxonomic Level Comparisons

APPENDICES

OVERVIEW OF THE NEW, OR GEOMETRIC MORPHOMETRICS

Leslie F. Marcus and Marco Corti

INTRODUCTION

The new morphometrics, or geometric morphometrics, is a rapidly evolving field. "Morphometrics" has been used to describe a number of fields that study measurements of organisms, and that is why the modifier "new" or "geometric" is necessary to set aside the special subject matter discussed here: the shape of biological organisms as it is studied using as data, point coordinates in two or three dimensions. This includes landmark and outline coordinates. Landmarks are specific points on an organism that correspond in a sensible way over the forms being studied, that is they are homologs; while outline points do not share this notion of homology. The data of morphometrics is now being extended to include tangent directions at coordinates as well (see Bookstein and Green, 1993; Little and Mardia, this volume). Shape is primarily concerned with properties of coordinates that are invariant to scale, location and orientation (see Appendix I, Glossary). The current status of morphometrics together with some future objectives are summarized in a review by Rohlf and Marcus (1993a).

A series of workshops has brought the methodology to the attention of research biologists. Publications from these workshops has produced books in a variety of colored bindings: Red—*Morphometrics in Evolutionary Biology* by Bookstein et al. (1985); Blue—*Proceedings of the Michigan Workshop in Morphometrics* edited by Rohlf and Bookstein (1990); Orange—*Morphometric Tools for Landmark Data: Geometry and Biology,* the treatise on the subject by Bookstein (1991); Black—*Contributions to Morphometrics* by Marcus et al. (1993); and White—this volume. Important contributions to morphometrics are to be found in these books and in collateral articles by Bookstein, Goodall, Reyment, Rohlf, and Rohlf and Slice. Another Black book, by Reyment (1991), *Multidimensional Paleobiology,* includes important material on the new morphometrics.

As Bookstein points out (this volume, "Combining" article) these developments are supported by a theoretical underpinning of shape studies in the United Kingdom. The principal definitions of "shape space" are discussed by Rohlf ("Morphometric Spaces" article, this volume). A series of meetings at Leeds over the past several years has brought researchers together to discuss theoretical statistical developments of shape analysis (for latest proceedings see Mardia and Gill, 1995, and MORPHMET archive described in Appendix III; and how to obtain announcements of meetings, see Appendix II).

Advances in Morphometrics, Edited by L. F. Marcus et al.
Plenum Press, New York, 1996

Traditional morphometrics has been concerned with the traditional caliper distances or length measurements of organisms, as well as areas and volumes—and traditional morphometrics does include some shape measurements such as angles. Distances may be constructed from point coordinates in two and three dimensions, but the reverse is not true unless an extensive series of measurements is recorded that includes the same landmarks as endpoints in several distance measurements. Landmarks archive the shape and size of an organism, an accomplishment that no simple set of distances can duplicate.

The statistics of the new morphometrics is the same classical multivariate statistics used for distance measurements, now applied to coordinates of suitably aligned landmarks (see Bookstein "Combining" article and Rohlf "Morphometric Spaces" article, this volume) so that the same principal components, factor analysis, or canonical variates may be applied to landmark data.

Therefore the "revolution" in morphometrics is more of a synthesis of: developments in shape theory; the recognition of coordinates and landmarks as manageable data; and analysis using multivariate statistics. Much of this has been made possible through the development of widely available software by Rohlf (1993 a,c,e) and Slice (1993b). Beginning with the Blue book, new and better software has become available to analyze coordinate data—especially for the IBM compatible personal computers (PCs using DOS). Much of this software is free and available by anonymous FTP (file transfer process; see Appendix II).

At the time that the Red book was published collecting coordinate data in two dimensions was just becoming relatively easy. Digitizing tablets were widely available for collecting two-dimensional (2D) coordinates from drawings or photographs or projecting images on the tablet by means of a camera lucida. Video digitizing could be used on similar images or by direct video input from actual specimens. However, calipers were used to take extensive series of measurements, and, if these were organized, as pointed out in the Red book, the coordinates could be reconstructed—though not always as easily as one might like. See Carpenter et al. (this volume) for procedures to transform distances to landmarks; see also Reig (this volume) and Demeter et al. ("Identification" article, this volume) for comparisons of distance and landmark data collected from the same specimens.

More recently three-dimensional (3D) data acquisition has begun to be affordable (see Dean, this volume) and 3D coordinates are now much easier to collect. Clever homemade 3D devices (for an example see Corti et al. this volume; Spencer and Spencer, 1995) have also been developed for a fraction of the cost of commercial devices for some special applications.

The synthesis is currently in a mature stage (Bookstein, 1993) and those who have used caliper distances in the past perceive the new developments as a real revolution. Coordinates finally brought a way to visualize shape and shape changes in 3D organism space. It is now possible to visualize in plots those parts of the organism encoded in the coordinates that are involved in shape change. This contrasts with traditional morphometric summary tables of columns and columns of numbers resulting from multivariate analyses and 2D plots of multivariate ordinations. It was popular to interpret the covariances among linear measurements, but visualization of shape change in 3D organism space was not possible. One had to rely on one's own multivariate experience and familiarity with the organisms measured to imagine the morphometric changes in a skull, a wing, a carapace or a leaf and to interpret biological processes usefully.

Landmarks as described by Bookstein (1991) reflect a biological homology in tissues, anatomical locations and coordinates, and these forms of data are very attractive to the evolutionary and systematic biologist. These ideas have built on the work of Galton and D'Arcy Thompson (see Bookstein, 1993; Reyment "History" essay, this volume). By superimposing landmarks according to a criterion (Galton), or by making them coincide, the differences in shape are made clearer.

At the NATO Advanced Study Institute the methods discussed and put into practice were for landmarks and outlines. Shape coordinates, Procrustes methods and thin-plate splines were applied to landmarks; elliptical Fourier analysis and some new techniques were applied to closely spaced coordinates on outlines.

A textbook or primer on new morphometrics is sorely needed, and at least one is currently being prepared. Until such is available, the various colored volumes and these proceedings will provide much of the basic literature.

OVERVIEW OF THE NEW MORPHOMETRICS

Here we briefly review new morphometrics—derived largely from Rohlf and Marcus (1993a,b) and Corti (1993).

Two-Point Shape Coordinates

A constellation of *n* landmarks in two dimensions are related to a fixed base defined by any two of the landmarks. All *x*- and *y*- coordinates are rigidly translated and rotated and rescaled to have one end of the baseline at (0, 0) and the other at (1, 0), by dividing the rotated and translated coordinates by the length of the base line . The other *n* - 2 landmarks now form apices of triangles, each with a base of length one. The data now consist of *n*-2 *x*- and *y*- coordinate pairs and are called two-point shape, or Bookstein coordinates .

It is possible to decompose overall shape differences among shape coordinates for a set of organisms into affine and nonaffine components (see Bookstein "Uniform" article, this volume) and map the shape differences (eg., see Corti, 1993). Shape coordinates may be analyzed by univariate or multivariate statistics (see Bookstein "Combining" article, this volume), and the statistics are relatively insensitive to choice of baseline. Bookstein coordinates are especially useful for simple production of mean forms and analysis or tests of relationships between landmark configurations and outside causes or effects, clines or allometry.

Rohlf and Marcus (1993b) pointed out that the method of two-point shape coordinates can be viewed as a kind of superimposition (Procrustes) analysis. The visualization and analysis, however, does not summarize covariation across all landmarks. Plots do provide useful visualization with respect to certain landmarks, but this method is probably most useful for relatively few landmarks or for relatively simple patterns of correlation among landmarks.

Although the multivariate statistics are nearly invariant to choice of baseline, it is important to choose the baseline carefully for visualization and analysis of individual landmarks (see Bookstein, 1991, for a discussion). Becerra et al. (this volume) give an application using Bookstein coordinates in a taxonomic context.

Superimposition Methods

Superimposition methods overlay a series of specimens so that corresponding landmarks match as closely as possible. Differences in shape are recorded as residuals from a reference shape. The residuals may be plotted at each landmark as point scatters or vectors relative to the reference. Least squares and resistant-fit are two common methods of superimposition available in software packages (Rohlf and Slice; 1990 a,b; Slice "Three Dimensional" article, this volume; Slice, 1993b).

The least-squares techniques are based on an optimality criterion and are very fast. Resistant fit may produce a more interesting alignment when shape differences are limited to a small proportion of the landmarks, because the algorithm is based on the median of differences. The computations may require considerably more time to converge to a consensus fit.

The statistical properties of least-squares superimpositional methods are relatively well understood. Both least-squares and resistant-fit programs can compute affine or uniform shape differences and produce residuals, but the resistant-fit residuals have a nonaffine component left in them. It is possible to show covariation of displacements at different landmarks, although this is more versatile in relative warp analysis.

Relative Warp Analysis

Relative warp analysis as pointed out by Rohlf and Marcus (1993a) is perhaps the most exciting new method. Based on the thin-plate spline transformation, relative warp analysis maps a reference configuration of landmarks, usually the mean of a sample, onto each specimen. The parameters of these thin-plate spline transformations can be used as new variables in multivariate statistical analysis, such as principal components analysis or canonical variates analysis. Because the new variables are linear combinations of the residuals from the reference for the specimens, normally distributed digitizing error on the specimen results in multivariate normally distributed variation.

One very important result from these statistical analyses is that the scores can be used to visualize shape differences through thin-plate spline displays. For example, variation along a canonical variate axis can be examined in this fashion (see Loy et al., 1993; Hugot and Baylac, this volume).

A parameter α in relative warp analysis adjusts the weights for the linear combinations. An α of 1 scales the principal warps of the reference to give large-scale changes greater weight. This may be appropriate in growth studies, for example. On the other hand, an α of 0 equally weights the principal warps. For exploratory studies in systematics this strategy is suggested (Rohlf, 1993d) to be more appropriate, because there are no a priori expectations that shape differences at large or small scale will be more important. Because this area is one of considerable interest to many evolutionary biologists and experimentation with α is strongly recommended.

The affine, or uniform components and nonuniform components of shape may also be regressed on size to analyze and display allometric shape changes. The software TPSREGR (Rohlf, 1993c) is available for this purpose.

Outline Methods

These methods received little attention at the Advanced Study Institute as compared to earlier workshops. No papers are devoted to the methods described earlier in Rohlf (1986, 1990c) and Rohlf and Archie (1984). Elliptical Fourier analysis is now routinely used for the study of complex shapes, and software is widely available, such as NTSYS-pc (Rohlf, 1993a) for 2D and 3D outlines. A new Windows based 2D program by Isaev (1994) is very easy to use and has excellent graphics—it is available by means of anonymous FTP (see Appendix II for details). Demeter et al. (this volume) have built the elliptical Fourier fitting algorithm into their television capture and image analysis program.

Newer outline methods are developed and discussed in the articles by Dean (this volume) and Sampson et al. (this volume). Dean combines landmarks and coordinates between landmarks in an analysis of curves in three dimensions, called ridge curves; Sampson et al. apply a form of Procrustes type eigenshape analysis after a Procrustes-type alignment to 2D coordinates of outlines of hearts in normal and diseased human patients.

Other Methods

Euclidean Distance Matrix Analysis (EDMA). This method compares forms—shapes before they have been scaled—based on landmarks. Basic to the method is the representation of form in terms of distances among all pairs of n landmarks. n landmakrs generate $n \cdot (n-1)/2$ distances. For example twenty landmarks would generate 190 distances. Lele (1993) has developed methods for estimating the mean and variance of a set of forms. For any two specimens, if the ratio of all the distances in the specimens' two lists of distances is a constant, then the forms have the same shape. The common ratio gives the difference in size. A statistical test to compare two shapes in two populations has been developed by Lele and Richtsmeier (1991). However to date, there is no method of providing a visualization of shape differences. One must examine long lists of distance ratios to interpret shape differences or changes. Because this method is not based on Kendall's shape space and has arbitrarily low statistical efficiency in some applications, it is not considered part of the synthesis (see MORPHMET discussion, April and November, 1995—Appendix III gives instructions on how to access the MORPHMET archive).

Finite Element Scaling Analysis (FESA). This method has been used in physical anthropology and development. The set of landmarks on the form serve as vertices of elements of small closed regions, usually triangles in two dimensions, and tetrahedra in three dimensions. Two forms are compared by computing an affine transformation for each small region. Results are usually displayed in terms of strain crosses inside each element (triangle or tetrahedron) or at each landmark. One problem with the technique is that a different partitioning of the specimen into different elements provides in some cases, rather different patterns of strain crosses. It is therefore difficult to partition shape differences into global effects such as affine or uniform shape changes. Bookstein (1991) describes further difficulties in using this technique.

THEORY COVERED IN THIS VOLUME

Because the body of theoretical development of the methods discussed in the previous section is well documented in the literature and various of the colored volumes, it was decided not to reproduce these developments here. However, a few articles in Part I of this volume augment or complement the literature available elsewhere, and the more important developments are briefly mentioned here.

Rohlf has outlined some of the terminology and concepts of shape space that underlay methodology used in landmark analysis (see that Section 2). This is Kendall shape space, which as Bookstein (see below) points out forms an important mathematical basis for practical shape analysis. A second important aspect of Rohlf's article is how the various transformations available in landmark analysis affect the results of the multivariate analysis undertaken.

Bookstein, in his article on "Combining the Tools of Geometric Morphometrics" develops the relation between Kendall shape space and the practical steps in an analysis of landmark data and as well provides a series of DOs and DON'Ts based on the properties of the methods.

The need for this article was realized at the first meeting of the editorial committee in May 1994, after drafts of all of the contributions had been submitted and read. It became clear to the editors that the researchers who had started to apply the methodology had common questions, had made some common errors in their applications and as well had some common misunderstandings. Consequently, Bookstein volunteered to write this article to incorporate the ideas and content agreed upon at the editorial conference. Obviously, the article's content would had been better presented at the Advanced Study Institute before any of the research reported here had been attempted. But hindsight is always better than foresight.

Bookstein makes a second important contribution in his article "Standard Formula for the Uniform Shape Component." This essay compares the variety of formulas and procedures available for estimating uniform components, and provides a standardized one.

Slice gives an up-to-date overview of superimposition methods. His paper nicely complements other superimposition and landmark papers and provides the background for using Slice's GRF-ND software (Slice, 1993b) for superimposition in two, three and higher dimensions. Three-dimensional arrays of landmarks may be rotated, and superimposed, with a number of analysis and display options. A dynamic visualization of principal components of residuals from a reference in 3D organism space is quite useful. Also residuals from other analyses can be imported into GRF-ND and displayed.

An extension of landmarks to landmarks with directions or edgels, is explored in three dimensions by Mardia and Little based on the work of Bookstein and Green (1993). This small contribution is an enticing sampler of new ideas yet to come.

APPLICATIONS IN THIS VOLUME

All of the editors and editorial consultants felt that the most important contribution of this volume is the variety of applications to biological problems, and in one case an archeological problem. Organizing these contributions was difficult because they covered a number of animal groups and analyzed geographic variation, sexual dimorphism and a variety of taxonomic problems at different hierarchical levels—from infraspecific to intergeneric.

For convenience the editors decided on an hierarchical organization for Part II based more or less on taxonomic level and a priori knowledge of the diversity being studied. Each section has an introductory essay, but the contents are briefly summarized here as well.

The first-level papers are in "Section 1. Geographic Comparisons," with an introductory essay by Loy. Topics include fluctuating asymmetry in mice (Auffray et al.); evolution of insect wing venation patterns and coevolution with host plant species (Baylac and Daufresne); recognition of coral species based on Procrustes analysis (Johnson and Budd); geographic variation in shrews (Sarà); geographic variation in 3D landmarks of mole rats in relation to climate and soils (Corti et al.); and ecomorphological variation in sticklebacks with some general methodological discussion (Walker).

Section 2, edited by Corti, discusses applications in which a priori categories were distinct but generally of low taxonomic rank. Demeter et al. attempted to discriminate two sympatric species of *Mus* and compare caliper and landmark data in two dimensions. Reig does a similar comparison of two sympatric species of weasels in two and three dimensions. Two studies, one by Ahlström and the other by Wood and Lynch, examine sexual dimorphism in human crania. Reyment applies geometric morphometrics to ostracod data sets. Loy et al. apply thin-plate spline regression analysis in a comparison of sea bass raised in fresh and salt water.

Section 3 is concerned with higher-level taxonomic comparisons and is introduced by Naylor. Both shape coordinates and caliper distances are used to ordinate water mite genera by Becerra et al. Berge uses the method of thin-plate splines to compare pelvic ilia shapes in *Australopithecus* and *Homo*. Bogdanovich and Owen analyze 2D skull shapes by means of landmarks in a subfamily of bats. Kent Carpenter looks at shape differences in a family of fish related to feeding mode. There are two parasite studies: one by Gubanyi who examines the diversity in outline and landmarks, both collected automatically, of tapeworm scolex hooks; the other by Hugot and Baylac who look at variation in pinworms in relation to the association and classification of their rodent and primate hosts. Van Dam analyzes a radiation of fossil murids based on molar cusp patterns by means of thin-plate splines to describe shape changes that accompany distinctive anatomical changes.

Finally, Naylor has attempted to evaluate the use of partial warps as characters in a simulation of a phylogenetic analysis of fish shapes. There has been a recent debate (especially aired over the on-line bulletin board, MORPHMET [see Appendix III]) as to the relevance of geometric morphometrics to phylogenetic, and specifically cladistic, analysis. There has been considerable frustration on the part of some cladist attendees at the earliest Michigan morphometrics workshop as to the seeming lack of attention to the quest for characters useful in cladistic analysis. Adding to this was the different meaning of homology attached to landmarks and the ideas of homology in cladistics. Both Bookstein and Rohlf (see MORPHMET archive for April 1994; Zelditch et al., 1995) have been rather negative about the use of the new morphometrics in cladistics (Bookstein, 1994). Others see some promise, and publications are beginning to appear (Swiderski, 1994; Zelditch et al., 1995; Fink and Zelditch, 1995). There is no doubt in our mind that the new morphometrics will provide us with some new characters for cladistic analysis. However, this is true for any careful comprehensive multivariate analysis (Laduke, 1991) Naylor provides a simulation that traces landmark evolution in a clade and then applies the thin-plate spline algorithm in an attempt to do a cladistic analysis on the partial warp scores.

DATA ACQUISITION

The section on data acquisition (Part I, Section 1) includes some new developments and a summary of 3D data acquisition equipment and software. Dean provides an up-to-date (at the time of submission) review of 3D hardware, software and model-building techniques including the newer laser scanners. However, he has not summarized the computer tomography (CT) and magnetic resonance (MR) hardware and software, which is just coming into use in comparative biology and is very expensive.

Becerra provides a description of Imagina—a free-ware package that runs on the PC Vision European OFG model, and Demeter et al. describe Budapest Video, which uses a Hungarian frame grabber. Both have some new interesting features. García-Valdecasas updates the article in the Black book by Becerra et al. (1993), and covers the newer high-resolution video equipment of special interest for museum archiving. Carpenter et al. explain software for transforming a truss of distances to landmarks—a practical problem for those who collected a lot of caliper data and would now like to take advantage of the newer landmark morphometrics.

SOFTWARE DEVELOPMENTS

The software for landmark and outline morphometrics continues to evolve. Although there are not many new programs or packages available, many of the software tools available along with the Blue book and Black book are free and continue to be updated. Perhaps the most important addition to the set of programs from the State University of New York (SUNY) at Stony Brook is Slice's GRF-ND (1993b) which is a 2D and 3D graphics and computation superimposition program. GRF-ND and all of the free-ware discussed in the

Blue and Black books is now available at the SUNY Stony Brook anonymous FTP site (see Appendix II for details).

Additional software is discussed in this volume, including the two video-digitizing programs and the truss-to-landmark conversion mentioned in the previous section. Rasskin-Gutman and Buscalioni describe software for visualization of interactive affine transformations in their article.

Edgewarp is graphical software for interactive thin-plate spline, edgel and other analyses and runs on a workstation. Instructions for obtaining this software from Michigan are given in Appendix II.

Walker (this volume) has written a comprehensive software program, Morphometrika, for landmark analysis for the MacIntosh family of computers. Contact him for further information.

Several commercial packages are of considerable interest in morphometrics: Rohlf's NTSYS-pc (1993a) is a relatively inexpensive comprehensive clustering and multivariate statistical package that is frequently updated. It contains elliptical Fourier analysis, shape coordinate and relative warp analysis in two and three dimensions. Additional features not widely available in other packages are partial least-squares (PLS), singular value decomposition (SVD) and common principal component analysis (CPCA).

SAS (Statistical Analysis System) is perhaps the most versatile of the large multiplatform computer packages because of its special matrix processor language PROC IML. Most classical multivariate and newer statistical routines are available in other PROCs, and the user can add programs written in matrix algebra by use of PROC IML. The advantage of doing this within a comprehensive package is that results can be passed back and forth between the classical PROCs and the matrix processor. Klingenberg has provided a CPCA program written in PROC IML (available at the Stony Brook FTP site, see Appendix II); Marcus has written a SAS IML Appendix for Reyment's (1991) *Multidimensional Paleobiology*, which perform most of the analyses discussed in that book. This group of programs is available as well from the Stony Brook FTP site (see Appendix II for details).

MATLAB is a stand-alone matrix processor program, currently available on IBM PCs (WINDOWS and DOS), MacIntosh and several Unix workstations (available from The MathWorks Inc.). This package is very much like PROC IML in SAS but easier to use, and it uses less hardware resources. Also, it has attractive easy to use graphics. A set of Toolboxes gives easy access to applications in a number of fields. A set of programs written in MATLAB by Marcus is an Appendix to the book *Applied Factor Analysis for the Natural Sciences* by Reyment and Jöreskog (1993). The programs are available by anonymous FTP from Stony Brook in the file REYBLUE. It is relatively easy to translate a routine from SAS's PROC IML to MATLAB and vice versa.

S-plus is a stand-alone function and object-oriented program with very powerful statistical and graphical processing. It is another preferred choice for those who want to be up to date in software and write their own techniques as the methodology advances.

Resampling techniques such as the bootstrap, jackknife and randomization tests are especially easy to write in SAS's PROC IML, MATLAB or S-plus. All perform relatively fast and are available on several platforms.

Other software of interest to morphometrics is discussed in some of the papers in this volume, where full references are given.

STEPS IN THE ANALYSIS

It was thought worthwhile to outline some steps required in a morphometric analysis in the form of a flow chart. This preliminary attempt is not derived from the Advanced Study

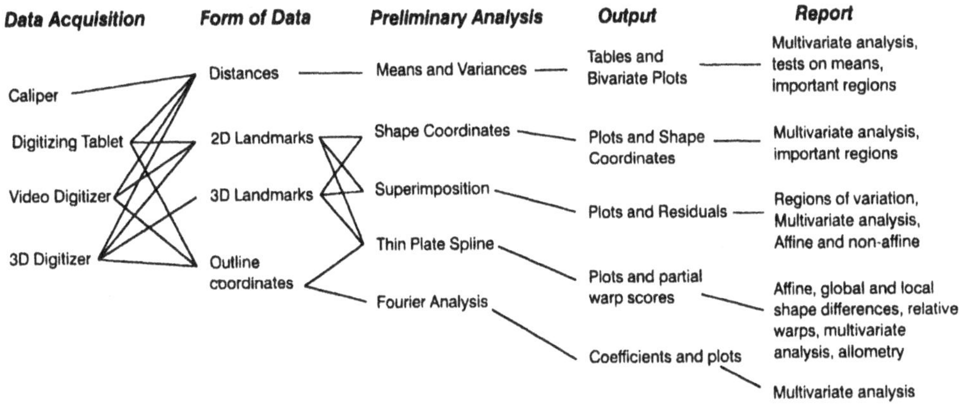

Figure 1. Steps in the Analysis. Note EDMA and FESA were left out of the diagram because they were not fully covered in the NATO Advanced Study Institute. They would appear under "Preliminary Analysis" of 2D and 3D coordinates.

Institute but from a small workshop held in Leiden in 1995 by the senior editor. As this overview was being written one new line had to be added to the figure because Green (1995) has developed a method for applying thin-plate splines to outline coordinates.

GUIDE TO MORPHOMETRICS

Finally it was decided to provide a brief list of recommended reading and software to get started with, as well as some additional books and papers to peruse.

The most comprehensive short article is the one by Rohlf and Marcus (1993a), with the added discussion by Corti (1993) and Rohlf and Marcus (1993b). This short article should wet the reader's appetite for further involvement and help the investigator decide whether to delve more deeply into the subject.

In a biological context, the chapters in Reyment (1991) cover geometric morphometics with real data from fossils. Additionally that book reviews many other interesting mathematical and statistical developments of a more traditional kind relevant to the systematist and evolutionary biologist.

Chapters in the Blue and Black books mentioned below are "must" reading. Of especial interest are the articles by Rohlf, Bookstein, and the application by Loy et al. in the Black book. These should be supplemented by various articles, many of which are published in Systematic Biology or its earlier incarnation, Systematic Zoology.

Bookstein's article on landmarks in the Blue book is a must as is his critique of landmark types and his discussion of their use in the articles in that book. Rohlf's articles in the Blue and Black books give an excellent overview and should be read prior to reading his article in this volume.

Klingenberg (this volume) is a comprehensive review of allometry. Bookstein's essay on size and shape (Bookstein, 1989a) is highly recommended.

Video data acquisition is explained in the articles by Fink (1990), MacLeod (1990) and Rohlf (1990b) in the Blue book, and a more practical article is given by Becerra et al. (1993) in the Black book.

In order to get a theoretical foundation for superimposition techniques, the articles by Rohlf, and Rohlf and Slice are musts. Statistical tests and confidence intervals are discussed in Goodall (1991). For thin-plate splines a shorter overview is given in Bookstein (1989b); his 1991 book remains the definitive treatise. The development of edgels is discussed in the paper by Bookstein and Green (1993).

GLOSSARY

Last, but not least, it is predicted that one of the most useful and lasting contributions of this volume will be Appendix I. This is the most comprehensive attempt to date to define those terms and concepts felt relevant to morphometrics at this time.

The glossary was shepherded by Slice and also contributed to by Rohlf, Bookstein and Marcus with solicitations to the other editors, authors and the MORPHMET bulletin board for additional terms.

It was not easy to reach a consensus as to the terms to include and their definitions in all cases. There was some dissension as to whether to include the more common statistical terms such as ANOVA. Because the target audience is dominated by biologists, we leaned in favor of being more comprehensive rather than more specific to morphometrics per se, with an attempt to reach clarity—perhaps in some cases we suffered some loss of precision and rigor.

This glossary will serve as a first generation glossary, as part of the World Wide Web. There, on the server at SUNY Stony Brook, it is for all to see and criticize—and is a dynamic thing—with entries to be added (or deleted) as required—and with discussion on the MORPHMET bulletin board. So feel free to contact any of us about commissions or omissions. Alternative definitions are also a possibility.

THE FUTURE

What the future holds, of course, is unknown and unknowable. So perhaps a set of hopes and wishes will set some of the directions in future research and applications. It has been the senior editor's claim for some time (Marcus, 1988) that data acquisition capability has moved at a much more rapid pace than has our ability to analyze that data. Rohlf (1993b) points out that in one illustration there may be a richness of data available to the biologist who does not yet know how to quantify that data let alone analyze it.

The rapid advances in speed of computing, memory storage capacity and lowering of costs has made many of the acquisition tasks available to the researcher on a desktop or small laboratory computer. The senior author's portable Polhemus 3DRAW, in his office, is a wonder to behold—making it now routine to collect rapidly a comprehensive set of landmarks on 3D objects which can be done as easily when visiting other institutions as at his resident institution. However, CT data is still too expensive and difficult to obtain and process for most biologists. Laser scanners are just becoming affordable and are installed in some museums and biology departments. Expect to see more laser scanners being used.

Software is a major limiting factor—visualization software for all platforms is not available, and some computational programs are limited by the number of objects or landmarks that they can handle. These limits will be rapidly moved outward as computer speed increases and costs decrease, as platform-independent packages develop and as more workers become conversant with general purpose packages such as MATLAB, SAS or S-plus.

The convergence of the MacIntosh, IBM PC and clones and workstations means that software such as Edgewarp, installed on UNIX-based workstation is able to be run on a network using such protocols as X-Windows with IBM PCs and Macintosh computers as terminals.

There is still a considerable way to go in analyzing landmark and outline data together, although Dean (this volume) and Bookstein and Green (1993) have made some progress. Surfaces are even more challenging, and our ability to quantify textures such as graininess, bumpiness and hairiness in our quantitative biological descriptions has yet to develop. We await these developments with great interest.

It is clear that the Michigan, Stony Brook, Valsain, Paris and Il Ciocco get-togethers have done a great deal to educate biologists about morphometric developments. Smaller get-togethers and workshops at national and international meetings have served similar educational ends. More national and international meetings will occur in the future, and meetings are relatively inexpensive and quick ways to train and inform biologists in morphometric theory and techniques and new developments. There, the developers can receive feedback on the relevance and modifications required for their theory to make it yet more useful. At some major universities courses are being taught at graduate level on shape analysis, and it is hoped that these will become more popular. In journal articles and at international and national meetings one sees more thin-plate spline applications and shape analysis based on coordinates in two and three dimensions. Applications are just beginning to become common in the literature, and as new generations are trained and take their places, we w'll really see what this revolution has wrought.

REFERENCES

Becerra, J. M., E. Bello, and A. García-Valdecasas 1993. Building your own machine image system: A user point of view. Pages 65–92 *in* L. F. Marcus, E. Bello, and A. García-Valdecasas, (eds.), Contributions to morphometrics. Monografias del Museo Nacional de Ciencias Naturales 8, Madrid.

Bookstein, F. L. 1989a. "Size and shape": A comment on semantics. Systematic Zoology 38:173–180.

Bookstein, F. L. 1989b. Principal warps: thin-plate splines and the decomposition of deformations. IEEE Transactions on Pattern Analysis and Machine Intelligence 11:567–585.

Bookstein, F. L. 1991. Morphometric tools for landmark data: Geometry and biology. Cambridge University Press: Cambridge

Bookstein, F. L. 1993. A brief history of the morphometric synthesis. Pages 15–40 *in* L. F. Marcus, E. Bello, and A. García-Valdecasas, (eds.), Contributions to morphometrics. Monografias del Museo Nacional de Ciencias Naturales 8, Madrid.

Bookstein, F. L. 1994. Can biometrical shape be a homologous character? Pages 197–227 *in* B. K. Hall, (ed.), Homology: The hierarchical basis of comparative biology. Academic Press: New York.

Bookstein, F. L., B. Chernoff, R. L. Elder, J. M. Humphries, Jr., G. R. Smith, and R. E. Strauss. 1985. Morphometrics in evolutionary biology: The geometry of size and shape change, with examples from fishes. Academy of Natural Sciences of Philadelphia Special Publication 15. [Red book]

Bookstein, F. L. and W. D. K. Green 1993. A feature space for edgels in images with landmarks. Journal of Mathematical Imaging and Vision 3:231–261.

Corti, M. 1993. Geometric morphometrics: An extension to the revolution. Trends in Ecology and Evolution 8(2):302–303.

Fink, W. L. 1990. Data acquisition for morphometric analysis in systematic biology. Pages 9–19 *in* F. J. Rohlf and F. L. Bookstein, (eds.), Proceedings of the Michigan Morphometrics Workshop. University of Michigan Museum of Zoology Special Publication 2.

Fink, W. L. and M. L. Zelditch. 1995. Phylogenetic analysis of ontogenetic shape transformations: A reassessment of the piranha genus *Pygocentris* (Telostei). Systematic Biology 44:343-360.

Goodall, C. R. 1991. Procrustes methods in the statistical analysis of shape (with discussion and rejoinder). Journal of the Royal Statistical Society, Series B, 53:285–339.

Green, W. D. K. 1995. Spline-based deformable models. Pages 290–301 *in* R. Melter, A. Wu., F. Bookstein, and W. D. K. Green, (eds.), Vision geometry IV. SPIE Proceedings.

Isaev, M. 1995. EFAWin: Window shell for Elliptic Fourier. Genesys Labs, 1111 BayHill Dr, Suite 180, San Bruno, CA 94066.

Laduke, T. C. 1991. Morphometric variability of the precaudal vertebrae of *Thamnophis sirtalis sirtalis* (Serpentes: Colubridae), and implications for interpretation of the fossil record. Ph. D. dissertation, City University of New York.

Lele, S. 1993. Euclidean distance matrix analysis (EDMA): Estimation of mean form and mean form difference. Mathematical Geology 25:573–602.

Lele, S., and J. T. Richtsmeier. 1991. Euclidean distance matrix analysis: A coordinate free approach for comparing biological shapes using landmark data. American Journal of Physical Anthropology 86:415–428.

Lele, S. and J. T. Richtsmeier. 1992. On comparing biological shapes: Detection of influential landmarks. American Journal of Physical Anthropology 87:49–65.

Loy, A., M. Corti, and L. F. Marcus. 1993. Landmark data: Size and shape analysis in systematics. A case study on Old World Talpidae (Mammalia, Insectivora). Pages 215–240 *in* L. F. Marcus, E. Bello, and A. García-Valdecasas, (eds.), Contributions to morphometrics, Monografías del Museo Nacional de Ciencias Naturales 8, Madrid.

MacLeod, N. 1990. Digital images and automated image analysis systems. Pages 21–35 *in* F. J. Rohlf and F. L. Bookstein, (eds.), Proceedings of the Michigan morphometrics workshop. University of Michigan Museum of Zoology Special Publication 2.

Marcus, L. F. 1988. Automated data acquisition in museums. Workshop on Computers in Museums, University of Mexico. Spectra.

Marcus L. F., E. Bello and A. García-Valdecasas. (eds.). 1993. Contributions to morphometrics. Monografías del Museo Nacional de Ciencias Naturales 8, Madrid. [Black book]

Mardia, K. V., and C. A. Gill (eds.). 1995. Proceedings in current issues in statistical shape analysis. Leeds University Press: Leeds, United Kingdom.

Reyment, R. A. 1991. Multidimensional palaeobiology. Pergamon Press: Oxford.

Rohlf, F. J. 1986. Relationships among eigenshape analysis, Fourier analysis, and analysis of coordinates. Mathematical Geology 18:845–854.

Rohlf, F. J. 1990a. Rotational fit (Procrustes) methods. Pages 227–236 *in* F. J. Rohlf and F. L. Bookstein, (eds.), Proceedings of the Michigan morphometrics workshop. University of Michigan Museum of Zoology Special Publication 2.

Rohlf, F. J. 1990b. An overview of image processing and analysis techniques for morphometrics. Pages 37–60 *in* F. J. Rohlf and F. L. Bookstein, (eds), Proceedings of the Michigan morphometrics workshop. University of Michigan Museum of Zoology Special Publication 2.

Rohlf, F. J. 1990c. Fitting curves to outlines. Pages 167–177 *in* F. J. Rohlf and F. L. Bookstein, (eds.), Proceedings of the Michigan morphometrics workshop. University of Michigan Museum of Zoology Special Publication 2.

Rohlf, F. J. 1990d. TPSPLINE: A program to compare two shapes using a thin-plate spline. Department of Ecology and Evolution, State University of New York, Stony Brook, New York 11794.

Rohlf, F. J. 1993a. NTSYS-pc - Numerical taxonomy and multivariate analysis system, version 1.80. Exeter Software: Setauket, New York.

Rohlf, F. J. 1993b. Feature extraction in systematic biology. Pages 375–392. *in* R. Fortuner, (ed), Advances in computer methods for systematic biology: Artificial intelligence, databases, computer vision. John Hopkins University Press: Baltimore.

Rohlf, F. J. 1993c. TPSREGR: A program for regression of partial warp scores. Department of Ecology and Evolution, State University of New York, Stony Brook, New York 11794.

Rohlf, F. J. 1993d. Relative warp analysis and an example of its application to mosquito wings. Pages 131–159 *in* L. F. Marcus, E. Bello and A. García-Valdecasas, (eds.), Contribution to morphometrics. Monografías del Museo Nacional de Ciencias Naturales 8, Madrid.

Rohlf, F. J. 1993e. TPSRW—Thin-plate spline relative warp. Department of Ecology and Evolution, State University of New York, Stony Brook, New York 11794.

Rohlf, F. J., and J. W. Archie. 1978. Least-squares mapping using interpoint distances. Ecology 59(1):126-132.

Rohlf, F. J. and J. W. Archie. 1984. A comparison of Fourier methods for the description of wing shape in mosquitoes (Diptera: Culicidae). Systematic Zoology 33:302–317.

Rohlf, F. J and F. L. Bookstein (eds.). 1990. Proceedings of the Michigan morphometrics workshop. University of Michigan Museum of Zoology Special Publication 2. [Blue book]

Rohlf, F. J., and L. F. Marcus. 1993a. A revolution in morphometrics. Trends in Ecology and Evolution 8(4):129–132.

Rohlf, F. J., and L. F. Marcus. 1993b. Geometric morphometrics: Reply to M. Corti. Trends in Ecology and Evolution 8(9):339.

Rohlf, F. J., and D. E. Slice. 1990a. Extensions of the Procrustes method for the optimal superimposition of landmarks. Systematic Zoology 39:40–59.

Rohlf, F. J., and D. E. Slice. 1990b. GRF: A program for generalized rotational fitting. Department of Ecology and Evolution, State University of New York, Stony Brook, New York 11794.

Slice, D. E. 1993a. Extensions, comparisons, and applications of superimposition methods for morphometric analysis. Ph.D. dissertation: State University of New York at Stony Brook.

Slice, D. E. 1993b. GRF-ND: generalized rotational fitting of N-dimensional data. Department of Ecology and Evolution, State University of New York, Stony Brook, New York 11794.

Spencer, M. A. and G. S. Spencer. 1995. Video-based three-dimensional morphometrics. American Journal of Physical Anthropology 96:443-453.

Swiderski, D. L. 1994. Morphological evolution of the scapula in tree squirrels, chipmunks and ground squirrels (Sciuridae): An analysis using thin-plate splines. Evolution 47:1854-1873.

Zelditch, M. L., W. L. Fink, and D. L. Swiderski. 1995. Morphometrics, homology and phylogenetics: Quantified characters as synapomorphies. Systematic Biology 44(2):179–189.

PART I: THEORY AND METHODS

AN IDIOSYNCRATIC HISTORY OF EARLY MORPHOMETRICS

Richard A. Reyment*

Institute of Earth Sciences
Department of Historical Geology and Paleontology
University of Uppsala
Uppsala, Sweden

INTRODUCTION

The desire to depict shape must be as old as cultivated thought itself. Figures on ancient Egyptian monuments provide early evidence of this wish. More widely known are the designs of Dürer, the fifteenth century German artist, who invented a system of mapping caricatures of faces. Even other artists have used this technique to vary physiognomies in mass scenes, using basically few faces as a starting point.

Scientific interest in the quantitative analysis of shape has quite a long history in biology, back almost 100 years, and owing to the activities of the early biometricians. Notwithstanding that little real progress was made in these first attempts, it is significant that there was a definite desire to come to grips with analysis of shape. It was in 1917 that D'Arcy Wentworth Thompson published his remarkable treatise on growth and form (Thompson, 1917). This volume is replete with captivating suggestions, indicative of deep insight into the fundamental concepts in the study of change in shape. Thompson's method of depicting such change in terms of coordinates is well known and has been cited, with copies of some of his figures, in many textbooks on theoretical biology. Intellectually significant as these examples may be, they were not constructed by any mathematical method but were produced freehand. Huxley (1932) seems to have been the first to note this, but the point has been overlooked by others.

Attempts have been made at different times to deduce a valid solution for the shape-coordinate concept, but all have failed to resolve the real issues involved, and for decades the concept was in limbo. Instead, an algebraic approach gained ascendancy. Beginning with some exploratory work by Teissier (1938), the algebraic mode of analysis in the form of principal components surged to the fore. Teissier's results involved what we would now call allometric size change and are quite forecastable, in that they are a natural outcome of the Perron-Frobenius theorem. This theorem of linear algebra states that if all

* E-mail: Richard.Reyment@pal.uu.se

the elements of a square, symmetric matrix are positive, then all the elements of the first latent vector extracted from that matrix will be positive.

Although the principal component solution has come to be attributed to Jolicoeur and Mosimann (1960), the real originator of the size-shape reduction of a matrix of covariances and variances is S. C. Pearce, a statistician employed by the East Malling Research Station. Early in the fifties, Pearce was making routine use of the principal component model for designing the shape of fruit trees. Mardia et al. (1979) reviewed Pearce's contributions and use one of his examples. There is a biometry text by Pearce (1965), which is an illuminating source of information. An even earlier reference to shape analysis of trees and human bodies is the book by Quenouille (1952), again in an algebraic garb.

ALGEBRAIC TREATMENT OF SIZE AND SHAPE

What was it that Georges Teissier actually did? He applied the relatively new concept of "factor analysis" to a set of multivariate measures made on the carapace of the crab *Carcinus maenas*. Although Teissier may have believed he was doing a factor analysis, he was, in fact, not. In 1938, the difference between the factor-analytical model of the psychometricians and the principal component analysis of educational proficiency tests (Hotelling, 1931) was not appreciated; in truth, nobody in those days was really clear on the distinction. Teissier extracted the first principal component from a matrix of correlations, and it was this that provided the basis of his factor analysis. Jolicoeur's (1963) resuscitation of Teissier's idea consists of the extraction of the latent roots and vectors of a covariance matrix of logarithmically transformed morphological distance measures (which was an innovation in relation to the original proposition and which was proposed as a direct generalization of the Huxleyan equation for allometric growth).

It must be realized that the size-shape justification attached to the algebraic solution is an a posteriori one and not the outcome of a specific biometrical model. The same algebra can, and is, applied to any number of problems in physics and chemistry. A feeling for what is involved can be obtained from a contrived example. Suppose that the analysis of a quadrivariate, log-transformed data set has given the following array of latent vectors, extracted from its covariance matrix:

	I	II	III	IV
x_1	0.5	0.5	-0.5	-0.5
x_2	0.5	-0.5	-0.5	0.5
x_3	0.5	0.5	0.5	0.5
x_4	0.5	-0.5	0.5	-0.5

The principal component interpretation here for morphometric variables would be that the first vector expresses equal growth in all elements, or, more precisely, isometric growth. Assume, as usually is the case, that the latent root associated with vector I is much greater than the sum of all the remaining roots; based on the Teissier-Pearce-Jolicoeur model, one would say that most variation is related to size variation and that this variation has been partialled out. What remains is taken to represent various linear combinations that express shape in relation to unspecified sources.

Imagine now that the first latent vector had contained unequal elements: (0.3, 0.8, 0.3, 0.4)'. The relationship would then no longer be isometric but differential allometric and hence provided with a shape component. Obviously, this situation must confound any attempt at a size-shape interpretation.

There is a natural explanation for the confounding of concepts. Distance measures—i.e., the usual measures made in morphometric work—lengths, heights and breadths on organisms—contain information on size as well as shape. One piece of essential information they do not contain is the relative location of each point in space, that is, the geometry of the form. As long as the caliper remains in place on the organism, we can see how the distance between two points is oriented. Once removed, this knowledge is lost forever.

The coordinates of selected points on an organism, the landmarks, retain all information on distances and geometry. Thompson's diagrams expressing shape comparisons as deformations of coordinate systems had to be brought into service, if the subject of morphometric analysis was going to advance. The important breakthrough was achieved by F. L. Bookstein (1991), who realized that the thin-plate spline technique could be invoked for producing a practical solution to the Thompsonian intercalation problem. In a way, Bookstein bridged the gap between the concepts of Albrecht Dürer and D'Arcy Thompson.

Before passing to the idiosyncratic part of my note, I wish to mention a few of the algebraists who have contributed to the analysis of distance measures. Hopkins (1966) provided an intelligent improvement to Jolicoeur's method of assessing multivariate allometry by means of a factor-analytical model of a reduced rank covariance matrix. Sprent (1968) and Rao (1964, 1971) tried to make the algebraic treatment more general, and Mardia et al. (1979), following Rao (1971), underscored the distinction between size and shape factors, on the one hand, and the size and shape vectors of Pearce, on the other. It is historically interesting to record that Mardia et al. (1979) thought that the good side of the method was the many size-shape possibilities it offered, and the bad thing about it was the arbitrary behavior of the principal component loadings, a point already noted by Rao (1964) in his monograph on principal component analysis.

Shape can be given different meanings by different people. This is not surprising, for there must always be semantic uncertainties when vernacular terms are forced into precise scientific dress. I have just learned that there is a well-developed field of shape analysis about which I knew nothing. This is expounded in a recent article by Girling (1993) on the shape analysis of anisotropic corpuscles, by means of Wicksell's size-shape independence condition.

SOME PERSONAGES OF EARLY MORPHOMETRICS

The origins of morphometrics lie with the early biometricians, Karl Pearson (1857-1936), Francis Galton (1822-1911) and W. F. R. Weldon (1860-1906). It is certainly thought provoking that the two main surges in mathematical statistics took place over the years 1890-1905 and 1915-1930, in answer to the need for solutions to very real biological problems. The incentive for founding the journal *Biometrika* around 1900, and the coining of the concept of "biometry," was an outcome of Weldon's appointment to the Chair of Zoology at the University of London, where Pearson was already installed as professor of Applied Mathematics and one of his main tasks was teaching mathematics to students in the engineering school. The way in which the two came into contact bears reflection. Pearson had just revised, and largely rewritten, Isaac Todhunter's "A History of the Theory of Elasticity" and was looking around for new fields. He was well equipped for such an enterprise, having been a student of such giants as Cayley, Maxwell, and Stokes while at Cambridge. He had become aware of biological problems requiring statistical solutions, and he began experimenting with the theory of distributions, problems of evolution, differential death rates and selection, the standard deviation, the coefficient of correlation, and the study of racial differences through measurements on human skulls. He was in a restless period, and it may come as a surprise that Pearson was an accomplished Germanist (it was while a student at Heidelberg and Berlin that Charles was changed to Karl). He devoted a considerable part of his research energy throughout the last two decades of the nineteenth century to erudite works on mediaeval and Renaissance German literature and folklore. It must have come as a welcome breath of air when he was contacted by his new colleague at London concerning some statistical problems Weldon had with his work. Pearson is said to have exclaimed "metron shall be applied to bios." Galton, an aged sage by that time, was valuable as a counselor, and it was he who first had the ideas of correlation and regression. The real input was to come from the collaboration between Pearson and Weldon. Pearson, born and bred in London, where his father was an accomplished jurist, actually began his professional career within the legal profession, but soon tired of that. Weldon had in 1890 produced the first paper in which quantitative methods were applied to biological types other than humans in his article dealing with the distributions of four different measurements made on the shrimp *Crangon vulgaris*, which he found closely followed the Gaussian law. Galton, as referee, perceived faults in the article, but, rather than hand it back, he kindly made the necessary improvements. However, the credit for making the vast number of measurements and formulating of the biological problem was Weldon's. The second paper dealt with correlations between organs in the same species contrasted over four demes. So intense became the collaboration between Weldon and Galton, that revisions of notes swifted back and forth over the day, with final versions being sent by the "midnight mail," a convenience that the telephone and telefacsimile seem to have made extinct.

Pearson was an enthusiastic computer and spent much of his time reducing vast sets of measurements. It was in connection with his work on skew frequency curves (the stimulus for which came from F. Y. Edgeworth, not Galton nor Weldon, as was once lightly assumed) that he felt he had to have access to a computing machine. After much trouble in obtaining the necessary finances, he was able to purchase a Brunsviga apparatus. To the end of his life in 1936, he used a Brunsviga calculator that was of the beginning-of-the-century vintage. Sir Maurice Kendall, the renowned English statistician, was still using Udny Yule's Brunsviga in 1965.

Pearson's great respect for Galton is well documented, not least in a three-volume appreciation. In 1930, he added some comments on Galton's ability, stating that Galton was no faultless genius and that he made slips of many kinds. However, Galton had the persistency of the truly great to redeem these errors and thereby gain greater insight. There is a tendency to downgrade the significance of Pearson's work these days; this, I think, is not constructive. He made errors and was often unwilling to accept justified criticism, but there is no doubt in my mind that without his contributions over a wide field, the development of statistics would hardly have advanced to the extent it did in the first half of this century.

We know a great deal about Pearson and his associates thanks to his son Egon, who succeeded him at London University. We know, for example, that in 1894 Weldon was concerned with a Royal Society Committee for the Statistical Analysis of Measurable Characteristics of Plants and Animals. Of the many experiments proposed and carried out, there were an astonishing number concerned with Pearson's asymmetrical frequency curves. There was some concern on the part of Weldon and Galton about Pearson's monograph on the subject, and they were looking for a physical explanation of the fundamental differential equation, which they could not discover. Pearson, for his part, with the enthusiasm of the creator of such an elastic system, was trying out his curves on anything he could get hold of, without bothering about the niceties of biological relevance.

In retrospect, we can see that the shrimp data so often used by Weldon to illustrate biometrical ideas had many defects, making them unsuitable for what he was trying to do—defects such as broken parts, lack of homogeneity, unknown effect of age on relative growth of parts and so on, but his work was important because it stimulated interest in many others to work in the morphometrical field. There was also the problem of relying upon Pearson's unsure handling of probability theory, which, to a certain extent, unsettled Weldon.

Pearson was not primarily biologically oriented, his main interests lying with applied mathematics and engineering. Evidence of his lack of insight into biological problems may be seen in his bitter refusal to accept the generalized statistical distance of P. C. Mahalanobis as being a more suitable way of tackling morphometrical relationships based on multiple measurements than the coefficient of racial likeness, to which, of course, the generalized distance reduces to zero correlations between all traits. What Pearson had to offer were solutions in search of a problem.

We need to backtrack a half decade now, to look at the role played by Francis Galton in setting the stage for later developments. Galton belonged to the economically sound British middle class. He was born near Birmingham on February 16, 1822, into a prosperous Quaker business family—sensible, practical, thrifty. His grandfather was a friend of such notables of his day as Priestley and Watt, and his father was an amateur scientist, as well as a successful banker. On his mother's side he was related to the Darwin family. Charles Darwin was his cousin and close friend. The Darwins were marked by exceptional longevity, a property Galton identified among his siblings—he himself lived to be 89, and it was this fact that may have interested Galton in his studies of inherited characteristics. His schooling and university training did little to suggest he had exceptional ability, and he finally ended up practicing medicine. At the age of 22 he inherited his father's entire estate and could then devote his life to leisure and being a gentleman scientist. He began publishing his observations on almost everything that attracted his attention and then left for a two-year stay in southwest Africa, where his explorations gained him renown. One of his most noteworthy achievements was the invention of the correlation coefficient (the mathematical details were later supplied by Pearson and others), which arose out of his preoccupation with the principles of inheritance. His influence on the subsequent development of biometry cannot be overstated—not only as a source of interesting problems to be solved by the more able but also as an organizer of research. He died on January 17, 1911.

No account of the early stirrings in biometrical thought would be complete without reference to Ronald Aylmer Fisher (1890-1962), who, when all things are considered, must be granted credit as being the originator of modern biometry. Fisher was one of eight children (his twin brother did not survive), and himself had nine children. His father was an auctioneer who took a strong interest in his children and their schooling. He seized on Ronald's obvious mathematical ability and saw to it that he received an appropriate education. When Fisher left school, it was to find that the family was in financial straits. Fortunately, he was able to win an entrance scholarship to Cambridge, which assured his future. He graduated in 1912 with distinction in optics. He put in a fourth year of study in mathematical physics. Although he had no formal background in biology, he is said to have shown interest in genetics, such as the subject was in those days. It was at about this time he began looking into statistical theory and, in 1915, produced his work on the distribution of the correlation coefficient. The fourth year of study at an end, he had to find employment and spent two years as an actuary. His extremely poor eyesight kept him out of the Army during the first world war and, having tired of the insurance business, he became a science teacher at a succession of schools. Pearson was impressed by the young schoolmaster and offered him a lectureship. Fisher chose, however, a post at the Rothamsted Agricultural Research Station, a fortunate choice, because it kept him away from teaching, for which he was singularly ill-suited. Rothamsted Station provided him with an endless source of problems that led to the large-scale development of biometrical theory and practice. Fisher's ability to master difficult problems, quickly, is legendary. There is no doubt that without Fisher the whole science of biometry would not be where it is today.

Not long after 1920, Fisher and Pearson (and associates by extension) became bad friends. This was inevitable. Pearson had a stormy and rather overbearing temperament, more than matched by Fisher's own deficiencies of character, likewise overbearing, vindictive and revengeful. Nobody really knows how the quarreling started, but Fisher, when pressed, would say it was because Pearson would not let him reproduce Eldertson's table of chi-square. Much time was wasted in petty feuding, although a positive aspect is that Fisher cleared up many of the murkier results of Pearson's as part of his vendetta. Fisher was unable to accept criticism, even if mild. He extended his dislike of Pearson to encompass Jerzy Neyman and anybody else working with Pearson.

The heroic period of multivariate biometry now took place: the discriminant function appeared in 1935 and 1936, applied to morphometrical problems, and the T^2 of Hotelling was presented in 1932, followed by principal components and canonical correlation. These were not isolated events, as one might be tempted to assume today. Hotelling belonged to the Rothamsted group in the late 1920s, as did also Wishart, and it was during this period many of the main concepts were thrashed out, largely under Fisher's guidance. In 1933, Pearson finally left his old department at London, and it was divided into two. Fisher took up the new chair of Eugenics, and Pearson's son, Egon, was later appointed to the chair of Statistics. In 1943, Fisher was offered the chair of Genetics in Cambridge and left the field of statistical theory. His later years were devoted to the statistical development of genetics. He left Cambridge on retirement and settled in Adelaide, Australia, where he had been offered an appointment by the Commonwealth Scientific and Industrial Research Organization. He died there from embolism in 1962. These final days of his life have been well described by a member of the family that rented him a room, a person who later made a name for himself in the field of mathematical genetics and, who as a student at the time, profited greatly from Fisher's willingness to solve his statistical problems. Fisher was experienced by his hosts as an enigma, a strange, often gratuitously rude, person who paid scant attention to the niceties of polite conversation and who suffered from chronic insomnia.

And now I come to D'Arcy Wentworth Thompson. Thompson was a Scot, born in Edinburgh in 1860, the grandson of a sea captain and the son of an able classical scholar who was born at sea off the coast of Tasmania, a dramatic event marred by the death of his mother. Thompson's father eventually passed from being a schoolmaster to professor of Greek at Galway. Thompson was given a first-rate classical education that led to his being remarkably fluent at Latin and Greek. He studied medicine at Edinburgh but after two years obtained a scholarship to Cambridge, where he devoted himself to natural science. His hopes for a research career at Cambridge were dashed when he failed to obtain a fellowship. He returned to his native Scotland and entered the university college at Dundee, later incorporated into the University of St. Andrews. In 1884 Thompson was appointed professor of biology at Dundee, which later became a chair of natural history at St. Andrews, a post he filled for 64 years. His early work was strictly taxonomic and, largely, bibliographical. He prepared extensive catalogues on protozoans, sponges, coelenterates and worms and, throughout his life, glossaries dealing with Greek birds and fishes. There were also translations of treatises (including a work of Aristotle) and morphological studies of tapeworms, fish, birds and cephalopods, among other taxa. In addition to his academic activities, Thompson indulged in consulting work. For 41 years he was scientific adviser to the Fishery Board for Scotland, editing reports and preparing notes on practical fishery problems.

Shortly before he died in 1948, he lectured in India on the skeletal structure of birds, holding a live, and unhappy, hen under his massive arm as a model. Thompson was a huge man, with an untrimmed, bushy beard that originally was flaming red. The popular prestige surrounding him, particularly in later life, was enormous. The first edition of *Growth and Form* actually appeared in 1915, issued in just 500 copies. It sold poorly but, once out of print, became much sought after in the secondhand market. This stimulated the interest of the publishing sector, always on the lookout for saleable material. It was frequently subject to minor revisions after that, and Thompson, himself, did not consider it finally ready until the edition of 1941. He was wont to say, rather dramatically, that his book needed little in the way of a preface, it being all preface. His aim, he proclaimed, was to reduce biology to physics and mathematics. Biology is no exception to the rule "Theos aei Geometrthei," God always geometrizes. Problems of growth are essentially physical problems, and the morphologist is, in truth, a student of physical science. Thompson's wide classical background and excellent knowledge of German, acquired while a student at Cambridge, opened fields to him that remained terra incognita to others. His idea for shape-coordinates came, as all morphometricians know, from his fascination with the work of the fifteenth century artist Albrecht Dürer. His generalization of the significance of the equiangular spiral, was no doubt inspired by the paper on the pearly nautilus by the Reverend H. Mosely in 1822. But there are many other remarkable observations in his treatise, many of which are now feeding current research in functional morphology. The flight of birds, stress diagrams, the secretion of echinoderm shell material, all bear witness to an uncommonly well developed sense of geometry. In comparison with some of the notables reviewed in this note, Thompson had an even temper and a generous nature.

Albrecht Dürer gets an honorable mention in many overviews of morphometrics, not forgetting Thompson's original treatise, but little more is provided to appease the curious. I started with Dürer and I think it is fitting to finish this note with a brief sketch of him—after all, it was perhaps he who was responsible for starting it all.

Albrecht Dürer was born in Nuremberg in 1471, the third of eighteen children of a goldsmith of Hungarian descent. After a desultory and disinterested start in the goldsmith's trade, he was allowed to follow his natural artistic bent. His father apprenticed him to an artist, Michael Wohlgemut, a fortunate choice that assured his learning the basics of drafting. He had the good fortune to marry well and, at a relatively early age, to gain immense fame, both as a teacher as well as an artist.

He was fascinated by geometry for intellectual and practical reasons, and there are good grounds for believing that recognition for the discovery of descriptive geometry should go to him. He published noteworthy treatises on practical geometry and he is usually credited with providing the first genuine mathematical proof in the German language in one of these (Dürer, 1525). He died on April 6, 1528. He is said to have been a handsome, well-favored man with a winning disposition.

REFERENCES

Dürer, A. 1525. Underweysung der Messung mit dem Zirckel un Richtscheyt, in Linien Ebnen und ganzen Corporen. Nuremberg.

Girling, A. J. 1993. Shape analysis for the anisotropic corpuscle problem. Journal of the Royal Statistical Society, B, 55:675–686.

Hopkins, J. W. 1966. Some considerations in multivariate allometry. Biometrics, 22:747-760.

Hotelling, H. 1931. The generalization of Student's ratio. Annals of Mathematical Statistics 2:360–378.

Huxley, J. S. 1932. Problems of relative growth. Methuen: London.

Jolicoeur, J. 1963. The degree of generality of robustness in *Martes americana*. Growth 27:1–27.

Jolicoeur, J. and Mosimann, J. E. 1960. Size and shape variation in the painted turtle, a principal component analysis. Growth 24:339–354.

Mardia, K. V., Kent, J. T. and Bibby, J. M. 1979. Multivariate analysis. Academic Press: New York.

Panofsky, E. 1945. Albrecht Dürer. Princeton University Press: Princeton, New Jersey.

Pearce, S. C. 1965. Biological statistics, an introduction. McGraw-Hill: New York.

Quenouille, M. H. 1952. Associated measurements. Butterworths: London.

Rao, C. R. 1964. The use and interpretation of principal components in applied Mathematics in the archaeological and historical Sciences. Edinburgh University research. Sankhya 26:329–358.

Rao, C. R. 1971. Taxonomy in anthropology. *In* F. R. Hodson,. F. R. Kendall, and P. Tautu, (eds.), Mathematics in the archaeological and historical sciences. Edinburgh University Press: Edinburgh, U. K.
Sprent, P. 1968. Linear relationships in growth and size studies. Biometrics 24:39–656.
Teissier, G. 1938. Un essai d'analyse factorielle. Les variants sexuels de *Maia squinada*. Biotypologie 7:73–96.
Thompson, D. W. 1917. On growth and form. Cambridge University Press: Cambridge.

MULTIVARIATE ALLOMETRY

Christian Peter Klingenberg

Department of Biological Sciences
University of Alberta
Edmonton, Alberta T6G 2E9, Canada

ABSTRACT

The subject of allometry is variation in morphometric variables or other features of organisms associated with variation in size. Such variation can be produced by several biological phenomena, and three different levels of allometry are therefore distinguished: static allometry reflects individual variation within a population and age class, ontogenetic allometry is due to growth processes, and evolutionary allometry is the result of phylogenetic variation among taxa. Most multivariate studies of allometry have used principal component analysis. I review the traditional technique, which can be interpreted as a least-squares fit of a straight line to the scatter of data points in a multidimensional space spanned by the morphometric variables. I also summarize some recent developments extending principal component analysis to multiple groups. "Size correction" for comparisons between groups of organisms is an important application of allometry in morphometrics. I recommend use of Burnaby's technique for "size correction" and compare it with some similar approaches. The procedures described herein are applied to a data set on geographic variation in the waterstrider *Gerris costae* (Insecta: Heteroptera: Gerridae). In this example, I use the bootstrap technique to compute standard errors and perform statistical tests. Finally, I contrast this approach to the study of allometry with some alternatives, such as factor analytic and geometric approaches, and briefly analyze the different notions of allometry upon which these approaches are based.

INTRODUCTION

Variation in size of organisms usually is associated with variation in shape, and most metric characters are highly correlated among one another. These associations are the subject of allometry (Huxley, 1932; Cock, 1966; Gould, 1966, 1975). Although allometry is often used to examine the consequences of size for ecological or physiological variables (Günther, 1975; LaBarbera, 1989; Reiss, 1989), this review deals only with measurements of traits used to characterize the morphological form of organisms.

Advances in Morphometrics, Edited by L. F. Marcus et al.
Plenum Press, New York, 1996

Unlike other approaches in morphometrics, which are built on geometric theory, allometry has a largely empirical basis. Huxley (1932) realized that scatter plots of two trait measurements in growing organisms often closely follow a curved line, and that this relationship usually becomes linear if both measurements are transformed to logarithms. From this, he derived his formula of simple allometry

$$y = bx^{\alpha}$$

or, in log-transformed notation,

$$\log y = \log b + \alpha \log x,$$

where x and y are trait measurements, and b and α are constants. The constant a, the slope in log-log plots of x and y, is often called the allometric coefficient (terminology is not uniform; some authors call b coefficient). The special case when $\alpha = 1$ is called isometry, and indicates direct proportionality between x and y. If $\alpha > 1$, there is positive allometry, whereas for negative allometry, $\alpha < 1$ (Huxley and Teissier, 1936). In humans, for example, the long bones of the limbs show positive allometric growth relative to overall stature, and the height of the head shows negative allometry.

In most morphometric data sets, measurements are positively correlated, i.e., x and y increase or decrease simultaneously. Even if there is negative allometry, a still is positive; negative allometry implies only that the relative variation in y is smaller than that in x, e.g., y grows by 10% for every 20% growth increment in x. If a is negative, however, there is an absolute reduction in y associated with an increase in x. This case is called enantiometry (Huxley and Teissier, 1936). Reduction of the absolute size of organs during growth is a real phenomenon, although it is not found commonly in morphometric studies. The most striking example is the shrinking of larval structures during metamorphosis, e.g. the gills and tail of anuran tadpoles; but in a subtler way, enantiometry even occurs in cranial growth of primates (Corner and Richtsmeier, 1991).

Huxley's approach is not restricted to pairs of measurements. In many multivariate data sets, log-log plots of all pairwise combinations of morphometric variables show approximately linear relationships. Therefore, Huxley's bivariate allometry can be generalized to multiple dimensions. Moreover, it is not confined to growth data, as straight-line relationships are also found in log-log plots of intra- and interspecific variation within one particular ontogenetic stage (most often adults). From this perspective of allometry, some major questions follow: How much variation is there? Are the data points clustered around a straight line and, if so, how closely? What is the direction of that line in multidimensional space? Do different groups of organisms share the same allometric relationship?

In this paper I review concepts and techniques used in studies of multivariate allometry. First, I introduce the main levels of variation that have been the subject of allometric studies within and between species,. Then, I present the multivariate generalization of allometry using principal component analysis and some more recent developments, such as the bootstrap and techniques dealing with multiple groups. Finally, I briefly contrast some alternative approaches to allometry.

LEVELS OF ALLOMETRY

Huxley (1932) devised allometry mainly as a tool to study the relative growth of parts in various organisms. Growth, however, is not the only origin of variation in overall size and

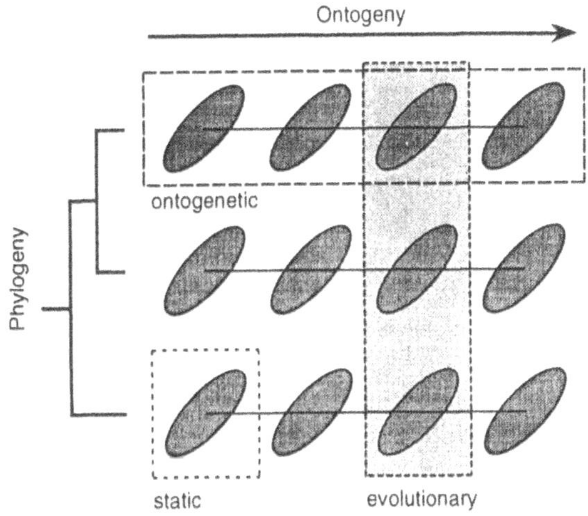

Figure 1. The three levels of allometry. The diagram shows three species, each with four different ontogenetic stages, that are considered to be homologous among species. Rectangles enclose the species and stage groups included in an analysis of allometry at each of the three levels. Ontogenetic allometry can be separately analyzed for all three species, evolutionary allometry for each of the four stages, and static allometry for each of the 12 species and stage groups.

associated variation in shape, because evolutionary changes and individual variation also can generate allometric relations. These levels or types of allometry have been included in elaborate classification schemes (see Cock, 1966; Gould, 1966, 1975). Because of its simplicity, I prefer the terminology proposed by Cock (1966) who distinguished static, ontogenetic and evolutionary allometry (Fig. 1). This classification has also been used in most empirical comparisons between levels of allometry (Cheverud, 1982; Leamy and Bradley, 1982; Boag, 1984; Gibson et al., 1984; Leamy and Atchley, 1984a; Shea, 1985; Klingenberg and Zimmermann, 1992a).

Static Allometry. Static allometry, which is also referred to as size allometry, results from variation among individuals of the same population and age group (intraspecific scaling, Gould, 1975). Static allometry is particularly easy to study in organisms with discrete growth stages, such as insects (Cuzin-Roudy, 1975; Klingenberg and Zimmermann, 1992a), or in adults of organisms with determinate growth, such as birds (Boag, 1984; Gibson et al., 1984). These studies, among others, found that the largest proportion of multivariate variation is contained in one dimension, and that the model of simple static allometry therefore is appropriate. This phenomenon has been termed morphometric or phenotypic integration (e.g., Leamy and Atchley, 1984b; Zelditch, 1987, 1988; Zelditch and Carmichael, 1989). Although there is an extensive literature describing static allometry and morphometric integration, relatively little is known about their developmental basis (but see Cowley and Atchley, 1990; Atchley and Hall, 1991; Shea, 1992; Paulsen and Nijhout, 1993). In theoretical models, Riska (1986) explored how developmental processes can affect static correlations among the traits they produce (see also Cowley and Atchley, 1992). Patterns of static allometry have sometimes been used to deduce underlying developmental processes (e.g., Zelditch, 1987; Wheeler, 1991). Such inferences, however, should be substantiated by direct observations.

Ontogenetic Allometry. Ontogenetic allometry or growth allometry deals with covariation among characters during growth. Simple allometry occurs if the ratio between the specific growth rates (percentage increment per time unit) of two different characters is constant (Huxley, 1932; Reeve and Huxley, 1945; Shea, 1985; Blackstone, 1987). Theoretical studies showed that various models of growth as a function of time can result in simple allometry (e.g., Laird, 1965; Laird et al., 1968; Katz, 1980). The rule of simple allometry holds often, but not always, as Huxley (1932) demonstrated with an impressive list of bivariate examples. Correspondingly, multivariate studies often find that one dimension contains the largest proportion of the total variation, sometimes more than 99% (e.g., Jungers et al., 1988; Solignac et al., 1990; Strauss, 1990b; Klingenberg and Zimmermann, 1992a). Some studies, however, show clear deviations from simple allometry in certain structures (Cuzin-Roudy and Laval, 1975; Boag, 1984; Cane, 1993) or subtle curvatures of growth trajectories in the space of log-transformed characters (Bookstein, 1991: Figs. 4.2.2 to 4.2.4; Klingenberg and Zimmermann 1992a; Klingenberg and Spence, 1993). Studies of plant growth showed particularly strong deviations from simple allometry (e.g. Kampny et al., 1993; McLellan, 1993). Three types of data are used to study ontogenetic allometry: longitudinal data based on measurements of the same individuals at several developmental stages, cross-sectional data with different specimens in several known stages, and mixed cross-sectional data collected without information on ontogenetic stage (Cock, 1966).

Evolutionary Allometry. Evolutionary allometry reflects covariation among changes in different traits along the branches of a phylogeny. It is concerned with character covariation among contemporaneous species sharing a common ancestor (Fig. 2a) or among fossil members of an evolutionary lineage (Fig. 2b). I do not distinguish separate levels for analyses that use these two types of data (for different terminology, see Gould, 1966). Evolutionary processes leading to the associations between trait changes presumably do not differ depending on whether these changes occur within one lineage successively or in different lineages giving rise to sister groups. It is important to use specimens in comparable ontogenetic stages to avoid confounding evolutionary and ontogenetic variation. This is straightforward in organisms with determinate growth, such as birds or insects (e.g., Livezey, 1989; Strauss, 1990a; Klingenberg and Zimmermann, 1992a), but it is more difficult in organisms with indeterminate growth, for which some studies assume that specimens are "typical" for the respective species (e.g., Strauss, 1985). In these studies, among others, the model of simple allometry fits the data fairly well, indicating that evolutionary variation is constrained in its dimensionality (Maynard Smith et al., 1985; Gould, 1989; Arnold, 1992). Some of this covariation among traits may be determined by developmental processes, as Riska (1989) showed with a simulation study.

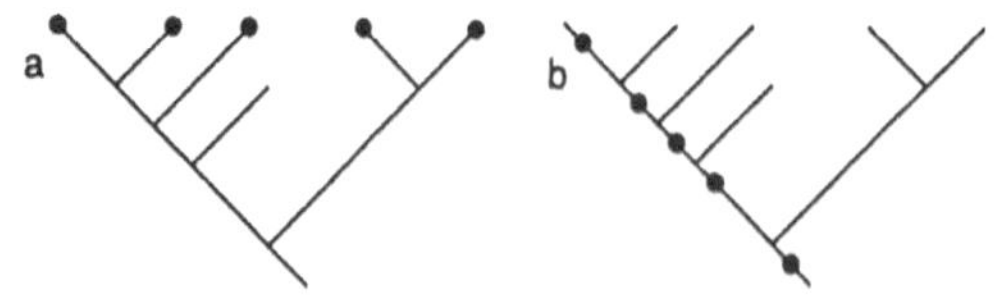

Figure 2. Types of data for evolutionary allometry. (a) In neontological studies, the data always are measurements taken on recent species from several lineages in a clade, that are related as sister groups rather than ancestors and descendants. (b) In paleontological studies, evolutionary allometry often refers to character covariation among members of a single evolutionary lineage. Because it is difficult to distinguish with certainty whether two fossil species are related as ancestor and descendant or as sister groups, many of these studies may in fact use designs that are mixtures of (a) and (b).

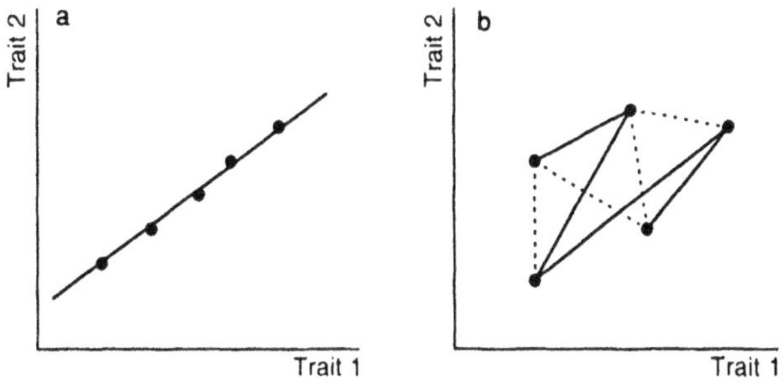

Figure 3. Influence of the correlation between characters on the robustness of evolutionary allometry. For simplicity, I assume that the species are members of a single evolutionary lineage (as in Fig. 2b). (a) If the traits are highly correlated, most morphometric variation is in a single direction. Estimates of evolutionary allometry, the "average" direction of evolutionary changes, will yield almost the same result regardless of the phylogenetic relations among species (i.e., how the points are connected). (b) If the correlation between traits is low, however, the "average" direction of changes depends strongly on the phylogenetic scenario, i.e., on how the points are connected. The solid and dashed lines represent two alternative hypotheses of ancestor–descendant relationships, which have drastically different directions of evolutionary changes.

Evolutionary allometry, like all interspecific comparisons, presents some statistical problems because the species are not independent of one another but are parts of a hierarchically structured phylogeny (e.g., Felsenstein, 1985; Pagel and Harvey, 1988). This interdependence is most evident for species presumed to be members of a single, unbranched lineage (Fig. 2b), but it also applies to comparisons among terminal taxa in a clade. A possible solution is the method of phylogenetically independent contrasts (Felsenstein, 1985; Martins and Garland, 1991; Garland et al., 1992), which analyzes character changes along the branches in the phylogeny instead of the measurements (character states) of terminal taxa. Changes are either directly measured, if actual ancestor–descendant series of fossils are available, or inferred by indirect methods such as parsimony or maximum likelihood. This approach requires knowledge of the phylogeny of the group under consideration and a multivariate theory for reconstructing state vectors of quantitative characters for internal nodes, which still needs to be developed (for univariate approaches, see Maddison and Maddison, 1992). For morphometric data, however, the problem of phylogenetic dependence may not be as severe as for other data types, because often most of the variation is in a single dimension. These high correlations make estimates of the direction of evolutionary changes relatively robust against errors in phylogenetic reconstruction (Fig. 3).

Further levels of allometry exist in organisms with modular organization, such as colonial animals and most vascular plants. In addition to the ontogeny of individual zooids, colonial animals have an additional level of colony wide development, which is called astogeny (e.g., Pandolfi, 1988). Buss and Blackstone (1991) showed that colony growth in a marine hydroid follows well-determined trajectories and that colonies react to experimental perturbations in an integrated manner (see also Anstey, 1987). Similarly, the structure of plant parts changes with the age of the entire plant (heteroblasty; e.g., McLellan, 1993). Jones (1992, 1993) studied correspondences between the development of individual leaves and the succession of leaves during whole-plant ontogeny. New methodological approaches, such as "process morphology" (Sattler, 1992; Jeune and Sattler, 1992), reflect the morphological flexibility of modular organization in plants but are only semiquantitative and cannot be directly related to allometry (a similar approach in zoology is the "skeleton space," Thomas and Reif, 1993).

The causes of allometric variation at different levels are mutually interrelated. Static variation, which is caused by variation in ontogenetic processes that produce the structures of interest, is the raw material upon which natural selection can act. Response to selection, in turn, generates evolutionary changes affecting the developmental processes. One way to study these interactions is to compare empirically the patterns of variation between different levels of allometry. Various such comparisons have been made (Cheverud, 1982; Leamy and Bradley, 1982; Boag, 1984; Gibson et al., 1984; Leamy and Atchley, 1984a; Shea, 1985; Klingenberg and Zimmermann, 1992a). Most of these studies found that patterns of allometry at different levels were similar but not equal. It is not possible, however, to make further generalizations of the results because the studies differ widely in the kinds of data and methods used. As an alternative to this observational approach, the mechanisms that are the basis of allometric variation can be investigated by experimental techniques, e.g., using genetically engineered organisms (Shea et al., 1990) or directly manipulating the size of eggs or embryos (Sinervo, 1993).

PRINCIPAL COMPONENTS AND ALLOMETRY

Under the model of simple allometry, bivariate plots of pairs of log-transformed morphometric variables are straight lines. If there are three variables, and all pairwise combinations satisfy this condition, then the data points follow a line in the three-dimensional space defined by the variables. This argument can be extended to more than three variables: data points still are arranged along a straight line under simple allometry (e.g., Teissier, 1955), but this line now is in the multidimensional space defined by all the variables. Therefore, dimensionality of morphometric variation is a prime concern of allometry (e.g., Hopkins, 1966; Sprent, 1972). As in bivariate allometry, points may be scattered around the line, rather than exactly lying on it, and one has to find a line that "optimally" fits the scatter of data points (Pearson, 1901). Jolicoeur (1963) proposed the first principal component, estimated from the covariance matrix of log-transformed measurements, as a multivariate generalization of simple allometry.

Many texts of multivariate statistics introduce principal component analysis (PCA) as a technique for summarizing most of the variation in a multivariate data set in fewer dimensions (e.g., Pimentel, 1979; Jolliffe, 1986; Flury, 1988; Flury and Riedwyl, 1988; Johnson and Wichern, 1988; Jackson, 1990; Jobson, 1992). The first principal component (PC1) is the linear combination that accounts for the maximum variance. Geometrically, it corresponds to the direction of the longest axis through the scatter of data points. Subsequent principal components take up maximal variance, subject to being orthogonal to all preceding component axes.

Fig. 4 shows a contour ellipse of the bivariate distribution of two variables X_1 and X_2 with its centroid (mean vector) at the point labeled 0. For simplicity, data are centered by subtracting the means of X_1 and X_2, which shifts the coordinate system to the new axes x_1 and x_2 each with a sample mean of zero. Therefore, the x_1 and x_2 values themselves are the deviations from their mean; their sample variances can be calculated as the sum of squared x_1 and x_2 values divided by $(n-1)$. In Fig. 4, one data point is labeled P, and its projection onto the x_1 axis is Q. The sum of squared x_1 values is the sum of the squared distances between 0 and Q for all data points. By the same argument, the sum of squared x_2 values corresponds to the sum of squared distances between P and Q. According to the Pythagorean theorem, the squared distance between 0 and P is the sum of the squared distances between 0 and Q and between P and Q. It follows that the sum of the squared distances of all data points from the sample centroid, divided by $(n-1)$, is the sum of the variances of x_1 and x_2, or total variance. Now, consider the same data set after rotating the coordinate system to the

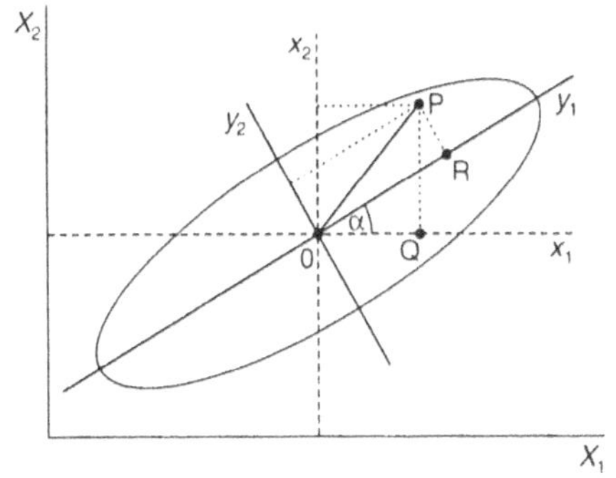

Figure 4. Principal component analysis. The diagram shows the contour ellipse of a bivariate distribution with its centroid at the point labeled 0. The centered coordinates x_1 and x_2 are derived from the original variables X_1 and X_2 by subtracting their mean values. The principal components y_1 and y_2 are the directions of maximal and minimal variance, respectively. See text for details.

directions of the principal component axes y_1 and y_2. Because the PC1 axis, y_1, is defined as the direction that has maximal variance, the sum of the squared distances between 0 and R, the projection of P onto the PC1 axis, is maximal. Because the rotation of the coordinate system does not change the distances between the data points and the centroid, maximizing the sum of squared distances between 0 and R also results in minimizing the sum of squared distances between P and R. The sum of squared distances between P and R, divided by $(n - 1)$, is the part of the total variance not accounted for by the PC1, i.e., the residual variance. This argument also holds for more than two dimensions: because the PC1 is the direction that has maximal variance, all other principal components taken together have minimal variance. Hence, the PC1 axis can be seen as a least-squares fit of a straight line to the scatter of data points in the space of log-transformed, bivariate or multivariate data. This justifies Jolicoeur's (1963) multivariate generalization of allometry (see also Hopkins, 1966).

PCA decomposes a covariance matrix **S** into eigenvectors and eigenvalues, so that **S** = **BLB'**. The matrix **B** of eigenvectors is used to transform the original data **X** into a set of new variables **Y** = **XB**, the principal components (PCs). The matrix **L** is the covariance matrix of the PCs, and as the PCs are uncorrelated among each other, all off-diagonal elements of **L** are zero. The diagonal elements of **L**, the eigenvalues, are the variances for which the associated eigenvectors account. They are difficult to interpret by themselves, because they depend on the measurement units and the base of the logarithm used to transform the data. However, the proportion of the total variance for which the PC1 accounts is important to assess how well the model of simple allometry fits the data.

Principal component analysis can be interpreted geometrically as a rotation of the coordinate system. The PC axes are aligned with the directions of the axes of the multidimensional "scatter ellipsoid" (in two dimensions, this is an ellipse, Fig. 4). The PC coefficients of the original variables can be interpreted as "direction cosines," i.e., the cosine of the angle between the PC axis and the coordinate axis of the respective variable (α for PC1 and x_1 in Fig. 4). Principal component axes are mutually orthogonal, and the vectors of PC coefficients are usually normalized to have unit length so that the squares of the coefficients sum up to unity (**b'b** = 1, where **b** is an eigenvector). As a result, the coefficient values depend on the number of variables. Nevertheless, translating PC coefficients to bivariate allometric coefficients (Huxley's α) is quite easy (Jolicoeur, 1963; Shea, 1985). The ratio of PC1 coefficients for two variables corresponds to the variables' bivariate allometric coefficient. For example, in a study of two species of voles, Airoldi and Flury (1988) found that the PC1 coefficients of skull length, width and height were approximately 2/3, 2/3, and 1/3, respectively. Thus, in allometric plots with skull length as the independent variable, skull width would be isometric with a slope of about 1, whereas skull height, with a slope of about 0.5, would show strong negative allometry. With p variables, isometry in all pairwise combinations of variables results in a PC1 in which all coefficients are equal, and have the value 1 divided by the square root of p, i.e., $p^{-0.5}$. Isometry can be assessed with Anderson's (1963) test, which is based on normal theory (Pimentel, 1979: 70; Flury, 1988: 34), or by comparison with jackknifed or bootstrapped confidence intervals (see below; e.g., Diaconis and Efron, 1983; Gibson et al., 1984; Klingenberg and Zimmermann, 1992a). Multiplying PC1 coefficients by the square root of p yields values that can be interpreted as bivariate allometric coefficients for each of the variables against a measure of overall size (a weighted geometric mean of all variables). These allometric coefficients or the principal component coefficients can be graphed as Huxley's (1932) growth gradients (e.g., Boitard et al., 1982; Solignac et al., 1990), or they can be displayed on diagrams of the measurements, such as the truss network (Strauss and Bookstein, 1982; Bookstein et al., 1985). Another type of graphical display for PC coefficients is the biplot (e.g., Marcus, 1993).

Empirical comparisons of bivariate and multivariate approaches found that both estimated corresponding patterns of allometry (Davies and Brown, 1972; Shea, 1985). Jungers and German (1981) criticized the multivariate approach because allometric coefficients derived from principal components of skeletal measurements did not match those from bivariate regressions on a known variable for size that was not included in the analysis, either body weight or length. Hills (1982) showed that these discrepancies disappear if one considers allometry between the traits and the size measure that is taken as a reference, e.g., between skeletal size and body weight.

It is possible to perform PCA by use of a correlation matrix instead of a covariance matrix (see also Pimentel, 1979; Bookstein et al., 1985; Johnson and Wichern, 1988). This corresponds to an analysis of standardized variables. Geometrically, it means that all the variables are adjusted to have standard deviations of 1 by stretching or shrinking their coordinate axes before the analysis. This maneuver can be used to remove scaling effects if variables are measured in different units. Standardizing may also be useful if one is interested in only ordination of specimens, as in some applications in systematics, where giving equal weight to all variables may be more important than scaling. For allometry, however, scaling is essential. After removing scale by standardization one still can determine whether the data points lie along a straight line, but it is impossible to estimate allometric coefficients because standardization changes the direction of the allometric axis. A simple hypothetical example, constructed from purely allometric variation without any residual scatter, can show this. Let the multivariate allometric coefficients (eigenvector) be 2/3, 2/3 and 1/3; these coefficients correspond to strong deviations from isometry. Then, the covariance matrix is a multiple of

$$\begin{bmatrix} 4 & 4 & 2 \\ 4 & 4 & 2 \\ 2 & 2 & 1 \end{bmatrix}$$

and the correlation matrix is

$$\begin{bmatrix} 1 & 1 & 1 \\ 1 & 1 & 1 \\ 1 & 1 & 1 \end{bmatrix}$$

The PC1 of this correlation matrix has coefficients that all take the value $3^{-0.5}$, and thus falsely indicate isometry, with no residual variation at all (see also Johnson and Wichern, 1988: 350). Likewise, in allometric interpretations of PCAs based on correlation matrices of real data (e.g., Teissier, 1955; Somers, 1986), all "non-isometric" variation that may be inferred from PC1 coefficients merely results from the residual scatter about an allometric relationship, but does not reflect allometry. Therefore, it is crucial to use the covariance matrix for allometry. Routines for PCA in many statistical software packages use the correlation matrix as the default option. Users of these programs should make sure to specify the option for PCA based on the covariance matrix.

A related point is the transformation of data to logarithms. There are numerous practical and theoretical reasons why it is often useful to transform data to logarithms (Pimentel, 1979; Reyment et al., 1984; Bookstein et al., 1985; Bookstein, 1991; Reyment, 1991). For ontogenetic allometry, Huxley (1932) justified the use of power functions, and therefore also of logarithms, by his rule of constant ratios among specific growth rates of different organs (see also Reeve and Huxley, 1945; Günther, 1975; Katz, 1980; Bookstein et al., 1985; Shea, 1985; Blackstone, 1987). Such a theoretical justification, however, is more difficult to find for static and evolutionary allometry. The multiplicative nature of growth processes also may be important for these levels because all variation in morphological structures is due to variation in the developmental processes that generate them. Mosimann (1970) and Mosimann and James (1979) pointed out statistical advantages of the log-normal distribution (but see Smith, 1993, for biases in predicted values from allometric regression). In practice, log transformation often renders relations among variables more linear and also can make variances more homogeneous. Finally, log-transformed data are independent of measurement units (e.g., millimeters or the units of an eyepiece micrometer) but retain the information about scale (e.g., lengths versus surfaces). It does not matter which base for the logarithms is chosen, as long as the same base is used consistently throughout a given analysis.

Results of PCAs are estimates of allometric patterns in the populations from which the study samples are drawn. To assess how reliable these estimates are, standard errors or confidence intervals should be calculated. Formulas for these statistics (e.g., Flury and Riedwyl, 1988) are based on the assumption of multivariate normal distribution and on large sample sizes. In most allometric studies, however, the distribution of data cannot be assumed to be multivariate normal. In ontogenetic allometry, for example, the distribution of measurements depends on the age composition in the sample, as well as on the growth dynamics of the structures investigated. In these cases, the bootstrap and jackknife procedures are helpful (an excellent introduction is Efron and Tibshirani, 1993; other useful references are Diaconis and Efron, 1983; Efron and Gong, 1983; Efron and Tibshirani, 1986; Manly, 1991). The bootstrap is a computer-intensive procedure that substitutes repeated sampling from the sample distribution for a theoretical model of that distribution. The only assumption that must be made is that the specimens have been sampled randomly. Applications of the bootstrap to PCA include Diaconis and Efron (1983), Stauffer et al. (1985), Daudin et al. (1988), and Efron and Tibshirani (1993). In multivariate allometry, Gibson et al. (1984) and McGillivray (1985) used the jackknife, whereas Klingenberg and Froese (1991) and Klingenberg and Zimmermann (1992a, b) used the bootstrap. Marcus (1990) compared the jackknife and the bootstrap with each other and with results from large-sample theory.

The fundamental idea of the bootstrap, and the procedures to apply it, follow immediately from the definitions of standard errors and confidence intervals for a statistic q (e.g., mean value or PC coefficients) estimated from a sample of n specimens. Both standard errors and confidence intervals provide answers to the same question: If the same study were repeated numerous times, estimating q from a sample with n specimens each time, how variable would the estimates be? The standard error of a statistic is the standard deviation of these estimates and a confidence interval is the interval containing a certain percentage of the estimates (e.g., the 95% confidence interval is delimited by the 2.5% and 97.5% quantiles). This is exactly what the bootstrap does, assuming that the sample distribution is representative of the totality of organisms about which statements are made (e.g., all members of a local population or all females of a species). Repeatedly, a "bootstrap sample" of n specimens is drawn randomly, with replacement, from the original sample, and q is estimated for each bootstrap sample. The standard deviation of these estimates is the bootstrapped standard error, and confidence intervals can be derived from the distribution of bootstrap estimates (for details, see Efron and Tibshirani, 1993). About 100 bootstrap samples usually are sufficient to establish standard errors, but at least about 1000 are necessary for confidence intervals (Efron and Tibshirani, 1993). The bootstrap can even be used for hypothesis tests by generating bootstrap replicates of a test statistic under a particular null hypothesis and then comparing the resulting distribution to the test statistic calculated for the observed data (Efron and Tibshirani, 1993; for a related topic, permutation tests, see also Manly, 1991). An advantage of the bootstrap is that it can be adapted to the particular design of a study. For instance, if there are discrete growth stages that can be identified unambiguously (see Fig. 1), there may be no sampling error in the stage composition of the data set (e.g., a design with equal numbers from each stage). The bootstrap procedure for ontogenetic allometry can be adapted by drawing "bootstrap subsamples" from these stages separately. These are then pooled into one bootstrap sample and principal components are calculated (Klingenberg and Zimmermann, 1992a).

ANALYSES OF MULTIPLE GROUPS

Many morphometric studies deal with several groups of specimens, e.g., different sexes, species or ecomorphs. In all these cases, variation within and between groups has to be separated. Otherwise, levels of allometry may be confounded, or within-group variability may invalidate discrimination between groups. Separation of size-related variation within groups from between-group differences has been a traditional topic in morphometrics (Burnaby, 1966; Gower, 1976; Reyment and Banfield, 1976; Pimentel, 1979; Humphries et al., 1981; Thorpe, 1983; Reyment et al., 1984; Bookstein et al., 1985; Rohlf and Bookstein, 1987; Marcus, 1990; Reyment, 1991).

Multivariate comparisons of allometric patterns often focus on the directions of the major axes of scatter ellipsoids in several groups. A straightforward measure for differences between two groups is the angle between their first principal components. For normalized principal components (i.e., squared coefficients sum up to unity), the angle a between components **b** and **c** in two groups is the arc cosine of the inner product of the two vectors,

$$\alpha = \arccos(\Sigma b_i c_i) = \arccos(\mathbf{b}'\mathbf{c})$$

($\mathbf{b}'\mathbf{c}$ is sometimes called vector correlation; note that this is not the correlation between corresponding elements of the two vectors; see also Pimentel, 1979; Bryant, 1984). Angles can even be calculated from published tables of PC coefficients. Applications of angular comparisons include Boitard et al. (1982) and Gibson et al. (1984). Cheverud (1982) and Klingenberg and Zimmermann (1992a) used Monte Carlo simulations of angles between random vectors to assess statistical significance. For the example in this paper, I used the bootstrap to test the more appropriate null hypothesis of equal PC vectors (0° angles).

Another method for comparison among multiple groups of organisms is based on a multivariate ordination of the directions of allometric axes (Solignac et al., 1990; Klingenberg and Froese, 1991; Klingenberg and Spence, 1993). The first principal component of each group is considered as a data point in the space spanned by the coefficients of the original variables. The vectors of PC1 coefficients are entered as observations in an ordination by a second PCA. The results of this analysis can then be displayed as plots of the "meta-PC" scores. Bootstrap estimates of allometric coefficients can be used to draw confidence ellipses (e.g., Owen and Chmielewski, 1985; Johnson and Wichern, 1988) as a visual indication of statistical accuracy (Klingenberg and Froese, 1991; Klingenberg and Spence, 1993).

Comparisons of allometry within several groups often, but not always, show that the coefficients of the PC1s differ only minimally. In these cases, it may be feasible to use one of the models of common covariance structure (Fig. 5; Airoldi and Flury, 1988; Flury, 1988). These models are based on the assumption that the groups share a common allometric pattern, i.e., that the major axes of their scatter ellipsoids are parallel. Therefore, the differences between the observed PCs of the samples are regarded as effects of sampling error. Principal component analysis, however, is a procedure for analyzing variation in a single sample, and therefore the method needs to be generalized for the context of multiple groups. Usually, the PC1 of the pooled within-group covariance matrix has been used to characterize this common allometric pattern, e.g., in multigroup PCA (Pimentel, 1979; Thorpe, 1983), Burnaby's procedure (Burnaby, 1966; Rohlf and Bookstein, 1987) and in the shearing procedure (Humphries et al., 1981; Bookstein et al., 1985; Rohlf and Bookstein, 1987). Airoldi and Flury (1988) criticized the use of the pooled within-group covariance matrix, because it implicitly assumes that the covariance matrices of all groups are identical (Fig. 5, right panel). They proposed an alternative procedure, common principal component analysis (CPCA), which only assumes that the PCs are common to all groups (see also Flury, 1988; Flury and Riedwyl, 1988). Whereas the directions of the principal axes are assumed to be the same, the amount of variation associated with each PC can vary between groups (Fig. 5, center). Common principal component analysis is available in the NTSYS software package, FORTRAN routines are contained in the IMSL/STAT library (routines KPRIN and DKPRIN), a MATLAB program was written by L. F. Marcus and a SAS/IML version is available from the author. In applications of CPCA to multivariate allometry, the first common principal component (CPC1) is interpreted as an allometric pattern shared by all groups (Airoldi and Flury, 1988; Klingenberg and Zimmermann, 1992a, b; Klingenberg and Spence, 1993).

Discrimination between groups is often difficult because of allometric variation within groups. Especially in organisms with indeterminate growth, the amount of within-group variation may far exceed between-group differences. For instance, fish can increase in size by several orders of magnitude during their life cycle. Other causes, such as nutrition, also contribute to variability within groups (Bernays, 1986; Patton and Brylski, 1987; Meyer, 1990; Smith and Palmer, 1994). Depending on the particular organisms of interest, within-group variation is mostly ontogenetic or mostly static allometry, or a mixture of both.

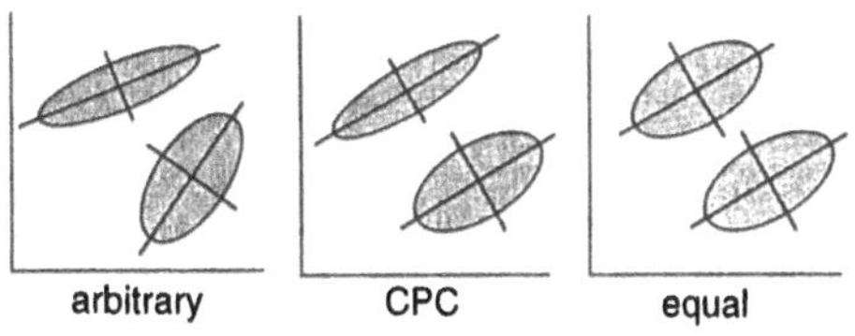

Figure 5. Three levels of similarity between covariance structures. Groups can have arbitrarily different covariance matrices; scatter ellipses then differ both in the directions and lengths of their principal axes. Under the common principal component model (CPC), groups share the same directions of principal axes but may differ in the amount of variation associated with each axis. Groups with equal covariance matrices have the same directions and lengths of principal axes. More details, including additional levels of similarity, are given by Airoldi and Flury (1988) and Flury (1988).

Because most of this variation is often confined to a single dimension, along the allometric axis, it can be removed from an analysis by eliminating the variation in that direction. This approach uses allometry as a criterion of subtraction (Gould, 1975), as it has been done traditionally in bivariate studies. The central assumption of all methods for "size correction," is that the groups share the same allometric vector. If the groups have different allometric patterns, size correction is not possible because all corrections that are suitable for one group will not work in other groups. Several methods have been developed following this principle (Burnaby, 1966; Gower, 1976; Humphries et al., 1981; Thorpe, 1983; Bookstein et al., 1985; Rohlf and Bookstein, 1987). None of these methods, however, can be a substitute for careful examination of the specimens: hidden heterogeneity within the groups, e.g., undetected sex dimorphism or cryptic species, may invalidate the entire analysis.

The procedure proposed by Burnaby (1966) eliminates the effects of growth from multivariate data by projecting data points onto a subspace that is orthogonal to the growth vector (see also Gower, 1976; Rohlf and Bookstein, 1987; Reyment, 1991). This growth-invariant subspace has one dimension fewer than the original space. With three variables, for example, the growth-invariant space is a plane, and for two variables, it is a line (Fig. 6). The growth-adjusted data are coordinates of the projected points, expressed in the coordinate system of the original variables. The growth-invariant data for an $n \times p$ data matrix $\mathbf{X}$ and a $p \times 1$ growth vector $\mathbf{b}$ can be obtained as

$$\mathbf{X}(\mathbf{I} - \mathbf{b}(\mathbf{b}'\mathbf{b})^{-1}\mathbf{b}'),$$

where I is an identity matrix of rank p. With a normalized vector, such as a principal component, the formula simplifies to

$$\mathbf{X}(\mathbf{I} - \mathbf{b}\mathbf{b}').$$

Usually, the PC1 of the pooled within-groups covariance matrix has been used as the growth vector in Burnaby's procedure (e.g., Reyment and Banfield, 1976; Riska, 1981; Rohlf and Bookstein, 1987). Because CPCA is based on less-stringent assumptions (see above),

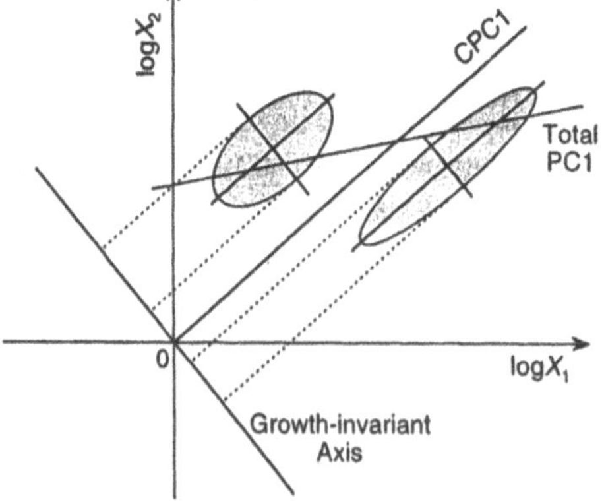

Figure 6. Burnaby's (1966) procedure in two dimensions. The model of common principal components is appropriate, because the principal axes of the scatter ellipses in the two groups are parallel. Therefore, the groups share a common axis of allometric growth, the CPC1. All the variation is projected onto an axis perpendicular to the CPC1 by setting the CPC1 scores to zero. Subsequent analyses focus on between-group differences in this "growth-invariant" axis. The PC1 of the combined samples ("Total PC1") confounds variation within and between groups.

the CPC1 seems preferable as an estimate of a common allometric pattern. This version of Burnaby's technique is equivalent to a procedure involving three consecutive steps: (1) a rigid rotation to the common principal components, (2) setting the CPC1 scores of each data point to zero, and (3) rigid rotation back to the original coordinate system.

For analyses of growth-adjusted data, it may be more convenient to omit step (3). First, common principal components are computed as an estimate of within-group variation, and the CPC scores, except for the CPC1, are used as variables in subsequent analyses. The results of these analyses, e.g., discriminant analysis or MANOVA, are identical to the results based on data adjusted by Burnaby's original procedure, but within-group covariance matrices are of full rank. This technique is almost identical to the one proposed by Thorpe (1983; the PC1 of the pooled within-groups covariance matrix as an estimate of the allometric axis), which has been used in numerous studies (e.g., Wiig, 1985; Thorpe and Baez, 1987; Corti et al., 1988; Lessa and Patton, 1989).

Interpreting the values resulting from Burnaby's procedure is somewhat difficult. With some caution, they can be seen as measurements adjusted to "unit size." This is possible for most morphometric data sets because the CPC1 has high positive correlations with all variables, and therefore can be interpreted as an overall size variable. Setting the CPC1 score for log-transformed measurements to zero corresponds to setting the untransformed value of overall size to one. The adjusted variables therefore allow one to compare organisms of different sizes by rescaling all measurements allometrically to unit size. For example, in Fig. 6 the group to the left has higher X_2 and lower X_1 values at corresponding sizes than does the group to the right. The difference between this approach and comparisons of ratios or geometric shape is that Burnaby's procedure takes into account the shape changes caused by allometric growth.

Another way to understand Burnaby's procedure is by analogy to the most familiar method of size correction in bivariate allometry, regression residuals. Regression residuals are an appropriate way of correcting for size if one is interested in the relationship of a dependent variable y (e.g., an ecological or physiological parameter) to a size variable x known a priori and measured independently (e.g. body weight). Then, the residuals are the deviations of actual measurements from the value expected for an "average" specimen of that particular size. These deviations are computed in the direction of the y axis by subtracting the y value estimated by regression from the observed value. Hence, the interpretation of allometry as a "criterion of subtraction" (Gould, 1975) can be taken literally. In multivariate allometry, however, there is no a priori size variable that can be measured independently from the other variables. All measurements are affected simultaneously by overall size, which only can be estimated from them. As I explained above, the PC1 is a good choice for such an estimate. The "residuals" are subsequent PCs, which are perpendicular to the PC1 (step [2] above). In bivariate regression, comparison of intercepts of several groups makes sense only when the slopes in all groups are equal. Analogously, Burnaby's procedure only works if all the groups share a common allometric pattern, as it is estimated by the CPC1. (For groups that differ in their growth vectors, Burnaby [1966] suggested removing all growth vectors from the data; after this, however, there may not be much meaningful variation left to study [e.g., Humphries et al., 1981].)

Because it includes only rigid rotations, Burnaby's procedure conserves the spatial relationships among data points in all directions except that of the growth vector. Therefore, data adjusted by this technique can be used to quantify variation perpendicular to the growth trajectories. For instance, lateral transpositions of growth trajectories (also called vertical transpositions) can be distinguished from group differences produced by shifts along the growth axis (ontogenetic scaling, Gould, 1975; Shea 1985, 1992). These two types of group differences, as well as within-group variation, are confounded in PCA of combined samples (Fig. 6; for discussion, see Voss et al., 1990; Voss and Marcus, 1992; Klingenberg and Spence,

1993). Burnaby-adjusted data also can be used to examine within-group deviations from simple ontogenetic allometry, such as curvatures of the growth trajectories. Klingenberg and Spence (1993) used a MANOVA of Burnaby-adjusted data to separate lateral transposition from nonallometric growth in a data set containing samples of six discrete ontogenetic stages from each of six waterstrider species (i.e., a data structure corresponding to Fig. 1). The PC1 scores of the between-species matrix can be used to display lateral transposition, and likewise the between-stage matrix for nonallometric growth.

Procedures for size correction in several groups all assume that the groups share the same allometric pattern. But what if that is not true? In this case, the formal procedures fail, and one has to find a visual and at best "semistatistical" way to assess group differences. A very useful technique of this kind is the "tomographic representation" introduced by Boitard et al. (1982). First, they calculated the PCs for the covariance matrix of the pooled data. Then, they plotted the second versus third PC scores separately for data points grouped by the data points PC1 scores. Finally, they combined these layers into a plot showing the scatter ellipsoids "suspended" by their within-group PC1 axes in a box representing the space of the first three PCs of the pooled data.

EXAMPLE: GEOGRAPHIC VARIATION

Some of the techniques described above are applied in a simple example, a morphometric data set taken from a larger study on geographic variation in the waterstrider *Gerris costae* (Klingenberg, 1992). Only adult specimens were measured, and because these bugs do not grow after they reach the adult stage, I made no attempt to correct for size in the original study. Here, using allometric techniques, I reanalyze a part of the data.

Three samples, each representing a different subspecies, are included here: the nominate subspecies *G. c. costae* is represented by a sample from the Swiss Alps ($n = 32$), *G. c. fieberi* is represented by a sample from northern Greece ($n = 33$), and *G. c. poissoni* by a sample from the eastern Pyrenees in France ($n = 28$). In this example, I consider only adult males. Allometric variation within the three samples is therefore purely static allometry.

Four measurements were chosen for the example: (1) total thorax length, (2) the length of the first antennal segment, (3) the middle femur length and (4) the hind femur length. The raw data are presented in the Appendix. Data were transformed to natural logarithms. For the convenience of presentation, variances and covariances are multiplied by 10^4 (this also applies to eigenvalues). The covariance matrices are

$$\begin{bmatrix} 12.67 & 8.39 & 8.75 & 11.09 \\ 8.39 & 14.34 & 9.04 & 11.61 \\ 8.75 & 9.04 & 12.05 & 13.63 \\ 11.09 & 11.61 & 13.63 & 17.83 \end{bmatrix} \text{ for the sample from the Alps,}$$

$$\begin{bmatrix} 5.55 & 2.07 & 2.53 & 4.06 \\ 2.07 & 9.66 & 4.03 & 4.67 \\ 2.53 & 4.03 & 7.82 & 9.02 \\ 4.06 & 4.67 & 9.02 & 12.98 \end{bmatrix} \text{ for the sample from Greece, and}$$

$$\begin{bmatrix} 4.93 & 5.97 & 3.90 & 5.16 \\ 5.97 & 18.65 & 6.78 & 8.35 \\ 3.90 & 6.78 & 6.94 & 7.56 \\ 5.16 & 8.35 & 7.56 & 10.70 \end{bmatrix} \text{ for the sample from the Pyrenees.}$$

As in many morphometric data sets, all covariances are positive and most of them are relatively high. There are, however, some differences between groups and among variables. The sample from the Alps is more variable than the other two samples. Whereas the two femur lengths are highly correlated in all groups, correlations involving thorax length and the first antennal segment tend to be lower and more variable (e.g., the correlation between these two measurements is only 0.28 in the Greek sample).

Principal components were computed for each sample, and parametric standard errors for the estimates of PC coefficients and eigenvalues were calculated using the formulas given by Flury (1988). Moreover, standard errors were also determined with a bootstrap procedure. For each group, 1000 bootstrap samples were randomly drawn (with replacement), and principal components were computed for each. (This number of replications is more than actually needed for standard errors; Efron and Tibshirani, 1993.) The standard deviations of these 1000 estimates of the PC coefficients or eigenvalues are their bootstrap standard errors.

The amount of variance (± parametric and bootstrapped standard errors) for which PC1 accounts is 46.1 (± 11.7, 12.5) or 81% of total variance in the sample from the Alps; for 24.1 (± 6.0, 4.5), or 67% in the Greek sample; and for 31.4 (± 8.5, 7.2), or 76% in the sample from the Pyrenees. Standard errors determined by the two approaches are similar in magnitude, although the bootstrap standard error for the Greek sample differs from the parametric estimate by about 25%. The percentages of total variance taken up by the PC1 are quite typical for static allometry (e.g. Cuzin-Roudy, 1975; Klingenberg and Zimmermann, 1992a). The estimated coefficients of the PC1 and their parametric and bootstrapped standard errors are

$$\begin{bmatrix}0.441\\0.471\\0.477\\0.597\end{bmatrix}, \begin{bmatrix}0.057\\0.061\\0.036\\0.034\end{bmatrix}, \text{and} \begin{bmatrix}0.060\\0.055\\0.032\\0.030\end{bmatrix} \text{for the sample from the Alps,}$$

$$\begin{bmatrix}0.269\\0.409\\0.527\\0.695\end{bmatrix}, \begin{bmatrix}0.085\\0.119\\0.045\\0.054\end{bmatrix}, \text{and} \begin{bmatrix}0.093\\0.113\\0.045\\0.060\end{bmatrix} \text{for the Greek sample, and}$$

$$\begin{bmatrix}0.315\\0.695\\0.400\\0.507\end{bmatrix}, \begin{bmatrix}0.050\\0.082\\0.057\\0.070\end{bmatrix}, \text{and} \begin{bmatrix}0.037\\0.077\\0.057\\0.078\end{bmatrix} \text{for the sample from the Pyrenees.}$$

The estimates of PC coefficients are fairly stable, as indicated by the relatively small standard errors. Bootstrap standard errors agree well with parametric estimates. The coefficient values clearly do not conform to the pattern for overall isometry, where all coefficients would be equal to $p^{-0.5}$, i.e., the vector [0.5, 0.5, 0.5, 0.5]'. This is confirmed by Anderson's test (Anderson, 1963; Pimentel, 1979:70; Flury, 1988:34), which was significant for all three groups (Alps: $\chi^2 = 11.52$, $P < 0.01$; Greece: $\chi^2 = 20.37$, $p < 0.001$; Pyrenees $\chi^2 = 23.82$, $p < 0.0001$; df = 3 in each case).

Coefficients vary considerably between samples. In the sample from the Alps, the middle femur (third measurement) is almost isometric relative to thorax length (first measurement), as the ratio of their coefficients is 0.477 / 0.441 = 1.08. In the Greek sample, however, the corresponding ratio is 0.527 / 0.269 = 1.96, which indicates strong positive allometry of the middle femur relative to the thorax. Whereas the lengths of the first antennal segment and of the middle femur (second and third measurements) are almost isometric in the sample from the Alps, the middle femur shows positive allometry relative to the first antennal segment in the Greek sample, and negative allometry in the Pyrenees. As an overall measure of these differences, I computed the angles between PC1 axes of different groups;

they are 12.22° between Alps and Greece, 16.29° between Alps and Pyrenees, and 21.19° between Greece and the Pyrenees.

If we expect a common allometric pattern, these angles seem quite large. But are the differences from zero statistically significant? Because there is no parametric test for the angles between PCs, I used the bootstrap approach to test the null hypothesis that the common principal component model holds. The bootstrap test procedure is straightforward: use the data to generate a modified data set that conforms to the null hypothesis, then repeatedly draw bootstrap samples and compute the test statistic, and finally compare the empirical distribution of the test statistic to the value calculated for the original data (Efron and Tibshirani, 1993). To produce a data set conforming to the CPC model, I rotated the data in each sample so that the within-group PC axes are aligned exactly with the CPC axes. This is easy to do because the matrix of PC coefficients **B** can be used to rotate the data points from the original coordinate system to the PC coordinates $\mathbf{Y} = \mathbf{XB}$, whereas the transpose of **B** performs the reverse rotation $\mathbf{YB}' = \mathbf{X}$. A data set conforming to the CPC model can be obtained by using each group's own PC coefficients for the first rotation, but the coefficient matrix from a CPCA for the reverse rotation. Thus, the modified data in the *i*-th group are $\mathbf{X}^*_i = \mathbf{X}_i\mathbf{B}_i\mathbf{B}'_{CPC}$, where $\mathbf{B}_i$ is the matrix of within-group PC coefficients, and $\mathbf{B}_{CPC}$ is the matrix of CPC coefficients (using the PC scores for each group, $\mathbf{Y}_i = \mathbf{X}_i\mathbf{B}_i$ is equivalent). For this test, 5000 bootstrap replications were performed. For each replication, bootstrap samples were drawn from the modified data of all three groups, and the angles between the PC1 axes of the three groups were computed. The 95% quantiles of the angles are 18.56° between the samples from Alps and from Greece, 16.32° between the Alps and the Pyrenees, and 20.59° between the Pyrenean and the Greek samples. Therefore, the two angles that involve the sample from the Pyrenees seem to indicate a significant difference from the Greek sample and a borderline significance for the difference from the Alps. Because there are three comparisons, however, the chance of rejecting a true null hypothesis by chance at an α of 0.05 is $1 - (1 - \alpha)^3 = 1 - 0.86 = 0.14$. The Bonferroni technique can be used to adjust the significance level for *m* individual comparisons to a new significance threshold at $(1 - \alpha/m) =$ 98.33%. None of the comparisons was significant at this adjusted level (98.33% quantiles are 22.23, 20.63, and 25.48, respectively). Therefore, the data seem to be consistent with a model of an allometric pattern that all three groups have in common.

This preliminary conclusion justifies using common principal components to estimate an allometric pattern for all three groups simultaneously. Again, standard errors were estimated both by using formulas based on large-sample theory (Flury, 1988) and by bootstrapping with 1000 bootstrap iterations. The estimates of the CPC1 coefficients with their parametric and bootstrapped standard errors are

$$\begin{bmatrix} 0.361 \\ 0.510 \\ 0.482 \\ 0.614 \end{bmatrix}, \begin{bmatrix} 0.034 \\ 0.047 \\ 0.024 \\ 0.028 \end{bmatrix}, \text{ and } \begin{bmatrix} 0.039 \\ 0.053 \\ 0.024 \\ 0.031 \end{bmatrix}, \text{ respectively.}$$

Although in a CPC model there is only one set of eigenvectors, which is shared by the groups, each group has its own eigenvalues. The variance taken up by the CPC1 (± parametric and bootstrap standard errors) in each sample is 45.8 (± 11.4, 12.2) for the Alps; 23.6 (± 5.8, 4.7) for Greece, and 30.0 (± 8.0, 7.4) for the Pyrenees. This corresponds to 80.4%, 65.7% and 72.8% of the total variance in the respective samples. These values are only a little lower than corresponding values from the separate one-group PCAs of each sample; one common allometric pattern can account for almost as much of the variation as the PC1 of each group separately. Further support for a common model comes from the log-likelihood ratio test of the CPC model (Flury, 1988; Airoldi and Flury, 1988), which does not show a significant difference from the unrestricted model in which every group has its own PCs ($\chi^2 = 10.55$, df = 12, $p = 0.57$). For the same χ^2 statistic, the bootstrap test with 5000 iterations yielded a 95% quantile of 20.90, which agrees nicely with the corresponding value of 21.03 in statistical tables (the same bootstrap replicates as for the angles between one-group PCs, above). The angles between the CPC1 axis and the one-group PC1 axes indicate that the CPC estimate is a "compromise" between the one-group PCs (5.2° for Alps; 9.4° for Greece; 13.5° for Pyrenees). In the bootstrap test, none of these angles exceeded the 98.33% quantiles (Bonferroni adjustment for $\alpha = 0.05$; quantiles are 9.85°, 18.25°, and 17.75°, respectively). Therefore, it is reasonable to assume that the three groups share the same allometric pattern and that the differences between the PC1 estimates of individual groups are due to sampling error.

To examine whether the differences between groups are simply extensions of within-group variation, I used Burnaby's approach. "Size-invariant" variation between groups was analyzed by a MANOVA of CPC scores, omitting the CPC1. The first eigenvector of the resulting between-groups matrix indicates the axis that contains the most variation among groups, subject to being perpendicular to the within-group allometric axis. Because there are three groups, the matrix of between-groups sums of squares (mean squares would yield equivalent results) has only two non-zero eigenvalues. The first eigenvector of this matrix accounted for 82% of the total between-group variation, and it therefore summarizes most of the differences between samples after adjusting for size.

Fig. 7 is a plot of this axis of group differences versus the CPC1 scores. The Greek specimens differ from the Alpine ones mainly by their higher CPC1 scores, which indicate greater overall size. Therefore, the bugs of the Greek sample can be seen as "scaled-up" versions of their counterparts from the Alps, corresponding to intraspecific scaling along the axis of static allometry (Gould, 1975; although these data deal with subspecies, and not with different species, I do not think it would be helpful to coin a new term). The differences between the samples from the Alps and Pyrenees are largely unrelated to within-group variation. These conclusions are also supported in a more quantitative way by the Mahalanobis distances between the three groups. The Mahalanobis D^2 value, computed from the second to fourth CPC scores, is clearly smaller between the samples from the Alps and from Greece

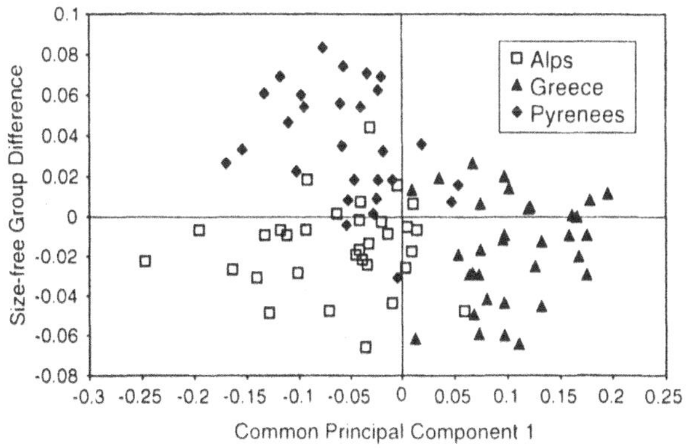

Figure 7. Geographic variation in the waterstrider *Gerris costae.* The first common principal component of the three samples (abscissa) can be interpreted as a measure of overall size. It indicates that Greek specimens are considerably larger than those from the Alps or the Pyrenees. The vertical axis is the PC1 of the between-group matrix from a MANOVA of "size-free" data (the scores of CPC2–CPC4) and summarizes differences between groups independent of within-group allometry. It shows that the samples from the Alps and Pyrenees are separated fairly well, although they are of similar overall size.

($D^2 = 2.38$) than between either of these and the sample from the Pyrenees (Alps versus Pyrenees $D^2 = 6.55$; Greece versus Pyrenees $D^2 = 4.92$). These values, however, also show that scaling does not account for all the difference between the specimens from the Alps and from Greece.

The bootstrap technique has several advantages for assessing statistical accuracy and even for hypothesis testing in morphometric analyses. First, it does not require that the data conform to any particular probability distribution as other techniques do. Nevertheless, assumptions about distributions can be incorporated in the simulations (parametric bootstrap; Efron and Tibshirani, 1993). For the present example, the parametric bootstrap for a multivariate normal CPC model gave results similar to the nonparametric results presented above. Second, the bootstrap can be used for any test statistic, even if its statistical properties are unknown. In the example, I extensively used the angles between PC axes, because angles are particularly intuitive as a measure of overall similarity for allometric vectors. Moreover, the bootstrap can be adapted easily to a variety of experimental designs or hypothesis tests. For the bootstrap test of the CPC model, rotation of the original data was sufficient to generate a data set conforming to the null model. With tools such as the singular value decomposition (e.g., Marcus, 1993), a variety of other null models could be simulated with real data.

These advantages have a cost, however, as the bootstrap technique is based on massive amounts of numerical calculation. As computers become faster and cheaper, this may not be a serious problem except for extremely large or complex data sets. For this example, all bootstrap analyses were done by a personal computer with a 486/50MHz processor, and using SAS/IML software (version 6.08). Although the number of bootstrap replications was substantial, the computation time was moderate. For standard errors in one-group PCA, with 3000 analyses (1000 bootstrap replications for three groups), the entire bootstrap procedure took less than 1.5 minutes. The 1000 bootstrap replicates for CPCA took about 45 minutes, much longer than for ordinary PCA, because the computational procedure for CPCA is more complex. The most effort was required for the bootstrap test: the 5000 iterations, each with three one-group PCAs and CPCA, took a little more than 3 hours. The high number of bootstrap replications for the test was necessary because one is interested in the tails of the test statistic's empirical distribution (Efron and Tibshirani, 1993). The computational effort used for this example shows that the bootstrap is a reasonable option, even with a personal computer and for relatively complex problems.

ALTERNATIVE APPROACHES IN ALLOMETRY

In the preceding sections, I presented allometry in a pragmatic way, extending the familiar logic of two-dimensional scatter plots to a multivariate context. This approach focuses on the patterns of variation by determining amount, dimensionality and direction of morphometric variation in the space of log-transformed variables.

Some readers may have noticed that I only used the words "size" and "shape" in a rather informal way, although they are the central concepts for other approaches in morphometrics (Bookstein et al., 1985; Bookstein, 1989, 1991, 1993; Rohlf, 1990; Rohlf and Marcus, 1993). In studies based on the approach described above, "size" and "shape" may appear in interpretations of the results, but they are not parts of the analyses themselves. The analyses are exploratory or they test simple hypotheses about the structure of variation, such as whether or not the scatter ellipsoids of several groups have major axes that are parallel. Principal components, used in many allometric studies, can "account for" or "take up" variation, but do not "cause" or "explain" it. Interpretation and explanation are extrinsic to the analyses, and they consist of arguments about biological processes producing the observed patterns of variation, e.g., growth dynamics or evolutionary constraints.

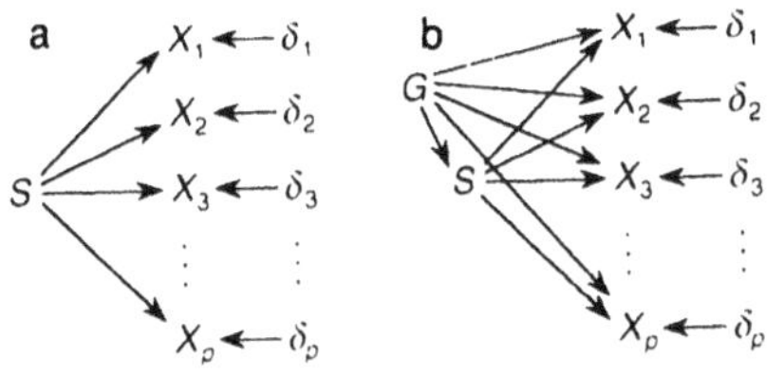

Figure 8. Path models in allometry. (a) Path diagram for simple allometry in a single group. All covariation among variables is caused by a general size factor *S*. Residual variation (δ) is uncorrelated among variables. (b) Allometric variation in two groups. Both groups share the same general size factor and, therefore, also the same within-group covariance structure. Group differences are determined by the group factor *G* in two different ways: directly as group shape differences (arrows from *G* to the variables), or as group size difference via the general size factor (arrow from *G* to *S*).

A very different framework underlies the factor analytic approach, which starts with an explicit model of the origin of variation in measurement variables (see Bookstein et al., 1985; Bookstein, 1989, 1991). Factors are formally included in the analysis as causes of covariation among several morphometric variables; explanation is therefore an intrinsic part of the analysis. A path model is constructed according to biological knowledge and specifies a hypothesis of relationships between factors and observed variables (Wright, 1968; Bookstein et al., 1985; Loehlin, 1987; Zelditch, 1987; Marcus, 1990; Bookstein, 1991). In a path model of allometry, general size is a factor, or latent variable, simultaneously affecting all morphometric measurements and causing the covariance among them (Fig. 8a). As a latent variable, size cannot be measured directly, but it can be estimated. This is usually done using the within-group PC1 (Bookstein et al., 1985; Rohlf and Bookstein, 1987).

Hopkins (1966) proposed a similar factor model of allometry. The observed covariance matrix **S** of log-transformed characters is composed of two parts, $\mathbf{S} = \mathbf{T} + \mathbf{D}$. The matrix **T**, which reflects systematic covariance, is of rank one. Therefore, it has only one principal component, which corresponds to the allometric axis, or to the general size factor in Fig. 8a. The PC coefficients of **T** (which cannot be observed) are proportional to the factor loadings of general size. The matrix **D** stands for the residual variation, which is assumed to be uncorrelated among variables; therefore, **D** is a diagonal matrix. The structure of **D** is crucial for the choice of methods to estimate the parameters of the model. The PC1 of **S** is only appropriate to estimate the size factor if the diagonal elements of **D** are equal (Hopkins, 1966); otherwise, the size factor should be estimated from the off-diagonal elements (see Bookstein, 1991). If all variables are highly correlated to each other and **D** therefore makes only a minor contribution to the total covariance matrix **S**, as in many morphometric data sets, the PC1 is a reasonable estimator.

If there is more than one group of specimens, one or more additional factors for group differences (*G* in Fig. 8b) explain differences between two or more groups (Bookstein et al., 1985; Rohlf and Bookstein, 1987; Bookstein, 1991). Group factors can affect the measurements through general size or in a size-invariant manner. Group size differences (arrow from *G* to *S* in Fig. 8b) cause shifts along the growth axis and correspond to ontogenetic scaling, whereas size-invariant differences (arrows from *G* to the variables in Fig. 8b) correspond to lateral shifts of trajectories (Shea, 1985, 1992; Klingenberg and Spence, 1993). This is the path model for the shearing procedure, which was originally introduced by Humphries et al. (1981) and later reformulated by Bookstein et al. (1985), Rohlf and Bookstein (1987) and Bookstein (1991). The main purpose of this procedure is to obtain factor loadings interpretable as path coefficients (Rohlf and Bookstein, 1987) rather than to obtain an ordination. The geometric basis for the shear is more complex than for Burnaby's procedure because it is not just a rigid rotation (see Humphries et al., 1981; Bookstein et al., 1985; Rohlf and Bookstein, 1987). As a consequence, it does not conserve the spatial relationships among data points and is difficult to use, e.g., to quantify lateral transposition of growth trajectories. Applications of the shear include Bookstein et al. (1985), Strauss (1985), Voss et al. (1990), and Voss and Marcus (1992).

Several studies have used factor analysis to investigate more complex models of correlation or covariance structure among morphometric variables (Bookstein et al., 1985; Zelditch, 1987, 1988; Zelditch and Carmichael, 1989; Marcus, 1990). In addition to general size, these models include factors explaining joint variation in groups of variables that are developmentally or functionally related. An alternative procedure (Cowley and Atchley, 1990; Paulsen and Nijhout, 1993) uses hypotheses about relations among characters to predict the pattern of a correlation matrix, and then compares these to observed correlation matrices by means of randomization tests (Cheverud et al., 1989; Manly, 1991). These models, however, are beyond the scope of allometry.

Yet another, more general method of characterizing size was introduced by Mosimann (1970), when he defined standard size variables. Any positive, real-valued function $G(\mathbf{x})$ of a vector of measurements $\mathbf{x}$ is a standard size variable, if multiplication of each measurement by a constant a results in an a-fold value of the size function, i.e., $G(a\mathbf{x}) = aG(\mathbf{x})$. This condition ensures that the variable scales as a linear dimension. A standard size variable transformed to logarithms is called a log-size variable. A class of log-size variables important for multivariate allometry is defined as linear combinations of log-transformed measurements, i.e., $\log G(\mathbf{x}) = \Sigma b_i (\log x_i)$, with $\Sigma b_i = 1$ (Mosimann and James, 1979; Darroch and Mosimann, 1985). Rescaling the PC coefficients for ontogenetic allometry so that they sum up to unity yields a log-size variable that indicates each specimen's position along the growth trajectory (Klingenberg and Zimmermann, 1992b; Klingenberg and Spence, 1993). Similar measures of size, but without rescaling, were used by Creighton and Strauss (1986), Strauss (1990b) and Voss and Marcus (1992), among others.

If size alone is of interest, the choice of a size measure often does not matter very much. In many morphometric data sets, the variables and the size measures derived from them are highly correlated among one another. Hence, different size measures may produce different scaling factors, due to allometry, but they will yield basically the same ordering from small to large specimens and similar size differences between them. In studies of allometry based on either PCA or factor analysis, "shape" often does not appear explicitly at all, or if it does, it is used in a sense very different from everyday language (e.g., Bookstein, 1989). In those cases, "shape" is usually (though rarely explicitly) defined as "everything that is not size." It is through this notion of "shape" that the choice of a size measure matters for morphometric studies: because "size" often takes up a large fraction of the total variation in a data set, relatively small changes in the size measure produce proportionally large changes in what remains after "size" is removed.

In multivariate allometry based on PCA, the second and subsequent PCs often are interpreted as "shape scores." Nevertheless, they do not reflect a geometric concept of shape. If two specimens, differing in size, have the same "shape scores," they can be interpreted as geometrically similar only if the corresponding size vector is isometric; otherwise, there are allometric changes in shape (see also Bookstein, 1989). To separate size and geometric shape, Somers (1986, 1989) proposed a size-constrained version of PCA in which variation in the direction of an isometric vector is removed first. Unfortunately, Somers used the correlation instead of the covariance matrix, thereby removing not only isometric size from the data but also all allometric variation (see above). As an alternative procedure to achieve Somers's original objective, Burnaby's procedure can be used to eliminate a vector representing isometric variation, which is mathematically equivalent to performing a PCA on the covariance matrix of doubly centered data (Somers, 1989) or on the "principal components of shape" proposed by Darroch and Mosimann (1985; see also Jungers et al., 1988). Notice, however, that removing isometric size adjusts for only variability in overall size itself, but not for any size- or age-related shape variation. For example, although Barbie dolls are smaller than are many other dolls clearly representing infants, it is easily recognizable that Barbie dolls are modeled after human adults.

Mosimann (1970) presented a definition of a shape space based on geometric similarity. Each vector of measurements $\mathbf{x}$, divided by a standard size variable $G(\mathbf{x})$, constitutes a measure of shape. Darroch and Mosimann (1985) developed principal components and canonical variates for the space of these shape measures and applied them to two examples. Further applications are found in Mosimann and James (1979) and Jungers et al. (1988). In this framework, allometry exists if variations in size and shape are associated; isometry means that variation in size and shape are statistically independent. Mosimann's theory of size and shape links some aspects of multivariate allometry to the landmark-based methods of geometric morphometrics (Bookstein, 1989, 1991, 1993). The underlying concept of allometry, however, differs fundamentally from the other frameworks presented here, as it abandons the straight-line relationship among log-transformed variables, which is the basis of allometry as devised by Huxley (1932). In this point, Mosimann's concept is closer to the much broader notion of allometry adopted by Gould (1966), who characterized it as "the study of size and its consequences."

Whereas Mosimann's (1970) approach, although based on considerations of geometric similarity, still uses vectors of length measurements, geometric morphometrics goes one step further and analyzes shape as geometric configurations of morphological landmarks (e.g., Rohlf, 1990; Bookstein, 1991, 1993; Rohlf and Marcus, 1993; other chapters in this volume). The strong emphasis on shape in geometric morphometrics is reflected in two recent definitions of morphometrics, characterizing it as "the quantitative description, analysis and interpretation of shape and shape variation in biology" (Rohlf 1990) and as "the geometrically reified description of effects on geometric shape" (Bookstein, 1993) without even mentioning size. Clearly, geometric morphometrics presents a dramatically different framework for allometry. Variation in size is removed from the data (by the two-point registration for Bookstein's shape coordinates or by standardizing for centroid size; Bookstein, 1991; Rohlf, 1993), and shape changes alone are included in the analysis. Allometry can be assessed by combining the results from shape analysis with additional information, either directly by nonlinear regression of "shape scores" (relative warps, shape coordinates, or Procrustes residuals) on a measure of size (Bookstein, 1991; Walker, 1993), or by subdividing specimens into size classes (MacLeod and Kitchell, 1990) or age groups (Reilly, 1990; Bookstein, 1991; Zelditch et al., 1992) and comparing their mean shapes. Allometric variation over an extended size range, as it occurs in many growth studies, often leads to highly nonlinear trajectories (Bookstein, 1991; Zelditch et al., 1992; Walker, 1993).

The choice of methods for a particular study depends on what questions a study is supposed to answer. The results obtained from analyses using the geometric methods can be interpreted directly in terms of shape. Here, "shape" is used in its intuitive sense, meaning a geometric configuration. The disadvantage of these methods, however, is the complexity of allometric relations. For example, a procedure analogous to Burnaby's technique to adjust for shape differences due to allometric growth would have to use nonlinear regression of shape measures on overall size. On the other hand, the results of analyses that use distance data are more difficult to describe in everyday language; graphical displays, like Fig. 7 are abstractions rather than pictures of real organisms. If the notion of "shape" is used at all, it denotes the relative sizes of parts of the organism. The advantage of methods using log-transformed distances is that these data often fit linear models due to their relationship to growth dynamics, which was used by Huxley (1932) to justify his formula for simple allometry.

From an extreme point of view, the configuration of morphological landmarks of an organism could be considered as merely an epiphenomenon of the growth processes affecting the tissues between the landmarks. Ideally, therefore, morphometric methods should be based on models of biological processes rather than geometrical or statistical considerations (e.g., Sattler, 1992). Although this view is correct in principle, our knowledge of the mechanisms involved in developmental processes is incomplete even for simple and well-studied experimental systems (e.g., Atchley and Hall, 1991). For less well-known organisms and for more complex problems, such as evolutionary comparisons, landmark configurations or length measurements must be used as the basis for our understanding of organismic form.

ACKNOWLEDGMENTS

I thank Michel Baylac, Marco Corti, Bernard Flury, Les Marcus, Rich Palmer, John Spence, Manfred Zimmermann and various participants of the Advanced Study Institute for discussions about multivariate allometry and constructive comments on earlier versions of this paper. Financial support was provided by NATO, a University of Alberta Ph.D. Scholarship, an Izaak Walton Killam Scholarship, and a Mary Louise Imrie Graduate Award (Faculty of Graduate Studies and Research, and the Vice-President [Research] of the University of Alberta) to the author and by grants from the Natural Sciences and Engineering Research Council of Canada to John R. Spence.

REFERENCES

Airoldi, J.-P., and B. K. Flury. 1988. An application of common principal component analysis to cranial morphometry of *Microtus californicus* and *M. ochrogaster* (Mammalia, Rodentia). Journal of Zoology (London) 216:21–36.

Anderson, T. W. 1963. Asymptotic theory for principal component analysis. Annals of Mathematical Statistics 34:122–148.

Anstey, R. L. 1987. Astogeny and phylogeny: Evolutionary heterochrony in Paleozoic bryozoans. Paleobiology 13:20–43.

Arnold, S. J. 1992. Constraints on phenotypic evolution. American Naturalist 140:S85–S107.

Atchley, W. R., and B. K. Hall. 1991. A model for development and evolution of complex morphological structures. Biological Reviews 66:101–157.

Bernays, E. A. 1986. Diet-induced head allometry among foliage-chewing insects and its importance for graminivores. Science 231:495–497.

Blackstone, N. W. 1987. Specific growth rates of parts in a hermit crab *Pagurus longicarpus*: A reductionist approach to the study of allometry. Journal of Zoology (London) 211:531–545.

Boag, P. T. 1984. Growth and allometry of external morphology in Darwin's finches (*Geospiza*) on Isla Daphne Major, Galapagos. Journal of Zoology (London) 204:413–441.

Boitard, M., J. Lefebvre, and M. Solignac. 1982. Analyse en composantes principales de la variabilité de taille, de croissance et de conformation des espèces du complexe *Jaera albifrons* (Crustacés Isopodes). Cahiers de Biologie Marine 23:115–142.

Bookstein, F. L. 1989. "Size and shape": A comment on semantics. Systematic Zoology 38:173–180.

Bookstein, F. L. 1991. Morphometric tools for landmark data: Geometry and biology. Cambridge University Press: Cambridge.

Bookstein, F. L. 1993. A brief history of the morphometric synthesis. Pages 15–40 *in* L. F. Marcus, E. Bello, and A. García-Valdecasas, (eds.), Contributions to morphometrics. Monografias del Museo Nacional de Ciencias Naturales 8, Madrid.

Bookstein, F. L., B. Chernoff, R. L. Elder, J. M. Humphries, Jr., G. R. Smith, and R. E. Strauss. 1985. Morphometrics in Evolutionary Biology: The geometry of size and shape change, with examples from fishes. Academy of Natural Sciences of Philadelphia Special Publication 15.

Bryant, P. 1984. Geometry, statistics, probability: Variations on a common theme. American Statistician 38:38–48.

Burnaby, T. P. 1966. Growth-invariant discriminant functions and generalized distances. Biometrics 22:96–110.

Buss, L. W., and N. W. Blackstone. 1991. An experimental exploration of Waddington's epigenetic landscape. Philosophical Transactions of the Royal Society of London B 332:49–58.

Cane, W. P. 1993. The ontogeny of postcranial integration in the common tern, *Sterna hirundo*. Evolution 47:1138–1151.

Cheverud, J. M. 1982. Relationships among ontogenetic, static, and evolutionary allometry. American Journal of Physical Anthropology 59:139–149.

Cheverud, J. M., G. P. Wagner, and M. M. Dow. 1989. Methods for the comparative analysis of variation patterns. Systematic Zoology 38:201–213.

Cock, A. G. 1966. Genetical aspects of metrical growth and form in animals. Quarterly Review of Biology 41:131–190.

Corner, B. D., and J. T. Richtsmeier. 1991. Morphometric analysis of craniofacial growth in *Cebus apella*. American Journal of Physical Anthropology 84:323–342.

Corti, M., R. S. Thorpe, L. Sola, V. Sbordoni, and S. Cataudella. 1988. Multivariate morphometrics in aquaculture: A case study of six stocks of the common carp (*Cyprinus carpio*) from Italy. Canadian Journal of Fisheries and Aquatic Sciences 45:1548–1554.

Cowley, D. E., and W. R. Atchley. 1990. Development and quantitative genetics of correlation structure among body parts of *Drosophila melanogaster*. American Naturalist 135:242–268.

Cowley, D. E., and W. R. Atchley. 1992. Quantitative genetic models for development, epigenetic selection, and phenotypic evolution. Evolution 46:495–518.

Creighton, G. K., and R. E. Strauss. 1986. Comparative patterns of growth and development in cricetine rodents and the evolution of ontogeny. Evolution 40:94–106.

Cuzin-Roudy, J. 1975. Étude de la variabilité et de l'allométrie de taille chez *Notonecta maculata* Fabricius (Insectes, hétéroptères), par les méthodes classiques et par la méthode des composantes principales. Archives de Zoologie Expérimentale et Générale 116:173–227.

Cuzin-Roudy, J., and Ph. Laval. 1975. A canonical discriminant analysis of post-embryonic development in *Notonecta maculata* Fabricius (Insecta: Heteroptera). Growth 39:251–280.

Darroch, J. N., and J. E. Mosimann. 1985. Canonical and principal components of shape. Biometrika 72:241–252.

Daudin, J. J., C. Duby, and P. Trecourt. 1988. Stability of principal component analysis studied by the bootstrap method. Statistics 19:241–258.

Davies, R. G., and V. Brown. 1972. A multivariate analysis of postembryonic growth in two species of *Ectobius* (Dictyoptera: Blattidae). Journal of Zoology (London) 168:51–79.

Diaconis, P., and B. Efron. 1983. Computer-intensive methods in statistics. Scientific American 248 (May):116–130.

Efron, B., and G. Gong. 1983. A leisurely look at the bootstrap, the jackknife, and cross-validation. American Statistician 37:36–48.

Efron, B., and R. Tibshirani. 1986. Bootstrap methods for standard errors, confidence intervals, and other measures of statistical accuracy. Statistical Science 1:54–77.

Efron, B., and R. J. Tibshirani. 1993. An introduction to the bootstrap. Chapman and Hall: New York.

Felsenstein, J. 1985. Phylogenies and the comparative method. American Naturalist 125:1–15.

Flury, B. 1988. Common principal components and related multivariate models. Wiley: New York.

Flury, B., and H. Riedwyl. 1988. Multivariate statistics: A practical approach. Chapman and Hall: London.

Garland, T., Jr., P. H. Harvey, and A. R. Ives. 1992. Procedures for the analysis of comparative data using phylogenetically independent contrasts. Systematic Biology 41:18–32.

Gibson, A. R., A. J. Baker, and A. Moeed. 1984. Morphometric variation in introduced populations of the common myna (*Acridotheres tristis*): An application of the jackknife to principal component analysis. Systematic Zoology 33:408–421.

Gould, S. J. 1966. Allometry and size in ontogeny and phylogeny. Biological Reviews 41:587–640.

Gould, S. J. 1975. Allometry in primates, with emphasis on scaling and the evolution of the brain. Contributions to Primatology 5:244–292.

Gould, S. J. 1989. A developmental constraint in *Cerion*, with comments on the definition and interpretation of constraint in evolution. Evolution 43:516–539.

Gower, J. C. 1976. Growth-free canonical variates and generalized inverses. Bulletin of the Geological Institutions of the University of Uppsala, N. S. 7:1–10.

Günther, B. 1975. Dimensional analysis and theory of biological similarity. Physiological Reviews 55:659–699.

Hills, M. 1982. Bivariate versus multivariate allometry: A note on a paper by Jungers and German. American Journal of Physical Anthropology 59:321–322.

Hopkins, J. W. 1966. Some considerations in multivariate allometry. Biometrics 22:747–760.

Humphries, J. M., F. L. Bookstein, B. Chernoff, G. R. Smith, R. L. Elder, and S. G. Poss. 1981. Multivariate discrimination by shape in relation to size. Systematic Zoology 30:291–308.

Huxley, J. S. 1932. Problems of relative growth. Methuen: London. Reprinted 1972, Dover Publications: New York.

Huxley, J. S., and G. Teissier. 1936. Terminology of relative growth. Nature 137:780–781.

Jackson, J. E. 1990. A user's guide to principal components. Wiley: New York.

Jeune, B., and R. Sattler. 1992. Multivariate analysis in process morphology of plants. Journal of Theoretical Biology 156:147–167.

Jobson, J. D. 1992. Applied multivariate data analysis. Volume II: Categorical and multivariate methods. Springer-Verlag: New York.

Johnson, R. A., and D. W. Wichern. 1988. Applied multivariate statistical analysis. 2nd edition. Prentice-Hall: Englewood Cliffs, New Jersey.

Jolicoeur, P. 1963. The multivariate generalization of the allometry equation. Biometrics 19:497–499.

Jolliffe, I. T. 1986. Principal component analysis. Springer-Verlag: New York.

Jones, C. S. 1992. Comparative ontogeny of a wild cucurbit and its derived cultivar. Evolution 46:1827–1847.

Jones, C. S. 1993. Heterochrony and heteroblastic leaf development in two subspecies of *Cucurbita argyrosperma* (Cucurbitaceae). American Journal of Botany 80:778–795.

Jungers, W. L., and R. Z. German. 1981. Ontogenetic and interspecific skeletal allometry in nonhuman primates: Bivariate versus multivariate analysis. American Journal of Physical Anthropology 55:195–202.

Jungers, W. L., T. M. Cole, III, and D. W. Owsley. 1988. Multivariate analysis of relative growth in the limb bones of Arikara Indians. Growth, Development and Aging 52:103–107.

Kampny, C. M., T. A. Dickinson, and N. G. Dengler. 1993. Quantitative comparison of floral development in *Veronica chamaedrys* and *Veronicastrum virginicum* (Scrophulariaceae). American Journal of Botany 80:449–460.

Katz, M. J. 1980. Allometry formula: A cellular model. Growth 44:89–96.

Klingenberg, C. P. 1992. Multivariate morphometrics of geographic variation of *Gerris costae* (Heteroptera: Gerridae) in Europe. Revue Suisse de Zoologie 99:11–30.

Klingenberg, C. P., and R. Froese. 1991. A multivariate comparison of allometric growth patterns. Systematic Zoology 40:410–419.

Klingenberg, C. P., and J. R. Spence. 1993. Heterochrony and allometry: Lessons from the waterstrider genus *Limnoporus*. Evolution 47:1834-1853.

Klingenberg, C. P., and M. Zimmermann. 1992a. Static, ontogenetic, and evolutionary allometry: A multivariate comparison in nine species of water striders. American Naturalist 140:601–620.

Klingenberg, C. P., and M. Zimmermann. 1992b. Dyar's rule and multivariate allometric growth in nine species of waterstriders (Heteroptera: Gerridae). Journal of Zoology (London) 227:453–464.

LaBarbera, M. 1989. Analyzing body size as a factor in ecology and evolution. Annual Review of Ecology and Systematics 20:97–117.

Laird, A. K. 1965. Dynamics of relative growth. Growth 29:249–263.

Laird, A. K., A. D. Barton, and S. A. Tyler. 1968. Growth and time: An interpretation of allometry. Growth 32:347–354.

Leamy, L. and W. Atchley. 1984a. Static and evolutionary allometry of osteometric traits in selected lines of rats. Evolution 38:47–54.

Leamy, L., and W. R. Atchley. 1984b. Morphometric integration in the rat (*Rattus* sp.) scapula. Journal of Zoology (London) 202:43–56.

Leamy, L., and D. Bradley. 1982. Static and growth allometry of morphometric traits in randombred house mice. Evolution 36:1200–1212.

Lessa, E. P., and J. L. Patton. 1989. Structural constraints, recurrent shapes, and allometry in pocket gophers (genus *Thomomys*). Biological Journal of the Linnean Society 36:349–363.

Livezey, B. C. 1989. Morphometric patterns in Recent and fossil penguins (Aves, Sphenisciformes). Journal of Zoology (London) 219:269–307.

Loehlin, J. C. 1987. Latent variable models: An introduction to factor, path, and structural Analysis. Lawrence Erlbaum Associates: Hillsdale, New Jersey.

MacLeod, N., and J. A. Kitchell. 1990. Morphometrics and evolutionary inference: A case study involving ontogenetic and developmental aspects of foraminiferal evolution. Pages 283–299 *in* F. J. Rohlf and

F. L. Bookstein, (eds.), Proceedings of the Michigan morphometrics workshop. University of Michigan Museum of Zoology Special Publication No. 2: Ann Arbor.

Maddison, W. P., and D. R. Maddison. 1992. MacClade: Analysis of phylogeny and character evolution. Sinauer Associates: Sunderland, Massachusetts.

Manly, B. F. J. 1991. Randomization and Monte Carlo methods in biology. Chapman and Hall: London.

Marcus, L. F. 1990. Traditional morphometrics. Pages 77–122 *in* F. J. Rohlf and F. L. Bookstein, (eds.), Proceedings of the Michigan morphometrics workshop. University of Michigan Museum of Zoology Special Publication No. 2: Ann Arbor.

Marcus, L. F. 1993. Some aspects of multivariate statistics for morphometrics. Pages 95–130 *in* L. F. Marcus, E. Bello, and A. García-Valdecasas, (eds.), Contributions to morphometrics. Monografias del Museo Nacional de Ciencias Naturales 8, Madrid.

Martins, E. P., and T. Garland, Jr. 1991. Phylogenetic analyses of the correlated evolution of continuous characters: A simulation study. Evolution 45:534–557.

Maynard Smith, J., R. Burian, S. Kauffman, P. Alberch, J. Campbell, B. Goodwin, R. Lande, and L. Wolpert. 1985. Developmental constraints and evolution. Quarterly Review of Biology 60:265–287.

McGillivray, W. B. 1985. Size, sexual size dimorphism, and their measurement in Great Horned Owls in Alberta. Canadian Journal of Zoology 63:2364–2372.

McLellan, T. 1993. The roles of heterochrony and heteroblasty in the diversification of leaf shapes in *Begonia dregei* (Begoniaceae). American Journal of Botany 80:796–804.

Meyer, A. 1990. Morphometrics and allometry in the trophically polymorphic cichlid fish, *Cichlastoma citrinellum*: Alternative adaptations and ontogenetic changes in shape. Journal of Zoology (London) 221:237–260.

Mosimann, J. E. 1970. Size allometry: Size and shape variables with characterizations of the lognormal and generalized gamma distributions. Journal of the American Statistical Association 65:930–945.

Mosimann, J. E., and F. C. James. 1979. New statistical methods for allometry with application to Florida red-winged blackbirds. Evolution 33:444–459.

Owen, J. G., and M. A. Chmielewski. 1985. On canonical variates analysis and the construction of confidence ellipses in systematic studies. Systematic Zoology 34:366–374.

Pagel, M. D., and P. H. Harvey. 1988. Recent developments in the analysis of comparative data. Quarterly Review of Biology 63:413–440.

Pandolfi, J. M. 1988. Heterochrony in colonial marine animals. Pages 135–158 *in* M. L. McKinney, (ed.), Heterochrony in evolution: A multidisciplinary approach. Plenum Press: New York.

Patton, J. L., and P. V. Brylski. 1987. Pocket gophers in alfalfa fields: Causes and consequences of habitat-related body size variation. American Naturalist 130:493–506.

Paulsen, S. M., and H. F. Nijhout. 1993. Phenotypic correlation structure among elements of the color pattern in *Precis coenia* (Lepidoptera: Nymphalidae). Evolution 47:593–618.

Pearson, K. 1901. On lines and planes of closest fit to systems of points in space. Philosophical Magazine and Journal of Science, 6th series 2:559–572.

Pimentel, R. A. 1979. Morphometrics: The multivariate analysis of biological data. Kendall/Hunt: Dubuque, Iowa.

Reeve, E. C. R., and J. S. Huxley. 1945. Some problems in the study of allometric growth. Pages 121–156 *in* W. E. Le Gros Clark and P. B. Medawar (eds.), Essays on growth and form presented to D'Arcy Wentworth Thompson. Clarendon Press: Oxford.

Reilly, S. M. 1990. Comparative ontogeny of cranial shape in salamanders using resistant fit theta rho analysis. Pages 311–321 *in*. F. J. Rohlf and F. L. Bookstein (eds.), Proceedings of the Michigan morphometrics workshop. University of Michigan Museum of Zoology, Special Publication 2.

Reiss, M. J. 1989. The allometry of growth and reproduction. Cambridge University Press: Cambridge, U.K.

Reyment, R. A. 1991. Multidimensional palaeobiology. Pergamon Press: Oxford.

Reyment, R. A., and C. F. Banfield. 1976. Growth-free canonical variates applied to fossil foraminifers. Bulletin of the Geological Institutions of the University of Uppsala, N. S. 7:11–21.

Reyment, R. A., R. E. Blackith, and N. A. Campbell. 1984. Multivariate Morphometrics. 2nd edition. Academic Press, London.

Riska, B. 1981. Morphological variation in the horseshoe crab *Limulus polyphemus*. Evolution 35:647–658.

Riska, B. 1986. Some models for development, growth, and morphometric correlation. Evolution 40:1303–1311.

Riska, B. 1989. Composite traits, selection response, and evolution. Evolution 43:1172–1191.

Rohlf, F. J. 1990. Morphometrics. Annual Review of Ecology and Systematics 21:299–316.

Rohlf, F. J. 1993. Relative warp analysis and an example of its application to mosquito wings. Pages 131–159 *in* L. F. Marcus, E. Bello, and A. García-Valdecasas (eds.), Contributions to morphometrics. Monografías del Museo Nacional de Ciencias Naturales 8, Madrid.

Rohlf, F. J., and F. L. Bookstein. 1987. A comment on shearing as a method for "size correction." Systematic Zoology 36:356–367.

Rohlf, F. J., and L. F. Marcus. 1993. A revolution in morphometrics. Trends in Ecology and Evolution 8:129–132.

Sattler, R. 1992. Process morphology: Structural dynamics in development and evolution. Canadian Journal of Botany 70:708–714.

Shea, B. T. 1985. Bivariate and multivariate growth allometry: Statistical and biological considerations. Journal of Zoology (London) 206:367–390.

Shea, B. T. 1992. Developmental perspective on size change and allometry in evolution. Evolutionary Anthropology 1:125–134.

Shea, B. T., R. E. Hammer, R. L. Brinster, and M. R. Ravosa. 1990. Relative growth of the skull and postcranium in giant transgenic mice. Genetical Research, Cambridge 56:21–34.

Sinervo, B. 1993. The effect of offspring size on physiology and life history. BioScience 43:210–218.

Smith, L. D., and A. R. Palmer. 1994. Effects of manipulated diet on size and performance of brachyuran crab claws. Science 264:710–712.

Smith, R. J. 1993. Logarithmic transformation bias in allometry. American Journal of Physical Anthropology 90:215–228.

Solignac, M., M.-L. Cariou, and M. Wimitzki. 1990. Variability, specificity and evolution of growth gradients in the species complex *Jaera albifrons* (Isopoda, Asellota). Crustaceana 59:121–145.

Somers, K. M. 1986. Multivariate allometry and removal of size with principal components analysis. Systematic Zoology 35:359–368.

Somers, K. M. 1989. Allometry, isometry and shape in principal component analysis. Systematic Zoology 38:169–173.

Sprent, P. 1972. The mathematics of size and shape. Biometrics 28:23–37.

Stauffer, D. F., E. O. Garton, and R. K. Steinhorst. 1985. A comparison of principal components from real and random data. Ecology 66:1693–1698.

Strauss, R. E. 1985. Evolutionary allometry and variation in body form in the South American catfish genus *Corydoras* (Callichthyidae). Systematic Zoology 34:381–396.

Strauss, R. E. 1990a. Patterns of quantitative variation in lepidopteran wing morphology: The convergent groups Heliconiinae and Ithomiinae (Papilionoidea: Nymphalidae). Evolution 44:86–103.

Strauss, R. E. 1990b. Heterochronic variation in the developmental timing of cranial ossifications in poecilid fishes (Cyprinodontiformes). Evolution 44:1558–1567.

Strauss, R. E., and F. L. Bookstein. 1982. The truss: Body form reconstructions in morphometrics. Systematic Zoology 31:113–135.

Teissier, G. 1955. Allométrie de taille et variabilité chez *Maïa squinado*. Archives de Zoologie Expérimentale et Générale 92:221–264.

Thomas, R. D. K., and W.-E. Reif. 1993. The skeleton space: A finite set of organic designs. Evolution 47:341-360.

Thorpe, R. S. 1983. A review of the numerical methods for recognizing and analyzing racial differentiation. Pages 404–423 *in* J. Felsenstein, (ed.), Numerical taxonomy. Springer-Verlag: Berlin.

Thorpe, R. S., and M. Baez. 1987. Geographic variation within an island: Univariate and multivariate contouring of scalation, size, and shape of the lizard *Gallotia galloti*. Evolution 41:256–268.

Voss, R. S., L. F. Marcus, and P. Escalante. 1990. Morphological evolution in muroid rodents I. Conservative patterns of craniometric covariance and their ontogenetic basis in the neotropical genus *Zygodontomys*. Evolution 44:1568–1587.

Voss, R. S., and L. F. Marcus. 1992. Morphological evolution in muroid rodents II. Craniometric factor divergence in seven neotropical genera, with experimental results from *Zygodontomys*. Evolution 46:1918–1934.

Walker, J. A. 1993. Ontogenetic allometry of threespine stickleback body form using landmark-based morphometrics. Pages 193–214 *in* L. F. Marcus, E. Bello, and A. García-Valdecasas, (eds.), Contributions to morphometrics. Monografías del Museo Nacional de Ciencias Naturales 8, Madrid.

Wheeler, D. E. 1991. The developmental basis of worker caste polymorphism in ants. American Naturalist 138:1218–1238.

Wiig, Ø. 1985. Multivariate variation in feral American mink (*Mustela vison*) from Southern Norway. Journal of Zoology (London) 206:441–452.

Wright, S. 1968. Evolution and the genetics of populations. Volume 1. Genetic and biometric foundations. University of Chicago Press: Chicago.

Zelditch, M. L. 1987. Evaluating models of developmental integration in the laboratory rat using confirmatory factor analysis. Systematic Zoology 36:368–380.

Zelditch, M. L. 1988. Ontogenetic variation in patterns of phenotypic integration in the laboratory rat. Evolution 42:28–41.

Zelditch, M. L., F. L. Bookstein, and B. L. Lundrigan. 1992. Ontogeny of integrated skull growth in the cotton rat *Sigmodon fulviventer*. Evolution 46:1164–1180.

Zelditch, M. L., and A. C. Carmichael. 1989. Ontogenetic variation in patterns of developmental and functional integration in skulls of *Sigmodon fulviventer*. Evolution 43:814–824.

APPENDIX

Table 1. Morphometric data (raw values, in millimeters) for the example in the text. Four measurements were made on male *Gerris costae* from three different locations in Europe (Klingenberg, 1992). The measurements are the lengths of the thorax (Tho), the first antennal segment (Ant), and the femora of the middle and hind legs (MF and HF), corresponding to numbers 35, 43, 49, and 53 of Klingenberg (1992)

Alps				Greece				Pyrenees			
Tho	Ant	MF	HF	Tho	Ant	MF	HF	Tho	Ant	MF	HF
5.11	1.51	6.07	6.13	5.74	1.72	6.65	7.05	5.15	1.45	6.05	6.36
5.10	1.53	6.15	6.25	5.40	1.59	6.39	6.76	5.05	1.46	5.86	6.11
5.08	1.60	6.28	6.41	5.34	1.67	6.47	6.79	4.97	1.52	6.12	6.35
5.16	1.55	6.18	6.3	5.51	1.72	6.65	7.05	5.03	1.43	6.01	6.21
4.78	1.54	5.79	5.86	5.44	1.66	6.46	6.76	5.34	1.53	6.13	6.40
4.64	1.41	5.55	5.46	5.49	1.69	6.27	6.48	4.99	1.41	5.81	5.89
5.27	1.60	6.20	6.29	5.64	1.75	6.75	6.95	5.31	1.59	6.39	6.67
4.71	1.43	5.74	5.67	5.45	1.71	6.70	7.12	5.29	1.54	6.23	6.62
5.42	1.55	6.28	6.35	5.36	1.71	6.84	7.12	5.33	1.59	6.35	6.60
4.91	1.48	5.70	5.71	5.44	1.72	6.31	6.71	5.18	1.46	6.25	6.54
5.03	1.48	6.07	5.82	5.51	1.69	6.34	6.55	5.22	1.52	6.08	6.49
5.11	1.48	6.29	6.38	5.71	1.69	6.44	6.88	5.23	1.47	6.02	6.46
5.01	1.56	6.10	6.23	5.42	1.65	6.44	6.59	5.16	1.51	6.06	6.20
5.25	1.56	6.06	6.13	5.35	1.67	6.36	6.61	5.07	1.42	5.89	6.27
5.15	1.57	6.03	6.21	5.45	1.57	6.18	6.62	5.13	1.54	5.91	6.25
5.18	1.57	6.20	6.3	5.28	1.56	6.20	6.49	5.14	1.55	6.06	6.35
4.96	1.53	5.92	5.93	5.43	1.66	6.28	6.58	5.26	1.47	6.18	6.49
5.02	1.48	5.87	6.17	5.44	1.63	6.28	6.52	4.96	1.40	5.81	6.15
4.82	1.51	6.02	6.08	5.48	1.75	6.44	6.61	5.10	1.42	5.97	6.07
5.22	1.62	6.13	6.21	5.51	1.71	6.17	6.52	5.17	1.54	6.11	6.35
5.16	1.63	5.98	6.07	5.32	1.67	6.52	7.02	5.06	1.48	6.02	6.35
5.34	1.58	6.19	6.41	5.58	1.62	6.51	6.80	5.08	1.53	6.22	6.38
5.18	1.55	6.04	6.14	5.46	1.61	6.45	6.66	5.08	1.39	5.92	6.15
4.94	1.59	6.36	6.46	5.44	1.66	6.59	6.92	5.25	1.43	6.08	6.39
5.05	1.48	5.83	5.92	5.63	1.61	6.46	6.80	5.06	1.46	6.18	6.57
4.87	1.48	5.79	5.93	5.61	1.70	6.75	7.16	5.21	1.61	6.05	6.34
5.10	1.55	6.28	6.44	5.42	1.74	6.30	6.65	5.05	1.40	6.19	6.33
5.16	1.69	6.42	6.50	5.67	1.71	6.75	7.27	4.96	1.40	5.75	5.82
5.13	1.53	6.15	6.20	5.15	1.66	6.24	6.24				
4.82	1.50	5.85	5.79	5.36	1.75	6.52	6.86				
4.97	1.58	5.90	6.06	5.57	1.65	6.29	6.48				
5.02	1.56	6.17	6.25	5.62	1.74	6.57	7.06				
				5.53	1.71	6.45	6.86				

PART I

Section 1. Data Acquisition

INTRODUCTION TO DATA ACQUISITION

Leslie F. Marcus

Data acquisition developments have gone hand in hand with new methods of morphometric analysis. The articles in this section by Becerra, Dean, Demeter et al., and García-Valdecasas address new needs and developments in data acquisition. The article by Carpenter et al. provides the explanation for software to convert distance data to coordinate data where possible.

Equipment for three-dimensional (3D) data acquisition is still expensive and not as widely available as we would like. This will hopefully change in the near future. Corti et al. (this volume) describes one home-made device for 3D coordinate acquisition from rodent jaws, and I am aware of some others for larger objects—though these devices have not been widely discussed or reviewed.

David Dean gives a comprehensive review of available 3D data acquisition and printing hardware and software—giving costs, specifications and evaluations where he has experience or, alternatively, information from the literature or supplied by manufacturers. There is an increasing number of handheld and automated acquisition instruments, and some of their costs are within reach of the biological community. The Polhemus 3SPACE and newer 3DRAW have been and are being widely used by biologists for larger objects and the REFLEX microscope for smaller ones. Hopefully, prices will come down for some of the more expensive automated 3D equipment and instruments such as laser scanners, and they will be more widely available in the not-too-distant future. Most of the development of this equipment is driven by medical and industrial applications, as well as interest in virtual reality, so that some of us should keep track of those areas and inform others through the Internet (see Appendix III).

For two-dimensional data acquisition we have three contributions describing various aspects of video image analysis. The article by García-Valdecasas is an update of the contribution by Becerra et al. (1993) in the Black book. We are given further information about features of video systems in comparison with photography and a review of the latest high-resolution digital cameras recently available. As García-Valdecasas points out, these new cameras are expensive and more attuned to specimen archives than to morphometrics, where the previous generation of medium-resolution systems was up to the job (see Becerra et al., 1993).

In separate articles Becerra and Demeter et al. describe video image analysis and data acquisition systems developed with the morphometrician in mind. Each system has new and useful features fully described in their respective articles. Imagina, written by Jose Becerra, is freeware (available on the Internet) that does not require a separate monitor for real-time data acquisition as do two of the most popular packages, MeasurementTV and MorphoSys. It can also handle gray image files in TIFF format collected from anywhere. One very good feature of Imagina is the ability to pan when dealing with images that are larger than can be displayed in VGA or super VGA modes. Another valuable feature is the ability to export landmark and distance data into formats directly compatible with morphometric software. For real-time data acquisition, Imagina has restrictive hardware requirements, because it captures images using only a monochrome AT-OFG frame grabber (European version only) from Imaging Technology. Becerra plans to generalize the program so that other frame grabbers may be used in the future. One should contact him for new developments or watch MORPHMET (see Appendix III).

Demeter et al. have developed Budapest Video, because the high cost of adequate frame grabbers in Hungary necessitated use of a locally available board. Along with the board, an innovative image capture, data acquisition and analysis package was developed by the Hungarian team. Like Imagina, IMAGOES does not need a separate monitor for image acquisition. An attractive feature that speeds up and eases repetitive data collection is the circling and highlighting of the area around each landmark and the prompting of the user for the next point to be taken. This can save time and effort when analyzing a series of similar specimens. Elliptical fourier analysis is built into the program, and fitted outlines can be displayed to see how well they capture the features of digitized outlines. Coefficients of fitted functions with up to 30 terms may be saved to a file for further data analysis.

Whereas it is clear that the majority of data being collected today is coordinate data in two or three dimensions, earlier, distances were the main data analyzed in traditional morphometrics. Calipers, originally mechanical and now digital, are still probably the most widely used measuring tools. An exhaustive collection of all interlandmark distances can be converted to coordinates easily, but the task is more difficult if the data were collected in the form of a truss or incomplete set of distances. Much data exist in the form of the truss from when that method was widely promulgated (Bookstein et al., 1985). Carpenter et al. provide a method for converting these data to coordinates so that the new morphometrics becomes available to those with an adequate set of interlandmark distances. The procedure and its use for actual data is discussed in detail in the article.

Yet more expensive and interesting acquisition equipment is used in medicine, such as Computer Aided Tomography (CAT), Magnetic Resonance Imaging (MRI) and Proton Emission Tomography (PET). They are primarily used for 3D imaging and structure reconstruction. The confocal microscope is another tool that allows us to see 3D structure at small scale. Although the emphasis to date has been on visualization of surface and internal structures only previously available by intrusive methods, if at all, these technologies will be important for acquisition of landmarks, space curves and surface data as morphometrics matures.

The richness of two-dimensional and three-dimensional data that can be collected by means of these tools may exceed our capability to analyze them in biologically important ways (Rohlf, 1993). We have yet to be able to adequately and fully analyze 3D surfaces and textures, let alone nonsurface textures and structures. Applications that make use of these newer techniques are discussed in this volume by Dean et al., Sampson et al., and Little and Mardia.

REFERENCES

Becerra, J. M., E. Bello, and A. García-Valdecasas. 1993. Building your own machine image system for morphometrics analysis: A user point of view. Pages 41–61 *in* L F. Marcus, E. Bello, and A. García-Valdecasas, (eds.), Contributions to morphometrics, Monografías del Museo Nacional de Ciencias Naturales 8, Madrid.

Bookstein, F. L., B. Chernoff, R. Elder, J. Humphries, G. Smith and R. Strauss. 1985. Morphometrics in evolutionary biology. Academy of Natural Sciences of Philadelphia Special Publication 15.

Rohlf, F. J. 1993. Computer vision needs in systematic biology. Pages 365–373 *in* R. Fortuner, (ed.), Advances in computer methods in systematic biology, artificial intelligence, databases, computer vision. The Johns Hopkins University Press:Baltimore, Maryland.

THREE-DIMENSIONAL DATA CAPTURE AND VISUALIZATION

David Dean*

Department of Anatomy
Case Western Reserve University
10900 Euclid Avenue
Cleveland, Ohio 44106-4930

The next great surge in morphometrics will involve the superb new technology of high-speed workstations. (Bookstein, 1991)

ABSTRACT

Three-dimensional (3D) digitization of biological specimens overcomes many of the problems of specimen registration attendant when using "standard" two-dimensional views or projections. External surfaces can be digitized directly by hand with electromagnetic, laser-light, sound, servo-mechanism and stereophotogrammetric devices, which are reviewed here. Single-point source laser-light devices are generally the most precise. Data may be captured directly from external or internal surfaces as adjoining tomographic slices (e.g., microscope; CT, and MR not reviewed here). The resulting stack of slices may be used to produce an isosurface image. There are a great many visualization environments available, some of which are briefly described, that facilitate the construction of isosurface images and measurement collection from Three-dimensional images. Three-dimensional models of Three-Dimensional image data are usually "printed" in plastic via a variety of new technologies briefly summarized.

INTRODUCTION

The hallmark of the new morphometrics is the use of landmark coordinate data. Current computer-based digitization, visualization and analysis applications in morphometrics are heavily geared toward two-dimensional (2D) data sets. Although some organisms are flatter than others, all are three-dimensional. We are accustomed to conflating these

*E-mail: dean@lucifer.cwru.edu

three-dimensional (3D) organisms to 2D despite the compromises involved. If coordinates are collected from pictorial representations (e.g., photographs, x-rays and frame-grabbers) it is important to attempt to register each organism in the same position before image or data capture. This is especially true for oblique views. The lateral view of "bilaterally symmetric" organisms is a very popular means to represent three-dimensional organisms in 2D. This approach ignores minor, but interesting, bilateral asymmetries—a popular topic of investigation (for example see Auffray et al., this volume). Problems of data capture, object registration or asymmetry generally do not affect the collection of data sets in three-dimensions.

Although organisms live in the fourth dimension, it is the task of morphometrics to describe three-dimensional organismal variation adequately. This variation is due to various 4D processes: ontogeny, phylogeny, pathology, injury and sexual differentiation. Most, if not all, 2D morphometric algorithms are easily extended to 3D. Until recently 3D data capture devices and data visualization software have not been readily available. This paper discusses the recent profusion of such devices and software.

Is it always desirable to undertake 3D morphometric analyses? There is no single answer to this question. Each morphometrician must perform his or her own cost-benefit analysis. As a lure Bookstein (1991) notes that

> In contrast to two-dimensional biological form, three-dimensional form may suggest a richness of local and regional geometric descriptors that often matches our intuitive sense of how biological forms are usefully compared.

Thus, whereas many measurements can be made on 2D views of, or cuts through, an organism, only the 3D surfaces give evidence of topology and texture. When possible, these important biological shape descriptors ought not be ignored.

The focus of this paper is tools that assist in the analysis of macroscopic biological surfaces and volumes (e.g., bones, skin and organ systems) that are readily visualized. When structures are not visible from the surface, often radiological techniques can be applied to visualize them in vivo. It should be mentioned that there are a host of problems unique to remote 3D visualization, especially those related to thresholding and isosurface construction, which are not discussed here (Fruhauf, 1991; Magnusson et al., 1991).

Data collection hardware and 3D image reconstruction software (i.e., from data slices) for morphometrics needs to have a high degree of precision, at least within the acceptable limits of direct data capture means such as calipers. There are lots of homemade devices (e.g., Corti et al., this volume) that we will not discuss here which obtain this level of accuracy. All of these devices are used to collect data directly from the surface of the study object. When the data are collected indirectly, i.e., by scanning, one must have a 3D visualization environment that allows quick and precise collection of landmark, surface or volumetric data. Of course, creating ones own visualization environment requires a great deal of expertise.

There are primarily two industrial applications driving the development of both direct and remote 3D digitizing devices and visualization environments: rapid prototyping and virtual reality modeling. Tools developed for rapid prototyping and virtual reality visualization have already been put to use in preparing fossil reconstructions (Kalvin et al., 1992). Medical imaging applications have also been the motivation for many new developments in 3D visualization (Coatrieux and Barillot, 1990).

The following is an annotated listing of 3D data capture and visualization hardware and software. I have firsthand experience with many of the devices and applications discussed below. Independent review of each device or package was requested, but, if not available, summaries were based on product literature: caveat emptor. There are sections on hand-operated data capture devices and remote laser-light and stereomicroscopic-surface imaging.

Additional sections cover a broad range of visualization software packages that have morphometric applicability. Finally there is a review of currently available rapid prototyping devices also referred to as CAMs (Computer-Aided-Manufacturing) or 3D printers.

DATA CAPTURE DEVICES: HAND DIGITIZING

These devices are most useful for capturing landmark-point and line-tracing coordinate data. However, as compared with the scanning devices discussed next, they are not as useful for precise capture of surface patches. Radiological slice data (e.g., CT and MR) or histological methods are the best way to gather information on the complete volume. Several of the handheld devices involve a source of electromagnetic radiation that creates a field through which a handheld sensor with a stylus moves. The tip of the stylus is tracked as it is applied to an object of interest. The location of significant surface features are recorded to disk at the operator's instruction.

These devices are in high demand in human kinematic studies, virtual reality applications, and industrial robotics applications, activities where 3D position tracking is a requirement. Several of these devices are designed to be lightweight and portable. Most offer devoted software interfaces or can be easily adapted to use data-recording software on a laptop computer. Several of these devices offer interfaces to common computer assisted design (CAD) programs so that the data can be visualized as it is collected. Many of these interfaces allow the user to change the point of view and zoom in and out of the image data. Most often the stylus tip controls the point of view, but in some interfaces the data lie within a cube and one may chose any one of the six faces as a view.

Multiple viewing options are critical when collecting large numbers of points or line tracings. Visualization becomes more and more difficult as the data come to overlie each other. When collecting large amounts of data the visual check on position accuracy allows mistakes to be corrected immediately rather than discovered at an inopportune time.

In most cases it is necessary to rigidly secure the object being digitized. If more than one view of an object is to be digitized (e.g., the part of the object that was anchored in the initial digitizing session), it is very helpful to take multiple redundant landmarks between the two sessions. A Procrustes fit can then be done between the landmarks that show "no" error, and one can throw away landmarks with noise in order to match up all the landmark data from each view of the object. Several of the various handheld digitizers that produce 3D coordinate data are reported below. Unlike laser light or other scanning digitizers, handheld stylus devices may be used for 2D digitizing of images. Note: Unless otherwise indicated, repeatability is given in inches (e.g., 0.05"). Monetary values are in U.S. dollars.

Electromagnetic Devices

DigiTracker ($3,395).
Visual Circuits, 3309 83rd Avenue North, Brooklyn Park, Minnesota 55443
Tel: 612-560-6205; FAX: 612-659-6629

As of this writing the device has not yet been released. A commercial version of the Ascension Flock of Birds (see below), it is reported to be a work-alike of the Polhemus 3DRAW. Like the Flock of Birds, its accuracy rating is a magnitude lower, 0.1 inch root mean square (RMS). It may have greater utility for biometric applications than does an unaltered Flock of Birds device, which Visual Circuits retails for $2,575, because of its well-designed stylus. It is supplied with a powerful digitizing software environment, Bit-

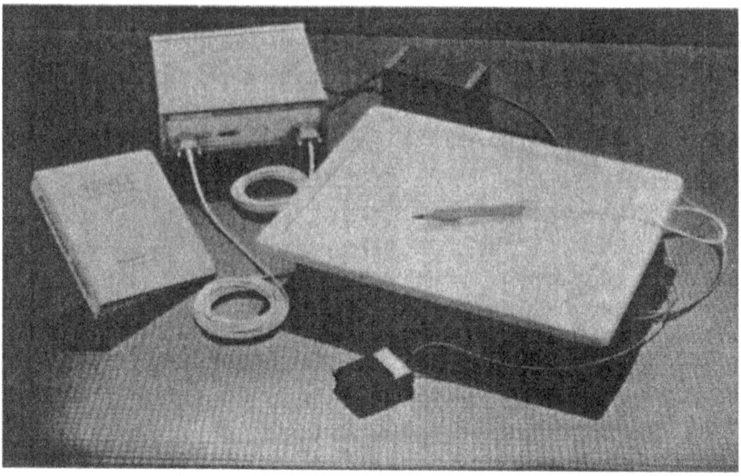

Figure 1. Polhemus 3DRAW. A magnetic source underneath the digitizing platform is detected by coils in the stylus. The 3D position of the stylus tip is reported through a serial port to a desktop or portable computer. An entire system, including a portable computer, can be assembled weighing less than 20Kg.

Blitzer. A number of mainstream Draw, Video and CAD software export filters are supplied with BitBlitzer.

Flock of Birds ($2,695).
Ascension Technology Corp., PO Box 527, Burlington, Vermont 05402
Tel: 802-860-6440; FAX: 802-860-6439

An early split-off from the Polhemus group. The same technology has been adapted to tracking multiple electromagnetic 6D (xyz coordinates and normals) sensors ("a flock") within the magnetic field emitted by a single stationary source. A popular device for biomechanical studies of joint movement. Transitional accuracy rated to 0.1 inch RMS. The above price is for single sensor device.

3DRAW ($6,500).
Polhemus, One Hercules Drive, PO Box 560, Colchester, Vermont 05446
Tel: 800-357-4777 or 802-655-3159; FAX: 801-655-1439

This fourth-generation device has been used by several investigators to study nonhuman primate and human crania (Vannier and Conroy, 1989; Corner and Richtsmeier, 1992; Lele and Richtsmeier, 1992; Richtsmeier and Walker, 1993; Dean, this volume). Originally developed to be mounted on a fighter pilot helmet to track head movements (Raab et al., 1979), this application was soon made obsolete by laser-based sighting devices. The device consists of a platform under which a fixed source emits a three-axis magnetic field. When a specimen is placed on the surface of this platform, it is digitized with the aid of a handheld stylus.

The stylus is equipped with a three-axis magnetic sensor that allows a small computer to determine the location of the tip in 3 space, including *xyz* normal information. Because the device assumes the magnetic field is constant, the work area should be kept reasonably clear of metal objects. None should be within a meter of the digitizing envelope, and other large metal objects such as steam-heated radiators and filing cabinets should be at least 2-3 meters away. Given a changing metal environment, the device should be recalibrated (it comes with calibration software) before each digitizing session. Unless the stylus is used very carefully, the normal information is not very important as part of the morphometric data collected. However, the viewpoint of the data collection interface can be tied to the stylus.

This allows easy perusal of the data from various points of view. This is a crucial error-checking step whenever potentially overlapping coordinates are collected.

There is firmware control of the stylus data report; it can be set in point or continuous mode interactively. Continuous mode is very useful for tracing edge features (see Dean, this volume). The accuracy of this device has been tested by Polhemus by means of widely accepted methods (Krieg et al., 1992) and has shown a 0.01 inch or 0.254 millimeter RMS. Although this is more than a two-fold improvement over the original 3SPACE device, the model commonly used with large, nonhuman primate and human skull material, the accuracy is still below the acceptable range for most small mammal, e.g. rodent, and human dental studies.

The 3DRAW is highly portable. The 3DRAW table, stylus, and data-handling device altogether weigh less than 12 pounds. There is a padded, hard-shell, airline check-in carrying case available.

An interface was recently released that allows the data collected by this device to be sent to the CADKEYS generic CAD program. This flexible PC program can be readily tailored to most biometric applications. The interface is included, gratis, with the device. Another long-standing interface for Polhemus, Metrecom, and Scientific Accessories (see the latter two below) devices is marketed by Mira Imaging, 2257 South East, Suite 1A, Salt Lake City, UT 84106. Tel: 801-466-4641; FAX: 801-466-4699. This interface, **Hyperspace**™, is primarily aimed at producing surface mesh representation of solid objects. The NATO morphometrics workshop attendees tested this software and found it difficult, but not impossible, to adapt to morphometric applications such as single-point or space-curve digitization.

Laser Light Stylus-Based Digitizers

Flashpoint Model 3000i ($14,900).
Pixsys, Inc., 5680-B Central Avenue, Boulder, Colorado 80301
Tel: 303-447-0248; FAX: 303-447-3905

A device like this or the much more expensive Optotrak (discussed next) is likely to be optimal for many morphometric studies. These devices do not have the echo or humidity artifacts of acoustic digitizers or the artifacts due to changing magnetic fields common with electromagnetic digitizers. These devices are, technologically, the newest; thus their high cost.

Both devices consist of one or more light emitting diodes (LEDs) or fiber optic light emitters. These light sources are detected by an array of three light sensors mounted in a bracket so that they are fixed in a noncollinear position. This fixed offset is used to triangulate the position of the light source. The light source can be attached to a stylus. Data can be

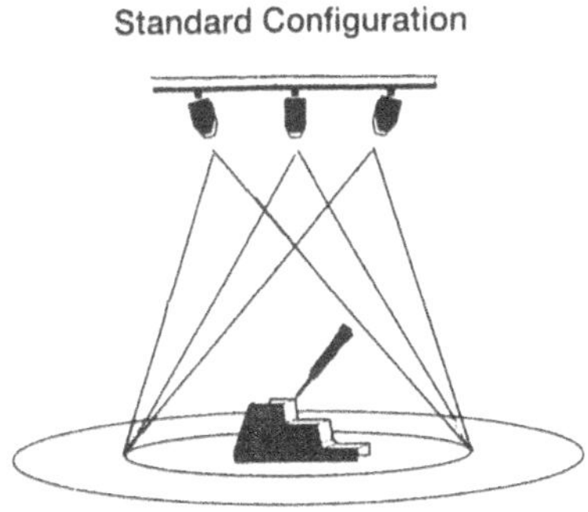

Figure 2. Laser-light digitizer. The stylus emits a light that is detected by three noncollinear sensors. Based on sensor position, offset location of the stylus is determined by triangulation. These expensive devices have high accuracy and low artifact noise (figure supplied by Pixsys, Inc.).

collected in point or continuous mode. There are device drivers for all the common CAD programs.

The rated accuracy of the current model is 0.5 millimeters (0.02 inches). Over the next few years I expect to see rapid increases in the accuracy of this device coinciding with price reductions. The entire device weighs 22.4 pounds and should be quite portable.

Optotrak ($57,500).
Northern Digital, Inc., 403 Albert Street, Waterloo, ON N2L 3V2, Canada
Tel: 800-265-2741 or 519-884-5142; FAX: 519-884-5142

This device has basically similar components and works in the same fashion as the Pixsys. Its accuracy is rated at 0.1 mm (0.004 inch) in the *x-y* plane, 0.15 mm (0.006 inch) in *z* when the stylus is within 2.25m (7 feet) of the sensors. Weighing approximately 90 pounds, not including a computer to record and visualize data, it is not easily transported.

Acoustic Stylus-Based Digitizer

GP-12 ES ($3,200).
Science Accessories Corporation, 2 Research Drive, PO Box 825, Shelton, Connecticut 06484
Tel: 203-925-1661; FAX 203-929-9636

The stylus in this device produces a clicking sound. This sound is differentially detected by three wall-mounted acoustic sensors. With the three separate readings of the click, the location of the source can be triangulated. The stylus should be at least as flexible as that of the 3DRAW or DigiTracker (I have not tried to use it), making complex concavities more reachable by this probe than, for example, those probes used in fixed-base, servo-mechanism digitizing arms (see next section). However this maneuverability is not without problems. Convexities may prevent sound from reaching the acoustic sensors without distortion. High humidity is also known to effect the quality of the data. The manufacturer's customers claim an accuracy of 0.01 inch(0.254 mm), in a cubic meter digitizing envelope. There is a real-time driver that works with a large number of CAD programs, including AutoCAD, Cadkey, and Mastercam. Scientific Accessories is also willing to develop a customized interface.

Rigid and Servo-Mechanism Arms

Cyclone ($52,000).
Renishaw, Inc., 623 Cooper Court, Schaumburg, Illinois 60173
Tel: 708-843-3666; FAX: 708-843-1744

This device comes from the rapid prototyping industry. It includes a computer terminal and digitizing base. This unjointed arm digitizer sits in place, retreating in the *z* direction as protrusions on the object are passed under its probe on a linearly moving platform. Successive cross sections are collected in this manner to build up an isosurface of the object. It moves slowly and gently enough during its slow traverse that there should be little chance for damage of specimens (stiff casts would be at less risk than would friable originals). However, complex convexities are ignored. A similar device, the **Retroscan** ($20,000) uses a drill bit to mill a model of a digitized object from solid plastic blocks.

Diagraph 202 ($1,472).
Seritex Inc., 450 Barell Avenue, Carlstadt, New Jersey 07072
Tel: 201-939-4606; FAX 201-939-3468

The Diagraph, designed by Swiss manufacturer GPM, is used to directly digitize objects (e.g., a bone) held rigidly in place. One end of the Diagraph's rigid arm points at surface features while the opposite end of the arm supports a pen or pencil on a piece of paper. The paper sits on a horizontal plane of tracing (or graph) paper set at a particular z-height. The Diagraph is useful for the collection of serial sections for 3D reconstruction.

Faro Arm ($14,400).
Faro, 125 Technology Park, Lake Mary, Florida 32746
Tel: 800-736-6063 or 407-333-9911; FAX: 407-333-4181

Morphometricians have attempted to use digitizing arms for at least the past 30 years. This attempt has not been successful in most craniofacial applications because the objects of study have been too topologically complex. These Faro Arm devices have four joints, giving them a great deal of flexibility. Even so it might be difficult to capture some points in the orbits, temporal fossa, posterior choanae, the palette and parts of the basicranium. Tracing lines on bony tori or elsewhere on the skull is likely to range between difficult and impossible. The accuracy of the Bronze Series 6 is 0.012 inch (0.3 mm), within a 6 foot (1.8 m) spherical diameter.

Faro manufactures a more expensive line of digitizing arms, the **Surgicom** series. These are used in human surgical planning, execution, and follow-up. A complete neurosurgical package, including visualization software, has recently been approved for use by the U. S. Food and Drug Administration and is marketed by ISG Technologies, 6509 Airport Road, Mississauga, ON L4V 1S7, Canada; Tel: 905-672-2100; FAX: 905-672-2307. The digitizing arm is primarily used as a virtual-reality tool in designing a surgical plan. The user digitizes landmarks to set up a reference frame on the patient that matches the one in the surgical plan, and then the reference frame is used as a position tracker for surgical instruments. Advances in the use of this and other technologies for frameless stereotactic surgery (Sandeman et al., 1992; Zinreich et al., 1993) will likely be furthered by the incorporation of newer morphometric techniques.

Romer Model 2000 ($105,000).
Romer Supratech, Inc., 5145 Avenida Encinas, Carlsbad, California 92008
Tel: 619-431-7944; FAX: 619-431-7940

This model is the smallest of a line of four-arm devices with digitizing envelopes of between 6.5 and 9.8 feet (2 and 3 m) in diameter. These arms have rotary joints making them more maneuverable than the Faro arm. These devices are also designed to be portable. The Model 2000 weighs 17 pounds. The small base includes a table clamp and an input/output processing unit with a serial port. It can thus be easily adapted to use with a portable computer. The accuracy rating is 0.00125 inch (0.032 mm) RMS, as tested by the NIST (National Institute of Standards and Technology).

DATA CAPTURE DEVICES: REMOTE DIGITIZATION DEVICES

The two types of digitizers discussed here, laser light and optical, are primarily used to digitize one surface, the external surface. Optical systems calibrate the x and y values from the power of magnification, and the z can be determined by the focus level. In this section we will not discuss multisurface, slice-based approaches such as CT or MR scanning.

Laser-light digitization has a reasonable possibility of providing data that can be used to construct highly accurate surface representations of the original object, i.e., an isosurface. Point resolution is a function of the width of the light source beam. Scan accuracy is determined not only by beam width, but also camera sensitivity, surface shininess and the

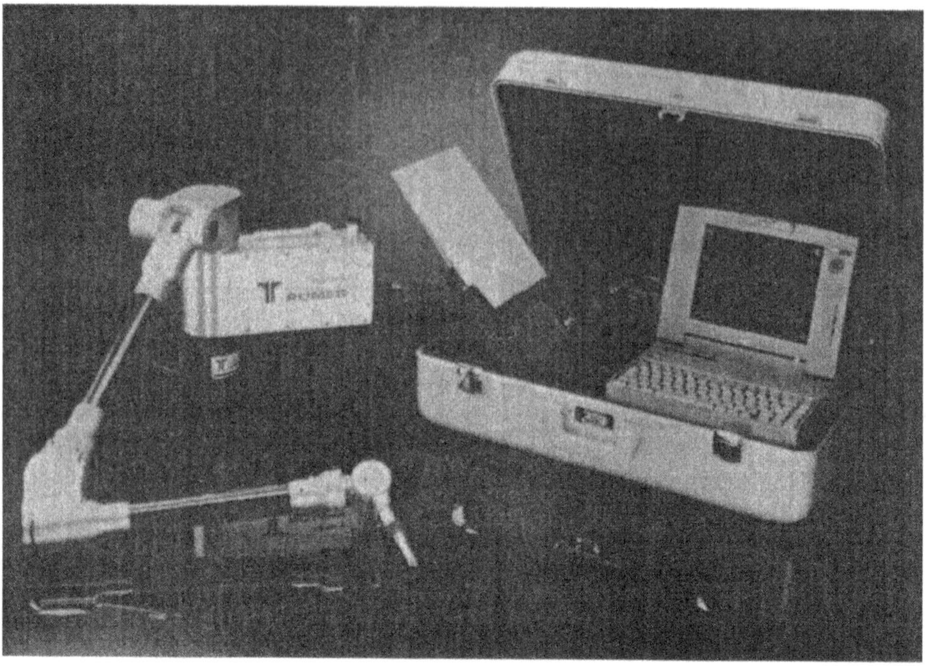

Figure 3. Romer Model 2000. This highly portable device provides high accuracy well within the range required for dental studies. Because of their limited mobility, digitizing arms may not be useful for complex surfaces. The hard-shell, airline check-in case on the right holds a portable computer, printer, and the digitizing arm.

system's ability to screen out artifactual reflections. In most cases users will benefit from importing the data into a visualization package and digitizing the required 3D landmarks or lines there.

Multiple-Point Source Laser-Light Scanning

Digibot II ($49,000).
Digibotics, 2800 Longhorn Boulevard, Suite 102, Austin, Texas 78758
Tel: 512-832-6544; FAX 512-832-1163

The Digibot II uses a four-axis laser-light scanner that checks sequential points on the surface. This allows some correction for artifacts. The four-axis array also obviates focusing problems that other systems may have.

Hirez Model 3030 ($46,400).
Cyberware Laboratory, Inc., 2110 Del Monte Avenue, Monterey, California 93940
Tel: 408-675-1440; FAX: 408-675-1494

This device consists of a moving source of laser light that is detected by a video camera at a known angle of offset. Like all of this first group of laser light digitizers, the Cyberware scanners use multiple laser-light sources. This allows the device to cover more surface area in less time. An entire human face can be scanned in less than 10 seconds. This is regularly done for a number of different morphometric applications (Cutting et al., 1988; Wohlers, 1992). However, diffuse light source artifacts, especially of complex surfaces, tend to corrupt the data. The recommendation of scanning a surface from different orientations and co-registering the scans can lead to compounded artifacts. Assuming no artifacts, landmark accuracy is approximately 0.02 inch (0.5 mm).

One version of this device also captures color information from the surface, which can be used to render a more realistic isosurface. A version of this device weighs less than 50 pounds.

Optica ($50,000).
3D Technology, Inc., 12 Cambridge Drive, Trumbull, Connecticut 06611-4764
Tel: 203-371-8500; FAX: 203-371-6300

Using a technology developed in England, but reminiscent of the Cyberware scanner, this vendor claims 0.0008 inch (0.02 mm) accuracy (Vinarub and Kapoor, 1992).

Single-Point Source Laser-Light Scanning

CyberScan ($30,000).
CyberOptics Corp., 2505 Kennedy Street NE, Minneapolis, Minnesota 55413
Tel: 612-331-5702; FAX: 612-331-3826

Sharnoa Corp., 45901 5 Mile Road, Plymouth, Michigan 48170
Tel 313-925-1661 FAX 313-454-7198

This single source laser sits at a preset "stand-off" distance from the surface to be digitized. It has a high accuracy, 0.001inch (0.0254 mm), but can fail on surfaces with convexities that block the path of the scan head. These devices would take several hours to scan an entire human face. Sharnoa uses the same scan head and provides integration with CNC machine milling tools for rapid prototyping.

Hyscan (n.a.).
Hymarc Ltd., G.A. Davis Associates Inc., PO Box 36240, 19959 Vernier Avenue, Suite 2B
Harper Woods, Michigan 48225
Tel: 313-86-4101; FAX: 313-86-1107

A single-point laser device with very high accuracy and relatively slow data acquisition.

Lasar and TriCam ($50,000 and $11,000, respectively).
Perceptron, Inc., 23855 Research Drive, Farmington Hills, Michigan 48335
Tel: 810-78-7710; FAX: 810-478-7059

These are single-point laser devices that use radar focusing. The digitizing envelope of the Lasar is commonly a cube of 2 - 3 meters (6.5 - 10 feet) with an accuracy of 0.1 mm (0.004 inch). The TriCam is more accurate in its smaller digitizing envelope. The TriCam weighs a portable 20 pounds.

Surveyor ($40,000).
Laser Design, Inc., 9401 James Avenue South, Suite 162, Minneapolis, Minnesota 55431
Tel: 612-884-9648; FAX: 612-884-9653

This device is scheduled to be available shortly. The developers claim digitizing speeds comparable with the Cyberware scanner and accuracy improvements of at least an order of magnitude, i.e., 0.05 mm (0.002 inch). The whole-body version of this scanner is scheduled to ship at $145,000.

Moire

CadEyes ($80,000).
Medar Company, 38700 Grand River, Farmington Hills, Michigan 48335
Tel: 810-477-3900; FAX: 810-477-8897

As with all Moire photography, light is passed through a screen grating. The resulting topographic pattern can be used for 3D digitization. Medar claims 0.0001 inch (0.00254 mm) accuracy in the *z* dimension with this device.

Mini-Moire ($39,000).
Electro-Optical Information Systems, 528 Euclid Street, Santa Monica, California 90402
Tel: 310-451-8566; FAX: 310-393-2453

The Electro-Optical Information Systems device works on the same principle as does the CadEyes scanner. It claims about the same accuracy level; however, it is much more portable.

Convergent Photogrammetry

Photomodeler (Windows) ($895).
Eos Systems, Inc., 2040 West 12th Avenue, Vancouver, BC V6J 2G2 Canada
Tel: 604-732-6658; FAX: 604-732-4716

This software uses standard photographs or 35 mm slides to produce a precise 3D model (AutoCAD 3D *.DXF file) of any object. The documentation shows you how to calibrate your camera's focal length, image aspect ratio, image position, and lens distortion for this application. There are a few more planning steps to take so that all views will be complementary.

Macroscopic Stereo-Microscopes

Reflex Microscope ($35,000).
Reflex Measurements Limited, Hadley House, Water Lane, Butleigh, Somerset BA6 8SP, England
Tel/FAX: 44-458-50332

This device has been used with great success on small biological specimens (MacLarnon, 1989). An accuracy of 0.002 mm (0.00008 inch) is claimed for the 110 x 110 x 50 mm digitizing envelope.

Tool-Makers and Digital Contracers Microscopes ($35,000-45,000).
Mitotoyu, M-R-O Industrial Supplies, Inc., 440 South Main Street, Manville, New Jersey 08835
Tel: 201-368-0525 or 800-345-4135; FAX: 908-707-1266

The Tool-Makers device has a 100 x 100 x 50 mm digitizing envelope with a claimed accuracy of .02 mm (0.0008 inch). The Digital Contracers Microscope has a 100 x 100 x 25 mm digitizing envelope with a claimed accuracy of .02 mm. These are standard optical digitizing devices designed for the tool and die industry but also readily adaptable to submacroscopic morphometrics, especially on micro-organisms and teeth. It has been widely used in morphometrics. A more expensive motorized version is available.

Video Comparator DVC110 ($8,900)
Deltronic, 3900 West Segerstrom Avenue, Santa Ana, CA 92704, USA
Tel: 714-545-0401; FAX: 714-641-0946

More like the Reflex in having a potentially larger digitizing envelope: 2 x 2 x 4 inch (51 x 51 x 102 mm). The manufacturer rates its accuracy at 0.0002 inch (0.05 mm), a level of accuracy acceptable for mammalian dental studies.

VISUALIZATION SOFTWARE

NATO conference attendees tested DSDIGIT, the shareware visualization environment created by Dennis Slice that runs under MS-DOS. It provides a view of a data "cube" from various sides and can be used to calculate distances between landmarks. Jeffrey Walker (SUNY-Stony Brook) demonstrated his Morphometrika package for the Mac. It provides sophisticated morphometric image-analysis tools.

The visualization environments discussed below have been developed for different primary audiences, but all have some application useful to morphometrics. Many of these packages offer a 3D "pick" tool to collect landmark and line-tracing coordinates from isosurface models. Many of these packages provide image enhancement tools such as isosurface shading and lighting options, to help the user better identify the morphology to be digitized.

Shareware and Freeware

Khoros 2.0 (Shareware Multiple UNIX Platforms, NeXT, PC-XWindows, Macintosh-XWindows).
e-mail: khoros-request@chama.eece.unm.edu.
Usenet: comp.soft-sys.khoros.
ftp: Khoros may be downloaded from the following ftp locations (read $KHOROS_ftp/release/install.ftp first). Given below are country, Internet host name, IP address in parentheses and the subdirectory where the software resides:

USA:	ftp.eecc.unm.edu	(129.24.24.119)	/pub/khoros
USA:	ftp.uu.net	(192.48.96.9)	/pub/window-sys/khoros
Canada:	popeye.genie.uottawa.ca	(137.122.20.3)	/pub/khoros
Germany:	ftp.rrz.Uni-Koeln.DE	(134.95.80.5)	/graph/khoros
Germany:	ftp.lrz-muenchen.de	(129.187.10.35)	/local/khoros
Italy:	ipifidpt.difi.unipi.it	(131.114.8.130)	/pub/khoros
Japan:	ftp.waseda.ac.jp	(133.9.1.32)	/pub/khoros
UK:	ftp.mcc.ac.uk	(130.88.203.12)	/pub/cgu/khoros
UK:	unix.hensa.ac.uk	(129.12.21.7)	/pub/uunet/window-sys/khoros

A powerful visualization environment written at the University of New Mexico. There are several tools for adjusting isosurface thresholds by use of image gradient information.

NIH Image 1.54 (Shareware Macintosh).
ftp site: USA: zippy.nimh.mih.gov \pub\image
Source Names: nih-image-154-fpu.hqx; nih-image-154-nonfpu.hqx; nih-image-154-docs.hqx

A plug-in (Plug-in Digitizer), written by Cyrus Daboo of Cambridge University, now allows NIH Image to grab images by use of the frame grabber built into the AV Macs or from any other Quicktime compatible digitizer with a 'vdig' component. NIH image has tools for measuring areas, distances and angles, and for counting things. The digitizer patch is available as a binhexed, self-extracting archive by anonymous ftp from zippy.nimh.nih.gov, in the /pub/nih-image/plug-ins directory. The archive also includes a report on the AV digitizer, which points out some its flaws, such as automatic gain control that can't be disabled. It also includes Apple's AV Digitizer Options extension that allows the user to switch the AV digitizer into gray-scale mode, which greatly improves the quality of images captured using gray-scale cameras.

QSH (Multiple Unix workstations).
Dr. Marilyn Noz, Department of Radiology, New York University School of Medicine
550 First Avenue, New York, New York 10016-6402

This is radiological image-handling software with a wide range of previsualization tools, including pixel editing, thresholding, and in-slice measuring. After the object of interest has been isolated, a simple image stack (*.qsh file) and header file (*.qhd file) can be built.

SGI Explorer (Gratis Software with Standard Workstation Purchase) (IRIX [SGI UNIX]).
Silicon Graphics Incorporated, 2011 North Shoreline Boulevard, Mountain View, California 94039
Tel: 415-390-1960 or 800-800-4744; FAX: 415-390-6153

Free visualization tool that comes with an SGI workstation. Piggybacking on the power of SGI's GL (graphics language), it provides several tools to display and explore 3D data.

Viewit freeware (Multiple UNIX Platforms, Macintosh).
National Center for Supercomputing Applications, Computing Applications Building
605 E. Springfield Avenue, Champaign, Illinois 61820

Viewit has a minimal user interface. It can be used to produce volumetric images from slice data. It is known to require a lot of RAM. There are also visualization tools. The Viewit electronic newsletter is available from the ftp site: viewit@ncsa.uiuc.edu. Clint Potter is the primary author. Contact cpotter@ncsa.uiuc.edu.

Commercial Packages

Alice ($1,495, $1,046.50 academic price; Windows, Windows NT, Macintosh, PowerPC).
Hayden Image Processing Group, 1919 14th Street, Suite 405, Boulder, CO 80302, USA
Tel: 303-449-3433; FAX: 303-449-3772

Originally released under the nomen: DIPstation. A medical imaging toolkit with help for segmentation, volumetric image rendering, measurements from slices, and volumes, and data analysis. The segmentation tools in Alice are stronger than many more expensive packages.

Analyze 6.2 ($16,000; Multiple UNIX platforms, Macintosh).
Mayo Medical Ventures, 200 First St. SW, Rochester, Minnesota 55905
Tel: 507-284-8452; FAX: 507-284-5410

Analyze offers a large set of slice-data processing tools and 3D image-building algorithms (Robb and Hanson, 1991). Analyze provides import/export facility for a large number of radiological, medical imaging and histological data formats. Linear measurements, landmarks and line tracings may be captured from the resulting isosurface models.

AVS ($6,500; Multiple Unix platforms, Macintosh, Windows NT, Windows).
Advanced Visual Systems, Inc., 300 Fifth Avenue, Waltham, MA 02154, USA
Tel: 617-890-4300; FAX: 617-890-8287

This visualization environment comes with a large number of image-handling primitives that assist in 2D and 3D image analysis. These primitives support several hardware-based graphics languages such as SGI GL, Xlib, GDI, XGL and PEX, speeding redraw of the user's private interface.

Data Explorer ($5900; Multiple UNIX Platforms; substantial academic discount)
IBM Data Explorer Group, Product Marketing, Department 590, 8 Skyline Drive, Hawthorne, New York 10532
Tel: 914-784-5089; FAX: 914-784-5082

Data Explorer is image manipulation software with applications similar to 3DViewnix. The isosurface-building algorithm, Alligator (Kalvin, 1991), derives its flexibility from the "Winged Edge" data structure (Kalvin et al., 1991). Recently, powerful data-reduction software, "**Superfaces**," was added to this package. The package also includes a powerful data-capture pick tool and image-handling primitives (Rogowitz and Treinish, 1993). These image-handling primitives greatly speed the creation of tools for image analysis.

IDL ($3,750; $1,500 academic price; Windows, Macintosh, Multiple UNIX Platforms)
Research Systems Inc., 2995 Wilderness Place, Suite 203, Boulder, Colorado 80301
Tel: 303-786-9900; FAX: 303-786-9909

Interactive Data Language provides tools for image data reduction, visualization, segmentation, measurement collection and analysis. Instead of canned image-processing tools, there is a high-level, user-friendly, image-processing language, IDL.

Image Volumes ($4,500; IRIX).
Minnesota Datametrics Corp., 1000 Ingerson Road, St. Paul, Minnesota 55126-8146
Tel: 612-482-7938; FAX: 612-490-9717

A package to assist collection of measurements from volumetric images and which includes a large image-processing tool kit. Provides facile collection of linear distances, surface areas and volumes. Also includes native tools for superimposition by landmarks, bounding boxes or other transformations.

MacStereology ($750; Macintosh).
Ranfurly Microsystems, 16 Derwent Avenue, Haddington, Oxford OX3 0AP, England
ftp demo copy: zippy.nimh.nih.gov /pub/nih-image/programs

This program includes tools to make measurements directly from images or to produce 3D reconstructions from stacks of images. MacStereology can also accept data from digitizing tablets or *.PICT files. The user can set magnification. Measurement tools include area, perimeter and center of gravity. Requires only 1 megabyte of RAM.

Mocha ($1,495; Windows).
Jandel Scientific, 2591 Kerner Boulevard, San Rafael, California 94901
Tel: 800-874-1888 or 415-453-6700; FAX: 415-453-7769

Primarily aimed at a histological and microbiological audience, this package provides a large number of general image-processing and data-capture tools. Analysis proceeds primarily on *x-y* slices of 3D subjects, but 3D reports can be generated.

Surfacer 3.1 ($4,000; Multiple UNIX Platforms, Windows).
Imageware, Inc., 313 North First Street, Ann Arbor, Michigan 48103
Tel: 313-994-7300; FAX: 313-994-7303

Surfacer imports data from a wide variety of standard formats. In addition to data rendering and digitization it offers a large complement of CAD reverse engineering tools.

3DViewnix ($1,000; Multiple UNIX Platforms).
Medical Imaging Processing Group, Radiology Associates, University of Pennsylvania
Brockley Hall, 4th Floor, 428 Service Drive, Philadelphia, Pennsylvania 19104
Tel: 215-662-6780; FAX: 215-898-9145

This software provides a large number of image-manipulation and data-capture tools (Udupa et al., 1993), primarily for radiological slice data (e.g., MR and CT) but adaptable to other 3D data sets. Isosurfaces are built from topological information encoded by a highly adaptable data structure known as the "Shell" (Udupa and Odhner, 1993).

VoxBlast ($2,500; Multiple UNIX platforms, Macintosh, PowerMac, Windows).
VayTek, Inc., PO Box 732, 305 West Lowe, Fairfield, Iowa 52556
Tel: 515-472-2227; FAX: 515-472-8131

VoxBlast produces isosurfaces from a wide variety of slice data. There are tools to take measurements from these isosurfaces and to calculate volumes.

VoxelView ($10,000; academic price $5,000, $1,295 Macintosh Version; Irix and Macintosh).
Vital Images, 505 N. Fourth St., Fairfield, IA 52556, USA
Tel: 515-472-7726; FAX: 515-472-1661

Produces 3D isosurface models from a variety of data sources. One of the few visualization environments (other than SGI Explorer) that takes advantage of SGI's IRIX GL (graphics language) and thus the hardwired graphics engines found in SGI workstations. This native hardware support speeds up image rendering. Reduced redraw times mean faster digitization from multiple views as well as when one changes visualization parameters (e.g. threshold, contrast or number of light sources).

RAPID PROTO-TYPING DEVICES

Generating an average landmark configuration, especially if the landmark density is high enough to warrant an isosurface model, is a major part of geometric morphometrics. Similarly, an average form or morphotype is best viewed as a real model. These 3D printers are still very expensive but have dropped enough in price to make requests for single copy output affordable. This copy can then be used as a positive in mold production if copies are needed.

FDM 1000 ($50,000).
Stratasys, Inc., 14950 Martin Drive, Eden Prairie, Minnesota 55344
Tel: 612-937-3000; FAX: 612-937-0070

You select a *z* axis within your 3D data set. The data are then divided into *z* slices of 0.05 inch (1.3 mm) thickness. An extruder begins at the center of each *z* slice and spins concentric circles emitting a 0.05 inch thick filament of plastic at just over the melting temperature. The plastic left wherever there were data in the original data set instantly cools. The extruder finishes that *z* slice and moves up 0.05 inch to the next *z* slice. If an overhang is expected in the data, a bridge is built that can be removed later. The models produced are almost seamless. It can interface with most of the aforementioned data visualization environments.

Helisys ($95,000).
Helisys, Inc., Building M10, 2750 Oregon Court, Torrance, California 90503
Tel: 310-782-1949; FAX: 310-782-8280

The volumetric data is converted to z slices along the preferred z axis. Each slice is cut into a sheet of paper. The sheets of paper are then laminated together. The models have hard-wood resilience and are virtually seamless. Unlike laser sintering devices, there is no warping in the part of the plastic model already produced due to heat in the part that is being formed. The Helisys can interface with most of the aforementioned data visualization environments.

QuickCast ($195,000).
3D Systems, 26081 Avenue Hall, Valencia, California 91355
Tel: 805-295-5600; FAX: 805-257-1200

A standard sterolithographic device (Jacobs, 1993) that uses a laser to etch the desired surfaces in an epoxy resin block. This technique is referred to as laser sintering. Built in a series of z slices, slice thickness can be as low as 0.005 inch (0.13 mm). The QuickCast can interface with most of the aforementioned data visualization environments.

SLS Sinterstation ($350,000).
DTM Corp., 1611 Headway Circle, Building 2, Austin, Texas 78754
Tel: 512-339-2922; FAX: 512-339-0634

This is a standard laser sintering device. DTM claims to have made more careful study of the materials used by other laser sintering vendors, especially for biomedical applications. It has been used to produce skeletal casts. The device can interface with most of the aforementioned data visualization environments.

Solider ($490,000).
Cubital America Inc., 1307F Allen Drive, Troy, Michigan 48083
Tel: 810-585-7880; FAX: 810-585-7884

A standard stereolithography device developed in Israel. Solider can interface with most of the aforementioned data visualization environments.

Z-mode ($800 per Model + Model Costs)
3D Medical Imaging Services, 1241 Garden Street, San Luis Obispo, California 93401
Tel: 805-542-9281; FAX: 805-542-9168

Uses proprietary volumetric-rendering software and the Cubital 3D printing technology from ISG (see Metrecom Arm description above). This company accepts a wide range of 3D CT and MR data for the production of 3D models. These models are typically used in surgical planning or prosthesis development. The z plane resolution of these models is claimed to be 0.002 inch (0.05 mm).

SUMMARY

Most morphometric investigations discussed at the NATO Advanced Study Institute derived their data from the external surfaces of organisms. However, the leading 3D data-capture technologies are laser-light based and these digitizers are currently the most expensive devices. Therefore advances in less expensive electromagnetic devices such as the Polhemus 3DRAW may be the near-term answer for landmark and space curve digitization. The Laser Designs device looks to be a breakthrough, combining speed and accuracy in a single-point source laser-light digitizer. Three-dimensional visualization software is now readily available from a variety of sources. High-quality three dimensional printers should start making their way into the mainstream shortly.

ACKNOWLEDGMENTS

Thanks to Jim Rohlf, Alan Kalvin (IBM TJ Watson), Chip Maguire (Columbia University), and John Russ (Univ. North Carolina) for careful comments on the manuscript. Much of the visualization information was obtained from "med.volviz.faq.94-08.2/3" and "med-volviz-faq.94-10" produced by Matti Haveri (mhaveri@cc.oulu.fi).

REFERENCES

Bookstein, F. L. 1991 Morphometric tools for landmark data: Geometry and biology. Cambridge University Press: Cambridge.

Coatrieux J. L. and C. Barillot 1990 A survey of 3D display techniques to render medical data. Pages 175–195 *in* K. H. Hoëhne, H. Fuchs, S. M. Pizer, (eds.) 3D imaging in medicine. Springer-Verlag: Berlin.

Corner, B. D., S. Lele, and J. T. Richtsmeier. 1992. Measurement error of 3D landmark data. Quantitative Anthropology 3:347–359.

Corner, B., and J. T. Richtsmeier 1992 Cranial growth in the Squirrel monkey (*Saimuri sciureus*): A quantitative analysis using 3D coordinate data. American Journal of Physical Anthropology 87:67–81.

Cutting, C. B., J. G. McCarthy, and D. B. Karron. 1988. 3D input of body surface data using a laser light scanner. Annals of Plastic Surgery 21:38–45.

Fruhauf, M. 1991. Volume visualization on workstations: Image quality and efficiency of different techniques. Comput. and Graphics 15:101–107.

Jacobs, P. 1993. Stereolithography: From art to part. Cutting Tool Engineering 45:1–3.

Kalvin A. D. 1991. Segmentation and surface-based modeling of objects in 3D biomedical images. Ph.D. dissertation: New York University.

Kalvin, A. D., C. B. Cutting, B. Haddad, and M. Noz. 1991. Constructing topologically connected surfaces for the comprehensive analysis of 3D medical structures. SPIE 1445:247–258.

Kalvin A. D., D. Dean, J.-J. Hublin, and M. Braun. 1992. Visualization in anthropology: Reconstruction of human fossils from multiple pieces. Pages. 404–420 *in* A. E. Kaufman and G. M. Nielson, (eds.). Proceedings of IEEE Visualization '92.

Krieg, J. C., H. R. Jones, A. G. Rodgers, and M. R. Schneider. 1992. A 4 millisecond, low latency, 120 Hz, electromagnetic tracker for virtual reality applications. Internal Application Note, Polhemus.

Lele, S., and J. T. Richtsmeier. 1992. On comparing biological shapes: Detection of influential landmarks. American Journal of Physical Anthropology 87:49–65.

MacLarnon, A. M. 1989. Applications of the Reflex instruments in quantitative morphology. Folia Primatologica 53:33–49.

Magnusson, M., R. Lenz, and P. E. Danielsson. 1991. Evaluation of methods for shaded surface display of CT volumes. Comput. Med. Imaging Graph. 15:247–256.

Raab, F. H., E. B. Blod, T. O. Steiner, and H. R. Jones 1979 Magnetic position and orientation tracking system. IEEE Trans. Aerospace and Elec. Sys. 15:709–718.

Richtsmeier, J. T., and A. Walker 1993. A morphometric study of facial growth. Pages 391–410 *in* A Walker and R Leakey (eds.), The Nariokotome *Homo erectus* skeleton. Harvard University Press: Cambridge.

Robb, A. R., and D. P. Hanson. 1991. A software system for interactive and quantitative visualization of multidimensional biomedical images. Austral. Phys. and Eng. Sci. in Med. 14:9–30.

Rogowitz, B. E., and L. A. Treinish. 1993. An architecture for rule-based visualization. Proceedings of IEEE Visualization '93:236–234.

Sandeman, D. R., N. Patel, C. Chandler, R. J. Nelson, H. B. Coakham, and H. B. Griffith. 1992. Advances in image directed neurosurgery: Preliminary experience with the ISG Viewing Wand compared with the Leksell G frame. Clinical MRI 2:91–92.

Udupa, J. K., and D. Odhner. 1993. Shell rendering. IEEE Comp. Graphics and Applic. 27:58–67.

Udupa, J. K., K. I. Goncalves, D. Narendula, D. Odhner, S. Samarasekera, and S. Sharma. 1993. 3DVIEWNIX: An open, transportable software system for the visualization and analysis of multidimensional, multimodality, multiparametric images. SPIE 1897:47–58.

Vannier, M. W., and G. C. Conroy. 1989. Imaging workstations for computer-aided primatology: Promises and pitfalls. Folia Primatologica 53:7–21.

Vinarub, E. I., and N. Kapoor. 1992. 3D digitizing hidden surface using borescopic sensor technology. Proc. App. Machine Vision Conf. '92: MS92 160:1–11.

Wohlers, T. T. 1992. 3D Digitizers. Computer Graphics World 2:73–77.

Zinreich, J. S., S. A. Tebo, D. M. Long, H. Brem, D. E. Mattox, M. E. Loury, C. A. Vander Kolk, W. M. Koch, D. W. Kennedy, and R. N. Bryan. 1993. Frameless stereotaxic integration of CT imaging data: Accuracy and initial applications. Radiology 188:735–742.

TWO-DIMENSIONAL IMAGING

An Update

Antonio García-Valdecasas*

Museo Nacional de Ciencias Naturales
José Gutierrez Abascal, 2
Madrid 28006, Spain

ABSTRACT

Digital image acquisition and display is compared with traditional photography. Some of the criteria used to evaluate the quality of images and tips for testing them in digital imaging are reviewed.

INTRODUCTION

There is an abundant literature on two-dimensional (2D) digital imaging, so that it is unnecessary to produce yet another paper on fundamentals. What is rare is to find presentations in which digital imaging is discussed in relation to other types of imaging. This is the main objective of this paper. In order to reach this goal, we will look a little further at what is customarily discussed in morphometrics: acquisition of images for the simple purpose of measurement and recording of landmark coordinates. Inoué (1986) and Russ (1992) are two good reference books. For introductions to digital image acquisition systems applied to morphometrics see Fink (1990) and MacLeod (1990). Meacham (1992) gives useful clues for working with digital systems, and Becerra et al. (1993) give guidelines for testing different morphometric hardware and software. All these papers include references to further literature.

This paper is organized into the following topics:

- general concept of imaging;
- criteria for the evaluation of the quality of imaging systems;
- low-resolution, medium-resolution and high-resolution imaging systems; and
- software for morphometrics.

* e-mail: mcnhv00@cc.csic.es

Advances in Morphometrics, Edited by L. F. Marcus et al.
Plenum Press, New York, 1996

GENERAL CONCEPT OF IMAGING

Imaging can be loosely defined as a process that includes the acquisition, storing and display of images. Any chemical or electronic systems of imaging include artifacts peculiar to each of these steps, although, as is the case with photography, the storage medium also functions as a display.

Electronic analog imaging includes acquisition by a video camera, recording on a videocassette using a recorder (VCR), and display on a television (TV) set. Electronic digital imaging can share with analog imaging a mixed acquisition step, whereby an analog video camera's video signal is digitally transformed before storage or display. More recently, the circuitry to produce a digital output has moved into the camera. This avoids the signal degrading process of producing a standard video signal by codification of the voltage variation produced by the charged couple device (CCD). Some digital cameras can store one or more frames in their own dedicated memory and even perform some image processing in their hardware. In any case, this does not question the generality of the three steps mentioned above.

In a more restricted sense, imaging refers to acquisition, storing and display of images by computers. This narrow sense was useful at the beginning of image analysis by computers. It now seems more fruitful to point out the similarities and limitations among the different kinds of imaging.

CRITERIA FOR THE EVALUATION OF THE QUALITY OF AN IMAGING SYSTEM

The usual criteria employed to evaluate the quality of photography can be used for electronic imaging as well. The parameters are the same. The criteria and indices to evaluate those parameters may be the same or include some new ones.

Three general parameters can be used to evaluate the quality of an imaging system: tone, sharpness and graininess (see Miyake et al., 1984, for additional quality parameters).

Tone

Tone is the different intensity of light reflected by objects. It is synonymous with brightness or luminance. A natural scene can have a range of brightness between black-and-white as high as 1:10,000 (Adams, 1981), although the average ratio has been found to be around 1:160 (Stroebel et al., 1986). This brightness or luminance ratio has a direct relation to the general contrast of the scene. Scenes with a luminance ratio higher than average have high contrast, whereas those with lower ratios are called flat.

In any imaging system there are two aspects of tone: (1) the ability to record the luminance ratio of the scene; and (2) the number of intermediate steps that can be distinguished between the extreme values (the number of shades of gray in monochrome electronic imaging).

For many image-processing routines the number of tones or shades distinguished by a system is an important variable. Typical digitizing systems distinguish 2^8 (256) shades of gray, or $2^8 \times 2^8 \times 2^8$ for red, green and blue in color systems. Hardware with 2^{10}, 2^{12}, 2^{16} and even greater numbers of shades are available for special applications. Note that being able to acquire images with 256 shades of gray is not the same as to be able to display these 256 shades.

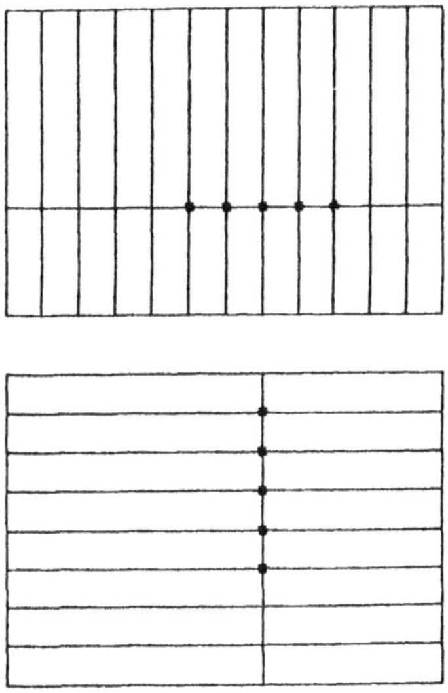

Figure 1. Resolution as measured with a standard TV chart.

The fidelity of a scene's tone reproduction for an imaging system can be easily tested with a special strip of gray densities manufactured by photographic companies (it is also usually included in the TV charts used to test resolution and other TV variables). This strip also tests the linearity of the system; different shades are spaced at regular intervals. A linear relation among shades of gray and voltage produced by a video camera leads to what is called a gamma value of 1.

Sharpness

Sharpness, or different luminance in boundary areas, is a subjective term that refers to a combination of resolution and acutance. Resolution is the ability to distinguish between two nearby details. Acutance is the physical measurement of the sharpness of an edge. Both are related to the contrast of the image, and this relation is used in what is known as the modulation transfer function (MTF).

Resolution of an electronic imaging system is easily measured with any standard TV chart (like the one included in Becerra et al., 1993) or other special resolution test targets (available from companies such as Edmund Scientific or the Rochester Institute of Technology). In the first case a relative measure is usually expressed as number of TV lines. Note that horizontal resolution is given by the number of vertical lines and vertical resolution by the number of horizontal lines (Fig. 1). Video systems have a 4:3 aspect ratio on a display.

Table 1. Some specifications of common TV standards

Monchrome	RS-170	CCIR-625	
Color	NTSC	PAL	SECAM
Interlace (fields per frame)	2:1	2:1	2:1
Aspect ratio	4:3	4:3	4:3
Lines per frame	525	625	625
Visible Lines per frame	484 (485)	575 (576)	575 (576)
Frames per second	30	25	25

Table 2. Kell correction factors for RS-170 and CCIR-635 standards

RS-170	$\frac{H}{484} = \frac{4}{3}$	H = 645 vertical lines for horizontal resolution
		Kell correction factor = 0.75
	$\frac{H}{484 \times 0.75} = \frac{4}{3}$	H = 482 (rounding to even number)
CCIR-625	$\frac{H}{575} = \frac{4}{3}$	H = 766 lines
		Kell correction factor = 0.75
	$\frac{H}{575 \times 0.75} = \frac{4}{3}$	H=573

Knowing that the number of horizontal lines (vertical resolution) is fixed for a video standard (Table 1), it is easy to calculate the maximum number of vertical lines (horizontal resolution). A correction factor—Kell factor—is needed in order to take into account some resolution degrading effects (Table 2). Another way to determine the video resolution is through the bandwidth used. One MHz is roughly equivalent to 80 vertical lines (Fig. 2).

Special resolution test targets give an absolute value of lines per millimeter. Number of TV lines can be converted to lines per millimeter by the following formulas (Edmund Scientific, 1994):

Horizontal resolution (lines/mm) = Number TV lines (H) · 1.3333 / (2 · horizontal sensing area)

Vertical resolution (lines/mm) = Number TV lines (V) / (2 · vertical sensing area)

The following equation allows one to calculate the resolution of a system, when the resolution of its components are known:

$$\frac{1}{R} = \frac{1}{r1} + \frac{1}{r2}$$

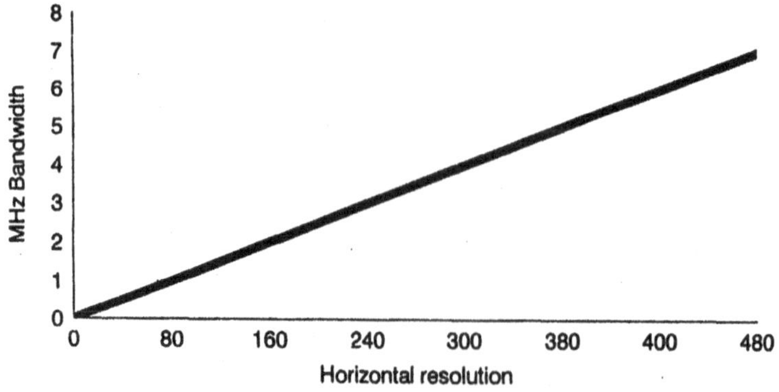

Figure 2. Relation between horizontal resolution and bandwidth in Megahertz (MHz).

where R is the resolving power of the system and $r1$ and $r2$ the resolving powers of the lens and film separately, or the lens and the CCD. For example, a film with 200 lines/mm and a lens with 200 lines/mm resolution gives an overall resolution of 100 lines/mm. In general, as the number of elements that take part in the acquisition of images increases, the resolution decreases.

The modulation transfer function (MTF) expresses the variation of the contrast in the system as the spatial frequency of the target varies. It has been traditionally tested with a sine wave target (a sequence of black-and-white lines progressively closer). A qualitative explanation of how the MTF can be measured follows.

For a pattern of black-and-white lines clearly spaced, the minimum brightness, corresponding to the center of a black line, and the maximum brightness, corresponding to the center of a white line, do not interfere and both are clearly distinguished in the image. As the spatial frequency of the target increases, light from the white lines will fall onto the dark area, due mainly to lens aberration and diffraction in the system. The general contrast or range of luminance of the system will be reduced. When the contrast falls to zero, the resolution limit of the system has been reached. There are different methods of measuring MTF than that given here.

A plot of MTF for a system can be calculated as follows (Stroebel et al., 1986) by finding separately for the object and the image

$$\text{Modulation object } Mo = (L_{max} - L_{min})/(L_{max} + L_{min})$$

$$\text{Modulation image } Mi = (L_{max} - L_{min})/(L_{max} + L_{min})$$

$$\text{Modulation factor} = Mi/Mo \text{ for each spatial frequency}$$

where L stands for Luminance.

Finally, one plots the spatial frequency in lines/mm or cycles/mm on the abscissa and the modulation factor on the ordinate. Modulation transfer functions are specially useful for comparing different performance of system components, like lenses or CCDs, or systems taken as a whole. Figure 3 displays the MTF of two hypothetical systems. The higher the MTF values the better.

Graininess

What is usually evaluated in photography as graininess, the detection of emulsion grain due to a high-speed emulsion or high enlargement, can be thought of as noise (Stroebel et al., 1986: 391). In this sense, the more usual signal to noise ratio of electronic equipment

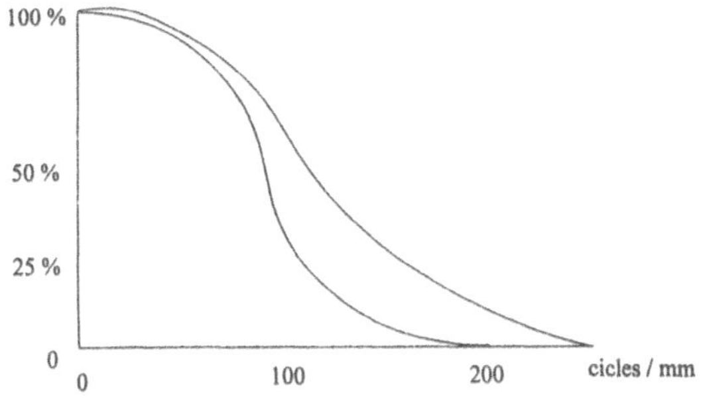

Figure 3. Modulation transfer function of two hypothetical devices.

can be equated to graininess in photography. Electronic imaging systems, especially those with analog circuitry, are especially prone to suffer from noise. The noise comes mainly from the camera itself and the analog to digital circuitry (Russ, 1992).

Some simple tests to gain an idea of the random noise level of the system can be (1) to acquire several frames with the objective of the camera completely covered and count the number of pixels different from the black value or (2) to use a standard photographic gray card (with an 18% reflection value) uniformly illuminated and calculate in the acquired images the dispersion around the predominant value. Frame averaging is a good procedure to improve the signal to noise ratio. It is better done through hardware than software, as frame averaging through software can be extremely tedious. The ratio improves as the square root of the number of averaged images (Russ, 1992).

These two tests are easy to perform and useful for comparative purposes. Imagina (Becerra "Imagina" article, this volume) includes a routine that, after giving a graphical display of the histogram of an image, writes an ASCII file with the number of pixels assigned to each gray level.

ELECTRONIC IMAGING SYSTEMS

Generalities

Following the three-step process presented above, I will summarize the more usual components of electronic imaging systems.

The acquisition step requires an optical system, a sensitive receptor and circuitry to produce a codified signal. The most common receptors are charged couple devices, CCD, which are usually of the following common sizes: 1/2 inch (6.4 mm x 4.8 mm), 2/3 inch (8.8 mm x 6.6 mm) and 1 inch (12.8 mm x 9.6 mm) (numbers in parenthesis may vary slightly). Knowing the CCD size is important, among other things, in order to choose the correct optical adapter for a microscope. There are four common CCD functional architectures (Janesic et al., 1987): (1) simple linear-shift register—the image is recorded by slowly scanning the scene; (2) area array as a full-frame imager—the scene is imaged onto the entire device; (3) integrated, frame-transfer structure—the scene is imaged onto half the array for an integration time (1/25 second or 1/30 second) and rapidly transferred to the optically masked lower half (this is the common architecture in commercial broadcast systems); and (4) interline transfer system—a parallel array of line sensors is separated by a transfer electrode from opaque CCD readout registers. The last was developed for electronic shuttering and eliminating image smearing common to the frame-transfer CCD.

Finally, apart from CCD pixel size, CCDs used in top-of-the-line cameras are classified in three or four classes of quality, based on the maximum number of point defects in the CCD (the numbers corresponding to each class varies from company to company).

The optical components that produce an image on the surface of the CCD do not need to be much different from those used for photography. In this sense, the classical criteria of aberrations, resolution or even MTF can be usefully applied.

The acquisition step ends with coding the voltage variation produced by the CCD into either a standard video signal, or a high-frequency signal or transforming it to digital format before leaving the camera.

For the storage step, a wide number of devices are available. Analog storing in VCRs suffers from a fixed resolution limit imposed by the recording standard. See Becerra et al. (1993) for common video recording formats. Once the analog signal is digitized a much wider offering of storage devices is available, from optical drives to permanent CD-ROM or the new KODAK CD-PHOTO, to more traditional hard drives and, more recently, digital audio tapes (DAT). Digital images use a lot of memory, so, at the least, a criterion for selecting a storage device should take into account transfer speed, total storage memory size and price.

In electronic imaging systems the display components consist minimally of a graphic board and a monitor. Resolution and number of shades (or colors) are the two parameters usually employed to evaluate these devices. The common VGA board can display only 64 shades of gray. Frame sizes of 640 x 480, 800 x 600 or 1024 x 768 are very common in so-called super VGA (sVGA). Enough video RAM is necessary to display true color (2^8 red, 2^8 green, 2^8 blue) for a certain frame size. Monitor dot pitch is a good indication of quality. Present good-quality monitors have a dot pitch of 0.26 or lower. More expensive monitors display more than 2^8 shades of gray, but they are not usually needed in morphometrics.

Low-Resolution Systems

I include these for completeness. They are not very useful for morphometrics. Low resolution can be the result of a low-resolution camera, like the FOTOMAN or the CANON ION RC-260 (XAPSHOT in the United States), or a limited recording system, like 8-mm amateur video or VHS (see Becerra et al., 1993 for more on this).

These systems play a useful role in the production and editing of newsletters, catalogs, etc. Curiously enough, they need not be cheaper than a medium-resolution camera.

Medium-Resolution Systems

Several configurations are possible here. The more traditional and still the recommended system for morphometrics includes the following.

Video Camera. Monochrome cameras achieve higher resolution than do similar color cameras. Good resolution for a monochrome video camera is around 570 TV lines of

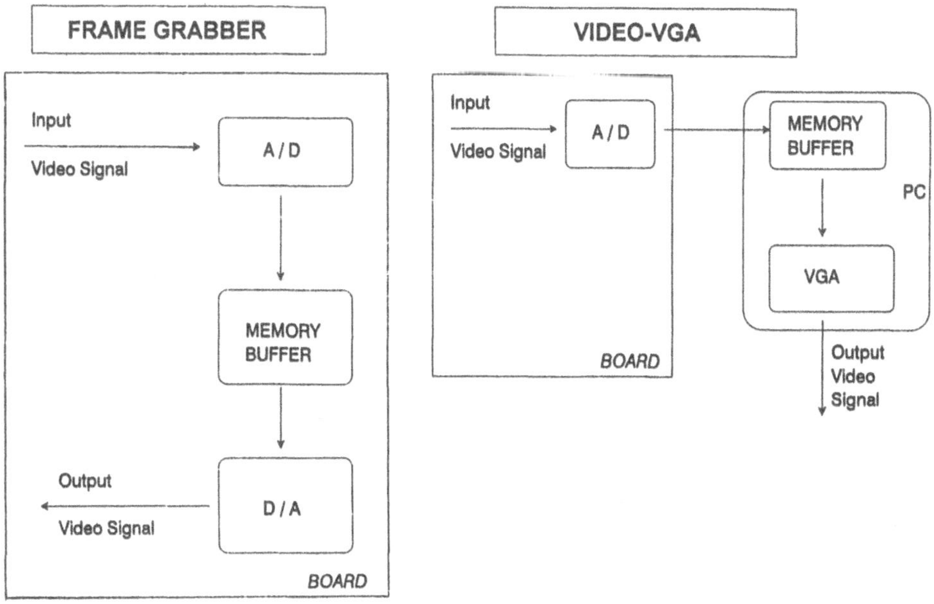

Figure 4. Two common types of analog to digital (A/D) converters.

horizontal resolution. High-quality camcorders are around 440 TV lines (horizontal,) that degrade to an upper limit of 400 if recorded on Hi 8 mm or SVHS and 200 if recorded on 8 mm or VHS. Alternative digital video cameras (25 or 30 frames per second) are available. They are more expensive and need a dedicated board but have better signal to noise ratios.

Analog to Digital Converter. To incorporate images for computer processing the analog signal must be digitized through a frame grabber or similar device. Figure 4 illustrates two common types of these analog to digital (A/D) converters. The process is graphically explained in Fig. 5. The voltage of the video signal is first sampled at a certain rate. Next, those voltages are constrained to some values (quantification). Each value represents a shade of gray. (Refer to the citations at the beginning of this chapter for a detailed treatment of analog to digital conversion.) Boards for A/D conversion can be found from US$150 (Reis, 1993). The more expensive ones include dedicated memory on the board and hardware-implemented image processing. Some of the tests suggested above are worth trying with the whole camera-board system before making a decision. European users are advised to take note of the aspect ratio of the digital signal output from the board. Some boards give 768 horizontal x 512 vertical output that leads to a 3:2 aspect ratio. It is more convenient to have a frame digitized at 768 horizontal x 576 vertical, which keeps the 4:3 aspect ratio of the computer screen.

Display. It has been mentioned that VGA boards cannot display more than 64 shades of gray. Depending on the video RAM, more or fewer colors will be available for display.

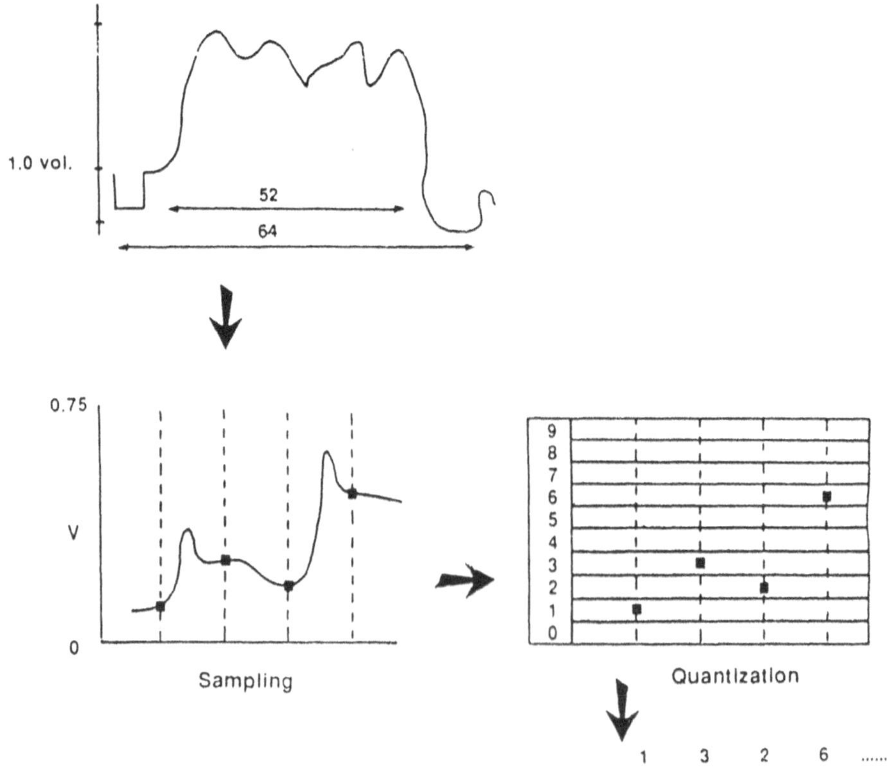

Figure 5. Sampling and quantification of analog to digital conversion.

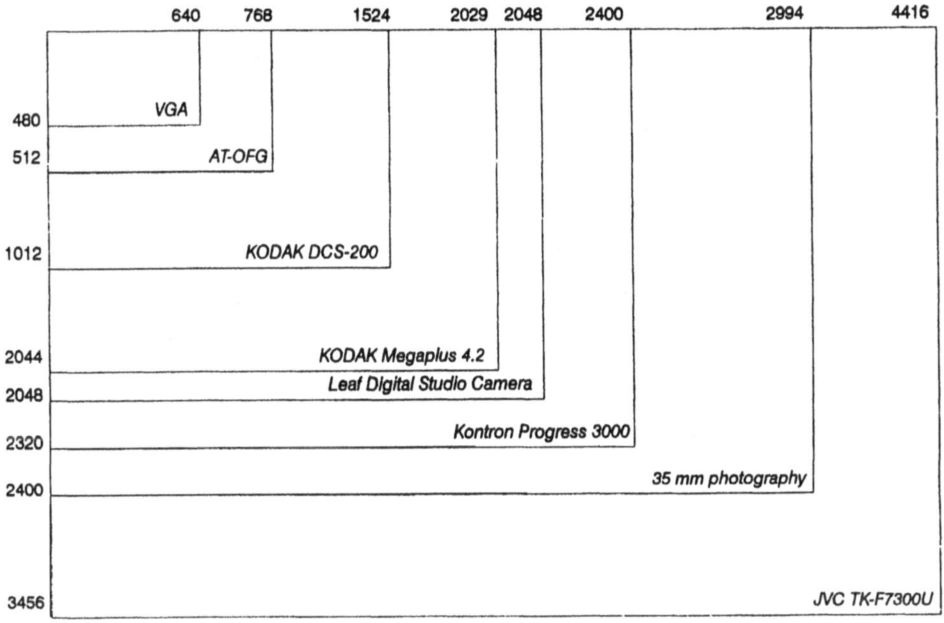

Figure 6. Horizontal and vertical resolution in pixels of some high-resolution systems.

The medium-resolution systems are the best choice for morphometrics in terms of quality and price. Adapters for microscopes are easily found, and there is inexpensive, available software ready to work at this resolution.

High-Resolution Systems

If the interest in the acquisition of images is beyond immediate use for morphometrics but instead, for example, for some sort of digital bank of very high-quality images for taxonomic and future reference work, then top digital quality systems should be employed. The high-resolution and image quality of these systems are not needed for measuring.

Figure 6 names some high-resolution systems. These systems can be arbitrarily delimited with a lower boundary of an array of 1000 x 1000 pixels. The main differences from the previous systems lie in the acquisition step, although they have high requirements for storage and display. No clear classification is available for these systems.

To be able to digitize at high-resolution it is not necessary for the camera to have a big CCD. Two different cameras illustrate the process that can be called "building image acquisition through integration." The Kontron Progress 3000 and 3012 have a 2/3-inch CCD of 500 x 582 pixels. By a special movement of the CCD, image multiples of this array can be built up to a resolution of 3000 x 2320. Acquisition is through three colors, red green and blue that can be transformed later to shades of gray by software. When the highest resolution is used, it can take more than 20 seconds to acquire an image. The signal leaves the camera in digital format to a dedicated board in the computer.

Similar in concept although working on a different mechanism is the JVC TK-F7300U. The basic image size is 736 x 576. A single image of 4416 x 3456 pixels is built up as a multiple of the active elements through the use of a piece of angled optical glass that lies between the CCD and the lens. The glass rotates in tiny horizontal and vertical steps to compile a full image. A digital board integrates the image. Signal to noise ratio is inferior to the Kontron, but the JVC is less expensive.

Integration cameras seems to have a promising future, specially taking into account how expensive it is to produce large, high-quality CCDs.

High-resolution cameras based on big, high-quality CCDs are illustrated by the KODAK Megaplus series. With pixels arrays from 1317 x 1035 to 2029 x 2044 and pixel size of 6.8 x 6.8 microns, the KODAK Megaplus series can be considered the state of the art at this resolution. The CCD manufactured by KODAK for these cameras is used by other companies to produce their own equipment, e.g., PHOTOMETRICS or XILLIX.

Apart from the time that these cameras take to acquire images, they pose a special problem in storage and display. Compression algorithms are limited to the "lossless" case (it does not make sense to acquire high-quality, high-resolution images to immediately degrade them). Display is limited by the size of the monitor and the power of the graphics board. Big images cannot be fully displayed at once.

High price and expensive additional equipment do not justify an investment in these systems for morphometrics. Only if the purpose is to archive images for future data retrieval can these systems play a role in taxonomic imaging.

The KODAK PHOTO-CD could be a good alternative to high-resolution systems. Images are first taken on film and then scanned by the company and loaded onto a compact disk. One-hundred images can be stored on a disk, and images can be displayed at different resolutions. PHOTO-CD can achieve higher resolution than can the JVC TK-F7300U.

Software

As for hardware, software for morphometrics can be divided into low-cost and high-cost systems. Becerra et al. (1993) review MeasurementTV, MorphoSys and Java. The first two, being strictly dedicated to morphometrics, are worth the investment; however, MorphoSys (as Java) will not run unless a PCVISION board is in the computer.

The trend seems to be for software to work under WINDOWS and to offer more diversified functionality than the pure morphometrics routines. Price consequently increases. An example of this tendency is MOCHA, the imaging software of JANDEL for WINDOWS.

Imagina (Becerra, this volume) is free software for image acquisition and production of some elementary morphometric functions, such as distances and landmark coordinates.

When one wants to go further into image treatment, it seems that highly priced software like OPTIMAS and others are not justified unless an equivalent investment is done in hardware.

Hardware and software must go together.

CONCLUSION

In order to use the morphometric tools described in this volume, the following recommendation is in order: use a high-resolution (570 lines horizontal resolution), monochrome, video analog camera with a reasonably priced Video VGA board (such as Screen Machine or Video Blaster). Take care to keep the same aspect ratio in acquisition and display. Use software such as Imagina, MeasurementTV, or the like to take distances and landmark coordinates.

ACKNOWLEDGMENTS

This work became possible thanks to a grant from the CSIC, project PB92-0082. Leslie F. Marcus made very useful comments on the manuscript. I also want to thank to José M. Becerra, Elisa Bello, Maria Soledad Vicente, James Rohlf and Fred Bookstein for all the comments they made.

REFERENCES

Adams, A. 1981 The negative. Little, Brown and Company, Boston.

Becerra, J. M., E. Bello and A. García-Valdecasas 1993. Building your own machine image system: A user point of view. Pages 65–92 *in* L. F. Marcus, E. Bello, and A. Garcia-Valdecasas (eds.), Contributions to morphometrics. Monografias del Museo Nacional de Ciencias Naturales 8, Madrid.

Edmund Scientific. 1994. Annual reference catalog. 101 E. Gloucester Pike, Barrington, NJ 08007-1380, USA.

Fink, W. L. 1990. Data acquisition for morphometric analysis in Systematic biology. Pages 9–19 *in* F. J. Rohlf and F. L. Bookstein, (eds.), Proceedings of the Michigan morphometrics workshop. University of Michigan Museum of Zoology Special Publication no. 2: Ann Arbor.

Inoué, S. 1986. Video microscopy. Plenum Press, New York.

Janesic, J. R., T. Elliot, S. Collins, N. M. Blouke and J. Freeman. 1987. Scientific charged-coupled devices. Optical Engineering 26: 692–714.

MacLeod, N. 1990. Digital images and automated image analysis systems. Pages 21–35 *in* F. J. Rohlf and F. L. Bookstein, (eds.), Proceedings of the Michigan morphometrics workshop. University of Michigan Museum of Zoology Special Publication no. 2: Ann Arbor.

Meacham, C. E. 1992. Practical machine vision for morphometrics data acquisition. *In* J. T. Sorensen and R. G. Footitt, (eds.), Ordinations in the study of morphology, evolution, and systematics of fishes: Applications and quantitative genetics rationales. Elsevier: Amsterdam.

Miyake, Y., S. Inoué, Y. Tamoto, and S.Kubo. 1984. Evaluation of image quality of digital pictures obtained by drum scanner and photo-printer. Pages 62-66 *in* H. M. Graves (ed.), Photographic and electronic image quality. Royal Photographic Society: Cambridge.

Reis, C. 1993. What a $150 PC imaging board can deliver right now. Advanced Imaging (May):32-34.

Russ, J. C. 1992. The image processing handbook. CRC Press: Boca Raton, Florida.

Stroebel, L., J. Compton, I. Current, and R. Zakia. 1986. Photographic materials and processes. Focal Press: Boston.

IMAGINA

A Direct Tool for Image Analysis in Systematics

José M. Becerra*

Museo Nacional de Ciencias Naturales
José Gutierrez Abascal, 2
Madrid 28006, Spain

ABSTRACT

Imagina is a program for the acquisition, treatment and quantitative analysis of images developed for morphometric analysis. The structure of the program is presented, and technical information for all options is discussed. Imagina is compared with other programs. Some planned future developments are discussed.

INTRODUCTION

Imagina is a program for the acquisition, treatment and quantitative analysis of images developed for morphometric analysis. There are many programs for image analysis, so, why Imagina?

I decided to develop Imagina with two purposes in mind. The first purpose is to have a program in which there is control and knowledge of what processes are applied to the images, from transformations to visualization. Too many programs process images like "black-boxes," i.e., an image goes in, a process is performed and a result is obtained at the other end, with no knowledge of what goes on between the input and output. Knowing what is done and precisely the way it is done is critical for scientific endeavor. The second purpose for developing Imagina is to obtain a platform on which to test algorithms and routines (new and old) in the field of systematics.

Imagina was developed in Borland Pascal 7.0 for IBM PCs and clones, using the protected mode option. This setting allows use of up to 16 megabytes of RAM memory, so that images as large as 4K pixels by 3K pixels can be used for transformation and display. The program is menu driven and can work with a mouse. A beta version of Imagina is available at http://albia.museo.csic.es/tools.html at the Museo Nacional de Ciencias Naturales.

* e-mail: mcnjm14@cc.csic.es

An important point is that although the program is capture-board dependent for image acquisition, it can perform all the other functions without a board being present.

Becerra (1994) classified three levels of image software in terms of routines included and price. In this classification, Imagina belongs to the first level, grouped with programs like MorphoSys, MeasurementTV, and Java (Becerra et al., 1993). Fig. 1 shows the main menu and the options for every selection.

STRUCTURE OF THE PROGRAM

The main choice of options, presented in the top-bar menu, follow:

1. **Board.** Allows one to capture images and extract basic information (coordinates and distances).
2. **Utilities.** Allows one to view or transform images or data files, to edit ASCII files, to temporarily exit to DOS, and to configure the program.
3. **Morphometrics.** Allows one to obtain shape coordinates or run programs from other authors.
4. **Image.** Allows one to perform several image treatment operations.
5. **Help.** Gives information on the program.

These options are discussed below in terms of logical modules that follow the kinds of processes applied to images.

Image Capture

Imagina has an acquisition module that works through a monochrome AT-OFG frame grabber (European version only) from Imaging Technology. The images can be stored in a file, in either TIFF or binary format, with 256 gray levels. If a video monitor is not available,

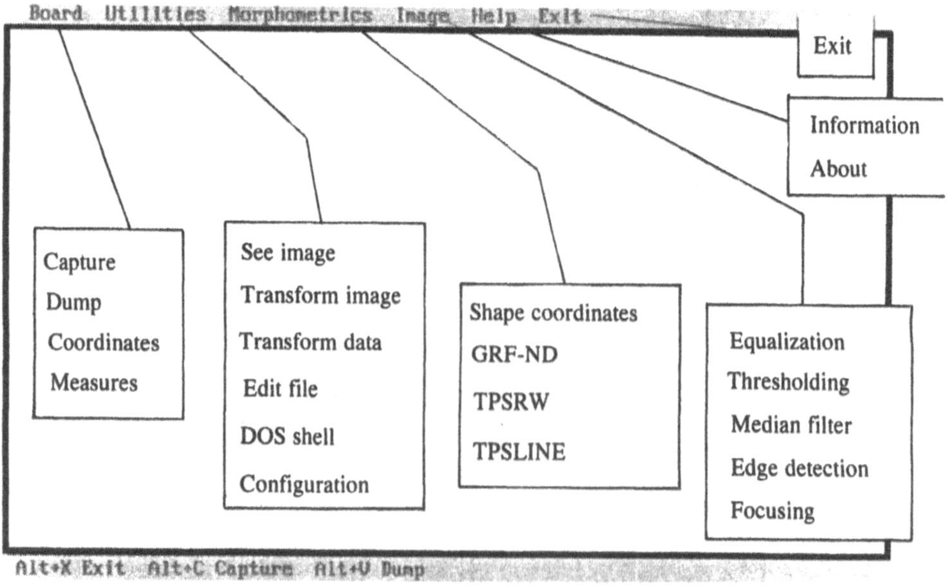

Figure 1. Main options in Imagina with their sub-options.

```
INFORM ABOUT FILE acaro.tif

ImageWidth (256)
  Type= SHORT
  Length= 1
  Value= 768
ImageLength (257)
  Type= SHORT
  Length= 1
  Value= 512
Bitspersample (258)
  Type= SHORT
  Length= 1
  Value= 8
Compression (259)
  Type= SHORT
  Length= 1
  Value= 1
PhotometricInterp (262)
  Type= SHORT
  Length= 1
  Value= 1
ImageDescription (270)
  Type= ASCII
  Length= 0
  desplazamiento a 0
StripOffsets (273)
  Type= LONG
  Length= 1
  Value= 98
```

Figure 2. Tags included in the header of the TIFF files generated by Imagina. This list has been obtained using the program Tags (Becerra, 1994).

one can focus the image on the computer screen. The screen is refreshed every 7 seconds with a superVGA (sVGA) resolution of 800 x 600 pixels. This is very useful for dead or fixed specimens, for example in paleontological work, as you do not need a video monitor. The TIFF (Kay and Levine, 1992) file generated in the capture option can be read by any imaging program because TIFF is a very well known standard for image files.

The tag fields that Imagina includes in the header of the TIFF file are shown in Fig. 2. This list of fields has been obtained using the program Tags (Becerra, 1994; see below).

Image Treatment

Imagina allows certain transformations of images, which are discussed below.

Histogram Equalization. Histogram equalization (Jensen, 1986:127) stretches the contrast of an image. This operation is the same as histogram flattening in MeasurementTV.

Thresholding. Thresholding (Lindley, 1991:321) converts an image into a binary image, given a threshold. It is useful for work with edges.

Median Filter. Median filter (Lindley, 1991:381), eliminates the random noise generated during capture. It also performs a smoothing operation.

Edge Detection. Edge detection (Lindley, 1991:377) combines a Sobel routine with a thresholding operation.

Rohlf (1990) gives useful insight to all of these algorithms. All these routines share certain properties in that they are performed via software instead of hardware. This implies that the operations take more time for large images, especially for the median filter operation.

Importantly, although an sVGA graphic card displays only 64 shades of gray, which is equivalent to 256 colors, the operations mentioned above are performed over the 256 shades of gray saved in the file. If at a later time we want to dump the image onto the frame grabber memory, we still retain the whole range of levels. Images can be seen at sVGA resolution (up to 1024 x 768 pixels with 256 colors or 64 shades of gray) thanks to a special driver for Borland Pascal (Hargrave, 1993).

Data Analysis

Once one has obtained an image, the program can extract quantitative data and then make an analysis of that data. Information can be extracted in two ways: coordinates, for landmark data, and distances between landmarks. The coordinates are taken from image files, again of any size and in TIFF or binary format. The coordinates are stored in ASCII files, in "Imagina format." One can convert this proprietary format into others (NTSYS, GRF-ND, MATLAB or SAS) with the data conversion option. If the image cannot fit on the screen for the chosen VGA resolution, only the portion of the image that fits on the screen is shown, and, using a menu (on the bottom of the screen), one can virtually move or pan the screen to see the hidden parts. In this way, images obtained with a system different from the AT-OFG (which acquires images with a size of 768 x 512 pixels), such as a high-resolution digital camera or a scanner, can be treated. The measurement option follows MorphoSys design (Meacham, 1993) in that a command file is applied to the coordinate file to obtain a file with the distances if desired. This distance file can be in any of the formats mentioned above.

Analysis of the extracted coordinates is done through the operations included in the option Morphometrics. These operations are of two classes: one routine obtains the shape coordinates of a set of landmarks, following Bookstein et al. (1985) and Bookstein (1991); the other class is a group of routines that interface with programs of other authors, namely TPSRW, TPSPLINE and GRF-ND.

Utilities

Some utilities have been added to the program. The important ones are the options transform data and transform image; which make textual or visual information available to other programs. There are two options for changing the data format. The image conversion routine is a subset of another program, called Tags, that can read the header information contained in TIFF, Targa and PCX files and write the header to an ASCII file in a readable format.

Edit File. A small text editor is available to generate the command files or to see data or other files.

See Image. An image display option to see an image quickly, before any process executed on it is included. This option can display TIFF or binary images of any size, although here panning is not yet available.

The tests in Becerra et al. (1993) were performed using Imagina. The results are given in Table 1. Figure 3 shows the geometric forms used to obtain the measures. We can see that Imagina is between MorphoSys and MeasurementTV with regards to precision. The high accuracy (low standard deviation) in acquisition may be due to the ease of placing the mouse-controlled cursor on the same point every time we record a landmark.

Table 1. Results (absolute mean error and standard deviation in parenthesis) of the test performed on different programs for image and morphometrics analysis. The results for Imagina are given for different VGA resolutions. Results for MorphoSys, Java and MeasurementTV are obtained from Becerra et al. (1993). Images measured are shown in Fig. 3.

		Scale on X axis			Scale on Y axis		
		1	2	3	1	2	3
Distance 1-2	*MorphoSys*	1.22 (0.39)	0.76 (0.18)	0.36 (0.11)	-	-	-
	MTV	-0.18 (0.33)	-0.52 (0.08)	-0.37 (0.0)	0.88 (0.06)	0.25 (0.09)	0.00 (0.00)
	Java	-0.78 (0.32)	-0.17 (0.06)	-0.02 (0.13)	0.75 (0.00)	0.06 (0.20)	0.02 (0.06)
	Imagina 640 x 480	-0.59 (0.0)	0.78 (0.0)	-0.39 (0.0)	-	-	-
	Imagina 800 x 600	-0.59 (0.0)	0.78 (0.0)	-0.39 (0.0)	-	-	-
	Imagina 1024 x 768	-0.59 (0.0)	0.63 (0.50)	-0.39 (0.0)	-	-	-
Distance 2-3	*MorphoSys*	1.20 (0.34)	1.11 (0.27)	0.42 (0.15)	-	-	-
	MTV	0.44 (0.23)	-0.02 (0.11)	-0.19 (0.00)	1.75 (0.13)	0.68 (0.16)	0.30 (0.10)
	Java	-0.14 (0.28)	0.04 (0.20)	0.04 (0.12)	1.39 (0.43)	0.36 (0.19)	0.17 (0.06)
	Imagina 640 x 480	0.77 (0.0)	-0.78 (0.0)	-0.39 (0.0)	-	-	-
	Imagina 800 x 600	-0.59 (0.0)	-0.27 (0.0)	-0.14 (0.0)	-	-	-
	Imagina 1024 x 768	-0.59 (0.0)	-0.27 (0.0)	-0.14 (0.0)	-	-	-
Distance 1-3	*MorphoSys*	2.04 (0.23)	1.47 (0.18)	0.68 (0.09)	-	-	-
	MTV	-0.13 (0.11)	-0.48 (0.09)	-0.52 (0.13)	1.80 (0.31)	0.69 (0.08)	0.15 (0.17)
	Java	-0.41 (0.43)	-0.02 (0.06)	-0.04 (0.14)	0.55 (0.65)	0.59 (0.09)	0.10 (0.13)
	Imagina 640 x 480	0.12 (0.0)	-0.01 (0.0)	-0.56 (0.0)	-	-	-
	Imagina 800 x 600	-0.83 (0.0)	0.36 (0.0)	-0.41 (0.0)	-	-	-
	Imagina 1024 x 768	-0.83 (0.0)	0.24 (0.35)	-0.38 (0.0)	-	-	-

IMAGINA AND OTHER IMAGE SOFTWARE AT THE SAME LEVEL

One important difference between Imagina and MorphoSys and Java is that the former is able to operate without a frame grabber in the computer. Imagina can use image files, obtained by other means, as input. In this way, one can have high-level image capture equipment at a central facility and some midlevel computers for image analysis elsewhere. At present, Imagina can only acquire images with the AT-OFG frame grabber. The program is being restructured to have several executable modules called from a principal module, i.e.,

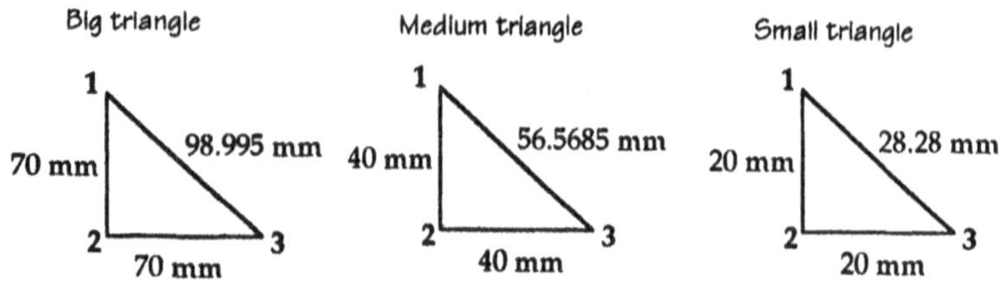

Figure 3. Real values for the measures shown in Table 1. The big triangle was 1, the medium was 2, and the small was 3 (taken from Becerra et al., 1993).

the main menu. It will then be easy to develop new versions for other frame grabbers by replacing the acquisition module with a new one, without altering the rest of the program.

Imagina can use a computer monitor as the sole image display hardware during acquisition. MeasurementTV can display images on the computer monitor, but only when using image files—that is, not live images. When using a computer monitor for Imagina, as García-Valdecasas (this volume) points out, be aware that the difference in the aspect ratio value between the AT-OFG and cards which digitize at 512 horizontal points instead of 576 (which is a ratio of 3:2) and the computer screen (which is a ratio of 4:3) can lead to errors when performing a specific operation. This does not affect measurements because of calibration but may present distorted images.

Imagina requires a specified sVGA graphics card, namely any of those using the common Tseng 4000 or Paradise chips. I am planning to allow the program to use a wider range of graphic cards. Through the sVGA driver it is possible to use the following extended VGA modes: 640 x 480, 800 x 600, and 1024 x 768 pixels, all of them with 256 colors (64 shades of gray). The high resolution modes give better visualization, because they combine a large number of points with a midlevel number of shades of gray.

MorphoSys can work only with images of the size of the frame grabber. MeasurementTV can display TIFF images of any size, but because the program fits the image on the computer screen area (640 x 480 or 320 x 200 points), when the size of the image is greater than the area of visualization, MeasurementTV removes pixels. Imagina maintains a 1:1 ratio for pixels in the file and pixels in the screen. This mean that sometimes one will not see the whole image on the screen at one time, but, as in the case of coordinate acquisition, one can move a virtual window over the image.

One important tool for studying images is the histogram (Lindley, 1991). With this tool the dispersion of the brightness values in the image can be studied and improved by means of one of the routines mentioned above. Imagina includes the possibility of extracting the histogram of an image and displaying the lowest and highest intensity values with associated pixels.

ACKNOWLEDGMENTS

Antonio Garcia-Valdecasas helped with many ideas during the development of the program and encouraged me to write this paper. Leslie F. Marcus made useful comments on the manuscript. J. Hargrave helped to set up his driver for our sVGA card. This work was possible thanks to a CAM grant and a grant from the CSIC under project PB92-0082.

REFERENCES

Becerra, J. M. 1994. La automatización de la información visual en Historia Natural: Aplicación a la morfometria. Ph.D. dissertation, Universidad Autónoma de Madrid.

Becerra, J. M., E. Bello and A. García-Valdecasas 1993. Building your own machine image system: A user point of view. Pages 65–92 *in* L. F. Marcus, E. Bello, and A. García-Valdecasas, (eds.), Contributions to morphometrics. Monografías del Museo Nacional de Ciencias Naturales 8, Madrid.

Bookstein, F. L. 1991. Morphometric tools for landmark data: Geometry and biology. Cambridge University Press: Cambridge.

Bookstein, F. L., B. Chernoff, R. Elder, J. Humphries, G. Smith, and R. Strauss, 1985. Morphometrics in evolutionary biology. The Academy of Natural Sciences of Philadelphia Special Publication 15: Philadelphia.

Hargrave, J. 1993. Graphics driver for Turbo Pascal. Jordan Hargraphix Software.

Jensen, J. R. 1986. Introductory digital image processing: A remote sensing perspective. Prentice-Hall: Englewood Cliffs, New Jersey.

Kay, D. C. and J. R. Levine 1992. Graphics file format. WindCrest/McGraw-Hill: New York.

Lindley, C.A. 1991. Practical image processing in C: Acquisition, manipulation, storage. Wiley: New York.

Meacham, C. 1993. MorphoSys: An interactive machine vision program for acquisition of morphometric data. Pages 393–402 *in* R. Fortuner, (ed.), Advances in computer methods for systematic biology. Artificial intelligence, databases, computer vision. The Johns Hopkins University Press: Baltimore, Maryland.

Rohlf, F. J. 1990. An overview of image processing and analysis techniques for Morphometrics. Pages 37–60 in: F. J. Rohlf and F. L. Bookstein, (eds.), Proceedings of the Michigan morphometrics workshop. University of Michigan Museum of Zoology Special Publication 2.

AN IMAGE-CAPTURE AND DATA-COLLECTION SYSTEM FOR MORPHOMETRIC ANALYSIS

András Demeter,[1,3] János Vámosi,[2] László Peregovits,[3] and György Topál[3]

[1] Secretariat of the Hungarian Academy of Sciences
Nádor u. 7
Budapest, H-1051 Hungary
[2] Institute of Nuclear Research of the Hungarian Academy of Sciences
Debrecen, Pf. 51
H-4001 Hungary
[3] Zoological Department
Hungarian Natural History Museum
Baross u. 13, Budapest, H-1088 Hungary

ABSTRACT

We describe a video data-acquisition system called Budapest Video and its software component IMAGOES, designed to collect two-dimensional landmark and outline data. Various calibration tests on linear scales and biological specimens show that the accuracy of the system may exceed the resolution of landmarks on biological specimens. The software was designed to help the user collect landmark data by developing training sets of landmarks. The results of fitting elliptical Fourier analysis to outlines may be visually inspected by plotting the captured and reconstructed outlines simultaneously.

INTRODUCTION

Biological morphometrics has undergone major development over the past fifteen years. Rohlf and Marcus (1993) have called this a revolution. Two classes of novel geometric morphometric techniques are now being used. They focus on either coordinates of landmarks (biologically homologous points) or the outlines of the objects being studied.

Whereas data collection can be done by hand, it is extremely laborious and cumbersome, especially with outlines. Initially, the development and application on a wider scale of the new morphometric techniques was hindered by lack of both data capture software and analytical programs. Progress in data-capture has been reviewed by Fink (1990) and Rohlf (1993a). A considerable advance was the development of interfaces between video cameras and computers (VDAS, video data acquisition systems). Ever better and more powerful hardware and increasingly more sophisticated software have been produced during the past decade to capture (digitize), manipulate and store images (see MacLeod, 1990 and Rohlf, 1990a). There is now a lot of commercial image-processing software available. However, there exists a relatively limited number of hardware products and software for data capture designed with the biological morphometrician in mind. Biomedical applications are too specialized and often prohibitively expensive to suit the needs of systematists wanting to use morphometric tools.

In 1990 the senior author embarked upon a project studying craniometric variation in house mice and their wild relatives (genus *Mus*, Rodentia). Because of the unavailability of appropriate hardware and software in Hungary (e.g. those reviewed in Fink 1990), we decided to build a complete hardware and software system of our own. There were several reasons for doing so besides the difficulty in obtaining commercial products, which might eventually have been overcome. A limited budget was one, and heeding Fink's advice, we were wary of turn-key systems. At the same time that we had just begun using the locally developed system, Becerra et al. (1993) offered some important "hindsights" relevant to our experience.

Below is a brief description of the Hungarian system, nicknamed "Budapest Video" (B-V) by Leslie F. Marcus. The software is called IMAGOES. This title should sound familiar to biologists: it refers to the adult stage of insects. It is also the plural of the Latin word for image, as well as an acronym for **IM**age **A**nalysis of **G**eometric and **O**utline data in **E**volution and **S**ystematics.

DESCRIPTION

Hardware

The heart of the system is a nonpatented frame grabber developed and manufactured by a small firm in Hungary. It was originally designed for a system analyzing X-ray images. It is an 8-bit, black-and-white, high-resolution (768 x 576 pixels) digitizing card that accepts composite video input (a color card is now available).

In contrast with several other systems, the image to be captured and analyzed is not displayed on a separate monitor but on the computer monitor. Super VGA (sVGA) cards with at least 512 Kb RAM are supported through VESA drivers. An sVGA monitor capable of at least 800 x 600 pixel resolution (preferably in noninterlaced mode) is recommended (although 640 x 480 pixel resolution is also supported for VGA monitors). The video signal may also be relayed through an RF modulator to a black-and-white or color television set so that positioning and adjusting the illumination of the specimen is made easier, but it is not an absolute requirement to run the system.

There are many cameras that may be used but greater than 350 scan line resolution is recommended. The prototype runs with a black-and-white Panasonic WV-BP100 CCD camera, which has a horizontal resolution of 380 lines. Currently, the camera is fitted with a custom-made macro lens that enables the user to work on images 3-50 mm in size; but by inverting the lens, one may view images more than 50 cm in size.

The minimal computer configuration is a 386 processor, 2MB RAM, with DOS 3.3 or higher, but a faster system, i.e., 486 processor and large hard disk capacity, is recommended if large numbers of image files are to be stored. The frame grabber card has been successfully installed in various IBM clones.

Features of IMAGOES

The software is coded in C++ with some assembly routines. The program is completely menu-driven and requires a mouse or track-ball pointing device. The user is guided by pull-down and nested menus, and on the graphics screen (when the captured image is displayed on the monitor), the user operates the program by clicking on push-buttons along a tool bar. After having carried out the necessary configuration commands, the user is ready to read a new image or recall a previously saved image. If there is no television set attached to the camera, pressing F10 will bring up a small graphic display of the digitized image in the upper-right corner of the screen in real-time mode, and adjustments can be carried out easily.

The following features were thought to be important when designing the software.

Landmark Data (Example Screen in Figure 1). The aspect ratio of the screen is corrected for by the program, therefore only one scale must be given before collection of landmarks begins, or after a change of magnification.

It is a mandatory function to set up an internal coordinate system for the object, unequivocally determined by two points and the order of entry. The first coordinate gives the origin, the second gives the direction of the x axis.

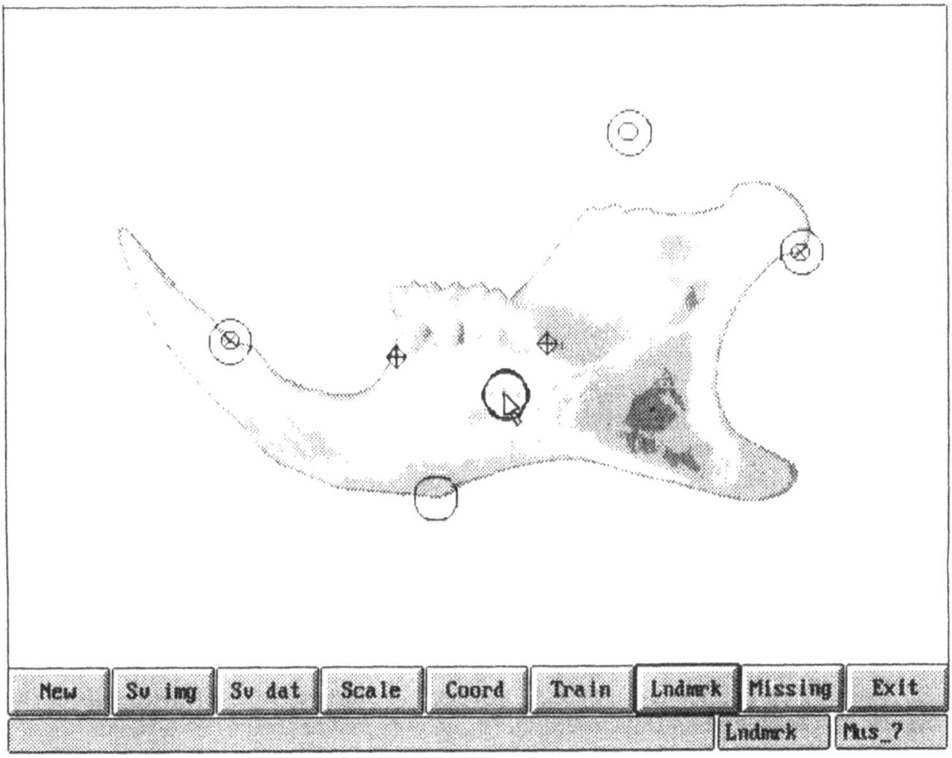

Figure 1. A sample screen of IMAGOES in landmark collection mode. The mandible of a mouse is viewed from the lingual side. The training sets are shown by circles, collected landmarks appear as crossed circle; and missing landmarks are empty circles. The next landmark to be collected is indicated by a thick training circle; the cursor is resting on that landmark. The system coordinates are crossed rectangles.

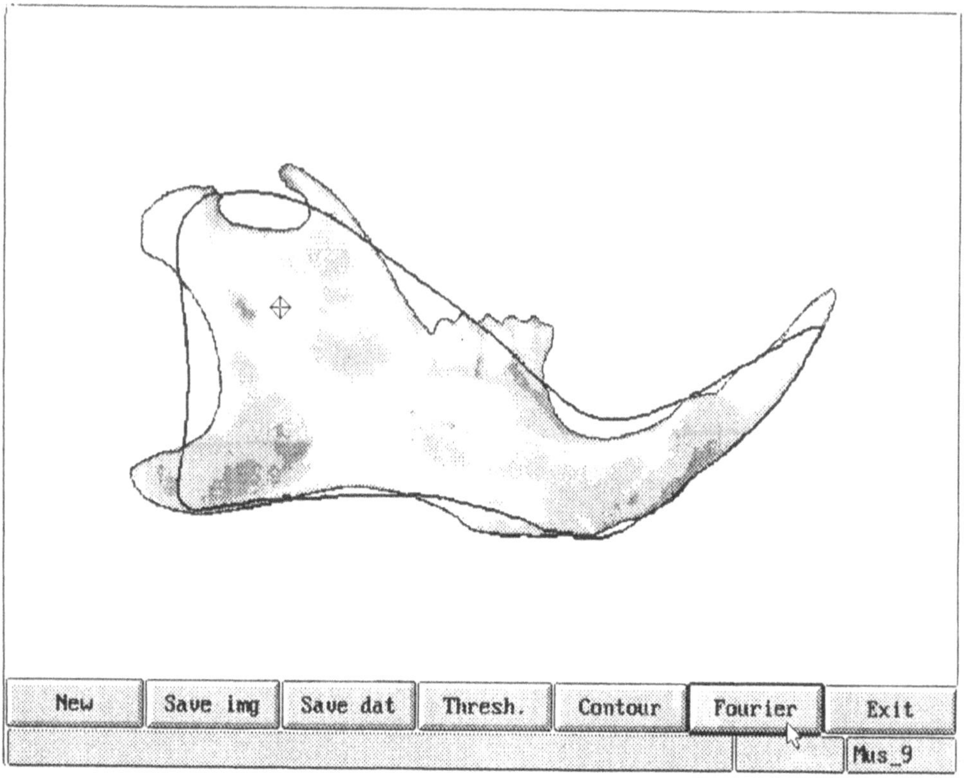

Figure 2. A sample screen from IMAGOES in Fourier analysis mode. The mandible of a mouse is viewed from the lingual side. The outline found by the program is superimposed in this example by the inverse Fourier analysis of eight harmonics, which gives only a rough approximation of the original shape. About 25 harmonics are necessary to describe this shape accurately.

The positions of landmarks may be recorded as a nondeformable template. This is called the "training mode," and "training circles" that indicate the approximate positions are placed on the object so that inexperienced or fatigued users are guided through the routine of data capture. The scaling and the coordinate systems are necessary so that the absolute coordinates of the landmarks can be calculated and the training set circles can be referenced to the object invariant to its size and position on the screen.

There is a choice of two types of coordinate data: raw (Cartesian) coordinates or shape variables (Bookstein, 1991) standardized to the baseline defined by the system coordinates.

Missing values are declared as such during data capture to avoid any uncertainty and possible error during subsequent analyses (coded 999999 in the data file).

Outline Data (Example Screen in Figure 2). Basic image processing (threshold adjustment) to sharpen contrast between object and background is provided.

Automated feature extraction is available: normalized (invariant to size and translation) elliptical Fourier coefficients (Kuhl and Giardina, 1982) can be calculated and saved to a data file. The terms A_0 and C_0 and for each harmonic i, the coefficients A_i, B_i, C_i and D_i are saved. An inverse Fourier transformation is then carried out, and the reconstructed outline overlaying the gray-tone image of the object is projected onto the screen. The number of harmonics necessary for describing the object to a visually ascertained level of precision may be determined interactively.

Common to Both Landmark and Outline Data. Information as regards type of measurement, image compression (if any), setting of display resolution, name of data file, and so on are saved in a configuration file. The user may switch between various configuration files.

The resolution of the computer monitor may be set from 640 x 480 through 800 x 600 to 1024 x 768 pixels.

Serial capture and recording of images and data is facilitated by the program supplying previously declared root names for identifiers and automatically incrementing serial numbers. This may be very helpful if data are collected from large series of specimens that have incremental numbers.

Image may be saved to disk in industry-standard TIFF 5.0 format (Aldus tagged image file format; choice of black-and-white, gray scale or 8-bit palette colors, with options of uncompressed, standard or differential LZW compressed formats). If no compression is chosen, files are accessed much more quickly but occupy considerably more disk space (approximately 440 Kb).

Data saved in a dBase III file format are easily imported by statistics packages (e.g. SYSTAT, SPSSPC+). Because of the limitation of this file format to 128 fields, up to 60 landmark coordinates, or 30 harmonics of an outline may be collected.

Detailed, key-sensitive, on-line help at most stages of the data collection session is provided.

TESTS OF THE SYSTEM

We have carried out a number of calibrations and tests of the system in order to measure its precision and the repeatability of the measurements (accuracy).

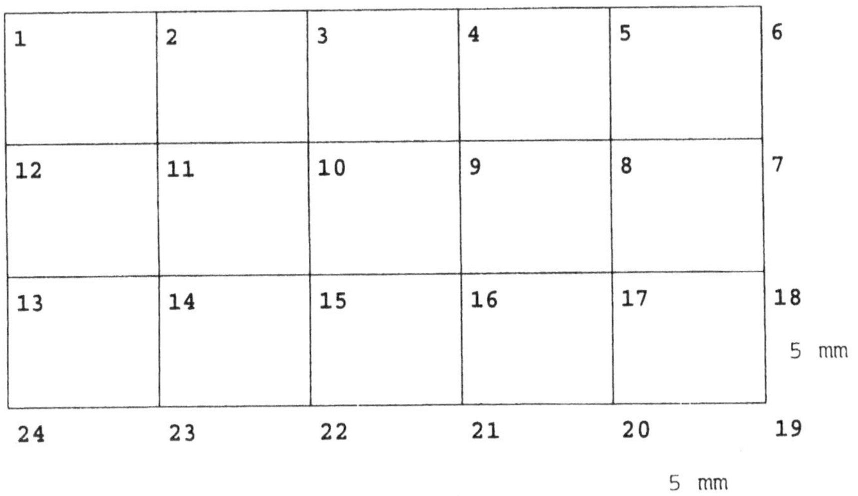

Figure 3. The grid system from graph paper used to test the accuracy of the VDAS.

Table 1. Basic statistics of distances in microns of digitized landmarks from calibration points (intersections of lines in a grid on graph paper; $n = 20$)

Point	Mean	S.D.	Minimum	Maximum
1	65.83	9.93	45.89	80.52
2	31.34	7.88	15.81	46.32
3	9.51**	5.57	0.00	16.49
4	29.84	6.77	13.00	39.20
5	63.69	10.87	41.44	79.16
6	93.63***	17.67	45.01	110.64
7	93.46	14.80	59.30	115.57
8	62.89	11.03	40.31	80.23
9	30.11	6.44	18.00	40.02
10	6.04*	5.12	0.00	16.64
11	31.48	7.11	14.87	40.26
12	63.36*	10.51	41.23	76.66
13	62.81*	10.08	36.35	78.64
14	29.63	6.33	18.44	40.02
15	3.88***	2.82	0.00	9.00
16	30.42	5.98	18.44	44.00
17	65.54	10.84	40.31	87.14
18	94.11	15.28	58.31	114.86
19	93.81	16.12	58.86	114.11
20	64.77*	11.85	40.31	79.43
21	30.37	8.05	17.00	45.54
22	8.29	4.01	0.00	16.64
23	32.22	5.75	18.03	44.18
24	62.19	11.55	32.02	83.15

* $p < 0.1$; ** $p < 0.05$; *** $p < 0.01$

Test 1

Following Becerra et al. (1993), we carried out calibration readings on graph paper. Because of our current interest in small objects, and our desire to test the custom-made lens, a grid of 24 points (6 along the horizontal axis and 4 along the vertical axis, 5 mm apart (Fig. 3) was digitized 20 times.

Table 1 shows the mean digitized distances from the intersections of the grid. It can be seen that in general, the differences are very small, hardly exceeding 0.1 mm on a screen viewing a surface of 15.5 x 25.5 mm. Given that some of the variation may be due to faults in the printing of the graph paper (some intersecting lines appeared quite thick when viewed through the macro lens), the accuracy of the system seems quite good. There is, however, a tendency for more accurate point repositioning near the center of the screen than at the edges. This is probably due to better perception of the pixels in the center of the screen or perhaps to a parallax effect.

A few of the distances from the bivariate centroids are nonnormally distributed, in three cases due to a very small range of variation and many values close to zero.

Table 2 shows that both vertical and horizontal distances across the screen are repeatedly measured with great accuracy. The consistent, though small degree of, underestimation of the horizontal distances must be due to a very slight error in positioning the camera vertically (the object table might have been slightly tilted).

These tests show that B-V is a very accurate data-capture system. The basic assumption of normally distributed measurement errors of the landmarks (Bookstein, 1991) is generally met.

Table 2. Basic statistics of percentage differences of distance measurements along the horizontal axis (25.00 mm) and the vertical axis (15.00 mm) ($n = 20$)

Line	Mean	S.D.	Minimum	Maximum
Vertical 1	-0.08	1.18	-1.47	2.73
Vertical 2	0.18	1.08	-1.20	2.73
Vertical 3	0.29	1.22	-1.53	3.33
Vertical 4	0.27	1.16	-1.20	3.00
Vertical 5	0.16	1.11	-1.27	2.67
Vertical 6	-0.05	1.17	-1.47	2.67
Horizontal 1	-0.67	0.96	-2.20	1.48
Horizontal 2	-0.52	1.03	-2.20	1.52
Horizontal 3	-0.36	1.04	-2.20	2.08
Horizontal 4	-0.37	1.00	-2.04	1.88

Test 2

To test the repositioning accuracy of the system on real biological objects, the coordinates of 46 landmarks on the wing of a butterfly *Parnassius mnemosyne* Linnaeus (Lepidoptera: Papilionidae) were captured 20 times without moving the specimen by two persons (10 repetitions each). Landmarks were intersections of venation, vein endings on the edge of the wing and the greatest length and width of the color patterns. The mean measurement errors (in microns) are tabulated in Table 3.

That table shows that most of the repeated measurements are very close. It is also instructive to know that whereas 47 of the 92 (2 x 46) samples (of the original x and y coordinates) had at least one outlier greater than ± 1.96 standard deviation from the mean, only 7 had two outliers and none more than 2. Of the 92 samples of measurement distribution, 8 ($7x$, $12x$, $14x$, $19x$, $29x$, $32y$, $38y$ and $41y$) were significantly different from normal at $p < 0.1$, 5 ($2x$, $22x$, $31x$, $35y$ and $37y$) at $p < 0.05$ and none at a greater level of significance. Most of these landmarks (12, 14, 29, 32, 38, 41, and 31, 35, 37) were difficult to study because they were located at the edges of color spots without a sharply defined margin. This point also illustrates the need for adequate landmark selection. For well-defined landmarks, the repeatability of the measurements is very good even when taken by different operators. In sum, the trial tests demonstrate that the measurement error of B-V is low, generally smaller than the visual definition of many biological landmarks.

Test 3

The results reported here are part of ongoing studies on morphometric variation in Pliocene bats (Topál, 1985). The purpose of the present exercise was to see whether having collected a large series of distance data by caliper, it would be advisable to switch over to B-V.

A large number of linear-distance measurements of the mandible were taken in a series of 36 specimens of *Myotis aemulus* from the Pliocene, Locality No. 9 of Osztramos, northern Hungary. Because the bones were fragmented, none of the specimens had all the characters, so 29 of those specimens with $n > 10$ are given here.

First, the distance measurements were recorded with the help of a Mitutoyo digital caliper to 0.05 mm accuracy. Then coordinates of landmarks representing the endpoints of the characters were digitized using IMAGOES. Inter-landmark distances were computed using a routine written in FoxPro.

Table 3. Basic statistics of distances (in microns) from bivariate centroids of 46 digitized landmarks in *Parnassius mnemosyme* ($n = 20$). Asterisks indicate significant deviations from normality (Shapiro-Wilks test)

Landmark	Mean Distance	S.D.	Minimum	Maximum
1	216.19	78.51	75.45	372.94
2	226.49	112.66	82.73	467.57
3	230.16	121.74	36.09	493.12
4	200.87	126.14	39.33	544.64
5	245.28**	136.49	69.59	542.15
6	237.44	97.99	76.50	436.17
7	238.58	74.71	94.58	385.03
8	238.36	126.02	36.18	533.74
9	204.72	91.61	46.74	402.44
10	227.08	110.74	15.44	497.84
11	296.18	171.60	33.15	657.14
12	187.45***	132.51	15.89	636.24
13	140.97	60.66	52.96	248.97
14	153.49	65.21	46.67	279.86
15	210.05	118.68	26.34	454.29
16	183.98	99.23	37.24	359.59
17	230.09	106.06	42.01	453.37
18	295.77	147.91	42.32	577.68
19	300.76	107.49	144.78	465.72
20	300.89	160.18	57.00	569.68
21	354.69	123.02	136.26	529.06
22	302.83	122.70	138.60	593.67
23	221.98***	125.17	79.59	438.01
24	185.92	103.80	22.67	431.50
25	219.95	123.74	43.31	477.70
26	246.13	113.02	11.18	434.03
27	231.56**	178.89	23.41	657.67
28	248.55	146.74	25.29	618.05
29	198.29***	179.34	37.83	792.89
30	194.80*	113.26	33.45	429.84
31	229.72**	148.90	46.29	671.06
32	186.14*	88.83	29.46	456.20
33	192.46	107.00	41.17	417.38
34	250.47	125.60	65.52	567.03
35	325.97	111.48	142.42	564.65
36	189.06*	123.55	20.67	405.40
37	244.18	93.44	82.29	415.55
38	151.95**	87.40	39.40	356.76
39	216.36	114.74	66.78	464.12
40	211.93	77.78	68.57	370.02
41*	204.59	119.78	20.93	507.97
42	215.35	90.55	24.33	377.43
43	277.34	140.41	78.11	528.89
44	221.49	134.68	20.58	467.77
45	250.25	120.57	69.31	489.52
46	195.17	80.83	95.10	388.66

$^{*}p < 0.1$; $^{**}p < 0.05$; $^{***}p < 0.01$.

Scatter plots of the caliper versus video measurements were screened for gros outliers, and 3% of the observations were rejected as obvious observer error.

Table 4 contains results of regressing the video measurements on the caliper meas urements. The caliper measurements serve as a yardstick for historical reasons, with whic

Table 4. Correlation and regression analyses between distance measurements of mandibles of fossil bats from caliper readings and landmarks digitized using Budapest Video

Variable	n	Mean percent difference	r	Constant	S.D.	Slope	S.D.
Alveolar length C-M^3	11	6.18	0.958***	-3.767***	1.343	1.753***	0.175
Alveolar length C-P_4	22	0.06	0.828***	0.632	0.414	0.813	0.123
Alveolar length P_4-M_3	15	2.54	0.818***	-0.193	1.362	1.290	0.252
Crown length M_1-M_3	15	1.14	0.917***	0.203	0.522	0.966	0.117
Alveolar length M_1-M_3	18	0.69	0.806***	1.599***	0.489	0.629***	0.116
Alveolar length lower C	23	31.88	0.554***	0.323	0.319	0.992	0.322
Alveolar width lower C	17	5.57	0.308	0.425*	0.206	0.398*	0.317
Cross-sectional length P_2	10	-0.90	0.738**	0.120*	0.211	0.837	0.271
Alveolar length P_2	26	24.53	0.456***	0.363**	0.141	0.615	0.245
Crown width P_2	11	-0.13	0.909***	0.023	0.110	0.968	0.148
Alveolar width P_2	17	3.77	0.163	0.441***	0.129	0.162***	0.254
Cross-sectional length P_3	12	6.07	0.808***	0.114	0.124	0.875	0.201
Alveolar length P_3	29	8.55	0.663***	0.51*	0.083	0.777	0.169
Crown width P_3	10	3.58	0.866***	-0.202	0.182	1.340	0.274
Alveolar width P_3	17	4.97	0.644***	0.079	0.118	0.870	0.267
Cross-sectional length P_4	21	0.00	0.708***	-0.679	0.407	1.621	0.371
Cross-sectional width P_4	20	-0.60	0.729***	0.094	0.162	0.881	0.195
Alveolar length P_4	29	10.08	0.376	0.621***	0.187	0.388***	0.188
Crown length M_1	20	-0.91	0.866***	-0.313	0.267	1.180	0.161
Talonid width M_1	20	-3.11	0.643***	0.339	0.213	0.671	0.188*
Crown length M_2	20	0.12	0.578***	0.875***	0.244	0.457***	0.152
Talonid width M_2	20	-3.70	0.645***	-0.153	0.348	1.098	0.307
Crown length M_3	18	-1.62	0.793***	0.011	0.267	0.976	0.187
Trigonid width M_3	17	-0.98	0.634***	-0.058	0.309	0.982	0.309
Talonid width M_3	13	-4.03	0.469	0.438***	0.178	0.399**	0.227
Height of mandible at M_1	23	-1.93	0.828***	-0.139	0.315	1.049	0.155
Height of mandible at M_3	11	-11.41	0.788***	0.110	0.437	0.832	0.217
Alveolar length P_2-P_3	26	28.18	0.908***	0.245*	0.120	1.073	0.101
Symphysis length	17	-1.83	0.937***	-0.304	0.318	1.082	0.104

* $p < 0.1$; ** $p < 0.05$; *** $p < 0.01$.

the digitized measurements are compared. Ideally, there should be no or little difference between the two sets of measurements, indicated by small mean difference, high correlation coefficient, zero intercept and unit slope.

Eighteen (62%) of the 29 variables are commensurate, with small mean percentage differences, high correlation coefficients, and slopes and intercepts not significantly different from the expected values. Some of the measurements even appear to be almost exact reproductions as indicated by the above-mentioned sampling statistics (e.g. crown width P_2).

There are notable differences, however. Some of the problematic measurements are various widths and lengths of single teeth (C, P_2, P_4), where the caliper might have been pushed to or below the lower limit of its applicability. There might have been problems in accurately positioning the caliper needle on teeth 0.5-0.7 mm across. Other considerable measurement discrepancies involve larger distances measured along the mandible. We suspect these errors were introduced by regions of the bone not lying perpendicular to the axis of the camera. An example is alveolar length of C-M_3, which was measured with a very high correlation coefficient, but, on average, the video measurements overestimated the caliper values by over 25%, also shown by a negative intercept and a steep slope. In summary, the B-V is a reliable device for capturing coordinates of landmarks for computing distances. However, if we want caliper and VDAS data to be comparable, caution should be exercised in selecting landmarks for endpoints and in positioning specimens.

DISCUSSION

As far as we know, Becerra et al.'s (1993) report is the only comparison of the performance of various other VDASs (cf. review on the Reflex Microscope by MacLarnon, 1989). Budapest Video was designed with specific objectives that distinguish it from other systems. The hardware (digitizing board) provides a resolution that is 45.3% greater than that of the board used by the systems tested by Becerra et al. (1993).

The training set feature of IMAGOES designed by one of the authors, J. V., greatly facilitates the collection of landmarks. Templates have been used in various software, and a similar feature for positioning the cursor has been programmed in the software driving the 3D Reflex Microscope.

Automatic feature extraction is an area for further improvement in video image analysis (Rohlf, 1993b). MorphoSys (Meacham, 1992, 1993) is strong in its command language structure. In that system once an outline is captured, fairly complicated feature extraction can be automated using the geometric commands. IMAGOES attempts nothing of that sort, but both the shape coordinates and the standardized elliptical Fourier coefficients are a step towards feature extraction. The outline collection feature of IMAGOES is limited to one descriptor which Rohlf and Archie (1984) found to be a very effective and convenient method for closed outlines. We have found the present limit of 30 harmonics sufficient to describe complex outlines, such as mouse mandibles accurately.

The precision of the system compares well or even exceeds those reported for other systems by Becerra et al. (1993). Precision is essentially limited by pixel size.

Further development of the software could include more flexible output file formats, including ASCII and the format of NTSYS-pc (Rohlf, 1990b). The latter format is used by several recently developed programs that do deformation analyses, so it would definitely be a step towards integrated analysis if the data-collection software produced files readable by analytical software.

Measurement errors may significantly affect the results of morphometric analyses, as shown, for example, by Francis and Mattlin (1986). The VDAS reported here provides good measurement repeatability.

Credit for this contribution is distributed among the coauthors as follows: A.D. initiated and oversaw the project, formulated the principal ideas of the software and carried out most of the writing of this paper; J. V. designed and coded the software; L.P. helped design the software-hardware system, brought the hardware-software team together, managed the project and carried out some testing. Finally, G.T. was the first to put the system to intensive use.

Budapest Video is available from Pars Kft., Budapest, Cinkotai ut 73/c, H-1141 Hungary.

ACKNOWLEDGMENTS

Funds for this work were in part provided by OTKA grant no. 3177/91 to A. Demeter and grant no. 3179/91 to L. Peregovits. We thank the participants of the NATO Advanced Study Institute for useful suggestions for improving version 1.0 of IMAGOES. Most of the writing of this contribution was carried out while A. D. was recipient of a Fulbright fellowship in the National Museum of Natural History, Smithsonian Institution, Washington, D.C., under the mentorship of Robert S. Hoffmann. Ralph Chapman of the Applied Morphometrics Laboratory of that museum is thanked for welcoming A. D. to facilities under his supervision, for discussions on morphometrics and for making useful comments on the manuscript.

REFERENCES

Becerra, J. M., E. Bello and A. Garcia-Valdecasas. 1993. Building your own machine image system for morphometric analysis: A user point of view. Pages 65–92 *in* L. F. Marcus, E. Bello and A. Garcia-Valdecasas, (eds.), Contributions to morphometrics. Monografias del Museo Nacional de Ciencias Naturales 8, Madrid.

Bookstein, F. L. 1991. Morphometric tools for landmark data. Geometry and biology. Cambridge University Press: Cambridge.

Fink, W. L. 1990. Data acquisition for morphometric analysis in systematic biology. Pages 9–19 *in* F. J. Rohlf and F. L. Bookstein, (eds.), Proceedings of the Michigan morphometrics workshop. University of Michigan Museum of Zoology Special Publication No. 2: Ann Arbor.

Francis, R. I. C. C., and R. H. Mattlin. 1986. A possible pitfall in the morphometric application of discriminant analysis: Measurement bias. Marine Biology 93:311–313.

Kuhl, F. P., and C. R. Giardina. 1982. Elliptic Fourier features of a closed contour. Computer Graphics and Image Processing 18:236–258.

MacLeod, N. 1990. Digital images and automated image analysis systems. Pages 21–35 *in* F. J. Rohlf and F. L. Bookstein, (eds.), Proceedings of the Michigan morphometrics workshop. University of Michigan Museum of Zoology Special Publication No. 2: Ann Arbor.

MacLarnon, Ann M. 1989. Application of the Reflex instruments in quantitative morphology. Folia Primatologica 53:33–49.

Meacham, C. A. 1992. Practical machine vision for morphometric data acquisition. Pages 113–126 *in* J. T. Sorensen and R. Footit, (eds.), Ordination in the study of morphology, evolution and systematics of insects: application of quantitative genetic rationales. Elsevier: Amsterdam.

Meacham, C. A. 1993. MorphoSys: an interactive machine vision program for acquisition of morphometric data. Pages 393–402 *in* R. Fortuner, (ed.), Advances in computer methods for systematic biology. Artificial intelligence, databases, computer vision. Johns Hopkins University Press: Baltimore, Maryland.

Rohlf, F. J. 1990a. An overview of image processing and analysis techniques for morphometrics. Pages 37–60 *in* F. J. Rohlf and F. L. Bookstein, (eds.), Proceedings of the Michigan morphometrics workshop. University of Michigan Museum of Zoology Special Publication No. 2: Ann Arbor.

Rohlf, F. J. 1990b. NTSYS-pc. Numerical taxonomy and multivariate analysis system. Version 1.6. Exeter: Setauket, New York.

Rohlf, F. J. 1990c. Fitting curves to outlines. Pages 167–177 *in* F. J. Rohlf and F. L. Bookstein, (eds.), Proceedings of the Michigan morphometrics workshop. University of Michigan Museum of Zoology Special Publication No. 2.

Rohlf, F. J. 1993a. Computer vision needs in systematic biology. Pages 365–373 *in* R. Fortuner, (ed.), Advances in computer methods for systematic biology. Artificial intelligence, databases, computer vision. Johns Hopkins University Press: Baltimore, Maryland.

Rohlf, F. J. 1993b. Feature extraction in systematic biology. Pages 375–392 *in* R. Fortuner, (ed.), Advances in computer methods for systematic biology. Artificial intelligence, databases, computer vision. Johns Hopkins University Press: Baltimore, Maryland.

Rohlf, F. J. and Archie, J. W. 1984. A comparison of Fourier methods for the description of wing shape in mosquitoes (Diptera: Culicidae). Systematic Zoology 33:302–317.

Rohlf, F. J. and L. F. Marcus. 1993. A revolution in morphometrics. Trends in Ecology and Evolution 8:129-132.

Topál, G. 1985. Pliocene species of *Myotis* from Hungary. Acta Zoologica Fennica 170:153–155.

CONVERTING TRUSS INTERLANDMARK DISTANCES TO CARTESIAN COORDINATES

Kent E. Carpenter,[1] H. Joseph Sommer III,[2] and Leslie F. Marcus[3]

[1] Department of Biological Sciences
Old Dominion University
Norfolk, Virginia 23529-0058
[2] Department of Mechanical Engineering
327 Reber Building
The Pennsylvania State University
University Park, Pennsylvania 16802
[3] Department of Biology
Queens College of the City University of New York
Flushing, New York, 11367

ABSTRACT

Coordinate-based landmark data are required to use many of the recent advances in morphometrics, although interlandmark distance measurements were used extensively in the past. In many cases the latter are still being used in morphometric research. The conversion of distance data to coordinate data is not straightforward because insufficient measurement redundancy among landmarks and measurement error obscures the possibility of direct geometric reconstruction. Iterative multivariate methods and redundant measurements, such as the truss protocol, allow for a reasonably accurate conversion of interlandmark distance data into coordinate landmark data. We examine two multidimensional scaling methods for making this conversion. One is based on the nonmetric method found in the statistical package, NTSYS-pc, and the other is a simplified weighted least squares method tailored for this type of conversion. The latter method is amenable to heuristic modification and found to be more accurate than the former, although the NTSYS-pc based method may be more convenient for some users.

INTRODUCTION

Rapid advances in morphometrics over the past dozen or so years have led to a bewildering progression of methods that have been newly developed, discarded, replaced, improved upon and redeveloped. Certain components of this time series survive today as

Advances in Morphometrics, Edited by L. F. Marcus et al.
Plenum Press, New York, 1996

relics of less successful methodological lineages. Their force of presence today is a testimony to their novel attraction in the past. An example of one particularly successful morphometric method "has-been" is the truss protocol for measuring specimens, which is based on distance measurements on specimens (Strauss and Bookstein, 1982). A number of participants of the Advanced Study Institute expressed an interest in analyzing their truss data using the new geometric morphometrics, but these methods are based on coordinate data. The purpose of this paper is to evaluate three methods that can be used to translate interlandmark truss distances into Cartesian coordinates.

Traditionally, size and shape analyses of biological specimens have employed multivariate statistics for scalar distance measurements such as length, width and thickness. This approach has also been used for landmark-based morphometrics with multiple scalar interlandmark distances (ILD) by Strauss and Bookstein (1982) and Lele and Richtsmeier (1992), among others. In this context, landmarks are specific locations on specimens that are replicable and identifiable across phylogeny, ontogeny or pathology. More recently however, new methods such as Procrustes analysis (Goodall, 1991) and thin-plate splines and relative warps (Bookstein, 1989, 1991) have been developed for landmark-based morphometrics using two-dimensional (2D) or three-dimensional (3D) Cartesian coordinates of landmarks rather than ILDs. Because both ILD and coordinate methods have relative overall advantages and disadvantages, investigators may wish to compare and contrast results from both. Unfortunately, conversion from ILD data to coordinate data is not as easy as the converse.

The truss protocol was originally developed to systematically and exhaustively measure body form based on ILDs (Strauss and Bookstein, 1982). This method was viewed as less biased than were classical methods of collecting morphometric data that usually tend to over emphasize the use of certain landmarks such as tip of the snout and end of the caudal skeleton in fish. In addition, the pattern of redundant measurements in the truss network allows reconstruction of Cartesian coordinates of the landmarks and compensation for measurement error.

The usefulness of the truss network was again mentioned by Bookstein et al. (1985) at the same time that coordinate landmark data became increasingly important as the basis for geometric morphometrics. The emphasis today is to collect coordinate data directly using digitizing boards, and 2D or 3D machine image systems (Becerra et al., 1993; Roth, 1993; and Dean "Three dimensional article", this volume). However, some truss distance data sets continue to grow because of the amount of time invested in them in the past and the advantage of continuing along a previous line of research, or because the equipment for collecting coordinate data are not readily available (as, for example, on board a fishing vessel where a set of calipers can often be more easily used). Distance data may also be more convenient in disciplines other than morphometrics, such as in ecology where an analogous version of the truss was developed by Rohlf and Archie (1978). Subsequent conversion of these data to coordinates can also be useful in ecological analyses and therefore the conversion methods described here may have wider application.

METHODS

The microcomputer programs reviewed here for converting interlandmark distances to coordinate data are based on multidimensional scaling. A mainframe computer program that uses the geometric approach described by Strauss and Bookstein (1982) is also available through Fred L. Bookstein. However, we were not able to locate the library of subroutines necessary to run this program on a mainframe nor were we successful in porting it to the microcomputer environment.

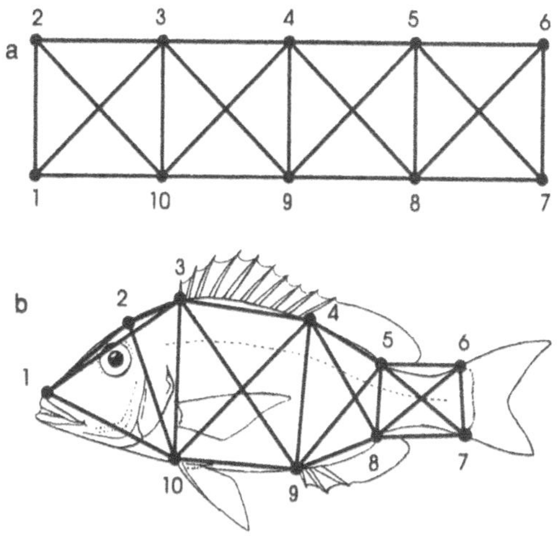

Figure 1. The truss network used in tests with landmarks labeled 1-10; (a) the basic truss configuration; (b) the truss configuration superimposed on a typical lethrinid fish.

Multidimensional scaling (MDS) is a method to assign relative values for linearly independent parameters (coordinates) that describe multiple elements in a given set (landmarks) based on different observations (ILDs) between the elements (Davison, 1983). We employ two different MDS methods, one based on the nonmetric MDS algorithm found in NTSYS-pc (Rohlf, 1993) and another, described here, is based on a simplified algorithm that is more amenable to heuristic modification. For both methods, however, the number of ILD measurements commonly available in morphometric data sets, while sufficient for traditional multivariate ILD analyses, are often mathematically insufficient for unambiguous coordinate reconstruction. Often, the reconstructed coordinates are accurate in terms of the ILDs, but the shape of the reconstructed specimen is distorted because its planer representation is folded upon itself during MDS. Therefore, some form of artificial flattening or unfolding during or after MDS must be employed to reconstruct both the coordinates and the specimens' true shape. The main comparisons in this paper are not based on the different MDS approaches but, rather, mostly on the three different approaches employed to ensure that the coordinates reflect the actual shape of the specimen.

The Truss and the Test Data Set

A typical truss configuration, and the one used for our test data set, is shown in Fig. 1a, with 21 ILDs and landmarks labeled from 1 to 10. This truss was measured on specimens of fishes (Fig. 1b) of the family Lethrinidae (Perciformes). Altogether 413 specimens from 28 species were completely measured for the truss, and all specimens were pooled for the tests presented here. For a more detailed explanation of this data set see Carpenter (this volume).

MDS Using NTSYS-pc and a Permanent Artificial Flattener

The algorithm used in Rohlf's (1993) NTSYS-pc was nonmetric MDS (Kruskal, 1964a, 1964b). We also tried linear MDS but our unpublished results were similar to the nonmetric method. Maximum iterations was set at 40 and minimum stress at 0.001. An initial matrix was used with *x,y* coordinates of the landmarks 1 to 10 (as in Fig. 1) set as 0,0; 0,1; 1,1; 2,1; 3,1; 4,1; 4,0; 3,0; 2,0; and 1,0. However, it was still necessary to include an artificial flattener throughout the MDS procedure in order to ensure that the truss network did not fold upon itself.

This was done by using the standard length of the fish (from the tip of the snout to the end of the caudal skeleton) as an ersatz ILD between landmarks 1 and 7. This "artificial flattener" forces the truss network to lay flat but will cause some distortion because the standard length is not equivalent to the distance between these landmarks. A more direct approach would have been to measure extra distances (as suggested by Rohlf and Archie, 1978) such as between landmarks 1 and 7 in order to avoid this distortion. However, because the original truss protocol did not call for this redundancy, truss data sets often do not include these distances.

MDS Using a Simplified Algorithm and a Temporary Artificial Flattener

The method described below is a simplistic MDS approach, more amenable to heuristic modification than to existing MDS algorithms in statistical packages such as NTSYS-pc (Rohlf, 1993). A computer program to compute this type of MDS is available from the second author (H. J. S.). The following development is based on 2D coordinates but it may be easily extended to 3D or higher dimensions. If coordinate locations have been measured for all landmarks on a specimen, all ILDs may be determined by equations 1 and 2. Conversely, if only ILDs are known, Equation 2 may be used to iteratively estimate landmark coordinate locations.

$$\max\_nild = nland \cdot \frac{(nland - 1)}{2}, \tag{1}$$

where max_*nild* is the maximum possible number of ILDs and *nild* is the actual number of ILD measurements available.

$$d_{ij}^2 = (x_i - x_j)^2 + (y_i - y_j)^2 \tag{2}$$

where d_{ij} is the ILD between landmark i and landmark j; x_i, y_i is the coordinate location for landmark i; and x_j, y_j is the coordinate location for landmark j.

Not all ILDs are required for reconstruction of coordinate locations. At a minimum *ncoord* (= number of coordinate dimensions), respective ILDs are required for each landmark. Further, each landmark should have at least (*ncoord* + 1) respective ILDs for unambiguous reconstruction. For example, only two ILDs for a given 2D landmark will allow that landmark location to be calculated, but it may be flipped into one of two alternate but mathematically valid locations as discussed below.

An iterative relaxation approach is employed for coordinate reconstruction. It estimates the location of one landmark at a time assuming the locations of all other landmarks are known. It then sequentially estimates new locations for all other landmarks in turn during that iteration. Subsequent iterations cycle through the landmarks one at a time in a similar manner. Assuming that landmark location j is known, equation 2 may be rewritten as shown in equations 3 and 4 to calculate the location of landmark i. Equations 5 and 6 are expanded assuming that previous estimates for all landmarks are known.

$$\begin{bmatrix} x_j & y_j & 1 \end{bmatrix} \cdot \begin{bmatrix} -2 \cdot x_i \\ -2 \cdot y_i \\ r_i^2 \end{bmatrix} = d_{ij}^2 - r_j^2 \tag{3}$$

$$r_i^2 = x_i^2 + y_i^2 \tag{4}$$

$$\begin{bmatrix} x_1 & y_1 & 1 \\ & \cdots\cdots & \\ x_j & y_j & 1 \\ & \cdots\cdots & \\ x_{nland} & y_{nland} & 1 \end{bmatrix} \cdot \begin{bmatrix} -2\cdot x_i \\ -2\cdot y_i \\ r_i^2 \end{bmatrix} = \begin{bmatrix} d_{i1}^2 - r_1^2 \\ \cdots\cdots \\ d_{ij}^2 - r_j^2 \\ \cdots\cdots \\ d_{i,nland}^2 - r_{nland}^2 \end{bmatrix} \tag{5}$$

$$X \cdot beta = y \tag{6}$$

where *beta* is the linearized least squares solution vector, nland is the number of landmarks, r_i is the distance of landmark *i* from the origin, *y* is the linearized least-squares right-hand side vector, and *X* is the matrix of current landmark coordinate estimates.

A weighted least squares solution may then be used to estimate the location of landmark *i* as shown in equations 7, 8, 9 and 10. A statistical weight of 1 should be associated with all available ILD measurements, and a statistical weight of 0 should be used to exclude those ILDs that are not available. A statistical weight of 1 is recommended to include the old estimate for location of landmark *i* when computing a new estimate for landmark *i* as shown in equation 10. This heuristic improves iterative convergence and prevents singularities if only (*ncoord*) ILDs are available for any landmark. This linearized solution numerically places slightly more emphasis on longer ILDs.

$$beta = inv(X' \cdot W \cdot X) \cdot (X' \cdot W \cdot Y) \tag{7}$$

$$W_{jj} = w_{ij}^2 \tag{8}$$

$$w_{ij} = 1 \text{ if } d_{ij} \text{ is available, otherwise } w_{ij} = 0 \tag{9}$$

$$w_{ii} = 1 \text{ for } d_{ii} = 0 \tag{10}$$

where *inv* is the matrix inverse, W is the diagonal weighting matrix, and w_{ij} is the statistical weight associated with d_{ij}.

The iterative relaxation algorithm uses linearized equation 7 to estimate sequentially new locations for all landmarks in turn within each iteration. Typically 10 to 20 iterations are required. Random initial landmark locations may be used for the first iteration if approximate landmark coordinates are not available from another prototypical specimen. Landmark coordinates should be shifted to place their centroid at the coordinate origin before each iteration to prevent numerical drifting and to improve convergence. Convergence is best monitored by computing ILD estimates after each iteration, subtracting them from respective measured ILDs and finding the maximum ILD error for that iteration. Maximum ILD error may be associated with a specific pair of landmarks for several successive iterations and will decrease as those landmarks are iteratively adjusted. Because maximum ILD error may later appear between a different pair of landmarks, maximum ILD error may not decrease monotonically during the iterative process.

Using variation in maximum ILD error over five sequential iterations as a convergence criterion is robust and independent of specimen size and ILD measurement noise. Moderate inconsistencies in ILD measurements, particularly for marginally defined landmarks with only (*ncoord* + 1) respective ILDs, can however cause band-limited oscillation near convergence. Consequently, a limit of 50 iterations is recommended. If a specimen exceeds the iteration limit, the iteration process should be restarted with new random initial coordinate estimates. No ILD data sets have required more than one restart.

This approach does not inherently enforce chirality of landmark coordinates. That is, ILDs from a dextral specimen may reconstruct landmark coordinates for an equivalent but sinistral reflection. Chiral inversion and rigid-body rotational alignment of landmark coordinates may be included before each iteration with no effects on convergence.

The above MDS is accurate for generating coordinates based on ILDs, however, the number of ILD measurements commonly available for these trusses are often mathematically insufficient for unambiguous coordinate reconstruction. The individual cells of the truss network can be reconstructed adjacent to each other or they can be folded on top of one another unless an ILD is specified that adequately spans both cells. The first method to prevent folding in the type of truss shown in Fig. 1 using the MDS approach described above, was to apply a temporary 'artificial flattener' (the extension of this method to encompass trusses with a terminal triangle is a little more complicated and is the subject of a paper currently in preparation by H. J. Sommer, et al.). As in the flattening approach using NTSYS-pc, the standard length was initially set as the distance between landmarks 1 and 7. After a number of iterations the truss lays flat and the artificial flattener is removed to prevent distortion. We found for the data set used here that 11 iterations was sufficient to force the truss to flatten without cells folding upon one another during subsequent iterations.

MDS Using a Simplified Algorithm and Geometric Unfolding

Another approach to ensure accurate reconstruction of truss coordinates and shape is to use the simplified MDS algorithm described above, allow the truss to fold upon itself during MDS, and later unfold the truss geometrically. This process involves aligning, testing, and unfolding guided by a double-ordered list of ILD pairs describing possible folding lines. It preserves geometric integrity within unambiguous truss segments, unfolds any single chain box truss no matter how badly folded, forces a specific chirality based on side ordering, and provides a simple framework for topological definition of folding. This method is described in more detail by Sommer et al. (in preparation).

Comparing the Accuracy of Reconstruction of Truss Coordinates

The method of testing the accuracy of the reconstruction of truss coordinates was simply to calculate ILDs from the reconstructed coordinates using the Pythagorean theorem and regress these against the actual measured ILDs. The degree of scatter about the regression line is an indication of the accuracy of the coordinate reconstruction method. In the case of the two simplified MDS approaches applied, the scale of the reconstructed ILDs is the same as the actual measurements and therefore the regression line should intersect close to the origin if the coordinate reconstruction method is accurate. The 21 measured versus reconstructed ILDs were each regressed using the 413 measurements for each of the three methods of truss flattening. All regressions were carried out using SAS for Personal Computers, version 6.03 (1988).

Table 1. Univariate statistics for intercepts from regression analyses of 21 measured versus reconstructed ILD's (n = 413 for each ILD regression) for simplidied MDS using a temporary flattener (Flattemer) and geometric unfolding (Unfolding)

	Mean	S. D.	Minimum	Maximum
Flattener	-0.1955	0.2530	-0.8354	0.0705
Unfolding	-0.2274	0.2566	-0.8875	0.0589

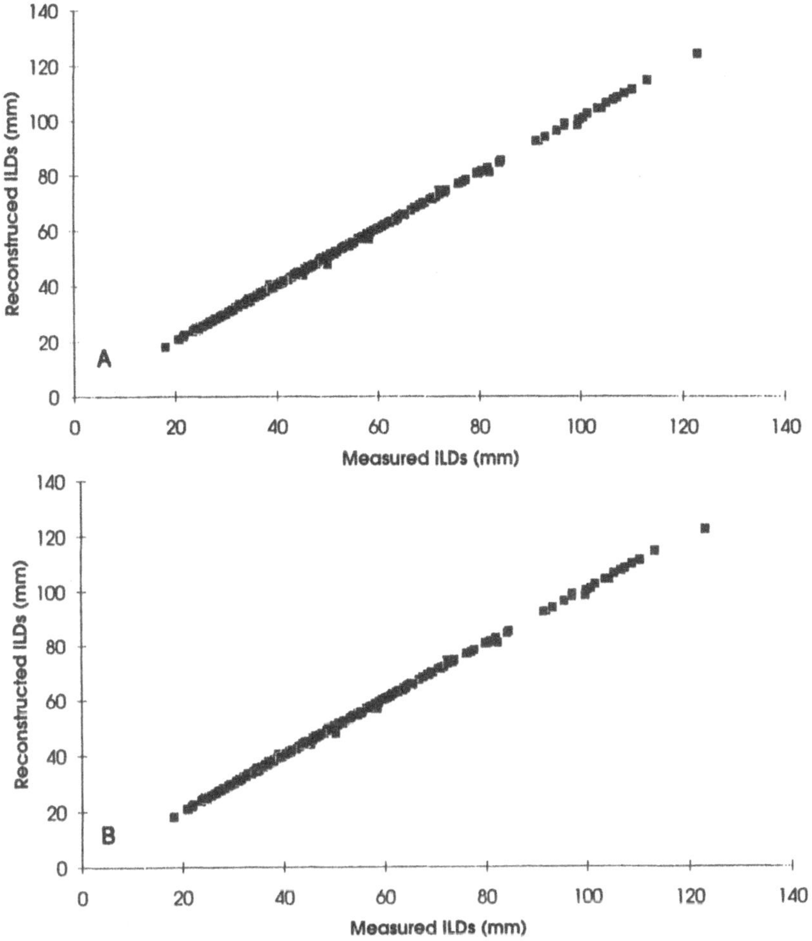

Figure 2. Scatter plots of measured versus reconstructed distances for the ILD between landmarks 1 and 2 (Fig. 1b); (A) scatter typical for simplified MDS using a temporary flattener (slope = 1.0012, model significant at $p = 0.001$); (B) scatter typical for simplified MDS using geometric unfolding (slope = 1.0020, model significant at $p = 0.001$).

RESULTS

The regression analysis of the 21 ILDs show that the simplified MDS approach using either the temporary flattener or geometric unfolding was more accurate in reconstructing the coordinates than the permanent flattener using NTSYS-pc. The regression model was significant ($p > 0.001$) for each of the 21 ILDs using the simplified MDS approach for both the temporary flattening and unfolding approaches, while only 10 of the 21 models were significant (at $p = 0.05$) for coordinate reconstruction using a permanent flattener under non-metric MDS (minimum stress was reached prior to 40 iterations in all of the 413 specimens analyzed under NTSYS-pc). The scatter of points typical for the simplified regressions are shown in Fig. 2a-b, using the ILD between landmarks 1 and 2 (Fig. 1). There was consistently more scatter for all 21 ILDs using the permanent flattener approach than either the temporary flattener (Fig. 2a) or the geometric unfolding (Fig. 2b) approaches. Both of the latter two approaches had similar, tight scatter for all of the 21 ILDs and their intercepts were close to 0 (Table 1). The mean intercepts for the temporary flattener was slightly closer to 0 than for geometric unfolding, but these differences were not significant.

DISCUSSION

Accurate reconstruction of Cartesian coordinates from truss distance data is possible using the simplified MDS approach presented here and either a temporary flattener or geometric unfolding. We do not recommend using a permanent flattener since the distortion apparently interferes with acceptable coordinate reconstruction. Of the computer programs for simplified MDS available, the one using geometric unfolding was considerably faster than the one using a temporary flattener (both available from the second author). Geometric unfolding was therefore the preferred method since the two approaches using simplified MDS did not differ significantly in their ability to reconstruct coordinates.

ACKNOWLEDGMENTS

We thank F. James Rohlf for making his latest version of NTSYS available for use at the Il Ciocco workshop and for initial guidance in using non-metric MDS for reconstruction of coordinates from truss distance data. We also thank Mauro J. Cavalcanti and John M. Lynch for e-mail discussions of truss coordinate reconstruction and computer programming.

REFERENCES

Becerra, J. M., E. Bello, and A. García-Valdecasas. 1993. Building your own machine image system for morphometric analysis: a user point of view. Pages 65–94 *in* L. F. Marcus, E. Bello, and A. García-Valdecasas (eds.), Contributions to morphometrics. Monografías del Museo Nacional de Ciencias Naturales 8, Madrid.

Bookstein, F. L. 1989. Principal warps: Thin-plate splines and the decomposition of deformations. I.E.E.E. Transactions on pattern Analysis and Machine Intelligence. 11: 567–585.

Bookstein, F. L. 1991. Morphometric tools for landmark data: Geometry and biology. Cambridge University Press: Cambridge.

Bookstein, F. L., B. Grayson, C. B. Cutting, H. C. Kim, and J. G. Mc Carthy. 1991. Landmarks in three dimensions: reconstruction from cephalograms versus direct observation. American Journal of Orthod. Dentofacial Orthop. 100(2): 133–140.

Bookstein, F. L., B. Chernoff, R. L. Elder, J. M. Humphries, Jr., G. R. Smith, and R. E. Strauss, (eds.), 1985. Morphometrics in evolutionary biology: The geometry of size and shape change, with examples from fishes. Academy of Natural Sciences of Philadelphia Special Publication 15.

Davison, M.L. 1983. Multidimensional Scaling. John Wiley and Sons: New York.

Goodall, C. 1991. Procrustes methods in the statistical analysis of shape. Journal of the Royal Statistical Society, B 53: 285–339.

Kruskal, J. B. 1964a. Multidimensional scaling by optimizing goodness of fit to a nonmetric hypothesis. Psychometrika. 29:1–27.

Kruskal, J. B. 1964b. Nonmetric multidimensional scaling: a numerical method. Psychometrika. 29:28–42.

Lele, S. and J. T. Richtsmeier 1992. On comparing biological shapes: detection of influential landmarks. American Journal of Physical Anthropology 87:49–65.

Rohlf, F. J. 1993. NTSYS-pc, Version 1.8. Exeter Publishing: Setauket, New York.

Rohlf, F. J. and J. W. Archie. 1978. Least-squares mapping using interpoint distances. Ecology. 59(1):26–132.

Roth, V. L. 1993. On three-dimensional morphometrics, and on the identification of landmark points. Pages 41-66 *in* L. F. Marcus, E. Bello, and A. García-Valdecasas, (eds.), Contributions to morphometrics. Monografias del Museo Nacional de Ciencias Naturales 8, Madrid.

SAS for Personal Computers, Release 6.03. 1988. SAS Institute: Cary, North Carolina.

Strauss, R. E., and F. L. Bookstein 1982. The truss: body form reconstruction in morphometrics. Systematic Zoology 31(2):113–135.

PART I

Section 2. Landmark Methods

INTRODUCTION TO LANDMARK METHODS

Dennis E. Slice

Most of the work in geometric morphometrics has focused on the analysis of landmark data—the two- or three-dimensional Cartesian coordinates of a fixed set of points identified as homologous across specimens (Rohlf and Marcus, 1993). The chapters in this section continue to address issues of the analysis of landmark data but differ from similar chapters in the proceedings volumes from earlier workshops (e.g., Rohlf and Bookstein, 1990; Marcus et al., 1993). In earlier workshops the focus was on the basic introduction and exposition of broad areas of landmark analysis (e.g., superimpositions, and spline-based analyses). The current chapters, on the other hand, emphasize the extension and refinement of the basic methods made familiar by those earlier works. This transition is made possible by the rapid maturation of the field in recent years — a maturation further evidenced by the high ratio of application to methodological chapters in the current volume. Although there is still much to be done, there is now a sufficiently rigorous, complete, and accessible foundation for landmark analysis upon which to base new and innovative investigations in evolution, development and numerous other areas. These latest contributions expand and strengthen that foundation.

Kendall (1984, 1985) provided the geometric foundation for the analysis of landmark data through his development of shape spaces, Σ_k^p for p points in a k-dimensional space. These spaces are non-Euclidean and curved and so do not directly support the analyses and intuitions derived from more traditional biometrics. Kendall's shape spaces, however, are very general constructs that can accommodate any k-dimensional configuration of p points. Most samples of configurations of biological interest range over a very small region of shape space, and analysis of shape variation, differences, covariances, etc., can be carried out adequately in linear tangent spaces by means of familiar statistical routines. To do this correctly, though, one must have a thorough understanding of the relationships between the various spaces and the implications of these relationships on the outcome of any statistical analyses or numerical or graphical summaries. In the first two chapters of this section, Rohlf and Bookstein provide the requisite information.

First, Rohlf presents a survey of the assorted spaces relevant to geometric morphometrics — physical space, shape space, tangent spaces, and various useful subspaces — and pays particular attention to the geometric relationships between these spaces and their dimensionality. He also covers the mathematics and geometry of the thin-plate spline, estimation of the uniform component, relative warps analysis, and the implications of various transformations on the reference configuration or sample specimens. In the course of this discussion, some seemingly counterintuitive points are brought to light. For example, why should the results of many standard statistical methods be invariant to an affine transforma-

Advances in Morphometrics, Edited by L. F. Marcus et al.
Plenum Press, New York, 1996

tion of the sample in certain relative warps analyses whereas a similar transformation of the reference produces different statistical results?

In the course of the Advanced Study Institute and subsequent discussions with workers beginning to apply the geometric methods, it became apparent that there are a number of procedures or analyses that, although individually quite correct and potentially useful, are not appropriate when used in certain combinations. Bookstein addresses this issue in his essay on "Combining the Tools of Geometric Morphometrics." He first traces the role of Procrustes distance throughout the three phases of morphometric analysis: the transition from landmarks to shape-space coordinates, the statistical analysis of shape, and the effective summary and display of the results. In some cases, e.g., the transition to shape-space coordinates, Procrustes distance is of fundamental importance and not to be tampered with, whereas in other situations, e.g., the display and interpretation of statistical results, it is part of a suite of possibly useful alternative methods. Following up with discussions of superimposition techniques, thin-plate spline and partial and relative warps analyses, Bookstein provides an extensive listing of DOs and DON'Ts invaluable to the applied morphometrician.

Together, the chapters by Rohlf and Bookstein present material of fundamental importance that must be mastered fully by anyone venturing into the field of geometric morphometrics.

Uniform transformations are global, nonlocalizeable transformations that have considerable appeal as simple models for shape variation. In the development of geometric morphometrics, several approaches have been proposed for estimating this component. Bookstein sorts these out in his second contribution to this section in which he develops a standard basis for the uniform subspace of shape variation and a method for estimating the uniform component with respect to that basis which preserves the Procrustes metric of shape space. That method is shown to be closely related to "alternative" formulations previously proposed. Of course, as discussed in Bookstein's earlier essay, if the uniform component is not associated with a significant portion of overall shape variation there is little justification in analyzing it separately from the nonlinear components.

Rasskin-Guttman and Buscalioni also emphasize the uniform component but take an approach different from the usual quest for "automated" analyses. They argue that an interactive, user-driven comparison of forms may sometimes be more insightful than comparisons based on mathematically optimized algorithms. Using the program D'ARCYGRAPH (see Appendix II for FTP information) they show how the interactive registration of forms based on affine models can suggest alternative explanations for shape differences. They also demonstrate how D'ARCYGRAPH can be used to generate affine morphospaces, in order to search for simple explanations of observed evolutionary sequences and provide a vocabulary, via "dispositional" variables, for describing such sequences.

The development of least-squares superimposition methods have involved general matrix formulations that can accommodate data of any dimension. Resistant-fit methods, on the other hand, have focused mainly on algorithms tailored to two-dimensional data sets. Slice presents a method for the resistant fitting of three-dimensional landmark data by exploiting a relationship between orthonormal and skew-symmetric matrices first suggested by Siegel and Pinkerton (1982). He also discusses several topics of relevance to geometric morphometrics such as the handling of missing data and the graphical representation of affine terms. The chapter concludes with a discussion and example of how to compare the results of least-squares and resistant fitting to isolate the kinds of atypical landmarks or landmark subsets the resistant approach is designed to identify. It is important to note that when comparing the results of the two methods one has different expectations of the relationships between them depending upon the level (individual or landmark) at which the comparison is being made. The methods discussed in the chapter are implemented in GRF-ND (see Appendix II for FTP information) or obtainable from the output of that program.

Finally, Rusakov tackles the problem of redundancy in the raw data. It may be, and probably often is, the case that in an attempt to sample a specimen adequately, the researcher will collect data from landmarks that provide little or no additional shape–form information. As such, variance–covariance matrices may not be of full rank even when adjustments for estimated parameters (see Bookstein's "Combining" article, this volume) have been made. If this situation is not carefully and adequately addressed, it could lead, in the worst case, to eloquent biological discussions of nothing more than rounding error associated with inappropriate matrix inversions. Rusakov suggests a way to address this problem by means of a principal components analysis of the residual coordinate covariance structure. He shows, using a simulated data set, how the sequential removal of landmarks along with an examination of the eigenvectors of the reduced covariance matrix can be used to identify landmarks whose coordinate variation can be expressed largely as linear combinations of variation at other landmarks.

ACKNOWLEDGMENTS

This work was supported by a grant (DEB 93-17572) from the Systematic Biology Program of the National Science Foundation. This is contribution number 933 from Graduate Studies in Ecology and Evolution, State University of New York at Stony Brook.

REFERENCES

Kendall, D. G. 1984. Shape-manifolds, procrustean metrics and complex projective spaces. Bulletin of the London Mathematical Society 16:81–121.

Kendall, D. G. 1985. Exact distributions for shapes of random triangles in convex sets. Advances in Applied Probability 17:308–329.

Marcus, L. F., E. Bello, and A. García-Valdecasas, (eds.), 1993. Contributions to morphometrics. Monografias del Museo Nacional de Ciencias Naturales, 8, Madrid.

Rohlf, F. J. and F. L. Bookstein (eds.) 1990. Proceedings of the Michigan morphometrics workshop. University of Michigan Museum of Zoology Special Publication 2.

Rohlf, F. J. and L. F. Marcus. 1993. A revolution in morphometrics. Trends in Ecology and Evolution 8(4):129–132.

Siegel, A. F., and J. R. Pinkerton. 1982. Robust comparison of three-dimensional shapes with an application to protein molecule configurations. Technical Report 217, Series 2, Department of Statistics, Princeton University.

MORPHOMETRIC SPACES, SHAPE COMPONENTS AND THE EFFECTS OF LINEAR TRANSFORMATIONS

F. James Rohlf

Department of Ecology and Evolution
State University of New York
Stony Brook, New York 11794-5245

ABSTRACT

The various types of mathematical spaces used in morphometrics are reviewed with emphasis given to the distinctions between the physical space of the organism, shape space and tangent spaces. The effects of linear transformations of the coordinates of the landmarks (on the specimens or the reference) on various statistical analyses and estimates of the uniform and nonaffine components of shape variation are presented. It is shown why statistical analyses of the estimates of the nonaffine components of shape variation are sensitive to affine transformations of the reference or the specimens (which may seem counterintuitive). Some implications of these results on the choices of methods for the study of shape are discussed.

INTRODUCTION

There has been considerable interest in recent years in geometrically based methods for the statistical study of shape variation. A short overview of the new field is given by Rohlf and Marcus (1993) and Rohlf (1990a). A comprehensive overview is given in the text by Bookstein (1991). Rohlf and Bookstein (1990) and Marcus et al. (1993) are proceedings of morphometrics workshops. In order to apply these new methods properly it is necessary that one understand some of the theory behind the methods. This paper discusses some aspects that are often misunderstood.

There has also been important theoretical work on various mathematical spaces that are useful for studying the statistics of shape change (e.g., Goodall, 1991). These spaces start with the assumption that information on shape has been captured by recording the coordinates of the positions of landmarks on an organism. The coordinates are two or three dimensional corresponding to the physical space in which the coordinates of the organism's landmarks were recorded. Although coordinate data contain some information not relevant

Advances in Morphometrics, Edited by L. F. Marcus et al.
Plenum Press, New York, 1996

to the study of shape (e.g., location and orientation of the specimen), such data also contain information sufficient to allow the computation of any possible mathematical index that might be proposed to describe the positions of the landmarks relative to one another. Thus working with coordinates is not a limitation. There has also been important work on the statistical distributions of points in these spaces (e.g., Kent; 1991, Goodall, 1992a).

Several different approaches are being used in biology to study shape. One of the purposes of the present note is to summarize some of the most important terms used in shape statistics in an accessible form to clarify some of the interrelationships among morphometric methods. It is not the intention of this paper to review all of the approaches currently being used (see Rohlf, 1990a, or Rohlf and Marcus, 1993, for more general reviews).

A major purpose of this paper is to point out some of the statistical consequences of different choices of reference configurations and linear transformations of the coordinates of the landmarks of the reference or the original organisms. In order to do this, it is necessary to review briefly some important equations used in the computations for each method. Knowing what manipulations of the data do or do not affect the results helps one interpret the results of a statistical method.

The notation used by other authors was adjusted in order to avoid duplication of symbols (especially the symbols α and **W**) within this paper and to be compatible with the notation used in Rohlf (1993a). The letter "t" when used as a superscript to a matrix indicates matrix transposition.

MORPHOMETRIC SPACES

Goodall (1991) presents a set of definitions for the various multidimensional spaces that are used in morphometrics. In all of these spaces the ith object (an individual specimen) consists of k coordinates in the physical space of the organism (x,y or x,y,z for two or three dimensions, respectively) at each of p landmarks, corresponding to the ith point in a multidimensional space. Although only some of these spaces will be used in most practical morphometric studies, an understanding of relationships between them and of their geometric properties is needed for understanding the new morphometric methods.

The original coordinates of the landmarks of an object (arbitrarily positioned in the digitizing plane or volume) characterize a **figure space**. This is a space of pk dimensions because the location of each of the p landmarks is described by k coordinates in the k-dimensional physical space of the organism.

If the n objects are translated in the plane or volume so that their centroids are superimposed, then the coordinates of a centered object corresponds to point in a **preform space**. This space is of $pk-k$ dimensions because the k coordinates of the centroid have been fixed for each object.

If the objects are translated and rotated (but not reflected or scaled) so that they superimpose optimally according to some criterion, then the resulting coordinates characterize a **form space** of $pk-k-k(k-1)/2$ dimensions. In two dimensions, one angle of rotation is fixed, and in three dimensions three angles are fixed. Note that Lele (1993) allows reflections in his definition of form space and gives a different number of dimensions (see below).

On the other hand, if one translates the objects to the origin and scales the objects to unit centroid size (the square root of the sum of squared distances of all landmarks to the centroid of the object, Bookstein, 1991) but does not perform any rotations or reflections, the resulting coordinates characterize what is called a **preshape space** of $pk-k-1$ dimensions. Preshapes that differ only by rotation are said to belong to the same fiber (the closed path of preshapes defined by all possible rotations of the object in the figure space).

Finally, if the objects are centered on the origin, scaled to unit centroid size and optimally rotated (but not reflected) so as to minimize the sum of the squared distances between homologous landmarks of a pair of objects (the square of the Procrustes distance between them), then an object corresponds to a point in Kendall's (1984, 1986) **shape space**. This space is of $pk-k-k(k-1)/2-1$ dimensions and is denoted by the symbol Σ_k^p (for p landmarks on a k-dimensional object). Each point in shape space corresponds to a fiber in preshape space. Shape space is non-Euclidean because Procrustes distance is used as the measure of distance between points in this space. The Procrustes distance between two shapes is the minimum distance between any two preshapes on the fibers to which the two shapes belong.

For three points in the plane, shape space, Σ_2^3 can be visualized as the surface of a sphere. Kendall (1986) gives an illustration and refers to it as a "spherical blackboard." Procrustes distances correspond to chord distances in this space. Some workers (e.g., Kendall, 1984) prefer to work with geodesic distances along great circles on the surface of the sphere. It is easy to convert back and forth between these two kinds of distances without any loss of information. In general, for p landmarks in the plane, shape space is the complex (p-2)-torus constructed from the Cartesian product of spheres with radius 1/2. The mathematical properties of this shape space are surprisingly complex given that it is generated as a consequence of pursuing the simple idea of using Procrustes distance between pairs of shapes. Shape spaces for $k = 3$ dimensions are "substantially more complicated" (Goodall, 1992b).

An alternative approach to the study of shape is to approximate the non-Euclidean shape space by a **tangent space** that has a Euclidean geometry. This space is made up of the projections of the objects in shape space onto a linear vector space that is tangent to shape space. The two spaces intersect at a point corresponding to the reference object, which can be an estimated mean (consensus) configuration from a Procrustes analysis. For three landmarks and two dimensions the tangent space is an ordinary plane that touches the surface of a sphere, corresponding to shape space, at the point which corresponds to the reference.

The shape variation captured in the linear tangent space is not limited to just linear changes in shape (uniform shape change, see below), as all shape variation is embedded in the tangent space. There is no loss of information as in a projection into a lower dimensional space. Every point in shape space corresponds to a point in the tangent space (and vice versa). For objects close to the reference, the Euclidean distances between pairs of objects in the tangent space will be good approximations to geodesic distances between points in shape space. Statistical analyses based on variation in the tangent space are expected to reflect accurately variation in shape space for only "small" variation in shape around the reference object. Bookstein (1991) gives some guidelines of what constitutes "small" variation in shape for triangles of landmarks. One can conjecture that when there are more than just a few landmarks, the usual magnitude of biological variation in shape should fill only a relatively small region of shape space and thus standard multivariate statistical analyses (such as described by Marcus, 1990, or Krzanowski, 1988) in the tangent space should be satisfactory.

THE THIN-PLATE SPLINE

Bookstein's (1989) use of thin-plate splines corresponds to one method for visualizing a tangent space for the statistical analysis of shape variation. The thin-plate spline function is a smooth function (it is twice differentiable) that maps all points in the physical space of the reference onto corresponding points in the space of the ith specimen. For completeness, basic aspects will be reviewed here. Bookstein (1991) gives a general description of thin-plate splines. The TPSPLINE program (Rohlf, 1990b) performs the necessary computations and plots the differences between two configurations of landmarks as a transformation grid based on the thin-plate spline.

For two dimensions the thin-plate spline function can be written as

$$\begin{bmatrix} x' \\ y' \end{bmatrix} = \mathbf{A} \begin{bmatrix} 1 \\ x \\ y \end{bmatrix} + \sum_{j=1}^{p} \mathbf{n}_j \, \mathrm{U}(r_j) , \tag{1}$$

where x and y are the coordinates of any point in the space of the reference, x' and y' are the resultant coordinates of a point in the space of the ith specimen, $\mathbf{A}$ is a $k \times (k+1)$ matrix of parameters that specify an affine transformation (translation, rotation, scale and shear), $\mathrm{U}(r_j)$ is the function $r_j^2 \ln r_j^2$, r_j is the distance between a point (x,y) in the space of the reference and the jth landmark in the reference, and the $\mathbf{n}_j$ are k-dimensional vectors of parameters that determine the nonaffine deformation of the space. For three dimensions the function $\mathrm{U}(r_j)$ is defined to be simply r_j. Points corresponding to the landmarks in the reference are mapped exactly to the locations of the homologous landmarks in the ith specimen. Intermediate points in other locations within the physical space of the reference are mapped to some mathematically homologous location within the physical space of the ith specimen. Because the only information used to compute a thin-plate spline is the locations of the landmarks and because this function is based on the physical properties of a thin sheet of metal rather than of biological structures, one should not take the implied homologies of intermediate points too seriously. The details of the transformation grids are very suggestive, but they are just a way of expressing the relative displacements of the landmarks. One of the most important properties is that the thin-plate spline function leads to convenient orthogonal vectors which span the tangent space and thus allows one to visualize all possible differences in shape that affect the positions of the landmarks. It is also very convenient that these vectors correspond to smooth functions. Thus any statistical analysis that expresses its results in terms of linear combinations of variables can be illustrated as a deformation in the style of a Thompson (1917) transformation grid.

The $k(k+1)+kp$ parameters needed to transform the coordinates of the landmarks in the reference into those of the ith specimen can be computed as follows:

$$\mathbf{A}_i = \mathbf{X}_i \, \mathbf{L}_q^{-1}$$

$$\mathbf{N}_i = \mathbf{X}_i \, \mathbf{L}_p^{-1\prime} \tag{2}$$

where $\mathbf{X}_i$ is the $k \times p$ matrix of coordinates of the ith specimen; L_p^{-1} is the upper left $p \times p$ block of the inverse of the $\mathbf{L}$ matrix (as defined in Bookstein, 1989, and Rohlf, 1993a) and is called the **bending energy matrix;** L_q^{-1} is the upper right $p \times (k+1)$ block of the inverse of $\mathbf{L}$; and $\mathbf{N}_i$ is a $k \times p$ matrix whose columns correspond to the $\mathbf{n}_j$ vectors in equation (1). Note that, for a fixed reference, the parameters, $\mathbf{A}_i$ and $\mathbf{N}_i$, are simply linear combinations of the coordinates of the ith specimen.

The different parameters correspond to differences between the reference and a given object at different geometric scales. The first $k(k+1)$ parameters, the elements of matrix $\mathbf{A}_i$, describe the differences in terms of an affine transformation. This includes translation, scale, rotation and shearing. The effects of an affine transformation are of infinite scale since the effects of these changes cannot be localized to any particular region of an organism. Only shearing is a component of shape. It corresponds to a uniform stretching or compression and possibly reflection, of an object in a particular direction. Bookstein (1991) refers to this as the **uniform shape component**. A shear can be specified in several ways. One of the easiest to visualize is in terms of **principal strains**—a set of orthogonal directions along which the object is uniformly stretched or compressed by some factor. For two-dimensional data, there is a direction of maximum stretching and a direction of minimum stretching at right angles to it. The ratio of the factors by which the object is stretched in these two directions is a dimensionless quantity called **anisotropy**. See Bookstein (1991) for a more detailed discussion with examples. For three-dimensional data, a shearing of the space can be described as stretching or compression along a set of three orthogonal directions.

The kp nonaffine parameters, $\mathbf{N}_i$, describe the differences between an object and the reference as the sum of nonlinear deformations of an object. These correspond to local regions of expansion, compression, bending, and so on. The subspace defined by these parameters is actually of rank $kp-k-k^2$ because translation, rotation, scaling and shearing have been fixed. This subspace can in turn be decomposed into $p-k-1$ k-dimensional geometrically orthogonal components (called partial warps, Bookstein, 1989, see below) corresponding to deformations along the x, y (and possibly z) coordinate axes at different geometric scales. Small-scale variation corresponds to changes in the relative positions of landmarks that are close together in the reference. Large-scale (but not infinite-scale) variation correspond to classic, evenly graded growth gradients.

The **principal warps** (Bookstein, 1989) are a set of eigenvectors that span the tangent space defined by the nonaffine components of the thin-plate spline. They are computed from the following decomposition of the bending energy matrix:

$$\mathbf{L}_p^{-1} = \mathbf{E}\Lambda\mathbf{E}' , \tag{3}$$

where $\mathbf{E}$ is a $p\times p$ matrix of eigenvectors and Λ is a diagonal matrix of eigenvalues (called bending energies, but it is better to interpret these simply as inverse measures of scale). The last $k+1$ eigenvalues are equal to 0 because their corresponding eigenvectors span the part of the tangent space pertaining to affine variation (which is of infinite scale). The principal warps correspond to the first $p-k-1$ column vectors of $\mathbf{E}$. The vectors are orthogonal because they are the eigenvectors of a real symmetric matrix.

One can express the thin-plate spline parameters of each specimen in terms of these principal warps rather than in terms of the original coordinates of its landmarks as in equation (2). This yields what are called **partial warps** (Bookstein, 1989). These are the principal warps applied separately along each coordinate axis and scaled appropriately so as to correspond to the nonaffine part of the thin-plate spline function that transforms the coordinates of the reference, $\mathbf{X}_c$, into those of a particular specimen, $\mathbf{X}_i$. The inner product of each row of $\mathbf{V}_i = \mathbf{X}_i - \mathbf{X}_c$ with the $p-k-1$ columns of $\mathbf{E}$ that correspond to eigenvalues greater than 0 yield what Rohlf (1993a) calls **partial warp scores**. They express the nonaffine shape differences between the reference and the ith specimen in terms of a new set of variables. They are computed as follows:

$$\mathbf{W}_i = \frac{1}{\sqrt{n}} \mathbf{V}_i \mathbf{E} \Lambda^{-\alpha/2} \tag{4}$$

where the division by $\sqrt{n}$ and the subtraction of $\mathbf{X}_c$ are for convenience in describing the variation in shape with a sample of n specimens; the multiplication by $\Lambda^{-\alpha/2}$ allows the possibility (depending on the value of α, see below) of different forms of weighting the principal warps according to geometric scale. Note that it is sometimes useful to retain all the columns in matrix $\mathbf{E}$ so that all variation captured by $\mathbf{X}_i$ is preserved in the $\mathbf{W}_i$ matrix. The additional columns of $\mathbf{W}_i$ are not partial warps because they describe affine differences not local deformations. Sometimes other information, such as an estimate of the uniform shape component, is appended to the matrix of partial warp scores.

The $\mathbf{W}_i$ matrix for the ith specimen has k rows and either p or $p-k-1$ columns depending on whether the last $k+1$ columns of $\mathbf{E}$ (the principal warps with eigenvalues equal to 0) are retained or not. If they are not retained then this equation can be simplified to

$$\mathbf{W}_i = \frac{1}{\sqrt{n}} \mathbf{X}_i \mathbf{E} \Lambda^{-\alpha/2} \tag{5}$$

because $\mathbf{X}_c\mathbf{E} = 0$ for the first p-k-1 columns of $\mathbf{E}$ (the principal warps were constructed to be orthogonal to the reference configuration). Although $\mathbf{W}_i$ is a linear function of $\mathbf{X}_i$, it specifies the nonlinear deformations needed to transform the physical space of the reference so that the locations of its landmarks are as in specimen i.

ESTIMATION OF THE UNIFORM COMPONENT OF SHAPE VARIATION

There are a number of methods to estimate the uniform component of shape variation. Bookstein and Sampson (1990) and Bookstein (1991) proposed two closely related methods for estimating the uniform component of shape change between a given specimen and reference specimen for two-dimensional data. Both methods use the following equation to compute the vector of parameters:

$$\mathbf{u} = (\mathbf{M}'\,\mathbf{S}^{-1}\,\mathbf{M})^{-1}\,\mathbf{M}'\,\mathbf{S}^{-1}\Delta\mathbf{V}\,, \tag{6}$$

where

$$\mathbf{M}' = \begin{bmatrix} v_{23} & 0 & v_{24} & 0 & \dots & v_{2k} & 0 \\ 0 & v_{23} & 0 & v_{24} & \dots & 0 & v_{2k} \end{bmatrix}$$

is a matrix of the y-coordinates of the shape coordinates (Bookstein, 1986) of landmarks 3 ... p in the reference (assuming a baseline defined by landmarks 1 and 2). The vector $\Delta\mathbf{V}$ contains the differences in x and y shape coordinates between the given specimen and the reference (strung out as a single column with $2k$–4 rows). The two methods differ in the estimation of the $(2k-4)\times(2k-4)$ variance–covariance matrix $\mathbf{S}$. For the method that minimizes Procrustes distance, $\mathbf{S}$ is the expected covariance among landmarks based on their geometrical position in the reference and assuming only circular digitizing noise with the same variance at each landmark. For their method that minimizes Mahalanobis distance, $\mathbf{S}$ is estimated from the observed variation in a sample of specimens. These methods are implemented in Bookstein's computer program PROJECT (distributed with Rohlf and Bookstein, 1990).

Rohlf (1993a) estimates the uniform shape component based on the affine parameters of the thin-plate spline. This has the advantage of extending the mathematically elegant decomposition of nonaffine differences in shape described in the previous section to include a decomposition of the affine differences into uniform shape change versus nonshape differences (translation, rotation and scale). Unfortunately, as pointed out by Bookstein ("Combining" article, this volume), the thin-plate spline computations weight differences inversely as a function of distances between landmarks in the reference. This means, for example, that shape differences involving a pair of landmarks that are very close together in the reference will be weighted very heavily compared with shape differences involving larger regions on the reference. This weighting seems inappropriate for estimating uniform shape differences (or the other affine components), for which larger scale differences should have the higher weight. The problem can be seen by the counterintuitive superimpositions one sometimes obtains using the "minimum energy superimposition" display option in the TPSRW program (Rohlf, 1992) on data in which some landmarks are very close together in the reference.

It is also possible to estimate the uniform component of shape variation as the "nuisance parameters" of affine superimposition methods. These superimposition techniques are described by Rohlf and Slice (1990a) and by Slice ("Three dimensional" article, this volume). In the least-squares approach (Generalized Affine Least Squares, GALS, superimposition) one estimates translation, rotation, scale and the uniform shape components so as to minimize the residual differences between two specimens (or between a specimen and an average consensus configuration of landmarks).

Bookstein ("Standard formula" article, this volume) proposes a linearized Procrustes method to estimate the uniform shape change component. He derives it to be equivalent to that obtained from GALS by Rohlf and Slice (1990a). For simplicity in notation, it is convenient to translate the reference (Generalized Least Squares, GLS, consensus configuration) to the origin, scale it to unit centroid size and then rotate it to its principal axes. Then in two dimensions $\Sigma xy = 0$, $\Sigma x^2 + \Sigma y^2 = 1$ and $\Sigma x = \Sigma y = 0$, where x and y are the x and y-coordinates of the reference after the centering, scaling and rotation operations described above are performed. Next use orthogonal least-squares Procrustes analysis (Sneath, 1967, Rohlf and Slice, 1990a, Goodall, 1991) to superimpose the ith specimen on the rotated reference and let Δx and Δy denote the differences between the new coordinates of the specimen and those of the rotated reference.

The uniform shape component for the ith object is then estimated as

$$u_1 = (\Sigma y_i \Delta x_i + \lambda_2 \Sigma (x_i \Delta y_i - y_i \Delta x_i)) / \sqrt{\lambda_1 \lambda_2}$$

$$u_2 = \Sigma y_i \Delta y_i / \sqrt{\lambda_1 \lambda_2} \tag{7}$$

where $\lambda_1 = \Sigma x_i^2$ and $\lambda_2 = \Sigma y_i^2$.

In two dimensions we can use u_1 and u_2 as uniform shape descriptors for statistical analyses. For a fixed reference, u_1 and u_2 are linear combinations of Δx's and Δy's. They will approximately follow a bivariate normal distribution if the deviations of the specimens from the reference are normally distributed landmark by landmark. The geometric meaning of u_1 and u_2 can be visualized by using the following equation. It describes the effect of a uniform shape transformation on the coordinates of landmarks in the reference.

$$\begin{bmatrix} x' \\ y' \end{bmatrix} = \begin{bmatrix} 1 & u_1 \\ 0 & 1 + u_2 \end{bmatrix} \begin{bmatrix} x \\ y \end{bmatrix} \tag{8}$$

where x and y are, as above, the centered, scaled and rotated coordinates of the reference and x' and y' are the coordinates of the reference after the shearing operation defined by the parameters u_1 and u_2 is applied.

RELATIVE WARPS ANALYSIS

In order to study variation in shape of a sample of n specimens, it is convenient to collect the $\mathbf{W}_i$ (partial warp scores, equation (4 above) into a single matrix $\mathbf{W}$. This new matrix has n rows corresponding to the n specimens and columns corresponding to the partial warp scores for each specimen strung out as a single row (e. g., all of the x projections followed by the y projections and, possibly, the z projections). There will be $k(p-k-1)$ columns (or more if additional columns are appended). This can be done by extension of equation (4) to

$$\mathbf{W} = \frac{1}{\sqrt{n}} \mathbf{V} (\mathbf{I}_k \otimes \mathbf{E}\Lambda^{-\alpha/2}) \tag{9}$$

where the symbol $\otimes$ denotes a direct (Kronecker) product of two matrices, $\mathbf{I}_k$ is the $k \times k$ identity matrix, and $\mathbf{V}$ is an $n \times kp$ matrix of the differences between each object and the reference strung out along each row (all of the differences in x-coordinates followed by those for the y-coordinates and, possibly, the z-coordinates). The Λ and $\mathbf{E}$ matrices and the scalar α are as defined in equation (4).

Relative warps analysis (Bookstein, 1989, 1991, Rohlf, 1993a) is a principal components analysis (PCA) of variation in shape in the tangent space as described by the **W** matrix. If $\alpha \neq 0$ then it is a weighted PCA with the weights a function of the geometric scale of each component of shape variation. If $\alpha = 0$ then relative warp analysis is consistent with the Procrustes metric for measuring the amount of difference between different shapes. Even though the principal warps are geometrically orthogonal, projections of a sample of specimens onto the principal warps are usually correlated. The purpose of a PCA is to describe the major trends of shape variation in as few statistically orthogonal dimensions as possible.

The principal component vectors of the **W** matrix are called **relative warps**. They are often expressed in terms of displacements of landmarks. They can also be visualized as a transformation grid showing deformations of the physical space of the reference configuration; this is done by expressing each relative warp as a thin-plate spline (examples are given in Rohlf, 1993a, and Rohlf and Marcus, 1993).

EFFECTS OF DIFFERENT CHOICES OF A REFERENCE

Different choices of a reference result in different physical distances between the landmarks in the reference. This leads to different **E** and **V** matrices and thus to different **W** matrices. However, if α is set equal to 0 in equation (9), as suggested by Rohlf (1993a) for exploratory studies, then it is possible to retain all p principal warps in matrix **E**. The **E** matrix will be an orthonormal matrix and different choices of a reference will result in **E** matrices that differ only by a rigid rotation. Thus any analyses of the distance relationships between objects (e.g., cluster analyses or principal coordinates analyses of distance matrices) based on the resulting **W** matrices will differ only as a result of the effect of subtracting off a different reference in computing the **V** matrix. Otherwise, the results will be identical because distances are invariant under rigid rotations.

If, as is usually the case, only the first p-k-1 principal warps are used (those with $\lambda_i > 0$) then the **W** matrix describes just the nonaffine part of the variation in shape among the objects (the part describable by local deformations). This can be visualized as a projection from the kp-dimensional tangent space (whose actual rank is $kp - k - k - k(k-1)/2 - 1$) into the $k(p-k-1)$-dimensional subspace of the tangent space that corresponds to local deformations. Different choices of a reference result in different principal warps and hence different partitionings of the tangent space into affine versus nonaffine components. Different partitionings lead to different distances among the objects. Distances in the subspace corresponding to nonaffine shape variation will be equal to or less than those obtained when all kp dimensions are used. If $\alpha > 0$ then different choices of a reference also affects what one considers to be large- versus small-scale differences in shape. This is because scale is measured by the reciprocals of the eigenvalues, λ_i. Thus, statistical analyses based on the partial warp scores (equation 9) will give numerically different results if different references are used. Analyses based on similar references should, however, give similar results and very similar statistical conclusions.

EFFECTS OF LINEAR TRANSFORMATIONS OF THE REFERENCE

If we consider only the nonaffine components of the thin-plate spline (i.e., matrix **E** is $p\times(p-k-1)$) then we can use equation (5) to compute the partial warp scores for specimen i. The effects of rotation, scaling and shearing of the reference can be represented as a premultiplication of $\mathbf{X}_c$ by a $k\times k$ matrix **T**. If **T** is orthonormal then using $\mathbf{TX}_c$ as the reference (but keeping the original $\mathbf{X}_i$) will have no effect on the $\mathbf{W}_i$ because the bending energy matrix is invariant to rotation of the reference. If **T** is not orthonormal then using $\mathbf{TX}_c$ shears the reference, which requires a recomputation of the bending energy matrix because distances between landmarks in the reference will change. This leads to new matrices **E** and **Λ**, which result in new $\mathbf{W}_i$ matrices that will differ in complex ways from those obtained originally. Thus an affine transformation of the reference affects the estimates of the non-affine components of the $\mathbf{W}_i$ through the effect of the transformation on the bending energy matrix.

Translation of the reference by a k-dimensional vector **t** will have no effect on the $\mathbf{W}_i$ because

$$\left(\mathbf{t}\mathbf{1}_p' + \mathrm{X_c}\right)\mathbf{E} = \mathbf{X}_c\,\mathbf{E} \tag{10}$$

Translation, rotation and scaling the reference will not affect the estimate of the anisotropy obtained from the linearized Procrustes method, equation (7), because the method eliminates the effects of these transformations initially. Bookstein and Sampson's (1990) and Bookstein's (1991) methods for estimating the uniform shape component by minimizing either Procrustes or Mahalanobis distances are not influenced by the alignment of the reference because this method is based on the use of Bookstein shape coordinates.

EFFECTS OF LINEAR TRANSFORMATIONS OF THE OBJECTS

Because a superimposition method such as a generalized Procrustes analysis is usually used to construct the reference as an average of the objects after alignment, a shearing of the objects leads to a change in the estimate of the reference (the consequences of which are described above). In this section we will, however, treat the reference as fixed.

Based on the relationship given in equation (5), the effect on $\mathbf{W}_i$ of a linear transformation of the ith object can be expressed as

$$\frac{1}{\sqrt{n}}\mathbf{TX}_i\mathbf{E}\Lambda^{-\alpha/2} = \mathbf{TW}_i \tag{11}$$

where **T** is a $k\times k$ transformation matrix. Thus any effects of the premultiplication by **T** (scaling, rotation, or shearing) are passed directly on to the matrix of partial warp scores. This affects both the affine and the nonaffine components. If **T** is an orthonormal matrix then the space defined by the partial warp scores will be rigidly rotated and the distances between the objects will be unchanged. Thus an ordination of the objects produced in a relative warps analysis will be unchanged if **T** is orthonormal. If **T** is not orthonormal then the results of a relative warps analysis will be affected because the specimens will have changed shape as a result of being sheared. On the other hand, statistical analyses such as MANOVA, canonical variates analysis, multiple regression analysis or the computation of Mahalanobis generalized distances are invariant to the effects of premultiplication by **T** as long as it is of full rank, *k*.

If instead of a constant matrix **T** one multiplies the *i*th specimen by $\mathbf{T}_i$ then the ith transformation effects will also be passed along to the corresponding rows of the matrix of partial warp scores. This will change the results that one expects to obtain in an ordination of the specimens or any other multivariate analysis applied to the **W** matrix. Therefore the specimens should be aligned in some consistent way (e.g., by an orthogonal Procrustes analysis) before the matrix of partial warp scores is computed. The use of superimposition programs such as GRF (Rohlf and Slice, 1990b), GRF-ND (Slice, 1993b), or Morphometrika (Walker, 1994) is recommended in order to align the specimens before using programs for relative warp analysis (e.g., TPSRW, Rohlf, 1992, or NTSYS-pc, Rohlf, 1993b), partial warp regression analysis (TPSREGR, Rohlf, 1993c) or for the estimation of the uniform component (Bookstein, "Standard formula" article, this volume). Different alignments of the specimens (even when the same reference is used) will lead to numerically different results for analyses based on the **W** matrix. Empirically (J. A. Walker, personal communication), the differences are usually minor and the biological and statistical conclusions unaffected between the usual methods of least-squares and resistant fitting.

The nonaffine components of **W** are not influenced by a translation of the objects by a vector **t** (or even $\mathbf{t}_i$) due to the same type of relationship shown in equation (10).

Translation, rotation and scaling of the specimens will not affect the estimates of anisotropy obtained from the linearized Procrustes method, equation, because that method first aligns the specimens to the reference by means of an orthogonal Procrustes analysis. Shearing the original specimens will of course affect any estimate of the uniform shape component. The estimated direction of the principal strain for each specimen will be affected because it is expressed relative to the initial orientation of each object. Bookstein and Sampson's (1990) and Bookstein's (1991) methods for estimating the uniform shape component by minimizing either Procrustes or Mahalanobis distances will not be affected because those methods are based upon the use of shape coordinates.

EFFECTS OF DIFFERENT CHOICES OF VALUES FOR α

The effect of different choices of values for α in the computation of the partial warp scores, **W**, is to determine how the matrix of eigenvalues, Λ, of the bending energy matrix is used to weight the principal warps. Because the diagonal matrix $\Lambda^{-\alpha/2}$ is the right-hand factor in equation (9), its effect is just to weight the principal warps. Although α is a continuous variable, it is most commonly set to either 0 (to weight the principal warps equally) or 1 (to weight them by the reciprocals of the square roots of their eigenvalues so that large scale features in the reference are given greater weight).

Multivariate statistical analyses using **W** as a "data matrix" will give numerically different results depending upon the value for α unless the analysis is invariant to scale. Statistical analyses that are invariant to scale (such as MANOVA, canonical variates analysis, discriminant functions and multiple regression analysis) will not be affected by different choices of values for α. On the other hand, principal components analysis is sensitive to the scaling of variables and thus the results of a relative warps analysis will depend on the value of α (Rohlf, 1993a).

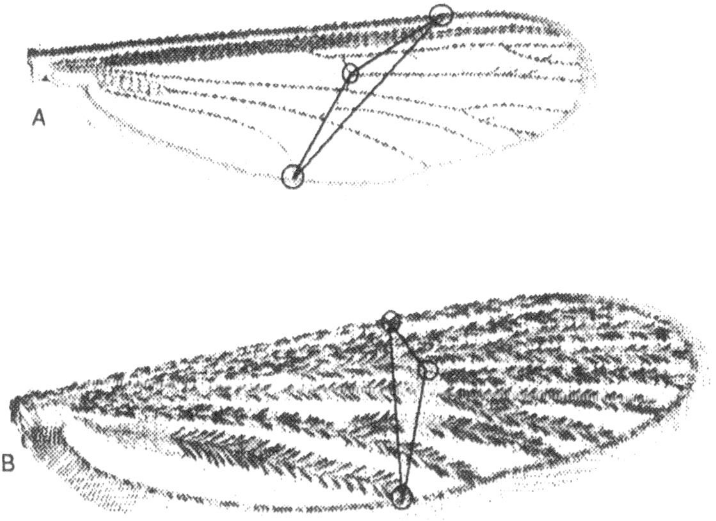

Figure 1. An example of a set of three landmarks whose relative positions in two species differ in part by a reflection. A. *Uranotaenia lowii.* B. *Orthopodomyia signifera.* Images of mosquito wings were scanned from illustrations in Carpenter and LaCasse (1955).

DISCUSSION

As mentioned above, Lele (1993) gives a different definition for form space. He uses a different definition because his methods are based on matrices of Euclidean distances among landmarks and cannot distinguish between a configuration of landmarks and its mirror image. Although invariance to reflection may sometimes be a convenience when working with incomplete specimens of bilaterally symmetric organisms, it can also be a serious limitation. Certain types of differences in shape will not be detectable, and nonsensical shape comparisons can result if reflections are ignored (see Fig. 1 for an example). Although it is much easier to find examples based on just three landmarks, such as shown in Fig. 1, examples can be constructed involving any number of landmarks. Differences due to reflection must be considered shape differences unless one knows that the organism itself has been reflected. Lele (1993) also states that form space is of $p(p-1)/2$ dimensions—because that is how many unique elements there are in a $p \times p$ distance matrix. However, these distances are partially redundant. The space corresponding to all possible physically realizable configurations of landmarks can occupy at most a pk-dimensional space. As discussed above, by fixing differences due to translation and rotation, the dimensionality is reduced to $pk-k-k(k-1)/2$.

It is interesting to compare statistical analyses based on principal warp scores with those based on the residuals from a superimposition analysis. If the reference used in equation (4) is the consensus configuration from an orthogonal generalized Procrustes analysis and if the $\mathbf{X}_i$ are the coordinates of the specimens after alignment with the reference (as is the recommended procedure, F. L. Bookstein, personal communication), then the elements of $\mathbf{V}_i$ are the residuals that are examined (usually graphically in a Procrustes analysis) to look for differences in shape. The square root of the sum of the squared elements of $\mathbf{V}_i$ is the Procrustes distance between the ith specimen and the reference. If $\alpha = 0$ and if all p principal warps are retained in matrix **E** in equation (4), then the $\mathbf{W}_i$ matrices are just a constant times a rigid rotation of $\mathbf{V}_i$. The $\mathbf{V}_i$ and $\mathbf{W}_i$ matrices contain the same information, and standard multivariate statistical analyses based on them will give equivalent if not identical results. Slice (1993a) gives examples of statistical analyses based on the coordinates of specimens aligned using superimposition methods. Slice (1993a) also contrasts the results with those based on partial warp scores.

Lele (1993) makes much of a supposed inconsistency of superimposition methods to estimate an average configuration of landmarks. Unfortunately, his analysis is based on the average of coordinates of landmarks in his definition of form space (specimens not size standardized), which does not correspond to any of the published methods. His example of a dataset for which Procrustes analysis yields an inconsistent estimate has a very extreme covariance structure. Interestingly, his own method, EDMA, also yields inconsistent estimates for that case (C. R. Goodall, personal communication). Procrustes analysis has been shown to yield consistent estimates for the case of circular isotropic error (Kent and Mardia, in prep). This paper is mostly concerned with shape statistics based on projections onto the tangent space (e.g., analyses based on partial warp scores) rather than operating directly in shape space. This is because the tangent space has a Euclidean metric and thus conventional multivariate statistical methods can be applied. For small variation in shape the approximation of shape space by a tangent space is very good. The issue of consistency does not seem critical for multivariate statistical analyses in the tangent space because small differences in the reference result in only slightly different orthogonal matrices to be used to determine the projection into the tangent space (Rohlf, in prep). Some researchers (e.g., Goodall, 1992a; Kent, 1991) are, however, working on the development of specialized techniques that operate in shape space. Such work is important because it will provide new methods capable of exact analyses when variation in shape is not small.

ACKNOWLEDGMENTS

The stimulating questions by Diego Rasskin, Stefan Revets, Jeff Walker and others at the NATO Advanced Study Institute "Advances in Morphometrics" at Il Ciocco, Italy, were very helpful in motivating the present note. Their persistence was very much appreciated. Fred Bookstein, Colin Goodall, Dennis Slice and Jeff Walker read a version of the manuscript and contributed many helpful suggestions that greatly improved this paper. This paper was supported by a grant (DEB 93-17572) from the Systematic Biology Program of the National Science Foundation.

This paper is contribution number 911 from the Graduate Studies in Ecology and Evolution, State University of New York at Stony Brook.

REFERENCES

Bookstein, F. L. 1984. Tensor biometrics for changes in cranial shape. Annals of Human Biology 11:413–437.

Bookstein, F. L. 1986. Size and shape spaces for landmark data in two dimensions (with discussion and rejoinder). Statistical Science 1:181–242.

Bookstein, F. L. 1989. Principal warps: Thin-plate splines and the decomposition of deformations. I.E.E.E. Transactions on Pattern Analysis and Machine Intelligence 11:567–585.

Bookstein, F. L. 1991. Morphometric tools for landmark data: Geometry and Biology. Cambridge University Press: Cambridge.

Bookstein, F. L., and P. Sampson. 1990. Statistical models for geometric components of shape change. Communications in Statistics: Theory and Methods 19:1939–1972

Carpenter, S. J. and W. J. LaCasse. 1955. Mosquitoes of North America (North of Mexico). University California Press: Berkeley.

Goodall, C. R. 1991. Procrustes methods in the statistical analysis of shape (with discussion and rejoinder). Journal of the Royal Statistical Society, Series B, 53:285–339.

Goodall, C. R. 1992a. Dynamic graphics in non-Euclidean spaces: The visualization and statistical analysis of shape. 1992 Proceedings of the Statistical Graphics Section American Statistical Association; Alexandria, Virginia.

Goodall, C. R. 1992b. Shape and image analysis for industry and medicine. Short course, XIIth Leeds workshop. University of Leeds, Leeds, UK.

Kendall, D. G. 1984. Shape-manifolds, procrustean metrics and complex projective spaces. Bulletin of the London Mathematical Society 16:81–121.

Kendall, D. G. 1986. Comment on F. L. Bookstein, Size and shape spaces for landmark data in two dimensions. Statistical Science 1:222–226.

Kent, J. T. 1991. The complex Bingham distribution and shape analysis. Technical report STAT 92/04, Department of Statistics, University of Leeds.

Krzanowski, W. J. 1988. Principles of multivariate analysis: A user's perspective. Oxford University Press: Oxford.

Lele, S. 1993. Euclidean distance matrix analysis (EDMA): Estimation of mean form and mean form difference. Mathematical Geology 25:573–602.

Marcus, L. F. 1990. Traditional morphometrics. Pages 77–122 *in* F. J. Rohlf and F. L. Bookstein, (eds.), Proceedings of the Michigan morphometrics workshop. University of Michigan Museum of Zoology Special Publication 2.

Marcus, L. F., E. Bello, and A. García-Valdecasas (eds.). 1993. Contributions to morphometrics. Monografías del Museo Nacional de Ciencias Naturales. 8, Madrid.

Rohlf, F. J. 1990a. Morphometrics. Annual Review of Ecology and Systematics 12,:299–316.

Rohlf, F. J. 1990b. TPSPLINE: A program to compare two shapes using a thin-plate spline. Department of Ecology and Evolution, State University of New York at Stony Brook, Stony Brook, New York, 11794.

Rohlf, F. J. 1992. TPSRW: A program for relative warps analysis. Department of Ecology and Evolution, State University of New York at Stony Brook, Stony Brook, New York 11794.

Rohlf, F. J. 1993a. Relative warp analysis and an example of its application to mosquito wings. Pages 131–159 *in* L. F. Marcus, L. F., E. Bello, and A. García-Valdecasas (eds.). 1993. Contributions to morphometrics. Monografías del Museo Nacional de Ciencias Naturales . 8, Madrid.

Rohlf, F. J. 1993b. NTSYS-pc, version 1.80. Exeter Software: Setauket, New York.

Rohlf, F. J. 1993c. TPSREGR: a program for regression of partial warp scores. Department of Ecology and Evolution, State University of New York at Stony Brook, Stony Brook, New York, 11794.

Rohlf, F. J., and F. L. Bookstein, (eds.), 1990. Proceedings of the Michigan morphometrics workshop. University of Michigan Museum of Zoology Special Publication 2.

Rohlf, F. J., and L. F. Marcus 1993 A revolution in morphometrics. Trends in Ecology and Evolution, 8:129–132.

Rohlf, F. J., and D. Slice. 1990a. Extensions of the Procrustes method for the optimal superimposition of landmarks. Systematic Zoology 39:40–59.

Rohlf, F. J., and D. Slice. 1990b. GRF: A program for generalized rotational fitting. Department of Ecology and Evolution, State University of New York at Stony Brook, Stony Brook, New York, 11794.

Slice, D. E. 1993a. Extensions, comparisons, and applications of superimposition methods for morphometric analysis. Ph.D. dissertation: State University of New York at Stony Brook.

Slice, D. E. 1993b. GRF-ND: Generalized rotational fitting of n-dimensional data. Department of Ecology and Evolution, State University of New York at Stony Brook, Stony Brook, New York, 11794.

Sneath, P. H. A. 1967. Trend-surface analysis of transformation grids. Journal of Zoology 151:65–122.

Thompson, D. W. 1917. On growth and form. Cambridge University Press: London.

Walker, J. A. 1994. Morphometrika. Geometric morphometrics for the Macintosh. Department of Anatomical Sciences, State University of New York at Stony Brook, Stony Brook, New York 11794.

COMBINING THE TOOLS OF GEOMETRIC MORPHOMETRICS

Fred L. Bookstein*

Institute of Gerontology
University of Michigan
Ann Arbor, Michigan 48109-2007

ABSTRACT

The greatest strength of the new geometric morphometrics is the system of interrelated multivariate and graphical procedures it offers for a variety of analytic questions involving landmark data. A typical analysis will begin with the conversion of landmark data into a multivariate statistical representation of shape, will continue with a series of broadly familiar multivariate matrix manipulations, and will conclude by inspection of a considerable variety of diagrams that represent the findings in both the space of shape coordinates per se and the space of the two- or three-dimensional image of the organism. The choices under the first heading, the passage to a multivariate representation of shape, include two-point shape coordinates, partial warp scores, and Procrustes residuals. Each of these except the partial warp scores is unsuitable for some subset of the reasonable matrix manipulations; for instance, shape coordinates do not supply sensible principal components analyses, and Procrustes residuals cannot lead to sound canonical variate analyses without modification. The modes of diagramming data include thin-plate splines, partial warp splines and scatters, Procrustes residual scatters, and resistant-fit scatters, among others. Most analyses benefit greatly from exploiting more than one of these.

Not every combination of shape coordinates, multivariate maneuvers, and diagram styles makes sense. Underlying the multivariate geometry of any sample is an a-priori Procrustes geometry of shape per se, and the three parts of any analysis must be mutually consistent in the use they make of this geometry as well as the usual linear geometry of multivariate analysis. Some formal structures serve two roles in this synthesis; for instance, the affine term of a generalized affine least-squares fit is equivalent to the uniform component of the relation between a specimen shape and the sample mean. Within the realm of multivariate computations per se, some analyses are automatically equivalent: for instance, relative warps analysis with $\alpha = -1$ is a principal coordinates analysis of bending energy. Familiar statistical tests (e.g. two-sample comparisons by Hotelling's T^2) come in several

* E-mail: fred@brainmap.med.umich.edu

Advances in Morphometrics, Edited by L. F. Marcus et al.
Plenum Press, New York, 1996

approximate versions for sample sizes smaller than those for which the standard procedure is appropriate. Some combinations are usually inappropriate; for instance, resistant-fit residuals are not shape coordinates and should not be used as input for any linear multivariate computation. And interpretation of certain computations requires the prior inspection of others; for instance, any relative warps analysis of the nonaffine space should be preceded by analysis of the full shape space. There are modifications of all the basic tools for applications to symmetric forms, and all come in one version for two-dimensional data and a different version for three-dimensional data.

This chapter explains the entire toolkit in terms of the logic by which diverse choices are combined into complete analyses that circumvent a variety of popular pitfalls.

INTRODUCTION

Geometric morphometrics is a new methodology for studying shape variation of landmark configurations. Its methods use all the information available about the locations of those landmarks while rigorously adhering to the geometric definition of shape. In particular, they systematically exploit the geometry of the mean shape to help characterize variation around that mean, or differences between one such mean and others, in biometrically and graphically effective ways. A good analytic strategy for this purpose must combine tools for generating a multivariate representation of shape from landmark coordinates and tools for more-or-less-familiar multivariate analysis of those shapes with a selection of graphical tools that bring to light many of the alternate interpretations that such findings might support.

More specifically, the new morphometric analyses comprise three phases: (1) rigorously converting one's data from digitized landmark locations into points of Kendall's shape space, and from there to their tangent space, a linearization customized to the vicinity of the mean shape of the sample; (2) carrying out a multivariate analysis in that tangent space appropriate to one's scientific goal and correctly matched to the way one got into that tangent space; and (3) getting back out of shape space — displaying the results of that multivariate analysis by an adequate variety of graphical methods — in order to visualize, interpret and understand. What this "shape space" is, and why its particular geometry is so important, are the concerns of the next section. The main variants of (1), means of getting to that geometry, are shape coordinates, partial warps, and Procrustes residuals; the methods under (2), multivariate analysis, are regressions, principal components analyses, canonical analyses, Partial Least Squares analyses, and multivariate analysis of variance; and the typical choices under (3), displays of shape findings on flat paper, include thin-plate splines, partial warps and scatters, uniform principal axes and scatters, Procrustes residual scatters, and resistant-fit scatters. Most of the intellectual effort is invested in alternating between phases (2) and (3), multivariate pattern construction and display.

Some of these triples make sense quite reliably (as long as one's landmarks are homologous and well spaced over the form, one's samples homogeneous, one's digitizing conscientious and precise, etc.). For example, when faced with a sample of two-dimensional landmark locations having no obvious heterogeneities, it is sensible to reduce a set of landmark configurations to shape space by GLS (Generalized Least Squares) superposition upon an estimated Procrustes mean shape, extract principal components of the covariance matrix of the residuals, and finally display the principal components — all but the last four — for inspection in two ways, as deformations and as Procrustes superpositions, along with scatters of scores. Allometry is displayed via the pattern of covariances with centroid size of those same residuals. (In fact, any basis for shape space can be adjusted for centroid size coordinate by coordinate, as long as those adjustments are not tested separately — see sections on generalized Procrustes analysis and on shape coordinates to a baseline.) But you can't carry out multivariate analysis of variance of the Procrustes residuals, adjusted or not: they are four too many. To explore group differences, a different shape basis is required—perhaps the weight matrix of partial warp scores and uniform component produced by Rohlf's TPSRW (Thin Plate Spline Relative Warp Analysis). Those MANOVAs are adjusted for allometry in the same way, using centroid size, in this case by conversion to MANCOVA.

To interpret findings like these, one can inquire unproblematically about the presence of large-scale aspects of the multivariate signal by inspecting the uniform component and the first few partial warps of the principal components, or the group mean differences or their centroid-size-adjusted versions. Or, circumventing TPSRW, one might compute a pair of group mean shapes by averaging their shape coordinates to any convenient baseline, test that mean difference for significance by Hotelling's T^2, and study the nature of the difference by partial warps and their scatters (to localize the processes under study) and also by GLS residuals and their scatters (a rearrangement emphasizing the roles of information at the landmarks separately). All these are standard three-phase sequences of computation and display.

But when one wishes to thrust more deeply into landmark data than is afforded by simple linear models and projections, these same tools must be recombined in novel arrangements. The fitness of arbitrary sequences of these tools is stringently constrained by the requirement that all incorporated references to the geometry of shape space, whether implicit or explicit, be coherent. In the last example, even though it is reasonable to compare within-groups principal components with group mean differences, one cannot pair such analyses if they begin from shape coordinates to a baseline. Those components change when one changes the baseline for shape coordinates, whereas the geometric descriptors of group differences, like most of the techniques of geometric morphometrics, are (virtually) baseline-independent. One must turn instead to an explicitly constructed variant of principal components, relative warps analysis, that produces findings with the necessary invariances. Similarly, there is probably no coherent "shape distance" to be derived from GRF (Generalized Resistant Fit) or GARF (Generalized Affine Resistant Fit) analyses, because these programs do not respect the Procrustes distance already locked into the algebra of all the other tools.

The commonest errors incurred in the course of this new praxis owe to inappropriate combinations of tools in ways like these. When first imagined (or when carried out by accident), these inappropriate combinations, or pitfalls, seem just about as reasonable as the combinations that we put forward elsewhere in this volume as "appropriate." Pitfalls are the sequences of operations that seem reasonable and familiar step by step but that, nevertheless, lose insight and result in more or less misleading "findings."

Some errors of this sort are committed by the toolmakers of this new morphometrics, others by tool-users. The oldest fragment of geometric morphometrics, the multivariate statistics of shape coordinates, dates only from the mid-1980s; of the graphical techniques we use, those deriving from thin-plate splines are quite a bit younger; and one of the tools (the uniform component of a shape scatter) dates from late 1993. Thus the field as a whole is new enough that no one's intuition is yet an adequate guide for discriminating the sensible sequences of computations from those that incorporate great gaping chasms where there should be rigor of biometrical inference. Nor is it easy to explain, in some cases, exactly why an error is, in retrospect, erroneous. The innocent question "Why shouldn't I do that?" usually has an answer, but it has taken the toolmakers a while to learn how to phrase these answers. Some of the sequences of computations I will characterize below as "errors" have already appeared in print in volumes refereed by these same editors, in theses read by them, or even, alas, in papers written by them before the moment of enlightenment. It is the task of this chapter to explain why some of these reasonable-looking procedures are sturdy even while other combinations of the same tools, combinations differing by what seem to be only insignificant changes of program option, are quite foolish.

This chapter is not intended as a scientific paper, only as an amalgam of useful advice. I have written it with only a handful of references, and those informal. A parallel version, "The Morphometric Synthesis," is Bookstein (1966). It will have a complete scholarly apparatus and will include two exhaustively worked examples.

THE CORE OF GEOMETRIC MORPHOMETRICS

What separates the reasonable combinations of tools from the unreasonable ones is the issue of underlying consistency with the shared geometrical substrate of all these methods. Morphometrics is unusual among multivariate statistical techniques in the ubiquity of this additional geometry. Irrespective of whether or not the "findings" make biological sense — indeed, regardless of whether there appear to be findings — many factors have to be taken into account in deciding whether a particular sequence of computations makes sense.

In brief, there is only one geometry of Euclidean shape for landmark data, and that is the geometry of Procrustes distance. In insisting that the geometry be of Euclidean shape we mean, among other things, that moving any one landmark by some specified distance in the digitizing plane gives rise to shapes that are all at the same distance from the starting shape regardless of the direction in which that landmark was moved. By "shape" we intend the ordinary scientific meaning of the word, referring to properties of figures that are invariant to changes in position, orientation, or geometric scale. Only one size variable, centroid size, is consistent with this geometry; it is the size variable divided out in the formula for Procrustes distance.* For small ranges of shape variation, there is essentially only one average shape for a sample, but many different algorithms (some of them rather elegant — Gower's principal components analysis, for example) for arriving at that average. Still for small ranges of shape, there is only one complete multivariate analysis of variation or covariation around that mean shape. But there are several different-looking ways of arriving at that shared multivariate analysis — baseline shape coordinates, partial warps, Procrustes residuals — and there is an even greater variety of ways of displaying those analyses, a variety that should never be reduced to any single choice.

Motivating these dogmas are fundamental theorems about the statistical geometry of shapes in the picture plane or ordinary (Euclidean) space, theorems first published by the probabilist David Kendall in 1984 for applications in astronomy and archaeology. Supposing throughout that we wish to talk about shape, by this let us agree to mean aspects of landmark configurations from which information about size, position, and orientation has been (very carefully) removed. (Techniques that do not explicitly parameterize or display shape variation, such as Euclidean Distance Matrix Analysis, are outside the bounds of this discussion.) It is easy enough to imagine removing information about position — just place each

*There is a subtlety associated with this business of the formula for Procrustes distance. That formula is usually taken as the root sum of squared distances between homologous landmarks when two configurations are each scaled to unit centroid size and then one is moved over the other until that sum of squares is least. Although this formula is graphically appealing, it does not give quite the right global geometry of shapes. The correct version includes an additional trigonometric transformation. For two-dimensional data in complex notation, for which up to rotation, shapes are sets of landmarks z_i with $\Sigma z_i = 0$ and $\Sigma z_i\bar{z}_i = 1$, the Procrustes distance between two shapes z and w is properly $\arccos|\Sigma z_i \bar{w}_i|$. The usual formula $\sqrt{\Sigma |z_i - w_i|^2}$ is, however, a perfectly adequate approximation for biometric applications.

configuration with its centroid at (0,0) — but the removal of "size" seems stickier, as we can't say just what size variable is to serve that function, and there is a similar problem of orientation that seems to require us to specify some angle to rotate by — again, apparently an arbitrary decision.

Kendall circumvented this problem in a manner remarkable for its elegance and mathematical simplicity. Instead of specifying the transformations that make a set of landmarks into a shape, he noticed, in effect, that adding just one further reasonable requirement leads to a unique shape distance, which, in turn, determines everything you could wish to know about the geometry of those shapes. The reasonable criterion Kendall added was that shape distance be an immersion of ordinary Euclidean distance in the digitizing plane; in other words, that if you vary one landmark in a small circle about its central position, while all the other landmarks remain exactly fixed at their positions, then the corresponding shapes all have to be exactly the same distance from the central shape.* The only shape distance function that satisfies this requirement is the arc-cosine version of the now-familiar Procrustes distance, brought to statisticians' attention just earlier by Peter Schönemann and John Gower for use in comparing different multivariate ordinations quite outside the realm of morphometrics.

The insistence upon Procrustes distance as the sole statistically meaningful shape distance for landmark data weaves together all the tools of geometric morphometrics. This dogma conceals an escape hatch for senses of shape that are different from the Euclidean one used here. One, of course, is the shape of outlines without landmarks, represented in other papers in this volume. Another, less often noted, pertains to studies of articulated linkages and other assemblages with internal mechanical degrees of freedom that cannot be controlled by a discipline of positioning. Although it is possible to wrestle the shape-space methods into compliance with problems of this sort, by intentionally suppressing shape dimensions that correspond to degrees of freedom at hinge joints and the like, the set of shape coordinates that results is usually statistically equivalent to the intuitive set of distance ratios and related biomechanical indices that could be written down by an investigator who had no acquaintance with geometric morphometrics at all. For data sets like those, these methods are overkill.

Models in which all landmarks vary around mean positions by circular Gaussian noise of the same small variance give rise to models in which all shapes vary around a mean shape by a very nearly spherical distribution in Procrustes distance; so the Procrustes distance between two mean shapes is often the pivotal statistic for testing whether they are significantly different. This same squared Procrustes distance, (very nearly) a sum of squares itself, is partitioned by several hypothesis-testing protocols (e.g., for uniform versus nonuniform shape change or fluctuating versus directional asymmetry) that look just like analyses of variance. The denominator of the Procrustes formula is the product of the centroid size factors by which the configurations should be scaled separately. That same centroid size is the size variable uncorrelated with all shape variables on that same null model of identical Gaussian noise at all the landmarks, and so on through several more theorems.

Procrustes distance plays another role in shape space: it serves as a criterion for deciding when sets of variables are orthonormal. In most applications of multivariate analysis, orthonormality is an empirical matter, the result of a construction (like principal components analysis) that enforces variances of unity and covariances of zero. But in geometric morphometrics, orthonormality is defined prior to any information about variability. Instead, orthonormal bases are determined purely by the shape of the average about which

*The squared radius of the shape circle, it turns out, is $1 - |z_i|^2$ times the squared radius of the Euclidean circle, where z_i is the landmark in the center of the circle according to the notation of the preceding footnote.

variation will obtain. That is, in morphometrics an orthonormal basis is rather more like the lines on a map than like the principal diameters of a city drawn superimposed over those lines.

The lines on a globe that represent latitude and longitude are at 90°. There are actually two sources of information involved here. One, the rule for the measurement of angle, arises from measurements of distance (e.g. by surveyors) on the sphere (or, more accurately, the surface of the earth at sea level) in exactly the same way that angles in Procrustes space arise from the formula for Procrustes distance in shape space. (The analogy is even tighter for triangles of landmarks: the Procrustes geometry of their shape space actually *is* the geometry of the surface of a sphere — Kendall calls this his spherical blackboard.) These lines specify the geometry of the globe quite independent of any feature — mountains, rivers, cities, roads — that will eventually be drawn on it; they describe the geometry of the space prior to data. On large-scale maps of small regions, these lines form a grid that is usually rectangular, degrees by degrees or minutes by minutes, but that could be made square if the cartographer spaced the longitude lines so their intervals were equal to the intervals of latitude. Once that correction is made, then those lines would form an orthonormal basis for the map of your neighborhood.

The second source of information is more arbitrary: it is the position of the equator, or the poles, which serves to select one set of these squares. The other orthonormal bases, in this simple analogy, arise by rotating this square of lines while leaving the contents of the map (houses, parks, etc.) unchanged. American cities often have street grids of squares like these, which are oriented with respect to major local geographical features (e.g., the direction of a river) rather than geographical north and east. As in Kendall's shape space, this analogy requires that neighborhoods be small: At the scale of any substantial patch of the whole earth, the grid lines must be curved, not straight, so that eventually either their spacing or their orientation has to become noticeably bent. Except in the vicinity of the mean shape — I mean, your house — these grid lines are either not quite "ortho" or not quite "normal". In biometric practice we usually ignore these curvatures.

If we choose shape variables carefully, they can be orthonormal in this geometry — they can share the a priori geometry of longitude and latitude — and still, like streets aligned with the river, make sense on the "map." In morphometrics, that means on the organism. The uniform component and partial warps mentioned over and over throughout this volume meet all the requirements and so inherit many nice properties. For instance, together they exactly partition the squared Procrustes distance between two forms, or two means, into components along each element of the basis. Principal components of shapes expressed in that same orthonormal basis decompose variance in commensurate units of Procrustes distance, so that any renormalization beforehand by "variance" is inappropriate — all principal components analyses in shape space apply to covariance matrices, not correlation matrices. Also, the graphics of partial warps and relative warps are already in Procrustes-optimal position, so that mean-squared displacements of landmarks in those plots are proportional to the net squared Procrustes distance accounted for by the contrast or component in question. In the particularly interesting subspace of uniform shape changes, changes that show no spatial gradient, Procrustes distance can be given a specialized graphical interpretation, as the ratio of axes of the ellipse into which a circle is deformed. There are likewise a variety of special scatterplots that describe sample variation in this subspace consistent with that distance function.

In this way, an underlying shape distance, Procrustes distance, ties together the characterization of single shapes, and sample means, the algebra of multivariate analysis of variation around those means and the geometry by which those findings are converted from vectors to meaningful diagrams on pictures of the typical form. Two quantities that would otherwise be relatively important in morphometrics are greatly overshadowed by this domination of Procrustes distance in weaving our tools together. The bending energy of our thin-plate splines plays no role in the geometry of shape space. In modifying the Procrustes sphere according to a flattened ellipsoid that expresses the spacing of landmarks in the mean form, it serves mainly as a geometric device by which to introduce the partial warps, a particular orthonormal basis for shape that seems consistent with biological modes of explanation. Bending energy serves a subsidiary role as a distance measure between shapes that emphasizes changes at small scale. (A principal-coordinates analysis of this distance measure is the version of relative warps analysis with $\alpha = -1$; see the section on "Partial Warps.") Whereas the formula for bending energy expresses the mean form, Mahalanobis distance replaces the Procrustes sphere by an ellipsoid expressing only the covariance structure of shapes in the sample, a matrix making no reference to the mean form. In geometric morphometrics, Mahalanobis distance is useful only when combined with another distance. The analysis of Mahalanobis distance with respect to Procrustes distance is a relative warps analysis with $\alpha = 0$; analysis of Mahalanobis distance with respect to bending energy is a relative warps analysis with $\alpha = 1$.

As far as I know, morphometrics is unique among contemporary biostatistical specialties in the pervasive presence of this shared geometry independent of any information about actual sample means or variances. In no other multivariate application is another metric "always there" alongside the usual statistical distance functions and empirical inner products. The pitfalls that often snag otherwise excellent morphometric analyses are pathways, usually inadvertent, by which constructions contradicting this ubiquitous underlying metric are inserted into chains of calculations and then propagated forward by tools that make sense only when the associated metric is the Procrustes distance.

For example, the undeniably evocative diagrams that arise from resistant fits produce graphical residuals the sums of squares of which disagree with Procrustes distance and that therefore must not be fed forward into subsequent steps of a multivariate analysis. They must be interpreted exactly where they are drawn, without further statistical processing. Because resistant fits are not algebraically consistent with the Procrustes projection that defines shape space, their "residuals" cannot serve as shape coordinates. Once we have looked at them, there is nothing further to do except to talk.

Part of the confusion that arises around Procrustes distance in geometric morphometrics owes to a subtle duality of roles that it plays in the toolkit: the same formula applies both in the original picture plane of the data and down inside Kendall's shape space. After all, the sum of squares of distances between corresponding landmarks varies as one varies the way one set of landmarks is put down over the other — as one varies the superposition rule. This sum of squares is the quantity being minimized across superpositions. The Procrustes distance between shapes, which is a distance computed between two shapes in shape space, is equal to the value of that minimum when computed between properly rescaled and repositioned pictures in the picture plane.

But Procrustes distance between shapes is not a function of superposition — it has a formula already that makes no explicit reference to superposition. (That is why plots of resistant fits have nothing to do with the Procrustes distance between shapes or with statistical analyses whether of distances or of shapes; they are only superpositions, however suggestive. These fits depend on the data, whereas the geometry of Procrustes distance derives ultimately from the semantics of the word "shape.") The Procrustes formula generates shapes from data and disciplines collections of descriptions of those shapes; it does not constrain the sample variation of those shapes in any way, but only our descriptions of that variation. For instance, the null model of circular noise of the same variance at every landmark corresponds to a probability distribution in shape space that is just about spherical. (That is, concentric Procrustes-spherical shells about a mean shape correspond very nearly to contours of equal probability of having arisen in samples from the null distribution about that mean.) But that sphericity has little to do with ordinary, normal-model multivariate analysis of shape data. It is the geometry of potential shape variables that is spherical, not the covariance structure of the data, for which any reasonable quadratic form (either of landmark location noise or of the shapes in their space) will suffice. One can make inferences about sets of shapes regardless of how they vary; but for any differences to be interpretable as collections of potentially separate features, sets of shape descriptors have to correspond to that spherical Procrustes geometry — i.e., they have to be orthonormal there. The Procrustes geometry constrains our descriptions of shape phenomena, not the phenomena themselves.

In my view, the errors and pitfalls of geometric morphometrics derive ultimately from the superb achievement of earlier multivariate statisticians in bridging the gap of imagination between real geometry and the abstractions of linear algebra. That they were able to make their students comfortable with notions such as regression and canonical variates analysis by means of diagrams constructed out of chalk lines and ellipses on a blackboard was a major pedagogical triumph. That pedagogy succeeded, however, because the blackboard was blank to begin with: there was no fear of confusion between real lines or curves and these algebraic lines and curves. The teacher's task was mostly to enhance the student's level of comfort with these metaphors. But when the study is of landmark data, so that the blackboard already has a sketch of a specimen on it, we must erase that vector geometry and then replace it by the distinctly non-Euclidean geometry of Procrustes superpositions and orthonormal shape bases.

The glory of geometric morphometrics is the way in which two geometries, the Procrustes and the multivariate, can be borne in the same diagrams. The responsibility of the toolmaker for this new morphometrics includes communicating patience, caution and conservatism in learning this new geometrical language of form. The articulations among tactics in the three phases of any analysis are extensive, delicate, and obscure enough that the applied researcher should not guess about new combinations of these techniques. Recombinants here can be lethals; don't experiment with them. If you don't fully understand the constraints placed upon these procedures by their shared reliance on Procrustes distance, that is probably because much of that reliance is cryptic (as in the methods of partial warps). You don't see it because it is concealed in mathematical identities lying far below the surface of the data. To verify that the computations that are so elegant separately are still coherent when run in series, there is no substitute for actually tracing the roles Procrustes distance is playing in full algebraic detail. This detective work has nothing to do with organisms, probabilities or blackboards; it is a matter of theorems in geometry.

In the remainder of this chapter, I will survey the combinations of tools that most directly lead to sensible data analyses in this new domain. I will also review the incoherences and errors of analysis that arise when these tools are mismatched — when the geometry of shape space is botched. The combinations, whether sound or unsound, fall under two convenient headings: superposition techniques and analyses of warps.

SUPERPOSITION TECHNIQUES AND THE ASSOCIATED BASES AND DISPLAYS

Superposition techniques are exploited in morphometrics at two different times: right at the outset, when landmark configurations have to be converted to points in shape space, and at the conclusion, when multivariate findings need to be interpreted via diagrams back in the picture plane of the forms under study. The rules applying in these two contexts are different, and the differences have not been stated clearly enough in our workshops or in earlier publications.

It is assumed throughout this section and the next that data were originally coordinate pairs. Corresponding discussions of three-dimensional data are set out in the section "In Three Dimensions."

Generalized Procrustes Analyses

As I explained above, only one shape distance plays any role in these techniques: Procrustes distance. When the residuals from a superposition sum in square to a Procrustes distance, those residuals serve as coordinates in shape space and may be used for many different multivariate maneuvers there. In the morphometric synthesis, GLS mainly serves in the first phase of the three-phase scheme, the production of a point in shape space from a set of landmark locations in the digitizing plane. Its role as an "analysis" is secondary to this geometric construction. For instance, although GLS computes a mean shape, in morphometrics that is of use mainly by way of supplying the formulas for the tangent space to shape space. When multiple group means are computed (see below), they are not computed separately by GLS but are computed jointly as ordinary multivariate means in the tangent space to the grand mean shape.

The GALS (Generalized Affine Least Squares) residual sum of squares is a component of the GLS sum of squares. The difference between them is the squared variance of the uniform component for the same sample, as is provided automatically by current versions of GALS. In practice, the GALS sum of squares — the "nonuniform distance" — is meaningful mostly as a fraction of the GLS sum-of-squares. The situation is analogous to one you already know from ordinary regression: for a residual from a model to be meaningful, first the model must be meaningful. The GALS configuration is the residual from a model of uniform shape variation, so for the GALS configuration to make sense, the fit to a uniform component should make sense first. Geometrically, GALS distance is the length of a projection onto a subspace of the tangent space. Just as the tangent space was characterized by the mean form, so this further subspace is defined by the mean form. GALS residuals are restricted to lie orthogonal to this particular plane of shapes through the mean form, whereas Procrustes distance per se has nothing to do with any such restriction. The geometry of the uniform subspace is not encoded into the geometry of Kendall's shape space in any meaningful way; it is an artifact of our search for certain convenient styles of biological explanation.

The resistant-fit versions of these techniques do not supply Procrustes distances. In general, different shapes are fitted to the consensus by reference to different subsets of landmarks, so that there is no shared formula for the separate residuals. Because GRF and GARF residuals are not homologous as variables, they cannot properly be compared among specimens, let alone squared and summed. Unlike GLS and GALS, GRF and GARF supply only diagrams to help search for simple interpretations of the spatial distribution of observed shape variations.

Hence, don't use GRF or GARF diagrams to produce residuals for later statistical analysis or distances for any sort of dendrogrammatical processing. Likewise, do not produce partial warp scores from resistant-fit superpositions; do not carry out analyses of their principal components or correlations; do not regress them on centroid size; and so on. None of these "summaries" mean anything unless they are identical to those deriving from analogous GLS fits, in which case nothing has been gained by the duplicate computation. Resistant-fit superpositions do not supply distances; in general, they do not even supply homologous shape variables. Their sole function is to yield interesting graphical hints that can simplify the description of form change. (This exception to the prohibition encompasses their application in Slice's work, "Three dimensional" article, this volume, in which inconsistency between GRF and GLS residuals serves as a startlingly effective filter for atypical landmark locations.)

Don't use all the Procrustes residuals in any procedure that involves inverting a covariance matrix (the list of procedures includes multiple regression of centroid size on shape, canonical variates analysis, canonical correlations analysis, Hotelling's T^2, and MANOVA.) The GLS residuals are subject to four almost exact linear constraints; the GALS residuals to six. Because these constraints are not exactly linear, the covariance matrix of the residuals appears to be invertible; but the inverse is almost wholly noise, and the resulting coefficients are meaningless. Analyses involving matrix inversions need access to bases of full rank. One suitable choice is the full set of shape coordinates to any baseline; another possibility, less symmetrical, is a sublist of all but two (any two) of the Procrustes residual vectors. Or one might use the orthonormal basis of partial warps and uniform component supplied by Rohlf's TPSRW, which is both nonarbitrary and nonredundant, at the cost (which is often a benefit) of being nonlocal. Principal components analyses, which must preserve Procrustes distance, may work through that same orthonormal basis or may use the GLS residuals themselves, of which they are a Procrustes rotation. If based on GLS residuals, the last four principal components express only the nonlinearity of the Procrustes transformation and have no biometric meaning.

If sample size is sufficient (roughly twice the dimensionality of our shape space — four times the number of landmarks), don't test selected Procrustes residuals separately for statistical significance. Using a full-rank subset of all but two of the residual vectors, for studies of group mean difference, carry out a T^2 analysis or MANOVA; for studies of adjusted mean differences, MANCOVA; for studies of associations with causes or effects of form, multiple regression of the full-rank subset or Partial Least Squares of all the residuals at once. (Partial Least Squares is not affected by rank of the set of shape variables per se.) Findings from every set of all but two should be substantively identical.

If sample size is not large enough, and if variation is nearly in accord with the null hypothesis except for known factors of shape that have been controlled by sample design, the sum of squares of the Procrustes residuals is often nearly as informative as the vectors of their values separately. For instance, in a two-group study (n specimens per group), if sample size is not large enough for the full-rank procedure of the preceding paragraph, do test the squared Procrustes distance between means against the mean squared within-group Procrustes distance from the group mean. In the null model their ratio, multiplied by $(n - 1)/2$, is distributed as an F-ratio with $2p - 4$ degrees of freedom for the numerator and $(2p - 4)(2n - 2)$ in the denominator.

If the null model seems untenable according to scatterplots of the GLS fit itself, or under the more powerful scrutiny of a relative warps analyses, instead test all the Procrustes-fitted residual vectors one by one in separate bivariate tests, after which apply an appropriate Bonferroni correction to the most significant p-value. (Or one can test the separate partial warp scores similarly; see "Partial Warps" below.) It is an error to test individual Procrustes residual elements (x- or y-components) one by one. Such tests are hopelessly confounded not only by orientation but also by the somewhat counterintuitive algebra of the Procrustes fit.

Make certain that all plots in Procrustes geometry are "squared," meaning that they have equal intervals on the two axes. This applies alike to Procrustes residuals, shape coordinates, partial warp scores, and the uniform component. Noncircularity matters in these plots: it announces the presence of possibly meaningful factors of shape. (Some of us would further extent this dictum to all plots arising from principal components analyses and canonical variate analyses regardless of subject matter.)

Don't expect specimens to be of centroid size exactly 1 after GLS fitting. The consensus or mean is set to centroid size exactly 1, to set up correctly the orthonormal bases for subsequent multivariate analyses. The centroid size of individual specimens as fitted to the consensus is (very slightly) less than 1 owing to the flattening of shape space that is required for graphics and multivariate statistics to go forward. Don't carry out more than one cycle of GLS fitting, no matter how many groups you have in your data set. All multivariate analysis must be in one consistent tangent space to shape space, usually associated with the grand sample mean. Subgroup mean shapes are computed as averages of shape coordinates in that single space and thus will have centroid sizes slightly less than 1 themselves.

Do treat GALS as a rearrangement of the GLS residuals that explicitly partitions out the uniform component. Don't use the GALS residuals for any later multivariate computation without first interpreting the uniform part that was sequestered.

Do feel free to think about the difference between the GLS and the GALS fit separately as the uniform component of whatever feature you're hypothesizing (see "Standardization of the Treatment of the Uniform Component"). You can also test to see if there is any information about group differences in the nonuniform part of shape variation or shape change; this is equivalent to a test of any $p - 3$ GALS residual pairs. See the Orange Book (Bookstein, 1991), Section 7.2.

Shape Coordinates to a Baseline

Shape coordinates to a baseline are the coordinates of the remaining landmarks when two are sent to fixed locations, usually (0,0) and (1,0). This construction is the simplest device for passing from a landmark configuration to a biometrically sensible representation of shape. Unlike GLS residuals, these shape coordinates have the correct count. Techniques that involve matrix inversion can proceed quite well using them, especially if the canonical vectors that result are visualized by splines or another baseline-free display. But baseline shape coordinates do not give forms at unit centroid size — they do not embody Procrustes distance in any graphically simple way; they cannot be submitted to principal components analysis; and they do not partition changes into components without special algebraic maneuvers (that actually convert them to an orthonormal Procrustes basis).

Do follow the advice about choosing baselines given in the Orange Book. When landmark reliabilities are fairly uniform, choose a baseline along the long axis of the typical form and that passes near the centroid. Don't use type-3 (extremal) landmarks at either end; if that's all you can get at one end, it is better not to use shape coordinates at all.

If sample size is sufficient (roughly four times the number of landmarks), don't test selected shape coordinate pairs separately for statistical significance. Use them all together in a full-rank multivariate test. If sample size is not that large, test the separate triangles — that is, the separate shape coordinate pairs — one by one in separate bivariate tests and then apply an appropriate Bonferroni correction to the most significant p-value. In the absence of a strong prior biomechanical justification, it is an error to test individual shape coordinates (x- or y-components) by univariate test. Such tests are strongly dependent on choice of baseline (an exception being shape coordinates with respect to an axis of symmetry; see "Symmetric Data" below).

If your shape variation is substantial, do check that crucial statistical findings are robust against sensible changes of baseline (e.g., substitution of one landmark by another nearby). But don't interpret differences between findings for different baselines, as these express nonlinearities in the basic shape-space formalism, not findings; and don't choose the baseline that gives you the most significant results!

Don't plot centroid size against your baseline length. The plot is hopelessly inefficient at communicating any insights about shape. (See Orange Book, Chapter 4.)

Do recognize baseline shape coordinates in various sets of ratios that may have appeared in earlier literature. For instance, as noted in Brande's chapter, this volume, archaeologists have long been interested in the height:width ratio of knives and in the relative location of the widest diameter. If height:width replaced by width:height, then the pair of variables (relative location, width:height) is a pair of shape coordinates: the widest point as moving landmark plotted against the two ends of the long axis as baseline. The appropriate statistics are thus those that apply to any other single pair of shape coordinates; for instance, a T^2 for group differences or a regression on centroid size for allometry.

Don't stop with a scatter plot of shape coordinates from a significant finding involving them. Do recognize shape coordinates as a superposition technique, sharing with all others the inability to visualize correlations of residuals across the form. Do supplement shape coordinate plots by relative warps in order to see if patterns of variation about mean differences are localized or global.

Adjustment of a shape-coordinate basis for centroid size or other covariates of shape is computed correctly by most MANCOVA routines. Do inspect the coefficients of "adjustment" shape coordinate by shape coordinate, but don't test those adjustments coordinatewise. The proper test for any such adjustment is a single test combining all the shape coordinates; for a single covariate, such as centroid size, this is a multiple regression based on the pooled within-group covariance structure. Prediction of shape from exogenous predictors, like these adjustments, proceeds shape coordinate by shape coordinate in terms of any basis at all.

Standardization of the Treatment of the "Uniform Component"

I explain elsewhere, ("Standard formula" article, this volume) how the uniform component of shape change or shape scatter arose separately in many different algebraic contexts of morphometrics and why it has now become desirable to standardize its definition. The cleanest way to achieve this goal (i. e., the way in which we wish we had explained the matter originally) is to couch the matter in terms of the Procrustes geometry that underlies the other paragraphs of this section, and so I place the discussion here.

For years we have measured the magnitude of uniform transformations — those that send squares to parallelograms — in terms of their anisotropy, the ratio of axes of the ellipse into which circles are transformed. Now Procrustes distance applies to pairs of forms related by a uniform transformation just as it applies to any other pair of homologously labeled forms. And, as it happens, in the vicinity of a mean form, anisotropy is a simple (mean-dependent) multiple of the Procrustes distances that describe the relations among the same shapes. Furthermore, these sets of uniform transformations are geometric constructions themselves: they are subspaces of shape space, one through every mean form, that can be defined prior to any inspection of variation.

It seems desirable, then, to characterize the uniform component of a change or scatter by recourse to this Procrustes geometry (except at the last stage of an analysis, the stage of visualization, when display of a single difference by principal axes remains the most effective). The necessary formulas had already appeared in the literature, as the "null-model generalized least squares estimates" of Bookstein and Sampson, but we had not realized that they were serving so geometrically cogent a role. The uniform component is best understood as a Procrustes-orthogonal projection of a point of shape space onto the plane of shapes that differ from the mean form by uniform transformations. Its uniform coordinates extract part of the information from shape somewhat as latitude and longitude extract part of the information that locates apartments in New York City. The Procrustes geometry enters into this construction by defining what we mean by "straight down." But do not push this analogy too far. As with shape coordinates, the two uniform coordinates have different formulas. One is a shear along the Procrustes horizontal, the other a dilation away from the Procrustes horizontal; they do not rotate like maps of cities.

Hence, do assign coordinates to the uniform component in a way consistent with the Procrustes geometry of projection onto the uniform subspace through the mean form. As of spring 1996, such projections are supplied by the GALS program, by the last two columns of the weight matrix from TPSRW, and (for two-dimensional data) by the exact shape-coordinate formulas of the Orange Book and my chapter here. Don't forget to show us the coordinate system (principal axes of the mean form) in which these components are supplied, or else we won't know how to interpret their mean differences, factors, and so on. Don't parameterize the uniform component of shape or shape change by the linear part of a thin-plate spline, by the two angular variables of "D'ARCYGRAPH" or by the strains and angles of a finite-element analysis. (But you may continue to use these descriptors to visualize findings.)

Don't use the uniform factor estimate of the Orange Book when you have access to the more recent formulas. (That formula is still the best available approximation for use in statistical packages that, lacking specialized machinery for geometrical morphometrics, restrict all but the most sophisticated users to manipulations of shape coordinates only.)

Don't forget that the uniform component is the "zeroth partial warp," to be included whenever one creates a set of visualizations of change or variation and to be correlated with any relevant causes, effects, or covariates of form.

Do treat the uniform components estimated by TPSRW as shape coordinate pairs. Make scatterplots of them. Compute their covariances with centroid size. If the mean component or regression representing your hypothesis is statistically significant, construct its principal directions and see how they lie upon the form. Don't forget that, like any other shape coordinate pair, these two numbers cannot be tested one by one. They represent one single entity, a tensor, and must remain paired throughout all multivariate manipulations.

Don't assume that the uniform subspace of shape space is interesting just because it has a convenient formula and is easy to diagram. Remember that to yield sensible interpretations, uniform findings have to be associated with conceivably uniform causes — biomechanical shears, actual energetics of stress and strain, length:width or height:width ratios, and the like. If there's no possibility of such a cause for the shape phenomenon you're studying, and if the uniform component has not much more than its fair share of shape difference, don't bother separating it out.

Symmetric Data

Most of these tools don't apply to symmetric forms without modification. The study of forms that are nominally symmetric (e.g., the mammalian skull from above) leads to special methodological decisions. None of the basic tools attend at all to asymmetry in the course of their ordinary operation. Procrustes superpositions, for instance, will cheerfully rotate the axis of symmetry as long as fit is improved; no warning message is generated if shape coordinates are computed to a nonsense baseline, like "left Gonion – right Condyle"; principal warps of symmetric configurations will wildly rotate features within the "planes" corresponding to reflections of the same underlying specific bending.

This will not matter if the usual Procrustes distance captures what you want to know about your forms regardless of asymmetry. Relative warps analysis, for instance, uses ordinary Procrustes distance and so needs no modifications for symmetric forms. However, no multivariate procedures that involves matrix inversion (see the list in "Generalized Procrustes Analyses") — including all tests for group differences, from T^2 to canonical variates analysis—can be expected to cope with the extreme ill condition of the distributions that represent almost-symmetric forms in shape space. The dimensions of nearly symmetry are of very small variance compared with the others, and these defective dimensions are numerous (two per bilateral landmark). Their collective effect is to torpedo any possibility of a sensible matrix inversion, that is to say, one that discerns a signal in the reliable symmetric part of shape.

Hence, in multivariate statistical tests of symmetric shapes, don't use the full-rank tests in any of the forms just reviewed in "Generalized Procrustes Analyses" or "Shape Coordinates to a Baseline" or discussed in "Partial Warps" below. All techniques must be modified to take nominal symmetry into account. (1) If you prefer to start from Procrustes residuals, construct a sum-and-difference basis of GLS residuals between antimeres. The easiest way to arrange for this is to duplicate your data set by pairing each original specimen with its reflection and then, by selecting one of the two completely equivalent sets of antimeres that result, to use half the sum of the paired residuals at the "same" landmark along with half the difference of the pairs. Landmarks along the axis of symmetry get no special treatment. (2) For shape coordinates, choose a baseline along the axis of symmetry and take averages of the x-shape coordinates of antimeres and half the differences of the y-coordinates for the paired landmarks, together with both coordinates of the unpaired landmarks along the baseline. (3) For partial warps, use a reference form that is half of the full form, along with landmarks on the axis of symmetry, and compute two sets of partial warps, one for sum and one for difference of half-forms. The uniform component of a symmetrized form has a special meaning (see the Orange Book).

Don't pool multivariate analyses over these two partitions; analyze effects on the symmetric part and on the asymmetric part totally separately. Do consider replacing the entire asymmetry component by a scalar summary, such as the mean-squared Procrustes distance between the whole form and its mirror image. This distance is the sum of two components, fluctuating and directional, as demonstrated in the Orange Book.

To date we have published no analyses of symmetries other than bilateral symmetry based on the new methods. The user who wishes to explore in advance of us toolmakers will need to construct analogues to these reduced bases by handwork in her favorite statistical package.

THIN-PLATE SPLINE INTERPOLATIONS AND THE ASSOCIATED STATISTICS AND DISPLAYS

The thin-plate spline is not a superposition technique. But, like the superposition techniques reviewed in the previous section, it plays multiple roles in the toolkit of geometric morphometrics. The spline supplies a visualization of any pair of forms in shape space as a Cartesian deformation taking one form to the other. It also supplies a visualization of any direction of shape space as the deformation of the mean form into another form some Procrustes distance out along that direction or its opposite. These images can be very powerful guides to the biological meaning of specific multivariate findings. In addition, the spline is solving an optimization problem to produce these maps — it has minimum bending energy of all deformations consistent with the change of landmark shape involved. Bending energy is proportional to integral quadratic variation, the sum of squares of all second derivatives of the map integrated over the whole picture. Think of it as a measure of how fast a local "uniform component" of the transformation has to vary across your picture to keep up with the possibly crazy things the landmarks are doing. The quantity being optimized itself plays only a modest role in morphometrics: it is a conversion of landmark displacements from the Procrustes metric (shifts scaled to centroid size) to a novel localized disproportion metric that is different from landmark to landmark (shifts scaled, in effect, to how tightly the landmark is packed with respect to its neighbors). More important than the value of this "bending" is its formula, which is a quadratic form in the Procrustes-fitted coordinates. The diagonalization of that formula is the only mechanism we have at present for supplying an interpretable orthonormal basis for shape space.

Don't confuse these two quite separate applications of the spline. The splines themselves are only a suggestive visual metaphor; don't talk about them as if they specified real changes of little bits of organism. They re-express the landmark data by lines drawn in-between the landmarks, where there isn't really any additional information. Likewise, the partial warps as warps need have nothing to do with the biology you are studying. But the partial warp scores, as shape variables, are the most useful basis for shape space that we have yet constructed—the only one that is orthonormal, with all the advantages that implies.

Don't report thin-plate spline visualizations using the language of superposition, in terms of how landmarks appear to have moved. The landmarks had already "moved" before the spline was displayed or even computed. A spline warps the picture plane in-between the landmarks by dilations or compressions of distances of various extents in various locations and directions. If that language seems not to suit the picture (e. g., if one landmark looks like it's moving on the background of all the others holding a more or less constant shape), try another visualization (in that case, GRF) and forget about the splines as graphics.

Do remember to convert from coefficients to loadings before plotting canonical vectors as splines. (That transformation won't affect relative warp loadings, which are already eigenvectors of the shape coordinate covariance matrix, and doesn't apply to Partial Least Squares analyses, the coefficients of which are already loadings.)

Partial Warps

If sample size is sufficient (roughly four times the number of landmarks), don't test selected partial warp scores separately for statistical significance. Use them all together, along with the uniform component, in a full-rank multivariate test. (To repeat, for studies of group mean difference, this test is T^2 or MANOVA; for studies of adjusted mean differences, MANCOVA; for studies of associations with causes or effects of form, multiple regression or Partial Least Squares.) If sample size is not that large, consider the single F-test of squared Procrustes distance (see "Generalized Procrustes Analyses" — that square is the sum of squares of all the partial warps — or test the partial warps and the uniform component pair by pair in separate bivariate tests and then apply an appropriate Bonferroni correction to the most significant p-value. (This test is more powerful than is the analogous approach to Procrustes residuals or shape coordinates whenever the hypothesis under investigation involves spatially correlated aspects of shape change, as in studies of growth or ecophenotypy.) Don't forget that partial warps are still orientation-dependent: they rotate with the coordinate system, and all aspects of the scores must be tested and described in pairs. The x-component of a partial warp does not have either a biological or a statistical meaning a priori, and p-values should not be assigned to hypotheses of differences on this component. But one component will often serve to help report findings and diagrams from analyses in which the components were properly paired throughout the computations.

Do use a sensible reference form for the computation of partial warp scores. In ordination studies, either within or between samples, and in studies of within-sample ecophenotypes or other correlations, this is best taken as the grand mean form. In studies of phylogeny, it is probably best to use this same grand mean, unless a good estimate of a hypothetical common ancestor is available. It makes no sense to use an outgroup for this purpose, as the uniform subspace and all the partial warps are biased without any compensating gain. The bias is an insidious one; the Procrustes distances to be approximated by shape coordinates have been silently foreshortened in a counterintuitive way that is a function of both the spacing of landmarks and the nature of the shape difference between the sample mean and the outgroup.

Don't believe any multivariate analysis unless its substantive import is independent of the choice of this basis vis-à-vis shape coordinates, Procrustes residuals or any other scheme. Any finding that requires the use of partial warps is erroneous.

Do inspect and test vector covariances of the partial warp scores with centroid size.

Do ransack the literature of your organism for characters and other patterns that may correspond to the geometry of the partial warps in particular orientations. In our experience, these characters and patterns have provided the most powerful way of communicating the partial warps' morphometric role to the naïve reader.

Don't forget that the uniform component changes differently from the other partial warps under change of coordinate system. It transforms like a tensor (see the Orange Book, Section 7.2), rotating twice as fast opposite to rotation of the reference direction, whereas the other warps rotate like vectors, exactly compensating for rotations of the reference direction. In the presentation of that component elsewhere, "Standard formula" article in this volume, I fixed the orientation of the mean form in order not to have to introduce notation for the more general case.

Don't use bending energy to order reports of the "importance" of partial warp scores unless you first make sure you understand how bending energy summarizes disproportion. Warp by warp, bending energy is the product of the squared Procrustes length of the partial warp score by the specific bending energy of the associated principal warp, which has no necessary biological meaning. Bending energy orders the geometry of the partial warps — their spatial localizability — whether or not it sensibly assorts their magnitude in particular shape changes or shape scatters. Bending energy as a magnitude is most consistent with style (5) of relative warps analysis, to be introduced below.

And don't use bending energy as a net "distance" between specimens, either, unless you understand its peculiarities. It weights relative displacements of closely spaced landmarks far more than it does shifts of the same Procrustes length between landmarks at greater separation, and it omits any reference to uniform changes, an omission that must be remedied by incorporation of the uniform component in all ordinations that result.

Relative Warps

Relative warps are eigenanalyses (principal components analyses) of the covariances of shape coordinates with respect to either Procrustes distance or some power of bending energy. Don't confuse their many different versions or consider them to be interchangeable. The versions are kept straight by a choice of subspace (the full shape space, the uniform subspace, or the nonuniform subspace) and, for the case of the nonuniform subspace, by a parameter α that can be negative, zero, or positive.

In the full shape space (for which, perforce, $\alpha = 0$), relative warps are a principal components analysis of shapes in the Procrustes geometry of shape space. Properly speaking, they are in the geometry of the tangent space. There are also versions of principal component analysis (PCA), appropriate to data sets of extremely large shape variation, that are situated in shape space itself. John Kent and Colin Goodall are responsible for these. When the uniform component is included, relative warps analysis is precisely equivalent to a principal coordinates analysis of Procrustes distance itself, should you prefer to circumvent any construction of a shape basis. The splines are a useful way of drawing the eigenvectors of such an analysis, as the drawing is at the same time a Procrustes superposition; but there is nothing about splines or bending energy in the actual multivariate formalism.

Any value of α other than zero converts the analysis to a relative principal components analysis (hence the "relative" of "relative warp") of shape covariances with respect to bending energy instead of Procrustes distance. All such computations apply only to the nonuniform subspace of shape space, because the bending energy of every change of shape within the uniform subspace is uniformly zero. For α greater than zero, warps of larger scale, excluding the uniform component, are assigned progressively more and more weight in the PCA; with α less than zero, they are more and more attenuated. The value $\alpha = 1$ is suitable for searching for growth gradients of the classical Huxleyan type (monotone across large regions of the form); the value $\alpha = -1$ is equivalent to a principal coordinates analysis of bending energy itself, appropriate if you are interested preferentially in phenomena at small scale.

Another way to think of this is by analogy with ordinary canonical variates analysis. Canonical variates analysis is a principal components analysis of group means after the ellipsoid of within-group variance has been "corrected" by shearing its shape to a sphere. Think of this as involving two Mahalanobis metrics: covariances within group and covariances between group. Similarly, relative warps analysis with α not zero is a principal components analysis inside the nonuniform subspace after the Procrustes geometry has been "corrected" by (1) discarding the uniform subspace entirely and (2) shearing (some power of) the bending energy matrix into a "disk" in the subspace remaining, the subspace of nonuniform variation. The confrontation of the matrices is very similar, although the interpretation is of course quite different inasmuch as a pattern of group means has been replaced by an a-priori geometrical structure.

Hence relative warps analyses come in five versions: (1) for the full shape space, with $\alpha = 0$, the routine prophylactic procedure recommended here; (2) for the uniform subspace (two-dimensional, for two-dimensional landmarks), as demonstrated in the Orange Book; (3) for the nonuniform subspace, still with $\alpha = 0$, as demonstrated by Rohlf in the Valsain proceedings; and finally, for the nonuniform subspace, (4) with $\alpha > 0$, as demonstrated in the Orange Book or (5) with $\alpha < 0$. I recommend that the five relative warps analyses be computed in this order and that the choice between the last two, the sign of α, depend on what is found in the first analysis of the complete Procrustes distance.

Don't carry out relative warp analyses of the nonuniform part of shape space, analyses (3), (4) or (5), unless some reason to do so has been supplied by prior morphogenetic theory, by biomechanical or other functional insight, or by an earlier step in the data analysis. In practice, this means finding a meaningful signal in an analysis of style (2), relative warps of the uniform subspace — a hint that there is some systematic aspect of uniform variability that makes sense to partial out, or using $\alpha < 0$ to search intentionally for small-scale structure. In either event, do compute a complete cross-covariance matrix for the two sets of scores, uniform and non-uniform, in the course of any examination.

Do begin with analyses of style (1), analysis of the full shape space (equivalently, analysis of Procrustes distance per se), to determine if any principal components seem to rise above the noise of the null model. (One ascertains this from the usual "scree plot" of eigenvalue against number.) If the first relative warp appears meaningful in view of its eigenvalue, and if it loads heavily on the uniform component or the first partial warp, do proceed with analyses (2) and (3), components of the uniform and nonuniform subspaces, separately. If analysis (3), of the nonuniform subspace, appears to show nonsphericity, sharpen its report by examining an additional analysis with positive α, style (4), and also carry out a principal components analysis of the first partial warp by itself (equivalent to an analysis of style (4) with infinitely large α). Large-scale phenomena are often detected satisfactorily using only the uniform component and the first relative warp; findings of this sort are easy to diagram (see the rat example in the Orange Book). When instead the first relative warp from analysis (1) is not significant or appears not to load heavily on the larger partial warps, consider a relative warps analysis with a modestly negative value of α, style (5), to see if consistent local remodeling of small subsets of landmarks emerges. In the presence of multiple groups, of course, all these analyses should be carried out within-group except for analysis (1), which supplies the "best" low-dimensional projection for ordination in the typical multivariate manner.

If you program your relative warps analyses for yourself instead of using Rohlf's TPSRW software, do respect all the necessary assumptions. Make sure you are using covariances (which are calibrated to Procrustes distance), not correlations; make sure all elements of the shape basis are included and that that basis is one that preserves Procrustes distance (GLS residuals or partial warp scores, not shape coordinates); and make sure you take the square root of the eigenvalues of bending energy and that you divide α by 2 before applying the eigenvalues to scale the principal warp scores. If you intentionally choose to compute any other eigenanalysis, please don't call it relative warps analysis.

Don't carry out a relative warps analysis and then report group mean differences. You wouldn't abuse a principal components analysis this way, would you? The components that result are grossly biased by the group mean difference, a difference that furthermore is not expressed in terms of the original shape coordinates but in terms of derived vectors that are a function of sample sizes and many other arbitrary factors. Group mean differences are to be reported by partial warps, not relative warps. Likewise, don't report the relative warps that "separate" groups of interest; not only is that not what they're designed to do, but it violates the assumption of a homogeneous sample under which they're computed in the first place. In particular, don't report an interesting relative warp way down in the stack — say, the eighth — without first successfully interpreting corresponding findings for the previous seven, to which the eighth was forced to be orthogonal.

Don't compute phenetic distances using only the first few relative warps, regardless of α. There is already a distance (Procrustes distance) underlying geometric morphometrics, and it corresponds to phenetic distance based on all the relative warps with $\alpha = 0$. Nothing is gained by restricting distance to only the dimensions that show substantial sample variability. The nonuniform distance that comes from the relative warps with $\alpha = 1$ weights large-scale gradients quite heavily; combined with the uniform distance, it may prove valuable in ordinations.

IN THREE DIMENSIONS

So far we have been discussing the case of Cartesian coordinate pairs, landmarks digitized in two dimensions only. But a steadily greater proportion of our raw information arises from digitizations in three full dimensions, and an increasing number of avoidable errors arise in the course of their analysis. What constitutes an appropriate sequence of computations is not altered: sequences that coherently keep straight how many quantities constitute a shape basis, bases that are orthonormal in Procrustes distance, and so on. But some of the algebraic details change and almost all of the graphical strategies. This section is a quick overview of some of the additional pitfalls that arise. Consult also Louise Roth's chapter in the Valsain proceedings for some very specific pitfalls that arise in pursuit of the raw data themselves, landmark locations in three dimensions.

Do consider the possibility of two-dimensional analysis of subsets of three-dimensional data: for example, analysis of the landmarks in a plane of bilateral symmetry, analysis of a single large triangle or a few triangles disconnected from one another, or analysis of nearly coplanar landmarks from standardized views. These do not supplant the complete analysis of your three-dimensional data but may provide succinct summaries and graphics that greatly ease the task of communicating and understanding findings. In particular, don't apply three-dimensional methods to the not-quite-three-dimensional data that arise from orthogonal photographs of opaque structures — the craniometrician's norma lateralis and her sisters. If your data are not actual points in three dimensions but only multiple two-dimensional views, do analyze them separately as two-dimensional data sets and then combine them by ordinary methods of covariance analysis or, better, by Partial Least Squares.

In three dimensions, shape still comprises the properties of landmark configurations that don't change under translation, rotation, or change of scale. Scale still is characterized by only a single parameter, centroid size, that is consistent with the geometry of shape to be imposed by Procrustes distance, and translation, quite reasonably, now has three parameters instead of two. But rotation now has three parameters instead of just one. Don't miscount this. For p landmarks, that leaves $3p-7$ degrees of freedom for shape. Within this new shape space, do persevere with the main geometrical thrust: statistical analysis in the tangent space, under control of this a priori Procrustes geometry. (Procrustes distance has the same verbal characterization in three dimensions as in two: the minimum of summed squared distances between landmarks when both forms are scaled to centroid size unity.) But don't try to visualize this new shape space — there is no longer a "spherical blackboard" to write on with chalk, as nothing in geological history has prepared the five-dimensional piece of slate on which the shapes of tetrahedra would need to be inscribed.

For spanning variation in this space, bases of "Bookstein shape coordinates" (really Galton's) — simple constructions of perpendiculars — are still possible. The easiest approach (see the Orange Book) is a basis of the same two-dimensional shape coordinates we've been accustomed to all along, shape coordinates of triangles that make up a rigid triangulation of your landmarks. (To get those coordinates, reconstruct selected triangles, each in its own separate plane, from sets of interlandmark distances in threes. See the Orange Book, Section 5.1.) These come in pairs, whereas the number of shape parameters we need to cover, $3p-7$, might be odd; if so, omit any one coordinate of any one triangle. Several of us have experimented with "true three-dimensional shape coordinates," the coordinates of the landmarks in a coordinate system with two landmarks set to (0,0,0) and (1,0,0) and a third restricted to the plane $(x,y,0)$. For landmarks off the base plane, this totals $3(p-3) = 3p-9$ shape coordinates; the last two of the $3p-7$ are the shape coordinates (x,y) of that third landmark in the base plane. Don't miscount this.

The GLS option in Slice's program packages continues to supply a shape basis assembled from the associated Procrustes-optimized superpositions. The residuals now total $3p$, so you must omit a full 7, either by explicit deletion or by computing principal components of all $3p$ and then dropping the last 7 eigenvectors. Again, don't miscount this. As for two-dimensional data, none of the resistant-fit algorithms supply shape bases, but they can be quite helpful at showing localized phenomena. For nearly symmetric data, the modifications of shape bases that sequester asymmetry prior to linear modeling may be copied over from the "Symmetric data" section without any essential change.

Thin-plate splines continue to make sense for three-dimensional data, retaining the same characterization: the maps that map landmarks onto their homologues with minimum integral quadratic variation. Don't try to visualize the "bending," however; it goes into dimensions we don't have. Imagine instead the general deformation of the sort of shipping crate in which fine drinking glasses are often packed in multiple layers, with three full sets of cardboard dividers separating them from neighbors in all three dimensions. Displacements of landmarks still push around the grid cells, just as in two dimensions, so that features of the transformation are propagated throughout the space of the grid with degrees of damping that are, as in two dimensions, inverse to the degrees of localization of the perturbations. The kernel function of the splines, that is, the actual function that is applied to deform grids, must now be $|r|$ instead of $r^2 \log r$, a change that makes no difference for statistical or biological interpretation. Bending energy still has eigenvectors, the principal warps, which supply an orthonormal basis of partial warps for the nonuniform part of shape space in the vicinity of a mean form. Don't forget that all these vectors — GLS residuals, partial warps, shape coordinates to a base plane — are now three-vectors, not two-vectors. A proper visualization requires stereo viewing, biplane projection or the simulation of tumbling. Mean differences can always be drawn as three-dimensional thin-plate splines, but no one has yet packaged suitable software that runs on PCs. Don't apply your familiar two-dimensional spline software to plane projections of your data; the grid lines do not compose a reasonable projection of any meaningful three-dimensional structure.

As in two dimensions, there are two approaches to orthonormal shape bases that support relative warps (principal components analyses) as well as linear models. One of these is the set of all but seven principal components of GLS residuals; the other is the set of partial warp scores augmented by a five-parameter uniform component. Do think through the ways in which the uniform component has been altered so as to apply to three-dimensional data. It now has five parameters, not two (see my other chapter), and so no longer can be represented as a simple scatter plot without loss of information. For ordinations, do parameterize the uniform component via the first five principal components of the six uniform coordinates that arise as three sets of two from the projections of the GLS fits onto their three cardinal planes. For visualization of findings relating to that component — factors within its subspace, covariances with centroid size, mean differences — revert to a basis of shape coordinates and then convert further to principal axes and principal strains of the implied tensor. Software for this purpose will shortly be supplied by the Slice–Rohlf team. As in two dimensions, don't assume that the uniform component bears any interesting information, and if it doesn't, don't sequester it from the rest of shape space in analyses of relative warps.

All these caveats notwithstanding, do follow the same three-phase system of computations that applies to two-dimensional data: (1) get from your landmark locations to shape space by GLS residuals, shape coordinates or uniform component and partial warps; (2) carry out multivariate analyses of those parameters; and (3) inspect and display the findings and the associated ordinations in a great variety of alternate ways.

CONCLUSION, WITH REASSURANCES

At this point some readers may be quite irritated — why didn't we explain these matters before the remaining chapters of this volume had matured to penultimate draft? — while others may be grateful not to have completed the contributions they'd planned. I have intentionally not commented on findings, only on methods. Findings that use our tools inappropriately may still be valid when verified by appropriate computations. My opinions that Euclidean Distance Matrix Analysis and finite-element scaling analysis are really nothing but pitfalls are well known and have been defended at length in the Orange Book; I will not go into them here. This essay is not aimed at reshaping through commentary what I was unable to affect as reviewer; nor is its purpose to set forth opinions about controversial matters in the guise of catechism. I have tried only to compensate for some unfortunate lapses in our collective style of teaching, lapses that probably could not be avoided, no matter how numerous the workshops, until they became visible in the analyses submitted for our review.

As our experience with geometric morphometrics continues to accumulate, countervailing processes will drive these directories of DOs and DON'Ts in opposite directions. The errors surveyed here, I trust, will decline in prevalence (although because many of them have already been published somewhere it is unreasonable to expect them to go extinct). Our customary packages will be rewritten to raise electronic eyebrows about particular combinations of options or to prohibit their computation entirely; other sequences will be welded into unified pipelines that do not bend the Procrustes geometry from one end to the other.

Opposing this tendency, I am sure, will be new pitfalls generated as these geometric techniques ramify from landmark data to include the other sources of information we are now striving to incorporate: outlines, pictures, geographic coordinates, and evolutionary history. Geometric morphometrics imposes a geometry; let us see how far we can travel using a chart given by that geometry. The tools reviewed in this volume pull out all the information contained in a set of landmarks, but we have no idea how much information that is, or how it relates to physiology, development, life history, or evolution. The fields to which geometric morphometric findings are just now coming to be juxtaposed are older than ours and have much larger, more carefully accreted archives of obviously successful developmental or evolutionary explanations. At this stage in the building of a new discipline, humility is rather more crucial than brilliance, and algebraic care a great deal more important than rhetoric.

ACKNOWLEDGMENTS

I am grateful to Paul Sampson, Jim Rohlf, Les Marcus, and Dennis Slice for detailed comments upon an earlier draft of this article. My work in morphometrics is supported by NIH grants DA–09009 and GM–37251 to F. L. Bookstein. The former of these is jointly supported by the National Institute on Drug Abuse, the National Institute of Mental Health, and the National Institute on Aging as part of the Human Brain Project.

STANDARD FORMULA FOR THE UNIFORM SHAPE COMPONENT IN LANDMARK DATA

Fred L. Bookstein*

Institute of Gerontology
University of Michigan
Ann Arbor, Michigan 48109-2007

ABSTRACT

In describing how the shapes of landmark configurations vary, an important role is played by the uniform component of shape variation. Geometrically, the uniform component parameterizes shape changes that leave parallel lines parallel throughout the form and so show neither a spatial gradient nor any spatial localization. Algebraically, the shapes produced from a mean form in this way make up a linearized subspace, *U*, of our multivariate shape space in the vicinity of that mean, so that the particular component that corresponds to a particular specimen or group difference can be computed as a pair of scores (for two-dimensional data) or a quintet of scores (for three-dimensional data) by a Procrustes projection onto *U*. The versions of this crucial projection that are offered by currently available program packages are inconsistent in several important aspects. This paper suggests a resolution of the contradictions among these approaches. I introduce a nearly exact Procrustes version of this component and show that it is identical to two separate versions previously suggested in the literature and that it represents a more precise estimate for yet another. A somewhat different version of this component is useful for multivariate statistical testing. The computations are demonstrated using the growing rat skull data set familiar from several other canonical demonstrations of the new geometric methods.

INTRODUCTION

The literature currently accumulating under the rubric of "geometric morphometrics" pursues two broad themes: fully efficient multivariate statistical tests of group differences and other effects on configurations of landmark points, and effective description of such effects when they are found to rise above biometric noise. One promising thrust for this second, descriptive task is the decomposition of observed shape changes into a hierarchy of

* E-mail: fred@brainmap.med.umich.edu

mixed geometrical–statistical components. The principal monograph in the field, Bookstein (1991), hereinafter referred to as "the Orange Book," puts forward two rather different suggestions for this hierarchy. One generalizes Huxley's notion of growth gradients by standardizing the polynomial trend-surface models of Sneath (1967). The other is the still-novel decomposition of changes into a series of partial warps, vector multiples of the eigenvectors of a certain quadratic form derived from the thin-plate spline interpolant. The eigenvectors, which are a function of the mean landmark configuration, describe a finite spectrum of geometric localizations according to which the picture around the landmarks can "bend."

Both these hierarchies begin with a term for uniform shape transformations, shape changes that are the same at all points. These are often called affine transformations or uniform shears. In D'Arcy Thompson's graphic of "Cartesian transformations," these are just the grid transformations that leave parallel lines parallel — transformations corresponding to taking the shadow of a rectangular mesh in the sun. Thompson's familiar figure (the transformation between two species of fishes, *Argyropelecus* to *Sternoptyx*) may be found on page 299 of the 1961 edition of his work (Thompson, 1961) — but see the example from Albrecht Dürer, 1524, on page 24 of Bookstein (1993). Many other examples are displayed in Chapter 7 of the Orange Book. Being free of all spatial gradients, the uniform shape transformations are the first term in the hierarchy of polynomial growth gradients; being wholly nonlocalizable — of "infinite" spatial scale — they are also the initial term (the "zeroth" partial warp) in the hierarchy of localized components of a thin-plate spline interpolant.

In contemporary practice the contemplation of the uniform and other components of these hierarchies has proved a fertile source of hypotheses about the ontogenetic, functional or evolutionary meaning of observed differences among forms. Uniform components are often interpretable even when most remaining variation is specific to individual landmarks. Their utility arises principally because they can be represented as single homogeneous tensors (Orange Book, Chapter 6) that arise in response to a great variety of biomechanical causes both ontogenetic and phylogenetic. The geometric operator such a tensor represents unambiguously takes circles to ellipses and triangles to sheared triangles (or spheres to ellipsoids and tetrahedra to sheared tetrahedra) in a spatially uniform fashion. Tensors and their variability may be visualized by their principal axes and principal strains in many ingenious graphical ways. The scope of this paper, however, is limited to the extraction of these single-tensor representations from real data in two dimensions — their geometric-statistical estimation, not their interpretation.

Formulas for the partial warps do not seem to be the subject of any disagreement. But the uniform component arose at many different junctures in the recent rapid development of geometric morphometrics in response to purposes that remain somewhat oblique to one another — statistical testing, feature extraction, visualization of deformations — and so the available programs and packages are inconsistent in the formulas that are used to represent this component and in several important details of the computation. There seem to be at least four different versions in print each claiming, not without diffidence, to be "the" uniform part of a group difference or of the variation of specimens about a reference form. All the versions share an approximation of the representation of shapes as a linearized space in the vicinity of a mean form (the tangent space to Kendall's shape space, see Bookstein, "Combining" article, this volume). In this approximation, each version is a projection of a normalized landmark shape onto the plane of exactly uniform shape changes; the algorithms differ, not necessarily subtly, in the normalization that precedes projection and in the direction of that projection. (In addition, there is a maximum-likelihood approach, which is necessarily nonlinear; see Mardia and Dryden, 1989.)

1. Projecting so as to minimize Procrustes distance.

- Fit of the vector $\hat{\alpha}$, Orange Book, Section 7.2, on the null model for shape covariances (Bookstein's program PROJECT, distributed with Rohlf and Bookstein, 1990).
- Nuisance parameter in the Rohlf–Slice GALS (Generalized Affine Least-Squares) fit (Slice's (1993) program GRF-ND).

2. Projecting so as to minimize Mahalanobis distance.
   - Fit of the vector $\hat{\alpha}$, Bookstein & Sampson (1990) or the Orange Book, using the sample shape coordinate covariance matrix instead of the null model (again, the program PROJECT, and also Paul Sampson's S code).
3. Projecting by the use of the last three columns of the matrix L^{-1} of the thin-plate spline formalism.
   - The weight matrix, option *A*, of Rohlf's programs TPSRW and NTSYS (Rohlf, 1993).
4. Hybrid in a Partial Least Squares spirit (easy to compute).
   - Uniform factor estimate to the long-axis baseline, the Orange Book (formula distributed by Marcus in his UNICOORD utility).

Now is a good time to try to bring some order into this roster. This note presents a new construction of the uniform component that is exact in the Procrustes geometry of the tangent space to shape space. Its simple formulas have already been distributed as an option in versions of Rohlf's TPSRW and NTSYS dated October 1993 or later. The new estimate has a simple relationship to approaches (1) and (4) of the previous list and helps explain why the values supplied by those earlier versions diverged in certain applications.

This note will explain the formula principally as it applies to two-dimensional data, for which complex arithmetic supplies a convenient notation easing some algebraic manipulations. In these expressions, the single complex number z, z', z_i, etc., will stand for the same geometric object as the usual Cartesian coordinate pair (x,y), (x',y'), (x_i,y_i), etc. The formulas for three-dimensional data cannot easily be expressed using this multiplicative notation; for data in the form of points (x,y,z) they will be expressed instead as a constrained superposition of three two-dimensional subproblems for the (x,y), (y,z), and (z,x) planar projections. The formulas that follow will all assume a registration of each landmark configuration upon its centroid, a scaling to unit centroid size, and an orientation to the principal axes of the mean shape (as computed by the usual Procrustes operations prior to the consideration of any basis for shape space).

EXPLICIT CONSTRUCTION OF THE UNIFORM SUBSPACE IN SHAPE SPACE

Let there be assigned a Procrustes mean configuration

$$z = \{z_i = (x_i, y_i)\}_{i=1}^{p}$$

of p landmarks centered at (0,0) and of unit centroid size

$$z\bar{z} = \sum_{j=1}^{p} z_j\bar{z}_j = \sum_{j=1}^{p} x_j^2 + \sum_{j=1}^{p} y_j^2 = 1.$$

(I am starting with two-dimensional data, so that these z's are complex numbers. In the context of complex arithmetic, the overbar, $^{-}$, represents complex conjugation, reflection in the x-axis, rather than averaging.) Note that z must be a true mean shape, not a mere reference configuration; it must have been produced by an algorithm like GLS (Generalized Least Squares) from Slice's GRF-ND. In order to simplify the formulas that follow, without loss of geometric generality, we rotate the mean shape to principal axes so that $\Sigma x_j y_j = 0$. (It does not matter which of the principal axes is set to horizontal by this operation, and if the principal moments α, γ of the formulas to follow are equal, the rotation makes no difference for the formulas.) The limits $1, p$ of the running index j in summations have been suppressed here and will be suppressed likewise in most of the formulas following; subscripts i will refer to the landmarks or their perturbations one at a time.

A general sketch of the geometry as we will construct it is as follows. The reader may have previously encountered David Kendall's shape space Σ_2^p, the space of all "shapes of configurations of p landmarks in two dimensions" quite abstracted from any considerations of graphics. (See Kendall, 1989; Bookstein, 1991; Goodall, 1991). The space Σ_2^p represents the set of configurations of p landmarks in two dimensions after position, orientation and scale are "quotiented out" by some delicate geometrical maneuvers. Shape space is a non-Euclidean metric space, with the squared distance between two forms z and z' given by the same Procrustes formula

$$\Sigma \mid z_j - z_j' \mid^2 = \Sigma \left((x_j - x_j')^2 + (y_j - y_j')^2 \right)$$

that we are already minimizing in the course of producing Procrustes fits.* It is important to note, however, that it is being used here as the shape metric, not the superposition metric that we use to produce pictures back in the space of the Cartesian data. (This distinction is from Goodall, 1991.)

Kendall's shape space is curved — that is, it has to be visualized as an ellipsoid of some high dimension, not as a flat Euclidean hyperplane of the same high dimension — but for sufficiently modest shape variation, we can assume it to be linear in the vicinity of a mean form (the reference form for all our superpositions). Properly speaking, this linear structure is the tangent space to shape space at the mean form (see Bookstein, 1996). Through this mean form there passes a plane, U, the collection of all shapes that can be gotten from the mean form by uniform transformations of shape (that is, by affine transformations of the original Cartesian plane in which the landmarks were located, followed by the obligatory Procrustes fit to the reference form, which takes us back down into shape space). In this linearized version of Kendall's space, any form of a sample, and any mean of a similar sample, can be related to the reference form by a vector. The principal concern of this paper is to construct the projection of any such vector onto the plane U, by means of a projection orthogonal to that plane based on the same Procrustes metric, and to express the result in a convenient basis for U.

We are used to learning about the Procrustes fit as an optimal superposition of the centered specimen form $z' = (z_i')$ over the reference. For small amounts of shape variation, this operation itself can be approximated by a projection of the original Cartesian coordinates. In terms of those coordinates, which are not shape coordinates, centering, scale change and reorientation are themselves projections that can be incorporated into the Procrustes fit by an additional matrix multiplication or two. The final Procrustes fit is the same whether the shape is Procrustes-fitted to the reference form before or after the shape change. That is, we can understand shape variation equally well whether the Procrustes fit is interpreted in shape space or in the original coordinate space (for which the landmark locations are actually

*Actually, true non-Euclidean distance in shape space is better taken as a slightly curved version of this, $\arccos \mid \Sigma z_j \bar{z}_j' \mid$. The sum-of squares metric applies only in the tangent space.

constrained by four separate approximate identities). This statement is not at all obvious, but depends on the profound fact that the Procrustes geometry is a "submersion" of the original Euclidean geometry of the digitizing plane (see Kendall, 1984).

When the result of the Procrustes fit is considered as a collection of locations in the plane of the original reference configuration, the vectors linking paired landmarks comprise a redundant set of shape coordinates for each specimen form; these will be called shape residuals here. Geometrically, we can imagine a shape change to apply either to the original landmark locations or to the locations after such a Procrustes fit. Either way, the "change" will be followed by a final Procrustes step bringing us back into shape space — back into superposition with the reference configuration. For small shape differences, the Procrustes fit is itself a projection, just as the extraction of the uniform component that is the subject of this paper is a projection. But the projection that is the Procrustes fit is onto Kendall's space, or its tangent space, whereas the projection that extracts uniform components for the forms of a sample is a projection within that tangent space, onto a special plane U inside it. In the text below, the kind of projection that is meant — onto shape space or onto the uniform subspace U — will be clear from the context.

For centered data in two dimensions, Procrustes fitting of z' to z may be approximated as complex multiplication of z' by the factor

$$P_z{}' = z\bar{z}'/z'\bar{z}' = (\Sigma\, z_j \bar{z}_j{}')/(\Sigma\, z_j{}' \bar{z}_j{}')$$

(the Orange Book, Section 7.1). The Procrustes shape residuals are then the values

$$P_{z'} z' - z = \{ \mathrm{P}_{z'} z_i{}' - z_i \}$$

We need a direct expression for these elements as a two-dimensional set of patterns of coordinated residuals over all i. This will be the space spanned by differentials of $P_{Az}(Az) - z$ as A ranges over the set of affine transformations near the identity. The expression $P_{Az}(Az)$ is just another way of writing out the shapes of the plane U of uniform changes of the mean shape z.

From this point the argument runs as follows. I will write down what looks like a remarkably arbitrary basis (the vectors S_1, S_2 to follow) for the set of affine transformations of the mean shape and then show that they are actually an orthogonal basis for this plane. It will follow that the characterization of the uniform component of any "nearby" shape as a change from the mean form is just the pair of scores of that shape on these two vectors, S_1, S_2. After a few rearrangements of those formulas, then, we will be done (for this two-dimensional case).

Because the subspace we seek is two-dimensional, it is spanned by any two independent affine transformations. Consider the following two one-dimensional subspaces:

$$S_1(\lambda) : z_i \to z_i' = z_i + \lambda(y_i, 0),$$
$$S_2(\lambda) : z_i \to z_i' = z_i + \lambda(0, y_i).$$

The first of these is a pure shear parallel to the x-axis; the second, pure dilation away from the x-axis. In pictorial terms (Fig. 1), S_1 takes horizontal rectangles to parallelograms of the same height by pushing the top edge of the rectangle to the left or right with respect to the bottom edge, without change of length in this horizontal direction. The parameter λ, in this example 0.56, is the horizontal shift for the top edge of a rectangle of height 1 (which this rectangle isn't). The second transformation, S_2, takes horizontal rectangles to horizontal rectangles that are taller or shorter, again without change of length in the horizontal direction.

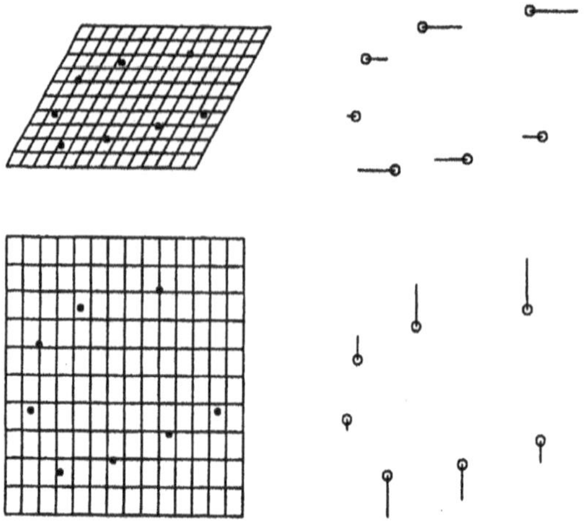

Figure 1. Typical basis elements S_1 (above) and S_2 (below) for the uniform subspace U. The data set for this and subsequent figures consists of the eight-landmark configurations for growing rats set out in the Orange Book, Appendix 4. Analysis is restricted to the 164 specimens with complete data. Illustrated are $S_1(0.56)$ and $S_2(0.56)$. Left column: Cartesian grids for the uniform shears specified. These grids are scaled separately by simple plotting software. Right column: changes at each landmark separately, horizontal or vertical, by multiples of the original y-coordinate of the mean shape (open circles). The starting grid for the shears at left is squared on this configuration of open circles, which is plotted at centroid size 1 and with major axis horizontal. $\alpha = 0.721$, $\gamma = 0.279$. This and all other figures were produced in Splus, a commercial statistical programming package.

The parameter λ is the change in the height of a rectangle that started at height 1; here it is seen in the increase in height of every little grid square by 56%.

The forms at the right in Fig. 1, even though their centroids haven't moved, are obviously not at Procrustes superposition; the formulas for the uniform component are produced by putting them back in Procrustes registration where they belong. The Procrustes fit will adjust both their centroid sizes and their orientations. For both pencils of transformations we treat only the limit of the Procrustes fits, residuals, etc., as $\lambda \to 0$. It may seem odd that we can cover all changes of shape by operations that appear to leave horizontal extents completely unchanged. It is because to each of these operations we append the obligatory step of projection "back down into shape space," that is, a final Procrustes fit back to the starting configuration. The rescaling in this fit will apply to the horizontal direction just as to every other.

The transformations S_1 and S_2 move shape around only in the plane U, but we would like to express them using a full basis for all of shape space. In fact, we will use the redundant basis of all $2p$ shape residuals (vectors at each of the original landmarks). The transformations S_1 have only one parameter, λ, and so will generate only one pattern of shape residuals, the pattern $(\alpha y_i, \gamma x_i)$ below. This is a formula for all the Procrustes residuals at the same time — "motion" of each landmark away from its mean location by a multiple λ of a landmark-specific vector determined by the mean configuration as a whole. Similarly, the transformations S_2 have only one parameter and will generate only multiples of the pattern $(-\gamma x_i, \alpha y_i)$, which is, again, a fixed vector of "movement" at each landmark with formulas determined by the mean shape as a whole. The next section derives these formulas, by means of complex arithmetic; other derivations are possible (Mordia, 1995).

For $S_2(\lambda)$, we have

$$z\bar{z}' = \Sigma\, z_j\,(\,\bar{z}_j - \lambda\,(0, y_j)\,) = 1 + \lambda\,(\,\Sigma\, y_j^2, -\,\Sigma\, x_j y_j\,),$$

$$z'\bar{z}' = \Sigma\,(z_j + \lambda(0, y_j))\,(\bar{z}_j - \lambda(0,y_j)) \approx 1 + \lambda\sqrt{-1}\,\Sigma\, y_j\,(z'_j - z_j) = 1 + 2\lambda\,\Sigma\, y_j^2$$

The multiplier $P_{z'}$ is the ratio of these:

$$\begin{aligned} P_{z'} &= \frac{1+\lambda\left(\sum y_j^2, -\sum x_j y_j\right)}{1+2\lambda\sum y_j^2} \\ &\approx 1+\lambda\left(-\sum y_j^2, -\sum x_j y_j\right) \\ &= 1-\lambda\sum y_j^2 \end{aligned}$$

a real number, because we set $\Sigma x_j y_j$ to 0 when we rotated to principal axes. The residual is then

$$\begin{aligned} P_{z'} z'_i - z_i &\approx \lambda\left((0, y_i) - \sum y_j^2 (x_i, y_i)\right) \\ &= \lambda\left(-x_i \sum y_j^2, y_i\left(1-\sum y_j^2\right)\right) \end{aligned}$$

Writing $\gamma = \Sigma\, y_j^2$, $\alpha = \Sigma\, x_j^2 = 1 - y$, we arrive at the shape residual vectors

$$\begin{aligned} \lim_{\lambda\to 0}\left(P_{z'} z'_i - z_i\right)/\lambda\Big|_{S_2} &= (-\gamma x_i, \alpha y_i) \\ &= (x_i, y_i)\left(\begin{pmatrix} 0 & 0 \\ 0 & 1 \end{pmatrix} - \begin{pmatrix} \gamma & 0 \\ 0 & \gamma \end{pmatrix}\right). \end{aligned}$$

Similarly, for $S_1(\lambda)$:

$$\begin{aligned} &z\bar{z}' = 1 + \lambda \sum y_j z_j, \\ &z'\bar{z}' \approx 1 + \lambda \sum y_j \left(z_j + \bar{z}_j\right) = 1 + 2\lambda \sum x_j y_j, \\ &P_{z'} \approx 1 + \lambda\left(-\sum x_j y_j, \sum y_j^2\right), \end{aligned}$$

which, for $\Sigma x_j y_j = 0$ and small λ, is nearly a pure turn (multiplication by a complex number of amplitude 1). Note, for later reference, that the magnitude of this turn is proportional to the total torque $\Sigma\, \lambda y_j^2$ of all of these changes around the centroid.

When the multiplication by this version of $P_{z'}$ is carried out, the shape residuals are

$$\begin{aligned} \lim_{\lambda\to 0}\left(P_{z'} z'_i - z_i\right)/\lambda\Big|_{S_1} &= (y_i, 0) + (x_i, y_i)\left(0, \sum y_j^2\right) \\ &= (\alpha y_i, \gamma x_i) \\ &= (x_i, y_i)\left(\begin{pmatrix} 0 & 0 \\ 1 & 0 \end{pmatrix} - \begin{pmatrix} 0 & -\gamma \\ \gamma & 0 \end{pmatrix}\right). \end{aligned}$$

Fig. 2, in panels parallel to those of Fig. 1, shows the resulting Procrustes superposition and the splined grids that represent our "shear" and "dilation" back in Procrustes space. The little vectors drawn may seem a bit more accurate if you keep in mind the values α = 0.721, γ = 0.279 that encode the elongation of the mean form.

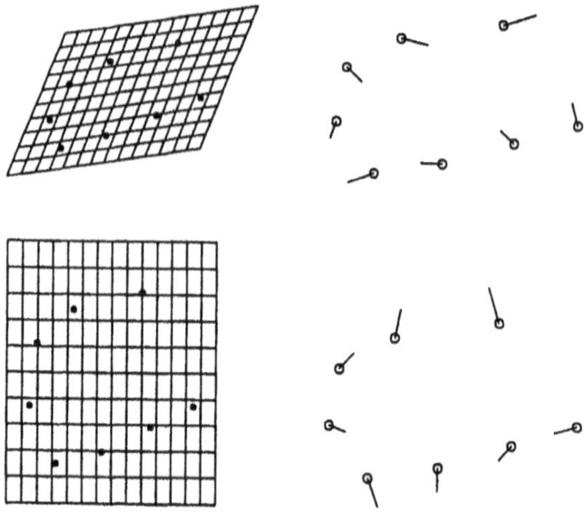

Figure 2. The same two transformations in Procrustes registration, with length $.56\sqrt{\alpha\gamma} \approx 0.25$. At the right are the outcome forms of Fig. 1 (ends of the vectors at the right) after Procrustes fit onto the same mean form. The grids at left are the same as those of Fig. 1 except for the consequences of the Procrustes fit: rotation for the upper shape change, rescaling for both. The vectors at right are exactly the vectors $(\alpha y_i, \gamma x_i)$ and $(-\gamma x_i, \alpha y_i)$ of the exposition in the text.

Geometrically, at each landmark of the reference configuration, the construction of this pair of residuals (the vectors in Fig. 2) is as follows. Draw the vector (x_i, y_i) from the centroid to the landmark and reflect it in the 45° line $x + y_i = y + x_i$ through the landmark to yield the direction (y_i, x_i). Then enhance what is now the horizontal term y_i with respect to what is now the vertical term x_i in the ratio $\alpha : \gamma = \Sigma x_j^2 : \Sigma y_j^2$, the ratio of the principal moments of the reference configuration (remember we have rotated so that the principal axes are horizontal and vertical). This set of vectors $(\alpha y_i, \gamma x_i)$ is proportional to the displacements of the Procrustes residuals as a whole corresponding to a small shear strain S_1 parallel to the x-axis. For instance, for landmarks on the x-axis, y_i is zero, and so the Procrustes-normalized change of landmark location under shears S_1 is wholly vertical, corresponding to the effect of a small rotation about the origin. The leftmost and rightmost landmarks of our mean shape are moved almost vertically in this way (Fig. 2, upper right). The rotation of the configuration as a whole is attenuated by a factor γ from the angle $\tan^{-1}\lambda$, expressing the original shear as it affects points on the y-axis; the attenuation owes to the failure of components of landmarks along the x-axis to have been "rotated" in Fig. 1.

The other set of vectors $(-\gamma x_i, \alpha y_i)$ (Fig. 2, bottom right), representing the basis for vertical dilation S_2 in terms of residuals landmark by landmark, is obtained by rotating each vector $(\alpha y_i, \gamma x_i)$ by 90° counterclockwise. These shape changes assign points on the x-axis purely real displacements and scale them back toward the centroid by the factor γ. But, even after the Procrustes step, points on the y-axis (e. g., the second from the left in the lower row here) end up farther from the x-axis than where they started, their original displacements attenuated by a factor α (which is less than 1). The value $\lambda = .56$ for these two figures was set so that the resulting length in Procrustes space, $.25 = .56\sqrt{\alpha\gamma}$, is both visually interesting and reasonable, even though it is larger (by some 25%) than the full range of the uniform component in the actual data set.

COMPUTING THE UNIFORM SHAPE COMPONENT

As the two vectors multiplying λ each have squared Procrustes length

$$\alpha^2 \Sigma y_i^2 + \gamma^2 \Sigma x_i^2 = \alpha^2 \gamma + \gamma^2 \alpha = (\alpha\gamma)(\alpha + \gamma) = \alpha\gamma ,$$

this basis is not only orthogonal landmark by landmark but also orthonormal in the Procrustes metric as a basis for the uniform subspace U, as long as each term is normalized by the factor $(\alpha\gamma)^{-1/2}$. We can absorb the factor into each coefficient separately to generate the orthonormal basis

$$\left(y_i\sqrt{\alpha/\gamma}, x_i\sqrt{\gamma/\alpha}\right), \left(-x_i\sqrt{\gamma/\alpha}, y_i\sqrt{\alpha/\gamma}\right).$$

Then the expression for the uniform shape coordinate from Procrustes residuals is just the dot product of these bases by the residuals at the landmarks separately. In the notation of Rohlf (1993 and "Morphometric spaces" article, this volume), we have arrived at the additional two columns that, concatenated with the matrix $\mathbf{I}_2 \otimes \mathbf{E}$ of partial warp scores, supply a Procrustes-orthonormal basis for all of Σ_2^p near the reference form. Notice that these additional terms do not have the symmetry of a tensor product, but instead encode the elliptic geometry of plane rotation. The maneuver fails if $\alpha\gamma = 0$, i. e., if the mean landmarks lie on (or, in practice, near) a straight line. In that case, none of the transformations S_1, S_2 have any effect upon shape.

Let Δx_i, Δy_i be the deviations of the landmarks from expected locations, usually a mean shape for their own group or another similar group. This notation intentionally mixes two spaces; whereas the "residual" relating one shape to the reference properly is construed only in shape space, we are expressing it instead back in the original digitizing plane, as if the landmarks were free to move about individually even after the Procrustes fit. They are not, of course: the $2p$ variables, Δx and Δy, satisfy four linear constraints, corresponding to the four degrees of freedom controlled in the course of a Procrustes superposition. Throughout this formalism, furthermore, the reference configuration $\{(x_i, y_i)\}$ is considered fixed, and likewise α and γ are fixed at the values Σx_j^2 and Σy_j^2 ; only the shape residuals Δx, Δy are allowed to vary.

Summing crossproducts of the shape residuals $(\Delta x_i, \Delta y_i)$ by the basis vectors $(\alpha y_i, \gamma x_i)/\sqrt{\alpha\gamma}$ and $(-\gamma x_i, \alpha y_i)/\sqrt{\alpha\gamma}$ just extracted, we arrive at the two scores of a specimen shape $(x_i + \Delta x_i, y_i + \Delta y_i)$ in U, the uniform subspace of shape space near the reference form. The estimate of the "horizontal" score, that aligned with the shears $S_1(\lambda)$, is

$$\left(\alpha \sum y_j \Delta x_j + \gamma \sum x_j \Delta y_j\right) / \sqrt{\alpha\gamma}$$

and that for the "vertical" score, the one aligned with the vertical strains $S_2(\lambda)$, is

$$\left(-\gamma \sum x_j \Delta x_j + \alpha \sum y_j \Delta y_j\right) / \sqrt{\alpha\gamma} .$$

These expressions are exact up to the linearization of the shape space itself. Because of the simplifying assumption $\Sigma x_j y_j = 0$, however, they apply only in this one specific coordinate system. More complex expressions in which that term is not suppressed can be generated to cover the general rotation but are of no interest here.

I propose this notation as the standard formula for the uniform component of shape variation. The component is expressed in the basis for the uniform subspace U of the pencils of transformations S_1 and S_2: shears along the principal axis of the reference configuration

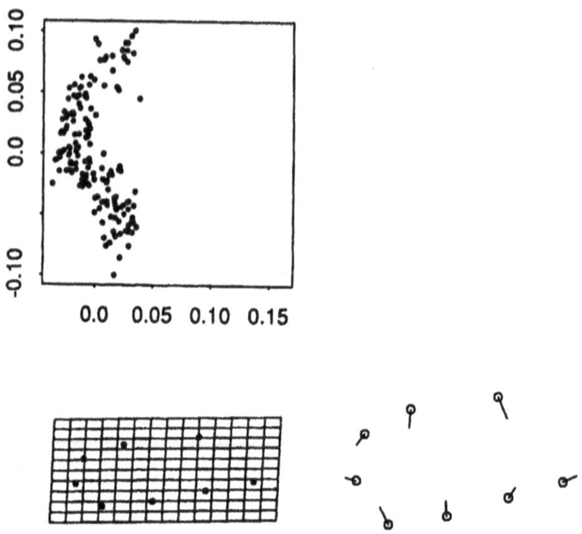

Figure 3. Scatter for the uniform component for the 164 rat skull shapes. Its Procrustes diameter is also its vertical range, about 0.2, and its total Procrustes variance (total variance explained by the uniform component of shape) is 0.00252, corresponding to a root mean square distance of about 0.05 of the dots from their centroid at (0,0). The comparable scatter in the Orange Book, Figure 7.6.5a, is taken to a baseline 45° from this one and hence is rotated about 90° on the page. Below, the first principal component of this uniform component, drawn as a sheared grid (left) and as a displacement in shape space (right) based on a multiple equal to the sample range.

and dilations perpendicular to that axis. These values are those supplied by the *U* option in Rohlf's programs TPSRW and NTSYS in all releases after September 1993 (see Rohlf, "Morphometric spaces" article, this volume). They are also supplied in versions of GRF-ND distributed after January 1995.

The scatter of the uniform component for the 164 rat skulls of our example here is shown at the top of Fig. 3. Although its factor structure is clearly nonlinear, it is useful nevertheless to extract its long axis, a "first uniform relative warp" (see Bookstein, 1995), which is aligned with the comparison of extreme forms (in this example, the 7-day-old rats versus the 150-day-old rats). This warp is shown at lower left as the spline of a grid squared on the usual mean form. The magnitude of the change shown here, which is also the length of the change vector shown at Procrustes superposition at lower right, is the Procrustes diameter of the scatter, about 0.2. Compare the lower grid, length 0.25, in Fig. 1 or 2.

Relation to the "Uniform Factor Estimate" of the Orange Book

Because $\alpha + \gamma = 1$, the core of the first formula for the "horizontal" (shear) score can be rearranged as

$$(1-\gamma)\,\Sigma\, y_j\,\Delta x_j + \gamma\Sigma\, x_j\,\Delta\, y_j = \Sigma\, y_j\,\Delta\, x_j + \gamma\Sigma\,(x_j\,\Delta\, y_j - y_j\,\Delta\, x_j)\,.$$

Also, recall that because these forms are being considered only after Procrustes superposition, centroid size is (almost exactly) constant. That is,

$$\Delta\,(\Sigma\,(x_j^2 + y_j^2)\,) \approx 0$$

(This is one of the four constraints on the Δx, Δy referred to above.) Then the core of the second equation, for the "vertical" score, can be simplified to yield

$$-\gamma\sum x_j\Delta x_j+(1-\gamma)\sum y_j\Delta y_j=-\gamma\left(\sum x_j\Delta x_j+\sum y_j\Delta y_j\right)+\sum y_j\Delta y_j$$
$$\approx-\gamma\left(\Delta\sum\left(x_j^2+y_j^2\right)/2\right)+\sum y_j\Delta y_j=\sum y_j\Delta y_j.$$

Written out explicitly as a vector in this basis, the rearranged formula now reads

$$\left(\sum y_j\Delta x_j+\gamma\sum\left(x_j\Delta y_j-y_j\Delta x_j\right),\sum y_j\Delta y_j\right)\ /\sqrt{\alpha\gamma}.$$

This pair of expressions is quite interesting. The combination of the first term of the S_1 element with the S_2 element results in a "partial estimate"

$$\sum_{1}^{p} y_j\,(\Delta x_j\,,\Delta y_j)\,.$$

(In the next few paragraphs I suppress the normalizing factor $\sqrt{\alpha\gamma}$, as the discussion deals only with proportionalities.) Save for the coordinate system in which the y's and Δ's are expressed, this fragment is proportional to the formula for a uniform factor estimate previously published in the Orange Book, page 279, for this same "shear, dilation" coordinate system for U. There the formula applied to shape coordinates, not Procrustes fits, so that, in particular, the sum went only "from 3 to p": the terms $(\Delta x_i,\Delta y_i)$ were taken as referring to a basis of $2p-4$ shape coordinates, not the larger count of $2p$ Procrustes residuals to which the formula applies here. That earlier discussion went on to recommend that, for greater accuracy, this estimate exploit shape coordinates to a baseline through the centroid of the mean shape and along the long axis of its intraspecimen scatter—precisely the same criteria as specify the x-axis of the Cartesian reference frame here.

The second term of the first element in this formula, the S_1 term, appears to be different between the approaches, but, actually, it is present in the shape-coordinate version as well. Recall that when we took care to choose a coordinate system with $\Sigma x_j y_j = 0$ (that is, to place the principal moments of the reference shape horizontal and vertical), then the horizontal shear $S_1(\lambda)$ was expressed after the Procrustes step as the pure turn, multiplication by $1+\gamma\lambda\sqrt{-1}$. The term $\gamma\Sigma(x_j\Delta y_j - y_j\Delta x_j)$ is the expression of another such shear, this time in a coordinate system prohibited from turning (because the Procrustes step is carried out only once per specimen).

It is exactly that rotation which the transformation to shape coordinates is carrying out. The term multiplying γ totals the torques of all the landmark changes around the centroid, just as in the expression for the multiplier $P_{z'}$ of the shears S_1. In the shape coordinate formalism, this is estimated using only two landmarks (the ends of the baseline), and the estimate is re-expressed after rotation to fix the arbitrary baseline instead of in the Procrustes frame. But the baseline back rotation of the shape-coordinate formalism and the additional "horizontal" term of the Procrustes formalism are estimating the same geometric quantity. (For instance, both methods are exact when transformations are exactly linear, regardless of presence or absence of this torque term.) If the formula of the Orange Book is to be applied directly to Procrustes residuals, this additional torque term must, of course, be computed and added in as a second term of shear along the principal axis of the reference shape. The factor of proportionality between the two versions of the formulas, $\sqrt{\gamma/\alpha}$, corresponds to the scaling of the shape-coordinate version by baseline length, that of the Procrustes version by centroid size.

For estimating transformations that are truly linear, up to digitizing noise, the two formulations (shape coordinates, no torque term, and Procrustes shape residuals, including torque term) have no relative bias, but differ only in precision. The Procrustes estimate is the more accurate, as the shape-coordinate formulation was forced to incorporate the error variance of the baseline landmarks explicitly, whereas the Procrustes approach attenuates those over all the landmarks of the configuration.

Partitioning Mean Squared Procrustes Length

Because the projection of shape residuals (Procrustes fits) onto the uniform subspace is orthogonal in the Procrustes geometry, the uniform component provides one term in an exact partition of mean-squared Procrustes distance of a sample of shapes around their mean. Here that mean square is 0.00509, representing an average Procrustes distance from the mean shape of 0.071. The left panel in Fig. 4 shows the usual graphic corresponding to that mean square (the Orange Book, Section 7.1); regardless of any correlations, Procrustes variance is proportional to the total variance of those "Gaussian" circles. At the right is the corresponding scatter of residuals after the uniform component is projected out. In this example, the partition is very nearly into equal halves: 0.00252 squared Procrustes length to the uniform component, 0.00257 to the residual. That residual is partitioned further by the partial warps as explained in principle in Bookstein (this volume); here, as usual in growth studies, it is mostly partial warp 1. Both terms, the uniform component and the nonuniform residual shape, are strongly correlated with age. That correlation is nonlinear for the uniform component but unidimensional for the residual (the Orange Book, Section 7.6.3), a unidimensionality reflected in all the separate scatters at the right of Fig. 4.

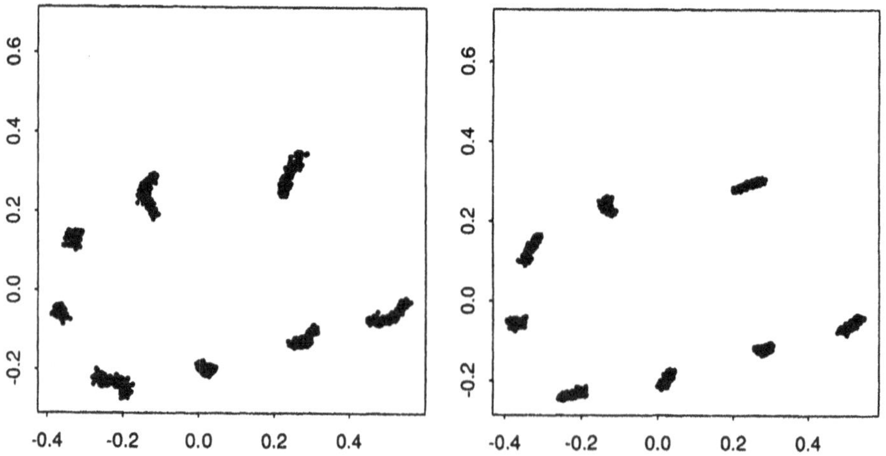

Figure 4. Procrustes fits of all 164 forms to their mean (left) and the fit of the forms after the uniform component of the previous figure is partialled out (right). This is equivalent to a GALS residual plot from Slice's (1993) GRF-ND.

OTHER APPROACHES

Recall the four different varieties of uniform component reviewed at the outset. We have just shown that a simple formula can be supplied under approach (1) that is much the same as that of approach (4), except for having a smaller standard error of estimate. The formula is already available in our most widely distributed software implementation, Rohlf's TPSRW, in all versions dated after September 1993. The "nuisance parameter" of the Rohlf–Slice GALS fit, which is also carrying out a minimization of Procrustes distance to the uniform plane U through the reference configuration, arrives at the same estimate (as always, up to nonlinearities of algorithm), as long as the reference configuration remains the GLS mean. The GALS scatter is of the residual shape variation after projection out of this estimated component in U, as demonstrated in Fig. 4.

Procrustes distance is not only the metric for Kendall's shape space but also the pivotal statistic for the distribution of shapes in that space under an assumption of identically distributed circular digitizing error in the original landmark locations separately. We can then project a sample landmark configuration perpendicularly back onto U starting from any basis for shape space, not just the shape residual vectors produced here, as long as we can construct an expression for Procrustes perpendicularity in the geometry of that more general basis. Up to the usual nonlinearities, the geometry of that projection is encoded in the covariance structure of the corresponding shape basis under the null model of identically independently distributed digitizing noise. Then the GLS fit of a uniform component to any basis will be exactly proportional to the Procrustes estimate as long as we use the correct covariance matrix to supply the geometry of the projection. Hence the formulas of the Orange Book, page 274, supply the equivalent of the present Procrustes estimate of the uniform component as long as the matrix S, regardless of actual shape coordinate covariances, is taken as the null model covariance matrix (Mardia and Dryden, 1989 or Orange Book, page 151).

The other two approaches can then be summarized quickly. To project along minimum Mahalanobis distance serves the same function here as in any other application: it provides the estimate of the uniform component that is likeliest on the multivariate normal model. In most other applications of the Mahalanobis technique, there is no metric other than minus log-likelihood to direct the investigator's choice of a projection. Here in shape space, the Procrustes metric leads to the more cogent visualizations of the geometry of this component. Whether or not the null model obtains, the Mahalanobis formula is the correct one for testing the significance of a group mean difference in this component and also for testing the significance of mean deviations from the uniform plane. The formulas are in the Orange Book, Section 7.2. But, issues of tail-probabilities notwithstanding, the Procrustes projection seems the more natural for decomposition of observed changes, cross-correlations and graphical displays — in short, for description of what has been found.

There remains the affine term of the thin-plate spline, projection orthogonal to bending energy as in approach (3). The decomposition of the interpolating spline by its partial warps is a very useful basis (the only one we have at present) for large-scale features of shape change in the complement of the uniform subspace. However, the uniform piece supplied by the spline, although guaranteed to result in an exact reconstruction of the observed reconfiguration of landmarks, does not optimize any geometric relation between the configuration's vector in shape space and the foot of its projection upon the uniform shape plane through the reference form. The direction along which this projection is taken is a very sensitive function of the spacing of the original landmarks and can be remarkably oblique to the direction of Procrustes orthogonality we're concentrating on here. (See, for instance, the way in which shears between closely spaced landmarks generate large affine terms that "balance" them: Bookstein and Green, 1993.) The A option of Rohlf's TPSRW should not be considered to supply estimates of the uniform component for subsequent statistical manipulation.

All these approaches, both the exact Procrustes formula presented here and the other Mahalanobis versions and approximations, apply in the tangent space to shape space, a linearization of shape variables "in the vicinity of the mean shape." Mardia and Dryden (1989) demonstrate the same computations without the linearization: true maximum-likelihood estimation of a uniform component within the non-Euclidean geometry of shape space itself. The formulas have been further extended to the case of nonisotropic digitizing noise (still the same at all landmarks, up to a rotation) in Mardia and Walder (1994).

IN THREE DIMENSIONS

For the extension to three-dimensional data, it is most convenient to proceed using redundant coordinates for the three-dimensionality of the scheme just as we have already used the redundant coordinates of the Procrustes fit itself to simplify the main formula in two dimensions. Assume a reference configuration of p points (x_i, y_i, z_i) in three dimensions, centered at (0,0,0) and oriented with principal axes along the x-, y-, and z-axes: $\Sigma x_i y_i = \Sigma x_i z_i = \Sigma y_i z_i = 0$. Write $\alpha = \Sigma x_i^2$, $\beta = \Sigma y_i^2$, $\gamma = \Sigma z_i^2$ and set centroid size $\alpha + \beta + \gamma$ to 1, as before.

We seek the uniform subspace U_3 of Kendall's shape space Σ_3^p in the vicinity of this mean form. For three dimensions, this subspace is of dimension 5. Consider three different versions of the preceding two-dimensional construction, corresponding to the xy-, yz- and zx-planes. For instance, in the xy-plane, we consider the limiting forms of the Procrustes projections down to Σ_3^p of the affine transformations $S_{1(xy)}(\lambda)$, which replace each point (x_i, y_i, z_i) by the point $(x_i + \lambda y_i, y_i, z_i)$, and of the transformations $S_{2(xy)}(\lambda)$, which replace each point by $(x_i, (1+\lambda) y_i, z_i)$. From an intermediate step in the preceding section, the $S_{1(xy)}$ projects down to Σ_3^p as the pure turn $(\alpha y_i, \beta x_i, 0)$. (This is a $(3p)$-vector, a 3-vector at each of the p landmarks of the reference form. In the same ambiguity as we exploited in the preceding section, the notation applies both in the linearization of Σ_3^p itself and to the "shape residuals" that emanate from the points of the reference configuration in the Cartesian space of the optimally superimposed data.) As for $S_{2(xy)}$, the change of scale induced by dilation along the y-axis is now restored by shrinkage along both of the other axes; the formula $(-\beta x_i, \alpha y_i)$ for the two-dimensional case is replaced by $(-\beta x_i, (1-\beta) y_i, -\beta z_i)$. These vectors are still orthogonal, but no longer have the same length.

Similarly, we have basis elements along $S_{1(yz)}$ and $S_{1(zx)}$ with formulas

$$(0, \beta z_i, \gamma y_i), \quad (\alpha z_i, 0, \gamma x_i)$$

and along $S_{2(yz)}$ and $S_{2(zx)}$ with formulas

$$(-\gamma x_i, -\gamma y_i, (1-\gamma) z_i), \ ((1-\alpha) x_i, -\alpha y_i, -\alpha z_i).$$

Assemble these in two sets, the S_1's and the S_2's:

$$S_1 = \begin{pmatrix} \alpha y_i & \beta x_i & 0 \\ 0 & \beta z_i & \gamma y_i \\ \alpha z_i & 0 & \gamma x_i \end{pmatrix}, \qquad S_2 = \begin{pmatrix} -\beta x_i & (1-\beta) y_i & -\beta z_i \\ -\gamma x_i & -\gamma y_i & (1-\gamma) z_i \\ (1-\alpha) x_i & -\alpha y_i & -\alpha z_i \end{pmatrix}.$$

Each of these matrices has three columns and $3p$ rows.

The columns of S_1 are orthogonal. For instance, the inner product of the first two columns is $\alpha\beta\Sigma x_i y_i$, but we set $\Sigma x_i y_i$ to 0 when we specified the Cartesian system for the reference shape. Likewise, the inner product of any column of S_1 with any column of S_2 is zero, as again every term involves $\Sigma x_i y_i$, $\Sigma x_i z_i$ or $\Sigma y_i z_i$.

The columns of S_2 themselves, however, are not orthogonal. In fact, by explicit cancellation, their sum is zero. After all, the effect of the combination of all three of these directional expansions by the factor λ is a simple change of scale of the entire configuration by λ, which of course has no effect on shape per se. It is in this linear dependency of the columns of S_2 that we locate the deficiency of rank in the (S_1, S_2) basis for the uniform subspace U_3: these six vectors span a five-dimensional subspace because the three vectors S_2 span only a two-dimensional subspace.

For either set of three, S_1 or S_2, the Procrustes lengths of the basis vectors must be set to 1 before the computation of the five elements of an empirical uniform component by inner product of these formulas with Procrustes residuals. The squared lengths of the columns of S_1 are $\alpha\beta(\alpha + \beta)$ and so on, and those of the columns of S_2 are $\beta(1 - \beta)$ and so on. The inner product of columns 1 and 2 of S_2 is $-\beta\gamma$, and likewise for other pairs. Written out in full, the Procrustes inner products of the basis vectors of S_2 are

$$\begin{pmatrix} \beta(1-\beta) & -\beta\gamma & -\alpha\beta \\ -\beta\gamma & \gamma(1-\gamma) & -\gamma\alpha \\ -\alpha\beta & -\gamma\alpha & \alpha(1-\alpha) \end{pmatrix}.$$

Because $\alpha + \beta + \gamma = 1$, the rows and columns of this matrix sum to 0 (as they must, because the columns themselves do) and so one of its eigenvalues is 0. Write the normalized eigenvectors that correspond to the other two eigenvalues as $(d_{1xy}, d_{1yz}, d_{1zx})$ and $(d_{2xy}, d_{2yz}, d_{2zx})$. The normalization here is one that is less common in matrix programming; the result of thc usual eigenroutine, which would normalize the d to sum of squares 1, must be divided by the square roots of the eigenvalues. The contribution of the shears S_1 to our uniform subspace is the set of three scores combining the $3p$ centered Procrustes coordinates $(\Delta x_i, \Delta y_i, \Delta z_i)$ according to the three $3k$-vectors of coefficients $S_{1(xy)}/\sqrt{\alpha\beta\,(\alpha+\beta)}$, $S_{1(yz)}/\sqrt{\beta\gamma\,(\beta+\gamma)}$, and $S_{1(zx)}/\sqrt{\gamma\alpha\,(\gamma+\alpha)}$, the normalizations of these already orthogonal shape variables. The contribution of the shears S_2 to the basis is the pair of scores $\Sigma(d_{1xy}S_{2xy} + d_{1yz}S_{2yz} + d_{1zx}S_{2zx}) \cdot (\Delta x_i, \Delta y_i, \Delta z_i)$ and $\Sigma(d_{2xy}S_{2xy} + d_{2yz}S_{2yz} + d_{2zx}S_{2zx}) \cdot (\Delta x_i, \Delta y_i, \Delta z_i)$. This amounts to five scores, as required, all orthogonal and each of Procrustes length 1 in its formula.

A basis for the uniform subspace U_3 of shape space Σ_3^p, then, may be gotten from the three columns of S_1 and thereafter in one of two ways: symmetrically and redundantly, by using all three columns of S_2 as well but constraining all subsequent multivariate analyses to the subspace orthogonal to the sum of the columns of S_2 or using generalized inverses; or unsymmetrically, augmenting the columns of S_1 by scores corresponding to the two nonzero eigenvectors of the Procrustes geometry of the S_2's as just set out.

The equivalents of these formulas for shape-coordinate bases for Σ_3^p go as in the two-dimensional case. For shape coordinates to a ground triangle [the baseline/baseplane orientation, landmarks 1, 2, 3 fixed at (0,0,0), (1,0,0), $(x_3, y_3, 0)$], a feasible skew basis for U_3 is the space U_2 of uniform transformations of the baseplane, applied identically in all planes parallel to the baseplane, crossed by a three-dimensional subspace each element of which translates all landmarks not on the baseplane by multiples $z_i\alpha$ of a single three-vector α. This will suffice for tests of the existence of uniform components in group differences or covariances of shape. For ordination, as in the two-dimensional case, we need the null model covariances of such a basis. I have not worked out the relevant formulas, nor their equivalents for the general rigid triangulations mentioned in the Orange Book, but I suspect they will prove a great deal more complicated than those for the S_1's and S_2's printed above, which are their joint re-expressions after Procrustes refitting.

ACKNOWLEDGMENTS

The research reported here was underwritten in part by NIH grants DA–09009 and GM–37251 to F. L. Bookstein. The first of these is sponsored by the National Institute on Drug Abuse, the National Institute of Mental Health and the National Institute on Aging as part of the Human Brain Project. I am grateful to W. D. K. Green for correcting an error in an earlier version of the three-dimensional formulas. The faculty and students of the NATO Advanced Study Institute "Advances in Morphometrics," Il Ciocco, Italy, July 1993, must take collective responsibility for insisting that the discrepancies among the earlier versions of the "uniform term" deserved a careful algebraic resolution. Jim Rohlf, Dennis Slice, and Paul Sampson studied an earlier draft of this paper and made several helpful suggestions.

REFERENCES

Bookstein, F. L. 1991. Morphometric tools for landmark data: Geometry and Biology. Cambridge University Press: New York. (The Orange Book.)

Bookstein, F. L. 1993. A brief history of the morphometric synthesis. Pages 15–40 *in* L. F. Marcus, E. Bello, and A. García-Valdecasas, (eds.), Contributions to morphometrics. Monografias del Museo Nacional de Ciencias Naturales 8, Madrid.

Bookstein, F. L. 1996. Biometrics, biomathematics, and the morphometric synthesis. Bulletin of Mathematical Biology 58.

Bookstein, F. L., and W. D. K. Green. 1993. A feature space for edgels in images with landmarks. Journal of Mathematical Imaging and Vision 3:231–261.

Bookstein, F. L., and P. D. Sampson. 1990. Statistical models for geometric components of shape change. Communications in Statistics: Theory and Methods 19:1939–1972.

Goodall, C. R. 1991. Procrustes methods in the statistical analysis of shape. Journal of the Royal Statistical Society, Series B, 53:285–339.

Kendall, D. G. 1984. Shape-manifolds, procrustean metrics, and complex projective spaces. Bulletin of the London Mathematical Society 16:81–121.

Kendall, D. G. 1989. A survey of the statistical theory of shape. Statistical Science 4:87–120.

Mardia, K. V. 1995. Shape advances and future perspectives. Pages 57-75 *in* Current Issues in Statistical Shape Analysis. Leeds University Press: Leeds, U.K.

Mardia, K. V., and I. Dryden. 1989. The statistical analysis of shape data. Biometrika 76:271–282.

Mardia, K. V., and A. N. Walder. 1994. Shape analysis of paired landmark data. Biometrika 81:185–196.

Rohlf, F. J. 1993. Relative warp analysis and an example of its application to mosquito wings. Pages 131–159 *in* L. F. Marcus, E. Bello, and A. García-Valdecasas, (eds.), Contributions to morphometrics. Monografias del Museo Nacional de Ciencias Naturales 8, Madrid.

Rohlf, F. J., and F. L. Bookstein, (eds.), 1990. Proceedings of the Michigan morphometrics workshop. University of Michigan Museum of Zoology Special Publican 2.

Sneath, P. H. A. 1967. Trend-surface analysis of transformation grids. Journal of Zoology 151:65–122.

Thompson, D'A. W. 1961. On growth and form (1917 edition abridged by John Tyler Bonner.) Cambridge University Press: London.

AFFINE TRANSFORMATIONS AS A MODEL OF VIRTUAL FORM CHANGE FOR GENERATING MORPHOSPACES

Diego Rasskin-Gutman and Angela D. Buscalioni

Unidad de Paleontología, B-101
Departamento de Biología, Facultad de Ciencias
Universidad Autónoma de Madrid, 28049
Cantoblanco, Spain

ABSTRACT

Affine change might be used to seek general patterns of variation or to envisage new ways of describing and evaluating form change. In this paper, we assume a theoretical perspective and show a strategy for generating affine morphospaces (ordered collections of affine-transformed images of a base form), by means of a new computer program called D'ARCYGRAPH. We propose the use of affine transformations as models of change that might induce the search for new explanations based on the constraints imposed by the properties of affine change, that is, their invariants.

We introduce a technique to elaborate affine morphospaces; as an example, we build the morphospace of pelvic girdles generated by using as a base form the pelvis of *Deinonychus antirrhopus* (Dinosauria, Theropoda). We explore possible transformation paths from this base form, which is considered to be the closest sister group of the major clade Aves. A discussion of the type of variables (called "dispositionals") that can be used and explained by the affine model is provided. We conclude that it is possible to build an affine trend that simulates the fossil record transition, Theropoda–Aves, according to these dispositional variables.

Also, we introduce an interactive strategy for the superimposition of forms, without any automatic optimality criteria. A comparison with other superimposition methods, based on tridactyl dinosaur footprints is discussed.

INTRODUCTION

Morphology, as a discipline of biological science, might be divided into three main groups: descriptive morphology, theoretical morphology and functional morphology. The first two groups are concerned with shape and form change, which is also the primary object

Advances in Morphometrics, Edited by L. F. Marcus et al.
Plenum Press, New York, 1996

of comparative anatomy, whereas the third one is more focused on shape in relation to function. Shape and form change constitutes the basis for most issues related to evolution and taxonomy at organismic levels. Recognizing different expressions of the same structure, that is, studying shape and form change, is the first step towards an appreciation of diversity. As expressed in Scott-Ram (1990), information provided by shape and form change can be addressed from either a descriptive point of view or a theoretical one. In the first approach, the methodology rests on the computation of different coefficients of similarity or dissimilarity that are based on a definition of morphological distance, and which are finally treated statistically (Sneath and Sokal, 1973). The descriptive attitude is concerned with shape, rather than form, as discussed in Benson (1982): "Shape is the description (or visual impression, if limited to that) of the relative proportions of distances to all points composing an outline or contour of an object. It must go no further than this conceptually or it becomes form. Form is shape plus a conceived or stated organizational, usually geometric, relationship among the parts that defines the whole. Philosophically form has no reality. It is specified as the set of abstract conditions that allows shape to exist." This difference between form and shape is purely conceptual. Thus, form is defined as having some properties of organization that might be identified either as (1) a set of landmarks which characterizes the geometry of the shape, a definition which encompasses the concept of form as composed of shape and size, where shape has itself a geometry and numerical description, or, more abstractly, as (2) a set of properties concerning the arrangement of the elements that constitute that shape—the idea that we will follow in this paper. The methodology of the theoretical approach rests on the formalization of a pattern, namely an idea of change; the concern here is with form, rather than shape (Benson, 1982). Both approaches can be employed on intra- and interspecific levels. The former methodology has been used in numerical Taxonomy (Sneath and Sokal, 1973), having ordination as a main goal: "this element belongs to this specific set." Thus, similarity or dissimilarity coefficients on specimens in a sample can be used as an entry to any ordination procedure such as factor analysis, principal components analysis or cluster analysis (see for instance Gordon, 1981; Reyment, 1991 for an exhaustive revision). In the theoretical perspective, form change is integrated as a pattern that, if expressed by a mathematical function, can be used to simulate forms which can account for diversity at different levels. This latter approach constitutes an extension of theoretical morphology, in which the main goal is the study of morphological constraints (Raup and Stanley, 1978).

A classic example of a theoretical perspective is Raup's morphospace of coiling shells (Raup, 1966), in which a model—in this case of growth—appears to be suitable for explaining the different forms that occur in nature. The model, as simple as the equation of a spiral, generates forms by means of just changes in the values of the parameters of that equation. Furthermore, Raup's approach has a causal explanation for form change, based on Sewall Wright's theoretical concept of adaptive landscape. In this paper we will assume a theoretical attitude and we will show a strategy to generate affine morphospaces by means of a new computer program without previous causal assumptions called D'ARCYGRAPH. This approach allows an a posteriori evaluation of any causal hypothesis about change within an internalist, functional or evolutionary context.

AFFINE TRANSFORMATION AS A MODEL OF FORM CHANGE

D'Arcy Thompson's 1917 book, *On Growth and Form*, introduced the idea of comparison of two biological forms by inscribing the outline of one form in a coordinate system and subsequently transforming the coordinate system in such a way that the outline would fit conveniently atop the outline of the other form. This idea of "outlines that are transformed" was later formalized by the definition of landmarks and became "landmarks that are transformed." Bookstein (1993) has summarized some of the several attempts to provide this idea with a suitable mathematical and statistical framework. Geometric morphometrics uses this landmark-based approach to compute numerical components of form change throughout a sample. Form change can be separated into two components. For instance, of the several ways to map two forms conveniently idealized by landmarks, the thin-plate spline method decomposes change into two incommensurate types: uniform and nonuniform. This method yields both components separately, providing the possibility of using a numeric value for each one: the uniform component as anisotropy (log ratio of principal strains from the affine matrix) and the nonuniform component as an energy (bending energy); (Bookstein, 1990).

Affine change, which is the mathematical name for a uniform change, is defined as a smooth transformation for which parallel lines remain parallel and proportions along a line remain the same after transformation (see, for example, Goodall and Green, 1986). Other features of the affine transformations, the different groups, and the invariant properties of their induced geometry may be found in Preparata and Samos (1985).

We propose the use of affine transformation as a model that simulates forms, rather than using it only as a component of deformation in a sample (Fig. 1). The affine model is used here as a descriptive model of change that might induce the search for new explanations of form change. These new questions would be based on the constraints imposed by the affine change, that is, their invariants. In the following section we shall introduce the technique to elaborate affine morphospaces (ordered collections of affine-transformed images of a base form), as well as provide a brief description of the program that generates these morphospaces along with an example of such a procedure. Also, we will introduce a strategy for the superimposition of forms, without any automatic optimality criteria.

BRIEF DESCRIPTION OF D'ARCYGRAPH

D'ARCYGRAPH is a program for PC-DOS compatible computers. Its aim is to perform different morphometric tasks. As a tool directly related to the generation of forms (1) from a base form, it generates uniform deformations leading to a transformed image of the original form (Fig. 1) or to a space of simulated forms (morphospace, see Fig. 3) and (2) it generates new forms by moving one landmark at a time, that is, by making local movements in specific landmarks, which results in a non-uniform transformation of the base form. As a morphometric tool (3) it superimposes two forms with equal numbers of digitized points and allows interactive scaling, translating and rotation of two forms (one at a time). This allows

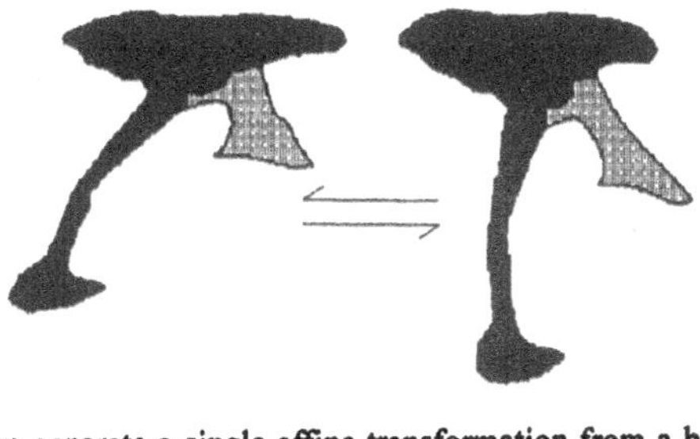

Figure 1. D'ARCYGRAPH can generate a single affine transformation from a base form. Here, the pelvic girdle of the theropod dinosaur, *Deinonychus* (right), is transformed. The transformation is biunivocal, as occurs in any affine transformation. The disposition, i.e., the orientation of the elements of the pelvis, has changed (see "Discussion").

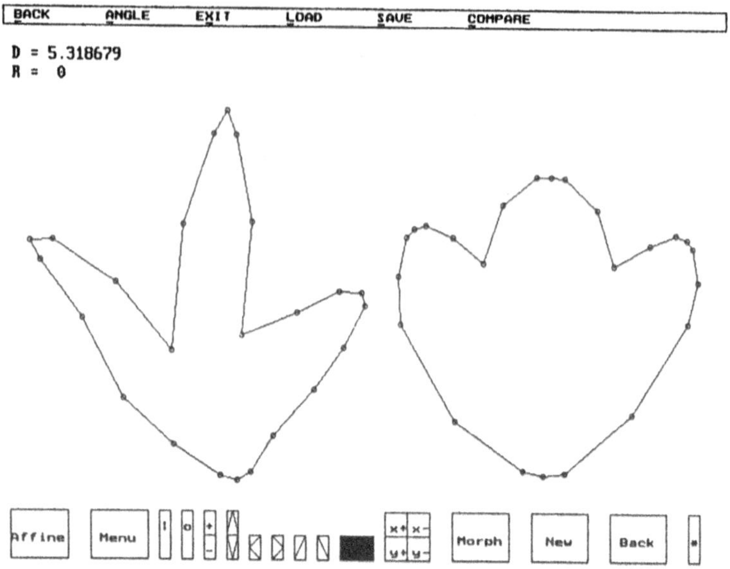

Figure 2. D'ARCYGRAPH's work screen, showing two different outlines in "Compare" mode. The outlines are dinosaur tridactyl footprints: theropod (left) and ornithopod, from the Spanish fossil record (Rasskin-Gutman et al., 1993). On the bottom of the screen are shown the options for manipulation of a second form over the base one. This manipulation is done interactively, by directly clicking with a mouse pointer. From left to right, a quick description of the most important options: affine box—allows a direct application of the **T** matrix over the base form (only on when not in "Compare" mode); manipulation buttons, from left to right—allow view of vectors that join homologous landmarks in "Compare" mode, view of the centroid and lines that join the landmarks with it, isometric scaling (changes the τ parameter), translation (changes ε and δ parameters), and rotation (changes ϕ parameter); the $x+$, $x-$, $y+$, $y-$ buttons—allow stretching in the x- and y-axes, respectively; morph box—generates a morphospace as shown in Fig. 3; and asterisk box—allows the user to choose a particular landmark and move it independently from the rest of landmarks (see text for an explanation of the different parameters).

different fits to be explored, which can be further added to an affine transformation or to local movements of specific landmarks. An estimate of the mean square distance between homologous digitized points (D) shows continuously as a similarity index of these two forms (Fig. 2). In addition, (4) it acts as a morphological visualizer, making isometric scalings by changing the baseline of the form under study as it is superimposed over a base form (this could be a consensus form, an ancestral form, an early ontogenetic stage or, simply, another form—the same form, as a matter of fact), highlighting interesting differences or similarities that might provide insights about how to handle the data or even point out where to locate landmarks.

Affine Morphospaces

The generation of simulated morphospaces is performed as follows: Let $\mathbf{C}$ be the $n \times 2$ matrix of the x,y coordinates of the n digitized points of the initial morphology that will be used as the base form for generating the simulated morphospace. Let $\mathbf{R}$ be the 2×2 rotation matrix and $\mathbf{T}$ be the 2×2 transformation matrix. Then the matrix that maps the x,y coordinates of $\mathbf{C}'$ onto $\mathbf{C}$ is $\mathbf{A} = \mathbf{R}^t\mathbf{T}$, and the final affine transformation of $\mathbf{C}'$ is $\mathbf{C}' = \mathbf{C}\mathbf{R}^t\mathbf{T}$. The matritial equation is:

$$\begin{bmatrix} x'_1 & y'_1 \\ x'_2 & y'_2 \\ \vdots & \vdots \\ x'_n & y'_n \end{bmatrix} = \left(\begin{bmatrix} x_1 & y_1 \\ x_2 & y_2 \\ \vdots & \vdots \\ x_n & y_n \end{bmatrix} \begin{bmatrix} \cos\phi & \sin\phi \\ -\sin\phi & \cos\phi \end{bmatrix} \tau + \begin{bmatrix} \varepsilon & \delta \\ \varepsilon & \delta \\ & \\ \varepsilon & \delta \end{bmatrix} \right) \begin{bmatrix} a & b \\ c & d \end{bmatrix} \quad (1)$$

where τ is the zoom term and ε, δ form an extra $n \times 2$ matrix of translation terms. The components of **T** can be defined algebraically as long as the determinant of **A** is not null. The equation employed for generating the morphospace is, in algebraic notation, as follows: $x' = x - y \tan(\alpha)$; $y' = y$. Then,

$$\mathbf{T} = \begin{bmatrix} 1 & 0 \\ -\tan\alpha & 1 \end{bmatrix} \quad (2)$$

The program automatically computes each affine image for generating the morphospace, applying the matrix **T** successively, and changing the orientation of the base form and the parameter angle of matrix **T**. The procedure is similar to that shown in Bookstein (1977; page 203, Fig. 11), although the stretchings were performed there in a direction that was parallel to the abscissa, as the form was rotating. With our program the generation process of the morphospace involves oblique stretchings, even though orthogonal stretchings might be added manually (see below). As discussed in Bookstein (1977), this procedure (rotation plus stretching) induces a complex change of the base form; however, this complex changc is easily explained by the affine matrix and can be described by the principal axes of deformation. (Bookstein, 1977, 1991).

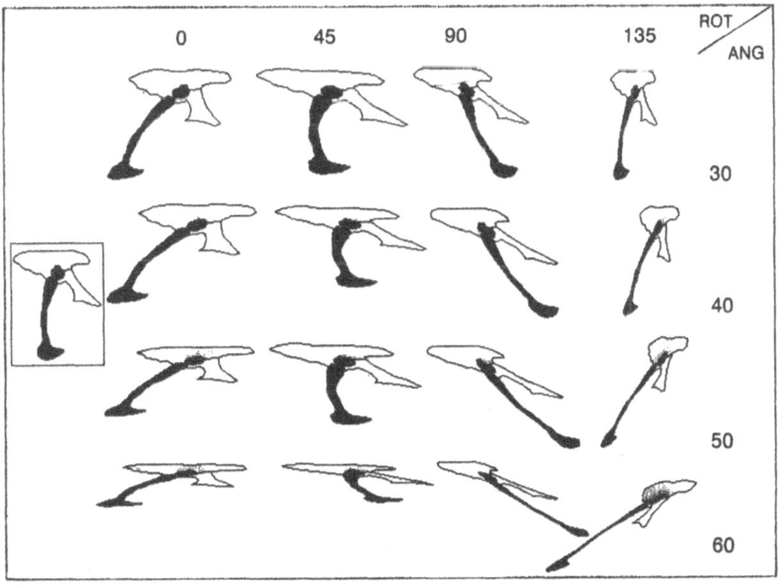

Figure 3. Generation of a morphospace with the affine model using D'ARCYGRAPH. The base form (*Deinonychus* pelvic girdle) is shown in the box. See the meaning of parameters α, ϕ (ANG, ROT) in the text (note that ϕ measures 0° from the x-axis, measuring counterclockwise). The series that simulates the dispositional trend of the dinosaur–avian transition is the column of 90° of ϕ (from Rasskin-Gutman and Buscalioni, 1992); see Fig. 5 for comparison.

As described above, the morphospace depends upon two parameters, ϕ and α (Fig. 3). The scaling and translating parameters vary conveniently as the needs of the morphospace vary. Also, it is possible to generate an independent stretching in either axis, because the scaling parameter is subdivided into an x component and a y component (not shown in the equation above, see Fig. 2). A singular value decomposition of the **A** matrix could be computed easily—although it is not an option of the program – leading to the canonical and the strain cross parameters of the affine matrix (Goodall and Green, 1986). This latter computation may be useful if an index of maximum and minimum change among the simulated forms needs to be used. To illustrate an affine morphospace we have generated 16 representative affine images (Fig. 3); we have used as a base form the pelvic girdle of the theropod dinosaur *Deinonychus antirrhopus*, modified and digitized from Ostrom (1976; see Fig. 3, box). These 16 images are ordered according to the generation process (rotation plus oblique stretching, illustrated in the figure by the entries ROT and ANGLE, respectively). Note that the affine morphospace is actually a continuum; α might vary continuously from 0 to 89°, while ϕ might vary continuously from 0 to 179°. Generating simulated forms does not require a careful definition or election of adequate homologous landmarks as would be necessary in a superimposition procedure, because no morphological distance or measure will be used in the evaluation of the results. However, the generated forms will strongly depend upon those digitized points. At the same time, because it is possible to perform statistical analysis with the simulated forms (Rasskin-Gutman et al., 1993), it would be better to define homologous landmarks. Finally, selection of an adequate base form is not arbitrary; it depends upon the framework of the problem being considered (see "Discussion").

Free Comparisons

When comparing two sets of landmarks, the different superimposition procedures use fitting algorithms that compute ϕ of the **R** matrix, τ, ε and δ, as are computed for the least-squares and resistant fit algorithms (Benson et al., 1982; Siegel and Benson, 1982; Chapman, 1990). Rohlf and Slice (1990) have developed an algorithm that also computes the parameters of the **T** matrix, a, b, c, and d. In these comparison methods, no consideration about the "biological" orientation of forms being analyzed is taken into account. Furthermore, the above-mentioned algorithms try to look for an arbitrary best fit that is considered to be the highest similarity between forms being analyzed (Rohlf and Marcus, 1993). With D'ARCYGRAPH, this algorithmic arbitrariness is bypassed; it is replaced by the biological criteria of the user who decides, for instance, which part of the form should be more conservative in the superimposing process (Fig. 4). To illustrate the use of the program for comparison of landmark-based forms without an automatic optimality criterion, we show two images of the work screen in "Compare" mode. The strategy is to play interactively with both forms, trying different superimpositions. This interaction—rotating, scaling, translating, stretching and moving specific landmarks separately in real time—might help to identify conservative and nonconservative areas or the behavior of the whole form when one region remains conservative. These identifications could result in very different conclusions regarding form change. In this example, three different fits for two dinosaur footprints are compared (Fig. 4). When the fit is based on a conservative area, the heel of the track (Fig. 4, left), changes in scale are dramatic - the most striking difference is the length of the digits, while the hypex maintains the same position in both footprints. If the fit is based on the election of a baseline, the extremal landmarks of the central digit (Fig. 4, center), the highest difference appears in the width of the digits and almost every landmark is moved. Finally, when the least-squares criteria is used (Fig. 4, right), differences are scattered all over the form. Which fit bears more biological information?

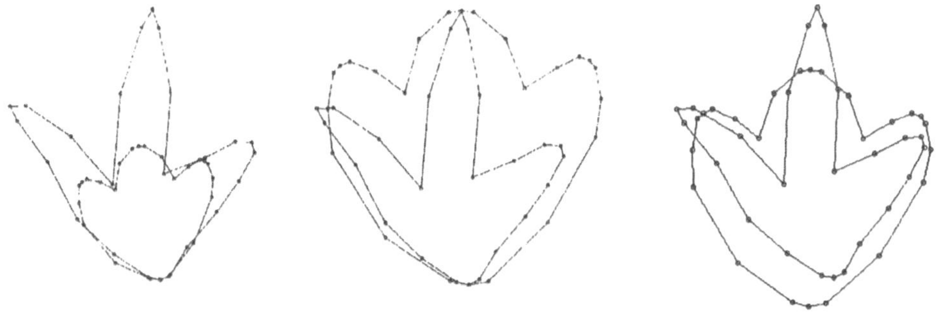

D = 1.610484 **D = 0.990411** **D = 0.659545**

Figure 4. Three different fits according to three different criteria. On the left, the fit is based on the proximal portion of the heel as a conservative area. In the center, the fit is based on the baseline defined by the extremal landmarks of the central digit. On the right, the fit is based on the least-squares optimality criteria. It is necessary to modify the τ, ϕ, ε and δ parameters in order to obtain these fits. Morphological distances are different for each fit, showing the dependence of this distance (see text).

DISCUSSION

The possibilities of use of D'ARCYGRAPH are focused on problems in which form plays a significant role. Playing around with forms—rotating, scaling, translating, stretching or moving landmarks separately—allows the recognition of features that might be employed for making different assessments and interpretations about the form being considered. Although no statistics are used in our approaches, saving any transformed form as a file of *x,y* coordinates allows those coordinates to be used as entries for quantitative evaluations of the transformations. The comparisons without automatic optimality criteria may help to highlight different zones of conservative and nonconservative behavior of the form that is being studied.

The model of affine change, as presented and used here, might indicate different interpretations about pattern variation. As noted by Tobler (1978), parsimony suggests that the affine model should be invoked for initial attempts to impute numerical values to influences that result in changes of form, even though biological materials would usually change in a nonuniform manner. Then, affine change might be used to seek general patterns of variation or, as indicated below, to envisage new ways of describing and evaluating form change. Thus, it is necessary to define variables that are related to the invariant properties of affine change to get useful insights from the results provided by the model.

When generating morphospaces with the affine model, the first aspect to be considered is which problem is going to be assessed. This issue will indicate the base form that will be used to build the morphospace. In the present example, *Deinonychus* belongs to the Deinonychosauria clade, which up to now has been considered the closest sister group of the major clade Aves (Novas, 1992). Thus, the selection of the base form is grounded on the consideration of "ancestral" form. Once the base form has been chosen and the morphospace generated, a comparison with the actual forms would indicate which variables should be selected to describe the whole spectrum of variation of the morphospace. These variables characterize the variation at a morphological level that we denominate "dispositional." Providing that form has an implicit idea of geometric arrangement of its elements; the disposition of the form "pelvic girdle" of dinosaurs might be characterized as the relative orientation of the elements that compose it, i.e., ilium, pubis, ischium and acetabulum. In lateral view, which is appropriate for evaluating patterns of variation of pelvic girdles, we consider that the disposition of the pelvis of all dinosaurs should be defined as follows: ilium anteroposterior and dorsal, pubis ventral and anterior, ischium ventral and posterior. and acetabulum bordered by all three rods. With these definitions, the position of the acetabulum and the orientation of the pubic and ischiatic rods with respect to the anteroposterior axis defined by the ilium are the variables that characterize the whole spectrum of variation of dinosaur pelvis. Thus, the disposition of the pelvic girdle of *Deinonychus* would be as follows: ilium centered with respect to the acetabulum, pubis vertical, and ischium retroverted.

When using the disposition of the pelvic girdle of *Deinonychus* as a set of initial conditions to generate an affine morphospace, there are just two options of change for the dispositional variable defined for the pubic rod. The vertical position of the ancestral condition allows only either an anteroversion or a retroversion when the ilium remains horizontal. These changes of state of the variable, "pubis orientation" are perfectly simulated by the affine model (Fig. 3). When $\phi = 45$ and $\phi = 90$, the simulated trend can account for the change of the disposition of the elements that takes place in the transition Dinosauria–Aves. Thus, as the parameter "angle" increases along both series, the ilium is progressively anterior with respect to the position of the acetabulum, while the pubis and ischium are progressively more retroverted. Note that in the other dispositional trends, the arrangement of the elements takes the opposite direction. For $\phi = 0$, the ilium remains without change, while both pubis and ischium are anteroverted; for $\phi = 135$, the ilium is progressively posterior with respect to the position of the acetabulum, while the pubis and ischium are anteroverted.

In conclusion, the affine model has generated a dispositional trend that might be used as a descriptive model of changes in disposition which occur in the fossil record of the transition Dinosauria–Aves, because it simulates the basal synapomorphy for the disposition of the pelvic girdle proposed for the clade Aves: "moderately retroverted pubis" (Fig. 5; Cracraft, 1988). The possibility of using the affine change as a descriptive model suggests that the mathematical properties of the affine change might be used as possible explanations of change in the above-defined concept of "disposition" of biological forms. Thus, for

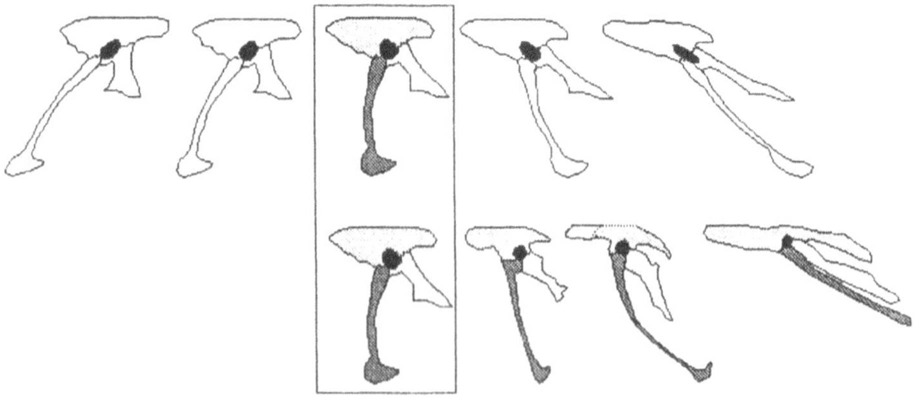

Figure 5. (Upper) Change of disposition, from an anteroverted to a retroverted state, of the pelvic girdle of *Deinonychus* (center box). (Lower) Sketch of the pelvic girdles of Mesozoic birds. Lower row, from left to right: *Archaeopteryx*, based on the proportions of the London specimen (sensu Wellnhofer, 1985); the Chinese bird, *Sinornis*, drawn from a cast of the specimen based on Sereno and Rao (1992)—the preacetabular area of the iliac blade (in white) has been reconstructed following the prediction of the affine model; the toothed bird, *Ichthyornis*, modified from Carroll (1988).

example, as all changes are precisely correlated, even though they follow different directions and extents, the model gives a description of change in disposition, suggesting a new view about the whole transition: a transition in which the disposition of all the elements changes in a correlated manner. This correlation might be employed further to provide new interpretations of form change. For instance, the model suggests the interpretation of the retroversion of the pubis as well as the fusion of the pubis and ischium as a correlated consequence of the anterior expansion of the ilium. This hypothesis would be an alternative to the traditional view, in which the retroversion of the pubic rod is explained either in terms of a functional requirement for the passage of the eggs or as a nonfunctional change (Gatesy, 1990). At the same time, the posterior expansion of the iliac blade would be interpreted as a de novo construction.

When used for the comparison of forms without an automatic optimality criterion, D'ARCYGRAPH assesses the superimposition of related forms from a morphodynamic perspective (Seilacher, 1991). The user is free to superimpose forms according to the researcher's criteria. In our example (Fig. 4), all superimpositions, based on different criteria, lead to different biological conclusions. These conclusions obey distinct a priori concepts or ideas underlying the comparison procedure and are motivated by different biological criteria. For instance, the fit based on a conservative area stresses a possible morphofunctional meaning of that area, that is, the tridactyl nature of these tracks shows that the plantar areas must be equivalent and, hence, convergent between theropod and ornithopods. On the other hand, when the fit is based on a baseline, the extremal landmarks of the central digit, the questions that we may ask are related to proportions and dimensions: how wide may the theropod track be while having the same length as the ornithopod ichnite? In the third fit, using the least-square algorithm for the procedure, we have the graphical representation of an independent similarity criterion useful for taxonomic purposes because an overall change is shown.

ACKNOWLEDGMENTS

The authors wish to express their thanks to Drs. Richard H. Benson, Ralph E. Chapman, Nieves López Martínez, and José Luis Sanz for stimulating valuable discussions and criticisms. We would also like to thank the Department of Paleobiology and the Automatic Data Processing Laboratory of The National Museum of Natural History at the Smithsonian Institution in Washington, D. C., for allowing the use of its facilities. This paper was supported with funds provided by the European Union, number CHRXCT930164, and D.G.I.C.Y.T., project APC930002.

REFERENCES

Benson, R. H. 1982. Deformation, Da Vinci's concept of form, and the analysis of events in evolutionary history. Pages 241–277 *in* E. M. Gallitelli, (ed.), Paleontology, essentials of historical geology. S.T.E.M. Mucchi: Modena, Italy.

Benson, R. H., R. E. Chapman, and A. F. Siegel. 1982. On the measurement of morphology and its change. Paleobiology 8(4):328–339.

Bookstein, F. L. 1977. The study of shape transformation after D'Arcy Thompson. Mathematical Biosciences 34:177–219.

Bookstein, F. L. 1990. Higher-order features of shape change for landmark data. Pages 237–250 *in* F. J. Rohlf. and F. L. Bookstein, (eds.), Proceedings of the Michigan morphometrics workshop. University of Michigan Museum of Zoology Special Publication 2.

Bookstein, F. L. 1991. Morphometric tools for landmark data: Geometry and biology. Cambridge University Press: Cambridge.

Bookstein, F. L. 1993. A brief history of the morphometric synthesis. Pages 15–40 *in* L. F. Marcus, E. Bello, and A. García-Valdecasas, (eds.), Contributions to morphometrics. Monografías del Museo Nacional de Ciencias Naturales 8, Madrid.

Carroll, R. L. 1988. Vertebrate paleontology and evolution. Freeman and Company: San Francisco.

Cracraft, J. 1988. The major clades of birds. Pages 339–362 *in* M. J. Benton, (ed.), The phylogeny and classification of the tetrapods, Volume 1. Clarendon Press: Oxford.

Chapman, R. E. 1990. Conventional Procrustes approaches. Pages 251–267 *in* F. J. Rohlf. and F. L. Bookstein (eds.), Proceedings of the Michigan morphometrics workshop. University of Michigan Museum of Zoology Special Publication 2.

Gatesy, F. G. 1990. Caudophemoral musculature and the evolution of theropod locomotion. Paleobiology 16(2):170–186.

Goodall, C. R. and P. B. Green. 1986. Quantitative analysis of surface growth. Botanical Gazette 147(1):1–15.

Gordon, A. D. 1981. Classification. Chapman and Hall: London.

Novas, F. 1992. La evolución de los dinosaurios carnívoros. Pages 126–163 *in* J. L. Sanz and A. D. Buscalioni, (eds.), Los dinosaurios y su entorno biótico. Actas del Segundo Curso de Paleontoloía en Cuenca, Instituto "Juan de Valdés." Ayto. de Cuenca, Spain.

Ostrom, J. H. 1976. On a new specimen of the Lower Cretaceous theropod dinosaur *Deinonychus antirrhopus*. Breviora 439:1–21.

Preparata, F. P. and M. I. Samos. 1985. Computational Geometry. An introduction. Springer-Verlag: New York.

Rasskin-Gutman, D. and A. D. Buscalioni. 1992. D'ARCYGRAPH. Primeros resultados: estudio de la deformación afín en la pelvis de *Deinonychus antirrhopus*. Pages 133–135 *in* VII Jornadas de Paleontología, Resúmenes, November, Barcelona, Spain.

Rasskin-Gutman, D., A. D. Buscalioni, and J. J. Moratalla. 1993. Theoretical morphospaces of tridactyl dinosaur footprints using D'ARCYGRAPH. Evolutionary implications. Page 364 *in* Evolution 93, IVth Congress of the European Society for Evolutionary Biology, 22–28 August, Montpellier, France.

Raup, D. M. 1966. Geometric analysis of shell coiling: General problems. Journal of Paleontology 40:1178–1190.

Raup, D. M. and S. M. Stanley. 1978. Principles of paleontology. W. H. Freeman and Company: San Francisco.

Reyment, R. A. 1991. Multidimensional palaeobiology. Pergamon Press: Oxford.

Rohlf, F. J., and D. E. Slice. 1990. Extensions of the Procrustes method for the optimal superimposition of landmarks. Systematic Zoology 39:40–59.

Rohlf, F. J., and L. F. Marcus. 1993. A revolution in morphometrics. Trends in Ecology and Evolution 8(4):129–132.

Scott-Ram, N. R. 1990. Transformed cladistics, taxonomy and evolution. Cambridge University Press: Cambridge..

Seilacher, A. 1991. Self-organizing mechanisms in morphogenesis and evolution. Pages. 251–271 *in* N. Schmidt-Kittler, and K. Vogel (eds.), Constructional morphology and evolution, Spinger-Verlag: New York.

Sereno, P. C., and C. H. Rao. 1992. Early evolution of avian flight and perching: New evidence from the Lower Cretaceous of China. Science 255:845–848.

Siegel, A. F., and R. H. Benson. 1982. A robust comparison of biological shapes. Biometrics 38:341–350.

Sneath, P. H. A. and R. R. Sokal. 1973. Numerical taxonomy. Freeman and Company: San Francisco.

Thompson, D. W. 1942. On growth and Form. Cambridge University Press: London. (From the Unabridged Dover (1992) republication of the complete revised edition of 1942. First edition, 1917.)

Tobler, W. R. 1978. Comparison of plane forms. Geographical Analysis 10(2):154–162.

Wellnhofer, P. 1985. Remarks on the digit and pubis problems of *Archaeopteryx*. Pages 113–122 *in* M. K. Hecht et al. (eds.), The Beginnings of birds. Proceedings of the International *Archaeopteryx* Conference, Eichstätt, 1984.

THREE-DIMENSIONAL GENERALIZED RESISTANT FITTING AND THE COMPARISON OF LEAST-SQUARES AND RESISTANT-FIT RESIDUALS

Dennis E. Slice*

Department of Ecology and Evolution
State University of New York
Stony Brook, New York 11794

ABSTRACT

The resistant-fit methods of superimposing configurations of landmarks are extended to three-dimensional data. Algorithms are described for fitting one configuration to another and for the fitting of a sample of configurations to a iteratively estimated consensus configuration. In addition, several topics relating to the practical use of the superimposition methods are discussed. These topics include missing data, visualization of results and the display and interpretation of affine components. Finally, an example is presented of how to use different graphical comparisons of resistant-fit and least-squares (Procrustes) residuals to identify subsets of landmarks contributing to localized shape differences. These procedures are shown to suggest considerable local variation in the posterior region of the carapace of certain individual yellow-bellied slider turtles, *Trachemys scripta scripta*.

INTRODUCTION

Much of the recent work in biological shape analysis has dealt with the extraction and analysis of shape information from the coordinates of points, or landmarks (Rohlf and Marcus, 1993)—part of what is now referred to as geometric morphometrics. Such data provide complete information about the shape of the configuration of landmarks including that information available in more traditional variables, such as interpoint distances, distance ratios or the angles of line-segment intersections. The least-squares and resistant-fit superimposition methods, frequently referred to as Procrustes techniques, have been developed

*E-mail: Dennis.Slice@sunysb.edu

Advances in Morphometrics, Edited by L. F. Marcus et al.
Plenum Press, New York, 1996

to provide the means to scale, rotate and translate sets of landmark coordinates into a common coordinate system in which shape differences are directly represented in differences between the homologous coordinates (Rohlf and Slice, 1990).

The least-squares procedures estimate the superimposition, or fitting, parameters to minimize the total sum-of-squared differences between homologous points on two or more configurations. The sum-of-squares minimization criterion on which these methods are based, however, may result in misleading representations of shape differences when one or a few of the landmarks manifest extreme displacements relative to the others in the configuration. The large contributions of the squared lengths of longer vectors to the fitting criterion result in the redistribution of larger local variation to other landmarks. The resistant-fit methods address this problem by using median-based parameter estimates that are less affected by the presence of larger differences in the relative locations of a few landmarks.

Both least-squares and resistant-fit methods have the desirable property that the relative physical arrangement of the data is retained throughout the superimposition process. This allows for the graphical display of individual configurations and samples after superimposition. Similarly, more abstract constructions, such as average, or consensus, configurations obtained from a generalized superimposition or hypothetical configurations based on a principal components, regression or other statistical analysis, can be drawn to provide intuitive and insightful presentations of these results. Such presentations are extremely important both for the interpretation of the results and as exploratory techniques to identify problematic data or patterns of variation worth more detailed examination.

Resistant-fit superimpositions are computationally expensive, and the time involved in their application can become prohibitive for samples involving larger numbers of landmarks or specimens. The least-squares superimposition methods, on the other hand, are relatively fast, and when certain modeling assumptions are met (Dryden and Mardia, 1991; Goodall, 1991; Goodall and Mardia, 1991), their results provide an optimal starting point for a more detailed study of shape such as multivariate analysis of variance, principal components analysis, relative warp analysis (Bookstein, 1989, 1991) or complex principal components analysis (Goodall, 1992). In other situations the resistant-fit methods may provide better results in the sense of more accurately reflecting the relative differences in landmark locations between individuals and the reference. Because it is not known at the outset of an analysis which method will be the best approach, the detailed comparisons of the results of the two methods can be very useful. If little or no differences in the results of the two methods are observed, subsequent analyses can proceed using the faster least-squares procedures. On the other hand, noteworthy differences between the results can be examined to identify individual specimens that are in some way atypical with respect to the rest of the sample or to identify individual or sets of landmarks behaving in a way deserving of special consideration.

This chapter provides a summary of methods for the least-squares and resistant fitting of three-dimensional data and describes methods for contrasting differences between the results of the two procedures. Most of the results included here are from larger studies carried out by Slice (1993) and will be reported in more detail elsewhere. All of the methods discussed here are, unless otherwise indicated, implemented in the program GRF-ND (Slice, 1994), which can be obtained as described in Appendix II.

THREE-DIMENSIONAL SUPERIMPOSITION

Much of the work on superimposition methods for morphometric analysis has focused on two-dimensional data (Sneath, 1967; Siegel and Benson, 1982; Mardia and Dryden, 1989; Rohlf and Slice, 1990; Dryden and Mardia, 1991). This emphasis is partly the result of the availability of such data and the addressing of problems associated with specific data sets. As the equipment for collecting three-dimensional data becomes more readily available, these methods must be extended to accommodate the increased dimensionality. In most cases, the development of the two-dimensional techniques has been sufficiently general to allow easy extension to higher dimensions. In a few cases, however, different approaches are required for three- or higher-dimensional data.

The model and notation used in the following development of three-dimensional superimposition will be

$$\mathbf{X}' = \rho\mathbf{X}\mathbf{H} + \mathbf{1}\tau, \quad (1)$$

where $\mathbf{X}'$ is the result of scaling, rotating and translating $\mathbf{X}$ by ρ, $\mathbf{H}$, and τ, respectively. Here, $\mathbf{X}$ and $\mathbf{X}'$ are p x k matrices of the k coordinates of the p points in a configuration before and after fitting, ρ is a scalar, $\mathbf{H}$ is a k x k transformation matrix, $\mathbf{1}$ is an p x 1 matrix of ones, and τ is a 1 x k matrix of translation parameters. Estimates of all of these parameters are made with respect to a reference specimen $\mathbf{Y}$, another p x k matrix that might represent another specimen or a mean configuration, so that the coordinates of points in $\mathbf{X}'$ can be used to assess shape differences relative to $\mathbf{Y}$. The matrix $\mathbf{H}$ is often constrained to effect only rigid rotations of the configuration, and methods using this constraint are referred to as orthogonal superimpositions. Removing the orthogonality constraint allows the least-squares procedure to reduce further the squared differences between configurations by stretching or shrinking the fitted configuration in orthogonal directions. The results of resistant fitting without the constraint similarly provides an improved fit, and in both cases, such an approach, called affine fitting, achieves a partitioning of shape differences between configurations into a global, uniform shape component (Bookstein and Sampson, 1987) and a separate nonaffine part.

Least Squares

The development of procedures for the least-squares superimposition of two or more sets of landmarks has used general matrix formulations. As such, the extension to three-dimensional data requires only the addition of an extra column to the end of the data matrices and the appropriate adjustments to the dimensions of the other matrices. This generality in the development of the least-squares techniques is a result of much of this work (Mosier, 1939; Gower, 1970, 1975) being done in areas of research in which the dimensionality of the data does not have the special physical meaning that it has in geometric morphometrics. Bookstein (1991) provides a useful discussion of this distinction.

Ordinary Resistant Fitting

Like the least-squares methods, many of the techniques for two-object resistant-fit superimposition can be carried over to three-dimensions with only minor modifications. Estimates of the scale parameter in the two-dimensional case are simply the repeated medians of ratios of Euclidean distances between pairs of landmarks. The same procedure can be used for three-dimensional data. The estimate of the translation used in the resistant superimposition of one two-dimensional configuration onto another is based on the simple medians of differences in the coordinates of landmarks taken independently for each coordinate dimension. One need only apply this approach to the third coordinate dimension in three-dimensional data sets to produce an estimate of the final translation vectors. As in the two-dimensional case, this must be done after scaling and rotation. The only real problem in developing resistant-fit procedures for three-dimensional data concerns the estimation of the rotation matrix, $\mathbf{H}$.

The differences between the method for estimating the rotation matrix for two-dimensional data and that to be described for three-dimensional data are not the result of any fundamental differences in the structure of the data or in the nature of the superimposition. Instead, the two-dimensional case just allows for an alternative, somewhat simpler, approach. In two dimensions, the elements of a rotation matrix are all functions of a single angle, θ:

$$\mathbf{H} = \begin{bmatrix} \cos\theta & -\sin\theta \\ \sin\theta & \cos\theta \end{bmatrix} \tag{2}$$

Any matched pair of line segments on the two configurations can be used to estimate this angle directly and their repeated median can be used to construct an estimate of $\mathbf{H}$ based on equation (2). The rotation matrix in three dimensions, however, is a function of the sines and cosines of three different angles:

$$\mathbf{H} = \begin{bmatrix} \cos\alpha\cos\gamma + \sin\alpha\sin\beta\sin\gamma & \sin\alpha\cos\beta & \sin\alpha\sin\beta\cos\gamma - \cos\alpha\sin\gamma \\ \cos\alpha\sin\beta\sin\gamma - \sin\alpha\cos\gamma & \cos\alpha\cos\beta & \sin\alpha\sin\gamma + \cos\alpha\sin\beta\cos\gamma \\ \cos\beta\sin\gamma & -\sin\beta & \cos\beta\cos\gamma \end{bmatrix}, \tag{3}$$

where α, β, and γ are angles of rotation around the *z*-, *x*-, and *y*-axes, respectively. These angles cannot be estimated independently, and so the entire rotation matrix must be determined directly. Line segments are clearly insufficient to estimate this matrix, so a minimum of three points must be used.

One must also define what is meant by "alignment" in three dimensions. For two-dimensional data, it is sufficient to specify the desired alignment as being that which points the two line segments in the same direction. Defining alignment for three-dimensional data based on three landmarks is a bit more complicated. One useful definition, assuming an ordered triplet of points, is to require that the vector defined by the first and second points in the triplet point in the same direction on both configurations and that the planes defined by the three points in both configurations be parallel. Furthermore, to distinguish between 180° rotations of the fitted configuration, one can also specify that the third landmark must be on the "same side" of the first two points in both configurations.

A rotation matrix that achieves such an alignment based on three points can be determined relatively easily (Siegel and Pinkerton, 1982). First, construct column vectors $\mathbf{x}_i$, $\mathbf{x}_j$, $\mathbf{x}_k$ and $\mathbf{y}_i$, $\mathbf{y}_j$, $\mathbf{y}_k$, which contain the coordinates of the points *i*, *j*, *k*, from the data and reference matrices, $\mathbf{X}$ and $\mathbf{Y}$. Next, use these vectors to construct unique sets of orthogonal vectors, $\mathbf{x}_{ij}$, $\mathbf{x}_{ijk}$, $\mathbf{x}_{ijk*}$ and $\mathbf{y}_{ij}$, $\mathbf{y}_{ijk}$, $\mathbf{y}_{ijk*}$ defined by the following equations:

$$\mathbf{x}_{ij} = \frac{(\mathbf{x}_j - \mathbf{x}_i)}{\|\mathbf{x}_j - \mathbf{x}_i\|} \qquad \mathbf{y}_{ij} = \frac{(\mathbf{y}_j - \mathbf{y}_i)}{\|\mathbf{y}_j - \mathbf{y}_i\|}$$

$$\mathbf{x}_{ijk} = \frac{(\mathbf{x}_k - \mathbf{x}_i) - [(\mathbf{x}_k - \mathbf{x}_i)\cdot\mathbf{x}_{ij}]\,\mathbf{x}_{ij}}{\|(\mathbf{x}_k - \mathbf{x}_i) - [(\mathbf{x}_k - \mathbf{x}_i)\cdot\mathbf{x}_{ij}]\,\mathbf{x}_{ij}\|},$$

$$\mathbf{y}_{ijk} = \frac{(\mathbf{y}_k - \mathbf{y}_i) - [(\mathbf{y}_k - \mathbf{y}_i)\cdot\mathbf{y}_{ij}]\,\mathbf{y}_{ij}}{\|(\mathbf{y}_k - \mathbf{y}_i) - [(\mathbf{y}_k - \mathbf{y}_i)\cdot\mathbf{y}_{ij}]\,\mathbf{y}_{ij}\|}$$

$$\mathbf{x}_{ij*} = \mathbf{x}_{ij} \times \mathbf{x}_{ijk}\text{, and}$$

$$\mathbf{y}_{ijk*} = \mathbf{y}_{ij} \times \mathbf{y}_{ijk}, \tag{4}$$

where × denotes the cross product, · denotes the inner product and $\|\,\|$ the vector norm (= length). The reader may recognize the first four of these equations as those of the Gram-Schmidt construction of orthogonal basis vectors. The final equations, those involving the cross product, are three-space shortcuts used to construct a unit length vector orthogonal to the first two and defining a "right-handed" three-dimensional coordinate system (Anton, 1984). These equations, of course, require that neither triplet of points be collinear.

The above vectors are used to construct the matrices $\mathbf{X}^*_{ijk} = (\mathbf{x}_{ij}, \mathbf{x}_{ijk}, \mathbf{x}_{ijk*})$, and $\mathbf{Y}^*_{ijk} = (\mathbf{y}_{ij}, \mathbf{y}_{ijk}, \mathbf{y}_{ijk*})$. Because both $\mathbf{X}^*_{ijk}$ and $\mathbf{Y}^*_{ijk}$ are sets of orthonormal basis vectors, $\mathbf{H}_{ijk}$, where $\mathbf{X}^*_{ijk}\,\mathbf{H}_{ijk} = \mathbf{Y}^*_{ijk}$, is the desired orthonormal rotation matrix. Premultiplication of both sides of the equation by $\mathbf{X}^{*t}_{ijk}$ gives the solution $\mathbf{H}_{ijk} = \mathbf{X}^{*t}_{ijk}\,\mathbf{Y}^*_{ijk}$.

Once $\mathbf{H}_{ijk}$ has been determined for a triplet of landmarks, one can use the median of all such estimates as the robust estimate of the rotation required to superimpose **X** onto **Y**. There are three levels over which the median can be taken so

$$\hat{\mathbf{H}} = \underset{i=1..p}{\text{med}}\left(\underset{j=1..p,j\neq i}{\text{med}}\left(\underset{k=1..p,k\neq i,j}{\text{med}}(\mathbf{H}_{ijk})\right)\right) \tag{5}$$

The above equation brings up the question, what is the median of a set of matrices? A general answer is not immediately apparent. One could use as each element of the median matrix the median value of that element across all matrices. This procedure is problematic because the elementwise median of orthonormal matrices need not be orthonormal. Siegel and Pinkerton (1982) suggest a solution that exploits the duality between orthonormal and skew-symmetric matrices.

A skew-symmetric matrix is one that satisfies the relationship $\mathbf{S} = -\mathbf{S}^t$. This definition requires that the matrix has zeros on the diagonal and elements such that $s_{ij} = -s_{ji}$. The elementwise median of a set of skew-symmetric matrices is a skew-symmetric matrix. Any special orthogonal matrix, $\mathbf{H}_{ijk}$, is uniquely associated with a skew-symmetric matrix, $\mathbf{S}_{ijk}$, through the relationship

$$\mathbf{S}_{ijk} = (\mathbf{H}_{ijk} + \mathbf{I})^{-1}(\mathbf{H}_{ijk} - \mathbf{I}), \tag{6}$$

where **I** is the identity matrix (Eves, 1966). Each $\mathbf{H}_{ijk}$ can be converted to $\mathbf{S}_{ijk}$, and the elementwise medians taken across levels i, j and k to obtain $\hat{\mathbf{S}}$. The required rotation matrix can then be determined using the inverse relationship

$$\hat{\mathbf{H}} = (\mathbf{I} + \hat{\mathbf{S}})(\mathbf{I} - \hat{\mathbf{S}})^{-1}. \tag{7}$$

For two-dimensional orthonormal rotation matrices this gives very nearly the same results as the elementwise median of the associated 2 x 2 rotation matrices, which does produce orthonormal estimates, for small variation. Differences between the results of these methods and from those using the repeated median of θ described previously are due only to the nonlinearity of the relationship between angles and their sines or cosines, which tends to be small over the range of angles usually encountered.

Generalized Resistant-Fit

For fitting in n objects in three-dimensions one need incorporate only the two-object, three-dimensional resistant estimates of the orthonormal rotation parameters with the iterative algorithm for generalized resistant fitting used for the two-dimensional case (Rohlf and Slice, 1990):

1. As in two dimensions, start with a least-squares fit of all the objects to some initial estimate of the consensus configuration. This achieves an approximate alignment that generally reduces the range of variability in rotation estimates. All configurations are then normalized to unit median size, and a new consensus configuration estimated.
2. For each configuration, repeated-median scale factors are estimated relative to the current consensus and applied to the configuration as for two-dimensional data.
3. For each point pair, $ij_{j \neq i}$, the orthonormal rotation matrices, $\mathbf{H}_{ijk}$, are determined using each of the other $p-2$ landmarks based on the construction described above.
4. The $\mathbf{H}_{ijk}$ matrices are transformed into their skew-symmetric counterparts by means of equation (6), and the elementwise median matrix, $\hat{\mathbf{S}}_{ij}$, determined for each pair of landmarks. For each i, the median skew-symmetric matrix, $\hat{\mathbf{S}}_i$, is calculated as the elementwise median of the $p-1$ matrices, $\hat{\mathbf{S}}_{ij}$. The overall median $\hat{\mathbf{S}}$ is determined using the p matrices, $\hat{\mathbf{S}}_i$.
5. The robust rotation estimate, $\hat{\mathbf{H}}$, is obtained from $\hat{\mathbf{S}}$ by means of equation (7).
6. Finally, translation for each configuration is estimated as the simple coordinatewise median of the differences in landmark locations between the configuration and the consensus *after* the estimated scale and rotation transformations have been applied.
7. After the individual parameter estimates are used to transform each configuration, a new consensus configuration is estimated and renormalized, and the appropriate measure of fit is calculated for each specimen and summed across the entire sample. Convergence is tested, and if judged satisfactory, the iterations are stopped. Otherwise, the process is repeated starting at step 2 and using the new consensus.

As with two-dimensional data, there is no guarantee that the results will converge or even monotonically decrease. However, experience with a variety of three-dimensional configurations has shown this procedure almost always converges, and even small samples generally stabilize after a few iterations. As a safeguard against ill-behaved data, one can limit the maximum number of iterations allowed. Convergence is, of course, a function of the critical value and statistic used to assess the change in the results.

The potentially important differences between the results of three-dimensional resistant and least-squares fitting are illustrated in Fig. 1. Here one specimen with eight landmarks was fit to a reference cube. The specimen was created by displacing two vertices of a cube and then scaling, translating and rotating the resulting configurations (top panel). The lower left plot shows how minimizing the least-squares criterion produces a general lack of fit at all vertices. The lower right plot shows a "correct" superimposition due to resistant fitting; the differences in the shape of the two configurations are isolated at the two vertices. For real data, additional information would be needed to ascertain whether or not the more parsimonious result of resistant fitting was more biologically reasonable than the result from the least-squares method.

Three-Dimensional Affine Resistant Fitting

Releasing the three-dimensional resistant-fit rotation matrices from the orthogonality constraint reduces the complexity of the computations in some ways while increasing it in others. The Gram-Schmidt orthonormalization steps are no longer required, and the transformation of the estimated rotation matrices to skew-symmetric form becomes unnecessary. Unfortunately, three points are not sufficient to estimate an affine transformation matrix—four are required.

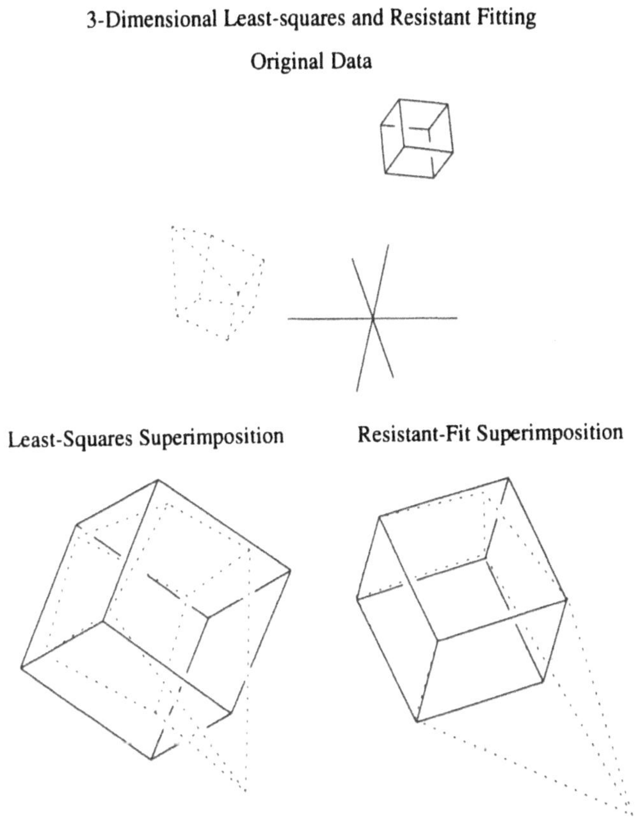

Figure 1. The top of the figure shows the original data to be fitted. Solid lines show the edges of a cubic configuration. Dotted lines are the edges of a configuration that is a cube within which two vertices have been displaced. The lower part of the figure shows the least-squares (left) and resistant-fit (right) superimposition of one onto the other. Note how the least-squares results indicate a general lack of fit at all vertices whereas the resistant-fit results correctly, in this case, isolates the differences (to the extent allowed by the precision of the original data) to the two perturbed vertices.

Each set of four, mean-centered landmarks, subscripted i, j, k, l in configuration $\mathbf{X}$, can be perfectly fit to the homologous set of similarly centered landmarks on configuration $\mathbf{Y}$ (actually, to any four mean-centered points) by the transformation matrix,

$$\mathbf{H}_{ijkl} = (\mathbf{X}'_{ijkl}\mathbf{X}_{ijkl})^{-1}(\mathbf{X}'_{ijkl}\mathbf{Y}_{ijkl}) \,. \tag{8}$$

If the set of points from configuration $\mathbf{X}$ is not of full rank, the inverse in the above equation is undefined. If the set from configuration $\mathbf{Y}$ is not of full rank, the transformation matrix collapses the set of points from $\mathbf{X}$ by one or more dimensions. Generally, neither is desirable, and in practice (e.g., GRF-ND) such results can be eliminated from consideration.

Medians of these matrices can be constructed directly by determining the univariate medians of individual matrix elements across the subsets of landmarks. This method avoids the computation of skew-symmetric matrices required by the orthogonal superimposition, but these computational savings are more than outweighed by the increase in the number of matrices that can be examined. For configurations of p landmarks in three dimensions, orthonormal fitting requires the estimation of $p(p-1)(p-2)$ rotation matrices. A complete affine resistant fitting, however, involves the estimation of $(p-3)$ times as many transformation matrices. In the case of the turtles discussed later, the orthogonal fitting of a single specimen estimates 388,944 orthonormal matrices whereas affine fitting requires the estimation of 27,615,024 matrices.

No specific scaling step is necessary because scale changes are incorporated into the affine transformations. The overall median affine transformation is then applied to the entire configuration, and the translation vector is determined just as in the orthogonal case.

For generalized fitting, consensus estimation and tests of convergence are the same as before. The modifications to specific steps outlined for the orthogonal fitting of three-dimensional configurations to allow for affine transformations are as follows:

Step 2. No scaling estimates are computed.

Step 3. Estimates of the rotation matrix now must involve quadruplets of landmarks, $\mathbf{H}_{ijkl}$, but these matrices can be determined directly without resorting to the Gram-Schmidt procedure.

Steps 4–5. The skew-symmetric transformation is not needed and the estimate, $\hat{\mathbf{H}}$, can be obtained directly as the repeated, elementwise median of the individual matrix estimates.

MISCELLANEA

It is worthwhile to discuss a few of the topics of immediate importance to someone seeking to use the superimposition methods before going on to the discussion of the comparison of least-squares and resistant-fit results. These topics include the problem of missing data, the visualization of data and results, which are key components of geometric morphometrics, and the interpretation and visualization of affine parameter estimates.

Missing Data

Of little interest to pure mathematicians but of potentially great importance to biologists, paleontologists or others dealing with applied problems is the problem of missing data. Despite the elegance of the algebraic formulations upon which the superimposition methods are based, they have no real provision for "blank spots" in the data matrices. Such occurrences can cause problems both in the mechanics of the superimposition and in subsequent statistical analyses.

The simplest way of addressing the problem of missing data in any analysis is to ensure that it doesn't arise at all. If data are plentiful, one can limit the collection or analysis of data to only those specimens for which complete information is available. This solution will not always be acceptable, especially in cases in which those individuals that would be excluded are the ones most interesting for the purposes of analysis or in which data may be extremely rare or difficult to obtain. Such filtering might also remove some of the natural and biologically relevant variation from the sample and produce misleading conclusions about shape differences and variation within and between populations.

The approach currently implemented in GRF-ND is to use the most information that is available. If a specimen has missing data values, it is fit to the reference using the coordinates of just those landmarks for which complete data are available. The same is true when landmarks are missing in the reference specimen. In the case of the iterative techniques, though, after the first iteration the estimates of the coordinates of the missing consensus landmarks are based on the subset of the sample for which that data is available. Similarly, the construction of landmark and configuration covariance matrices are based on the maximum set of available data with denominators adjusted accordingly.

The results of any of these superimposition methods can be treated as a multivariate data set and subjected to the usual analytical techniques available for such data. Such analyses can be carried out by using one of the commonly available statistical packages. These packages, however, may be less tolerant of missing data, and there are several approaches that might prove useful. The simplest of these is simply to substitute the mean or median values for the missing data. A more complicated, but possibly more satisfactory, solution is to estimate the coordinates of the missing data based on those of the available landmarks. This could involve a multivariate multiple regression of the form

$$\mathbf{x}_m = \mathbf{x}_p \, \beta, \tag{9}$$

where $\mathbf{x}_m$ is a vector representing the coordinates of the landmarks for which there are missing data in a particular specimen, $\mathbf{x}_p$ is a vector of coordinates of the points for which data are available, and β is the vector of regression coefficients. The regression coefficients can be estimated using either (1) the largest subset of points for which there are no missing values across the entire sample and those individuals for which data are available for the landmark coordinates to be estimated or (2) the subset of points common to the largest set of individuals with complete data for the missing landmarks. Such a method was used by Hartman (1989) in his analysis of the hominoid molars. As of this writing, none of these methods has been implemented in GRF-ND but are planned for future versions.

Visualization

Possibly the most important feature of the landmark-based approaches to morphometric analysis is the retention of all of the spatial relationships of the data throughout the fitting process and subsequent analyses (which includes the extraction of affine parameters during fitting). By maintaining the relative locations of the landmarks, results can be used to construct graphical presentations of the superimposed data and to visualize the results of more abstract analyses such as principal components analysis, multivariate analysis of variance, relative warp analysis and others. Such presentations provide a means for relating the results of an analysis to the actual specimens and serve as an exploratory tool to suggest biological explanations for a particular set of results or to direct further investigations.

A number of different graphical presentations are provided by GRF-ND for two- and three-dimensional data. The most basic of these is the plotting of the results of a superimposition. The mean configuration is plotted as a set of points and the deviations of the fitted data from the consensus are represented by vectors originating at the landmarks of the consensus. This is illustrated in Fig. 2 (top) which shows the results of the generalized resistant-fit superimposition of wing-vein landmarks from 127 species of North American mosquitoes (Rohlf and Slice, 1990). One can also plot these residuals in a more familiar manner as points to indicate the scatter of the fitted landmarks around those of the consensus. The means of different groups can be plotted in similar fashions with respect to a grand mean, though this is not currently automatic in GRF-ND.

The scatter of residuals around the consensus can be further summarized by the superimposition of scaled principal component axes onto the consensus landmarks. Such a plot is shown at the bottom of Fig. 2, where it can be seen that the deviations tend to have a major axis of variation parallel to wing edges or veins. This could indicate a biological effect of points of intersection tending to slide along existing boundaries or may reflect the more mundane explanation that the positions of intersections parallel to edges or veins are more difficult for the researcher to locate than are the positions of the bounding structure itself. In either case, the graphical summary of this variation facilitates the identification of the effect allowing the investigator to seek an explanation.

Individual residual vectors

Principal axes of landmark variation

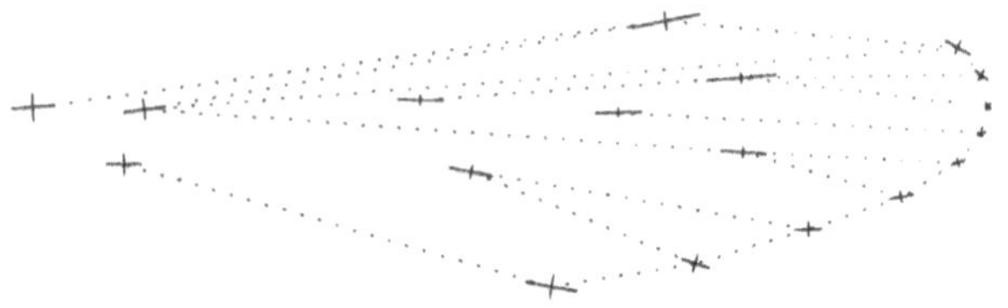

Figure 2. Graphical representations of the superimposition results. Results are from the generalized resistant fitting of 18 landmarks from the right wings of 127 species of North American mosquitoes. Top plot shows individual plots of residual vectors. Bottom plot shows principal axes of intra-landmark variation with points linked by straight lines (dotted) to visualize the wing better.

Another useful, and entertaining, presentation made possible by landmark-based methodology is the construction of hypothetical specimens. One such application is the visualization of sample principal component axes (as opposed to the individual landmark principal component axes discussed in the previous paragraph). After superimposition, the residual deviations of the fitted landmark coordinates from those of the consensus can be used to calculate the directions of major sample variation in the pk space of the fitted data by a principal components analysis of the $pk \times pk$ coordinate covariance matrix. (The actual shape space of these configurations is of lower dimension owing to the parameters estimated during fitting; see Rohlf, "Morphometric spaces" article, this volume). One can then visualize a specimen at any point along one of the principal axes by multiplying the coefficients of the axis by some appropriate factor, adding them to the reference (each coefficient corresponds to one coordinate on the reference) and plotting the result. I and others have found animated versions of this presentation, in which the hypothetical specimen is deformed as if it is sliding back and forth along a shape axis, extremely useful. Such a dynamic presentation facilitates the understanding of the shape change associated with a particular axis.

These displays need not be limited to the results of principal components analysis. Any set of linear coefficients from an analysis can be treated in the same way. For example, landmark-based regression results can also lead to effective graphical summaries. The program Morphometrika for the Apple Macintosh (Walker, 1994) plots estimated landmark trajectories based on linear or polynomial regression of shape onto size or other variables. These plots result in suggestive and intuitive presentations of how the shape of an organism changes with size, growth or variation in other extrinsic factors.

Other information can also be included in graphical presentations to enhance the interpretation of the results. GRF-ND offers two such options. The first involves simply drawing connecting lines between points in the data set, and the second allows the user to specify pairs of points between which lines are to be drawn. Both presentations are invaluable in screening the data to check the consistency of its ordering, which is a crucial assumption in all of the superimposition procedures. In addition, these lines help in visualizing the actual specimens from which the data were taken (e.g., see Fig. 1 and 2), which can be difficult when viewing a plot of unconnected points.

One could also include other real data digitized at the same time as the landmark coordinates. Although such data need not be involved in the superimposition process themselves, they can be carried along by the estimated transformation parameters and included in the graphical output. Such data might include other landmarks, digitized outlines or digitized surfaces. In the latter case, one might use raw data to generate interpolated surfaces, and, in the near future, I suspect hardware will be readily available with sufficient resolution to provide textural information. I would caution, however, against the overzealous inclusion of information that is not directly related to the analysis of the data involved in the superimposition. Deviations at the landmarks must be evaluated first, and only if these results warrant it, should one begin to consider data outside the fitting process.

Affine Parameters

Once one has invested the computational effort in estimating the affine transformation matrices, their geometric aspects can be further examined using the singular value decomposition

$$\hat{\mathbf{H}} = \mathbf{U}\Sigma\mathbf{V}^t , \tag{10}$$

where **U** and **V** are orthonormal matrices and Σ is diagonal (Goodall and Green, 1986). The matrix **U** simply rotates the configuration to align the major axes of affine stretching with the coordinate axes. This rotation will be largely a function of the original location of the configuration relative to the axes of the device used to collect the data and generally not of interest. The diagonal elements of Σ are the factors by which the axes are scaled after multiplication by **U**. Values less than 1.0 for these elements indicate a contraction along the associated axis whereas those greater than 1.0 indicate expansion. The overall change in scale due to the affine transformation can be evaluated as the product of these diagonal elements. This scaling is why a separate scaling step is not required in an affine fitting. The matrix **V** rotates the scaled configuration to align it with the reference. It will differ from configuration to configuration depending upon the orientation of the major axes of the affine deformation. Unlike the transformation implied by **U**, this transformation may actually contain relevant information about the nature of shape differences among configurations. If for instance, the major shape differences align with the long body axis of the organisms, then **V** would differ between relatively shorter, wider configurations and relatively longer, narrower ones by 90°. If there is no real affine component to the shape differences between a configuration and the reference, the directions derived from the singular value decomposition of the affine transformation matrix will be arbitrary. For a more detailed discussion of affine parameters see Bookstein (1984, 1991, and "Standard formula" article, this volume).

These relationships can be visualized by plotting strain crosses that indicate both the direction and magnitude of the uniform shape change effected by the affine transformation. This is done by simultaneously plotting the fitted data and the columns of $\pm\Sigma\mathbf{V}^t$. The comparison of these crosses across specimens, though, is complicated by the fact that they indicate both uniform shape changes and changes in scale. This problem can be addressed by scaling the vectors so that the product of their lengths is unity. Thus, the crosses will then represent uniform shape change alone. Note that this scaling constraint now introduces redundant information into the cross, and one need plot only k - 1 arms of the crosses because the direction of the last is orthogonal to the others and its length is the inverse of the product of the lengths of the other vectors. I and others have also found it preferable to use not Σ, but Σ^{-1}. Doing so means that the length of the vectors now indicates the stretching and shrinking that would be required to return the specimen to its original shape.

Other properties can be used to increase the information content of the plots of the affine strain crosses. For example, one can superimpose the vectors on a unit circle for scale and plot those with lengths equal to or greater than 1.0 as dotted lines and others as solid lines. I prefer to plot only one side of the arms of the strain cross because for moderate samples, especially for three-dimensional data, the plots quickly become uninterpretable. One can study these plots for one individual at a time to see how that individual's uniform shape component differs from that of the consensus or view all of the strain vectors simultaneously, which, with interactive rotation of the plot in the case of three-dimensional data, can help identify clustering of strain directions indicative of patterns of uniform shape differences. Figure 3 shows such a case in which the clustering of strain vectors indicates the tendency of the shapes of the turtle carapaces to differ in their relative width and height — an observation confirmed by other analyses and in other data sets (Jolicoeur and Mosimann, 1960; Slice 1993).

COMPARISON OF LEAST-SQUARES AND RESISTANT-FIT RESIDUALS

If it is reasonable to believe that the sample(s) adhere to the models used in the development of the least-squares methods (e.g., Goodall, 1991), one can treat the least-squares landmark residuals as a multivariate data set and analyze them by means of familiar methods such as multivariate analysis of variance, principal components analysis or multivariate regression techniques. Also, newer shape-space methodologies (e.g., Mardia and Dryden, 1989; Dryden and Mardia, 1991; Goodall, 1992) or the geometrical decomposition of shape variation by means of partial or relative warps (Bookstein, 1989, 1991) can be applied. Prior to the application of these methods, however, one should contrast the results of the least-squares and resistant-fit methods to scan the data for outlying specimens and to identify variation within specimens that are inconsistent with the basic model of shape

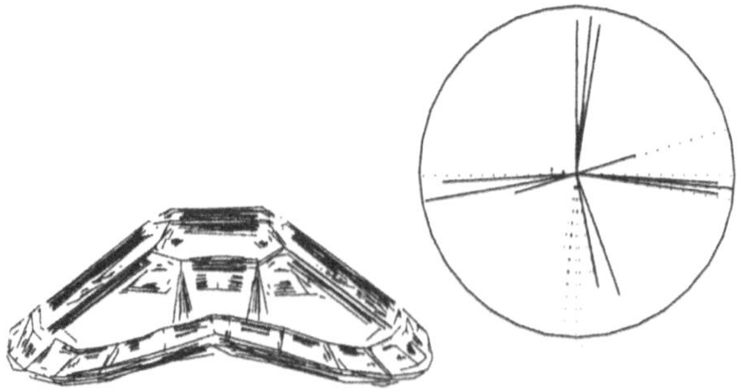

Figure 3. Results of a generalized affine least-squares superimposition of data from 10 male turtles showing fitted data and the affine strain vectors. The view is oriented to show the highly structured uniform shape variation that indicates differences in relative width and height. The strain vectors are shown relative to a unit circle. Those vectors with lengths greater than or equal to unity are drawn as dotted lines.

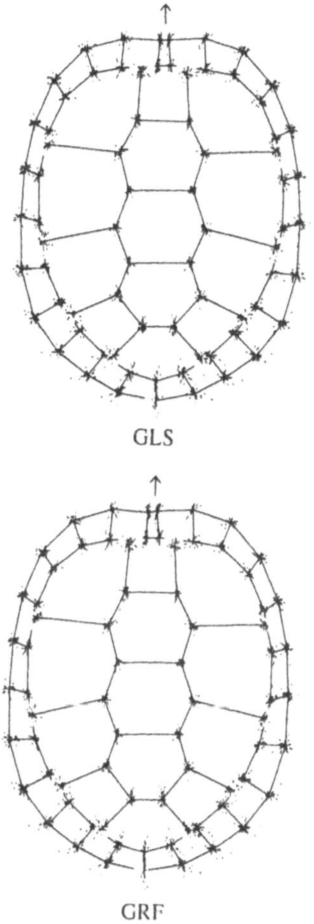

Figure 4. Dorsal view of consensus configurations and residuals for generalized least-squares fitting (top) and generalized resistant fitting (bottom) of male carapaces of *Trachemys scripta scripta*. Arrows indicate anterior direction.

differences. In this section, real data obtained from turtle carapaces are used to illustrate this type of analysis and the kinds of information that might be revealed.

The data used here consist of sets of three-dimensional coordinates of landmarks on the carapaces of 31 male and 23 female yellow-bellied slider turtles, *Trachemys scripta scripta*. The set includes data from undissected, alcohol-preserved individuals in the collection of the Carnegie Museum, Pittsburgh, Pennsylvania, USA from throughout the subspecies range and from live specimens from the site of the Savannah River Ecology Laboratory, Aiken, South Carolina, USA. The landmarks were defined by the intersection of seams between the epidermal shields that are found in most turtle species and points where seams intersect the edge of the carapace (see Fig. 4). The pattern and arrangement of these shields, at least on the carapace, are quite conservative across many species of turtles and developmentally conservative enough (Zangerl, 1969) to be used as descriptors of carapace shape. The coordinates for a total of 74 landmarks were collected using a SAC acoustic digitizer (see Dean "acquisition" article, this volume) and the DS-DIGIT software package (Slice, 1991). All individuals were superimposed using generalized orthogonal least-squares (GLS) and resistant-fit (GRF) methods described above and implemented in GRF-ND (Slice, 1993). For purposes of comparison, the results of both methods were scaled to have a consensus configuration of unit centroid size.

The results of the superimpositions are shown graphically in Fig. 4. The basic structure of the carapace is indicated in the figure by lines connecting the consensus (average or median) locations of the landmarks and reflecting the pattern of interscute seams. Deviations of individual shells from the consensus are drawn as residual vectors emanating from the consensus location of each landmark. The figure shows that similar results are obtained by both techniques and the residual variation around landmarks is fairly small compared with the size of the shell. There is little indication of structured, landmark-specific patterns of variation, but although not apparent in these figures, deviations at landmarks within individual shells are highly structured and are not uncorrelated random displacements from a mean form. The least-squares fitting converged after two iterations with a final average landmark residual of 2.69×10^{-5}. The resistant fitting stopped at four iterations without meeting the convergence criterion and with a final average residual (normalized to unit centroid size) of 2.82×10^{-5}, or about 4.8 % higher than in the least-squares result. Using the mean, instead of the median, configuration as the consensus estimate of the resistant-fit superimposition decreases the average residual to 2.78×10^{-5}, or 3.3% higher than that of the least-squares fitting.

The further comparison of these results follows two approaches, each with different expectations and providing different types of information. First, one can compare the average

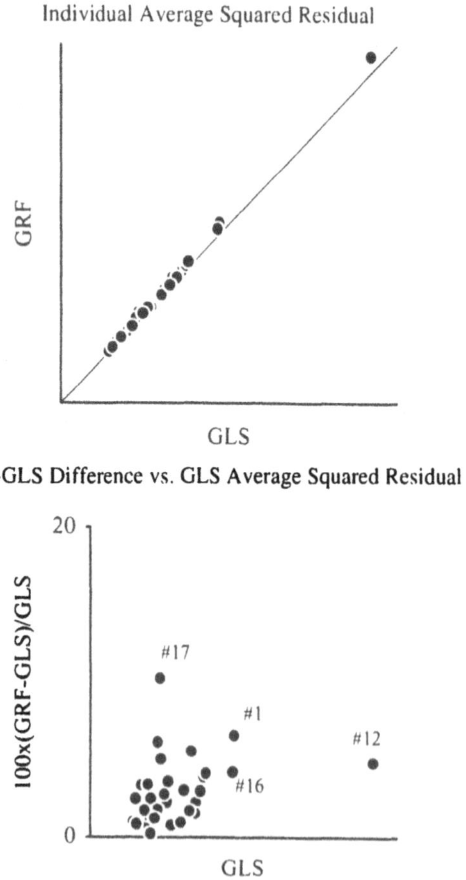

Figure 5. Comparison of GRF and GLS average squared residuals for the 31 male carapaces. Top plot shows residuals from GRF fitting plotted against GLS residuals. Bottom plot shows percent increase in GRF residual relative to GLS. Numbered points in bottom plot represent individual shells discussed in the text.

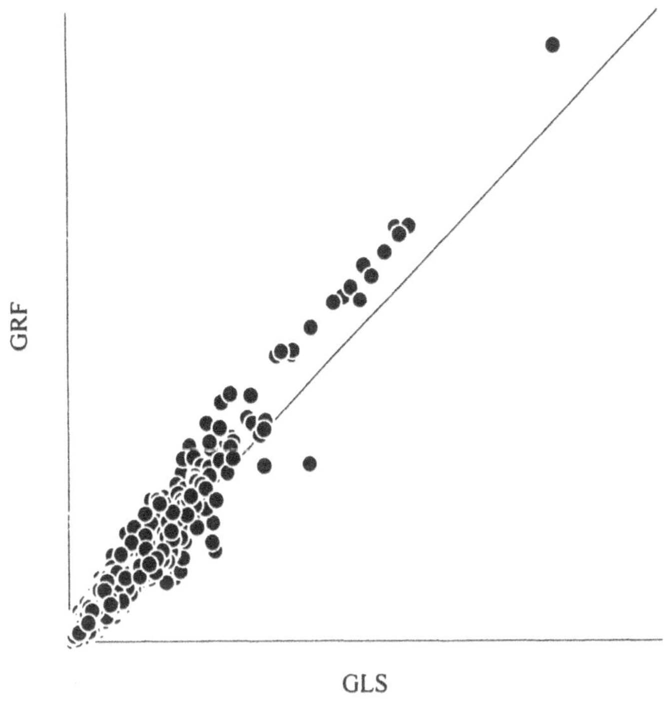

Figure 6. The squared landmark residuals from GRF fitting plotted against similar values from GLS fitting. Data are for 2294 landmarks from 31 male turtles.

squared residual for each individual shell after least-squares and resistant-fit superimposition. In this case, the average GRF residual must be greater than or equal to the corresponding GLS residual because this value is a function of the criterion minimized by the least-squares procedure. Exceptionally large discrepancies in such a comparison would indicate an individual specimen within which the residuals of a subset of one or more landmarks has been allowed to assume larger values by the resistant method. Such a subset might represent either a localized difference in shape in the region of the subset of landmarks or the landmarks might be more spatially dispersed in the configuration but developmentally linked (or not).

Such a comparison for the turtle data is illustrated in Fig. 5 where the abscissa indicates the average squared residual of each shell obtained from GLS fitting and the ordinate shows that from GRF fitting (least-squares normalized). If the results of the fittings were identical, the points would fall along the diagonal line shown in the plot. Although there is close agreement between the two sets of results, some GRF residuals have obviously larger magnitudes than do the GLS results. A more informative presentation of these differences might be that at the bottom of the figure, where the abscissa is as before but the ordinate now indicates the percent difference between the GRF and GLS results. Exceptionally large differences can more easily be identified in this presentation.

The differences seen at the bottom of Fig. 5 indicate a slight positive association between the discrepancy between GLS and GRF residuals and the magnitude of the GLS residuals. This pattern is rather consistent except for a shell (#17) that shows a larger difference than other shells with GLS residuals of comparable magnitude. Also, in both plots, each fitting indicates an extremely large average residual associated with specimen #12. This individual's average residual is farther from the next highest value by an amount greater than

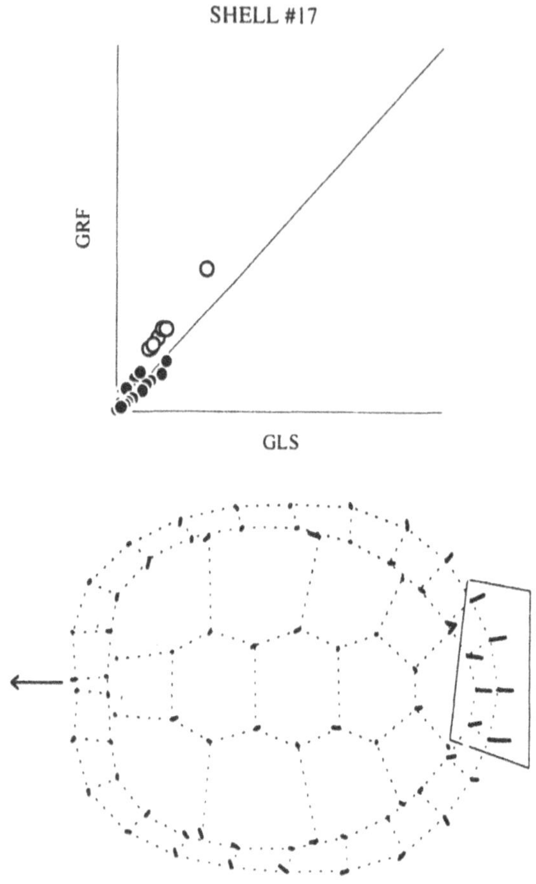

Figure 7. Individual deviations for shell #17. Top plot shows squared residuals for each landmark from GRF and GLS fitting. Open circles correspond to landmarks that are "boxed" in the lower plot. The lower plot shows GRF residual vectors from the consensus configuration. Arrow indicates anterior direction.

the entire range of the residuals of the other shells. Shell #1 shows a relatively large disagreement in the results of the fitting methods, but this difference is consistent with the pattern of increasing discrepancies with increasing GLS average residual. Shell #16 has the third largest average GLS deviation but shows relatively little disagreement between the two procedures.

The second step in contrasting the results of the least-squares and resistant-fit superimpositions involves the examination of the residuals at individual landmarks within a specimen. Here the expected pattern is different from that of the earlier plots. One now expects either agreement in the results of the two methods or a pattern of both relatively larger and smaller GRF residuals for individual shells. This expectation is because the GLS will minimize the total squared residuals by reducing large differences at certain landmarks at the expense of smaller residuals at others. GRF, on the other hand, will allow larger individual residuals in the final fitting and hence leave the naturally smaller ones untouched. Such a result for an individual shell would indicate the presence of individual or constellations of landmarks moving differently from the others with respect to the consensus configuration.

Figure 6 shows a plot of the individual landmark residuals for the entire data set of 74 landmarks × 31 shells = 2,294 points. Not much information can be gained from such a plot owing to the number of data points and the lack of information about which points belong to which shells, but one can see that most of the landmark residuals are

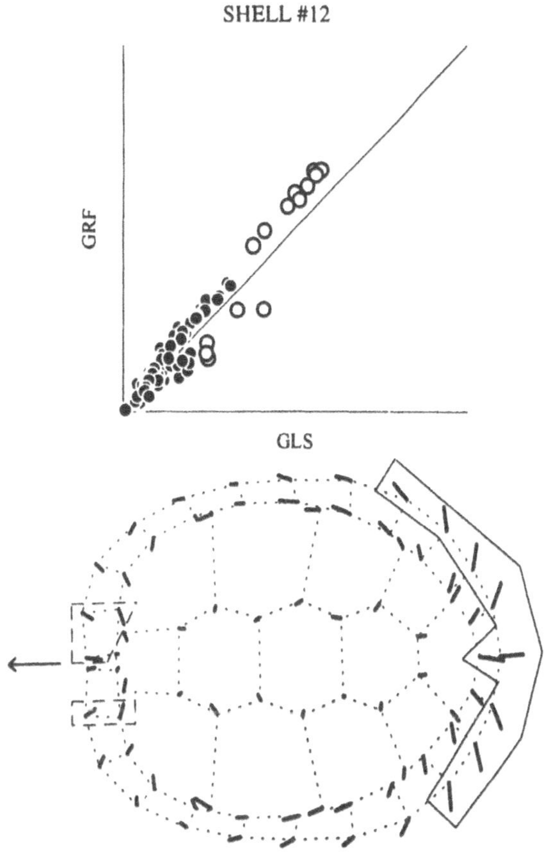

Figure 8. Individual deviations for shell #12. Top plot shows squared residuals for each landmark from GRF and GLS fitting. Open circles correspond to landmarks that are "boxed" in the lower plot. The lower plot shows GRF residual vectors from the consensus configuration. Dashed box indicates landmarks with lower GRF residuals. Arrow indicates anterior direction.

rather tightly clustered near the origin and many appear to have similar magnitudes for both least-squares and resistant fitting. On the other hand, there appear to be groups of landmarks with larger residuals offset above the diagonal line by a constant amount. Such an occurrence, should these points be from the same shell, is indicative of a subset of landmarks that has been allocated larger residuals by the resistant-fit superimposition. This may or may not be offset by a similar pattern of residuals below the diagonal because to reduce the larger values, least squares must shift some of the variation to other landmarks. Such a shift might be directed toward a particular subset of landmarks or the entirety of the other points depending upon the nature and geometry of the differences and configurations in question.

To really understand what patterns, if any, are indicated in Fig. 6, one must examine individual shells. Ideally, the pattern of residual variation in each shell should be investigated, but a likely group of individuals with which to begin are those whose overall residuals seemed atypical in Fig. 5. For the current illustrative purposes, the landmarks residuals from shells #17 and #12 are shown in Fig. 7 and 8, which are similar in scale and content to Fig. 6 but show only the residuals for a particular shell. Figure 7 (top) shows the squared residuals of the landmarks for shell #17. These data show the typical pattern of offset residual differences that suggest the presence of a relatively independent set of landmarks within the configuration. Note that here the discrepancy between the GLS and GRF residuals is due primarily to a few landmarks (plotted as open circles) having larger residuals in the GRF results. The

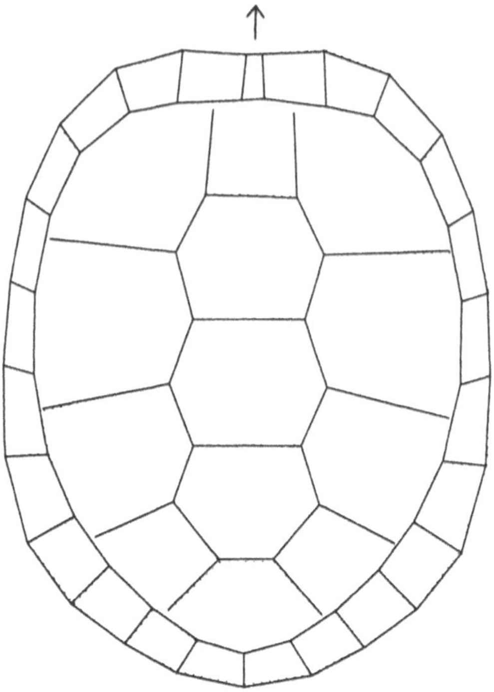

Figure 9. Dorsal view of least-squares superimposition of the consensus configuration from GRF (dashed lines) of male *Trachemys scripta scripta* carapaces onto the consensus configuration from GLS analysis (solid lines) of the same data. Lines are straight connections between landmarks for better visualization. Arrow indicates anterior direction.

compensatory reduction in the residuals of other landmarks appears slight and is spread over the rest of the configuration. The locations on the carapace of the landmarks associated with these larger GRF residuals are indicated in the lower plot, where it can be seen all are associated with the posterior-most region of the carapace. The direction of the residuals associated with these landmarks clearly indicate that this shell has the central and right portion of the posterior flange displaced away from the body of the shell relative to the consensus configuration.

The individual squared landmark residuals for shell #12 are shown in Fig. 8. This shell had the largest average squared residual in both the GLS and GRF results, and it is not surprising that the plot shows this to be the most deviant shell in the sample (compare with Fig. 6). Most of the largest landmark residuals associated with this shell are those for which GLS assigned smaller residuals relative to the GRF procedure (open symbols above the line). Identifying the extreme landmarks on the shell (Fig. 8, bottom) shows that largest residuals are, once again, associated with the posterior flange of the shell, which in this case is extremely extended and laterally expanded. A few landmarks with much lower squared residuals in the GRF results (open symbols below the line) are located on the anterior portion of the shell.

The extreme nature of the total shell residual for shell #12 was judged sufficient to warrant removing that shell from the sample as an outlier in the original study. Removing shell #12 reduced the average squared residual in the least-squares fitting of the GRF consensus to the GLS consensus by 12.8% from 4.7×10^{-7} to 4.1×10^{-7}. The closeness of these two estimates can be seen in Fig. 9, in which the two consensus estimates have been superimposed and are nearly indistinguishable. The least-squares difference between these two estimates is only about 1.7% of the average variation of the total sample.

Other shells with less extreme overall residuals showed similar patterns of localized variation in the posterior flange of the carapace. These patterns, combined with empirical observation of the affected area and the probability of an anteroposterior gradient of decreasing developmental control in the carapace (Cagle, 1950) suggests the variability identified in this region of the shell to be biological and not just an artifact of the interaction between configuration geometry and the superimposition process.

For comparison, individual average squared residuals for the female data used in the original study are shown in Fig. 10 (top); the same scaling as the comparable plot for the males is used. Here there is closer agreement between the results of the two methods and no obvious outliers, such as specimen #12 in the male data set. The range of these values falls roughly within the range seen for the males, excluding male specimen #12.

The lower plot in the figure shows the GRF-GLS percent difference plotted against the GLS average squared residual. The magnitudes of these differences are comparable to those of the male data but with less apparent association with the magnitude of the GLS residuals. Several specimens, #3, #4, #13 and #14, have a relatively larger GRF residual with respect to the GLS residual. The individual landmark deviations of these and all other shells were examined as with the male data (Fig. 11), but no obvious or extreme patterns suggested

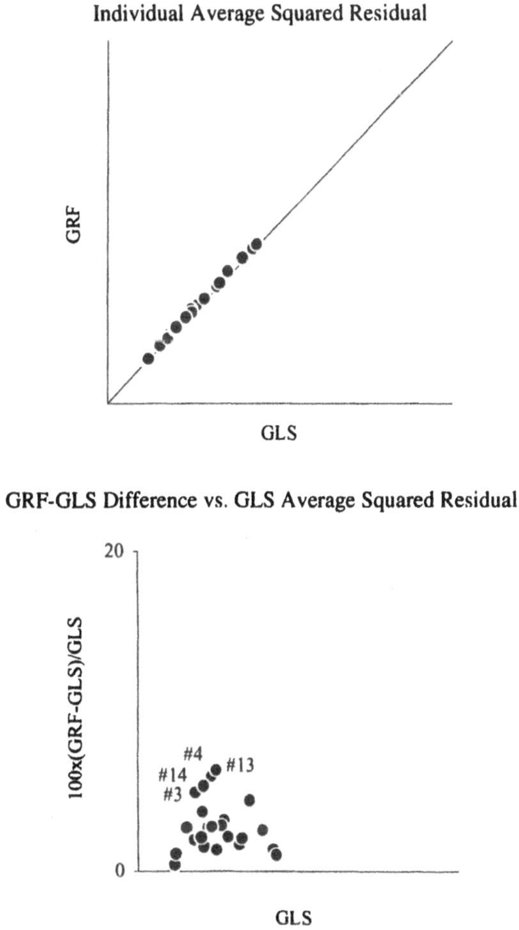

Figure 10. Comparison of GRF and GLS average squared residuals for the 23 female carapaces. Top plot shows residuals from GRF plotted against GLS residuals. Bottom plot shows percent increase in GRF residual relative to GLS. Numbered points in bottom plot represent individual shells discussed in text.

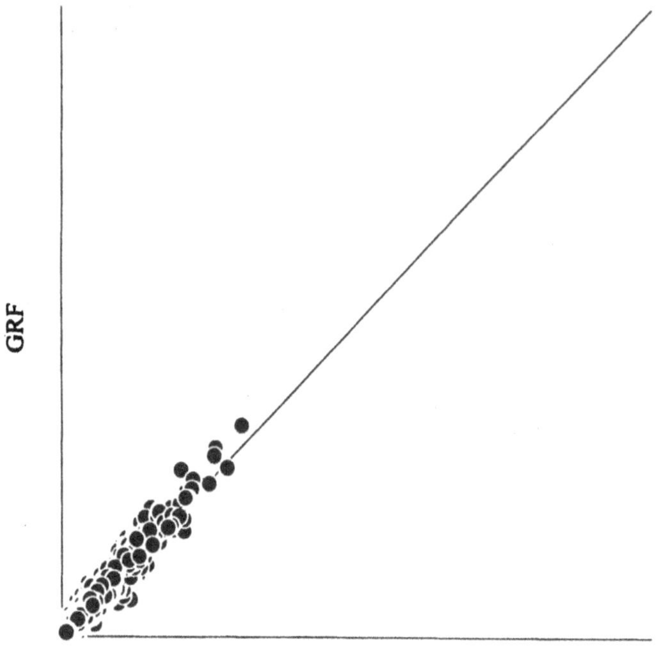

Figure 11. The squared landmark residuals from GRF fitting plotted against similar values from GLS fitting. Data are for 1,702 landmarks from 23 female turtles.

the removal of individual shells as outliers. Many of these differences were due to slight variation in the posterior region of certain shells as was found in the male shells.

ACKNOWLEDGMENTS

Early versions of this paper benefited from comments by F. James Rohlf. Fred L. Bookstein and the other members of the editorial committee provided valuable suggestions and criticisms on later versions. The work was supported in part by grants (BSR-89-18630 and BSR-90-15837) from the Systematic Biology Program of the National Science Foundation and through the hospitality and assistance of the staff of the Carnegie Museum, Pittsburgh, Pennsylvania and the Savannah River Ecology Laboratory, Aiken, South Carolina. This paper is contribution number 927 from Graduate Studies in Ecology and Evolution, State University of New York at Stony Brook.

REFERENCES

Anton, H. 1984. Calculus, with analytic geometry. John Wiley & Sons: New York.

Bookstein, F. L. 1984. Tensor biometrics for changes in cranial shape. Annals of Human Biology 11:413–437.

Bookstein, F. L. 1989. Principal warps: Thin-plate splines and the decomposition of deformations. I. E. E. E. Transactions on Pattern Analysis and Machine Intelligence 11:567–585.

Bookstein, F. L. 1991. Morphometric tools for landmark data: Geometry and biology. Cambridge University Press: Cambridge.

Bookstein, F. L. and P. Sampson. 1987. Statistical models for geometric components of shape change. Pages 18–27 *in* Proceedings of the Section on Statistical Graphics, 1987 Annual Meeting of the American Statistical Association. American Statistical Association: Alexandria, Virginia.

Cagle, F. R. 1950. The life history of the slider turtle, *Pseudemys scripta troostii* (Holbrook). Ecological Monographs 20(1):32–54.

Dryden, I. L., and K. V. Mardia. 1991. General shape distributions in a plane. Advances in Applied Probability 23:259–276.

Eves, H. 1966. Elementary matrix theory. Allyn and Bacon, Inc.: Boston.

Goodall, C. R. 1991. Procrustes methods in the statistical analysis of shape (with discussion and rejoinder). Journal of the Royal Statistical Society, Series B, 53(2):285–339.

Goodall, C. R. 1992. Dynamic graphics in non-Euclidean spaces: the visualization and statistical analysis of shape. 1992 Proceedings of the Statistical Graphics Section. American Statistical Association: Alexandria, Virginia.

Goodall, C. R., and P. B. Green. 1986. Quantitative analysis of surface growth. Botanical Gazette 147:1–15.

Goodall, C. R., and K. V. Mardia. 1991. A geometrical derivation of the shape density. Advances in Applied Probability 23:496–514.

Gower, J. C. 1970. Statistical methods of comparing different multivariate analyses of the same data. Pages 138–149 *in* F. R. Hodson, D. G. Kendall, and P. Tautu (eds.), Mathematics in the archaeological and historical sciences. Edinburgh University Press: Edinburgh.

Gower, J. C. 1975. Generalized Procrustes analysis. Psychometrika 40:33–51.

Hartman, S. E. 1989. Stereophotographic analysis of occlusal morphology of extant hominoid molars: phenetics and function. American Journal of Physical Anthropology 80(2):145–166.

Jolicoeur, P., and J. E. Mosimann. 1960. Size and shape variation in the painted turtle, a principal component analysis. Growth 24:339–354.

Mardia, K. V. and I. L. Dryden. 1989. Shape distributions for landmark data. Advances in Applied Probability 21:742–755.

Mosier, C. I. 1939. Determining a simple structure when loadings for certain tests are known. Psychometrika 4:149–162.

Rohlf, F. J. and L. F. Marcus. 1993. A revolution in morphometrics. Trends in Ecology and Evolution 8(4):129–132.

Rohlf, F. J., and D. E. Slice. 1990. Extensions of the Procrustes method for the optimal superimposition of landmarks. Systematic Zoology. 39:40–59.

Siegel, A. F., and R. H. Benson. 1982. A robust comparison of biological shapes. Biometrics 38:341–350.

Siegel, A. F., and J. R. Pinkerton. 1982. Robust comparison of three-dimensional shapes with an application to protein molecule configurations. Technical Report 217, Series 2, Department of Statistics, Princeton University.

Slice, D. E. 1991. DS-DIGIT: Basic digitizing software. Department of Ecology and Evolution, State University of New York at Stony Brook, Stony Brook, New York 11794.

Slice, D. E. 1993. Extensions, comparisons, and applications of superimposition methods for morphometric analysis. Ph. D. dissertation. State University of New York at Stony Brook.

Slice, D. E. 1994. GRF-ND: Generalized rotational fitting of N-dimensional data. Department of Ecology and Evolution. State University of New York at Stony Brook, Stony Brook, New York 11794.

Sneath, P. H. A. 1967. Trend-surface analysis of transformation grids. Journal of Zoology 151:65–122.

Walker, J. A. 1994. Morphometrika. Geometric morphometrics for the Macintosh. Department of Anatomical Sciences. State University of New York at Stony Brook, Stony Brook, New York, 11794.

Zangerl, R. 1969. The turtle shell. Pages 311–339 *in* C. A. Gans, A. d'A. Bellairs, and T. S. Parsons (eds.), Biology of the Reptilia. Volume 1. Academic Press: New York.

DIMENSION REDUCTION AND SELECTION OF LANDMARKS

A Monte Carlo Experiment

Dmitri A. Rusakov

The Open University
Milton Keynes
MK7 6AA, United Kingdom

ABSTRACT

A methodology of discarding landmarks that convey little shape and size information and is based on principal component analysis is considered. A sample of 100 two-dimensional "forms" described by five landmarks (three independent and two linearly dependent, with an independent perturbation component) is simulated. It is shown that the essential dimensionality of such data sets can be revealed by comparing eigenvalues of the data correlation matrix (PC values). A simple criterion to select the optimal landmark subset that uses raw data and minimizes the variance between PCs values is suggested.

INTRODUCTION

Classical statistical consideration of the shape variables deduced from landmark data (Bookstein, 1984, 1986) would assume that landmark perturbations are normally and independently distributed (in a set of forms). This paradigm has been studied in fine theoretical detail (Goodall, 1991; Goodall and Mardia, 1991; Dryden and Mardia, 1992). However, landmarks are usually placed on the form of an organism in accordance with biological criteria and, therefore, may often indicate statistical dependence and non-isotropic errors in their perturbations. In addition, in order to retain all available information about form, one usually considers more landmarks than are necessary.

Such a situation has long been known in multivariate analysis. A standard approach to the problem of statistical dependence is principal component (PC) analysis, which linearly transforms the observed variates into orthogonal PCs, thus maximizing the contrast between contributing variances (Mardia et al., 1979). A similar methodology which orthogonalizes the observed variates was developed for landmark data using canonical variables (Reyment, 1991) or, in an elegant manner, the "principal warps" being PCs in a space of the bending energy of thin splines (Bookstein, 1991). Such analyses allow, in principle, estimation of the essential data dimensionality by excluding PCs that have a negligibly small contribution (Mardia et al., 1979). However, all the original variables are still necessary to carry out PC analyses, and interpretation of both neglected and essential PCs is often subjective (McCabe, 1984).

This prompted development of a procedure that uses a PC analysis to select the most influential subset in the set of original (untransformed) variables by means of an optimizing technique (Joliffe, 1972; McCabe, 1984). To select a subset that maximally preserves data structure, Krzanowski (1987a,b) has suggested a direct Procrustes least-square cross-validation criterion, which compares PC scores of the complete data set and those of the candidate subset of interest.

This methodology is believed to have a potentially important effect on analyses of raw landmark data (Reyment, 1990). Indeed, the selection of an optimal landmark subset may be important when the biological versatility (within or between species or both) of the form of interest is related to a limited number of landmarks, whereas positions of the remaining landmarks are determined via common physiological or biomechanical causes. For instance, dimensions or proportions of one part of a skeletal bone may strictly determine dimensions or proportions of the other part, thus complying with the laws of mechanics (regardless of biological diversity).

The simplicity of examining the essential dimensionality of landmark data is that the multidimensional landmark scatter has the same Euclidean metric as individual landmark coordinates. Therefore, if the thickness of this scatter is negligibly small in one particular dimension (which can be examined by the PC analyses), one may assume that the space coordinates of one of the landmarks are fully determined by coordinates of the others. This trivial interpretation is similar to the (mechanical) degree of freedom and suggests the feasibility of landmark selection using the aforementioned selection technique (McCabe, 1984; Krzanowski, 1987b).

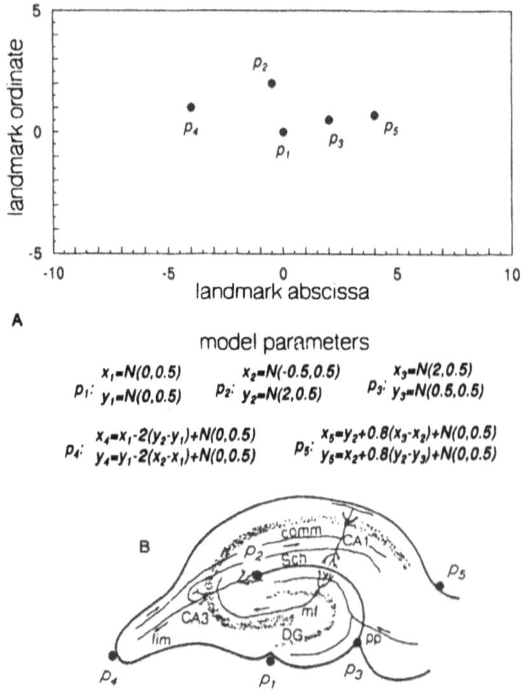

Figure 1. Simulated set of landmarks with model parameters (A) and an example of the underlying shape, a rat hippocampus section (B) (adopted from Bliss and Collingridge, 1993). Notation: p_i, x_i, y_i — ith landmark with its abscissa and ordinate; N(m, 0.5) — One Dimensional (1D) normal independent perturbation with center m and standard deviation 0.5.

However, the naturally persistent independent noise in landmark coordinates may affect the sensitivity of revealing the essential dimensionality and the optimal landmark subset by means of PC analyses. Moreover, given such highly structured data as landmark coordinates, it is difficult to foresee criteria that would ensure the best landmark subset selection. Therefore, the present study has used Monte Carlo simulations of designed forms to approach this problem.

MODEL

For simplicity, simulated two-dimensional forms included only five landmarks (Fig. 1A). Such a landmark set might be related, for instance, to the anatomy of a rat hippocampal section (illustrated in Fig. 1B), a widely accepted experimental model for the neurobiological studies of synaptic plasticity (see Bliss and Collingridge, 1993). Landmarks 1, 2, and 3 were set up to bear essential shape and size information being independent and normally distributed. The coordinates of landmarks 4 and 5 were set up (1) to be the linear functions of the coordinates of landmarks 1, 2, and 3, and (2) to produce an independent normal perturbation component as well (see Fig. 1A). Note that the model is designed so that the shape and size information can be retrieved using not only the apparently optimal landmark subset 1, 2, and 3, but also any other subset consisting of three landmarks (exceptions are subsets 1, 2 and 4; and 2, 3, and 5) because positions of the remaining landmarks are still determined by such subsets (see model parameters in Fig. 1A).

Computations were performed using a Maple V medium (Waterloo Maple Software). A sample of 100 forms was simulated (illustrated in Fig. 2A). The form size parameter was chosen in accordance with Kendall (1989):

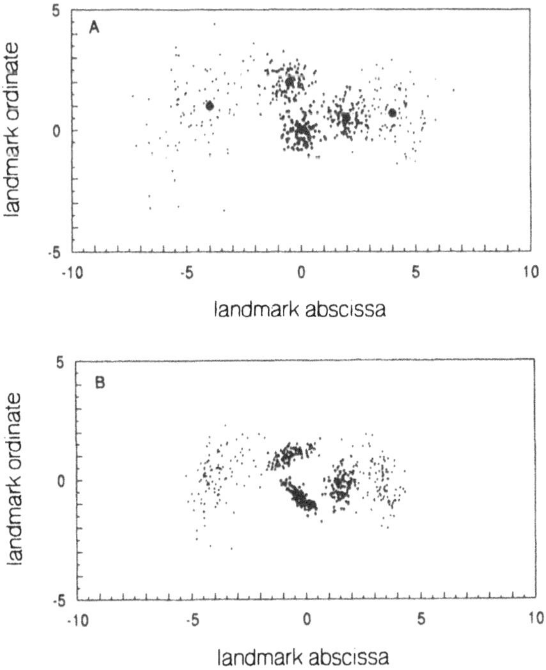

Figure 2. Original landmark coordinates of 100 simulated forms (A) and landmark coordinates of the centered and scaled forms (B).

$$\left(\frac{1}{5}\sum_{i=1}^{5} R_i^2\right)^{1/2}$$

where R_i denotes the distance between the ith landmark and the form centroid. To eliminate the factor of size, original forms (Fig. 2A) were fitted to the same centroid and scaled to the same size, thus giving the sample illustrated in Fig. 2B.

RESULTS

Dimension Reduction

Essential dimensionality of the simulated landmark data was examined using eigenvalues of the correlation matrix of the data sets, thus comparing contribution of PCs. In

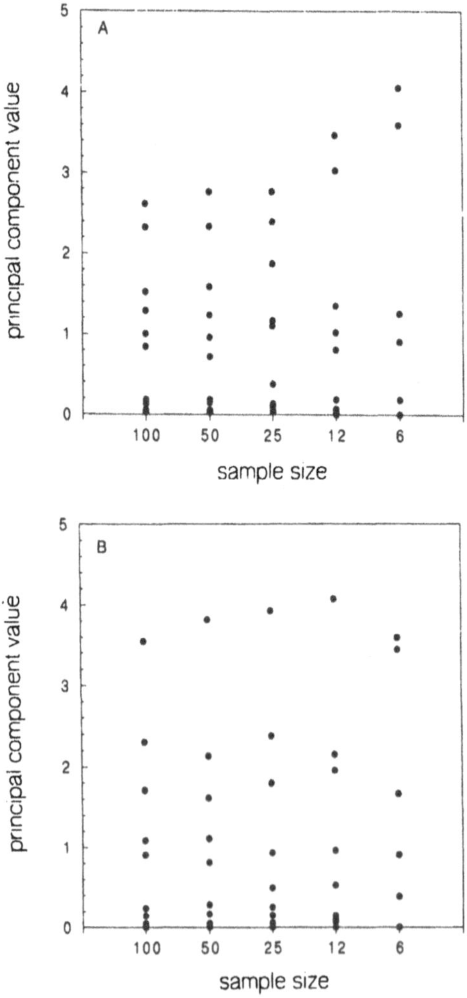

Figure 3. Principal component values (eigenvalues of the correlation matrix) for different sample sizes of the simulated original forms (A) and the centered and scaled forms (B).

addition, the effect of sample size upon the estimated dimensionality was tested in these experiments by taking not only 100, but also 50, 25, 12 or 6 first forms in the sample.

Fig. 3A shows PC values for the original data sets (Fig. 2A) at six different sample sizes. As one may expect, the landmark coordinates reveal four fairly negligible PCs (together having less than 4.1% of the total contribution), thus confirming six degrees of freedom in the data set. Note also that the sample consisting of six forms revealed only five degrees of freedom, whereas the other five PCs were indistinguishable from zero. This illustrates a trivial phenomenon: for the sample of n forms, dimensionality of the landmark data scatter is at most $n-1$, regardless of the number of landmarks.

The fitted and scaled forms revealed five degrees of freedom, with the remaining five PCs having together less than 5.1% of the total contribution, as shown in Fig. 3B; two degrees of freedom were "lost" when the original forms (Figs. 2A and 3A) were fitted at the same centroid and one degree of freedom was "gained" as a result of scaling.

These results suggest that dimensionality of such landmark data sets (or, in other words, the presence of direct inner connections between landmark positions) can be revealed by comparison of the eigenvalues. Data in Fig. 3 also indicate that changing the sample size may alter the estimated PC values so that the values become more dispersed in smaller samples.

Landmark Subset Selection

The estimated dimensionality of the model allows an attempt to find a subset of three landmarks that would preserve data structure. A sample of 100 original forms (Fig. 2A) and that of 100 fitted forms (Fig. 2B) were used in this experiment. Ten possible combinations of three of five landmarks gave the corresponding correlation matrices.

Fig. 4A (upper panel) shows the PC contribution scores for 10 landmark subsets taken from the original untransformed data. First, the plot indicates that each of two "noninformative" landmark subsets, indicated by minus signs, reveal another two negligible PCs having less than 4% of total contribution. This reflects (correctly) four degrees of freedom in the corresponding landmark sets. Second, as illustrated in the bottom panel (Fig. 4A), the optimal landmark subset (1,2 and 3) shows the least dispersion amongst the six PCs, thus suggesting that the shape and size information is fairly evenly distributed between landmarks 1, 2 and 3. This, in turn, permits the suggestion of a simple criterion to select a shape-bearing landmark subset from raw data: such a subset minimizes dispersion amongst eigenvalues of the correlation matrix.

Similar data for 100 fitted forms, as shown in Fig. 4B, indicate at least one negligibly small PC, thus confirming five degrees of freedom for these landmark subsets (see also Fig. 3B). However, in this sample the selection of an optimal landmark subset via the least dispersion criterion is not as definite as that in the raw data shown in Fig. 3A. One may assume that the procedure of fitting and scaling has altered the relative contribution of all the landmarks to the shape and size information. In some cases, this phenomenon may represent a potential flaw: fitting and scaling may increase an apparent importance of highly dependent (i.e., noninformative) but remote (marginal) landmarks.

CONCLUDING REMARKS

There are elegant methodologies that consider statistical inference with respect to shape variables. One combines Procrustes methods and PC analysis (see Rohlf, 1990; Reyment, 1991; Rohlf and Marcus, 1993), and another is principal warp analysis (see Bookstein, 1991). However, interpretation of results from these analyses that would account for the transformation of landmark variables to shape variables and then orthogonal shape variables, is not always easy.

A useful approach may lie in discarding those original variables (i.e., landmark positions) that convey little additional information about shape and size (Joliffe, 1972; McCabe, 1984). The corresponding "direct" cross-validation criterion, which optimizes landmark subset selection, has previously been described (Krzanowski, 1987a,b).

The present study has examined a similar methodology that uses simulated two-dimensional forms which have both independent and dependent landmarks. For these data, estimation of the essential dimensionality was unambiguous (Fig. 3A). In the raw landmark data the optimal subset was found to minimize dispersion among PC values (Fig. 4A), thus

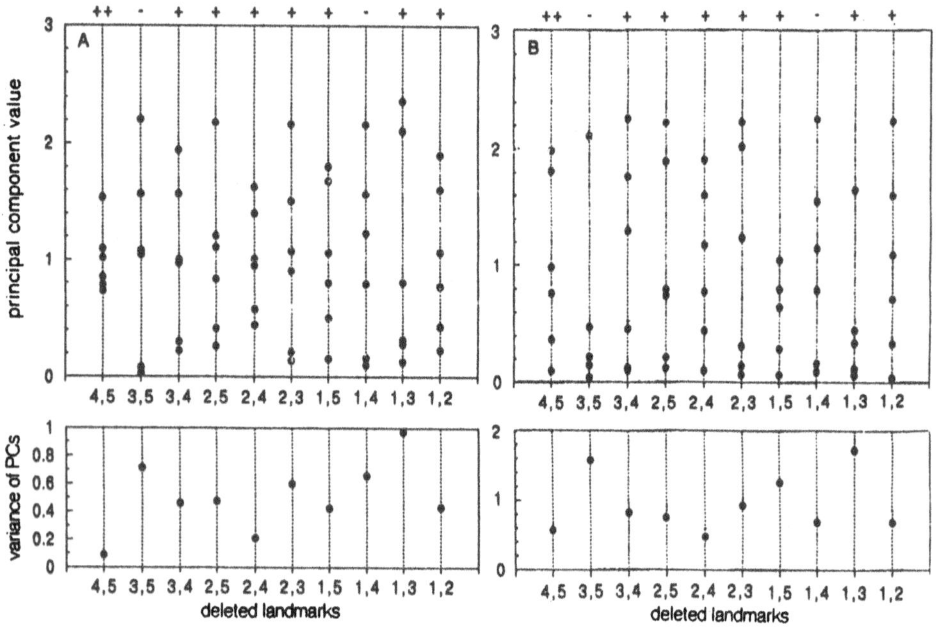

Figure 4. Principal component values (eigenvalues of the correlation matrix) for 10 selected subsets of the simulated original forms (A, upper panel) and of the centered and scaled forms (B, upper panel). Lower panels show the corresponding variance within PC sets. Notations: ++, the a priori optimal landmark subset; +, a

suggesting a simple criterion — even contribution of essential PCs — to reveal the structure-bearing landmark subset. Although this indicator seems to depend on the sample size (Fig. 3), it may be worthwhile to examine it further.

The same conclusions were drawn in a preliminary study where simulations of one-dimensional forms were performed (data not shown), and further extension of the method to three-dimensional landmark data seems to be straightforward. In addition, such a methodology has promising perspectives in allowing estimation of the essential degrees of freedom in the parameter space of dynamic biological systems (Goodwin et al., 1993; Rusakov, 1993a) or in the shape of cellular organelles (Rusakov, 1993b).

ACKNOWLEDGMENTS

To Professors Brian Goodwin and Michael Stewart for fruitful discussion. In part supported by INTAS-93-246 Grant.

REFERENCES

Bookstein, F. L. 1984. A statistical method for biological shape comparison. Journal of Theoretical Biology 107:475–520.

Bookstein, F. L. 1986. Size and shape spaces for landmark data in two dimensions (with discussion and rejoinder). Statistical Science 1:181–242.

Bookstein, F. L. 1991. Morphometric tools for landmark data: Geometry and biology. Cambridge University Press: Cambridge.

Bliss, T. V. P. and G. Collingridge. 1993. A synaptic model of memory: long-term potentiation in the hippocampus. Nature 361:31–39.

Dryden, I. L. and K. V. Mardia. 1992. Size and shape analysis of landmark data. Biometrika 79:57–68.

Goodall, C. 1991. Procrustes methods in the statistical analysis of shape (with discussion and rejoinder). Journal of the Royal Statistical Society, Series B, 53:285–339.

Goodall ,C. R. and K. V. Mardia. 1991. A geometrical derivation of the shape density. Advances in Applied Probability 23:496–514.

Goodwin, B. C., S. Kauffman, and J. D. Murray. 1993. Is morphogenesis an intrinsically robust process? Journal of Theoretical Biology 163:135–144.

Jolliffe, I. T. 1972. Discarding variables in a principal component analysis. I: Artificial data. Applied Statistics 21:160–173.

Kendall, D. G. 1989. A survey of the statistical theory of shape. Statistical Science 4:87–120.

Krzanowski, W. J. 1987a. Cross-validation in principal component analysis. Biometrics 43:575–584.

Krzanowski, W. J. 1987b. Selection of variables to preserve multivariate data structure, using principal components. Applied Statistics 36:22–33.

McCabe, G. P. 1984. Principal variables. Technometrics 26:137–144.

Mardia, K. V., J. T. Kent and J. M. Bibby. 1979. Multivariate analysis. Academic press: London:

Reyment, R. A. 1990. Reification of classical multivariate statistical analysis in morphometry. Pages 123–144 *in* F. J. Rohlf and F. L. Bookstein, (eds.), Proceeding of the Michigan morphometric workshop. University of Michigan Museum of Zoology Special Publication 2.

Reyment, R. A. 1991. Multidimensional palaeobiology. Pergamon Press: Oxford.

Rohlf, F. J. 1990. Morphometrics. Annual Review of Ecology and Systematics 21: 299–316.

Rohlf, F. J. and L. F. Marcus. 1993. A revolution in morphometrics. Trends in Ecology and Evolution 8:129–132.

Rusakov, D. A. 1993a. Estimation of the size distribution of closed cell elements from analysis of their plane random sections. Biometrics 49:141–149.

Rusakov, D. A. 1993b. Quantal behaviour of synaptic transmission can be statistically examined using the Fourier line spectrum of the histogram of synaptic potentials. Neuroscience Letters 163:231–234.

PART I

Section 3. Outlines with and without Landmarks

INTRODUCTION TO OUTLINES

F. James Rohlf

The articles in this section demonstrate that despite the emphasis in recent years on landmark-based morphometric methods (e.g., Bookstein, 1991; Rohlf and Marcus, 1993) there have also been important advances in methods for the analysis of outline data. This is important because in many cases there are not enough landmarks, or not enough in the right places to capture the variation in shape of the biological structure of interest adequately. The methods Rohlf and Bookstein (1990) discussed consisted simply of fitting various functions (usually weighted sums of sine and cosine terms) to an outline curve—or else deriving an "empirical function" (actually, just weighted linear combinations of tangent angles) to describe an outline as in eigenshape analysis. The four papers in this section represent distinct new approaches or important extensions to existing approaches.

A limitation of eigenshape analysis as originally defined is that it uses a rather arbitrary metric (sums of squares of differences in tangent angles) to measure the difference in shape between two outlines. The chapter by Sampson et al. presents a modification of eigenshape analysis that is based on a Procrustes metric (sum of squared distances between least-squares aligned points). Their new approach is expected to give results more compatible with the landmark-based methods that are based on the same metric. This is important because there is then the potential for combining outline and landmark information in a nonarbitrary way. The principal limitation of their proposed method seems to be that the outline curves must be fairly smooth and the shapes nearly convex; otherwise, the method can imply rather counterintuitive assignments of homology between points along the curves being compared. Except for some extreme cases, their method worked well for the analysis of left ventricular outlines, and, by using discriminant functions, they were usually able to distinguish among the normals and the groups of diseased cases.

In recent years there has been considerable interest in methods for the analysis of three-dimensional (3D) data. This is due in part to the fact that it is now more practical to collect such data, but it is also due to the recognition that important information can be lost if one analyzes only two-dimensional projections (Roth, 1993). The chapter by Sampson et al. also contains a number of suggestions about possible ways in which their approach could be generalized to handle space curves or surfaces in 3D. The chapter by Dean, Marcus and Bookstein discusses methods for the comparison of space curves such as digitized 3D line tracings along the brow ridge on a human skull. In their approach, curves for different specimens are aligned and compared with each other without fitting any function to the curves. The landmarks at the end points of the space curves are superimposed and then rotated to a position of best fit–unlike the unconstrained superimposition method used by Sampson et al. in which the starting and ending landmarks were not forced to superimpose. Dean, Marcus and Bookstein also show how to average curves within a sample and propose a test

Advances in Morphometrics, Edited by L. F. Marcus et al.
Plenum Press, New York, 1996

for differences between space curves based on a statistic whose distribution is expected to be proportional to a chi-square distribution.

The chapter by Guéziec is a review of some of the basic properties of splines and the relationships between several splines commonly used in computer graphics. He shows how Bezier curves based on triplets of points in a plane can be pieced together to form B-splines, which can, in turn, be joined together in three dimensions to form B-spline surface patches. An example is given of modeling the surface of a human head (from an MR scan). Such information can be used to find regions of high curvature that can be used to locate automatically such features as the ridge curves analyzed in the chapter by Dean, Marcus and Bookstein.

Little and Mardia present an example that shows how one can compare two outlines using the thin-plate spline when edgels (the slope of a tangent line at a landmark, Bookstein and Green, 1993 a,b) are included. The slopes can be taken as those of an outline contour at the point where it passes through a landmark. Little and Mardia then generalize the idea of edgels to the adding of information on the slope of a tangent plane when a 3D surface passes through a landmark.

ACKNOWLEDGMENTS

This work was supported by a grant (DEB 93-17572) from the Systematic Biology Program of the National Science Foundation. This is contribution number 966 from the Graduate Studies in Ecology and Evolution, State University of New York at Stony Brook.

REFERENCES

Bookstein, F. L., and W. D. K. Green. 1993a. A thin-plate spline for deformations with specified derivatives. Pages 14-28 *in* Mathematical methods in medical imaging II. S. P.I.E. Proceedings, vol. 2035.

Bookstein, F. L. and W. D. K. Green. 1993b. A feature space for edgels in images with landmarks. Journal of Mathematical Imaging and Vision 3:231–261.

Bookstein, F. L. 1991. Morphometric tools for landmark data: Geometry and Biology. Cambridge University Press: Cambridge.

Roth, L. 1993. On three-dimensional morphometrics, and on the identification of landmark points. Pages 41–61 *in* L. F. Marcus, E. Bello, and A. García-Valdecasas, (eds.), Contributions to morphometrics. Monografias del Museo Nacional de Ciencias 8, Madrid.

Rohlf, F. J., and F. L. Bookstein (eds.) 1990. Proceedings of the Michigan morphometrics workshop. Universiy of Michigan Museum of Zoology Special Publication 2.

Rohlf, F. J., and L. F. Marcus. 1993 A revolution in morphometrics. Trends in Ecology and Evolution 8:129–132.

EIGENSHAPE ANALYSIS OF LEFT VENTRICULAR OUTLINES FROM CONTRAST VENTRICULOGRAMS

Paul D. Sampson,[1] Fred L. Bookstein,[2] Florence H. Sheehan,[3] and Edward L. Bolson[3]

[1] Department of Statistics, Box 354322
University of Washington
Seattle, Washington 98195-4322
[2] Institute of Gerontology
University of Michigan
Ann Arbor, Michigan 48109-2007
[3] Division of Cardiology, Box 356422
School of Medicine
University of Washington
Seattle, Washington, 98195-6422

ABSTRACT

The left ventricle of the heart functions by contraction. From digitized outlines we analyze its function by describing its shape, shape change, and size change (or "ejection fraction") over the cardiac cycle, from end diastole (ED) to end systole (ES). For this purpose we introduce a new variant of eigenshape analysis for the morphometric analysis of outline data. The method begins with a mean outline defined by pointwise averages of a sample of outlines after they have been oriented in a Procrustes superposition by means of an "iterative closest point" algorithm. Individual outlines are then represented by vectors of deviations normal to the mean outline, and variation in shape is analyzed in terms of a singular value decomposition (SVD) of a sample matrix of such deviations. Principal modes of variation in shape are given by so-called "eigenshapes"—the left singular vectors of the SVD.

In application to the analysis of left ventricular outlines we compute an SVD for the joint representation of the outline shapes at both ED and ES. The results are discussed in terms of shape change. We use the scores on a subset of the principal eigenshapes to demonstrate a discriminant analysis distinguishing samples of "normals" from groups of clinical cases having either cardiomyopathy or infarcts associated with one of three types of coronary artery disease. We then discuss proposals for the morphometric analysis of two-dimensional outlines and three-dimensional surfaces that also include landmarks. These

Advances in Morphometrics, Edited by L. F. Marcus et al.
Plenum Press, New York, 1996

proposals integrate an eigenshape analysis with thin-plate spline based analyses of configurations of landmarks.

INTRODUCTION

Morphometric Methods for Landmarks and Outlines

Statistical methods for the morphometric analysis of landmark data are now quite well established. For a large body of problems researchers have access to the analytic tools to analyze landmark data most efficiently, both from the perspective of statistical analysis (e.g., the power of tests and the ability to discriminate groups) and from the perspective of biological and geometric interpretation. This is not to say that any analysis is straightforward, just that there is a body of statistical and data analytic tools on which there is reasonable consensus among the developers and practitioners of morphometric analysis. The series of workshops on morphometrics over the past five years, leading most recently to the NATO Advanced Study Institute "Advances in Morphometrics," provides the clearest documentation of this statement (see also Rohlf and Bookstein 1990, Marcus et al. 1993).

By contrast, there is far less certainty about the "best" methods of analysis for curving outline (or surface) data for which the available landmarks, if any, are inadequate to characterize the variation in shape of the outlines. A number of methods have been proposed for the analysis of two-dimensional (2D) outlines, typically with at most one landmark for orientation (see Rohlf and Bookstein, 1990, part III.A.). Perhaps best known are the elliptical Fourier analysis methods (Kuhl and Giardina, 1982; Rohlf and Archie, 1984; Rohlf 1990) and "eigenshape" methods (Lohmann, 1983). A basis for relating these two types of analysis was presented by Rohlf (1986). Unfortunately, in practice, these different types of analysis need not agree; that is, they may result in different multivariate ordinations of the sample outlines being analyzed. Both are characterized by certain arbitrary choices which influence the results of our analyses in ways that are difficult to describe.

In this paper we propose a new variant of eigenshape analysis that begins to address certain deficiencies in the current methods of analysis. Perhaps the most fundamental way in which these outline methods differ from the landmark-based methods is that they are not homology-based. Analyses that do not incorporate biological correspondences between points on different curves or specimens cannot offer findings discussed in terms of homology maps—regardless of the extent to which the pictures we draw of mean shapes and shape differences suggest such interpretations. In the case of both elliptical Fourier analysis and the usual approach to eigenshape analysis, calculations are based on a computed homology function that associates points on different curves linearly in arc length (measured from some starting point, ideally a landmark). As noted in Bookstein (1991a), in some cases one may consider homologies linear in radial angle measured from a suitably defined centroid. However, neither of these computed homologies is biologically based.

A second issue concerns the notion of "distance" between two shapes. Underlying all traditional multivariate statistical analyses is the specification of a distance between observations (specimens), either explicitly or implicitly. For a simplification of Lohmann's eigenshape analysis based on Zahn and Roskies' shape function (Zahn and Roskies, 1972) or, equivalently, the tangent angle function (Bookstein 1978, 1991a), distance is the summed squared difference between the tangent angles of points corresponding in arc length—a distance that is difficult to visualize. Bookstein points out that eigenshape analyses may be based also on representations of outlines by their curvature, or by their radial distances from a centroid, each of which corresponds to a different measure of the difference in shape for two outlines. Elliptical Fourier analysis represents another metric, summed squared distances between points of corresponding arc length (Rohlf, 1986). As in the choice of a homology function, these metrics are rather arbitrary. The latter choice is close to what we introduce in our variant of eigenshape analysis.

The eigenshape and Fourier methods are also related by the fact that their respective metrics lead to familiar orthogonal function analyses. Eigenshape analyses utilize an empirical orthogonal function decomposition of the observed outlines whereas the Fourier methods use geometric orthogonal functions that are Fourier components.

The morphometric methods for landmark data are, by contrast, quite specific in the metrics used for measuring differences in shape. For statistical purposes these are based on either Procrustes distance (Goodall, 1991) or Mahalanobis distances in shape coordinate space (Bookstein 1986). Two metrics describe differences or variation in shape geometrically: an anisotropy metric for uniform transformations, and the incommensurate bending energy metric for nonlinear changes of shape (Bookstein 1991a).

More recently, Bookstein has introduced an extension of the landmark methodology that accommodates information about the tangent directions of curves through landmarks (Bookstein and Green 1993). In this approach, homology is taken as known at some set of landmark points, and the outlines through those landmarks are taken to specify correspondence of additional landmarks that are arbitrarily close to those observed and that are constrained to the outline arcs in the vicinity. The distance function that results incorporates information about outline directions through landmarks in the form of a bending energy akin to that for landmarks, measured as an integral over the entire picture. This energy function thus includes terms for the interaction of outline changes at points widely distant along the outline. It is different from the version of bending energy treating the outline as an elastica (e.g., Amini et al. 1991).

Motivated by a problem from cardiology—the analysis of left ventricular function from 2D contours—we propose a new variant of eigenshape analysis that uses a different computed homology function (perhaps closer to, although still not truly, a biological homology function) and a different metric, easier to visualize for differences or variation in shape. This method computes an eigen decomposition or singular value decomposition (SVD) of distances measured normal to mean curves. Like current eigenshape analyses, it does not use information about landmarks; unlike current methods, it computes a mean curve based on a Procrustes-style superposition algorithm that helps to align certain geometric features of the outlines. The eigenanalysis is computed simultaneously for the pair of outline representations of the left ventricle at the extremes of the cardiac cycle, thus providing a joint analysis of the pair of shapes; we will show how analysis of change in shape is pulled out explicitly as a statistical feature of the resulting findings.

The features (imperfectly) aligned by the Procrustes superposition may indeed be homologous, but the alignment is only by rigid translation, rotation and change of scale of the outlines. For cases where (some of) the homologous features can be extracted as landmarks, we discuss a proposal to carry out a landmark analysis that maps the common landmarks to a mean configuration by means of the thin-plate spline methodology (Bookstein 1991a). An eigenshape analysis can then be applied to the corresponding segments of outline between the landmarks in the mean standardized registration. Calculated in this way, the eigenshape analysis describes only the information about outline shape that is not explained by the landmarks, for which multivariate statistical techniques are already available. We return to this point in our Discussion.

An Application in Cardiology

The left ventricle (LV) is the principal pumping chamber of the heart; its function is contraction. The LV's function may be reduced regionally due to coronary heart disease or globally in response to other disease processes. Ventricular shape may then change in compensation to achieve a more favorable distribution of stress around the LV. These shape changes are apparent to the trained observer but have been difficult to quantify, perhaps because of unsuitable geometric assumptions embedded in the models and analytic methods most often used to date. Interest in assessing shape derives from recent studies suggesting that even though volume change over the cardiac cycle is a useful predictor of survival and quality of life post infarction, LV shape holds additional information relevant to this prediction (Douglas, et al., 1989; Pfeffer and Braunwald, 1990; Kono et al., 1991).

The present investigation takes advantage of a database developed by two of us (F.H.S. and E.L.B.) that contains digitized LV contours from more than 20,000 patient studies. Each contour is based on a contrast ventriculogram, a projection image of the LV at end diastole (ED), when the LV is maximally filled, and at end systole (ES), when the LV has contracted to its minimum volume. These are 30º right anterior oblique (RAO) views. Comparison of the ED and ES volumes yields the ejection fraction, $EF = (V_{ED} - V_{ES})/V_{ED}$, a widely used parameter of LV function. Three groups were selected for the present analysis:

1. patients with normal hearts (data collected from six medical centers in three countries);
2. patients with recent myocardial infarction (heart attack) due to occlusion of one of the three main coronary arteries—LAD (left anterior descending), RCA (right coronary artery), CX (left circumflex)—who would be expected to have regional hypokinesis in the corresponding part of the LV; and
3. patients with cardiomyopathy, a diffuse disease process resulting in inadequate cardiac contraction that cannot be localized to any single region of the ventricle.

Figure 1 presents the pair of LV contours at ED and ES for one of the normal cases used in the analysis below. The contours were digitized between the anterior and inferior aspects of the aortic valve plane. The figure labels regions along the border of the contour.

The current investigation is based on a sample of 149 normal cases, 122 myocardial infarctions due to coronary artery disease, and 70 cases of cardiomyopathy. These numbers

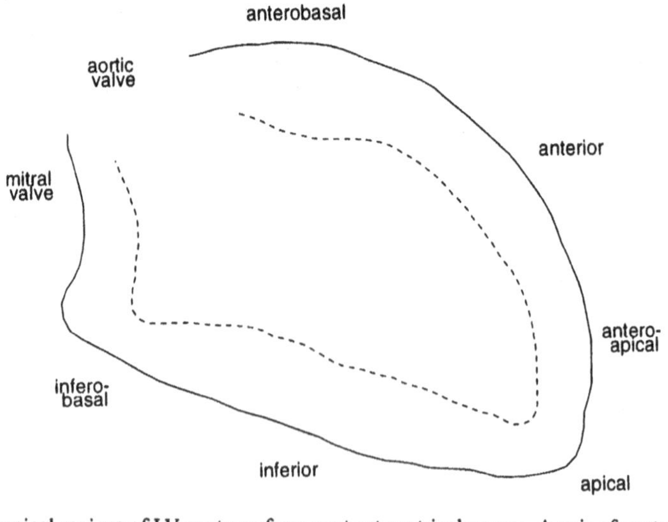

Figure 1. Anatomical regions of LV contours from contrast ventriculograms. A pair of contours taken at end diastole (ED) and end systole (ES) are shown superimposed to a common centroid. These are drawn with solid and dashed lines, respectively. Regions are aortic valve, anterobasal, anterior, anteroapical, apical, inferior, inferobasal, and mitral valve.

do not typify the clinical practice of the catheterization lab but were selected to provide a large sample for assessing normal variation as well as adequate samples of what were expected to be relatively extreme clinical deviations in LV shape. We used a training subsample for calculating the eigenshapes and developing discriminant rules, reserving the remaining subsample for validation of the discriminant rules.

EIGENSHAPE ANALYSIS OF OUTLINES WITHOUT LANDMARKS

Conventional Eigenshape Analysis of Tangent Angles

For completeness, we provide here a brief summary of one form of eigenshape analysis. This is a simplification of the eigenshape analysis introduced by Lohmann (1983), eliminating a difficult-to-justify "normalization" for the amplitude of the Zahn and Roskies shape function. A set of corresponding outlines are represented by a number, say 100, of equally spaced points. The outlines may or may not be closed. A single point (one of the endpoints if the outline is not closed) is chosen as the starting point, and the outlines are oriented with the tangent at the starting point in a specified orientation, say horizontal. The set of 100 points, because they are equally spaced, can then be represented completely by the vector of tangent angles as one circumscribes the outline. This is the first key point: we begin with a vector of numbers allowing us to reconstruct the actual shape as represented in the discretized approximation by 100 points. One could, alternatively, consider the local curvature rather than tangent angle and still integrate these to reconstruct the shape. Analysis of these tangent angles is indeed of "shape" only: position and orientation have been standardized explicitly, whereas geometric scale is wholly irrelevant to the tangent-angle descriptor.

Let the vector of tangent angles for an individual case i be denoted by z_i. A set of n outlines is then represented by a matrix $\boldsymbol{Z}$, $n \times 100$, one row for each outline, and one column for each of the 100 points. It would appear that there are 100 "variables" for each case. As noted in the introduction, the observations on a given variable are associated not by a biological homology function across the cases but by the computational device that associates points according to their relative arc length from a starting point.

A mean outline is defined to be represented by a vector of tangent angles that best approximates each of the outlines (rows) in a least-squares sense as explained below. Denote the mean by $\bar{\mathbf{z}}$. Variation about the mean is represented in terms of the patterns of deviations from the mean, the vectors of length 100, $\mathbf{z}_i - \bar{\mathbf{z}}$, $i=1,2, \ldots, n$. A convenient way of computing both the mean vector and a representation of the primary patterns of deviation from the mean is provided by the SVD. Let $m = \min(n,100)$. Then the SVD is written

$$\boldsymbol{Z} = \boldsymbol{U}\boldsymbol{D}\boldsymbol{V}',$$

where $\boldsymbol{U} = [\boldsymbol{u}_1 \ldots \boldsymbol{u}_m]$ is an $n \times m$ matrix of orthonormal columns called the left singular vectors, $\boldsymbol{D}$ is a diagonal matrix of singular values, d_i, and $\boldsymbol{V} = [\boldsymbol{v}_1 \ldots \boldsymbol{v}_{100}]$ is an $m \times 100$ matrix of orthonormal columns called the right singular vectors. The matrix $[\boldsymbol{d}_1\boldsymbol{u}_1\boldsymbol{v}_1']$ provides the best (least-squares) rank one approximation to $\boldsymbol{Z}$. It follows from this fact that $\boldsymbol{d}_1\boldsymbol{v}_1'$ is the vector of 100 tangent angles defining the mean shape as explained above. This is referred to as the first "eigenshape" (since the SVD can also be described as an "eigenanalysis" of the data matrix $\boldsymbol{Z}$). Subsequent columns or eigenshapes of $\boldsymbol{V}$ represent vectors orthogonal to the mean pattern of tangent angles and thus represent patterns of deviation from the mean.

There is a strong analogy here with the multivariate method of principal components analysis. However, in a principal components analysis the mean is computed first by simple averaging, and then all the principal components, including the first, represent patterns of multivariate deviation from the mean. In eigenshape analysis, we obtain a "mean" in the first eigenvector or singular vector because the data matrix Z is not first centered by subtraction of column means. In contrast to most principal components analyses, each of the components can be visualized by depicting the mean (component 1) plus or minus a multiple of the particular eigenshape (component 2, 3, ...).

The matrix of left singular vectors, U, is a matrix of "scores" on the eigenshapes, including the first. The 100 tangent angles representing each individual outline can be reconstructed exactly as

$$z_i = \sum_{j=1}^{m} d_j\, u_{ij}\, v_j' \,.$$

The outlines in the matrix Z can be approximated by retaining generally a small number, $k<m$, of eigenshapes, and statistical analyses such as discriminant analyses or tests of group differences can be based on the subset of score vectors, u_j , $j = 1, \ldots , k$.

Procrustes Averaging of Outlines and Eigenshape Analysis of Normal Deviations

For a class of outline shapes that are sufficiently smooth and nearly convex (as are contours of the LV from conventional ventriculograms) but lacking landmarks, we propose to put all the outlines of a sample in Procrustes superposition for calculation of a mean curve. Variation in shape can then be described by the deviations of individual sample curves measured along normals to the outline of the mean curve. This provides, perhaps, a more natural metric for variation in shape; certainly it is easier to visualize.

The Procrustes method of superposition is well known for the analysis of landmark data, but it seems not to have been considered for morphometric analysis of outlines. Consider an outline X represented by a set of n equally spaced points. The equivalence class of all outlines having the same shape is the set of n-point outlines Y that can be obtained by translating, rotating, and/or rescaling X. However, the points digitized along a curve are "deficient" in one coordinate (Bookstein, 1991a) because by virtue of the constraint that they are equally spaced along a curve, only variation in the direction normal to the curve is informative. Thus a procedure for putting different curves into superposition must allow the specific points representing a curve to slide or shift along the curve.

Consider first the Procrustes superposition of one curve on another assuming that the 100 points representing the two curves are fixed and corresponding. Let X and Y denote the 100×2 coordinate matrices representing the two curves. The solution to the problem of translating, scaling, and rotating the outlines to match with least sum of squared errors of fit can be expressed as follows (Gower, 1975; Rohlf and Slice, 1990; Rohlf, 1990):

- center both X and Y at the origin by subtracting their centroids;
- scale X and Y each to have centroid size one by dividing all the coordinates by the square root of the summed squared deviations from the centroid (now located at the origin after the first step); and
- compute the rotated coordinate matrix $X^* = XH$, where the orthonormal matrix H is computed as $H = VU'$ with V and U being the right and left singular vector matrices from the SVD $Y'X = UDV'$.

Figure 2 a–b shows the result of this calculation for the superimposition of a LV contour at ED, "X" in the notation above, on an estimate of a mean contour "Y" (which was

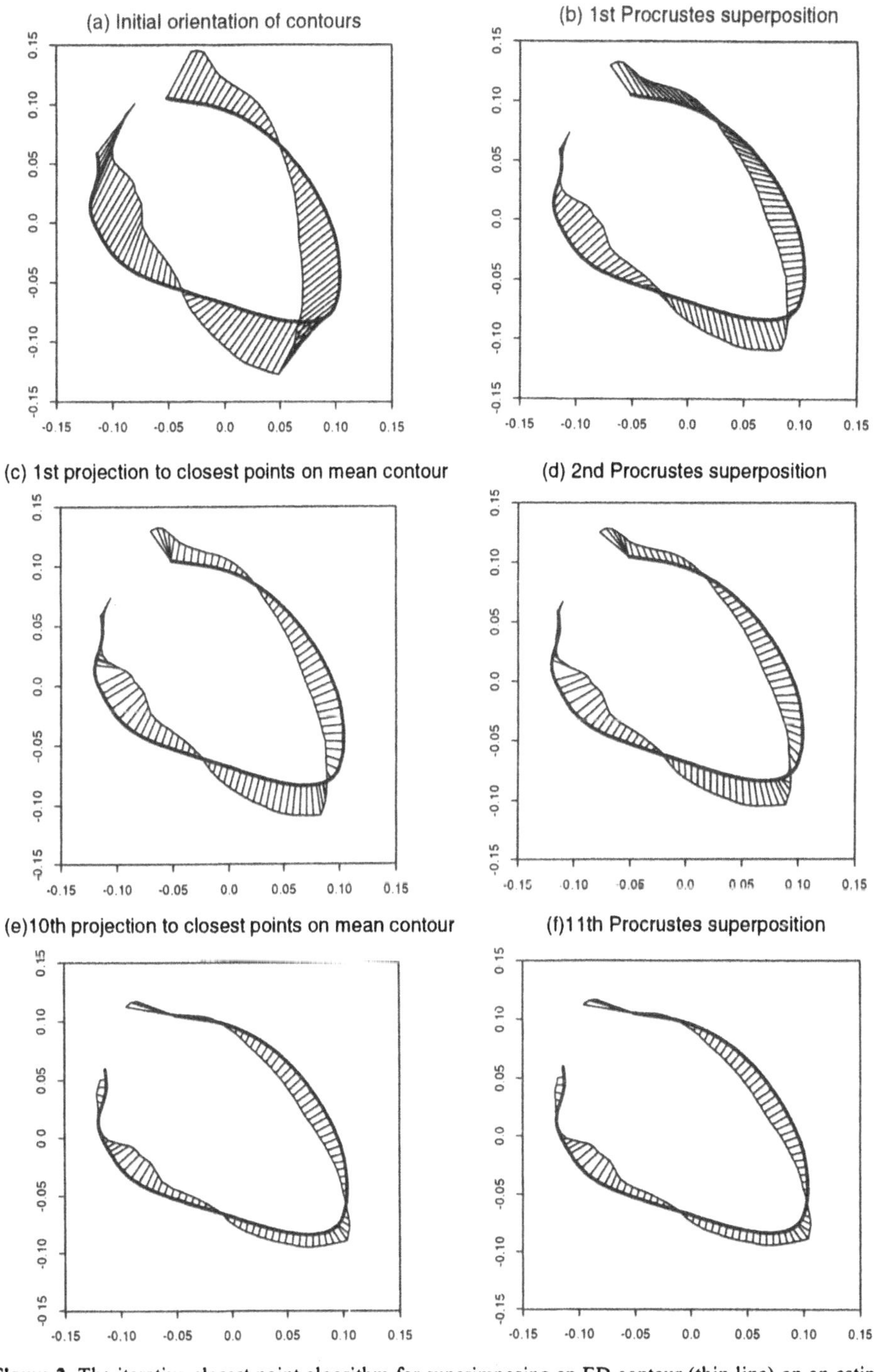

Figure 2. The iterative closest point algorithm for superimposing an ED contour (thin line) on an estimated mean contour (thick line). In each of the figures "corresponding" points are connected by line segments. (a) Initial orientations of the individual and mean contours. (b) Procrustes superposition of ED contour on the mean. (c) Projection of the (Procrustes superimposed) ED contour onto the mean for a new set of corresponding points. (d) Procrustes superposition of the ED contour relative to the corresponding points on the mean curve in (c). (e) Projection of the ED contour onto the mean for a new set of corresponding points after 10 iterations of the algorithm. (f) Final Procrustes superposition on the corresponding points of the mean curve.

computed by Procrustes averaging as described below). Line segments connect the corresponding points. The Procrustes superposition has partially corrected an obvious misorientation of the individual contour (thin line) with respect to the average (thick line), but it appears that the outline could be made to align better with the average by further rotation. In fact, the sum of squared segment lengths is a minimum for the superposition shown in Fig. 2b. It is clear that the correspondences of the equally spaced points provide a misleading measure of the distance between these two outlines. A more natural measure of distance between the curves is the sum of squared distances measured along normals to one or the other of the two curves.

We compute points on Y that correspond "best" to the equally spaced points of X^* by projecting these points orthogonally onto Y. The projections are taken as points on the line segments connecting the points originally defining Y. We then apply the Procrustes algorithm again to these more reasonably corresponding points to compute a further rotation of X^* to match Y. The result of this calculation is demonstrated in Fig. 2 c–d, where a somewhat better matching of the curves is seen. This calculation can obviously be iterated. Fig. 2 e–f shows the result after 10 iterations of the algorithm. We observe that this calculation appears to do a good job of matching regions of high and low curvature, notably the apex.

This algorithm is in fact the iterative closest point algorithm (ICP) proposed by Besl and McKay (1992) for the registration of 2D and three dimensional (3D) shapes. Convergence of the algorithm is easily proved because the sum of squared lengths of the line segments connecting the points of X (or X^*) to Y is decreased by each of the steps in the algorithm, first the simple Procrustes superposition, and then the calculation of closest points by projection of the points of X onto Y. Depending on the apparent misregistration of the original outline with respect to the target, convergence of this iterative algorithm can be slow. We have found results from 10 iterations to be adequate in our analyses of ventriculogram outlines.

Figure 3 provides a second example of the operation of the algorithm to match an ES outline (of a cardiomyopathy case) to an estimate of a normal ES mean contour. In this example we see that the initial Procrustes superposition, Fig. 3b, achieves only modest improvement in alignment due to the misspecification of corresponding points. The initial projection of points onto the mean curve, Fig. 3c, results in some unrealistic correspondences, but gradually, iteration of the algorithm matches the outline and the mean curve reasonably well, as shown in Fig. 3 e–f, although the projections leave gaps along the mean contour. For this example, note that we are left with a segment of the mean contour near the mitral valve that is unmatched with the specified outline. We return to this problem below. The contours of Figs. 2 and 3 are among those in our database most deviant from the mean curves.

The matching of outlines here is not too different from one version of a suggestion of Amini et al. (1991), which fits by a weighted average of this squared distance together with changes in curvature at corresponding outline points. Their measure was intended to be used for frame-by-frame tracking of ventriculograms, not the ED–ES analysis of the present application.

Assume now n outlines, X_i, $i = 1, \ldots, n$. The algorithm for computing a Procrustes mean curve is as follows (Gower, 1975; Rohlf, 1990):

1. Take as an initial estimate of the mean curve, Y, one of the original outlines, say X_1, or an initial pointwise average of the outlines X_i;
2. translate, scale, and rotate each of the outlines X_i into Procrustes superposition on Y by the ICP algorithm described above;
3. compute new corresponding points on each of the outlines X_i by projecting equally spaced points on Y along normals to the Y outline;
4. compute a new estimate of a mean curve Y as the pointwise average of these corresponding points on the superimposed X_i; and
5. return to step 2 and iterate until Y has converged.

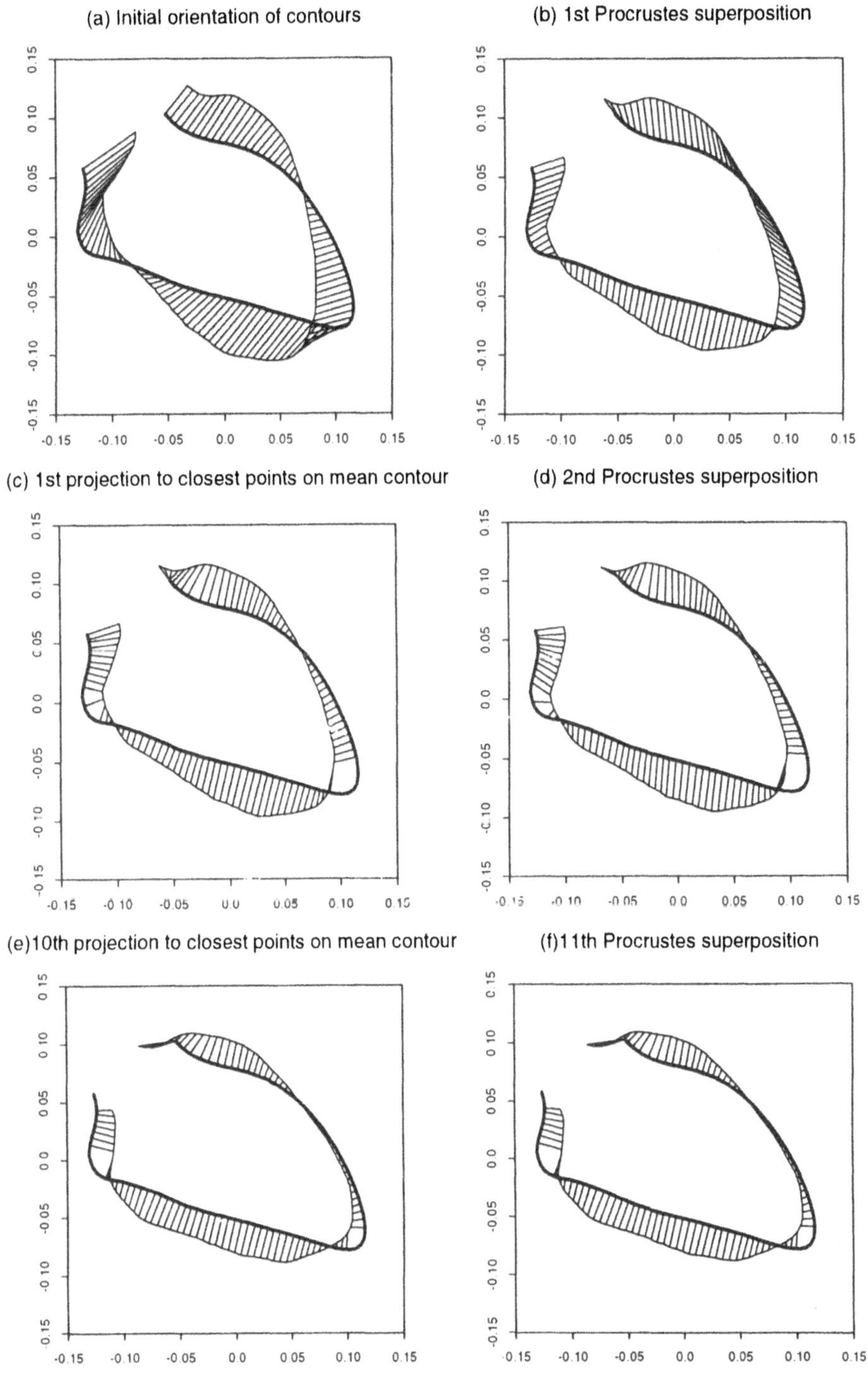

Figure 3. The iterative closest point algorithm for superimposing an ES contour on an estimated mean contour. This figure presents a second example with plots a–f as in Fig. 2.

Note step 3 in this averaging algorithm. Following the ICP algorithm, each of the outlines $\boldsymbol{X}_i$ is registered on $\boldsymbol{Y}$ according to the original points defining these outlines; these correspond to different projections on $\boldsymbol{Y}$. Before averaging it is necessary to identify corresponding points across all the $\boldsymbol{X}_i$, and we do so in terms of a set of equally spaced points on the (current) estimate of $\boldsymbol{Y}$. (see Dean et al., this volume; Cutting et al., 1993.) In practice we must sometimes compute corresponding points that are not found on the normals to the mean contour. For example, normals to the mitral valve segment of the mean contour in the final frame of Fig. 3 do not intersect the individual outline represented there. In such cases our implementation finds the nearest point on the target outline under the constraint that line segments to corresponding points do not cross. Near the region of high curvature at the apex, projections to the target outlines of diseased cases are often not exactly perpendicular to the mean curve of the normals because they are constrained not to intersect. This is demonstrated in Fig. 4 for the data of Fig. 3, one of our more extreme examples. The projections are slightly constrained near the apex and near the mitral valve as well. These departures from normal do not affect the statistical analysis of the lengths of these segments, but reconstructions of outlines would not be precise without accounting for the actual angles of the segments.

Following this calculation we can represent each individual outline by its set of (signed) normal deviations from the computed mean contour together with its scale factor. These can be assembled in a matrix $\boldsymbol{Z}$ for an SVD analysis exactly as in the case of Lohmann's eigenshape analysis as described above. As in a principal components analysis, but in contrast to Lohmann's eigenshape analysis, all of the singular vectors, from the first on, represent variation around the mean.

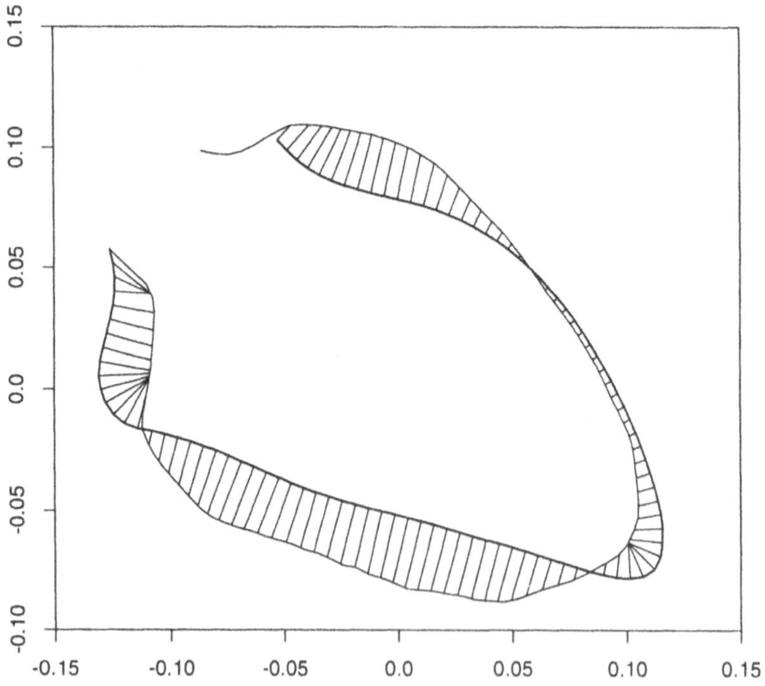

Figure 4. Projections from equally spaced points on the mean contour to the ES outline of Fig. 3. These projections are normal to the mean contour except in the vicinity of the apex and the mitral valve, where they were constrained not to intersect, and near the inferior limit of the aortic valve, where normals would not intersect the target outline.

APPLICATION

Eigenshape Analysis of LV Outlines from Ventriculograms

In this section we demonstrate an analysis of the ventriculogram database. First, using a training sample of 75 normal cases (half of the normal cases in the entire sample), we compute Procrustes mean outlines, separately for the ED and ES data. The remainder of the training sample for discriminant analysis was taken as a random sample of 2/3 of the diseased cases: 47 cardiomyopathy (CM) and 82 infarct cases—36 LAD, 36 RCA and 10 CX. (We selected more than half of the remaining clinical cases in order to have reasonably large sample size for "training" the discriminant procedure.) The LV contours for each of the 204 cases of the training sample were superimposed on the normal mean outlines at ED and ES by means of the Procrustes procedure described above, and eigenshapes were computed to summarize the joint shapes of the pair of LV borders at ED and ES. Our analysis is of the joint representation of shape at ED and ES, rather than a representation of shape change per se, because the change in shape is too great to permit a reasonable representation of all ED and ES outlines by superposition on a fixed set of 100 points lying on a midcontraction mean.

We will graphically represent the mean LV shapes at ED and ES for each of the groups in our training sample. Then, using a subset of the scores on the eigenshapes (the first few scores of the *m* score vectors calculated in the SVD), we carry out a discriminant analysis resulting in a classification rule to apply to both the training sample of 204 cases and the sample of 137 cases that was set aside for validation. This validation sample consists of 74 normals, 23 cases of cardiomyopathy, and 40 cases of coronary artery disease (CAD): 18 RCA, 17 LAD and 5 CX.

Let $\boldsymbol{Z}_d$ and $\boldsymbol{Z}_s$ represent the $n \times 100$ matrices of 100 normal deviations from the ED and ES Procrustes mean outlines, respectively, where $n = 204$. We combined these into a 204×200 matrix, $\boldsymbol{Z} = [\boldsymbol{Z}_d \mid \boldsymbol{Z}_s]$,which was decomposed by the singular value calculation as described above. The score matrix $\boldsymbol{U}$ was decomposed into partial score matrices $\boldsymbol{U}_d$ and $\boldsymbol{U}_s$ representing the contribution of the ED and ES outlines separately. Partitioning

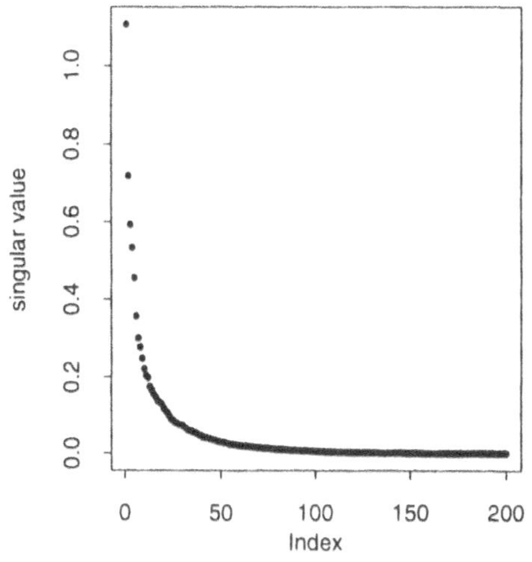

Figure 5. Plot of singular values from the analysis of 204 superimposed ED and ES ventricular contours.

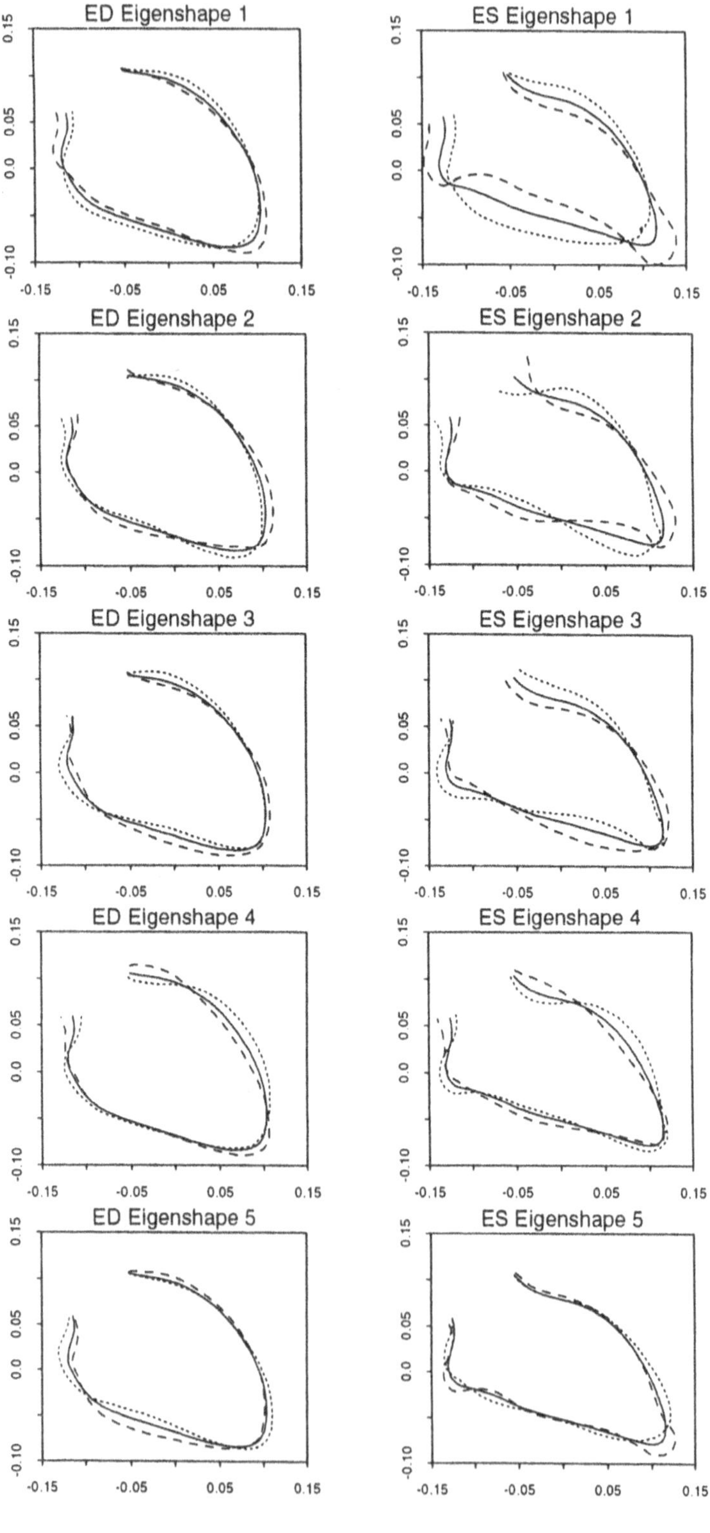
ED Eigenshape 1
ES Eigenshape 1
ED Eigenshape 2
ES Eigenshape 2
ED Eigenshape 3
ES Eigenshape 3
ED Eigenshape 4
ES Eigenshape 4
ED Eigenshape 5
ES Eigenshape 5
-0.15 -0.05 0.05 0.15
-0.10 0.0 0.05 0.15

$$V = \left[\frac{V_d}{V_s}\right],$$

then

$$U_d = Z_d V_d D^{-1},\ U_s = Z_s V_s D^{-1}.$$

Figure 5 plots the singular values (the elements of the diagonal matrix D), and Fig. 6 illustrates the patterns of shape variability represented by the first five eigenshapes: the first five columns of V_d and V_s, respectively. The mean outlines (ED and ES) are depicted by the solid curves; the dotted and dashed curves represent the mean plus and minus a multiple of the corresponding eigenshape. The multiple of the eigenshape added and subtracted from the mean outline is scaled to the 5th and 95th percentiles of the scores on the eigenshape in order to represent the empirical range of the shapes associated with the eigenshape. (These extreme percentiles result in slightly unrealistic reconstructions at ES for some eigenshapes because the largest deviations in the region of the apex are not always normal to the mean contour, as noted above.) The diagrams show that there is considerably greater variation in shape at ES than at ED.

At the outset it should be noted that the third through fifth singular values are not widely separated. They are best described jointly as approximately equipotent patterns of shape variation that, together, describe 25% of the total variation observed. The first five eigenshapes account for 76% of the variability of the normal deviations from the mean outlines (ED and ES); the first ten, 88%. We cannot readily judge these two numbers, 76% and 88%, as we do not yet have sufficient experience with the Procrustes analysis of these normal distances.

It appears that the first eigenshape, at both ED and ES, controls the relative sharpness of the ventricle at the apex, the curvature of the inferior wall and the orientation of the mitral valve. The major influence of the second eigenshape is to displace the apparent location of the apex and (at ES) the curvature and orientation of the anterobasal wall. The third eigenshape influences mainly the inferobasal wall and also the curvature at the apex.

The ED segment of the first eigenshape and the ES segment of the first eigenshape appear to describe similar patterns of deviation from their respective means (keeping in mind that they are defined with respect to 100 equally spaced points on distinct mean outlines). In fact, the ED–ES difference in mean outlines (computed by superimposing the ES mean on the ED mean) looks very much like the pattern of shape deviation explained by the ED part of the first eigenshape (V_d). Similarly, the ES–ED difference in mean outlines looks very much like the pattern of shape deviation explained by the ES part of the first eigenshape (V_s). These comparisons are illustrated in Fig. 7. This result is not happenstance.

The principal explanatory factor for change in shape is the contraction of the ventricle. The greater the contraction (i.e., the greater the ejection fraction) the greater the change in shape. The eigenanalysis describes the principal modes in which the contours of individual cases deviate from the mean contours. Those with small ejection fractions can be expected to have contours at ES that are closer, on average, to the mean contour at ED (and, similarly, contours at ED that appear closer to the mean contour at ES). Cases with high

Figure 6. Patterns of shape variability represented by the first five eigenshapes of the singular value decomposition of 204 superimposed ED and ES ventricular contours. Components of shape variability for the ED and ES contours are represented in the first and second columns, respectively. The mean contours are given by solid curves, and the dotted and dashed curves representing the mean plus and minus a multiple of the corresponding eigenshape (see text).

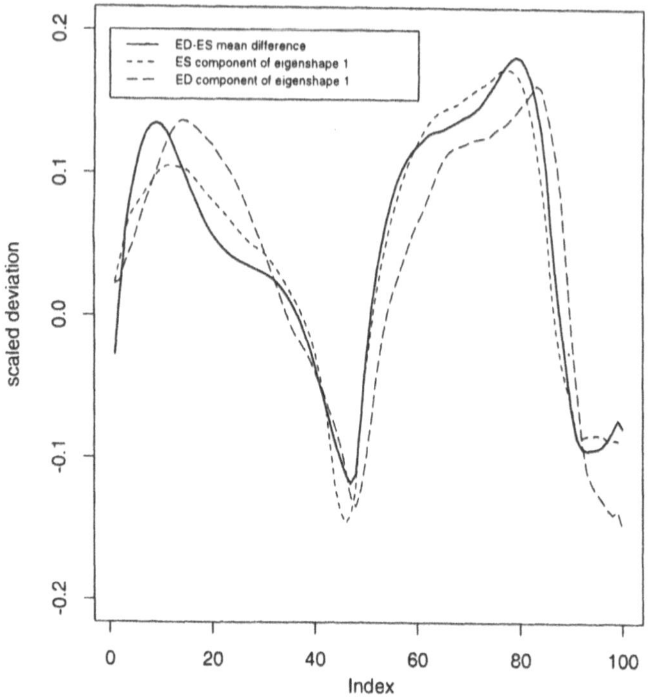

Figure 7. The difference in mean contours at ED and ES compared with the ED and ES components of eigenshape 1. The mean difference ED–ES is represented by the average of the normal deviations for the superposition of ES on ED and for the superposition of ED on ES. These deviations are computed at 100 equally spaced points indexed along the horizontal axis. They, along with the eigenshape components, have been scaled to sum of squares 1.

ejection fractions can be expected to have shapes at ES that deviate from the mean at ES further in the direction of the mean difference ED to ES. That is, the principal mode of variation in shape at ES is most likely explained by ejection fraction and should therefore correspond with the mean difference in shape, ES to ED (defined relative to the ES mean). By the same argument, the principal model of variation in shape at ED should correspond with the mean difference in shape, ED to ES (defined relative to the ED mean).

Consistent with this argument, we find that the ejection fraction (measured as relative change in area) is well correlated with the scores for the ES component of eigenshape 1 ($r = 0.60$). It is somewhat less correlated with the ED component of eigenshape 1 ($r = 0.31$), but there is less variation in shape at ED, and ejection fraction is still more correlated with eigenshape 1 than with any other. These terms being correlated are both "shapes," but they represent ratios of two quite different sorts. The eigenshape scores represent differences in geometric shape, whereas the ejection fraction is a change of geometric sizes. The correlation coefficient of 0.60 thus represents a strong degree of allometry.

A Discriminant Analysis

Figure 8 displays the mean shapes separately for each of the clinical groups. (These were computed not by separate Procrustes averaging but by reconstructing the mean curves as deviations from the mean contours for the normal cases based on the appropriate average scores—the complete set of column averages of the "*U*" matrix from the singular value

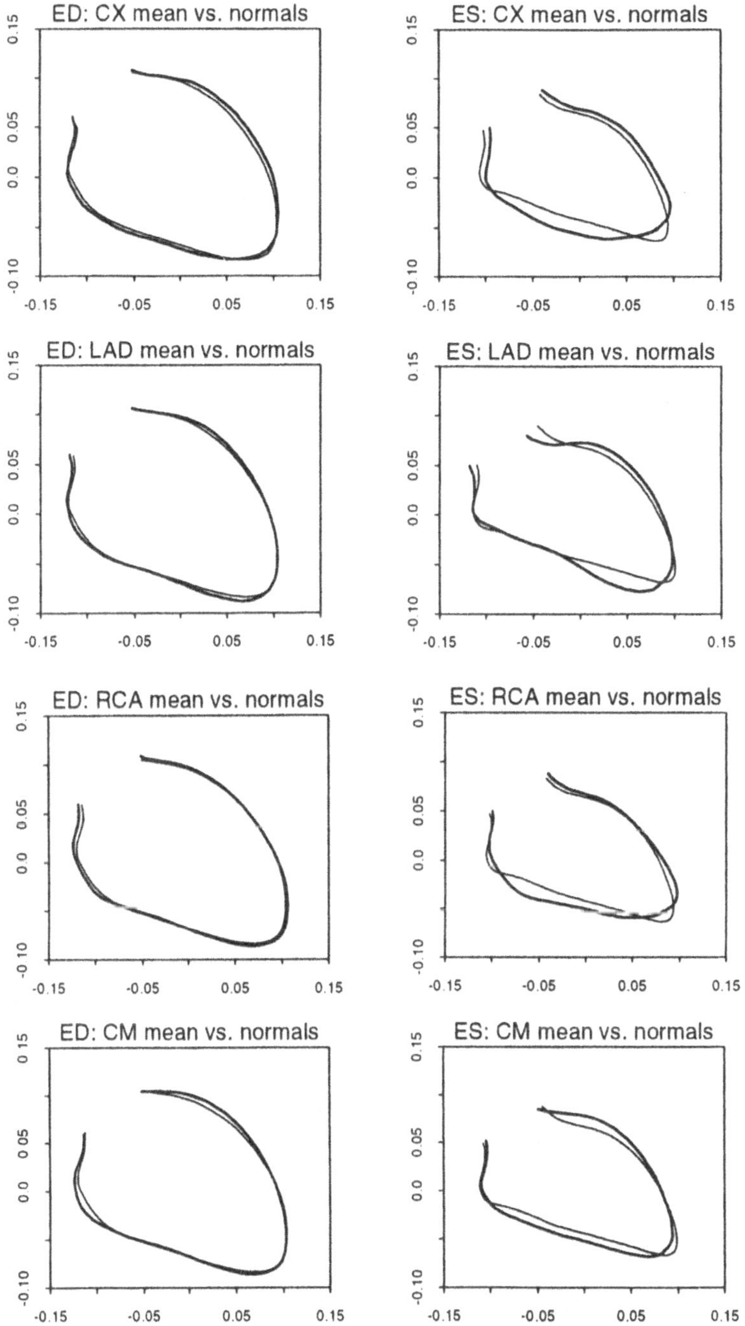

Figure 8. Mean ventricular contours at ED and ES for each of the clinical subgroups— CM, LAD, RCA and CX—represented by thick lines and drawn over the corresponding mean contours for the sample of normal cases, represented by thin lines.

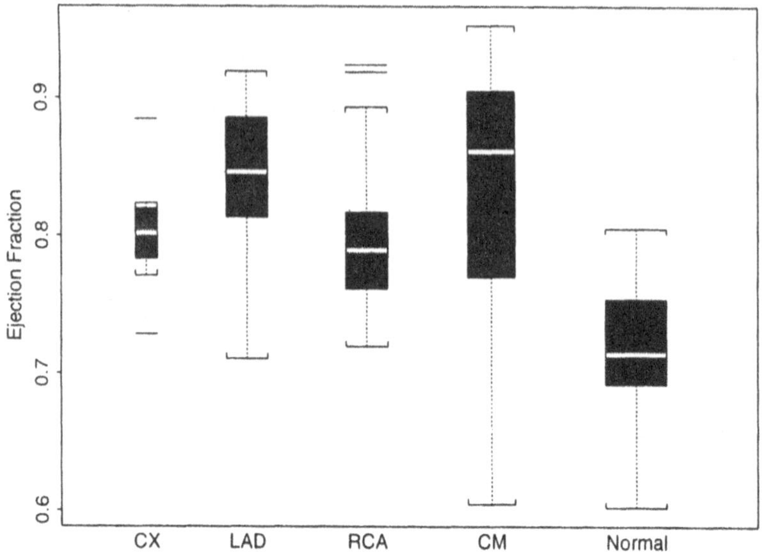

Figure 9. Boxplots of areal ejection fraction for samples from four cardiac disease groups and normals. Boxplot widths are proportional to square root of sample size.

decomposition.) It is immediately apparent that the means at ED for each of the diseased samples are almost identical with the mean outline for the sample of normal cases. The differences between these clinical groups are manifest almost exclusively in their shapes at ES, or equivalently, in their changes in shape from ED to ES.

We demonstrate here the results of a classical linear discriminant analysis (with discriminant axes computed as canonical variates) based on the eigenshape scores together with change in size, or equivalently, the global ejection fraction. This analysis should be considered primarily illustrative as we do not scrutinize the assumptions underlying the linear discriminant analysis nor do we consider alternatives.

For the five clinical groups (including normals) we can examine at most four discriminant axes in a linear discriminant function analysis. We considered calculations based on 5 and 10 eigenshape scores. These were partitioned into partial eigenshape scores, $\boldsymbol{U}_d$ and $\boldsymbol{U}_s$, representing the contributions of the ED and ES outlines separately, as described above. Classifications of the validation sample indicated that our best discrimination is achieved using the first 5 partial eigenshape scores for the ED and ES outlines (a total of 10 scores) together with the ejection fraction. We found nearly equivalent misclassification rates with three and four discriminant axes. The fourth axis helped to classify some of the small sample of CX cases correctly, but at the expense of more errors in other categories. We present here the results based on only three axes.

The ejection fraction alone distinguishes reasonably well the normal cases from all the diseased groups as shown in Fig. 9, but both the ejection fraction and eigenshape scores are necessary for discrimination of all the groups.

Figure 10 displays scatterplots of scores of the training sample cases on the first three discriminant axes. It can be seen that the first axis primarily separates the normals (solid diamonds) from the rest and the second axis distinguishes the LAD cases (open circles) from the RCA and CM cases (triangles and asterisks, respectively). The third discriminant axis helps to distinguish the RCAs.

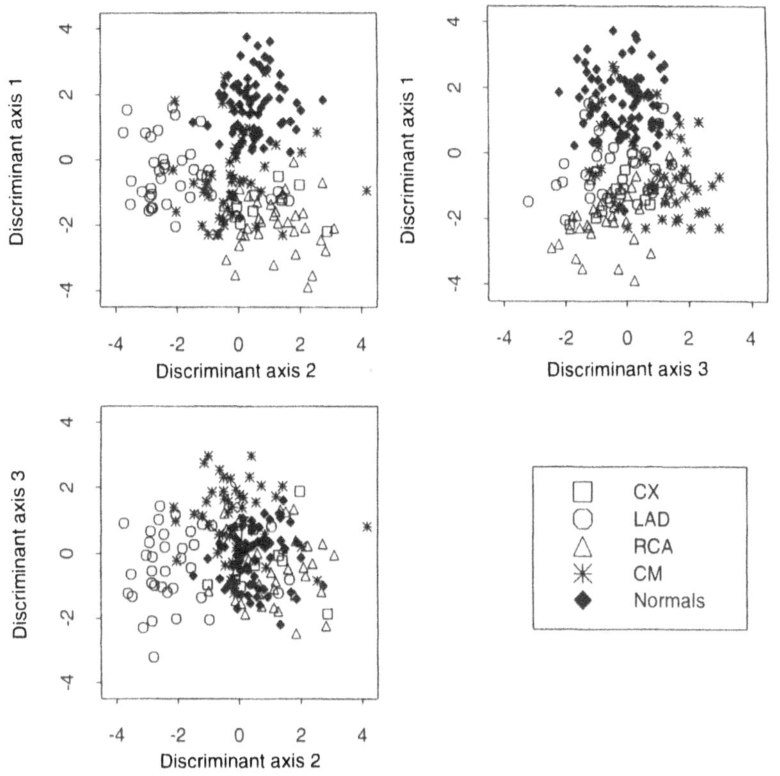

Figure 10. Scatterplots of scores of the training sample cases on the first three linear discriminant axes.

The discriminant axes reflect the fact that there are minimal differences between the groups at ED; discrimination is primarily via LV shape at ES. Scores on discriminant axis 1 are correlated most highly with the ES component of eigenshapes 1 and 5 ($r = 0.58$ and 0.80, respectively). Recall that eigenshape 1 correlates most highly with ejection fraction ($r = 0.60$). Eigenshape 5 is also correlated with ejection fraction ($r = 0.41$), and, like eigenshape 1, it also influences the curvature at the apex. Discriminant axis 1 correlates with ejection fraction ($r = 0.62$). Axis 2 (separating LAD from RCA) is correlated most highly with the ES component of eigenshape 2 and somewhat less correlated with eigenshapes 3 and 4. It also correlates well with ejection fraction ($r = 0.52$). Axis 3 (pulling out the CM cases) correlates mostly with eigenshape 3, both the ED and ES components.

The discriminant functions, linear combinations of the eigenshape scores, can be pictured like the individual eigenshapes (Fig. 6) as patterns of deviation from the mean contours. The illustrations in Fig. 11 are based on the corresponding linear combinations of the eigenshape vectors. Comparisons with Fig. 8 confirm our interpretations of these three discriminant functions.

Classification of observations in the training sample of 204 cases yielded the results shown in Table 1(top). The classification rule applied to the 137 cases in the validation sample yielded the results shown in Table 1(bottom). In both cases, the classification rules were based on "prior probabilities" set equal to the proportions of cases in the clinical categories in each of the samples.

In the training sample 80% of the cases (163 of 204) were correctly classified. In the validation sample 81% (111 of 137) were correctly classified. The greater accuracy in

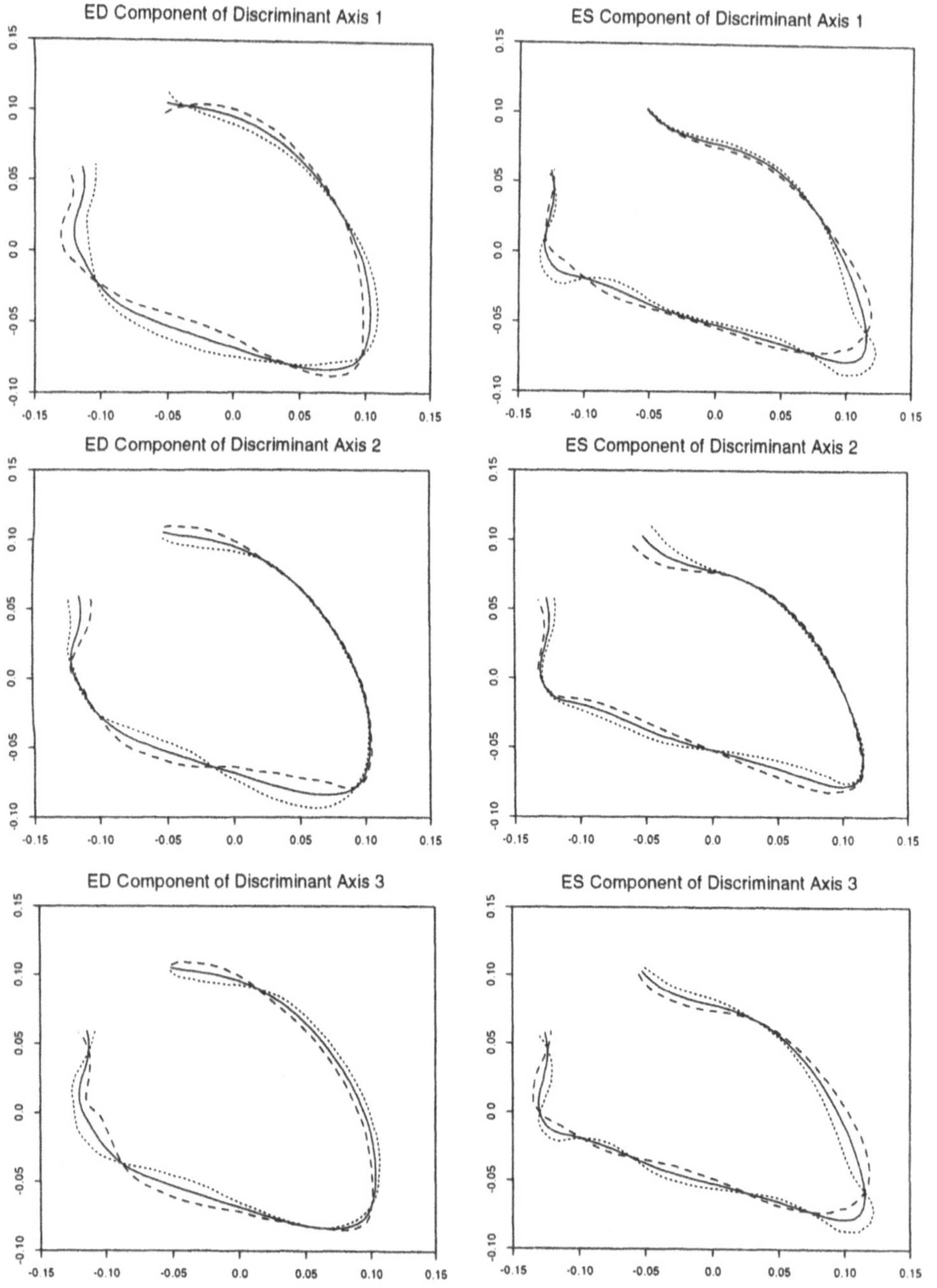

Figure 11. The first three discriminant functions represented as patterns of deviation from the mean ventricular contours, as in Fig. 6. The mean contours are given by solid curves, and the dotted and dashed curves represent the mean plus and minus a multiple of the corresponding discriminant axis.

classification of the validation sample is due to the validation sample having a greater proportion of Normals. The correct classification rate for the diseased cases drops, as expected, from 71% (92 of 129) in the training sample to 63% (40 of 63) in the validation sample.

Table 1. Discriminant analysis of normals and four groups of diseased cases based on five eigenshape scores for ED and ES outlines and relative size

Predicted classification	True classification				
	Normals	CM	LAD	RCA	CX
Training sample					
Normals	71	9	1	0	0
CM	2	32	3	5	4
LAD	1	2	30	1	1
RCA	1	4	2	30	5
CX	0	0	0	0	0
Total	75	47	36	36	10
Validation sample					
Normals	71	5	2	1	1
CM	2	13	1	1	0
LAD	1	4	14	3	2
RCA	0	1	0	13	2
CX	0	0	0	0	0
Total	74	23	17	18	5

DISCUSSION

Interpretation of the Procrustes Superpositions and Discriminant Results

Readers of the current literature of geometric cardiography, which refers throughout to regional dysfunction, would not find some of these results too surprising. For instance, in our plots the ES shape anomaly of the CX coronary artery disease cases is greatest at the inferior wall; in reality, that portion of heart muscle is the one supplied by the artery whose infarct characterizes this diagnosis. Likewise, the LAD group ES shape mean shows greatest difference from the normals along the anterior wall and apex. Often there is aneurysmal bulging at the apex. These anomalies, too, are what one would expect from the locations of the arteries infarcted.

But it would be unwise to proceed too far in this vein [sic]. Reports of "bulging" and other regional anomalies summarize only the present diagrammatic form of shape representation. There is no size information present in these pictures (although we retain size information for analysis). The only definition of size in these diagrams, total arc length, intentionally confounds contraction along the ventricular outline with the contraction across that the analysis intends to represent. Hence one could arrive at shape changes like these by changes of lengths around the outline in the absence of any contraction at all. For instance, the apical "blunting" seen in patients with aortic regurgitation, actually arises from physical lengthening of the inferior wall (Barletta et al., 1993). Such ambiguity of biological interpretation is universal among analyses of form change involving superposition—see Bookstein (1991a, section 7.1). The problem is inherent in the Procrustes step of our calculations, not the SVD that follows, and cannot be evaded without recourse to superposition-free methods.

Incorporation of Landmarks

An understanding of the discussion in this section presumes familiarity with the landmark methods presented in Bookstein (1991a). In the analysis of ventriculograms just presented, the end points of the border of the left ventricle as pictured above are the limits of the aortic valve plane in this projected view; they are not true anatomical landmarks. In fact, our analysis does not treat these endpoints as landmarks; they are not forced to correspond following the ICP–Procrustes method for calculation of a mean curve and corresponding points for the eigenshape analysis. Nonetheless, we should ask how landmarks may be incorporated in an analysis when they are available, in both 2D and 3D applications. In the following we briefly outline the analytical strategy we propose. We expect this strategy to be particularly useful in the analysis of 3D left ventricular images from 3D transesophageal echocardiography (Bashein et al., 1993) where a number of anatomical features of the left ventricle can be identified.

In this 2D illustration, there is only one other candidate for a landmark to supplement the endpoints of the border, the apex of the ventricle. The apex is an anatomical feature, but it cannot be identified unambiguously from these ventriculograms. The apex exists in the neighborhood of the extremum of curvature evident almost halfway around the border in all the images shown here, but it is not known whether or how well such an extremum corresponds with any "true biological apex." Nonetheless, for analytical purposes we may consider using this extremum of curvature, a "type 2" landmark in the classification scheme proposed by Bookstein (1991a).

We calculated the extrema of curvature for the data presented here. In fact, in about 5% of the cases the apex is too blunt to identify a unique extremum in the neighborhood of the apex. In these cases we simply took the midpoint of a pair of candidate extrema. Considering this additional landmark together with the limits of the aortic valve plane we have a triangle of landmarks, the simplest landmark configuration to which morphometric methods can be applied (Bookstein 1991a). An analysis of shape coordinates for this triangle showed that the normals and diseased cases have significantly different mean shapes at ES, but there is little power to discriminate normals and diseased cases based on the shape formed by these landmarks alone.

Two of the approaches that might be considered for utilizing the landmarks in an analysis of the shape of the contours are as follows. First, if the extrema of curvature do in fact correspond well with the anatomical apex, then we may expect better results by recognizing this homology and partitioning the outlines at the apex, associating points separately on the anterior and inferior segments. The Procrustes–ICP algorithm demonstrated above appeared to be indirectly aligning the apexes in most cases, but we should make this correspondence explicit if possible. We would need make only minor modifications in the details of the Procrustes–ICP algorithm described above. In effect, we would carry out a joint analysis of outline shape, as demonstrated above, for both segments.

This analysis, although straightforward, has (at least) two drawbacks. First, although it uses the landmarks to inject some degree of information about biological homology into the calculation of corresponding points, it still does not permit us to use the language of deformations, and it does not allow us to integrate a landmark analysis with an analysis of outline shape (except statistically in the sense that we may include both landmark shape coordinates and eigenshape scores in discriminant analyses). Second, this approach does not generalize readily to 3D surfaces.

An alternative approach is to base the analysis of outline shape on a landmark analysis that maps the common landmarks to a mean configuration by means of the thin-plate spline methodology (Bookstein 1989, 1991a). More specifically, we propose the following calculations.

(a) Compute a mean landmark configuration based on either a shape coordinate or Procrustes superposition. (The choice of method matters little for mean form; see Goodall, 1991.)

(b) Standardize each of the individual contours in the sample by unwarping (i.e., deforming) them by means of the thin-plate spline that maps the mean landmark configuration into the corresponding individual configuration. This requires numerical calculation of an inverse to the spline warping.

(c) Average the set of unwarped contours in order to define a mean contour. This average would be based on a brief iteration incorporating steps (3) and (4) of the Procrustes averaging algorithm described above (see also Cutting et al., 1993).

(d) Compute equally spaced points on the mean contours and project along normals in order to define new corresponding points on the individual (unwarped) contours. The individual contours will be represented by the vectors of (signed) deviations normal to the mean contour, just as they were in the calculations of this paper.

(e) Compute an eigenshape analysis of the contour deviations in this landmark-standardized mean registration.

There are two primary distinctions between this proposal and the calculations demonstrated above. First, the mean contour is to be computed by pointwise averaging of a set of deformed (unwarped) contours standardized to a mean landmark configuration. This is in contrast to pointwise averaging of a set of rigidly superimposed contours. Second, we propose to calculate the eigenshape analysis on these unwarped contours. Calculated in this way, the eigenshape analysis will effectively describe only the information about variation in outline shape that is not explained by the landmarks—i.e., that is not explained by a thin-plate spline warping of the mean configuration into individual specimen landmarks. We may think of this as a geometric adjustment for the effects of variation in the landmarks. Variables to be considered in discriminant analyses would include both components of variation in shape derived from the thin-plate spline-based analysis of the landmarks and the eigenshapes.

In the case of the 2D ventriculograms, the eigenshape analysis would be combined with a simple analysis of shape coordinates for the aortic valve–apex triangle. More generally, for larger sets of landmarks the eigenshapes may be combined with shape coordinates, scores, or both, describing uniform and non-uniform components of shape change as discussed, for example, in Bookstein (1991a), Rohlf and Bookstein (1990), or Reyment (1991). For an example (without eigenshape analysis) for an organ, the brain, whose shape is somewhat more static than that of the heart, see Bookstein (1991a, Appendix 1.3).

Extensions for Three-Dimensional Analysis in Cardiology

This work was motivated originally by the prospective task of analyzing left ventricular function from 3D data generated using 3D transesophageal echocardiography (TEE; Bashein et al., 1993). With 2D images we are limited primarily to consideration of "type 2" landmarks—extrema of curvature. This applies to 2D echocardiography as well as to contrast ventriculograms. Although there are a number of anatomical structures within the left ventricle (apex, mitral valve, aortic valve, interventricular septum and papillary muscles), these cannot be identified in plane images. We do expect, however, to be able to identify these anatomical structures by means of a new generation TEE probe now under development, and so expect to take advantage of the power of landmark-based methods. In fact, most of these structures do not define simple point landmarks. The interventricular septum, for example, traces out a curve along the 3D wall, and the valves and papillary muscles are 3D structures themselves. Research is ongoing on how best to analyze these features. One possibility is to augment landmarks by derivative information, as in Bookstein and Green

(1993); another is to represent landmarks as fuzzy points of some circular or elliptical shape, as in Bookstein (1991b).

It is not likely that the landmarks or anatomical structures alone will carry all the information necessary for clinical diagnosis. We must still model and analyze the shape of the ventricular surface. Arc length played a fundamental role in conventional 2D eigenshape and Fourier methods for analyzing outlines. Unfortunately, arc length does not generalize readily to 3D surfaces (although it can be used for 3D space curves). This led to our investigation of other approaches to eigenshape analysis.

Cutting et al. (1993) present a sketch of a composite method that may supply the necessary regularization of coordinate systems for this task. It applies to whole closed surfaces, such as the LV after it is closed by the aortic valve ring. The analysis registers shapes at some number of landmark points and splines all forms so that those points are at their averaged locations—as in steps (a) and (b) proposed above for 2D contours. (Information is not lost in this step but preserved in the form of shape coordinates for the original configuration, along with the size information already sequestered in any such standardization.) After such a registration it is found that several curves lie in reliable locations in the standardized space. In the present application some of these would be anatomical structures in their own right, like the base of the papillary muscles and the outline of the septum; others may be artificial traces, "geodesics" (shortest paths), connecting landmarks across long reaches of the myocardium. These curves can be extracted from surficial data by a variety of algorithms and then averaged in place in the standardized space, without any Procrustes step intervening. The curves parcel the surface in question into a map of patches, and ordinary surface points can then be averaged by normal distance over a grid on these patches exactly as we are constructing an average shape by averaging normal distances in one step of the present algorithm. Cutting et al. (1993) show an averaged skull form arrived at by this means. The analysis assumes, just as in the 2D example here, that point correspondences along normals to the mean surface are "reasonably homologous"—that is, that the mean surface can be used as a template for the description of variation by eigenshapes of distance around it. The calculation of the eigenshapes for such a 3D representation proceeds in a hierarchy of three stages—the first for landmark averaging, a second for curves around landmarks, and a third for surfaces around curves; the second and third of these are each exactly analogous to the 2D procedure sketched for ventriculograms in this paper. Only the visualization challenge increases.

REFERENCES

Amini, A. A., R. Owen, P. Anandan, and J. Duncan. 1991. Non-rigid motion models for tracking the left ventricular wall. Pages 343–357 *in* A. Colchester and D. Hawkes, (eds.), Information processing in medical imaging. Lecture Notes in Computer Science, No.511. Springer-Verlag: New York.

Barletta, G., M. Di Donato, M. Baroni, A. Fantini, F. Fantini. 1993. Left ventricular remodeling in chronic aortic regurgitation. International Journal of Cardiac Imaging 9:185–194.

Bashein, G., F. H. Sheehan, M. L. Nessly, P. R. Detmer, and R. W. Martin. 1993. Three-dimensional transesophageal echocardigraphy for depiction of regional left ventricular performance: Initial results and future directions. International Journal of Cardiac Imaging 9:121–131.

Besl, P. J., and N. D. McKay. 1992. A method for registration of 3-D shapes. IEEE Transactions on Pattern Analysis and machine Intelligence 14:239–256.

Bookstein, F. L. 1978. The measurement of biological shape and shape change. Lecture Notes in Biomathematics, Volume 24. Springer-Verlag: Berlin.

Bookstein, F. L. 1986. Size and shape spaces for landmark data in two dimensions (with discussion and rejoinder). Statistical Science 1:181–242.

Bookstein, F. L. 1989. Principal warps: thin-plate splines and the decomposition of deformations. IEEE Transactions on Pattern Analysis and Machine Intelligence 11:567–85.

Bookstein, F. L. 1991a. Morphometric tools for landmark data: Geometry and biology. Cambridge University Press: Cambridge.

Bookstein, F. L. 1991b. Four metrics for image variation. Pages 227–240 *in* D. Ortendahl and J. Llacer (eds.), Proceedings of the XI International Conference on Information Processing in Medical Imaging. Progress in Clinical and Biological Research, Volume 363. Wiley-Liss: New York.

Bookstein, F. L., and W. D. K. Green. 1993. A feature space for edgels in images with landmarks. Journal of Mathematical Imaging and Vision 3:231–261.

Cutting, C. B., F. L. Bookstein, B. Haddad, D. Dean, and D. Kim. 1993. A spline-based approach for averaging three-dimensional curves and surfaces. Pages 29–44 *in* J. N. Wilson and D. C. Wilson, (eds.), Mathematical methods in medical imaging II. S.P.I.E. Proceedings, Volume 2035.

Douglas, P. S., R. Morrow, A. Ioli, and N. Reichek. 1989. Left ventricular shape, afterload and survival in idiopathic dilated cardiomyopathy. Journal of the American College of Cardiology 13:311–315.

Goodall, C. R. 1991. Procrustes methods in the statistical analysis of shape (with discussion and rejoinder). Journal of the Royal Statistical Society, Series B 53:285–339.

Gower, J. C. 1975. Generalized Procrustes analysis. Psychometrika 40: 33–51.

Kono, T., H. N. Sabbah, P. D. Stein, J. F. Brymer, and F. Khaja. 1991. Left ventricular shape as a determinant of functional mitral regurgitation in patients with severe heart failure secondary to either coronary artery disease or idiopathic dilated cardiomyopathy. American Journal of Cardiology 68:355–359.

Kuhl, F. P., and C. R. Giardina. 1982. Elliptic Fourier features of a closed contour. Computer Graphics and Image Processing 18:236–258.

Lohmann, G. P. 1983. Eigenshape analysis of microfossils: A general morphometric procedure for describing changes in shape. Mathematical Geology 15:659–672.

Marcus, L. F., E. Bello, and A. García-Valdecasas (eds.) 1993. Contributions to morphometrics. Mongrafias del Museo Nacional de Ciencias Naturales 8, Madrid.

Pfeffer, M. A., and E. Braunwald. 1990. Ventricular remodeling after myocardial infarction: Experimental observations and clinical implications. Circulation 81(4):1161–1172.

Reyment, R. A. 1991. Multidimensional Palaeobiology. Pergamon Press: New York.

Rohlf, F. J. 1986. Relationships among eigenshape analysis, Fourier analysis, and analysis of coordinates. Mathematical Geology 18:845–854.

Rohlf, F. J. 1990. Fitting curves to outlines. Pages 167–177 *in* F. J. Rohlf and F. L. Bookstein, (eds.), Proceedings of the Michigan morphometrics workshop. University of Michigan Museums of Zoology Special Publication 2.

Rohlf, F. J., and J. Archie. 1984. A comparison of Fourier methods for the description of wing shape in mosquitoes (Diptera: culicidae). Systematic Zoology 33:302–317.

Rohlf, F. J., and F. L. Bookstein (eds.) 1990. Proceedings of the Michigan morphometrics workshop. University of Michigan Museum of Zoology Special Publication 2.

Rohlf, F. J. and D. E. Slice. 1990. Extensions of the Procrustes method for the optimal superimposition of landmarks. Systematic Zoology 39:40–59.

Zahn, C. T., and R. Z. Roskies. 1972. Fourier descriptors for plane closed curves. IEEE Transactions on Computing 21:269–281.

CHI-SQUARE TEST OF BIOLOGICAL SPACE CURVE AFFINITIES

David Dean,*[1] Leslie F. Marcus,†[2] and Fred L. Bookstein‡[3]

[1] Department of Anatomy
Case Western Reserve University
10900 Euclid Avenue
Cleveland, Ohio 44106-4930
[2] Department of Biology
Queens College of The City University of New York
and
Department of Invertebrates
American Museum of Natural History
Central Park West @ West 79 Street
New York, New York 10024
[3] Institute of Gerontology
University of Michigan
Ann Arbor, Michigan 48109-2007

ABSTRACT

This project attempts a discrimination of craniofacial shape between Pleistocene and recent humans. In this study three-dimensional craniofacial line tracings, referred to here as ridge curves, were surveyed. By means of a chi-square model, ridge curves taken from fossil specimens considered "transitional" were compared with homologous "average" ridge curves generated from *Homo erectus* and modern humans, *H. sapiens sapiens*. Results showed that only two of the five ridge curves, the medial brow and the nuchal torus, were useful in discriminating between these forms and determining the transitional specimens' affinities. For these two ridge curves the Middle Pleistocene forms showed overwhelming similarity with modern humans. The nature of the separation for the two significant ridge curves was studied with a principal coordinates analysis. This study supports the assignment of Middle Pleistocene transitional specimens to "early" *H. sapiens* rather than to *H. erectus* (Dean, 1993).

* e-mail: dean@lucifer.cwru.edu
† e-mail: marcus@amnh.org
‡ e-mail: fred@brainmap.med.umich.edu

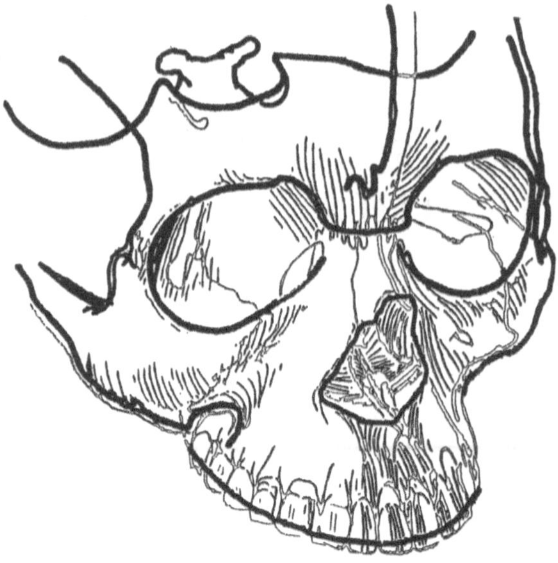

Figure 1. Ridge curves of the skull: Because they are the foci of curvature in their surface neighborhood, the edges of many structures can be captured and collected as coordinate data and compared as space curves. The heavy lines are ridge curves. Modified from Bookstein and Cutting (1988).

INTRODUCTION

Three-dimensional (3D) ridge curves (Bookstein and Cutting, 1988) representing major surface features of the skull (Fig. 1) summarize a great deal of bony skull morphology. However, only recently have such features been shown to be diagnostically useful (Dean, 1993; Cutting et al., 1995).

The ridge curves surveyed here were (1) the glabellar or midsegment of the brow, (2) the lateral brow, (3) the temporal line, (4) the coronal suture, and (5) the superior nuchal line.

The "transitional" sample exhibited overwhelming affinities with the average modern human medial brow and nuchal torus ridge curves. There was no clear association between

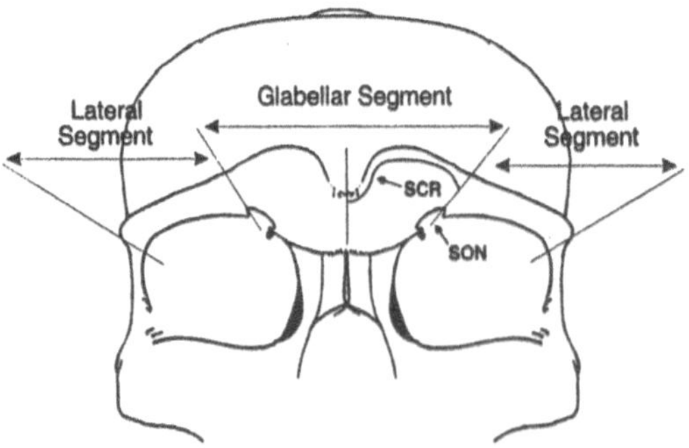

Figure 2. The supraorbital torus: The glabellar segment consists of both superciliary ridges [SCR] joined at glabella, and is separated from the lateral segment, the supraorbital torus proper, at the supraorbital notches (SON). Modified from Russell (1983).

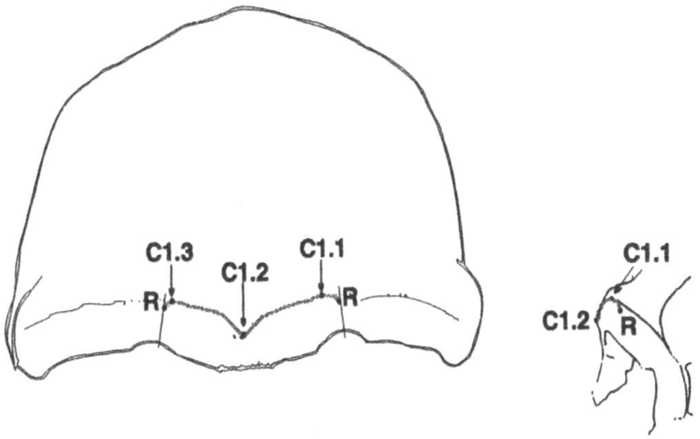

Figure 3. The mid-glabellar segment in *H. erectus*. Frontal view at left is Solo 11, Java, Indonesia. Lateral view at right is Zhoukoudian 12, "Peking Man" site. Both modified from Santa Luca (1980).

the transitional forms and the *Homo erectus* or *H. sapiens* samples for ridge curves 2, 3 and 4. Details are given in Dean (1993).

The Brow Ridge

The brow is typically divided into two segments (Fig. 2): the medial, or glabellar, segment and the lateral brows separated by lines positioned at the supraorbital notches (SON). The brow in *H. erectus* (Fig. 3) has been described by many workers as a continuous bar (Weidenreich, 1943; Le Gros Clark, 1955; Macintosh and Larnach, 1972; Jacob, 1976; Wood, 1984). This is true for the segments of the brow found over the orbits but in virtually every case where they are preserved, the superciliary ridges grade inferiorly to glabella. Glabella itself is usually somewhat posteroinferiorly depressed in *H. erectus*.

The brow is noticeably different from this condition in earliest *H. sapiens*. It has a glabellar segment, including both superciliary ridges, and two lateral brow ridge segments,

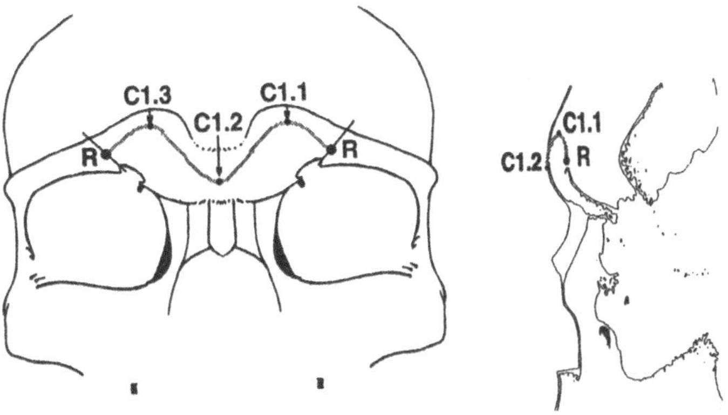

Figure 4. The glabellar segment in *H. sapiens sapiens*. Abbreviations represent the registration point (R); curvature maximum landmark (C); and glabella (C1.2). Figures are not actual size but approximately to scale. Both views are of an aboriginal Australian. Frontal view at left modified from Russell (1983); lateral view at right modified from Brown (1989).

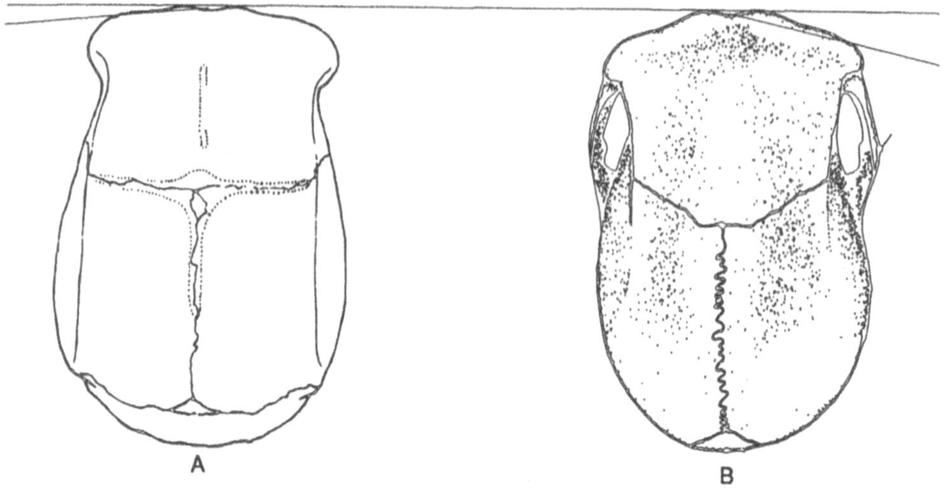

Figure 5. Superior view of skull: (A): *H. erectus*, Solo 11, modified from Santa Luca (1980); (B) *H. sapiens*, aboriginal Australian. Modified from Brown (1989). The angle is drawn between a line representing the frontal plane and a line connecting glabella and the superolateral corner of the brow ridge.

the supraorbital torus proper (Fig. 4). From a frontal view the superciliary ridges of *H. sapiens* begin laterally, above the supraorbital notches, and, in most cases, they rise well above the highest point of the lateral segment.

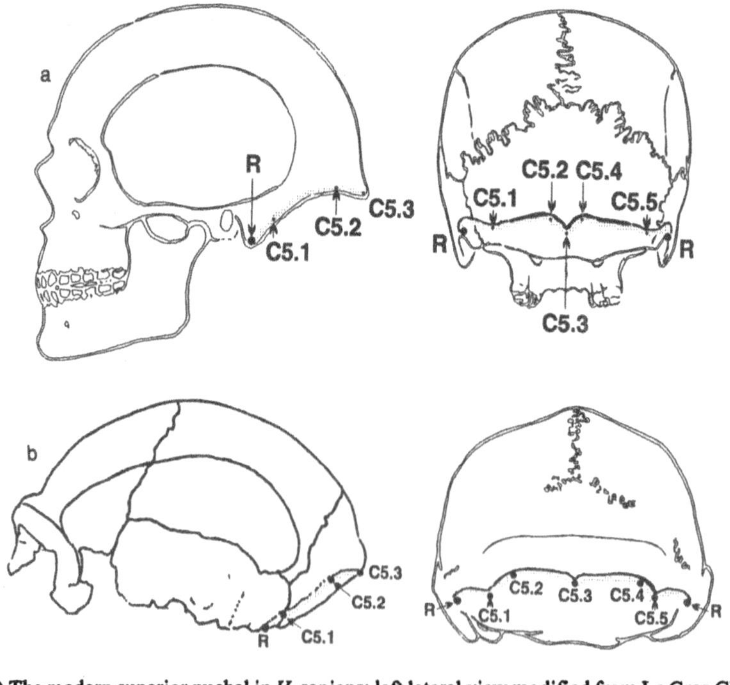

Figure 6. (a) The modern superior nuchal in *H.* sapiens: left lateral view modified from Le Gros Clark (1967); posterior view modified from Le Gros Clark (1955). (b) The superior nuchal line in *H. erectus*: left lateral view is Zhoukoudian 12 (modified from Santa Luca, 1980); posterior view modified from Weidenreich (1951).

Viewed from above, the superciliary ridges of the glabellar segment are in the frontal plane, whereas the entire lateral segment has rotated posteromedially. Thus, the angle between the lateral segment and the frontal plane (Fig. 5) in *H. sapiens* is greater than in *H. erectus*.

The Superior Nuchal Line

In lateral view inion (C5.3) for *H. sapiens* (Fig. 6a) is coincident with the external occipital protuberance, a configuration rarely found in *H. erectus* (Fig. 6b).

The superior nuchal line indicates the superior extent of the neck musculature. Tracing began at the tip of the mastoid process, registration point R, and continued medially along the muscle scar to inion, landmark C5.3. In *H. erectus,* inion and opisthocranion are usually coincident. Rightmire (1990) notes that from the earliest *H. sapiens* on we find a true inferiorly oriented and localized external occipital protuberance.

STUDY SPECIMENS

The *Homo erectus* Sample

Ashton (1950), Mayr (1951), Santa Luca (1980), Turner and Chamberlain (1989), and Rightmire (1990), amongst others, have all strongly advocated the specific unity of African and Asian *H. erectus* samples. This study used the taxon of Rightmire (1990) and Santa Luca (1980), *H. erectus* sensu lato (Stringer 1984). Among these specimens (Table 1) there is, perhaps, most disagreement over the taxonomic assignment of specimens from Solo (Java, Indonesia).

Table 1. The *H. erectus* sample. Specimens are either original (o) or cast (c). The curves measured are (1) the glabellar or midsegment of the brow, (2) the lateral brow, (3) the temporal line, (4) the coronal suture and (5) the superior nuchal line. Full citations for dating sources can be found in Dean (1993)

Specimen	Ridge curves	Date	Dating source(s)
KNM-ER 3733 (c)	1-5	1.78 Ma	Feibel et al., 1989
KNM-ER 3883 (c)	1-5	1.57 Ma	Feibel et al., 1989
OH 9 (c)	1-5	ca. 1.2 Ma	Hay, 1990
Sangiran 2 (o)	1-5	900 ± 200 Ka	Shizumu et al., 1985
Sangiran 4 (o)	4, 5	900 ± 200 Ka	Shizumu et al., 1985
Sangiran 17 (c)	1-5	900 ± 200 Ka	Shizumu et al., 1985
Solo 4 (c)	1, 2	450 ± 350 Ka	Aziz et al., 1991
Solo 5 (c)	1-5	450 ± 350 Ka	Aziz et al., 1991
Solo 7 (c)	1-5	450 ± 350 Ka	Aziz et al., 1991
Solo 8 (c)	4	450 ± 350 Ka	Aziz et al., 1991
Solo 10 (c)	1,2,3,5	450 ± 350 Ka	Aziz et al., 1991
Solo 11 (c)	1-5	450 ± 350 Ka	Aziz et al., 1991
Solo 12 (c)	1-5	450 ± 350 Ka	Aziz et al., 1991
Lantian (c)	1, 2	1,100 ± 125 Ka	An and Ho, 1989
Zhoukoudian 3 (c)	1-5	500 ± 250 Ka	Chen and Zhang, 1991
Zhoukoudian 10 (c)	1,2,3	500 ± 250 Ka	Chen and Zhang, 1991
Zhoukoudian 11 (c)	1,3,4,5	500 ± 250 Ka	Chen and Zhang, 1991
Zhoukoudian 12 (c)	1-5	500 ± 250 Ka	Chen and Zhang, 1991
Zhoukoudian 1966 (c)	1,2	200 ± 50 Ka	Chen and Zhang, 1991

Table 2. Transitional specimens are either original (o) or cast (c). The curves measured are (1) the glabellar or midsegment of the brow, (2) the lateral brow, (3) the temporal line, (4) the coronal suture and (5) the superior nuchal line. Full citations for dating sources can be found in Dean (1993)

Specimen	Ridge curves	Date	Dating source(s)
Bodo (c)	1,2	350 ± 150 Ka	White, 1986
Ndutu (c)	5	300 ± 100 Ka	Clarke, 1990; Hay, 1990
Salé (o)	3,4,5	300 ± 100 Ka	Hublin, 1985
Broken Hill (o)	1-5	400 ± 200 Ka	Delson, pers. comm. (estimate)
Eliye Springs (c)	4,5	200 ± 100 Ka	Bräuer and Leakey, 1986
Saldanha (c)	1, 2	550 ± 150 Ka	Klein and Cruz-Uribe, 1991
Petralona (o)	1-5	350 ± 150 Ka	Latham and Schwarcz, 1992
Arago (c)	1-3	350 ± 50 Ka	Yokoyama et al., 1991
Swanscombe (o)	5	325 ± 50 Ka	Bowen, et al., 1989
Steinheim (o)	1-5	300 ± 50 Ka	Adam, 1985; cf. Bowen et al., 1989
Reilingen (o)	4,5	200 ± 50 Ka	Dean et al., in prep.

The Transitional Sample

Fossils from the Middle Pleistocene are the earliest to present characters diagnostic for *H. sapiens*. Thus they are the earliest potentially "transitional" forms. Specimens are identified in Table 2.

Table 3. The Modern Human Sample

Museum	Catalog Number	Age	Sex	Race
CMNH	514	30	F	Caucasian
CMNH	560	23	M	Caucasian
CMNH	630	22	M	Caucasian
CMNH	667	24	M	Caucasian
CMNH	681	28	F	Caucasian
CMNH	856	23	M	Caucasian
CMNH	881	38	F	Caucasian
CMNH	883	39	F	Caucasian
CMNH	919	42	M	Caucasian
CMNH	920	42	M	Caucasian
CMNH	939	50	M	Caucasian
CMNH	962	44	M	Caucasian
CMNH	1157	25	F	Caucasian
CMNH	1253	31	F	Caucasian
CMNH	1273	34	F	Caucasian
CMNH	1759	40	F	Caucasian
CMNH	2048	52	F	Caucasian
CMNH	2401	22	M	Caucasian
CMNH	2584	23	M	Caucasian
CMNH	2860	46	F	Caucasian
AMNH	1439	e 40	M	S. Chinese
AMNH	1440	e 40	F	S. Chinese
AMNH	2382	e 25	M	W. African
AMNH	2386	e 25	F	W. African
AMNH	8173	e 25	M	Australian
AMNH	8175	e 25	F	Australian

The Modern Human Sample

A sample of 10 male and 10 female Caucasian specimens, aged 18 to 55, from the Laboratory of Physical Anthropology, Cleveland Museum of Natural History (CMNH), Hamann-Todd morgue collection, were used in this study (see Table 3). Additionally, one male and one female skull of each of three groups (Australian Aborigine, southern Chinese, and West African Black) from the von Luschan collection of the American Museum of Natural History (AMNH) were digitized.

DATA COLLECTION: HARDWARE AND SOFTWARE

Polhemus 3SPACE

Ridge curves were digitized with a handheld electromagnetic digitizer, the Polhemus 3SPACE. Multiple tracings and views proved to be very useful for error correction and confirmed that the same curve could be found on repeated attempts. Operator error (D. Dean) was tested by comparing multiple ridge-curve tracings of several of the modern human specimens obtained at greater than one month intervals.

Ridge Curve Data

The endpoints of each ridge curve serve as registration points and are considered type II landmarks (Bookstein, 1991) because they end at local curvature extrema. The homologous curvature maxima (Hilbert and Cohn Vossen, 1990:173) are also type II landmarks (Bookstein, 1991). These type II landmarks are used here as a basis for ridge-curve superposition and comparison (Cutting et al., 1993).

The intermediate points along each ridge curve, between a registration point and a curvature maximum and between curvature maxima, were splined and resampled at preset percentiles of total arc length.

RIDGE CURVE STATISTICS

Averaging Ridge Curves

A two-step process was used to produce the average ridge curve for each species, as described in Dean (1993) and Cutting et al. (1993). The first step begins after all of the ridge curves in each species sample have been splined and resampled to have the same number of equidistantly spaced points by arc length. Then, in each species sample, all of the ridge curves are aligned (translation, rotation and scale) to best fit (Cutting et al., 1993). The *xyz* coordinates of all of these fitted ridge curves are averaged. The points in the resulting average ridge curve are then splined to produce what is termed the arc-length average.

The second step begins by using each species' arc-length average as a baseline along which the original ridge curves are aligned. Perpendicular bisecting planes are positioned at equidistantly spaced intervals along the arc-length average ridge curve (Fig. 7). The intersected points are averaged. Use of the arc-length average insures that the perpendicular bisectors are positioned roughly perpendicular to the normal direction of the original ridge curves being sampled. This positioning of the bisecting planes avoids multiple and missed hits obtained by other sampling techniques (Dean, 1993).

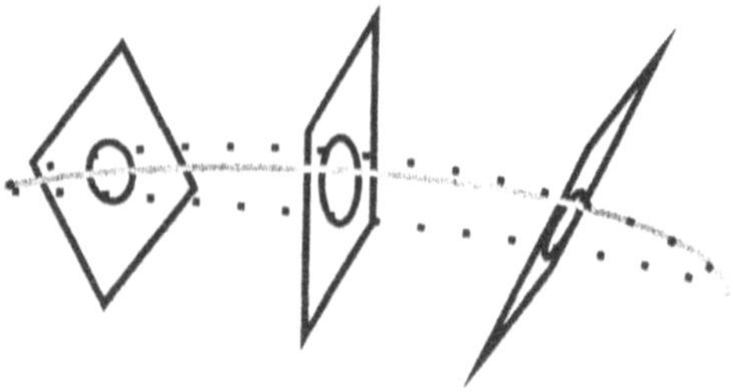

Figure 7. The Two-step average. The gray line is the arc-length average. It was determined by averaging points found at equal arc lengths along the other two curves. It is used to position perpendicular bisectors at specified arc lengths along the fitted samples of ridge curves to be averaged. The ridge curves to be averaged are represented as splines in order to calculate points intersected by the bisecting planes. The averages of the bisected points are used to produce the final average ridge curve.

Chi-Square Statistics

Each transitional ridge curve was compared with both species after matching endpoints to a baseline. This "fit" used the Procrustes least squares algorithm (Rohlf, 1990; Rohlf and Slice, 1990). The transitional specimen curve was then rotated around the common baseline to have minimum sum of squared residuals over curvature maxima landmarks (Fig. 8). Then the distances between intermediate points on both curves were squared and summed and added to the curvature maxima residual sum of squares. The smaller the sum of squares along this "ribbon", the more similar the two ridge curves. This squared Procrustes distance between the two curves at optimal position is used as the squared distance T between the curves in the analyses that follow.

Each specimen within each sample was compared to its species average. The distances were squared and summed to produce a statistic with chi-square (χ^2)-like properties. Converted to probabilities, these statistics were used to measure affinities between the transitional specimen ridge curves and the two species average curves.

The squared distances T sum distances between curves all along their length in a registration given by the position of the "homologous" landmarks. The χ^2 model assumes

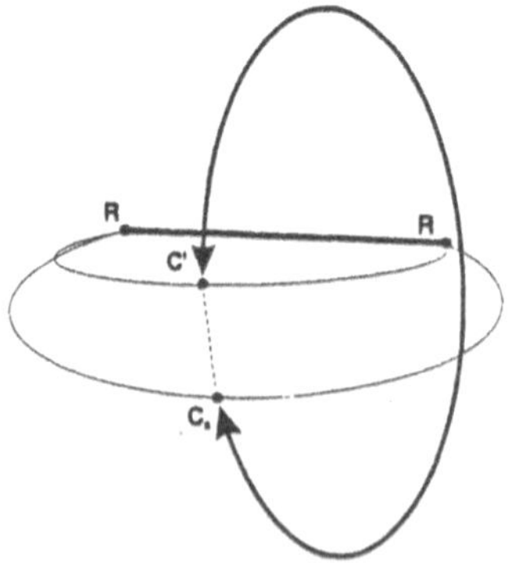

Figure 8. Ridge curve comparison. The two ridge curves are registered at their endpoints (R). The transitional curve is then rotated around their shared chord (bold straight line) to minimize the distances (dashed straight line) between one or more of its curvature maxima (C') and the homologous points on the average curve (Ca).

$$T = c\chi_d^2 \tag{1}$$

where T is the overall sum of squares comparison statistic , c is a coefficient to be estimated, and d represents the degrees of freedom, also to be estimated.

Let x_{ijkl} be the distance between the ith specimen ($i = 1,\ldots,n_k$) and the kth species mean ($k = 1,2$; i. e., *H. sapiens* or *H. erectus*) for the jth intermediate ridge curve point or type II landmark ($j = 1,\ldots,n_l$), for the lth ridge curve, here brow ridge or nuchal torus. We compute the within species sample sum of squares:

$$T_{ikl} = \sum_{j=1}^{n_l} x_{ijkl}^2 \tag{2}$$

Using the method of moments T/c is fitted to a χ^2 . This means that T is modeled as having mean

$$\mu_T = cd \tag{3}$$

and variance

$$\sigma_T^2 = 2c^2 d \tag{4}$$

Setting the sample moments of T equal to theoretical moments gives sample mean and variance estimates

$$\overline{T} = \hat{c}\,\hat{d} \tag{5}$$

and

$$s_T^2 = \hat{c}^2\, 2\hat{d} \tag{6}$$

so that the estimates are now:

$$\hat{c} = \frac{\overline{T}}{\hat{d}} \tag{7}$$

and

$$\hat{d} = 2\frac{\overline{T}^2}{s_T^2} \tag{8}$$

Then

$$\frac{T_{kl}}{\hat{c}}$$

will have an approximate χ^2 distribution with $\hat{d}$ degrees of freedom for the transitional specimens if they typify one or the other species.

The goodness of fit of the χ^2 model may be evaluated by a *p–p* plot (Wilk and Gnanadesikan, 1968). This plot compares the empirical cumulative distribution function F of T_{ikl} for each species to the cumulative distribution function of a χ^2 distribution with $\hat{d}$ degrees of freedom, $\hat{F}$. The plots were created in MATLAB. The expected F produces a line from 0,0 to 1,1.

The Kolmogorov–Smirnov statistic D was used to test for goodness of fit (Wilk and Gnanadesikan, 1968; Sokal and Rohlf, 1981). The K-S test statistic D is:

$$D = \max | F - \hat{F} | \tag{9}$$

If D was larger than the tabled 0.01 D value (Rohlf and Sokal, 1981) the χ^2 model was rejected. A significance level of α equal to 0.01 was used throughout this study. Values of D were small in this study, and the fit was adequate for both ridge curves for both species.

The chi-square model was appropriate: thus we compared the transitional curves (each C_t) to the sample averages for each ridge curve. We used the computed χ^2 and probability to make species assignments.

Species Attribution

A major assumption underlying this study was that there were only two hominid species present in the Middle Pleistocene. If this is true, then the transitional fossils should show unambiguous similarities with one of the two species. If the features measured are not similar, or if the measure is too crude, it is possible that a transitional specimen will not be similar to either species average, as occurs in a number of cases. Species assignment is justified only if the distribution of the T statistic has a χ^2 distribution.

Thus each of the transitional specimens and six non-Caucasian skulls (labeled AMNH in Table 3) were compared with the averages of both species. The right-tail probabilities for each species χ^2 is given in Tables 4 and 5. Greater affinity to either average ridge curve is expressed by a higher probability and disaffinity by a lower probability.

The AMNH non-Caucasian specimens give an estimate of the level of variability for each character within anatomically modern *H. sapiens*. It was initially hypothesized that for a given ridge curve the AMNH specimens would show greater affinities with the CMNH Caucasian average ridge curve than with the *H. erectus* average. In the case of ridge curves 2–4 there was significant overlap, with some AMNH specimens showing higher affinity with the *H. erectus* average. Therefore, ridge curves 2 to 4 were not considered further.

In addition a probability ratio, *PR* =(*H. sapiens* empirical probability)/(*H. erectus* empirical probability) was calculated as an affinity index. If there is little or no overlap of the AMNH specimens for this statistic, it provides a numerical diagnosis. This statistic is consistent in most regards with a direct examination of the probabilities. However it may be misleading when both probabilities are very small, and attribution is made ignoring this fact.

A *PR* of 1.0 or larger indicated *H. sapiens* affinities and less than 1.0 indicated *H. erectus* affinities. Based on all of the *PR*s in this study, a stronger allocation rule for *H. sapiens* is a *PR* greater than 2.0 and for *H. erectus* a *PR* less than 0.5. Results where the probabilities are low ($p < 0.01$) for both or neither species are a problem. Either would indicate that species affinity assignment cannot be made for that ridge curve.

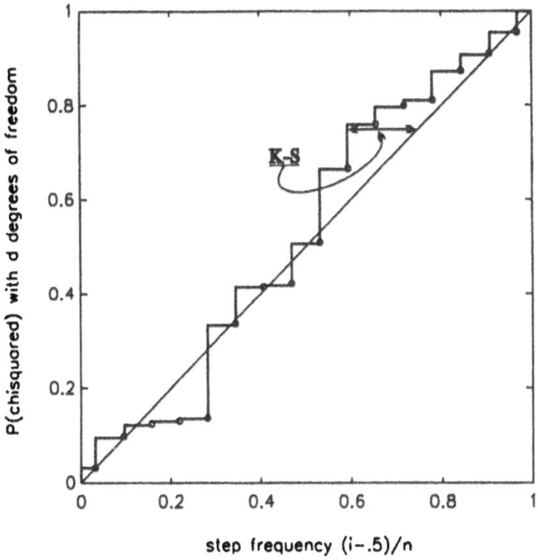

Figure 9. *H. erectus* glabellar ridge curve sample *p–p* plot. Length of line K-S is value of Kolmogorov-Smirnov *D* statistic.

RESULTS

Glabellar Ridge Curve

The *p–p* plot and *D* statistic for *H. erectus* is given in Fig. 9 and for *H. sapiens* in Fig. 10. The *T* values, empirical probabilities, and χ^2 model probabilities for *H. erectus* are given in Tables 4 and for *H. sapiens* in Table 5.

There is no sample overlap in this case. All of the non-Caucasian specimens show clear-cut affinities with the Caucasian average ridge curve and none exhibit significant

Table 4. *H. erectus* glabellar ridge curve intrasample data ($n = 16$). The mean of $T = 4.2988$, variance = 4.5620, $c = 0.7284$, and $d = 3.1314$

Specimen	*T*	Empirical probability	χ2 probability
Zhoukoudian 1966	0.6555	0.0312	0.0316
Solo 7	1.2058	0.0938	0.0978
Solo 12	1.3899	0.1562	0.1251
Solo 10	1.4276	0.2188	0.1309
KNM-ER 3883	1.4646	0.2812	0.1366
KNM-ER 3733	2.6401	0.3438	0.3358
Sangiran 17	3.1096	0.4062	0.4146
Zhoukoudian 3	3.1578	0.4688	0.4224
Zhoukoudian 11	3.7181	0.5312	0.5096
Zhoukoudian 12	4.9144	0.5938	0.6655
Lantian	5.8749	0.6562	0.7600
Solo 5	6.3739	0.7188	0.7995
OH 9	6.5446	0.7812	0.8116
Solo 11	7.6130	0.8438	0.8737
Solo 4	8.4966	0.9062	0.9102
Zhoukoudian 10	10.1948	0.9688	0.9544

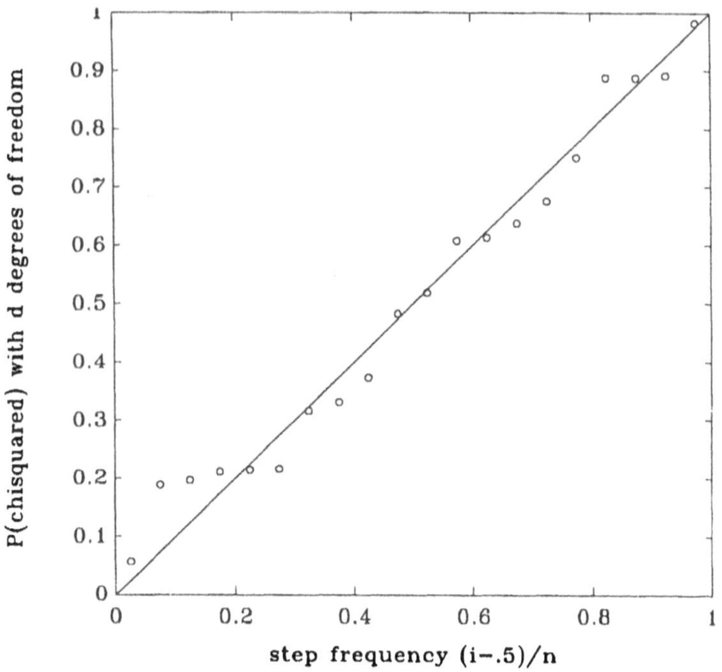

Figure 10. *H. sapiens* glabellar ridge curve sample *p–p* plot.

Table 5. *H. sapiens* glabellar ridge curve intrasample data ($n = 20$). The "unknown" non-Caucasian (AMNH) are listed by type and the transitional fossils are listed by site in Table 6. The mean of $T = 7.4307$, variance = 44.6095, $c = 3.0017$, and $d = 2.4755$

Specimen	T	Empirical probability	χ^2 probability
CMNH 681	0.2282	0.0250	0.0568
CMNH 1157	0.6595	0.0750	0.1884
CMNH 2584	0.6880	0.1250	0.1970
CMNH 920	0.7327	0.1750	0.2105
CMNH 1273	0.7458	0.2250	0.2144
CMNH 514	0.7492	0.2750	0.2154
CMNH 560	1.0998	0.3250	0.3164
CMNH 2401	1.1556	0.3750	0.3317
CMNH 881	1.3141	0.4250	0.3736
CMNH 1759	1.7736	0.4750	0.4834
CMNH 667	1.9402	0.5250	0.5190
CMNH 919	2.4102	0.5750	0.6080
CMNH 883	2.4394	0.6250	0.6130
CMNH 1253	2.5908	0.6750	0.6380
CMNH 630	2.8362	0.7250	0.6754
CMNH 856	3.4240	0.7750	0.7508
CMNH 2860	5.1651	0.8250	0.8880
CMNH 2048	5.2365	0.8750	0.8882
CMNH 962	5.1674	0.9250	0.8917
CMNH 939	9.1540	0.9750	0.9830

Table 6. Glabellar ridge curve. The He $d = 4.2988$, $c = 0.7284$. The Hs $d = 2.4755$, $c = 3.0017$

Specimen	Un-He T Statistic	Un-He χ^2 Probability	Un-Hs T Statistic	Un-Hs Emp. χ^2 Probability	PR^2
AMNH 1439 M Chin	8.5966	0.0235	11.3079	0.2133	9.1
AMNH 1440 F Chin	10.3836	0.0083	13.5849	0.1506	18.1
AMNH 2382 M WAfr	10.7346	0.0068	14.1234	0.1386	20.4
AMNH 2386 F WAfr	9.5150	0.0139	17.0494	0.0881	6.3
AMNH 8173 M Aust	8.3077	0.0278	17.6854	0.0797	2.9
AMNH 8175 F Aust	8.4822	0.0251	12.0233	0.1913	7.6
Arago	2.6209	0.5084	23.4688	0.0321	0.1
Bodo	14.2666	0.0008	16.9223	0.0898	112.3
Kabwe	15.3827	0.0004	19.3948	0.0610	152.7
Petralona	13.3719	0.0014	3.1874	0.6941	495.8
Saldanha	2.2553	0.5876	18.3423	0.0720	0.1

affinity with *H. erectus*. Among the transitional specimens, Arago and Saldanha show stronger affinities with *H. erectus*. The other transitional specimens, Bodo, Kabwe and Petralona, show stronger affinities with *H. sapiens*.

The *H. sapiens* sample was, unexpectedly, more variable than was the *H. erectus* sample (see the variances in Tables 4 and 5). Judging from the other ridge curves, and the fact that it was drawn from a larger geographic and temporal span, it seemed that the *H. erectus* sample should be more variable. However, despite this unexpected pattern of variability, all five transitional specimens showed greater affinities with the modern sample than any of the non-Caucasian specimens showed with the *H. erectus* average.

Superior Nuchal Line or Nuchal Torus Ridge Curve

The "unknown" non-Caucasians (AMNH) are listed by type and the transitional fossils are listed by site in Table 7.

As with the glabellar ridge curve there was little overlap. Three of the non-Caucasian specimens (AMNH 1439, 2382 and 8175) show χ^2 probabilities of greater than 0.1 of falling within the *H. erectus* sample; however, all three of these specimens also show greater affinity

Table 7. Findings: Superior Nuchal Line. The He $d = 4.3421$, $c = 62.6186$. The Hs $d = 6.9551$, $c = 66.2133$

Specimen	Un-He T Statistic	Un-He χ^2 Probability	Un-Hs T Statistic	Un-Hs Emp. χ^2 probability	PR
AMNH 1439 M Chin	428.50	0.1724	157.9591	0.9336	5.4
AMNH 1440 F Chin	953.20	0.0057	170.7828	0.9188	161.2
AMNH 2382 M WAfr	467.60	0.1366	274.8989	0.7579	5.6
AMNH 2386 F WAfr	862.20	0.0106	53.3090	0.9973	94.1
AMNH 8173 M Aust	700.60	0.0313	61.4035	0.9957	31.9
AMNH 8175 F Aust	326.80	0.3071	155.0844	0.9367	3.1
Eliye Springs	131.00	0.7645	720.7054	0.1411	0.2
Ndutu	214.40	0.5416	485.5790	0.3902	0.7
Petralona	434.00	0.1669	230.2105	0.8342	5.0
Reilingen	303.60	0.3477	393.7638	0.5409	1.6
Salé	1457.1	0.0002	269.8942	0.7668	3834.0
Steinheim	1803.0	0.0000	248.9834	0.8031	>8031.0
Swanscombe	1545.0	0.7000	497.4702	0.3728	0.5

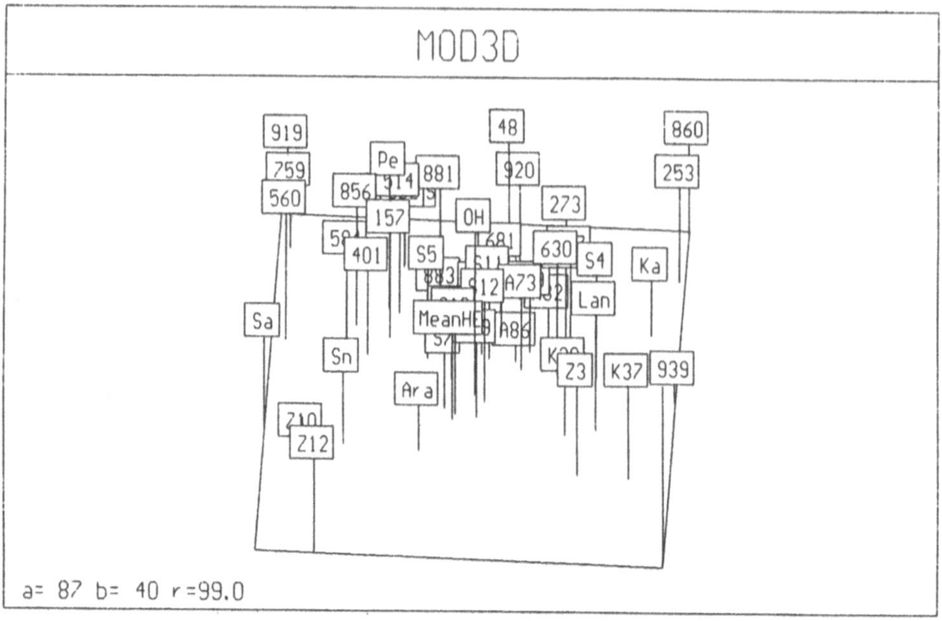

Figure 11. Principal coordinates plot of the glabellar ridge curve results. Numbered specimens correspond to modern human specimens listed in Table 3; unlabeled numbers are from the CMNH; other specimens are given as A (AMNH), HS (*H. sapiens)*, HE (*H. erectus)*, and from left to right; Pe (Petralona), Bo (Bodo), Ka (Kabwe), Sa (Saldanha), OH (OH9), S (Sangiran or Solo), K (Kenya National Museum), Ara (Arago), Z (Zhoukoudian). The second principal coordinates were multiplied by 2 to produce more equal length axes on the principal coordinate 1 and 2 plane.

with the *H. sapiens* average as all *PR*s are greater than 3.1. The Salé and Steinheim specimens show very strong similarity to *H. sapiens* with a *PR* greater than 3800 for both. However, the Salé specimen may exhibit a congenital deformation in this area (Hublin 1985, personal communication).

The Ndutu *PR* ratio is equivocal. The Swanscombe and Eliye Springs *PR*s demonstrate *H. erectus* affinities. Judging from the cast of KNM-ES (Eliye Springs) 11693 that we used, and photographs of the original, it seems possible that this cranium has been plastically deformed in this area. Therefore, interpretation has to be tempered. The Ndutu specimen may also be diagenetically deformed, through both plastic warping and breakage. There may be errors in the reconstruction (Clarke, 1990).

Reilingen and Swanscombe show the most affinity with *H. erectus*. Perhaps the incipient "en bombe" morphology in these two specimens is an evolutionary reversal? Petralona, Salé and Steinheim show clear affinities with *H. sapiens*.

Another Look: Principal Coordinates

An ordination of the transitional, *H. sapiens* and *H. erectus* specimens for the glabellar and nuchal torus ridge curves was produced to display the associations of the specimens. Principal coordinates (Gower, 1966) was used for the ordination.

Figures 11 and 12 show the principal coordinates based on the glabellar and nuchal torus ridge curve distance matrices, respectively. All distances were computed in the same way as the χ^2 test. However, in this case all ridge curves were scaled to have the same distance between the first and last point. The principal coordinate analysis was done in NTSYS-pc

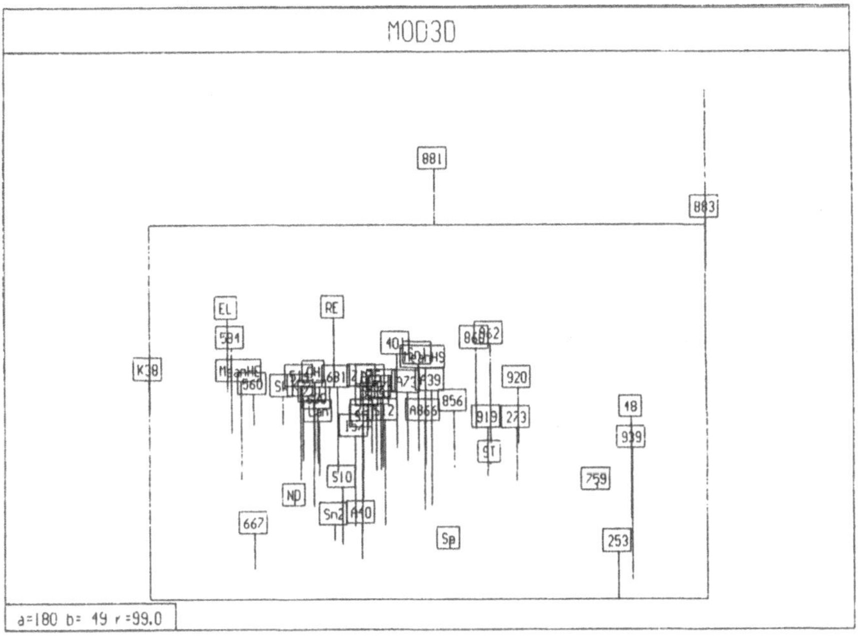

Figure 12. Principal coordinates plot of the nuchal torus ridge curve results. Key: Numbered specimens correspond to modern human specimens listed in Table 3; unlabeled numbers = Cleveland Museum; A = American Museum; HS = *H. sapiens*; HE = *H. erectus*, from left to right: K = Kenya National Museum; EL = Eliye Springs; Lan = Lantian, OH = OH9; Z = Zhoukoudian; S or Sn = Sangiran or Solo; RE = Reilingen; SW = Swanscombe; ND = Ndutu; Ka = Kabwe; Sa = Saldanha.

(Rohlf, 1993), and 3D graphic tools in that package were used to display the first three principal coordinates.

Whereas the χ^2 probabilities were used to assign the unknowns by looking at the distances from the *H. erectus* and *H. sapiens* average ridge curves separately, a principal coordinates ordination for all specimens displays a projection of the specimens in two or three dimensional space. The results show good separation of the two groups, with the transitional forms falling within the *H. sapiens* data point clouds. In both cases the *H. sapiens* specimens are more dispersed than are those for *H. erectus*.

CONCLUSIONS

The χ^2 model allowed an assignment of the transitional specimens to either *H. erectus* or *H. sapiens*. For ridge curves 1 and 5 there were clear affinities with *H. sapiens*. The transitional specimens and the AMNH non-Caucasian modern human specimens did not show clear-cut affinities with the two species for characters 2–4 which suggests those characters are of little or no diagnostic value.

The principal coordinate analysis showed that the *H. erectus* dispersion was less than that for *H. sapiens*. The 3D projection of the multidimensional scatter of all the specimens showed where each transitional ridge curve could be found within the modern human sample data cloud. These analyses reinforce the broad conclusion that Middle Pleistocene transitional specimens present the earliest evidence for *H. sapiens* (Dean, 1993).

ACKNOWLEDGMENTS

Court B.Cutting, New York University School of Medicine, coded the ridge-curve averaging program (discussed in Dean, 1993). Eric Delson supervised the dissertation work partially summarized here. Bruce Latimer, CMNH, kindly arranged several loans of the Caucasian human skulls used in this study. Ian Tattersall, AMNH, graciously provided access to the von Luschan collection. Original fossil material used in this study was provided by Jens Franzen, Senckenberg Museum, Frankfurt; Jean-Jacques Hublin, Musee de l'Homme, Paris; George Koufos, Aristotle University, Saloniki; Christopher Stringer, British Museum (Natural History), London; and Reinhard Zeigler, Naturkunde Museum, Stuttgart. Monetary support for many pilot projects, travel expenses, data collection and other activities surrounding the thesis project reported here was generously provided in grants directly to D. Dean from the Sigma-Xi Society; Senckenberg Museum, Frankfurt, Germany; Wenner-Gren Foundation; L.S.B. Leakey Foundation; and especially a City University of New York Dissertation Fellowship. The New York Consortium for Evolutionary Primatology provided D. Dean with access to a 3SPACE digitizer and a Silicon Graphics workstation during the data analysis. Finally, special thanks to colleagues Russell Taylor and Alan Kalvan, Computer Assisted Surgery Group, IBM T. J. Watson Research Center, Yorktown Heights., New York for providing D. Dean with an IBM PS/2 model 70 to interface with Dr. Cutting's Polhemus 3SPACE 3D digitizer during travels to museums in Europe during the summer and fall of 1991.

REFERENCES

Ashton, E. H. 1950. The endocranial capacities of the Australopithecinae. Proceedings of the Zoological Society of London 120:715–721.

Bookstein, F. L. 1991. Morphometric tools for landmark data: Geometry and biology. Cambridge University Press: New York.

Bookstein, F. L. and C. B. Cutting. 1988. A proposal for the apprehension of curving craniofacial form in three dimensions. Pages 127–140 *in* K. Vig and A. Burdi (eds.), Craniofacial morphogenesis and dysmorphogenesis. Center for Human Growth and Development: Ann Arbor.

Brown, P. 1989. Coobool Creek: A morphological and metrical analysis of the crania, mandibles and dentitions of a prehistoric Australian human population. Terra Australis 13:1–205.

Clarke, R. 1990. The Ndutu cranium and the origin of *Homo sapiens*. Journal of Human Evolution 19:699–736.

Cutting, C. B., F. L. Bookstein, B. Haddad, D. Dean, and D. Kim. 1993. A spline-based approach for averaging three-dimensional curves and surfaces. Soc. Pages 29–44 *in* I. N. Wilson and D. C. Wilson (eds.), Mathematical methods in medical imaging II. Society of Phto-Optical Institute Engineers 2035.

Cutting, C., D. Dean, F. Bookstein, B. Haddad, D. Khorramabadi, F. Zonnefeld, and J. McCarthy 1995. A three-dimensional smooth surface analysis of untreated Crouzon's disease in the adult. Journal of Craniofacial Surgery 6:444–453.

Dean, D. 1993. The Middle Pleistocene *Homo erectus/Homo sapiens* transition: New evidence from space curve statistics. Ph.D. dissertation; The City University of New York.

Gower, J. 1966. Some distance properties of latent root vector methods used in multivariate analysis. Biometrika 53:325–338.

Hilbert, D., and S. Cohn-Vossen. 1990. Geometry and the imagination. Chelsea: New York.

Hublin, J-J. 1985. Human fossils from the North African Middle Pleistocene and the origin of *Homo sapiens*. Pages 283-288 *in* E. Delson (ed.), Ancestors: The hard evidence. Alan R. Liss: New York.

Jacob, T. 1976. Solo Man and Peking Man. Pages 87–104 *in* B. A. Sigmon and J. S. Cybulski (eds.), Papers in honor of Davidson Black. University of Toronto Press: Toronto.

Le Gros Clark, W. E. 1955. The fossil evidence for human evolution, 3rd edition, University of Chicago Press: Chicago.

Le Gros Clark, W. E. 1967. Man apes or ape-men. Robert E. Krieger: New York.

Macintosh, N. W. G., and S. L. Larnach. 1972. The persistence of *Homo erectus* traits in Australian Aboriginal crania. Oceania 7:1–7.

Mayr, E. 1951. Taxonomic categories in fossil hominids. Cold Spring Harbor Symposia in Quantitative Biology 15:109–117.

Rightmire, G. P. 1990. The evolution of *Homo erectus*. Cambridge University Press: Cambridge, U. K.

Rohlf, F. J. 1990. Rotational fit (Procrustes) methods. Pages 227–236 *in* F. J. Rohlf and F. L. Bookstein (eds.), Proceedings of the Michigan morphometrics workshop. University of Michigan Museum of Zoology Special Publication 2.

Rohlf, F. J. 1993. NTSYS-pc. Numerical taxonomy and multivariate analysis system, version 1.80. Exeter Software: Setauket, New York.

Rohlf, F. J., and D. E. Slice. 1990. Extensions of the Procrustes method for the optimal superimposition of landmarks. Systematic Zoology 39:40–59.

Rohlf, F. J., and R. R. Sokal. 1981. Statistical tables. W. H. Freeman: New York.

Russell, M. D. 1983. The functional and adaptive significance of the supraorbital torus. Ph.D. dissertation, The University of Michigan.

Santa Luca, A. P. 1980. The Ngandong fossil hominids. Yale University Publications in Anthropology 78:1–175.

Sokal, R. R., and F. J. Rohlf. 1981. Biometry: The principles and practice of statistics in biological research, 2nd edition, W. H. Freeman: San Francisco.

Stringer, C. B. 1984. The definition of *Homo erectus* and the existence of the species in Africa and Europe. Cour. Forsch. Inst. Senckenberg 69:131–143.

Turner, A., and A. Chamberlain. 1989. Speciation, morphological change and the status of African *Homo erectus*. Journal of Human Evolution 18:115–130.

Weidenreich, F. 1943. The skull of *Sinanthropus pekinensis*; a comparative study on a primitive hominid skull. Palaeontologica Sinica Series D, 10.

Weidenreich, F. 1951. Morphology of Solo man. Anthropological Papers of the American Museum of Natural History 43:205–288.

Wilk, M. B., and R. Gnanadesikan. 1968. Probability plotting methods for the analysis of data. Biometrika 55:1–17.

Wood, B. A. 1984. The origin of *Homo erectus*. Cour. Forsch. Inst. Senckenberg 69:99–111.

SPLINE CURVES AND SURFACES FOR DATA MODELING

André Guéziec*

Computer Assisted Surgery Group
IBM T. J. Watson Research Center
P. O. Box 704
Yorktown Heights, New York 10598

ABSTRACT

We provide a brief introduction to spline functions. In modeling data, splines are associated with the notion of smoothness. They really are the analog to the draftsman's wooden spline. They are useful for curve and surface interpolation or approximation.

Our goal is to give a simple and original development of the basic material needed for understanding spline curves and the notion of curvature associated with them. The advantages of using splines for analyzing data are that splines are computationally simple and they satisfy the minimum curvature property. A review of algorithms involving splines as well as a comprehensive theory can be found elsewhere.

We begin by quickly tracing the ancestry of splines, which dates back to the eighteenth century. Then, we consider a quadratic Bézier curve, defined with only three points. In order to construct complex curves, we piece together Bézier curves, and, when specific continuity conditions are satisfied, we obtain B-spline curves. We provide a general definition of spline curves, comprising both Bézier and B-spline curves. We finish by combining B-spline curves to form tensor product B-spline surface patches.

INTRODUCTION

Spline functions have long been studied by mathematicians. Their origin dates back at least to the eighteenth century work by Leonard Euler (1755), and the modern theory makes ample use of the polynomials invented by Bernstein (1912).

After the 1940s, splines were investigated in the field of approximation theory (Schoenberg, 1946), and the minimum curvature property (Holladay, 1957), or minimum norm property, was discovered (Ahlberg et al., 1967). The minimum norm property can be

* e-mail: gueziec@watson.ibm.com.

roughly stated as follows: among all curves that pass through (i.e. interpolate) a set of points, the cubic spline has minimum curvature The curvature is a local quantity that measures how much a curve or surface is bent. In two dimensions it is computed by considering locally the best approximation of the curve with a circle and taking the inverse radius of that circle (Hilbert and Cohn-Vossen, 1990 is an accessible textbook). A formula is provided in the Appendix to this article. We shall see in the Appendix that only an approximation to curvature is meant. Thus, there is an analogy with the draftsman's wooden spline, made to pass through specified locations while being as fair as possible.

Indeed, it is through industrial design that the splines became popular as a computer-compatible description of three dimensional shapes for the purpose of machining the shapes. In the 1960s, the pioneering research was performed in the French automobile industry by Bézier (1966, 1967) at Renault and by De Casteljau at Citroën. De Casteljau was not granted permission to publish his work. Thus, independently from mathematicians engineers developed a set of mathematical tools because they needed them.

There is an extensive literature on splines for computer-aided design, now a well-established field. This includes comprehensive textbooks (de Boor, 1978; Bartels et al., 1987), as well as a surveys (Bohm et al., 1984). We generally use the same notation as Bartels et al. (1987). Farin (1988) is notable for using numerous examples and postponing the theory until it is essential. We take inspiration from this method.

In the section "Bézier Curves", we begin by introducing the simplest Bézier curves and the De Casteljau algorithm for constructing the curves. B-spline curves encompass a larger class of curves. Every Bézier curve is also a B-spline curve. Further, we show in the section "B-Spline Curves as Piecewise Bézier Curves" that B-spline curves can be obtained by piecing Bézier curves together. We then explain in the section "Curve Approximation Using B-Splines" how B-splines can be used for the approximation of an outline. Also in this section, we complement the contribution of Rohlf (1990) by giving some reasons why splines are useful for analyzing curve and outline data. Then, in "General Definition of a Spline Curve," we show that both B-spline and Bézier curves are in fact different representations of the same spline curves. For some applications, one representation is preferable to the other. Finally in the section "Surfaces as Tensor Products of B-Spline Curves," we introduce tensor-product B-splines, which are based upon B-spline curves and provide a popular description for surfaces.

We give an example of use in the analysis of medical data.

BÉZIER CURVES

We start with an ordered set of points, which we call control vertices. The simplest Bézier curve is a line segment joining two control vertices V_0 and V_1. By adding one more control vertex, V_2, we can define a quadratic Bézier curve (see Fig. 1, left). Here is a simple way to construct the curve. Take the middle of V_0 and V_1, $V_{0,1}$, then take the middle of V_1 and V_2, $V_{1,2}$, and finally take the middle of $V_{0,1}$ and $V_{1,2}$, $V_{0,1,2}$, also denoted by $V_{0,2}$. $V_{0,2}$ belongs to the quadratic Bézier curve defined with V_0, V_1, V_2. Then apply exactly the same process to the triples of points formed by V_0, $V_{0,1}$, $V_{0,2}$ and $V_{0,2}$, $V_{1,2}$, V_2. Four triples of points are created, which can be split again in eight triples, and so on. This process of refinement is known as recursive subdivision. Stop when the first and last points in any triple are close enough to each other.

Further, if instead of taking the middle of the segment V_i, V_{i+1}, we consider the point defined by $V_{i,k+1} = (1-\lambda)V_i + \lambda V_{i+1}$, where λ varies between 0 and 1, then we can write:

$$V_{0,2}(\lambda) = (1-\lambda)V_{0,1} + \lambda V_{1,2} = (1-\lambda)^2 V_0 + 2(1-\lambda)\lambda V_1 + \lambda^2 V_2 \tag{1}$$

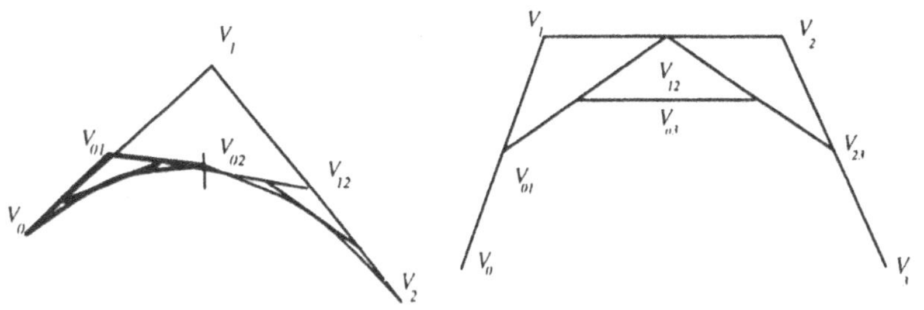

Figure 1. A quadratic Bézier curve (left) is determined with three control vertices V_0, V_1, V_2. Recursive subdivision—V_0, V_{01}, V_{02} determine the first half of the curve. A cubic Bézier curve (right) is determined by four control vertices, V_0, V_1, V_2 and V_3. We first construct V_{01}, V_{12}, V_{23}, which define a quadratic curve. We then apply the De Casteljau algorithm to construct the vertex V_{03} using V_{01}, V_{12}, V_{23}.

Equation (1) is that of a quadratic Bézier curve. Note that V_0 and V_2 are the two endpoints of the curve. However, V_1 does not belong to the curve. In the previous paragraph, we computed $V_{0,2}$ for $\lambda = 0.5$. The polynomials of the variable λ: $(1-\lambda)^2$, $2(1-\lambda)\lambda$, and λ^2 are the Bernstein polynomials of degree two. They are illustrated in Fig. 3 (left). This process, of constructing a curve point $V_{0,2}$ for each real value λ between 0 and 1, is known as the De Casteljau algorithm for evaluation of Bézier curves.

If we had started with four control vertices, V_0, V_1, V_2, V_3, instead of three, then we would have obtained a cubic curve. Cubics are easily deduced from quadratics. Consider the two quadratic curves defined by V_0, V_1, V_2 and V_1, V_2, V_3. For a given value of the parameter λ, two points are obtained, $V_{0,2}(\lambda)$ and $V_{1,3}(\lambda)$, located respectively on the first and second quadratic Bézier curve. The cubic curve can be obtained by a weighted sum of these two points: $V_{0,3}(\lambda) = (1-\lambda)\, V_{0,2} + \lambda\, V_{1,3}$. Another method is to form first the triple, $V_{0,1}$, $V_{1,2}$, $V_{2,3}$, by using our classical $(1-\lambda)$, λ weighted sum as shown in Fig. 1(right), and to use this triple as control vertices for a quadratic, thus resulting in (by means of Equation 1): $V_{0,3}(\lambda) = (1-\lambda)^2 V_{0,1} + 2(1-\lambda)\lambda V_{1,2} + \lambda^2 V_{2,3}$. Finally, we obtain the equation of a cubic Bézier curve, in terms of the original control vertices V_0, V_1, V_2, V_3, as

$$V_{0,3}(\lambda) = (1-\lambda)^3 V_0 + 3(1-\lambda)^2 \lambda V_1 + 3(1-\lambda)\lambda^2 V_2 + \lambda^3 I V_3 \qquad (2)$$

Cubic Bézier curves are more often used than are quadratic Bézier curves, for the following reasons: quadratics are necessarily planar whereas cubics are not (we encourage the reader to verify this) and cubics satisfy the minimum curvature property (see Appendix), which make them a much more suitable tool for data approximation or interpolation.

Five control vertices define a quartic curve, six control vertices define a quintic curve, and so on. For many applications a high number of control vertices is needed. In order to avoid increasing the degree of the Bézier curve, we need to piece together Bézier curves. We will denote these pieces by curve segments. We shall see that when continuity conditions are satisfied, the result is a B-spline curve.

B-SPLINE CURVES AS PIECEWISE BÉZIER CURVES

Let us consider two Bézier curves that share a vertex and are defined by the following control triples: V_0, V_1, V_2 and V_2, V_3 V_4. Without special attention to positioning V_0, V_1, V_3 and V_4 there will be a crease at vertex V_2, as visualized in Fig. 2. A crease will mean that the joint

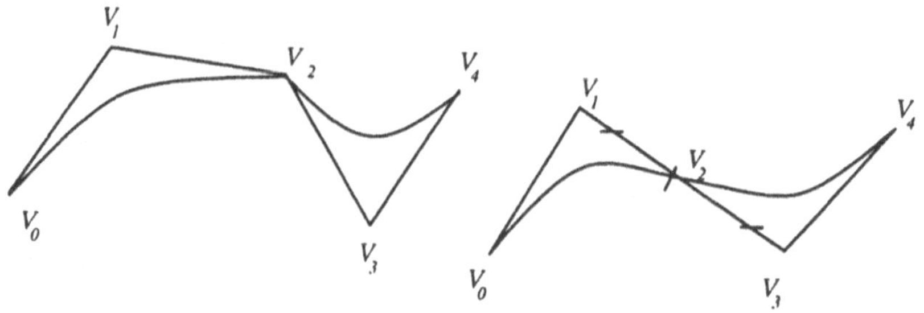

Figure 2. A C^0 (left), or angular joint, and a C^1 (right), or smooth joint, at V_2 between two quadratic Bézier curves. In the C^1 case, V_0, V_1, V_3 and V_4 define a quadratic B-spline curve.

between the two curves is simply continuous, or C^0. A curve defined with a function $f(u)$ on an interval $I \supset u$, as the loci of the points $(u, f(u))$, is said to be continuous or C^0 if and only if $f(u+h) - f(u)$ has a limit for each u when h approaches zero. The derivative curve $(u, f'(u))$, exists if and only if $(f(u+h) - f(u))/h$ has a limit for each u when h approaches zero. If it exists, the limit is noted $f'(u)$, and the curve is said to be one time differentiable. If $(u, f'(u))$ is continuous, then we say that the curve is C^1. The curve $(u, f'(u))$ can also have a derivative $(u, f''(u))$. If this curve is continuous, then $(u, f(u))$ is said to be C^2, and so on. The book on calculus by Smith (1983:278–282) provides a good review on continuity and differentiability. We may require that the derivatives of the two curves (Fig. 2) coincide at V_2, and thus obtain a C^1 joint.

By examining the derivatives of the two curves at V_2, we can see that for the C^1 condition the vectors V_1 to V_2 and V_2 to V_3 are the same, i.e.,

$$\overrightarrow{V_2 - V_1} = \overrightarrow{V_3 - V_2}.$$

If this condition is satisfied, V_0, V_1, V_2, V_3 and V_4 determine a quadratic B-spline curve (by definition C^1, and further, V_2 can be dropped. Similarly to equation 1, by eliminating V_2 in the equations of the Bézier curves, we can write the equation of a quadratic B-spline curve S with four control vertices V_0, V_1, V_3 and V_4. Traditionally, the parameter used for B-spline curves is denoted u. This is why we switch from λ to u in this section. The parameter u varies between 0 and 2 because we juxtapose two curves for which the parameter u varies between 0 and 1. Thus, in the second portion, we translate the parameter from [0,1] to [1,2]:

$$S(u) = \begin{cases} 0 \leq u < 1: & (1-u)^2 V_0 + u(2 - \frac{3}{2}u)V_1 + \frac{u^2}{2}V_3 \\ 1 \leq u \leq 2: & \frac{(2-u)^2}{2}V_1 + (2-u)(\frac{3}{2}u - 1)V_3 + (u-1)^2 V_4 \end{cases} \tag{3}$$

The polynomials in u involved in equation 3 are called B-spline functions. There are four piecewise polynomials associated respectively with V_0, V_1, V_3 and V_4. Piecewise means that the formula changes between the intervals [0,1] and [1,2], whereas the function is a polynomial in both intervals. However, the polynomials are all continuous with a continuous derivative (C^1) at $u = 1$, which means that they are C^1 across intervals. The four B-spline functions $\{B_{j,3}\}$ are:

$$B_{0,3} = (1-u)^2, u \in [0,1[, B_{0,3} = 0, u \in [1,2]; B_{1,3} = u(2-\tfrac{3}{2}u), u \in [0,1[, \tfrac{(2-u)^2}{2}, u \in [1,2]; B_{2,3} = \tfrac{u^2}{2}, u \in [0,1[, (2-u)(\tfrac{3}{2}u-1), u \in [1,2]; B_{3,3} = 0, u \in [0,1[, (u-1)^2, u \in [1,2].$$

They are illustrated in Fig. 3 (right).

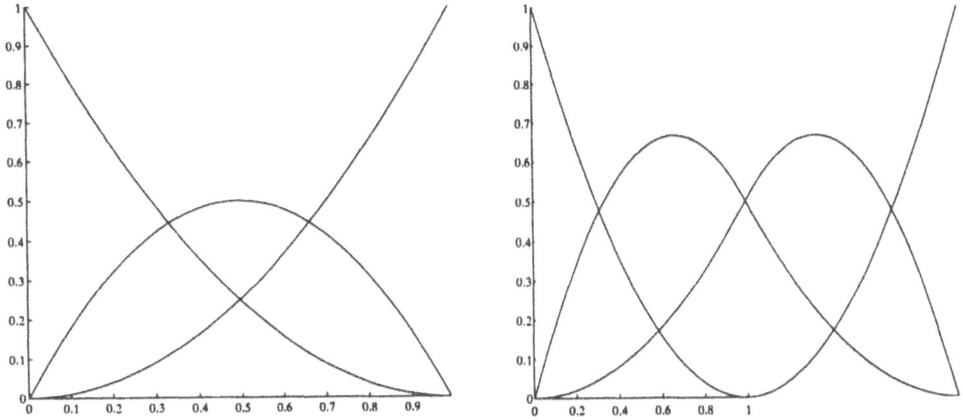

Figure 3. Three Bézier (left) and four B-spline (right) functions used, respectively, in the definition of a quadratic Bézier and B-spline curve.

The technique of piecing Bézier curve segments together is useful for data interpolation, i.e., when it is required to pass through each data point while minimizing the wiggling between points. A Bézier curve segment is created between each pair of consecutive data points. When this is done, the curve segments are pieced together in a C^1 or C^2 fashion. Data approximation is generally performed when the number of data points is too high or when the data is corrupted with noise. Contrary to the interpolation problem, it is not required to pass exactly through the data points. A fitness criterion is derived and the parameters of the curve are optimized in order to satisfy the fitness criterion. We develop data approximation by use of B-spline curves in the next section.

CURVE APPROXIMATION USING B-SPLINES

To complete the description given by Rohlf (1990) of cubic spline interpolation, we now show how B-splines can be used for approximating a discrete data curve. We use the quadratic B-spline curve with four control vertices obtained in the previous section. We assign variable coordinates to the four control vertices:

$$\left(V_0 \begin{pmatrix} X_0 \\ Y_0 \end{pmatrix}, V_1 \begin{pmatrix} X_1 \\ Y_1 \end{pmatrix}, V_2 \begin{pmatrix} X_2 \\ Y_2 \end{pmatrix}, V_3 \begin{pmatrix} X_3 \\ Y_3 \end{pmatrix} \right).$$

Given a sequence of $n + 1$ points $P_i(x_i, y_i)$, $i = 0 \ldots n$ in two dimensional space, we wish to minimize the sum of squared distances between the B-spline and the data (Pi). The notion of distance between a spline $S(u)$ and a data point P_i is based on the parameter value $\tilde{u}_i$ where the curve comes close to P_i. Bartels et al. (1987) suggest using for $\tilde{u}_i$ the current total length of the polygonal curve from P_0 to P_i:

$$\tilde{u}_i = \frac{\sum_{p=0}^{p=i-1} \| P_{p+1} - P_p \|}{\sum_{p=0}^{p=n-1} \| P_{p+1} - P_p \|}$$

Thus the criterion is to minimize

$$\begin{aligned}\Lambda &= \sum_{i=0}^{n} ||S(\tilde{u}_i) - P_i||^2 \\ &= \sum_{i=0}^{n}[(\sum_{j=0}^{3} X_j B_{j,3}(\tilde{u}_i) - x_i)^2 + (\sum_{j=0}^{3} Y_j B_{j,3}(\tilde{u}_i) - y_i)^2] \end{aligned} \quad (4)$$

Differentiating equation 4 with respect to the eight unknowns (X_j, Y_j), $j = 0,1,2,3$, yields the two following linear systems which are solved using a Choleski decomposition (Press et al., 1988):

$$\begin{aligned}\{\frac{\partial \Delta}{\partial X_j} &= 0 = 2\sum_{i=0}^{n} B_{j,3}(\tilde{u}_i)(\sum_{j=0}^{3} X_j B_{j,3}(\tilde{u}_i) - x_i)\}_{j=0,1,2,3}, \\ \{\frac{\partial \Delta}{\partial Y_j} &= 0 = 2\sum_{i=0}^{n} B_{j,3}(\tilde{u}_i)(\sum_{j=0}^{3} Y_j B_{j,3}(\tilde{u}_i) - y_i)\}_{j=0,1,2,3}.\end{aligned}$$

When analyzing curve data, performing such an approximation is useful for:

1. Removing noise from the data, i.e., smoothing the data. This is illustrated in the Fig. 4 .
2. Compressing the data; in our case, from $n + 1$ points down to four control vertices.
3. Using derivatives of the data, computed from the spline. The tangent is obtained using the first derivative of the spline. The curvature is obtained using the second derivative of the spline. When no obvious landmark is a priori available on the curve, the maxima of curvature can serve as landmarks. When working with biological shapes, some of the maxima of curvature do correspond to well-established landmarks. Dean (1993) performs space curve statistics based upon maximum curvature landmarks.

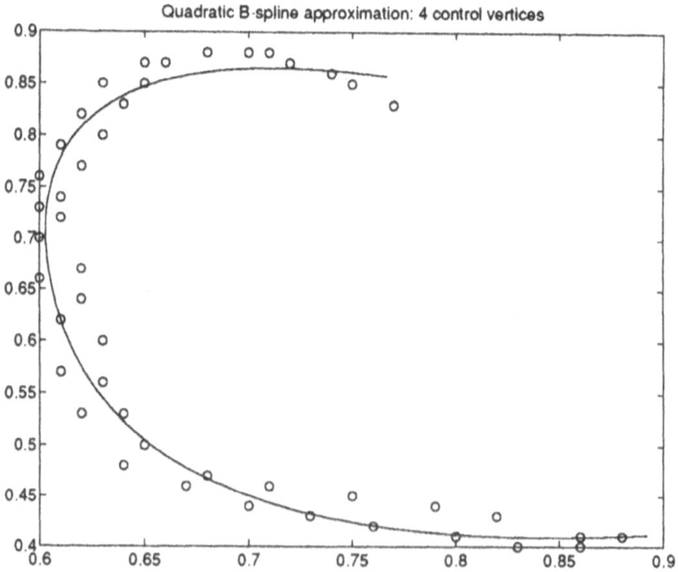

Figure 4. Approximation of a noisy set of points, represented with the circles, by a quadratic B-spline curve with four control vertices.

4. Resampling the data, so as to obtain the same number of samples in different data sets. Performing statistics on these samples, as well as performing statistics on the control vertices might produce poor results, because there is no guarantee that the samples nor the control vertices will be homologous. However, it is possible to resample the curves in a manner similar to arc length sampling.

GENERAL DEFINITION OF A SPLINE CURVE

For completeness's sake, we present the general formulation, but it is not easily produced from what precedes. However, to merge the material of the Sections on Bézier curves and B-splines with this theory is a good exercise. We define a B-spline curve of degree $K - 1$, with $m + 1$ control vertices as the formal sum

$$S = \sum_{j=0}^{m} V_j \, B_{j,K} \, .$$

When evaluated at a parameter value u, the equation gives the position of the current curve point from the evaluations of the B-spline functions $\{B_{j,K}(u)\}$: $S(u) = \Sigma_{j=0}^{m} \; V_j B_{j,K}(u)$. No new concept has been introduced yet. The generality of the definition comes next. The B-spline functions are specified by $m + K + 1$ knot values (u_j), with the relation

$$\begin{aligned} B_{j,1}(u) &= \begin{cases} 1 & u_j \leq u < u_{j+1} \\ 0 & \text{otherwise} \end{cases} \\ B_{j,K}(u) &= \frac{u - u_j}{u_{j+K-1} - u_j} B_{j,K-1}(u) + \frac{u_{j+K} - u}{u_{j+K} - u_{j+1}} B_{j+1,K-1}(u) \end{aligned} \tag{5}$$

In the example in the section "B-spline Curves as Piecewise Bézier Curves," the knot values were {0,0,0,1,2,2,2} which can be checked by deriving equation 3 starting with equation 5. Similarly, when using the tuple of knot values given by {0,0,0,1,1,1}, we obtain equation 1 when using {0,0,0,0,1,1,1,1} we obtain equation 2. Thus equation 5 also defines Bézier curves because B-spline curves form a more general family of curves.

More generally a function S is called a spline function of degree d if there exists a sequence of knots $u_0 \leq u_1 \leq \ldots \leq u_n$, such that inside each interval $[u_i, u_i{+}1]$, S is given by a polynomial of degree d at most, and such that the function S and its derivatives up to order $d - 1$ are continuous on $[u_0, u_n]$. Clearly B-spline functions and Bézier functions are examples of spline functions.

SURFACES AS TENSOR PRODUCTS OF B-SPLINE CURVES

We now use the B-spline curves to produce surfaces. We consider a rectangular mesh of control vertices $\{V_{ij}\}$ as in Fig. 5. A B-spline patch S of degree $K - 1$ can be defined by the tensor product, sometimes called exterior product (the $B_{i,K}$ and $B_{j,K}$ are the B-spline functions defined as in the section "B-spline Curves as Piecewise Bézier Curves"):

$$S\,(u, v) = \sum_{ij} V_{ij} \, B_{i,K}\,(u) \, B_{j,K}\,(v). \tag{6}$$

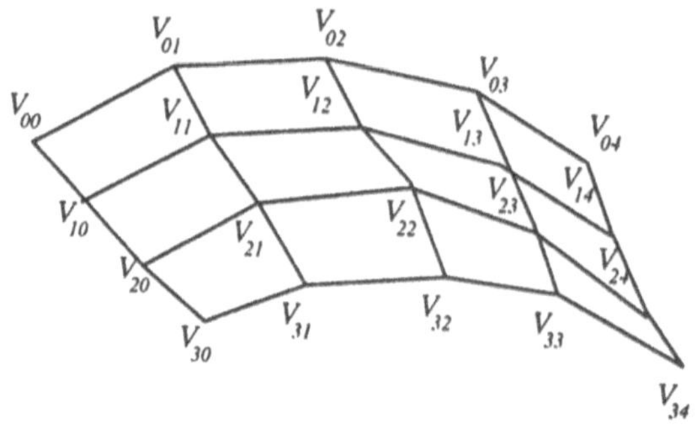

Figure 5. A rectangular mesh of control vertices (V_{ij}), defining a B-spline surface patch.

If we set u or v as a constant value in equation 6 then we obtain a B-spline curve as in the section, "B-Spline Curves as Piecewise Bézier Curves" or in the section "General Definition of a Spline Curve." Intuitively, the patch is obtained by blending two sets of B-spline curves.

Splines lend themselves easily to the evaluation of the curvature of a surface of curve because polynomial functions can be differentiated easily and also because the minimum curvature property (see Appendix) helps avoiding extraneous surface wiggles. Figure 6 illustrates a B-spline patch representing a human face, obtained after processing a 3D magnetic resonance (MR) scan. An abrupt intensity change occurs in the MR scan for the epidermis. We exploit this property by computing the gradient of the intensity, then extracting 3D points of high gradients, and computing the distance transform of these points (Borgefors, 1986). A cylindrical B-spline patch is initialized inside the volume of the distance transform. The B-spline patch then deforms under the influence of attractive elastic forces directed towards the points of high gradient. Internal forces allow the surface to maintain smoothness.

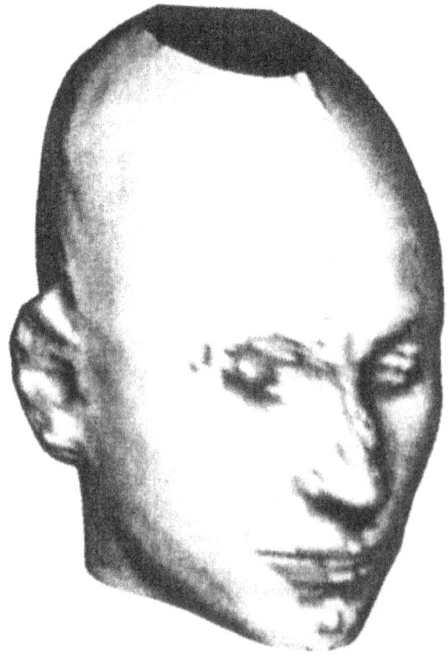

Figure 6. B-spline surface model consisting of 256 × 256 control vertices and obtained by "shrink-wrapping" inside an MR scan (Guéziec, 1995). The areas in darker gray show higher curvature loci on the face: ear lobes, mouth, nose and eye sockets.

The particular measure of curvature that we plot in Fig. 6 (sum of squared principal curvatures) represents the amount of bending energy accumulated locally by the surface during the deformation process.

CONCLUSION

We developed some features about spline functions. Spline curves and surfaces are very useful for modeling data because, as discussed in the Appendix, they have some optimal approximation properties (particularly cubic splines). They offer smooth representations of data, which reduce the error associated with data capture. The smoothness of splines is measured by the order of continuity of derivatives across the joints between different surface or curve elements.

APPENDIX

Minimum Curvature Property

Most of the present-day interest for splines was probably motivated by the following theorem.

Proposition 1 (Holladay, 1957). Let $u_0 < u_1 < \ldots < u_n$ and a set of real numbers (y_i), $i = 0, \ldots n$ be given. Of all functions f having a continuous second derivative on $[u_0, u_n]$ such that $f(u_i) = y_i$, $i = 0, \ldots, n$, the cubic B-spline S with knots sequence (u_i), (u_0 and u_n are repeated, see "General Definition of a Spline Curve", such that $S''(u_0) = S''(u_n) = 0$, minimizes the integral

$$\int_{u_0}^{u_n} (f''(u))^2 du. \tag{7}$$

Note that the curvature of a curve $(u, f(u))$ is given by

$$\text{eq. } \frac{f''(u)}{\sqrt{1+(f')^2}},$$

rather than $f''(u)$. This is why the name minimum norm property is often preferred.

Proof of the Proposition: We note e the error function such that $f = e + S$.

$$\int_{u_0}^{u_n} (f''(u))^2 du - \int_{u_0}^{u_n} (S''(u))^2 du = \int_{u_0}^{u_n} (e''(u))^2 du + 2\int_{u_0}^{u_n} e''S''(u)du. \tag{8}$$

Because S is a cubic spline, its third derivative is constant between the knots (u_i), $S'''(u) = m_i$, $u_i \leq u < u_{i+1}$, and the last integral vanishes:

$$\int_{u_0}^{u_n} e''S''(u)du = \sum_i \int_{u_i}^{u_i+1} e'S'''(u)du - \sum_i (e''S''(u_{i+1}) - e''S''(u_i)).$$

Finally, provided $S(u_0) = S''(u_n) = 0$, what is left is $\Sigma_i m_i(e(u_{i+1}) - e(u_i)) = 0$. Thus, both sides of equation 8 are positive and equal to zero if and only if $f = S$. This means that the "curvature" of f, written in equation 7, is always greater or equal to the curvature of S. S satisfies the minimum curvature property ■

Thin-plate splines provide a generalization of this property to surfaces.

REFERENCES

Ahlberg, J. H., and E. N. Nilson, and J. L. Walsh. 1967. The theory of splines and their applications, Academic Press: New York.

Bartels, R., J. Beatty, and B. Barsky 1987. An introduction to splines for use in computer graphics and geometric modeling. Morgan Kaufmann.

Bernstein, S. 1912. Démonstration du théorème de Weierstrass fondé sur le calcul des probabilités. Harkov Soobs. Matem ob-va 13:1–2.

Bézier, P. 1966. Définition numérique des courbes et surfaces I. Automatismes 11:625–632.

Bézier, P. 1967. Définition numérique des courbes et surfaces II. Automatismes 12:17–21.

Bohm, W., G. Farin, and J. Kahmann. 1984. A survey of curve and surface methods in CAGD. Computer aided geometric design. Elsevier Science Publishers: Amsterdam.

Borgefors, G. 1986. Distance transformations in digital images. Computer Vision, Graphics and Image Processing 34:344–371.

Dean, D. 1993. The Middle Pleistocene *Homo erectus/Homo sapiens* transition: New evidence from space curve statistics. Ph.D. dissertation. City University of New York.

de Boor, C. 1978. A practical guide to splines. Springer-Verlag: New York.

Euler, L. 1755. Institutionis calculi differentialis II. Petersburg.

Farin, G. 1988. Curves and surfaces for computer aided geometric design. Academic Press: New York.

Guéziec, A. 1995. Surface representation with deformable splines: using decoupled variables. IEEE Computational Science and Engineering 2(1): 69-80.

Guéziec, A. P., and N. Ayache. 1994. Smoothing and matching of 3-D space curves. International Journal of Computer Vision 12(1):79–104.

Hilbert, D. and S. Cohn-Vossen. 1990. Geometry and the imagination. Chelsea: New York.

Holladay, J. C. 1957. Smoothest curve approximation. Mathematical Tables Aids Computation 11:233–243.

Press, W. H., and S. A. Teukolsly, W. T. Vetterling, and B. P. Flannery. 1988. Numerical recipes in C. Cambridge University Press: New York.

Rohlf, F. J. 1990. Fitting curves to outlines. Pages 167–177 *in* F. J. Rohlf and F. L Bookstein, (eds.), Proceedings of the Michigan Morphometrics Workshop. University of Michigan Museum of Zoology Special Publication 2.

Schoenberg, L. J. 1946. Contributions to the problem of approximation of equidistant data by analytic functions. Quarterly of Applied Mathematics 4:45–99.

Smith, K. T. 1983. Primer of modern analysis. Springer-Verlag: New York.

EDGELS AND TANGENT PLANES IN IMAGE WARPING

John Little[1] and Kanti Mardia[2]

[1] Imaging Processing Group
Division of Radiological Sciences
3rd floor, Guy's Tower, Guy's Hospital
London Bridge, London SE1 9RT, United Kingdom
[2] Department of Statistics
University of Leeds
Leeds, LS2 9JT, United Kingdom

ABSTRACT

In this paper we first describe generally the uses of thin-plate splines for image deformation. We then go on to demonstrate why we feel that incorporating edgel information into a warp has great potential. An extension to the edgel method of Bookstein and Green (1993a,b) is outlined for use in three dimensions. Finally we give an example of the results of using this formulation for three-dimensional magnetic resonance images of the human head.

INTRODUCTION

In recent years there have been numerous methods used for deforming and transforming images. In this discussion we concentrate on thin-plate splines and recent extensions to the standard theory (Bookstein and Green, 1993a,b). Due to lack of space we will not be comparing the method discussed here to alternative registration methods, but for a recent review see Brown (1992).

Recent advances in computer hardware have enabled large data sets to be manipulated and displayed very quickly. These advances have lead a number of researchers to look at ways of warping images so that these images can be matched to another image and "data fusion" can take place, if desired. If a patient's image could be transformed to be of the same size and shape of an anatomical atlas image then the images could be compared and regions of dissimilarity highlighted automatically. In image fusion, images of the same view of a person's anatomy, from two or more different imaging modalities (e.g., Xray and nuclear medicine (NM) scan; see Mardia and Little, 1994) are combined (or fused) into one single image. In the case of Xrays and NM scans the two modalities show different things, the Xray shows anatomical information, and the NM scan shows functional information. Combining the two gives the position of a functional abnormality in the anatomy of the patient. It is expected that these image-matching methods will find increasing use in diagnostics.

WHY USE THE THIN-PLATE SPLINE?

For the above reasons, many researchers are looking at the problems of accurate image deformation but the majority of those who use landmarks appear to be using linear methods, with the exception of Bookstein and Green (1993a,b). We also feel that because of the enormous variation in "normal" human anatomy a nonlinear method would be far more relevant.

In trying to develop a method for image matching, we require that a template image can be deformed to cover the whole size and shape space for a given image type. This means that we require a very flexible method. The thin-plate spline (TPS) proves to be a very powerful tool for performing this class of deformation (Bookstein, 1989). Thin-plate spline image deformation is a landmark-based method that consists of exactly matching two sets of corresponding landmarks, by use of two thin-plate splines and interpolating throughout the plane in order to construct the warped image. There are already a very large number of anatomical landmarks that are well defined, and any suitable subset of these could be used to define the deformation.

TWO-DIMENSIONAL IMAGE DEFORMATION AND EDGELS

In terms of medical imaging, landmarks can be defined as points with an anatomical meaning, and hence they can be compared between images of the same anatomical region. Landmarks are, however, just points and contain no information about their immediate surroundings. If a method of incorporating extra information in the neighborhood of a landmark were available, this would enable a more accurate deformation to be performed.

An edgel is such a way of defining extra information at a landmark. Quite often landmarks lie on boundaries between objects within the image and so there is extra information at these boundaries—such as the direction of the boundary and its rate of change or curvature at the landmark. Bookstein and Green (1993a,b) were the first researchers to consider this type of information, and they have extended standard thin-plate spline theory to incorporate this derivative information.

Why is this called derivative information? A thin-plate spline for images consists of two coordinate surfaces (one for the x-direction and one for the y-direction; see Bookstein, 1989), obtained by using two sets of constraints (or heights) with a common deformation matrix. The edgel constraints taken from the two-dimensional image manifest themselves as a constraint of the gradient of each coordinate surface separately in the direction of the edgel. This is called the directional derivative of the surface and hence the name derivative information.

Edgels can, in fact, be defined to be in any direction and not just along the boundaries. For example, we can use the direction from one landmark to another or from a landmark to an anatomical feature in the image. Edgels can be anything the user thinks is important, and hence they are very flexible.

In practice, edgels may well represent a tangent to a boundary curve or outline because these are quite often easily found, Also, the shape change due to edgel direction change can be more easily explained in terms of boundary shape change. Figure 1 shows

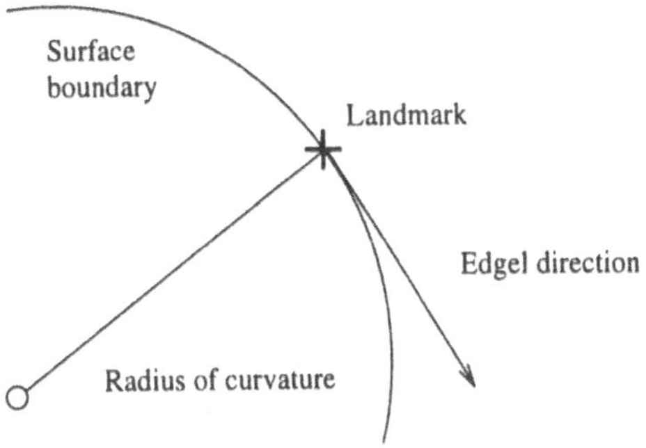

Figure 1. An edgel at a landmark.

how an edgel would be defined in this case. The landmark lies on a boundary, and the edgel represents the direction of the boundary as it passes through the landmark.

We have talked here in terms of deforming images but edgels could indeed also prove to be a very useful extension to the standard landmark-based analysis of within-and between-group shape variation. The extended deformation matrix can be decomposed into principal and partial warps in a similar manner to the standard thin-plate spline deformation matrix (Bookstein and Green, 1993a,b). These can then be used to describe the shape variation of a data set in a standard manner.

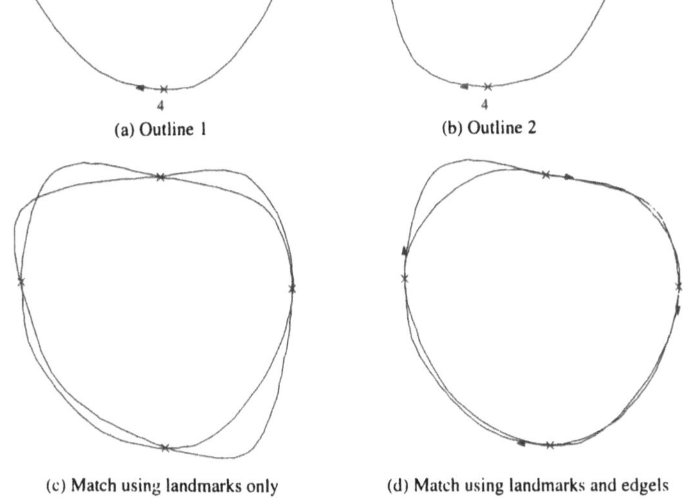

Figure 2. Improvement of an outline match by using edgels.

IMPROVEMENT DUE TO USING EDGELS

We now give an example to illustrate that an attempt to match two objects based only on landmark information may prove to be ineffective. Figure 2 (a and b) shows the outlines of two apple like objects that we wish to match. The results of matching only the landmarks of the two outlines is shown in Fig. 2c. Here the landmarks of outline 2 were matched to those of outline 1; the resulting planar deformation was constructed and then superimposed onto outline 1. It is obvious from Fig. 2c that although the landmarks are matched exactly, the outlines are not well matched. If, however, the tangents to the outlines at the landmarks (shown by the arrows) were also matched, then the outlines are matched far better, as shown in Fig. 2d.

In general we would expect that using edgels would significantly improve a required image or outline match.

THREE-DIMENSIONAL IMAGE WARPING

With recent advances in medical imaging hardware, it is now relatively easy to obtain three-dimensional images of human anatomy. Both computed tomography (CT) and magnetic resonance (MR) imaging systems now have the capability to acquire three-dimensional data sets in a relatively short period of time. This has lead to the need for a method of transforming these data sets. For example, CT images show hard bone much more clearly than does MR whereas MR is much better for viewing the soft tissue of a subject. Ideally if the two could be combined, a clearer indication of the subject's anatomy would be obtained.

The thin-plate spline in three dimensions is well known and requires the use of three splines, as opposed to two in two dimensions, to define the image deformation. Note that the kernel function will, of course, be different.

We have recently extended the edgel work of Bookstein and Green to three dimensions. As we are now looking at deforming three-dimensional structures, instead of specifying a tangent to a curve as in two dimensions, we have to specify a tangent plane to a surface. Still we can define a single direction vector at a landmark. At a landmark where we wish to define a tangent plane we take two perpendicular tangent vectors at the landmark and deform both of these to lie in the new tangent plane orientation. The landmark deformation may cause a compression of the data set within the tangent plane and so the required directions of the tangents must be carefully chosen so that this compression is not "unwarped."

We now state our three-dimensional extension of the Bookstein and Green (1993a,b) result for the two-dimensional case. We follow their notation.

Let $P_i = P(x_i, y_i, z_i)$, $i = 1,\ldots, k$ be k landmarks with respective heights $h_1,\ldots,h_k$ ($\mathbf{h} = (h_1,\ldots,h_k)^T$) and let there be m tangent planes, which without any loss of generality are at the first m landmarks.

In order to constrain the tangent plane we use two perpendicular three-dimensional edgels that lie in the tangent plane. Let these vectors be $\mathbf{t}_{j1}, \mathbf{t}_{j2}$, $j=1,\ldots, m$.

The thin-plate spline kernel function in three dimensions $U(r)$ for any point $P = (x,y,z)$ may be defined appropriately as (Mardia et al., 1994)

$$U(r) = |r|^3, \tag{1}$$

where $r = |P|$.

Using the above definitions we can now state the transformation for a general point P as follows:

$$f(P) = f_0(P) + f_1(P) + O(\delta), \tag{2}$$

where

$$f_0(P) = \sum_{i=1}^{k} a_{i,0}\, U(P-P_i) + c_{1,0} + c_{2,0}\, x + c_{3,0}\, y + c_{4,0}\, z, \tag{3}$$

and the coefficients of $f_0(P)$ are defined by

$$\begin{pmatrix} \boldsymbol{a}_0 \\ \boldsymbol{c}_0 \end{pmatrix} = L^{-1} \begin{pmatrix} \boldsymbol{h} \\ \boldsymbol{0}_4 \end{pmatrix}, \boldsymbol{a}_0{}^T = (a_{1,0}, \ldots, a_{k,0}),\ \boldsymbol{c}_0{}^T = (c_{1,0}, \ldots, c_{4,0}),$$

$$L = \begin{pmatrix} K & Q \\ Q^T & 0 \end{pmatrix},\ L : (k+4) \times (k+4),$$

$$(K)_{ij} = U(P_i - P_j),\ Q^T = \begin{pmatrix} 1 & 1 & \ldots & 1 \\ x_1 & x_2 & \ldots & x_k \\ y_1 & y_2 & \ldots & y_k \\ z_1 & z_2 & \ldots & z_k \end{pmatrix}.$$

And

$$f_1(P) = \sum_{i=1}^{k} a_{i,1} U(P - P_i) - \sum_{j=1}^{m} b_{j,1} U_{\boldsymbol{t}_{j1}}(P - P_j) - \sum_{j=1}^{m} b_{j,2} U_{\boldsymbol{t}_{j2}}(P - P_j) + c_{1,1} + c_{2,1} x + c_{3,1} y + c_{4,1} z, \tag{4}$$

where

$$\begin{pmatrix} \boldsymbol{a}_1 \\ \boldsymbol{c}_1 \end{pmatrix} = L^{-1} M \begin{pmatrix} \boldsymbol{b}_1 \\ \boldsymbol{b}_2 \end{pmatrix},$$

and,

$$a_{11} = (a_{1,1}, \ldots, a_{k,1}),\ b_1 = (b_{1,1}, \ldots, b_{m,1}),\ b_2 = (b_{1,2}, \ldots, b_{m,2}),\ c_1 = (c_{1,1}, \ldots, c_{4,1}),$$

$$M = \begin{pmatrix} -K_1 & -K_2 \\ V_1^T & V_2^T \end{pmatrix},\ M : (k+4) \times 2m,\ K_1 : k \times m,\ K_2 : k \times m\,.$$

The above matrices are defined as follows,

$$(K_1)_{ij} = U_{t_{j1}} (P_i - P_j)\ i = 1, \ldots, k, j = 1, \ldots, m$$

$$(K_2)_{ij} = U_{t_{j2}} (P_i - P_j)\ i = 1, \ldots, k, j = 1, \ldots, m$$

$$V_1^T = \begin{pmatrix} 0 & 0 & \dots & 0 \\ t_{11} & t_{21} & \dots & t_{m1} \end{pmatrix},$$

$$V_2^T = \begin{pmatrix} 0 & 0 & \dots & 0 \\ t_{12} & t_{22} & \dots & t_{m2} \end{pmatrix},$$

and

$$\begin{pmatrix} b_1 \\ b_2 \end{pmatrix} = (D - M^T L^{-1} M)^{-1} \Delta .$$

Where

$$D = \begin{pmatrix} -K_{22} & -K_{23} \\ -K_{23}^T & -K_{33} \end{pmatrix} + \begin{pmatrix} \frac{-2}{\delta^2} U(\delta) I_m & \frac{1}{\delta^2}(U(\sqrt{2}\delta) - 2U(\delta) I_m \\ \frac{1}{\delta^2}(U(\sqrt{2}\delta) - 2U(\delta) I_m & \frac{-2}{\delta^2} U(\delta) I_m \end{pmatrix}$$

and

$$(K_{22})_{ij} = U_{t_{i1}, t_{j1}} (P_i - P_j) \; i,j = 1, \dots, m,$$

$$(K_{33})_{ij} = U_{t_{i2}, t_{j2}} (P_i - P_j) \; i,j = 1, \dots, m,$$

$$(K_{23})_{ij} = U_{t_{i1}, t_{j2}} (P_i - P_j) \; i,j = 1, \dots, m,$$

and

$$\Delta = \begin{pmatrix} s_1 \\ s_2 \end{pmatrix} - M^T L^{-1} \begin{pmatrix} h \\ 0 \end{pmatrix}, \Delta : 2m \times 1, s_1 = (s_{11}, \dots, s_{m1})^T, s_2 = (s_{12}, \dots, s_{m2})^T .$$

Finally the derivatives were evaluated by using the spherical polar coordinates ($P = (x,y,z) = (r \sin\theta \cos\phi, r \sin\theta \sin\phi, r \cos\theta)$, $r = |P|$) and are given below; Mardia and Little, 1994)

$$U_{t_{jk}}(P) = \frac{U_r'(r)}{r}(t_{jk}.P) \tag{5}$$

and

$$U_{t_{jk}, t_{il}}(P) = \frac{U_r''(r)}{r^2}(t_{jk}.P)(t_{il}.P) + \frac{U_r'(r)}{r^3}$$

$$\left[\left(t_{jk}.\frac{\partial P}{\partial \theta}\right)\left(t_{il}.\frac{\partial P}{\partial \theta}\right) + \frac{1}{\sin^2\theta}\left(t_{jk}.\frac{\partial P}{\partial \phi}\right)\left(t_{il}.\frac{\partial P}{\partial \phi}\right)\right], \tag{6}$$

where

$$U_r'(r) = \frac{\partial U(r)}{\partial r}, U_r''(r) = \frac{\partial U_r'(r)}{\partial r} . \tag{7}$$

Figure 3 shows the results of using the above formulation in order to deform a three-dimensional MR image of the human head. The original data set is shown in Fig. 3a, the two images are of the same data set at different angles. In Fig. 3b a landmark only deformation has been performed, and the chin of the data set has been elongated. This was achieved using only nine landmarks. Figure 3c shows the result of changing the orientation

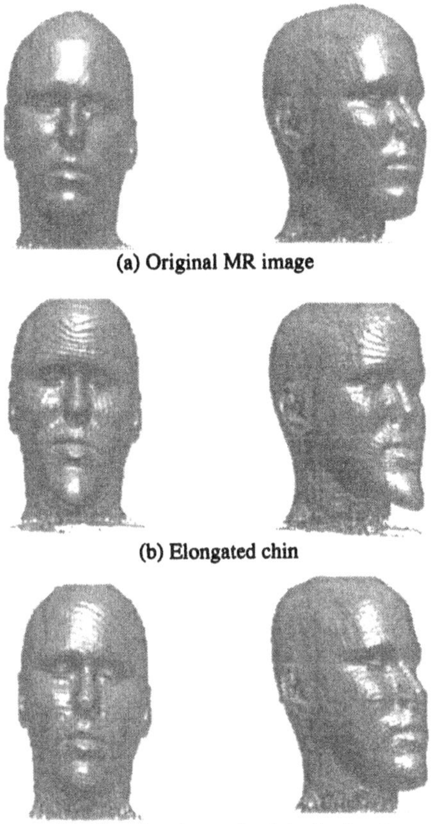
(a) Original MR image

(b) Elongated chin

(c) Changing the angle of the forehead

Figure 3. Three dimensional thin-plate spline deformations.

of a tangent plane that has been positioned in the center of the forehead. The tangent plane has been tilted towards the back of the head in order to "flatten" the forehead.

In all of these images the detail of the face is maintained, and despite the deformations using only a very small amount of information a reasonable form for the rest of the image is obtained. This, we feel, indicates that the three-dimensional extension of this work will be valuable in the field of medical image deformation.

DISCUSSION

In this paper we have talked of a possible method for image deformation. We feel that the added accuracy that can be gained by defining extra constraints such as edgels and tangent planes will prove to be very useful. In recent years there have been many researchers looking at methods of accurately matching and modeling anatomical images and structures. Matching and modeling images could be used for planning surgical procedures or even in virtual reality training systems (see, for example, Cover et al., 1993).

The accurate deformation of images and structures will be of great use in these methods and may be a landmark-based approach, such as that outlined above, or something similar that may prove helpful in achieving these aims.

ACKNOWLEDGMENTS

We are grateful to Fred Bookstein and William Green for the preprints of their edgel work and for Fred Bookstein's constant encouragement. The work of John Little was supported by an SERC Research Studentship, and the work of Kanti Mardia was supported by the Complex Stochastic Systems initiative of the SERC. We are also grateful to Liz Berry, John Ridgeway and COMIR (the Centre of Medical Imaging Researching Leeds) for supplying us with the MR images and for their helpful comments.

REFERENCES

Bookstein, F. L. 1989. Principal warps: thin-plate splines and the decomposition of defonmations. IEEE Transactions on Pattern Analysis and Machine Intelligence 11: 567–585.

Bookstein, F. L. and W. D. K. Green. 1993a. A thin-plate spline for deformations with specified derivatives. Pages 14–28 *in* Mathematical methods in medical Imaging II. S.P.I.E. Proceedings 2035.

Bookstein, F. L. and W. D. K. Green 1993b. A feature space for edgels in images with landmarks. Journal of Mathematical Imaging and Vision 3: 231–261.

Brown, L. G. 1992. A survey of image registration techniques. ACM Computing Surveys 24(4):325–376.

Cover, S. A., N. F. Ezquerra, J. F. O'Brien. R. Rowe. T. Gadacz. and E. Palm. 1993. Interactively deformable models for surgery simulation. IEEE Comp. Graph. and Appl. 13(6): 68–75.

Mardia, K. V., J. T. Kent, C. R. Goodall, and J. A. Little. 1996. Kriging and splines with derivative information. Biometrika, 83(1): 207-221.

Mardia, K. V. and J. A. Little. 1994. Image warping using derivative information. Pages 16-31 *in* Mathematical methods in medical imaging III. S.P.I.E. Proceedings 2299.

PART II: COMPARING SHAPES OF ORGANISMS

Section 1. Geographic Comparisons

AN INTRODUCTION TO GEOMETRIC MORPHOMETRICS AND INTRASPECIFIC VARIATION

A Fascinating Adventure

Anna Loy

It is quite peculiar that since its renaissance, geometric morphometrics has been little associated with the study of variation among closely related taxa. This is true despite the fact that the potential of landmark-based morphometrics has been widely explored during earlier morphometric workshops (see Rohlf and Bookstein, 1990; Marcus et al., 1993), where they proved to be powerful tools for investigations in systematics (for example, Loy et al., 1993). One important contribution from this NATO Advanced Study Institute volume is the relatively large number of papers addressing intraspecific variation, the subject of this section. These papers confirm that geometric morphometrics is maturing as one of the more powerful techniques for the description and interpretation of patterns of variation below the species level. The analysis of closely related groups, such as populations, demes and subspecies, often implies the comparison of very similar shapes, differing only in slight relative displacements of landmarks. This means that landmarks are more likely to be homologues when comparing different shapes. The absence of homology among landmarks for more distantly related taxa is a major problem for landmark morphometrics.

The integration of traditional multivariate approaches with a methodology that maintains close attention to the geometric relation among landmarks, before and after statistical analysis, allows a functional interpretation of results based on a direct correlation between variation of anatomical features and specific adaptive or historical factors. Further, the new methods offer the opportunity to analyze and visualize "pure" shape factors independent of size, as measured by centroid size. Size may often represent noise and be a nongeographic source of variation. Size, as centroid size, is extracted before the analysis as a separate component of morphological variation, and it can be summarized and correlated with patterns of shape changes in order to assess its role in morphological variation both within and among populations. Finally, the analysis of shape changes at different levels as linear and nonlinear transformations, expressed respectively by the uniform or affine and the nonuniform or nonaffine components, as well as large-scale and small-scale shape differences, may be coordinated with the different roles of genetic and phenetic plasticity involved. However, one must recognize that the affine and nonaffine partition is mathematical and may not be directly interpretable biologically. Graphic animations like the thin-plate spline deformation grids allow a visual amplification of shape changes that would be hardly

Advances in Morphometrics, Edited by L. F. Marcus et al.
Plenum Press, New York, 1996

recognized and difficult to describe using traditional approaches. The paper by Baylac et al. in this section is an example in which deformation grids associated with extremes of variation along canonical axes help the reader gain a quick understanding of the anatomical features that characterize the geographic pattern of variation.

On the other hand some problems are still encountered in the application of these methods in biogeography and intraspecific variation. These problems are related to the limits imposed by the algorithms for specific data sets. But a long way has been traveled to reach the present level by both parties in the discourse: the "parents" of the new geometric morphometrics on the one hand, trying to translate and modify software and formulations to link multivariate statistics to geometry and biology; and the "children," the evolutionary biologists, on the other, trying hard to understand the methods and raising problems to be solved in order to accomplish their objectives.

The papers of this section offer quite a wide spectrum of problems related to intraspecific variation—from sexual dimorphism to fluctuating asymmetry to variation along geographic and ecological transects. The papers are often supplemented with a comparison between the new and a more traditional approach, witnessing a rich and dialectical effort toward the definition of the most suitable instruments of investigation.

This section includes the paper of Auffray et al., an analysis of fluctuating asymmetry across a hybrid zone between two subspecies of *Mus musculus*. The authors compare the discriminant power of a traditional character set with data derived from Procrustes superposition of landmark configurations. Results suggest greater power for the landmark methods to detect lateral asymmetry in the populations studied. Landmark analysis also offers a simpler statistical treatment of the data. Baylac et al. analyze the morphometric variation among populations of the box midge *Monarthropalpus buxi*, an insect whose life cycle is strictly related to plants of the genus *Buxus*. They analyzed correlations between geographic variation of the insect and the distribution of the plant in France. Sexual dimorphism is also investigated among localities. The authors point out some problems related to the biological interpretation of their results when considering separately the uniform and the nonuniform shape variation. This illustrates a problem of which investigators must always be aware: results must be carefully checked for both their statistical and biological significance. The work of Corti et al. uses three-dimensional data, which will be very important in the future evolution of the methodology. Their analysis attempts to solve some of the intriguing evolutionary questions involved in the recent speciation of the fossorial rodent *Spalax ehrenbergi* in Israel. Walker applies landmark-based morphometrics to answer some puzzling questions posed to interpret diversification of the threespine stickleback *Gasterosteus aculeatus*, a fish distributed in both marine and freshwater in North America. Different methods of superposition of landmark configurations are tested for their effectiveness in relating body shape variation to the adaptive features of feeding and defense. Two papers attempt to deal with difficult species identification and classification problems. Sarà uses landmark-based analyses of the mandible to test for phenetic relationships among endemic taxa of *Crocidura*, insectivorous mammals that are widespread throughout the Mediterranean region. Their systematics are still controversial. Johnson and Budd successfully apply the geometry of landmarks in three dimensions to the distinction among different populations of the Caribbean reef-corals of the genus *Porites*. At this time, there is no reliable quantifiable technique for defining and classifying species.

ACKNOWLEDGMENTS

Special thanks go to Leslie F. Marcus, who gave friendly support for my participation in this project. Despite various "technical" impediments that prevented my physical presence at the NATO ASI, I was certainly there in spirit.

REFERENCES

Loy, A., M. Corti, and L. F. Marcus. 1993. Landmark data: size and shape analysis in systematics. A case study on Old World Talpidae (Mammalia, Insectivora). Pages 215-240 *in* L. F. Marcus, E. Bello and A. Garcia-Valdecasas, (eds.), Contributions to morphometrics, Monografias del Museo Nacional de Ciencias Naturales 8, Madrid.

Marcus, L. F., E. Bello and A. García-Valdecasas (eds.). 1993. Contributions to morphometrics, Monografías del Museo Nacional de Ciencias Naturales 8, Madrid.

Rohlf, F. J. and F. Bookstein 1990. (eds.) Proceedings of the Michigan morphometrics workshop. University of Michigan Museum of Zoology Special Publication 2.

FLUCTUATING ASYMMETRY IN *Mus musculus* SUBSPECIFIC HYBRIDIZATION

Traditional and Procrustes Comparative Approach

Jean-Christophe Auffray,[1] Paul Alibert,[1] Sabrina Renaud,[1] Annie Orth,[2] and François Bonhomme[2]

[1] Institut des Sciences de l'Evolution, CC064
Université Montpellier II, 34095
Montpellier Cedex, France
[2] Laboratoire Génome et Populations, CC063
Université Montpellier II
34095 Montpellier Cedex, France

ABSTRACT

The traditional approach to fluctuating asymmetry (FA) is compared with an application of the Procrustes superposition method. This comparison was performed on wild-derived and random bred strains of two house mouse subspecies, *Mus musculus domesticus* and *M. m. musculus*, originated from Denmark, and their hybrids in a controlled experiment. Results obtained with the Procrustes method show a decrease of FA for the hybrid sample, which is consistent with what was expected from previous FA studies on wild populations. However, no trend was detectable using the traditional approach in that specific case. Additionally, the observed levels of FA compared with those found in wild populations of the hybrid zone in Denmark suggest that the latter are under higher environmental stress than are laboratory-reared animals.

INTRODUCTION

It has been clearly established that fluctuating asymmetry (FA) which expresses the random difference occurring during development between the right and the left sides of a symmetric bilateral character, increases under environmental or genetic stresses (for a review see Parsons, 1992). This morphological expression is related to reduced developmental homeostasis, suggesting that an intense stress would be sufficient to impair the developmental stability of organisms.

Lerner (1954) formerly hypothesized that heterozygosity may be related to developmental homeostasis and predicted that heterozygosity allows organisms to buffer develop-

Advances in Morphometrics, Edited by L. F. Marcus et al.
Plenum Press, New York, 1996

mental processes (for a review see Mitton and Grant, 1984). Consequently, highly homozygous individuals (e.g., submitted to systematic inbreeding) should exhibit more phenodeviant or asymmetrical characters than do more heterozygous ones. A positive correlation between heterozygosity and developmental stability has often been noted both in strains or natural populations, e.g., in rodents, fishes and lizards (Bader, 1965; Soulé, 1979, Leamy, 1984; Leamy and Atchley, 1985; Leary et al., 1985; Mitton and Grant, 1984). To explain such a correlation, authors have invoked mechanisms of dominance, overdominance or effects due to fortuitous combinations of particular genes (Vrijenhoek and Lerman, 1982; Mitton and Grant, 1984) but, actually, the determinism and mechanism of FA are far from being resolved.

Practically, FA is detectable on meristic or metric characters. FA studies, which are *a fortiori* comparative studies between populations under different levels of environmental or genetic stress, are commonly performed on a set of paired characters. For a single population, each of these characters is treated independently. When combined together, FA levels for all characters give a more or less homogeneous image of FA for the population considered. To test this homogeneity, Soulé (1967) has proposed applying a Kendall test of concordance on FA levels among characters within populations. However, this method remains unsatisfactory because so little is known about the allometric constraints that could mask clear asymmetry patterns.

Bookstein (1991) has proposed a global approach to FA using Procrustes distances between landmark configurations of right and left sides. Using the Procrustes superposition method described by Rohlf and Slice (1990), a single index of FA is obtained per population (instead of one per population per character in the traditional approach), drastically simplifying the statistical synthesis of the FA pattern. The traditional approach to FA may not be performed on scaled material (this depends on the correlation between net asymmetry (*NA*) and the size of the character considered [see Palmer and Strobeck, 1986]) as it is focused on size asymmetry. On the other hand, Bookstein's method clearly measures shape asymmetry by previously scaling all configurations.

The purpose of this paper is to provide an application of the Bookstein Procrustes method and to compare it with the traditional approach. For this study we have chosen a model related to a phenomenon of high evolutionary interest.

According to the current nomenclature of Auffray et al. (1990a), the house mouse (*Mus musculus*) in Europe is subdivided into two subspecies, the Western (*M. m. domesticus*) and Eastern (*M. m. musculus*) house mice. These subspecies have colonized Europe following two different pathways (Auffray et al., 1990b; Auffray and Britton-Davidian, 1992) and have been allopatrically distributed long enough to diverge genetically but only slightly morphologically (Gerasimov et al., 1990; Boursot et al., 1993; Auffray et al., in press). Today, the two subspecies interact along a narrow hybrid zone that crosses all of Europe from Denmark to Bulgaria (Boursot et al., 1984). Although there is no evidence of a reduction in fitness for hybrids in nature, the quasi-absence of introgression for the sex chromosomes (Vanlerberghe et al., 1988; Tucker et al., 1992; Dod et al., 1993) and the higher susceptibility, genetically determined, to parasite infection in hybrid populations (Sage et al., 1986; Moulia et al., 1991, 1993) suggest that hybrid genomes are counter-selected as it is commonly described in narrow hybrid zones (Barton and Hewitt, 1989). However, a morphometrical study across this hybrid zone showed an unexpected reduction of FA towards the center of the hybrid zone in Denmark, suggesting better developmental stability of hybrids than of pure subspecific populations (Alibert et al., 1994). Theoretically, developmental stability in hybrid populations is related to the genetic proximity between hybridizing taxa and should result from a compromise between the stabilizing effect due to increased heterozygosity and the disruptive effect caused by breakdown of genomic coadaptation (Graham, 1992). Such discrepancies between the apparent benefit of hybridization on morphological development and counter-selection of house mouse hybrids could be explained by the decoupling of different biological features (e.g., developmental stability, reproduction and resistance to parasites). However, due to recurrent crossing between various hybrid combinations, natural hybrid populations are mixtures of complex hybrid recombinants. To evaluate the role of a hybridization between these two subspecies on the F1 hybrid developmental stability, we performed controlled crosses between parental subspecies in the laboratory, hopefully restricting also that way postnatal selective effects on hybrids.

MATERIAL AND METHODS

Mice tested were wild-derived, randombred and maintained for several generations in the laboratory. *Mus musculus domesticus* was represented by the DDO strain, randombred (13th generation) from wild animals of Ødis (Denmark), and *M. m. musculus* by the MDH strain (5th generation) originated from Hov (Denmark). The hybrid sample (F1) was obtained by crossing DDO males with MDH females as well as by the reciprocal crosses. To avoid any parental pair effect, samples were composed of one to three individuals per litter, and one to three litters for nine pairs. We used only males in the three samples to avoid potential sex-related confounding factors. Ages of animals were from 7 to 30 weeks for DDO and MDH, and from 7 to 10 weeks for hybrids.

FA was evaluated on the basis of the lower teeth row. Bader (1965) has already shown that in the house mouse lower molar widths clearly react to strain hybridization through reduction of FA levels. For the traditional approach, four measurements were taken (length and width of M/1 and of M/2, respectively named LM/1, WM/1, LM/2, WM/2); 20 landmarks were determined for the Procrustes method (Fig.1). Sample sizes were respectively 30, 28, and 29 for DDO, MDH and F1.

Data Acquisition

Measurements were taken on a NIKON Measuroscope (1/1000 mm). Landmarks were digitized in two dimensions directly through the microscope using BioScan Optimas version 4.0.

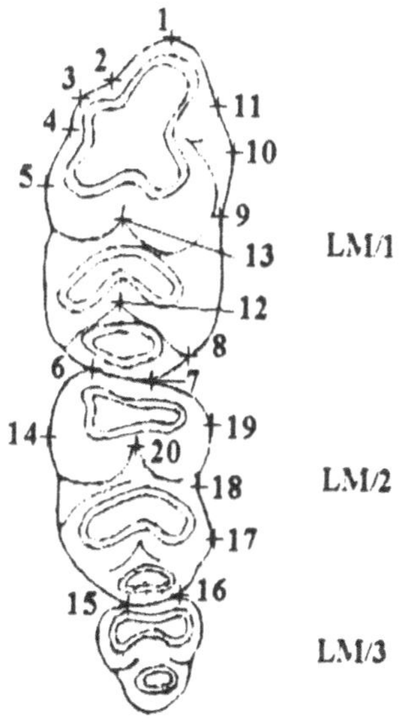

Figure 1. Location of landmarks on the left lower teeth row and on the mirror image of the right.

Traditional Approach

The traditional approach to FA is dependent on the type of characters used (meristic character or measurement) and on the sample size. All traditional approaches are exhaustively described in Palmer and Strobeck (1986). In our case, i.e., metric characters with a high sample size (n > 25), the procedure was as follows.

First normality had to be tested for the distribution of the Right - Left (*Ri-Li*) values for each character in the three samples to detect the possibility of antisymmetry (systematic nondirectional deviation from bilateral symmetry). The independence between ($|Ri-Li|$) and the size of the character expressed by (*Ri+Li*) was tested by calculating the coefficient of correlation, *r*, between the two sets of data. An eventual directional asymmetry (systematic, directional deviation from bilateral symmetry) was tested by the application of a mixed-model ANOVA (sides right and left, fixed; genotypes, random). Finally, FA was estimated by the variance of (*Ri-Li*) for each sample.

To detect whether the levels of asymmetry calculated for each parameter are randomly distributed among populations or clearly show a common pattern, it is necessary to apply a Kendall's coefficient of concordance test. Additionally, the main purpose of this study being the comparison between hybrid and parental groups, for any characters for which the parental FA levels were not statistically different, DDO and MDH samples were pooled and compared with the hybrid sample by a *t*-test. Software used for the traditional approach are Statworks version.1.1 and Biomeco version.4.0 (Lebreton et al., 1992).

Procrustes Distances Method

The method is based on the Procrustes superposition (Rohlf, 1990; Rohlf and Slice, 1990), i.e. the best-fitting superposition of several landmark configurations based on Procrustes distances for the calculation of directional and fluctuating asymmetry (Bookstein, 1991).

Briefly, the two sets of landmarks considered here correspond to the left teeth row and the mirror image of the right. The least squares Procrustes superposition method requires that both configurations be previously scaled to unit centroid size. Following the translation and the rotation of one of the sides to a position of best fit to the other, the individual net asymmetry (NA_i) is obtained by the calculation of the Procrustes distance between the two configurations. Afterwards, the population net asymmetry score (*NA*) is given by averaging individual net asymmetries (NA_i). The directional asymmetry (*DA*), which is considered by Bookstein (1991) as being a component of the *NA*, corresponds to the Procrustes distance between the population mean configuration of the left side and that of the right one. The remainder of (*NA-DA*) is the population fluctuating asymmetry (*FA*).

The significance of differences of *NA* and *FA* between populations was tested by one-way ANOVA on NA_i and on FA_i ($= NA_i - DA$). Similarly to the traditional approach, because the means of *FA* and *NA* were not different in the two parental groups, the parental *DA* and *FA* were calculated after having pooled the two samples and were compared with the corresponding values in the hybrid sample by a *t*-test.

As most of the landmarks taken on the lower teeth row of mice belong mainly to type 2 (maxima of curvature) and 3 (extreme point; for further information see Bookstein, 1991), it seemed important to evaluate the part of asymmetry that could be attributed to the lack of accuracy in the determination of landmarks. This has been done by considering only the left sides for one of the populations (hybrids). Landmarks were independently acquired twice, *NA*, *DA* and *FA* being estimated on this set of data.

Generalized Resistant Fit (GRF-ND) version 2.1 (see Slice, "Three-Dimensional" article, this volume) was used for this approach.

RESULTS

Traditional Approach

The independence between (|*Ri-Li*|) and the size of the character estimated by (*Ri+Li*) is shown by the lack of significance of all correlation coefficients calculated for each variable among and within samples (no significant *p*-values with a sequential Bonferroni test [Rice, 1989]). According to Palmer and Strobeck (1986), because there is an independence between (*Ri-Li*) and (*Ri+Li*), no scaling is necessary.

Antisymmetry. The preliminary treatment indicated that all (*Ri-Li*) samples considered here were normally distributed (Lilliefors, sequential Bonferroni test; no significant *p*-values), removing any antisymmetric pattern from our sampling.

Directional Asymmetry. Two-way ANOVAs showed that the means of right and left sides were not significantly different whereas the size of characters differed among samples (Tab. 1). However, because there is no correlation between (|*Ri-Li*|) and (*Ri+Li*), no scaling is required (see above). The non-significant difference of the right and left measurement means expresses the absence of directional asymmetry in our samples.

Fluctuating Asymmetry. According to preliminary analyses, FA may be simply expressed by the variance of (*Ri-Li*) for each population and measurement (Fig. 2). Comparisons of variances, population by population for the four parameters by applying a sequential Bonferroni test per line (pairs of populations compared for all variables) show that only WM/1 exhibits a significant difference between DDO and MDH (Tab. 2). Moreover,

Table 1. Statistical determination of directional asymmetry and size differences between populations by a two-way mixed-model ANOVA (side, fixed; sample, random)

Variable	Sum of squares	df	Mean square	*F*
LM/1				
Side	0.0007	1	0.0007	0.58
Sample	0.2686	2	0.1343	108.85**
Interaction	0.0003	2	0.0002	0.12
Residual	0.2073	168	0.0012	
WM/1				
Side	0.0000	1	0.0000	0.040
Sample	0.0442	2	0.0221	26.85**
Interaction	0.0001	2	0.0000	0.05
Residual	0.1384	168	0.0008	
LM/2				
Side	0.0000	1	0.0000	0.01
Sample	0.1264	2	0.0632	60.00**
Interaction	0.0003	2	0.0001	0.14
Residual	0.1769	168	0.0011	
WM/2				
Side	0.0005	1	0.0005	0.54
Sample	0.0155	2	0.0077	8.77**
Interaction	0.0003	2	0.0002	0.17
Residual	0.1481	168	0.0009	

**$p < 0.01$

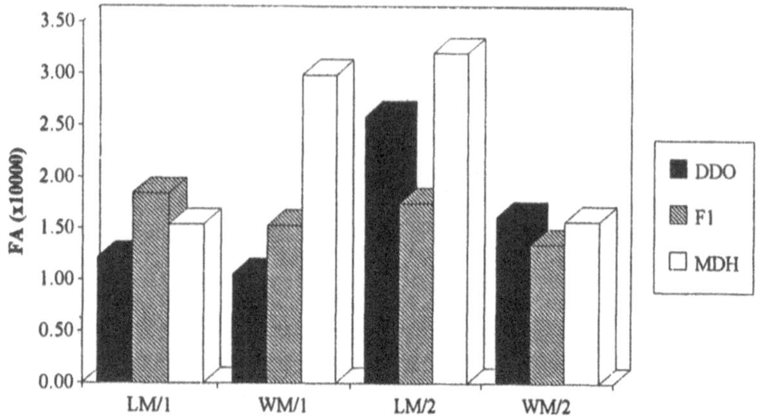

Figure 2. Traditional fluctuating asymmetry (*FA*) levels for the two strains, DDO and MDH, and their F1 hybrids represented for the four measurements considered here (length and width of the two first lower molars M/1 and M/2).

the Kendall test shows independence of rank ($W = 0.344, p > 0.05$), indicating no clear pattern of asymmetry within groups. For all characters except WM/1, a new *FA* value was calculated after having pooled the two parental samples. Comparisons between parental and hybrid *FA* do not provide any significant differences (Table 2).

Although the homogeneous pattern for M/2 indicates a decrease in asymmetry for the hybrid sample, no statistical test could corroborate that result.

Procrustes Superposition

The asymmetry patterns are given in Fig. 3. These results show a lower net asymmetry score for the F1 sample. The lower score also suggests that both *DA* and *FA* are reduced for that sample. ANOVA performed on NA_i distribution within samples shows that the F1 sample presents a reduced net asymmetry ($F = 3.51$, df 2,86; $p = 0.034$). When population *DAs* are respectively subtracted from individual NA_i, the population distributions of the remainders can be considered as the FA_i distribution. In this case, although the p-value is low, there are no significant differences between population means ($F = 2.83$, df 2,86, $p = 0.064$). Both *NA* and *FA* means do not exhibit significant differences between DDO and MDH (*NA*: $t = 0.13$; df = 57; $p = 0.90$; *FA*: $t = 0.042$, df = 57, $p = 0.97$). The two samples were then pooled and *FA* and *NA* recalculated globally for the parental group. Significant differences between hybrid and parental means are shown for *NA* and *FA* (*NA*: $t = 2.7$, df = 85, $p = 0.009$; *FA*:

Table 2. Differences between fluctuating asymmetry (*FA*) levels (i.e., between variances of *Ri-Li*) among populations in the traditional approach

		FAa/FAb		*(FAb/FAa)*	
Groups a-b compared	(Na/Nb)	LM/1	WM/1	LM/2	WM/2
DDO-F1	(30/29)	*1.52*	*1.44*	1.48	1.20
DDO-MDH	(30/28)	*1.27*	*2.82**	*1.24*	1.03
F1-MDH	(29/28)	1.20	*1.95*	*1.83*	*1.17*
<DDO-MDH>-F1	(58/29)	*1.31*	-	1.69	1.24

* $p < 0.05$ sequential Bonferroni test per line

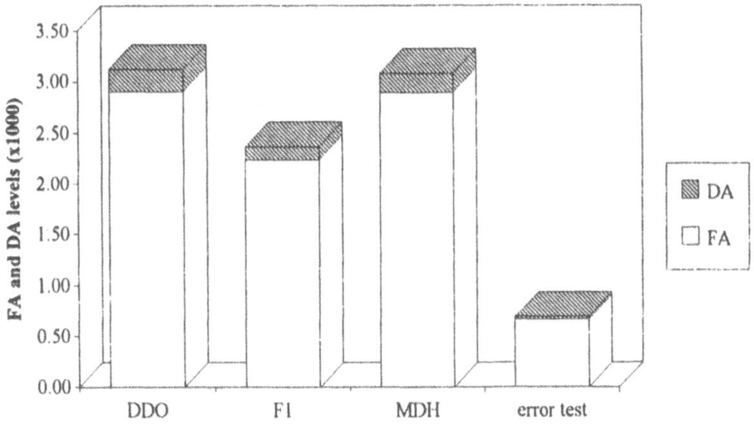

Figure 3. Fluctuating and directional asymmetry (*FA* and *DA*) levels by the Procrustes superposition method for DDO and MDH samples and their F1 hybrids (see text for error sample).

$t = 2.5$, df = 85, $p = 0.013$), clearly expressing a decrease of asymmetry parameters for the hybrid sample as compared with the parental one.

The levels of *NA* and *DA* determined in the error test are relatively important but remain nevertheless low compared with the true asymmetry values.

DISCUSSION

According to the results of Bader (1965) and Alibert et al. (1994) a significant decrease of *FA* was expected for the F1 hybrid sample. In the traditional approach, although not statistically significant, only M/2 measurements suggest this pattern. Using Procrustes superposition, both *NA* and *FA* scores show a significant decrease for the hybrid sample compared with the parental one. Consequently, the Procrustes approach gives results closer to those expected.

Although *DA* is insignificant using a traditional approach, the homologous values in the Procrustes approach are far from negligible; they represent between 5.4 to 7.1% of *NA* values. However, no tests are yet available to evaluate the homogeneity of *DA* among samples.

Practically, in the traditional approach, when four measurements and three populations are considered and compared, 22 tests are, in the simplest case, necessary to perform a statistical treatment for *FA* patterns (four correlations, four normal distributions, 1 two-way ANOVA, 12 variance comparisons and 1 Kendall test). In the Procrustes approach, only two one-way ANOVAs are required to get the same information.

From a biological point of view, the decrease of the *FA* level for hybrids is clearly shown by the application of the Procrustes method. Several experimental studies have already concluded that the developmental stability in strain hybrid mice is increased as compared with that of the parental strains (Leamy, 1984, 1992; Leamy and Atchley, 1985). However, most of these experiments have been performed on inbred strains. A certain heterotic effect in hybrids is usually invoked to explain this pattern, mitigating the destabilizing effect of inbreeding in parental strains. In the present study, the parental samples are randombred and belong to two different subspecies. The higher developmental stability observed in F1 hybrids is similar to the pattern observed in the study by Alibert et al. (1994) on wild populations of the Danish hybrid zone (see Introduction). The main difference is that the experimental hybrid sample is exclusively composed of F1 individuals whereas the natural hybrid sample is composed of complex recombinant individuals from populations for which the mean hybrid index ranged between 40 and 60%. Moreover, no F1 hybrid between the two pure subspecies has ever been found in the Danish hybrid zone. To explain the discrepancy between the benefit of hybridization for developmental stability and the apparent counter-selection on these natural hybrids, Alibert et al. (1994) invoked a decoupling of consequences of hybridization on different biological functions. Although some of the latter might profit from the heterotic effect of hybridization, others might suffer from the disruption in coadapted genomes. Whereas the disruption of coadapted genomes may not be clearly expressed or detected in F1, there is no evidence according to which heterotic effect may not be expressed in F1. Consequently, our results confirm the presence of heterosis on developmental stability in subspecific hybrids from F1 to further recombinant individuals in the house mouse.

Alibert et al.(1994) performed their study by applying the traditional approach for the calculation of *FA*. *FA* values found in wild populations are higher than those found in the present study (Wilcoxon signed rank test; $Ts = 0$, $n = 12$, $p < 0.001$). However, whereas the decrease of *FA* from wild mice (*M. m. domesticus* and *M. m. musculus*) to their laboratory homologues (DDO and MDH) is respectively of 66.4 and 46%, that decrease from wild hybrids to the F1 sample is only 18%. Bader (1965) did not find any *FA* differences between wild and randombred populations. The lower levels of *FA* found in the laboratory samples compared with those in wild populations may be explained by the environmental stress to which natural populations are subjected. Consequently, the reduction of *FA* due to hybridization is proportionally more important in wild populations than it is for the laboratory-reared animals. This may suggest that the stabilizing effect of heterozygosity is powerful enough to restore the developmental stability of hybrids in nature, despite the apparent environmental stress supported by wild populations.

ACKNOWLEDGMENTS

To Sarah Lenington and Janice Britton-Davidian for stimulating discussions, Guila Ganem for technical assistance and Alfred Zaegel for maintenance of animals. Pub. ISEM N°94-067.

REFERENCES

Alibert, P., S. Renaud, B. Dod, F. Bonhomme, J.-C. Auffray. 1994. Fluctuating asymmetry in the *Mus musculus* hybrid zone: a heterotic effect in disrupted co-adapted genomes. Proceedings of the Royal Society, London, Biological Sciences 258: 53-59.

Auffray, J.-C., P. Alibert, C. Latieule, and B. Dod. Relative warp analysis of skull shape across the hybrid zone of the house mouse (*Mus musculus*) in Denmark. Journal of Zoology, London (in press).

Auffray, J.-C., and J. Britton-Davidian. 1992. When did the house mouse colonize Europe? Biological Journal of the Linnean Society 45:187–190.

Auffray, J.-C., J. T. Marshall, L. Thaler, and F. Bonhomme. 1990a. Focus on the nomenclature of European species of *Mus*. Mouse Genome 88:7–8.

Auffray, J.-C., F. Vanlerberghe, and J. Britton-Davidian. 1990b. The house mouse progression in Eurasia: a palaeontological and archaeozoological approach. Biological Journal of the Linnean Society 41:13–25.

Bader, R. S. 1965. Fluctuating asymmetry in the dentition of the house mouse. Journal of Mammalogy 46:384–388.

Barton, N. H., and G. M. Hewitt. 1989. Adaptation, speciation and hybrid zones. Nature 341:497–503.

Bookstein, F. L. 1991. Morphometric tools for landmark data: geometry and biology. Cambridge University Press: Cambridge.

Boursot, P., J.-C. Auffray, J. Britton-Davidian, and F. Bonhomme. 1993. The evolution of house mice. Annual Review of Ecology and Systematics 24:119–152.

Boursot, P., F. Bonhomme, J. Britton-Davidian, J. Catalan, and H. Yonekawa 1984. Introgression différentielle des génomes nucléaires et mitochondriaux chez deux semi-espèces de souris. Comptes Rendus de l'Académie des Sciences, Paris. 299:365–370.

Dod, B., L. S. Jermiin, P. Boursot, V. M. Chapman, J. T. Nielsen, and F. Bonhomme. 1993. Counterselection on sex chromosomes in the *Mus musculus* European hybrid zone. Journal of Evolutionary Biology 6:529–546.

Gerasimov, S., H. Nikolov, V. Mihailova, J.-C. Auffray, and F. Bonhomme. 1990. Morphometric stepwise discriminant analysis of five genetically determined European taxa of the genus *Mus*. Biological Journal of the Linnean Society 41:47–64.

Graham, J. H. 1992. Genomic coadaptation and developmental stability in hybrid zone. Acta Zoologica Fennica 191:121–131.

Leamy, L. 1984. Morphometric studies in inbred and hybrid mouse. V. Directional and fluctuating asymmetry. American Naturalist 123:579–593.

Leamy, L. 1992. Morphometric studies in inbred and hybrid mouse. VII. heterosis in fluctuating asymmetry at different ages. Acta Zoologica Fennica 191:111–120.

Leamy, L., and W. Atchley. 1985. Directional selection and developmental stability: evidence from fluctuating asymmetry of morphometric characters in rats. Growth 49:8–18.

Leary, R. F., F. W. Allendorf, and R. L. Knudsen. 1985. Inheritance of meristic variation and the evolution of developmental stability in rainbow trout. Evolution 39:308–314.

Lebreton, J.-D., J.-P. Roux, G. Banco, and A.-M. Bakou. 1992. Biomeco (Biometry-Ecology), v. 4.2. Statistical Ecology Software. Centre d'Ecologie Fonctionnelle et Evolutive, CNRS: Montpellier

Lerner, I. M. 1954. Genetic homeostasis. Wiley: New York.

Mitton, J. B. and M. C. Grant. 1984. Associations among protein heterozygosity growth rate, and developmental homeostasis. Annual Review of Ecology and Systematics 15:479–499.

Moulia, C., J.-P. Aussel, F. Bonhomme, P. Boursot, J. T. Nielsen, and F. Renaud. 1991. Wormy mice in a hybrid zone: a genetic control of susceptibility to parasite infection. Journal of Evolutionary Biology 4:679–687.

Moulia, C., N. Le Brun, J. Dallas, A. Orth, and F. Renaud. 1993. Experimental evidence of genetic determinism in high susceptibility to intestinal pinworms infection in mice: a hybrid zone model. Parasitology 106:387–393.

Palmer, A. R., and C. Strobeck. 1986. Fluctuating asymmetry: measurement, analysis, patterns. Annual Review of Ecology and Systematics 17:391–421.

Parsons, P. A. 1992. Fluctuating asymmetry: a biological monitor of environmental and genomic stress. Heredity 68:361 364.

Rice, W. R. 1989. Analyzing tables of statistical tests. Evolution 43:223–225.

Rohlf, F. J. 1990. Rotational fit (Procrustes) methods. Pages 227–236 *in* F. J. Rohlf and F. Bookstein, (eds.), Proceedings of the Michigan morphometrics workshop. University of Michigan Museum of Zoology Special Publication 2.

Rohlf, F. J., and D. Slice. 1990. Extensions of the Procrustes method for the optimal superposition of landmarks. Systematic Zoology 39:40–59.

Sage, R. D., D. Heyneman, K. C. Lim, and A. C. Wilson. 1986. Wormy mice in a hybrid zone. Nature 324:60–63.

Soulé, M. E. 1967. Phenetics of natural populations. II. Asymmetry and evolution in a lizard. American Naturalist 101:141–160.

Soulé, M. E. 1979. Heterozygosity and developmental stability: another look. Evolution 33:396–401.

Tucker, P. K., R. D. Sage, J. Warmer, A. C. Wilson, and E. M. Eicher. 1992. Abrupt cline for sex chromosomes in a hybrid zone between species of mice. Evolution 46:1146–1163.

Vanlerberghe, F., P. Boursot, J. T. Nielsen, and F. Bonhomme. 1988. A steep Cline for mitochondrial DNA in Danish mice. Genetical Research, Cambridge 52:185–193.

Vrijenhoek, R. C., and S. Lerman. 1982. Heterozygosity and developmental stability under sexual and asexual breeding systems. Evolution 36:768–776.

WING VENATION VARIABILITY IN *MONARTHROPALPUS BUXI* (DIPTERA, CECIDOMYIIDAE) AND THE QUATERNARY COEVOLUTION OF BOX (*BUXUS SEMPERVIRENS* L.) AND ITS MIDGE

A Geometrical Morphometric Analysis

Michel Baylac* and Tanguy Daufresne

Muséum National d'Histoire Naturelle
Laboratoire d'Entomologie
45, rue Buffon
F-75005 Paris

ABSTRACT

The variability of the wing venation of the box midge, *Monarthropalpus buxi*, here characterized by six landmarks, is analyzed by geometrical morphometrics. Four localities in France were sampled, covering the northern disjunction in the distribution of box shrubs in Europe. Within- and between-locality analyses indicate that the southern locality is completely separated from the central, western and northern ones. This supports the hypothesis of Parent (1980), who considered the isolated box localities of the northern part of France and Germany to be relicts from the ice age, and is consistent with a close coevolution of box and its midge.

No differences were found between the parti- and univoltine populations of the midge. This result confirms that the dual life cycle of the species is a simple adaptation, originating perhaps during the ice age, but without taxonomic implications.

Sexual dimorphism involves a strong uniform component which corresponds to a stretching of the wing in the male sex, together with a nonuniform expansion of the medial area.

* e-mail: baylac@mnhn.fr

Advances in Morphometrics, Edited by L. F. Marcus et al.
Plenum Press, New York, 1996

INTRODUCTION

Box, *Buxus sempervirens* L., is an evergreen shrub with small, rounded and coriaceous leaves. The range of this Palearctic species is from Europe up to the Himalayas. It has been further spread by man to many continents for ornamental and also religious purposes.

The distribution and ecology of box in Europe have been intensively studied because of their numerous peculiarities (details and references in Christ, 1913 and Parent, 1980). The northern localities (Belgium, England, North of France and Germany) are in general small and isolated and restricted to stream or river valleys. In contrast, in the southern part of its range (South of France, Spain and Greece up to Iran), box sites cover large areas. As box seems to rely exclusively upon water for its large-scale dispersion, it has been suggested that the northern localities are possible relicts from the ice age (Vanden Berghen; 1955; Godwin, 1975). Palynological and hydrogeographical evidence, together with faunistic considerations, led Parent (1980) to substantiate this hypothesis in the case of the French and German Mosel valley localities. Additionally, Parent showed that this isolation process could be dated from the Eemian period, between 80,000 and 120,000 years BC. This hypothesis appears to apply equally well to most of the northern localities.

The entomofauna associated with box is relatively poor (Barnes, 1948), and the most frequently encountered species is a gall midge, *Monarthropalpus buxi* Laboulbène, commonly known as the box leaf midge. *M. buxi* is a monophagous species whose larvae live in an irregular blister gall within the leaf epidermis of box, *Buxus* spp. Up to 16 larvae may live in a single leaf, and the species has been considered as a pest of considerable importance (Barnes, 1948). The species is known from Europe and North America, where it was introduced during colonial times (Gagné, 1989) together with its host plant.

Recently, it was shown (Baylac, 1989) that *M. buxi* has two types of life cycles: a one-year univoltine (normal) and a two-year partivoltine cycle. In the latter case, the larvae spend two years within the galls before pupation, rather than a single year as in the former case. There are relatively few examples of partivoltinism among insects and particularly among Diptera. This condition has been generally considered to be an adaptation to harsh environmental conditions, such as those in deserts or high mountains (Waldbauer, 1978). This phenological adaptation may have allowed survival of the midge during the last ice age (Baylac, 1989), but this hypothesis remains to be tested.

The present study analyzes the degree of morphological diversification of the midge within and between geographical regions selected to represent northern and southern localities. Two other factors were investigated: voltinism, to assess possible divergence between parti- and univoltine midges, and sexual dimorphism of the wing venation, as very little information is available. It is impossible to sex the larvae, and additional morphological variability may be induced by endoparasites during the larval stage (Barnes, 1948, Baylac; 1989). It was therefore decided to restrict the analysis to adults. Wing venation was studied because many previous morphometric studies have shown the sensitivity and discriminating power of this indicator, in both the Diptera (Rohlf and Slice, 1990; Rohlf, 1993a) and other insects taxa (bees: Cornuet, 1975; Ruttner et al., 1978, Smith et al. in Bookstein, 1991; butterflies: Guillaumin, 1972).

MATERIAL AND METHODS

Galls were collected in the field in 1990, sorted into parti- and univoltine types, as assessed from their position on the branches, and reared until adult hatching. The four

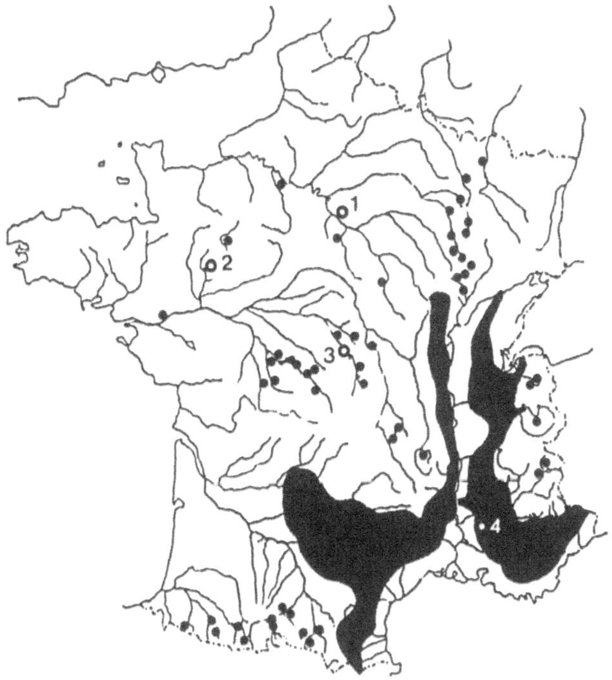

Figure 1. Distribution of *Buxus sempervirens* L. in France (modified from Lenoble and Broyer, 1945). The four localities sampled were (1) Poligny; (2) Saulges; (3) Boissereau; and (4) Sérignan.

localities (Fig. 1) represent the northern (Poligny), central (Boissereau), western (Saulges) and southern (Sérignan) parts of France. The northern and central localities correspond to isolated box sites, associated with the Seine basin and the Cher (tributary of the Loire) basin respectively. The Saulges location is completely isolated, located near a small river, without direct connection to the Loire and thus recent backward recolonization is unlikely. The last station, Sérignan, is representative of one of the two large areas of box located on either side of the Rhône, extending from Burgundy to Italy and the eastern border of Spain.

Two hundred and forty adults were investigated (Table 1). The samples were not balanced, and Poligny was the only location where the two voltine types of life cycles were sampled. The effect of voltinism was therefore tested using only the Poligny sample.

Left and right wings were stained overnight in a solution of fuchsin in lactic acid and temporarily slide mounted. Images of the left and right wings were obtained using a video camera mounted on a microscope with a 10X objective. The coordinates of six landmarks (Fig. 2) were recorded using MeasurementTV software (Updegraff, 1990). The first landmark was the lower end of the arculus; the second, third and forth landmarks were the intersections with the wing margin, or the costa, of, respectively, the veins R1, R4+5 and CuA1. The sixth landmark was the intersection of the radial and cubital sectors. It was not possible to include the distal end of CuA2, which never reaches the wing margin, at least in the females, even after overstaining.

Each wing was measured five times, and the results were averaged. The right side landmarks were superimposed onto those of the left by a Procrustes least-squares algorithm and averaged. Size was estimated by the logarithm of the mean over the left and right centroid sizes (square roots). Two additional parameters were calculated, the wing perimeter and the wing surface, by means of landmarks 1,2,3,4 and 6. All these calculations were performed using the XY program (Baylac, 1993) or specially devised PASCAL programs .

Table 1. Characteristics of the samples (n = 240)

	Boissereau		Poligny		Saulges		Sérignan		Total	
Localities	Males	Females	Males	Females	Males	Females	Males	Females	Males	Females
Partivoltine	8	8	12	16	5	8	-	-	25	31
Univoltine	-	-	29	39	-		56	60	85	99
Total	7	8	41	55	5	8	56	60	109	131

The coordinates of the whole sample were superimposed using Procrustes methods (Rohlf and Slice, 1990): Generalized Resistant Fit (GRF) and Generalized Least-square Fit (GLS) with the GRF-ND program (version 07/10/93; Slice, 1993). The two solutions appear to be equivalent: the normalized Mantel statistic, Z, between the GRF and GLS coordinates correlation matrices equals 0.911 (p = 0.001; none exceeded this value among the 1000 random permutations performed). Similarly, a canonical correlation analysis of the GRF and GLS coordinates matrices, furnished eight canonical axes explaining 100 % of the variation. The canonical correlations ranged from 0.9998 to 0.9916. The results were rotated to align coordinates with major axes of the reference (option REFPCA of GRF-ND). The partial warps—which parameterize the nonuniform part alone (Bookstein, 1991)—were calculated using the TPSRW package (version 09/27/93; Rohlf, 1993b) with the GLS coordinates and

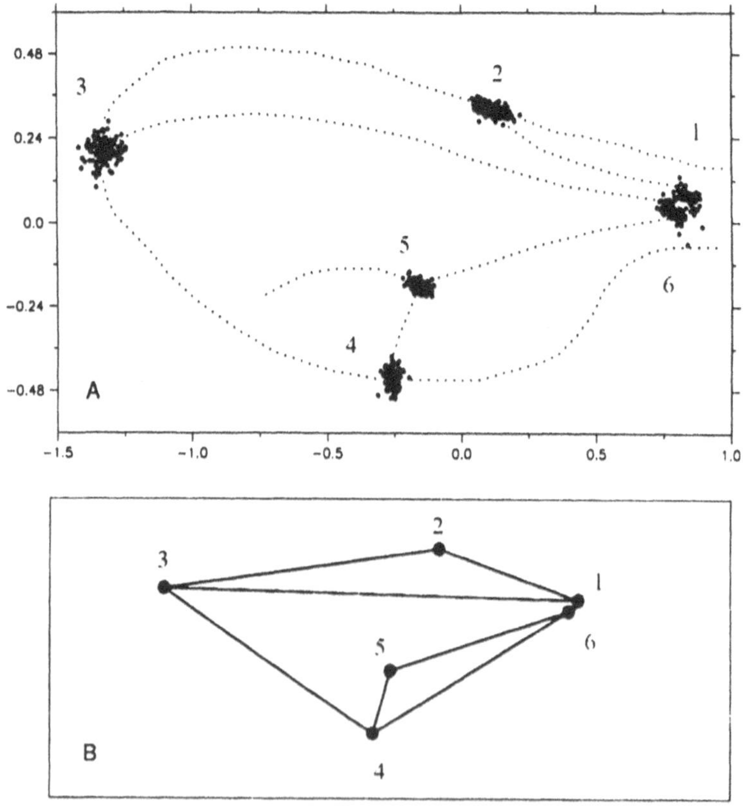

Figure 2. A Outline and landmarks of a left wing of *Monarthopalpus buxi*. Dots delineate the variability of the landmark location after a Generalized Resistant Fit superimposition (GRF). B simplified outline linking the landmarks used for the splines.

reference and setting $\alpha = 0$ (Rohlf, 1993a). The uniform parameters U1 and U2 were calculated following the formula of Bookstein ("Standard formula" article, this volume) as applied to GLS residuals. The substitution of the GRF residuals in the formula gave identical results (correlation coefficients $r_{U1} = 1.00$ and $r_{U2} = 1.00$).

The partial warps, uniform parameters and the log of centroid size were analyzed separately or together with ANOVA, MANOVA, and discriminant analyses using SAS 7.04. The splines illustrating the deformations along the canonical axes were calculated using the TPSREGR program (version 12/11/92; Rohlf, 1992). Bootstrap estimates of the discriminant functions were obtained using a specially written MATLAB function (in the case of resampling with unequal sample sizes) or the statistical package SPAD.N (CISIA, 1991). Between and within group allometry (Mosimann and James, 1979; Bookstein, 1991) were tested by simple and multiple correlation coefficients or directly by the multivariate approach (Marcus, 1990; Klingenberg, this volume) with principal component analyses (PCA) of the log of the 15 possible distances between the six landmarks. In the latter case the bootstrap estimates were obtained using a MATLAB function derived from Marcus (1990). The relationships between the uniform and the nonuniform parameters were investigated by canonical correlations and canonical redundancy analysis (SAS 7.04).

RESULTS

Size Differences

Centroid size statistics for each level are summarized in Table 2. Parti- and univoltine types from Poligny were pooled as there was no size differences between them, either for males (Student $t = 0.097$, $p = 0.923$, $df = 39$) or for females ($t = 0.856$, $p = 0.4$, df = 53), and the variances were statistically not different. Two-way ANOVA indicated that both the sex ($p < 0.015$) and the locality ($p < 0.001$) effects were significant. The interaction was not significant ($p = 0.27$). On the average, males tended to have larger wings than did females, and the differences were higher for Saulges and Boissereau. A Tukey multiple range test classified the localities into two groups, Sérignan and Poligny into one group with larger wings and Saulges and Boissereau into the other with smaller wings.

Shape Differences

Uniform Shape. There were no differences between the uniform parameters for the voltine types in the Poligny sample. The MANOVA was not significant, the probability of

Table 2. Centroid size (mean, standard deviation (Std), minimum and maximum) within locality, sex and voltine type

	Localities									
	Boissereau		Poligny				Saulges		Sérignan	
			Males		Females					
	Males	Females					Males	Females	Males	Females
Size	Partiv	Partiv	Univ	Partiv	Univ	Partiv	Partiv	Partiv	Univ	Univ
Mean	550.8	538.4	591.6	592.5	580.8	590.9	555.9	518.8	579.9	578.3
Std	27.3	27.0	28.9	20.9	37.1	45.4	8.1	38.0	26.2	25.5
Min	525.5	482.7	516.8	571.5	505.8	483.7	548.3	453.6	522.0	520.6
Max	612.0	568.1	635.0	625.8	635.3	644.7	568.5	558.1	629.8	632.5

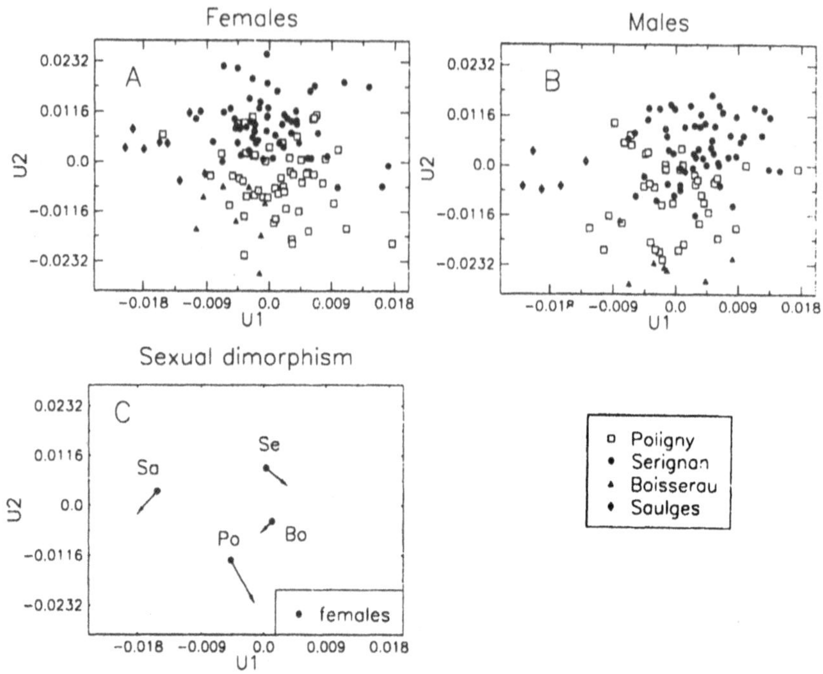

Figure 3. Plots of the affine components U1 and U2 for the males (A) and the females (B). (C) Vectors of the differences between mean males and females locations by localities; circles represent females and arrow heads, males.

the Wilks' Lambda parameter being equal to 0.51 (F test with df = 2, 93). The corresponding subsamples were pooled in the subsequent analyses.

On the plots of U1 versus U2 (Fig. 3A and 3B; see also Fig. 9), the discrimination is almost complete only in the case of Saulges. The effects of sex and locality are highly significant (MANOVA, $F = 7.14$, df = 2, 231, $p = 0.001$). Univariate analyses indicate that U1 did not differ between sexes, whereas, U2 differed between sexes and localities. The misclassification percentages, given by discriminant analyses of U1 and U2, ranged from 16 to 25 % in most cases, with the exceptions of Poligny (males) and Boissereau (females; Table 3). The patterns displayed within each locality were similar for the two sexes: a relative stretching of the wing in Poligny and a shear mainly apparent in Saulges. Although the overall size is similar for these localities, the uniform part is very different and cannot be related to a simple allometric effect.

The plot contrasting the mean configurations of males and females (Fig. 3C) shows that in all localities the males had stretched wings (U2). The misclassification percentages between the sexes were always lower for the females (Table 4).

Total Nonuniform Shape. MANOVA of the voltine types for Poligny were not significant either for the males (Wilks' Lambda = 0.0163, $F = 1.402$, $p = 0.24$, df = 6, 34) or for the females (Wilks' Lambda =0.959, $F = 0.338$, $p = 0.91$, df = 6, 40). The two types were therefore pooled. MANOVA comparing the effects of sex and locality was highly significant (sex factor: $F = 15.73$, df 6, 227, $p = 0.0001$; locality factor: $F = 29.95$, df = 18, 642.5, $p = 0.0001$). The interaction term was also significant ($F = 1.75$, df = 18, 642.5, $p = 0.028$), and thus all the analyses were conducted separately for each factor.

Table 3. Between-locality discriminant analyses. Misclassification percentages as function of size and shape parameters

	Uniform		Total nonuniform		Shape (uniform and nonuniform)		Shape and size	
Localities	Male	Female	Male	Female	Male	Female	Male	Female
Boissereau	14.3	62.5	14.3	25.0	14.3	0.0	14.3	0.0
Poligny	46.3	23.6	12.2	9.1	0.0	3.6	4.9	5.5
Saulges	0.0	25.0	60.0	87.5	0.0	0.0	0.0	0.0
Sérignan	16.1	16.7	3.6	3.3	0.0	5.0	0.0	3.3
Total	26.6	22.9	10.1	12.2	0.9	3.8	2.8	3.8

The importance of the nonuniform part differed substantially between localities. Between-locality misclassification percentages were as high as 60 % (males) and 75 % (females) for Saulges but only 3 to 25 % for the other locations (Table 3).

Only the first two canonical axes of the between-locality analysis were significant and explained 80.0 % and 17.1 % (cumulative 97.1 %) for the males against 88.6 % and 9.4 % (cumulative 98.0 %) for the females, respectively. The results were roughly the same for the two sexes. Nevertheless, the discrimination between Poligny and Sérignan as well as that between Boissereau and Poligny + Saulges (Fig. 4 and 5) was more pronounced for the males. On the third axis (not shown) Boissereau and Saulges were not clearly separated for either sex, and Saulges and Poligny overlapped substantially. This is a clear indication that the main difference between these two localities is of the uniform type. The differences between localities are summarized by the splines shown in Fig. 4 and 5. Sérignan was characterized by a smaller basal part of the wing, as is delimited by the polygon 1, 2, 5, 4 and 6. This corresponds to a horizontal displacement of the line delimited by the landmarks 4, 5, and 2 towards the base of the wing in Sérignan or towards its apex in the other localities. This backward shift was more marked for the dorsal (landmark 2) than for the ventral (landmarks

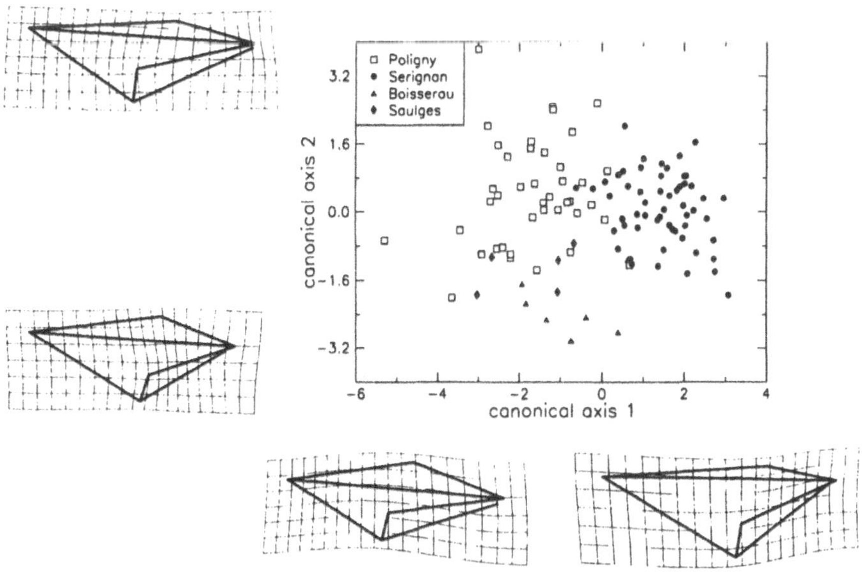

Figure 4. Between-locality canonical discriminant analyses. Total nonuniform warps, males.

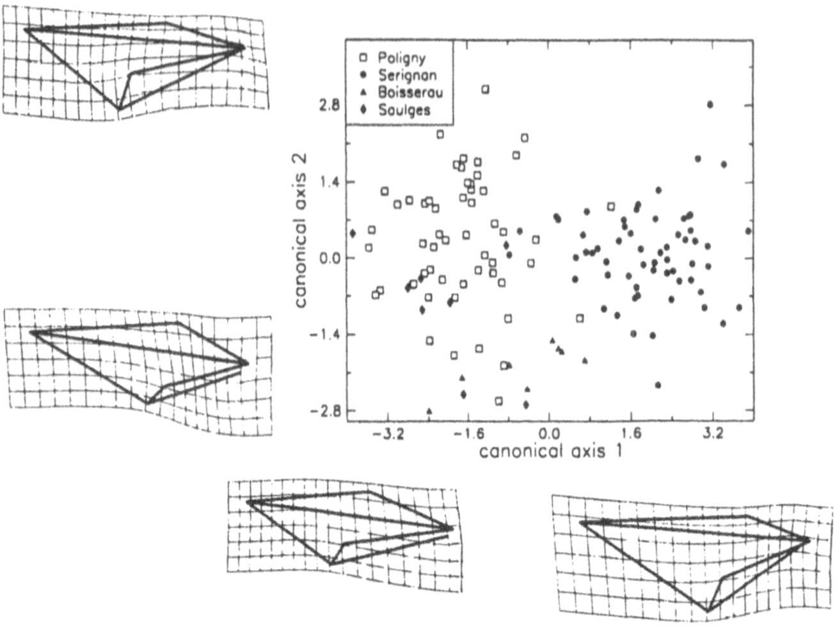

Figure 5. Between-locality canonical discriminant analyses. Total nonuniform warps, females.

4 and 5) margin. On the second axis, the Poligny locality was characterized by a distal (males) or distal and dorsal (females) shift of the fifth landmark together with a large cubital area (triangle 4, 5, 6) and a reduction of the triangle delimited by the landmarks 1 to 3. This can also be summarized by an increase of the distance between landmark 1 (or 6) and landmark 5, and a simultaneous decrease of the distance between the landmarks 2 and 5. Saulges and Boissereau were characterized by a small cubital area.

Although the misclassification percentages for sexes are lower than with the uniform part alone, the discrimination was complete only for Saulges. The misclassification rate was again higher for the males, with the exception of Sérignan (Table 4). The splines illustrating the discriminant functions (Fig. 6) depict very few differences between localities. The main difference was an anterior (males) or posterior (females) shift of the whole median region (landmarks 2, 5 and 4), which appeared more marked for its central part than for the margins of the wing. Correspondingly, the position of the fifth landmark was nearer the wing apex in males and nearer its base in females. The differences in the height of the triangle defined by the landmarks 1, 2 and 3 were more pronounced in Boissereau and Saulges populations.

Table 4. Between-sex discriminant analyses (F = females, M = males). Misclassification percentages as function of size and shape parameters. The results for shape alone are detailed in Table 9

	Uniform		Total nonuniform warps		Shape and size	
Localities	M to F	F to M	M to F	F to M	M to F	F to M
Boissereau	14.3	12.5	14.3	0.0	0.0	0.0
Poligny	73.2	18.2	17.1	14.6	14.6	1.8
Saulges	40.0	0.0	0.0	0.0	0.0	0.0
Sérignan	33.9	23.3	14.3	18.3	7.1	13.3

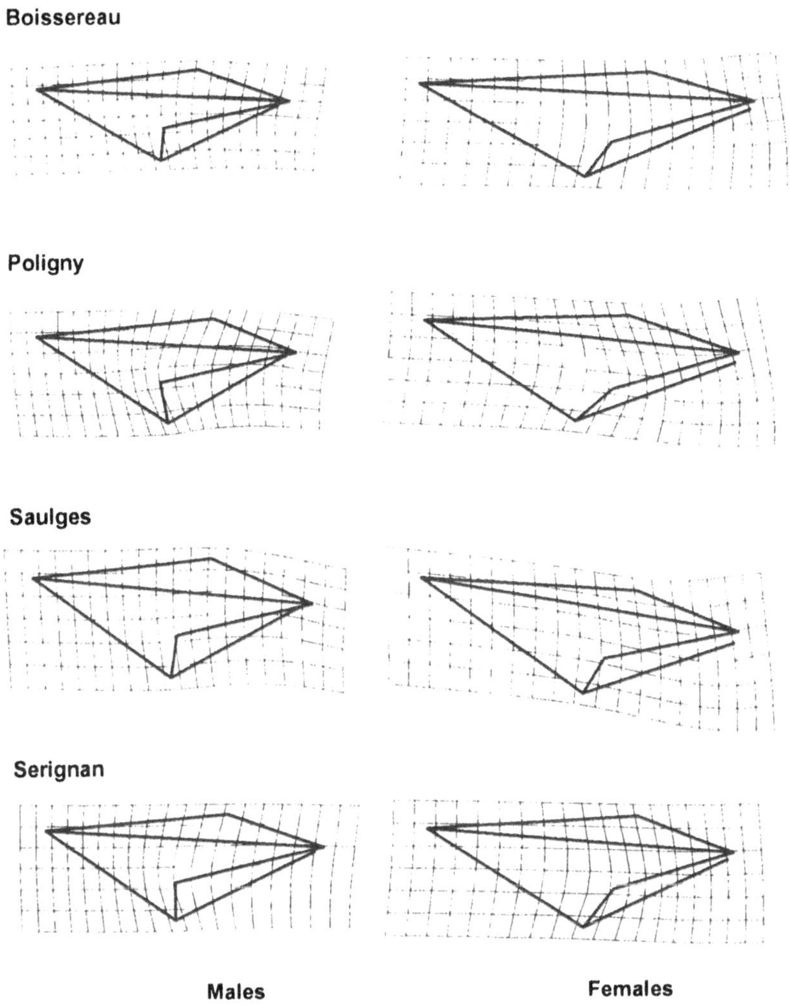

Figure 6. Sexual dimorphism within localities. Splines illustrating the discriminant functions calculated with the total nonuniform warps.

Relationships between Size, Uniform Shape and Nonuniform Shape

Relationships between Uniform and Nonuniform Components. Canonical redundancy analysis (Table 5) shows that Boissereau and Saulges differed completely from the other two localities. The percentages of variance of the uniform part explained by the nonuniform part were as high as 100 % in Boissereau and Saulges but account for less than 50 % in Poligny and Sérignan. Inversely, the percentages of variance of the nonuniform part were less variable, between 10 and 42 %. The small numbers of specimens from Saulges and Boissereau prevent further investigations. The canonical structures (not shown) differed from one locality to another, even between Poligny and Sérignan, for which there is no bias due to sample size.

Table 5. Canonical redundancy analysis of the uniform and the nonuniform parts within sex and locality; proportions of variance of one set explained by the second one

	Location			
Component	Boissereau	Poligny	Saulges	Sérignan
Females				
% Uniform	85.08	49.87	95.11	34.08
% Non-uniform	38.65	23.86	37.65	13.95
Males				
% Uniform	100.00	42.08	100.00	35.45
% Non-uniform	37.97	10.74	41.59	20.16

Relationships Between Size, Uniform and Nonuniform Components: Allometry. Multiple regression analyses were possible only in the cases of Sérignan and Poligny. For Boissereau and Saulges the model was not of full rank due to the small sample sizes; allometric patterns were therefore estimated by the simple correlation coefficients.

For the males, there was no significant relationship between size and shape; no regressions were significant either globally or separately for the uniform or the nonuniform components.

For the females (Table 6) neither the simple nor the multiple correlation coefficients were significant in Poligny. In Sérignan the multiple correlation was low, explaining only 30 % of the size variability. The simple correlations were on the whole higher in Boissereau and Saulges, but the allometric patterns for these two localities were very different.

The angles between the first axis of the within-groups PCA were between 4.12 (Poligny) and 12.03° (Boissereau) in the females and 4.03° (Poligny) and 63.07° (Saulges) in the males. None of theses axes appear statistically different from isometry. Bootstrap results with 1000 replications indicated that the lowest probability level was for the females of Saulges ($p = 0.101$), whereas in the other localities the probabilities were all above 0.20. Furthermore, the percentages of variance explained by these PCA were relatively low, both for the males, 56.4 (Sérignan), 57.4 (Poligny), 62.5 (Saulges), and 63.1 (Boissereau), and for the females, 58.8, 77.8, 93.8 and 83.2, respectively. According to Sprent (1972), most of these percentages fall out of the application range of the Jolicoeur generalization.

Table 6. Allometric patterns within locality for the females. Simple and multiple correlation coefficients of centroid size upon the uniform (U1S, U2S) and nonuniform (PW1 to PW6) components

Variates	Poligny	Sérignan	Saulges	Boissereau
U1S	0.047	0.326*	-0.561	-0.715*
U2S	0.164	-0.130	0.927***	0.241
PW1	0.022	0.077	0.534	-0.354
PW2	-0.063	0.346**	-0.449	0.101
PW3	-0.243	-0.323*	-0.522	-0.025
PW4	0.130	0.357**	0.787*	-0.402
PW5	-0.074	0.055	-0.534	-0.218
PW6	-0.010	0.011	0.450	0.587
Rmult	0.319	0.548*	-	-

*$p<0.05$; **$p<0.01$; ***$p<0.001$

However, many correlations between the interlandmark group means and the centroid size were significant at the 0.05 level or less: 1-3, 1-4, 1-5, 3-6, 4-6 and 5-6 in both sexes, and 2-4 and 2-5 for the females. These results confirm that the between-group differences are not purely isometric and do not correspond to simple extension (Poligny and Sérignan) or reduction (Boissereau and Saulges) of the size range.

Form Differences between Sexes and Localities

The effects of size were generally small or absent and will be addressed only briefly. The effect of voltinism was not statistically significant when all the components including size were pooled.

Geographical Variability. Pooling the uniform and the nonuniform parameters allowed a significant increase of the discriminations. The misclassification percentages never exceeded 16 % for the males and 5.5 % for the females (Table 3). The same holds if size is included, but with minor and insignificant differences that are probably due to variations in the heteroscedaticity of the covariances matrices.

For the males, canonical discriminant analysis of the uniform and nonuniform parameters (Fig. 7) almost completely individualized Sérignan onto the first axis (74.5 %). The second (14.4 %) and the third (10 %, not shown) axis partly separated the other localities. Nevertheless, the discriminations were less clear than for Sérignan, and one specimen from Boissereau falls into the Poligny group. The correlations between the canonical variates and the log of the centroid size were significant only for the second and the third axes. They do not exceed 0.35 and explain a maximum of 12 % of variance corresponding mostly to the separation of Poligny from Saulges and Boissereau.

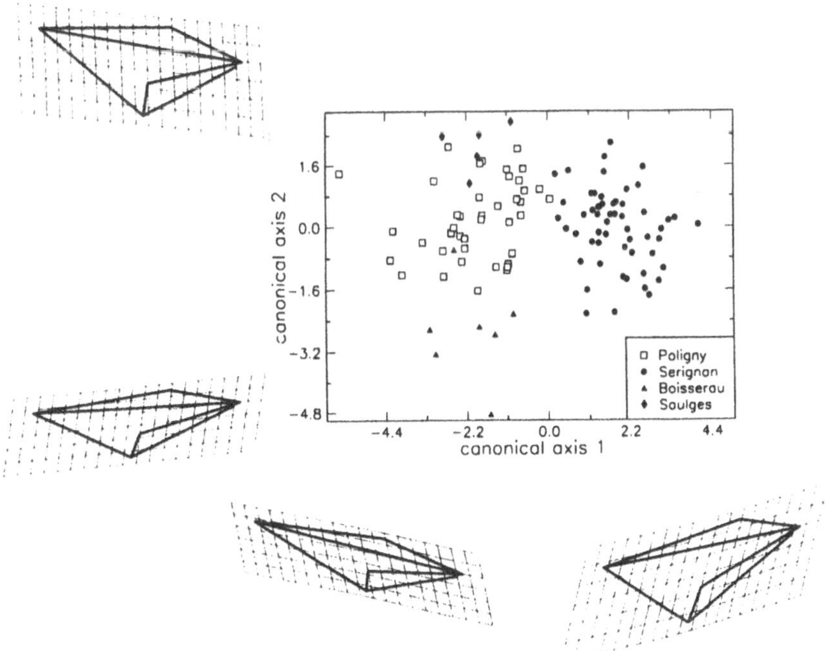

Figure 7. Between-locality canonical discriminant analysis. Uniform + total nonuniform parts, males.

Table 7. Mahalanobis distances (D^2) between localities. Females (upper diagonal, roman), males (lower diagonal, italic). (Shape: total nonuniform warps and uniform parameters)

	Boissereau	Poligny	Saulges	Sérignan
Boissereau	0	**10.63**	**13.47**	**21.48**
Poligny	*11.58*	0	**15.66**	**19.42**
Saulges	*24.99*	*13.62*	0	**25.70**
Sérignan	*23.90*	*14.32*	*25.53*	0

The deformations along the axes 1 and 2 are summarized in Fig. 7. Clearly, the uniform component overdominates the discriminations, particularly for the second axis, where no nonuniform component is apparent on the splines. The exact weights of the uniform and the nonuniform parts can be best determined by comparison between the Mahalanobis distances including or excluding the affine parameters (Table 7 and 8). The nonuniform part alone represents less than 19 % of the overall D^2 between Saulges and Boissereau and 72 % between Boissereau-Poligny. The second canonical axis is therefore merely representative of an opposition between Boissereau and Poligny rather than between Saulges and Boissereau, although the latter may appear more distant.

For the females, Sérignan segregates on the first canonical axis, which explains 77.47 % of the variance (Fig. 8). The second (9.35 %) and the third (6.50 %, not shown) axes clearly differ from the corresponding analyses for the males. Examples include the comparison of the splines on the second axes in Fig. 7 and 8, as well as the relative importance of the nonuniform part between Saulges, Boisserau and Poligny (Table 8). The main discrepancies are between Saulges and Boissereau (19 % in the males, 44% in the females) and between Poligny and Saulges (35 % in the males, 18% the females). This may be the origin of the significance of the interaction term between sexes and localities. Correspondingly, the second axis completely differs from that of the males analysis. It is no longer a cline but a clear discrimination between Poligny on one hand and Saulges and Boissereau on the other. This discrimination does not weight the relative displacement of the fifth and sixth landmarks so heavily. The third axis, which involves mainly an increase of the distance between the landmarks 1 and 6, partly separates Boissereau from Saulges. It appears, therefore, that the females from Boissereau and Saulges were much more similar than were the males. This is confirmed by the Mahalanobis distances (Table 7). Female size was correlated with the second and the third canonical axes (respectively 0.20, $p = 0.05$, and 0.30, $p = 0.01$), but as for males a correction for allometry does not seem to be justified. The separation between Poligny and Sérignan appeared to be free of any allometric component.

Sexual Dimorphism. Discrimination between the sexes (Table 9) was complete only in the cases of Saulges and Boissereau. The calculated misclassification percentages were

Table 8. Percentages of the between-locality Mahalanobis distance (D^2) corresponding to the total nonuniform part alone. Females (upper diagonal, bold) and males (lower diagonal, italic)

	Boissereau	Poligny	Saulges	Sérignan
Boissereau	0	**67.4**	**43.4**	**49.7**
Poligny	*71.9*	0	**17.5**	**70.0**
Saulges	*18.6*	*34.4*	0	**70.0**
Sérignan	*44.5*	*68.4*	*51.9*	0

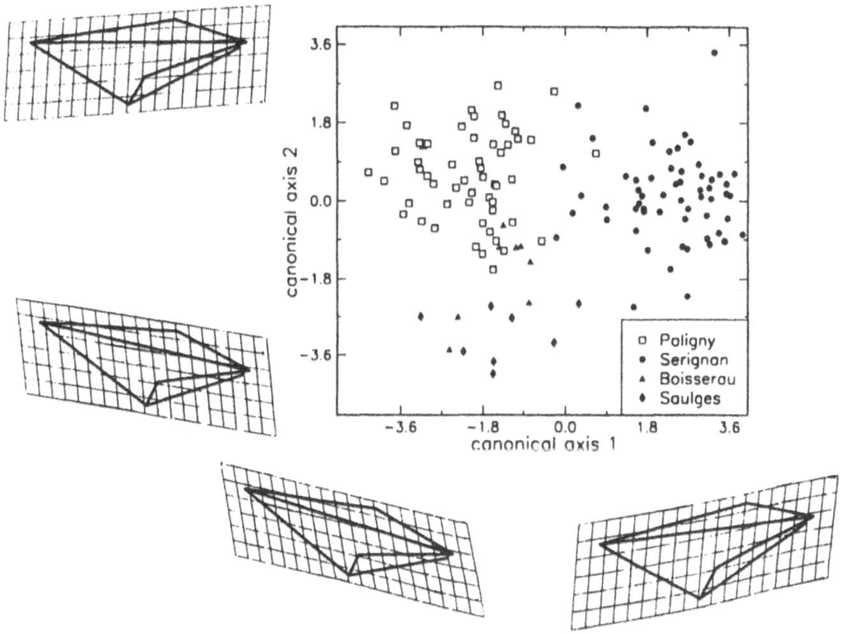

Figure 8. Between-locality canonical discriminant analysis. Uniform + total nonuniform parts, females.

slightly lower in Sérignan than in Poligny (Table 4 and 9) but are inversely distributed (Table 9). Bootstrap estimates (Table 9) are between 2 and 3 points above the calculated values, the highest misclassification percentage being 17 %.

Generalized distances between sexes differed also considerably between localities (Table 9). The influence, if any, of the sample sizes was assessed by bootstrapping the results of the discrimination in the large samples of Sérignan and Poligny. Subsamples of identical sizes to those of Boissereau and Saulges were randomly generated and the corresponding Mahalanobis distances were calculated. This procedure was repeated 10,000 times for each

Table 9. Sexual dimorphism. Discriminant functions of shape (total nonuniform warps and uniform parameters): Mahalanobis distances and misclassification percentages. The size of the nonuniform part is expressed as a percentage of the total D^2

	Mahalanobis D^2		Misclassification percentages			
Localities	D^2 (Shape)	Percent Nonuniform	From M to F	From F to M	Total	*n*
Boissereau	**24.26**	28.6 %	**0.00** -	**0.00** -	**0.00** -	15
Poligny	**6.70**	71.9 %	**14.63** *16.73 (4.97)*	**3.64** *6.68 (3.52)*	**8.33** *10.97 (3.35)*	96
Saulges	**82.69**	24.9 %	**0.00** -	**0.00** -	**0.00** -	13
Sérignan	**7.69**	70.7 %	**7.14** *9.27 (3.56)*	**13.33** *15.73 (4.08)*	**10.34** *12.61 (2.94)*	116

Calculated values are in bold; bootstrap estimates with standard deviations between parentheses are in italics (Serignan and Poligny); and abbreviations are M (males) and F (females).

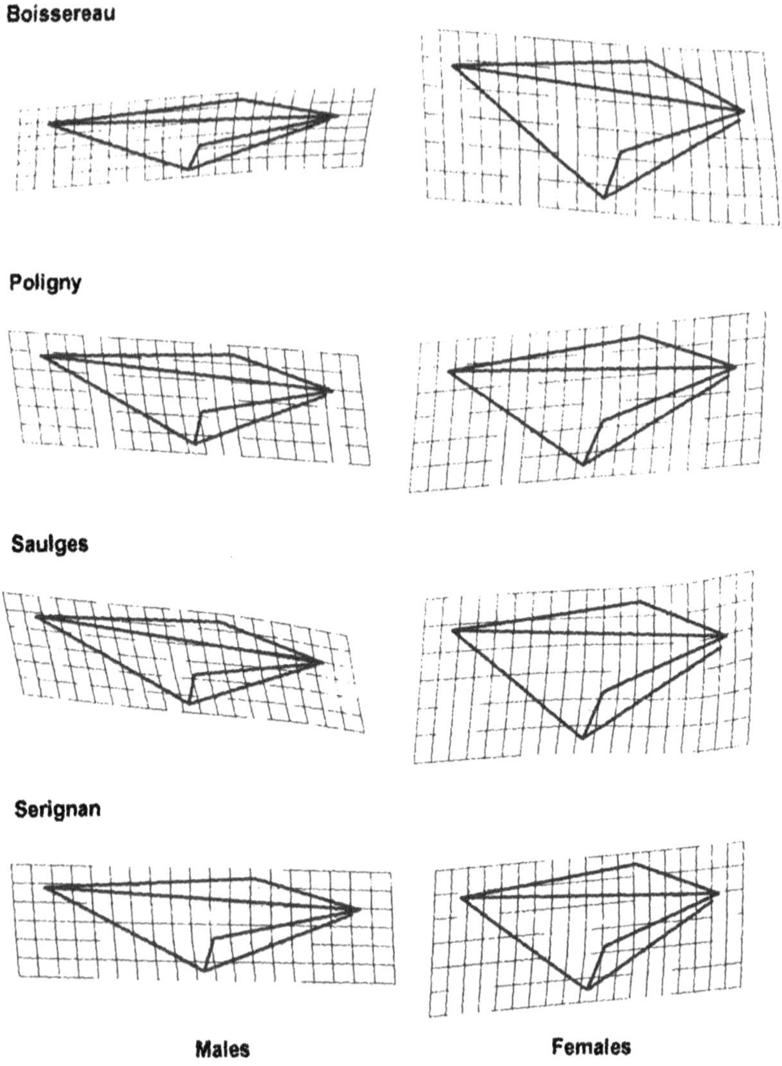

Figure 9. Sexual dimorphism within localities. Splines illustrating the discriminant functions calculated over uniform and nonuniform parts.

locality. Values greater or equal to the D^2 observed in Boissereau were found in 49.4 % of the bootstrap replicates of Sérignan and 10.8 % of those of Poligny. D^2 values greater or equal to those in Saulges were obtained in 24.4 % of the bootstrap replicates of Sérignan and 10.0 % of Poligny. If size was included, these percentages remained similar, increasing or decreasing by less than 2 %. Therefore, the D^2 differences can be explained by sample size inequalities only, and thus sexual dimorphism does not differ between localities.

The splines (Fig. 9) depict a pattern common to all localities: a relative stretching of the wing in the males with a nonuniform distal shift of the medial part. The only exception is Boissereau, for which the shear appeared reversed.

This sexual dimorphism of shape appeared free of any allometric component, at least for Poligny and Serignan, for which the correlation between centroid size and the discriminant scores within each locality explained no more than 1.6 % of variance. These percentages were 7.7 in Boissereau and 26.9 in Saulges, but the correlation coefficients are not significant in either case.

Wings of males were both larger and narrower in form than those of females. Differences in the perimeter lengths and the wing surfaces were tested. The perimeter was statistically longer in males (2725.5 pixels versus 2688.7, $p = 0.043$), but the surface was similar (291,561 pixels^2 for males and 288,472 for females, $p = 0.44$). Therefore, in males the length of the wing compensates for its narrowness leading to an equivalent surface of the wing membrane in both sexes. This compensation appears to be a between- and not a within-sex feature. Correlation between the uniform parameter U2 and the square root of the wing surface was not statistically significant within locality or sex but was highly significant (0.286, $p = 0.0001$) for the total sample.

CONCLUSIONS

The contrast between the low variability of the landmarks (see Fig. 1) and the clear discriminations achieved between localities and sexes demonstrates the ability of geometrical morphometrics to handle subtle form differences. Furthermore, the decomposition into affine and nonaffine parts (Goodall and Green, 1986; Bookstein, 1991) appears to be a valuable approach, helping both to dissect more precisely the differences or similarities between forms and to delineate large-scale deformations. The latter are sometimes difficult to trace, without the aid of grids, as with Procrustes residuals that focus on landmark by landmark. Richtsmeier et al. (1992), among others, asked about the biological significance of the uniform component. In the present case, the anisotropic uniform part appears to be a rough estimate—a simple model—of a complex biological process involving the differentiation of wings from imaginal discs. As pointed out by Bookstein (1991), the uniform part is merely a geometrical model of deformation that can be seen as the first degree of a polynomial decomposition of the form transformations. The purpose is descriptive, not causal, because the model is too general. In that sense, the issue is less about the biological significance of the uniform part per se, than its ability to depict a significant part of the deformations (but see the discussion about modeling in Rohlf, 1993c). The sexual wing dimorphism in *M. buxi* clearly illustrates the interest of the uniform part, which in this case explains as much as 85 % of the total differences.

The effect of voltinism appears negligible both on size and shape, albeit this was established using a single locality. This result is consistent with the apparent lack of genetic differences between voltine types and the fact that adults of the two types are on the wing at the same period of the year (Baylac, 1989), precluding any temporal isolation. This suggests that the alternative life cycles in the species are an adaptation without taxonomic implication. The similarity in size is inconsistent with previous findings (Baylac, unpublished data), which indicated that partivoltine larvae body lengths are on the whole greater than are univoltine larvae ones. This may result either from a difference in variability between stages or from a greater variability of the size of the wings. Although the length of the wing has been frequently used as an univariate size estimation, its ability to represent size has not been clearly established.

The sexual dimorphism is worthy of comment. There was no difference in wing surface area, contrasting with body mass differences between the two sexes : the abdomens of the females are twice as big as those of the males. This difference is not solely due to the presence of eggs (between 50 and 100; Barnes, 1948); there is a clear indication that the female larvae are bigger than the males (Baylac, 1989). A smaller wing size to body weight ratio, even partly compensated by broader wings, implies poorer flight ability for the females.

This hypothesis is supported by the low vagility of the females, exemplified by the high aggregations of attacks of the midge on box, which have been reported by most authors (Barnes, 1948).

These patterns of sexual dimorphism in *M. buxi* differ from what is known for other nematoceran groups. Krzeminska (1992), using a simple graphical superimposition, showed that sexual wing dimorphism is very important in the Trichoceridae. Females have larger wings because of an enlargement of the wings' ventral and distal parts. There remains much work to be done before a clear overview of the sexual dimorphism of wing form and wing venation is possible. Such work is of considerable importance, particularly owing to phylogenetic interest in wing characters, because many species, past or present, are known by only one of the two sexes or remain unsexed.

The main result of this study is a complete differentiation, at least for the males, between the northern localities – Poligny, Boissereau and Saulges – and the southern one, Sérignan. The differences between the populations in the former localities, although evident, are much less pronounced than are the differences they exhibit with Sérignan. The overall similarity between the northern midges is probably greater than is depicted by the canonical plots or by the Mahalanobis distances due to the heterogeneity of sample sizes; this is substantiated by the results of the bootstrap in the case of sexual dimorphism. The main difference between localities is in accordance with the hypothesis of Parent (1980) and suggests further investigations. For the northern range the limits of the local differentiations should be ascertained to determine whether they correspond with the basin limits (Seine, Loire and Mosel). For the southern range, the relationships between the populations in the South of France and those in Italy and Spain should be investigated. Spain and Italy were the two main postglacial centers of radiation for many species, both plants and animals. In the case of box, such a study would be particularly interesting as Italy does not appear to constitute an important potential center of recolonization in contrast to Spain. Christ (1918) described the Italian populations of box as relicts, in view of their scarcity and the fact that the eastern and the western borders of Italy constitute two major gaps within the range of box. Our results demonstrate that geometrical morphometric investigation allows fine analysis of intraspecific divergence and show the value of the particular coevolution model offered by the box and its midge in Europe.

ACKNOWLEDGMENTS

We thank the NATO referees, and particularly Jim Rohlf and Dennis Slice, for careful readings and constructive comments. Special morphometric and nonmorphometric thanks to Leslie Marcus for what he did, what he does and will do.

REFERENCES

Barnes, H. F. 1948. Gall midges of economic importance. IV. Ornamental Plants and shrubs, Crossby Lockwood, London.

Baylac, M. 1989. Coexistence d'individus univoltins et partivoltins dans les populations de *Monarthropalpus buxi* (Lab.) du Nord de l'Europe (Diptera, Cecidomyiidae). C.R. Acad. Sci. Paris 308 :337–340.

Baylac, M. 1993. XY: Utilities for 2D coordinates. MNHN. Lab. Entomologie 45, rue Buffon, F75005 Paris.

Bookstein, F. L. 1991 Morphometric tools for landmark data, geometry and biology. Cambridge University Press: Cambridge.

Christ, H. 1913. Über das vorkommen des Buchsbaumes (*Buxus sempervirens*) in der Schweiz und weiterhin durch Europa und Vorderasien. Vehr. Naturf. Ges. Basel 24:46–123.

CISIA 1991. Introduction à SPAD.N intégré version pc. CISIA, 1 av. Herbillon, F-94160 Saint-Mandé.

Cornuet, J. M. 1975. Discrimination et classification de populations d'abeilles à partir de caractères biométriques. Apidologie 6(2):145–187.

Gagné, R. J. 1989. The plant-feeding gall midges of North America. Comstock Publ. Assoc.: Ithaca, New York.

Godwin, H. 1975. The history of the British flora. A factual basis for phytogeography. Cambridge University Press: Cambridge.

Goodall, C. R. and P. B. Green 1986. Quantitative analysis of surface growth. Botanical Gazette 147:1–15.

Guillaumin, M. 1972. Analyse des différences de taille (T) et de forme (d_h de Sneath): Comparaison entre individus et entre caractéristiques moyennes de populations de *Pyrgus carlinae* Rbr. et *P. cirsii* Rbr. (Lep. Hesperiidae). Arch. Zool. exp. gén. 113:465–488.

Krezminska, E. 1992. Morphometric study of wing venation in the recent Trichoceridae—an application to the fossils? Acta Zool. cracov. 35(1):53–65.

Lenoble, F. and Ch. Broyer 1945. Sur la distribution du *Buxus sempervirens* en France. Bull. Soc. Bot. Fr. 92:118–131.

Marcus, L. F. 1990. Traditional morphometrics. Pages 77–122 *in* F. J. Rohlf and F. Bookstein, (eds.),: Proceedings of the Michigan morphometrics workshop. University of Michigan Museum of Zoology Special Publication 2.

Marcus, L. F. 1993. Appendix : Computer programs. Pages 289–350 *in* R. E. Reyment. Applied factor analysis in the natural sciences. Cambridge University Press: Cambridge.

Mosimann, J. E. and F. C. James 1979. New statistical methods for allometry with application to Florida red-winged blackbirds. Evolution 33(1):444–459.

Parent, G. H. 1980. Les buxaies mosellanes (France, G. D. de Luxembourg, Allemagne occidentale). Mém. Soc. Roy. Bot. Belg. 8:1–72.

Richtsmeier, J. T., J. M Cheverud. and S. Lele 1992. Advances in anthropological morphometrics. Annual Review Anthropology 21:283–305.

Rohlf, F. J. 1992. TPSREGR—Thin-plate splines regression analysis. Department of Ecology and Evolution, State University of New York, Stony Brook, New York.

Rohlf, F. J. 1993a. Relative warps analysis and an example of its application to mosquito wings. Pages 131–159 *in* L. F. Marcus, E. Bello and A. García-Valdecasas, (eds.), Contributions to morphometrics. Monografias del Museo Nacional de Ciencias Naturales 8, Madrid.

Rohlf, F. J. 1993b. TPSRW - Thin-plate spline relative warp. Department of Ecology and Evolution, State University of New York, Stony Brook, New York.

Rohlf, F. J. 1993c. Feature extraction in systematic Biology. Pages 375–392 *in* R. Fortuner (ed.), Advances in computer methods for systematic Biology. Artificial intelligence, databases, computer vision. John Hopkins University Press: Baltimore, Maryland.

Rohlf , F. J. and D. Slice 1990. Extensions of the Procrustes method for the optimal superimposition of landmarks. Systematic Zoology 39(1):40–59.

Ruttner, F., L. Tassencourt and J. Louveaux 1978. Biometrical-statistical analysis of the geographic variability of *Apis mellifera* L.. Apidologie 9(4):363–381.

Slice, D. 1993. GRF-ND Generalized rotational fitting of n-dimensional landmark data. Department of Ecology and Evolution. State University of New York. Stony Brook, New York.

Sprent, P. 1972. The mathematics of size and shape. Biometrics 28:23–38.

Updegraff, G., 1990. MeasurementTV, version 1.3. Data Crunch, 304 avenida Adobe, San Clemente, California.

Vanden Berghen, C. 1955. Etude sur les irradiations de plantes méridionales dans la vallée de la Meuse wallone. Bull. Soc. Roy. Bot. Belg. 87:29–55.

Waldbauer, G. P. 1978. Phenological adaptations and the polymodal emergence patterns of insects. Pages 127–144 *in* H. Dingle, (ed.), Evolution of insect migration and diapause. Springer-Verlag: New York.

SIZE AND SHAPE VARIATION IN THE MANDIBLE OF THE FOSSORIAL RODENT *Spalax ehrenbergi*

A Procrustes Analysis of Three Dimensions

Marco Corti,*[1] Carlo Fadda,[1] Shimon Simson,[2] and Eviatar Nevo[2]

[1] Anatomia Comparata
Dipartimento di Biologia Animale e dell'Uomo
Università di Roma 'La Sapienza'
via Borelli 50
00161 Roma, Italy
[2] Institute of Evolution
University of Haifa
Mount Carmel
Haifa 31905, Israel

ABSTRACT

Three-dimensional Procrustes analysis was used to study the variation of mandible size and shape in the fossorial rodent superspecies *Spalax ehrenbergi* across the four chromosomal species in Israel (2n = 52, 54, 58 and 60) and the Egyptian species (2n = 60). Because the animals use their incisors to dig their underground tunnel systems, we selected the mandible as a potentially rich source of information on phylogenetic and adaptive processes that characterized the evolution of the superspecies during Pleistocene and Holocene times. Eleven landmarks were recorded from the mandible in three dimensions as *x*, *y* and *z* coordinates. Differences in landmark position among populations and species, based on a 3D graphic visualization of the landmarks, were studied after Procrustes Generalized Least Square (GLS) fitting through uni- and multivariate statistical analysis. Sexual dimorphism was found only for size. Size also changes in a consistent pattern for species, geography, soil type and other ecological descriptors. A principal component analysis of the "shape" GLS residuals and the Mahalanobis distances between populations shows a pattern consistent with species differences in chromosomes. This favors a phylogenetic interpretation for the observed pattern of variation. However, Partial

* e-mail: corti@giannutri.caspur.it

Least Squares indicate that the change is also related to geography and current ecology and is associated with the increase in diploid number, suggesting that ecological factors affected speciation.

INTRODUCTION

The superspecies *Spalax ehrenbergi* is a subterranean rodent occurring in the Near East from southeast Turkey to Israel and North Africa. It is a remarkable example of recent ecological speciation driven by chromosomal rearrangements (reviewed by Nevo, 1991). The superspecies comprises four known chromosomal species in Israel with diploid numbers 2n = 52, 2n = 54, 2n = 58 and 2n = 60 and a newly described species with 2n = 60 in northern Egypt (which displays different chromosomal rearrangements, isozyme profiles and behav-

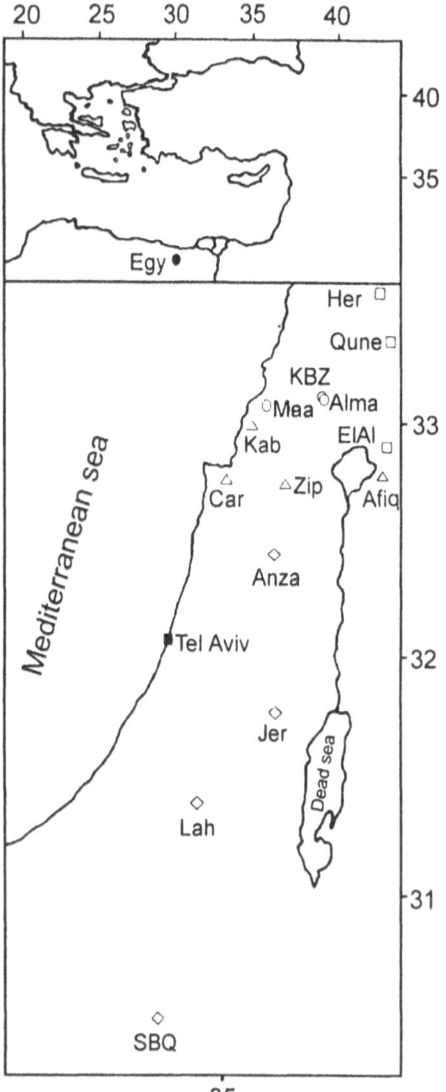

Figure 1. Distribution of the four chromosomal species of *S. ehrenbergi* in Israel and the locations of the populations studied with the acronyms used. The upper part of the figure shows the location of the Egyptian population. Chromosome species are represented by circles, 2n = 52; squares, 2n = 54; triangles, 2n = 58; diamonds, 2n = 60.

ioral patterns); (Nevo et al., 1991, 1992). Speciation commenced during the Pleistocene and is not yet fully completed, as evidenced by the permanence of parapatric hybrid zones.

Chromosomal species have progressively adapted to the deforestation and savannization process that distinguished the Pleistocene geological history of the Near East (Tchernov, 1968). The ranges of the species coincide grossly with four different climatic regimes that are characterized by increasing aridity southward and eastward and increasing temperature southward. The $2n = 52$ species occurs in humid and cool mountains of the Upper Galilee; the $2n = 54$ species in the semidry and cool heights of the Golan; the $2n = 58$ species in the warm and humid Lower Galilee; and the $2n = 60$ in the dry and warm coastal plains, Samaria, Judea, Jordan Valley and northern Negev (Fig. 1).

Multiple studies in the last 30 years explored the adaptations of the four Israeli chromosomal species to the different climatic regimes, in terms of physiology, genetics, behavior and morphology (see Nevo, 1991). The Egyptian species deserves further study; it differs from the Israeli complex in genetics and behaviour (Nevo et al., 1991, 1992) and little has been done on patterns of adaptation to the local sub-Saharan climate.

Animals dig their tunnel systems entirely with the upper and lower incisors, which have a peculiar development among rodents. The skull and the mandible are deeply modified to allow broader insertions of muscles supporting chisel-tooth digging, which is performed mainly by up- and downwards movements of the head. The head and the limbs are then used to move the soil dug out by incisors. In the upper part of the skull there is an evident nuchal crest formed by the squama occipitalis, os interparietale and os parietale and a protuberant sutura sagittalis, providing a wider surface for the muscles of the neck. In comparison with a typical rodent mandible which is almost a flat bone, the *Spalax*'s jaw is characterized by a distinct structure in three dimensions, with a developed processus coronoideus and a tuber massetericum that forms a broad lateral expansion (Fig. 2) which are peculiar to *Spalax*. These structures provide an increased surface for the insertion of temporal and masseter muscles respectively, and their main function is in digging. The traits are therefore modified as an adaptation for digging, i.e., for fossorial activity and a subterranean habitat.

Populations of the four species are scattered across diverse soils, from a very hard type, clay and basalt, to sandy loam and sand. Thus, populations may display a modification of the form of the mandible as an adaptation to the particular soil type where they occur.

We therefore chose the mandible architecture as a distinctive source of morphometric traits, in an attempt to relate the variation in the form—size and shape—to the phylogenetic (speciation) or adaptive processes that gave rise to the populations and species of the superspecies *Spalax ehrenbergi*. We studied the mandible form variation across the entire range of the superspecies, including populations from the main species ranges and a long-term isolate from the $2n = 60$ species. Our studies included also a population from the Egyptian $2n = 60$ species. We attempted to recover the main components of the functional design of the shape of this bone in three dimensions.

Geometric Morphometrics (Reyment, 1991) was used as it allows the extraction of variation in size and shape in the geometric space of the organism. By contrast, Traditional Morphometrics (Marcus, 1991) uses linear distances that cannot preserve the geometric relationship among characters. The form of the mandible was recorded using a set of "homologous" three-dimensional landmarks (Bookstein, 1991), i.e., points that can be identified unambiguously across specimens. The shape change from individual to individual is then described in terms of modification of the relative positions of these landmarks, after superimposition of specimens through rotation, translation and scaling to a common consensus form. Differences within and between samples and trends in shape change are estimated through the use of conventional univariate and multivariate statistics to these new coordinates.

Table 1. Populations, diploid number, acronyms and the location of collecting sites with soil types coded from a hard (basalt = 1) to soft (loess = 5) texture

Population	2n	Acronym	Geographic location	Soil type	Soil type code	Number of individuals
Alma	52	Alma	35° 47' N - 33° 03' E	Rendzina	3	5
Kerem Ben Zimra	52	KBZ	35° 28' N - 33° 02' E	Basalt	1	5
Maalot	52	Maa	35° 16' N - 33° 00' E	T. Rossa	2	9
El Al	54	ElAl	35° 45' N - 32° 48' E	Basalt	1	12
Hermon	54	Her	35° 44' N - 33° 18' E	T. Rossa	2	9
Quneitra	54	Qune	35° 49' N - 33° 07' E	Basalt	1	12
Afiq	58	Afiq	35° 42' N - 32° 46' E	Basalt	1	4
Carmel	58	Car	35° 00' N - 32° 44' E	Marl	4	8
Kabri	58	Kab	35° 09' N - 33° 01' E	Rendzina	3	11
Zippori	58	Zip	35° 17' N - 32° 44' E	Rendzina	3	11
Anza	60	Anza	35° 13' N - 32° 21' E	Rendzina	3	11
Jerusalem	60	Jer	35° 14' N - 31° 47' E	T. Rossa	2	8
Lahav	60	Lah	34° 52' N - 31° 23' E	Loess	5	6
Sede Boqer	60	SBQ	34° 47' N - 30° 52' E	Loess	5	6
Egypt	60	Egy	29° 28' N - 30° 21' E	Loess	5	16

There have been only a few studies on size and shape variation of organisms in three dimensions using geometric morphometrics. For example, Richtsmeier et al. (1993) used finite element scaling analysis for a three-dimensional description of facial growth patterns in three primates species. Slice ("Three dimensional" article, this volume) provides an exhaustive review of Procrustes methods to analyze shape variation in three dimensions. We adopted these methods to superimpose the mandibles of *Spalax* optimally.

MATERIAL AND METHODS

Animals, Populations, and Species

One hundred and thirty-three adult specimens were used in this study, which involved 15 populations representing the entire range of the four chromosomal species in Israel and one population from the Egyptian 2n = 60 species (Fig. 1, Table 1). The sample is composed of three populations for each of the species 2n = 52 and 2n = 54, and four populations for each of the species 2n = 58 and 2n = 60. The population from Sede Boqer (2n = 60) (Fig. 1,

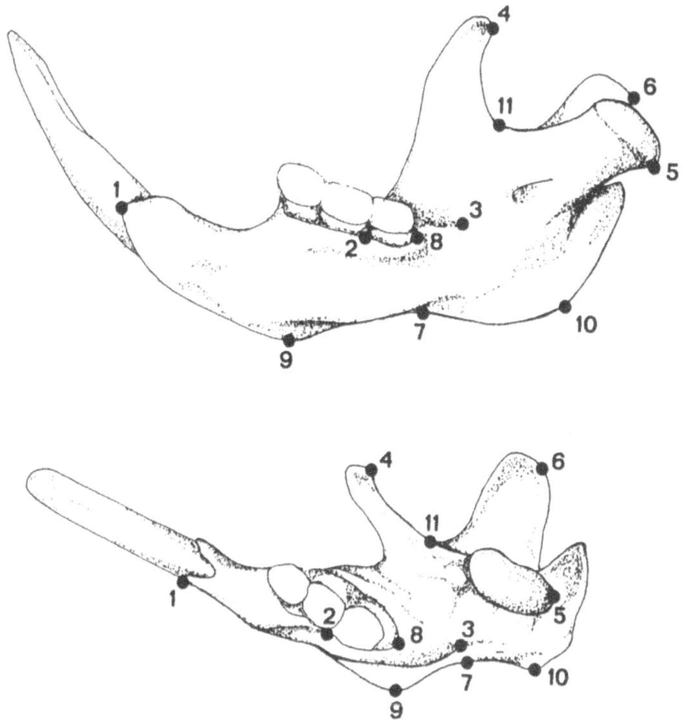

Figure 2. Drawing of the left mandible, lingual side and upper view. See text for landmark description.

Table 1) is an isolate with a small number of individuals (Nevo, 1989) located in the Negev desert 40 kilometers south of the southern border of the 2n = 60 species (isolation is estimated to have occurred in the last 10,000 years). The populations of Alma and Kerem Ben Zimra belong to the 2n = 52 chromosomal species and are separated by a distance that is less than 1 kilometer. They, however, occur on different soil types, i.e., rendzina and basalt respectively (Table 1).

Mandible Three-Dimensional Architectonics: Landmarks

Eleven homologous landmarks (L) were collected on the left mandible as x, y and z coordinates (Fig. 2): (1) extreme anterior part of the incisor alveolus, labial side; (2) extreme posterior part of the second molar alveolus; (3) maximum curvature of the mandible between the third molar and the foramen mandibulae dorsalis; (4) tip of the processus coronoideus; (5) tip of the processus condylaris; (6) tip of the tuberculum massetericum; (7) maximum curvature of the incisura vasorum facialum; (8) extreme posterior point of the alveolus of third molar; (9) maximum curvature point of the facies symphysalis; (10) maximum curvature in angulus mandibulae; and (11) maximum curvature of the incisura mandibulae.

A simple device was developed to record landmarks (Fig. 3). The mandible was placed under a slide on which transparent millimeter graph paper had been reduced 10 times. The x and y coordinates were recorded while looking through a binocular stereomicroscope. A second and third identical transparent slide of millimeter graph paper were placed perpendicularly along the x and y axes, and the third dimension was observed as reflected by mirrors placed at 45°. The z coordinate was also recorded while looking through the

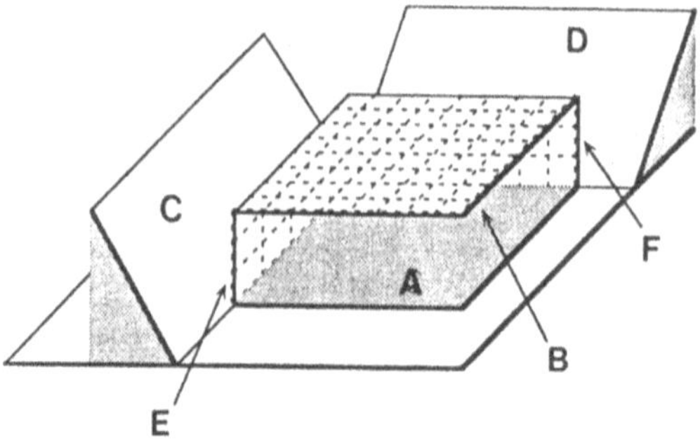

Figure 3. Schematic drawing of the x-y-z coordinates recording apparatus. The apparatus is placed under a binocular microscope and x, y and z coordinates recorded as follows: The mandible lies over the plane A and is observed through the transparent reticule B (coordinates x and y are recorded on the lingual side); two mirrors C and D are placed at 45° to the plane A along its x and y axes, together with two transparent reticules at 90° (E and F); the images of the lower and of the back sides of the mandible are reflected through the reticule by mirrors C and D and the z coordinates are then recorded.

microscope. The mandible was fixed in place by aligning L9 and L10 along the x axis and L1 along the *y* axis, which were constructed of fixed slides.

We will consider the mandible as an integrated structure consisting of three main parts: the diastema (L1, L2, L9); the molar or middle (L2, L8, L3, L7, L9) part; and the posterior (L10, L4, L11, L5, L6) part.

Size was registered as centroid size, the square root of the sum of the squares of the distances between each landmark from the centroid (Bookstein, 1989). In this way, centroid size is an estimate of size in the three dimensions and in all directions rather than being represented by a unidimensional measure such as, for example, total length.

Fitting Configurations and Statistics

Raw data were translated to locate a common origin at landmark 1, with *x*, *y* and *z* coordinates set equal to 0; then landmarks were rotated to have landmark 6 aligned along the *x* axis and *y* and *z* equal to 0. A further rotation of the other *n*-2 landmarks along the axis passing through L1 and L10 was prevented by the fixed position of the mandible in the plane L1, L9 and L10 provided by the recording apparatus. This new data set was used to compute the means for each population.

The possible effect of sexual dimorphism in the data was investigated by means of multivariate analysis of variance (MANOVA) landmark by landmark and for each species. To this purpose, we selected a subset of populations including a representative number of individuals for both sexes, and one population per chromosomal species: Kerem Ben Zimra (2n = 52), Quneitra (2n = 54), Carmel (2n = 58), Jerusalem (2n = 60) and Egypt (2n = 60); (Table 1).

A two-way analysis of variance (ANOVA, unbalanced design) was used to evaluate size differences between the sexes in the subsamples, testing for sex, population and their interaction. A GT2 test was performed to test for differences among the means (Sokal and Rohlf, 1981).

Landmark configurations of both the means and the set of all individuals were then superimposed by using a Procrustes fitting procedure (Rohlf and Slice, 1990) that uses Generalized Least Square (GLS) in the program GRF-ND (Slice, 1993). These two new data sets were used for subsequent analyses.

Eigenvectors were extracted from the covariance matrix of residuals from fitting the fifteen population consensuses with GLS. Residuals were then studied graphically landmark by landmark as vectors from the overall consensus in GRF-ND. Consensus populations were then projected onto the eigenvectors to investigate their ordination and to relate this distribution with the shape differences displayed by GRF-ND.

The 55 possible distances between landmarks were computed for all individuals after GLS fitting. ANOVA and GT2 tests between the means were then performed to show in detail certain aspects of shape variation and differences in trend between populations.

Mahalanobis distances between populations were computed, and their significance was tested by Hotelling T^2 after Bonferroni adjustment (Marcus 1991; Morrison 1976). The total number of comparisons are $c = k(k-1)/2$ for k groups, and the significance level that we have chosen, 0.05, was divided by c (in our case 105) to guarantee that overall significance. Moreover, since sample sizes vary a great deal over populations (Table 1), we also computed the unbiased Mahalanobis distances (Marcus, 1993).

Centroid size and principal component scores were related to ecogeography, i.e., population geographic locations and altitude and a set of geoclimatic variables (described below) including soil type (Table 1), through multiple regression and Partial Least Squares (PLS; Streissguth et al., 1993). PLS is a multivariate technique based on the singular value decomposition of the correlation matrix between two arrays or blocks of variables ("left" and "right" blocks), so that the predictive interrelations between the two is summarized by two sets of latent variables, one for each block.

The ecogeographical variables were selected as representative of the progressive increase of temperature and aridity and of the change in vegetation and plant cover occurring from north to south and from west to east. The geoclimatic variables are mean annual temperature, mean August temperature, mean January temperature, difference in temperature between the warmest and the coldest seasons, difference in temperature between day and night, mean annual rainfall, mean number of rainy days, climax plant community and percentage of plant cover. Data were collected from the Survey of Israel (1970) and from multiple-year records of the Meteorological Services of Israel. The Egyptian population was excluded from these tests as no information on ecogeography was available to us. Soil type was recorded from Butler et al. (1993) and coded as shown in Table 1, from hard (score 1) to soft (score 5) texture.

In order to compare shape variation with the environment, in the PLS analysis the left block was represented by population scores on the first three principal components from the covariance matrix of the GLS residuals, and the right block by the ecogeographic variables.

Computations were performed using the SAS (1993) statistical package (PC version 6.08), with various routines written in the IML language by L. F. Marcus (appendix in Reyment, 1991) and modified by Corti (1992), and the NTSYS-PC package (Rohlf, 1993, version 1.8). A program written in SAS IML by L. F. Marcus (see appendix III for instructions to obtains programs)was used to compute the PLS.

RESULTS

The ANOVAs testing for sexual dimorphism in size for the five selected populations (Kerem Ben Zimra, Quneitra, Carmel, Jerusalem and Egypt) did show significant differences. A plot of means with 95% confidence intervals based on the GT2 method is presented in Fig. 4. For Quneitra and Egypt the male and female means differ significantly ($p < 0.05$). There is no significant sex difference for Jerusalem and Carmel; the difference for Kerem

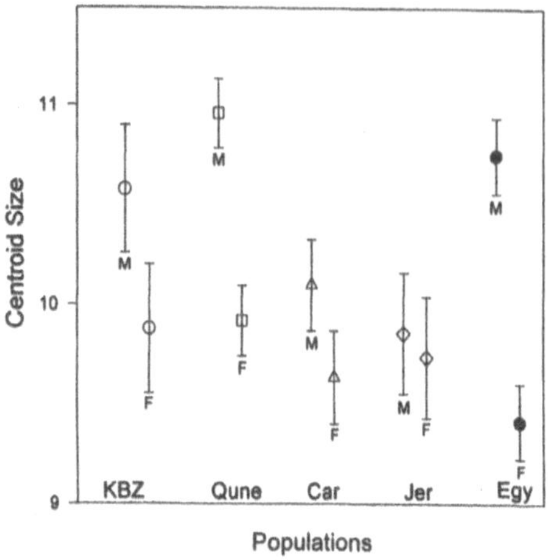

Figure 4. Sexual dimorphism: plot of the means of centroid size with 95% confidence intervals from the GT2 method for the selected subsample of Kerem Ben Zimra (KBZ, 2n = 52), Quneitra (Qune, 2n = 54), Carmel (Car, 2n = 58), Jerusalem (Jer, 2n = 60), and Egypt (Egy, 2n = 60). M = males; F = females.

Ben Zimra is larger and almost significant (Fig. 4). In general, males have a bigger mandible than do females, corresponding to their general bigger size (Nevo et al., 1986).

Multivariate analysis of variance on each landmark for the selected populations did not show any sign of significant difference between the sexes, except for landmark 8 in the Carmel population. When present, sexual dimorphism is expressed mainly as size. Because the landmarks were scaled after fitting with GLS, and size is not represented in the new data set, data were pooled over of sexes ignoring the single exception for the Carmel population..

Male and female mandible size differs significantly among the populations (males: $F = 2.5$, $p < 0.05$, df = 14, 42; females: $F = 3.97$, $p < 0.01$, df = 14, 89). Some of these differences are plotted in Fig. 4 where it is possible to appreciate size differences for each sex among the selected samples for Kerem Ben Zimra, Quneitra, Carmel, Jerusalem and Egypt. We see here the general tendency for decreased size over the chromosomal species, from 2n = 52 to 2n = 60 (as earlier shown for body size by Nevo et al., 1986) and Egypt.

Table 2. Correlations of ecogeographic variables and soil type with centroid size (* = p <0.05). The F-statistics associated with multiple regression on centroid size is significant ($p < 0.0001$). Data have been collected from the Survey of Israel (1970)

Ecogeographic variables	Correlation coefficient
Latitude	0.42235
Longitude	0.47935
Altitude	0.56879*
Mean annual temperature	-0.56691*
Mean August temperature	-0.64811*
Mean January temperature	-0.65841*
Difference in temperature between the warmest and coldest season	0.24025
Difference in temperature between day and night	-0.74456*
Mean annual rainfall	0.59048*
Mean number of rainy days	0.51058
Soil type	-0.56257*
Climax plant community	-0.29539
Plant cover (%)	-0.45620

The multiple regression of centroid size on ecogeographic variables is significant. There is a tendency for a size decrease from north to southwest. This is reflected in the correlations of size with latitude and longitude (Table 2). In fact, centroid size is significantly correlated with 7 out of 13 ecogeographic variables (Table 2).

UPGMA phenogram was computed using Mahalanobis distances to show population similarities.

The ordination of populations on PC1 and PC2, and PC1 and PC3 of the population consensuses is shown in Fig. 5. The first three components account for 25.76%, 17.97%, and 12.08% of variation respectively, summing to 55.83% of the total variation.

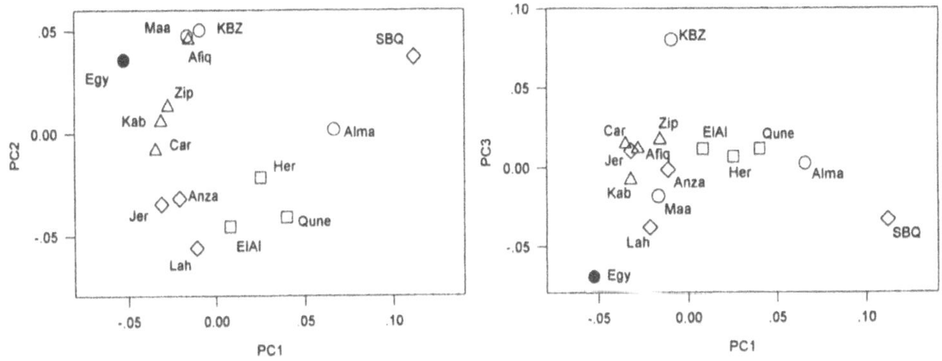

Figure 5. Principal Component Analysis of the 15 population consensuses based on residuals from a GLS fit to a common consensus form. Left: Scatter plot of PC1 and PC2; right: Scatter plot of PC1 and PC3. Chromosomal species are represented by circles, 2n = 52; squares, 2n = 54; triangles, 2n = 58; diamonds, 2n = 60; filled circles, Egyptian 2n = 60.

These principal components of the covariance matrix of population consensuses should be considered from two points of view: 1) as an ordination, which suggests a pattern in the similarity of shapes among populations and species; and 2) as a graphical representation of the major trends in shape change, i.e., the eigenvector displacements at each landmark of the consensus. In this sense, the positive and the negative directions at each landmark vector show how that landmark modifies its position (therefore affecting mandible shape) from the population with the highest (positive) to that with the lowest (negative) score for that given eigenvector.

The first three eigenvectors suggest shape changes that, apparently, are mostly related to phylogeny, e.g., populations of the same chromosomal species share very similar PC scores (Fig. 5). There are two main exceptions: the two populations of Alma and Kerem Ben Zimra (KBZ), both 2n = 52, and the desert isolate of Sede Boqer, 2n = 60.

Fig. 6 shows the plots of the consensus and the residual vectors for the first three eigenvectors in three dimensions from a lateral point of view. Notably, the direction of shape change associated with the first eigenvector involves a reduction of the diastema and of the molar parts of the mandible (L2, L3, L8) and a prolongation of the processus coronoideus (L4) and of the condilus (L5; Fig. 6). To this reduction there is a corresponding rotation occurring in the posterior part of the mandible: L3 moves forward with a corresponding

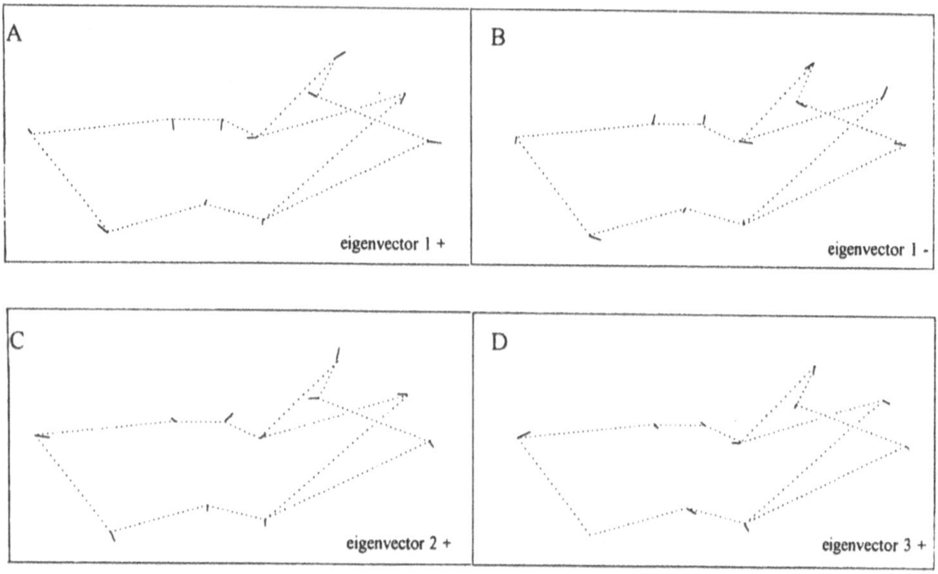

Figure 6. Labial view. Plots of consensus and residual vectors for the first three eigenvectors. The positive and negative displacements are presented for eigenvector 1 (A and B). Eigenvectors 2 and 3 are represented by the positive displacements only (C and D). Vector displacements have been intentionally exaggerated by a factor of 5.

rotation of 90° of L6, and an extension of L5 and L4, respectively backwards and upwards (Fig. 6). These landmark movements represent the shape differences between the Egyptian (negative) and the Sede Boqer (positive) populations that, in Fig. 5, have the most extreme scores for PC1. It should be noted that many populations share a common negative pattern, as they are scored between 0.00 and -0.05 (Fig. 5).

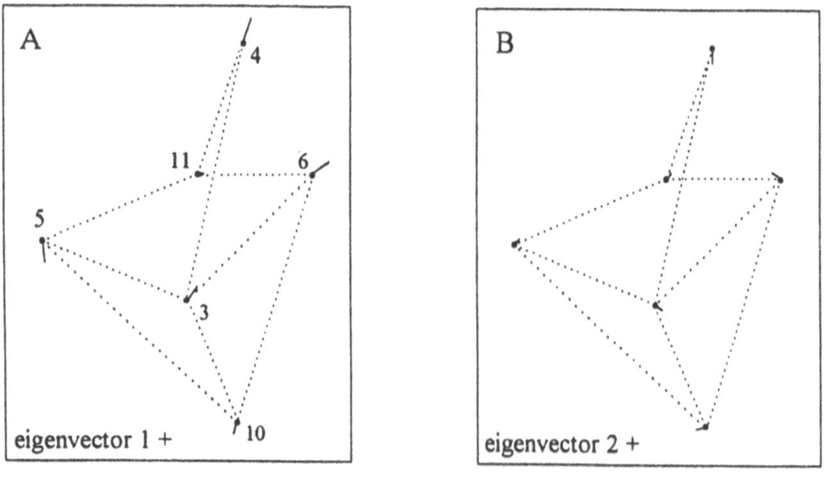

Figure 7. Plots of consensus and residual vectors for the first two eigenvectors (A and B respectively). View from the back, excluding landmarks of the anterior and middle part. Vector displacements have been intentionally exaggerated by a factor of 5. Vectors are in the positive direction. In A the landmark numbers are also given.

Eigenvector two shows an extensive lengthening of the mandible (from L1 to L6) accompanied by a contraction occurring in the diastema and molar parts (L2, L8, L3, L9, L7, L10) and a short coronoid (L4; Fig. 6).

The positive displacements are common to Kerem Ben Zimra, Afiq, Maalot, Sede Boqer (excluding the other 2n = 60 populations) and Egypt, whereas the negative ones are those for 2n = 54 and 2n = 60 (excluding Sede Boqer) populations (Fig. 5).

When the mandible is observed from the back (Fig. 7), the residual vectors associated with the first two eigenvectors show contrasting processes: following the first one, the three processes, i.e., condilaris (L5), tuberculum massetericum (L6) and coronoideus (L4) tend to diverge from each other, whereas the second eigenvector suggests convergent movements. If this tendency is seen as a combination of the first two eigenvectors in Fig. 5, Egypt, Kerem Ben Zimra and Maalot (2n = 52), and Afiq (2n = 58), are the populations in which the convergent movements of the three processes are at their maxima, whereas the opposite tendency is shown by all four populations with 2n = 60.

A GT2 comparison of means for all populations was performed for the distances among L4, L5 and L6. The means are less, on the average, in Egypt, whereas they are larger (often significantly) in the Sede Boqer and 2n = 60 populations. Plots of the means for these distances with 95% confidence intervals by the GT2 method are in Fig. 8.

This pattern is also shown in Fig. 9 for the populations from Maalot (2n = 52), Hermon (2n = 54), Lahav (2n = 60), Sede Boqer (2n = 60) and Egypt, which have the most different tendencies. The figure shows the triangles whose vertices are formed by landmarks 4, 5 and 6, as seen from the back, i.e., in the *y*, *z* plane. The positions of the vertices and their triangles

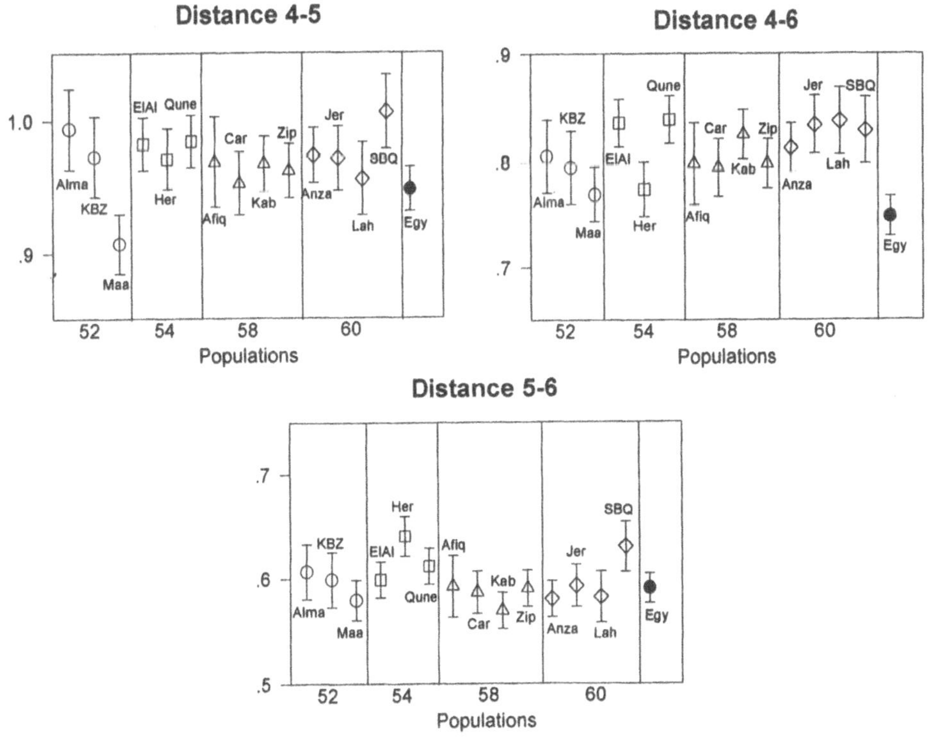

Figure 8. Plot of the means of all populations for the distances between landmarks 4-5, 4-6 and 5-6, i.e., between the tips of the processus coronoideus, condylaris and tuberculum massetericum, with 95% confidence intervals based on the GT2 method.

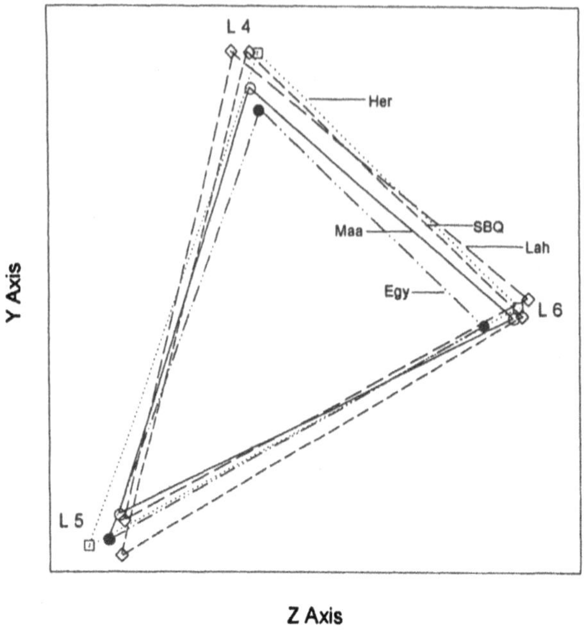

Figure 9. Plot of the consensus of Hermon (Her, 2n = 54), Maalot (Maa, 2n = 52), Lahav (Lah, 2n = 60), Sede Boqer (SBQ, 2n = 60) and Egypt (Egy, 2n = 60) for landmarks 4, 5 and 6 on the *z* and *y* axes. These landmarks are the tips of the processus coronoideus, condylaris and tuberculum massetericum viewed from the back of mandible. Egypt shows the maximum process of convergence of the three landmarks, whereas the opposite tendency is shown by Sede Boqer and Lahav.

highlight this pattern of divergence and convergence. Although it might be observed that the position of the vertices is a function of three-dimensional fitting, with the third dimension x missing here, it gives a very good approximation given the previous observation of the inter-landmark distances (Fig. 8).

Note that Sede Boqer exhibits the same pattern as the other 2n = 60 populations at the back of the mandible only for distance 4-6 (Fig. 8 and 9). The main differences in shape within this chromosome species involve mainly the diastema and the molar parts.

The upper view of the mandible is presented in Fig. 10. Ideally, the mandible is in the form of an arch, where an increase of the radius corresponds to a decrease in curvature. The first two eigenvectors show a straightening process evident in the diastema and molar parts. This involves an increase of the distance between the posterior parts of the left and

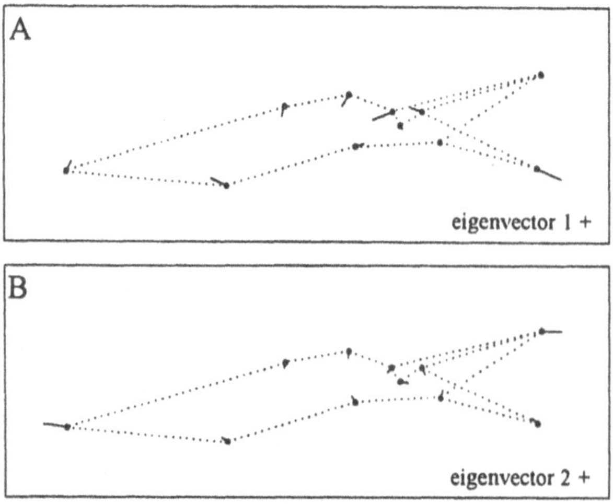

Figure 10. Upper view. Plot of the consensus and residual vectors for the first two eigenvectors (A and B, respectively). Vector displacements have been intentionally exaggerated by a factor of 5. Vectors are in the positive direction.

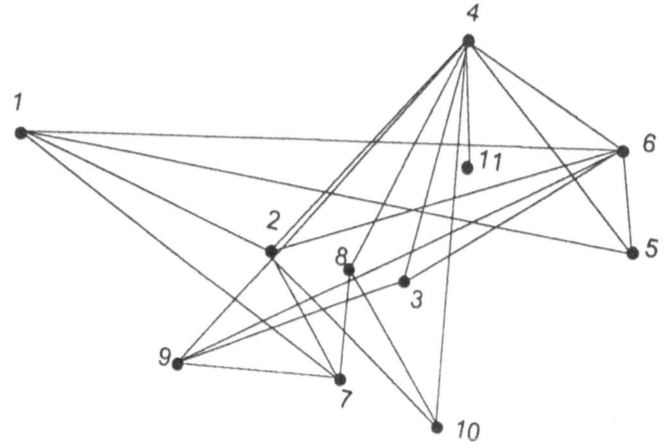

Figure 11. Analysis of variance on 21 inter-landmark distances. Lines represent only those distances which are significant ($p < 0.05$). Numbers correspond to landmarks as in Fig. 2.

right mandibles. In other words, the populations that have positive scores on the first two PCs (Fig. 5) have a wider angle between the left and right mandibles at the symphysis.

We performed ANOVAs on the 55 inter-landmark distances. A graphical representation of the significant inter-landmark distances is presented in Fig. 11. The great majority of significant differences concern the posterior part of the mandible and its principal axis. Eight of these concern landmark 4, and six landmark 6. Changes in shape concerning landmark 4 are the most distinctive.

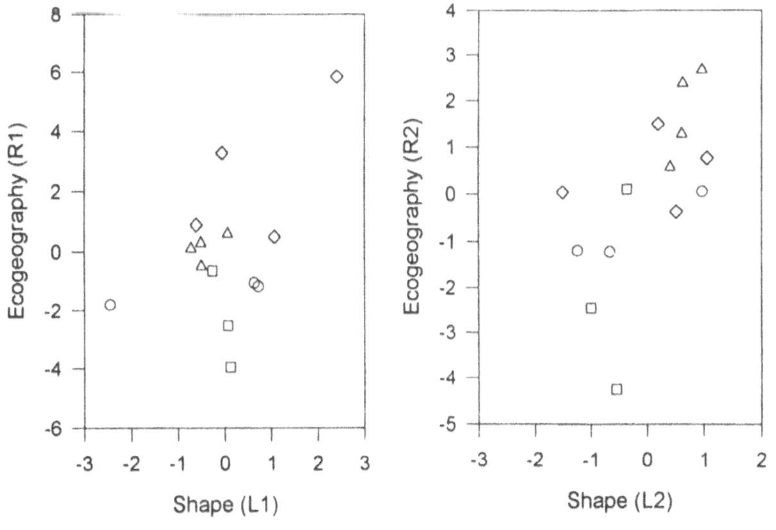

Figure 12. Partial Least Squares plots for the first (left) and second (right) latent variables for mandible population shape and ecogeography. Chromosomal species are represented by circles, 2n = 52; squares, 2n = 54; triangles/ 2n = 58; diamonds, 2n = 60.

Table 3. Partial Least Squares. Correlations of variables with the first three left and first three right vectors. The left vectors represent shape variation (they are the PC scores from the covariance matrix of the residuals after GLS fitting) and the right vectors are the 13 ecogeographic variables plus centroid size. Note that the second 'shape' PC is mainly represented by the third left and right vectors. Soil type is associated with the first vectors. Data have been collected from the Survey of Israel (1970)

Ecogeographic variables	Left 1	Left 2	Left 3	Right 1	Right 2	Right 3
Principal Component 1	0.61686	0.88763	0.15665	0.22759	0.48231	0.04558
Principal Component 2	-0.05468	0.02687	0.99088	0.03943	-0.07897	0.45260
Principal Component 3	-0.91411	0.19378	0.29206	-0.46778	0.22800	0.06032
Latitude	0.33766	-0.23349	0.02334	0.83423	-0.63325	-0.18027
Longitude	0.44105	0.11219	-0.15191	0.91752	-0.30650	-0.14301
Altitude	0.06007	-0.52429	0.13036	0.57503	-0.94230	0.59736
Mean annual temperature	-0.08396	0.39675	-0.06762	-0.71763	0.93471	-0.54115
Mean August temperature	-0.03854	0.31450	-0.03601	-0.62829	0.79921	-0.61231
Mean January temperature	-0.03261	0.50176	0.02615	-0.68047	0.95046	-0.40212
Difference in temperature between the warmest and coldest season	0.01210	-0.65993	-0.12206	0.42780	-0.84446	-0.01399
Difference in temperature between day and night	-0.31220	-0.03861	-0.05966	-0.78439	0.54558	-0.51061
Mean annual rainfall	0.19876	0.01037	0.10853	0.85704	-0.64266	0.46200
Mean number of rainy days	0.53727	0.05995	0.01379	0.96392	-0.40935	0.06975
Soil type	-0.51088	0.01898	0.14965	-0.76376	0.47200	0.32269
Climax plant community	-0.37620	-0.12209	-0.37773	-0.59334	0.45366	-0.79668
Plant cover (%)	-0.42330	-0.11356	0.15757	-0.90759	0.29075	0.15983

Correlations of right (ecogeography) and left (shape) vectors in PLS are 0.481, 0.618 and 0.445 for the first, second and third latent variables. The first three eigenvectors are equal to 1.563, 1.002, and 0.284, and account for 54.85%, 35.17% and 9.98% of total variance. A plot of the first two left (shape) and first two right (ecogeography) latent variables is given in Fig. 12.

The two plots show that there is a relation of shape change with geography and this is consistent within the species, i.e., populations of the same species tend to cluster, and this is consistent with an increase in chromosome number. The correlation of the variables (3 shape and 13 ecogeographical) with left and right vectors is presented in Table 3. The first left vector (shape) is mostly related to the mean number of rainy days, soil type, percentage of vegetation cover and latitude. This vector is essentially expressing PC1 and PC3. The second left vector expresses mainly PC1 and is influenced by difference in temperature between the warmest and coldest season, altitude, mean annual temperature and mean August temperature (Table 3).

Biased and unbiased Mahalanobis distances among group centroids were computed on the entire data set (Table 4). These distances were tested using Hotelling T^2 with a Bonferroni adjustment. After adjustment, only 23 of the possible 105 pairwise distances are not significant. Ten of these are for comparisons of populations within the same species, and 13 are interspecific comparisons (Table 4). The nonsignificant comparisons concern all those within the species $2n = 58$, 2 out of 3 within the species $2n = 54$, and 2 out of 6 in the species $2n = 60$. Distances within the $2n = 52$ species are all significant (Table 4).

The UPGMA computed on unbiased Mahalanobis distances between centroids is shown in Fig. 13. The most distinct populations are those from Egypt and Sede Boqer (2n

Table 4. Unbiased Mahalanobis distances between group centroids. Asterisks (*) denote $p < 0.05$ after Bonferroni's adjustment. Location abbreviations given in table 1

	Afiq	Alma	Anza	Car	Egy	ElAl	Her	Jer	Kab	KBZ	Lah	Maa	Qune	SB	Zip
Afiq	—														
Alma	4.00*	—													
Anza	3.24*	3.89*	—												
Car	2.65	3.39*	2.81*	—											
Egy	4.56*	5.13*	4.15*	3.09*	—										
ElAl	2.92	3.56*	2.91*	2.49*	5.16*	—									
Her	3.67	3.17*	3.43*	2.56	4.90*	2.77*	—								
Jer	2.65	3.96*	2.27	2.60	5.14*	2.21	3.36*	—							
Kab	2.70	3.53*	2.62*	1.94	4.28*	2.78*	3.57*	1.92	—						
KBZ	2.86	3.43*	4.23*	2.29	4.82*	3.43*	3.64*	3.60*	3.14*	—					
Lah	3.63*	3.36*	3.24*	1.95	5.17*	2.33	2.95*	2.52	2.68*	3.54*	—				
Maa	3.36*	4.88*	4.20*	3.44*	4.89*	4.40*	4.46*	4.02*	4.01*	3.79*	4.12*	—			
Qune	3.72*	3.66*	3.96*	2.87*	5.87*	2.00	1.93	3.00*	3.37*	3.75*	2.33	4.77*	—		
SBQ	4.30*	3.74*	4.53*	4.38*	5.10*	4.09*	4.39*	4.84*	4.59*	4.52*	4.21*	5.00*	4.53*	—	
Zip	2.22	2.93*	2.44*	2.14	3.87*	2.62*	3.12*	2.23	1.86	2.62	2.92*	3.28*	3.46*	4.23*	—

= 60). The 2n = 54 populations are clustered together, whereas 2n = 58 and 2n = 60 show less congruence with chromosomal number in their clustering. The remarkable distinction of the 2n = 52 populations should be noted. They do not form a cluster and have low similarities when compared with the other chromosomal species.

DISCUSSION

The populations and species display four main patterns of mandible morphology modification during their evolution. There are size differences between sexes and species and three kinds of shape change: (1) an extension along the principal axis (Fig. 7); (2) a lengthening of the three processes, coronoideus, condylaris and massetericum that form the posterior part, the former two providing insertions for masseters (Fig. 8); (3) a bending or stretching of the sagittal plane, the arc subtended by the anterior tip of the mandible and the processus articularis (landmarks 1 and 5; Fig. 10).

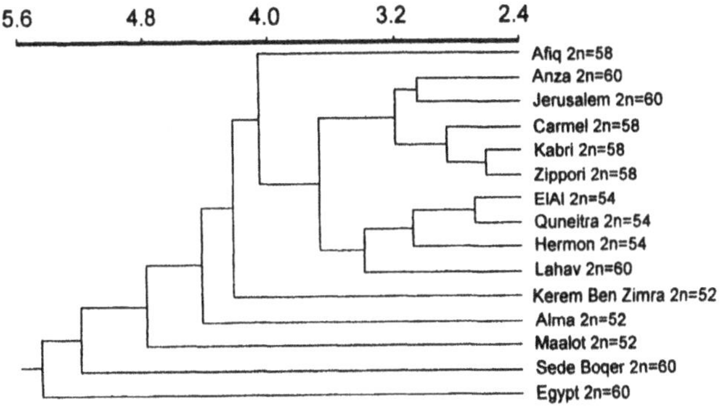

Figure 13. UPGMA phenogram computed on the unbiased Mahalanobis distances between centroids (Table 4).

Size is the main, if not the unique, component of sexual dimorphism, as shape components do not indicate differences between the sexes. Compared with the Israeli complex, the Egyptian species shows the greatest sexual dimorphism in size; the males' mandibles are much bigger than are the females (Fig. 4). It would be of interest to test whether higher sexual dimorphism in size is related to the trend towards the social structure shown by the Egyptian species (Nevo et al., 1992) compared with the solitary one of the Israeli species (Nevo et al., 1992).

Mandible size, as body size (Nevo et al., 1986), differs across the species in relation to chromosome numbers (an increase of diploid number corresponds to a decrease in size) and to ecogeography (Table 2), with a trend towards smaller size southwards and eastwards. The progressive decrease in size is significantly associated with the set of thirteen ecogeographic variables, and these change are concordant with karyotype number, i.e., the mandible gets smaller over geography and with the increase in chromosomal number.

Shape differences are mainly summarized by landmarks 1, 4 and 6 (Fig. 2 and 11). The latter two landmarks are the tips of the coronoid and of the processus massetericum; these tend to enlarge and to diverge from each other. The mandible tends to elongate to a maximum at Sede Boqer (landmarks 1, 4, 6; Fig. 6 and 10, by a combination of the first two eigenvectors as seen in Fig. 5) and reaches its minimum in the Egyptian population. A similar extreme opposite tendency between Sede Boqer and Egypt is shown in the divergence of the three processes (Fig. 5 and 8).

The function of these shape modifications in terms of biomechanics deserves further study in conjunction with the description of the masseters' insertion and extension. The relative enlargement of the three processes and their progressive divergence may indeed involve major differences in ability of digging and foraging. However, this should also be considered in relation to mandible articulation with the skull, which, among the other peculiar traits, is characterized by a particularly elongated fossa mandibularis, which should provide extended back and forth movements.

The consistency between population similarities and chromosomal number (see the UPGMA in Fig. 13) strongly suggests phylogeny as a factor, if not the major one, in the different shape features observed. The Egyptian species is the most distinct, together with the isolated $2n = 60$ population of Sede Boqer (Figs. 6 and 13; Table 4). Of the Israeli complex, the $2n = 52$ species is the most different. Within the group of the $2n = 54$, 58 and 60, there is a strong tendency to cluster according to chromosomal type and to form a $2n = 54$ group separated from the $2n = 58$ and $2n = 60$.

Within this description of differentiation, there are two remarkable exceptions: the two populations of Alma and Kerem Ben Zimra (KBZ), both $2n = 52$, less than 1 kilometer apart (Table 1). They are distinctively different (Fig. 5). Sede Boqer (SBQ), the desert isolate living in fluvial loess, is distinct from the other $2n = 60$ populations of Lahav, Anza and Jerusalem, which live in loess, rendzina and terra rossa, respectively (Table 1).

The population of Sede Boqer is a long-term isolate and its distinctive shape features may derive from a combination of isolation and soil type in the desert. This population is also characterized by a high degree of polymorphism in chromosome 1 (Nevo et al., 1988) and in mitochondrial DNA (Nevo et al., 1993) as well as allozymes (Nevo, 1989), and could be regarded as a potential for a future speciation event (Nevo, 1991).

Alma and Kerem Ben Zimra have no barriers to prevent gene flow. They might even be considered as one single population. The cause of such a divergence in their mandible shape might be an adaptation to very different soil types (Table 1). However, to verify that indeed soil selection affected mandible shape a larger sample size is needed.

There is also a good relation between shape and ecogeography. Shape changes across geography and ecology are consistent with species type and with the increase in chromosomal number (Fig. 12). In particular, the second left (shape) and right (ecogeography) latent vectors express a significant correlation. This ecogeographic vector represents mainly PC 1, reflecting latitude and the temperature variables (Table 3). Soil type and the other ecogeographic variables affecting soil texture, as well as vegetation type and amount of rainfall, are expressed mainly by the first latent vectors (3) and they show the same species consistency (Fig. 12), although correlation between these first left and right latent vectors is just below significance. Soil type alone cannot explain the trend in shape change, and a cline is rather evident from a combination of all ecogeographic variables together.

If size and shape have modified independently as adaptations to progressive increase in aridity and to different soil types, there is a unique pattern for each species. Soils become lighter, on average, southwards. The 2n = 52 and 2n = 54 occupy primarily hard soils whereas the 2n = 58 and 2n = 60 occupy lighter soils; this is correlated with a thinning of the outer enamel of the incisors (Flynn et al., 1987). Although this pattern is reflected in species specificity, soil differentiation within each species may account for part of the interpopulation mandible differentiation, as for Alma and Kerem Ben Zimra.

There is no doubt about the superiority of geometric morphometrics in 3D for analysis of form. However, there are still limitations in terms of precision and high cost for 3D data acquisition devices (reviewed by Dean, this volume). Also landmark methods for analyzing and comparing variation in 3D have not been explored or experimented with enough detail. Most of current applications of landmark-based morphometrics have dealt with 2D data, ignoring the effect of the third dimension in landmark position change (see Dean et al., Reig and Ahlstrom for other 3D analyses in this volume). For example, applications in previous books on morphometrics (Bookstein, 1991; Rohlf and Bookstein, 1991; Marcus et al., 1993) reported 3D morphological problems in terms of data and analysis in two dimensions, and this may lead to misinterpretation of the shape changes.

ACKNOWLEDGMENTS

The Editorial Committee and the Editorial Consultants provided helpful suggestions for new analyses and comments to improve the manuscript. Part of this project was supported by MURST, Fondi di Ateneo Università di Roma 'La Sapienza', quota 60% "Meccanismi di speciazione in ambiente tropicale" (M. Corti).

E. Nevo thanks the Israeli Discount Bank Chair of Evolutionary Biology and the Ancell-Teicher Research Foundation for Genetics and Molecular Evolution established by Florence and Theodore Baumritter of New York, for financial support.

REFERENCES

Bookstein, F. L. 1989. "Size and shape": a comment on semantics. Systematic Zoology, 38:173–180.

Bookstein, F. L. 1991a. Morphometric tools for landmark data: Geometry and biology. Cambridge University Press: Cambridge.

Butler P. M., E. Nevo, A. Beiles and S. Simson. 1993. Variation in molar morphology in the *Spalax ehrenbergi* superspecies: adaptive and phylogenetic significance. Journal of Zoology, London, 229:191–216.

Corti, M. 1992. Data analysis in systematics: a workshop and a manual to introduce geometric morphometrics. (Document accessible through anonymous ftp from life.bio.sunysb.edu - see Appendix II for instructions)

Flynn, L. G., E. Nevo and G. Heth. 1987. Incisor enamel microstructure in the blind mole rats: adaptive and phylogenetic significance. Journal of Mammalogy 68:500–507.

Marcus L. F. 1991. Traditional morphometrics. Pages 95–130 *in* F. J. Rohlf and F. L. Bookstein, (eds.), Proceedings of the Michigan morphometrics workshop. University of Michigan Museum of Zoology Special Publication 2.

Marcus L. F. 1993, Some aspects of multivariate statistics for morphometrics. Pages 95–130 *in* L. F. Marcus, E. Bello and A. García-Valdecasas, (eds.),Contributions to morphometrics. Monografias del Museo Nacional de Ciencias Naturales 8, Madrid.

Marcus L. F., E. Bello and A. García-Valdecasas (eds.). 1993. Contributions to morphometics, Monografías del Museo Nacional de Ciencias Naturales 8, Madrid.

Morrison, D. F., 1976. Multivariate statistical methods. McGraw Hill: New York.

Nevo E. 1989. Modes of speciation: The nature and the role of peripheral isolates in the origin of species. Pages 205–236 *in* L. V. Giddings, K. Y. Kaneshiro and W. W. Anderson, (eds.),: Genetics, speciation and the founder principle. Oxford University Press: London.

Nevo E. 1991. Evolutionary theory and processes of active speciation and adaptive radiation in subterranean mole rats, *Spalax ehrenbergi* superspecies, in Israel. Pages 1–125 *in* M. K. Hecht, B. Wallace and R. J. MacIntyre, (eds.), Evolutionary Biology, Vol. 25.

Nevo E., A. Beiles, G. Heth, and S. Simson. 1986. Adaptive differentiation of body size in speciating mole rats. Oecologia, 69:327–333.

Nevo E., M. Corti, G. Heth, A. Beiles, and S. Simson. 1988. Chromosomal polymorphism in subterranean mole rats and its evolutionary significance. Biological Journal of the Linnean Society, 33:309–322.

Nevo E., R. L. Honeycutt, H. Yonekawa, K. Nelson and N. Hanzawa. 1993. Mitochondrial DNA polymorphism is subterranean mole-rats of the *Spalax ehrenbergi* superspecies in Israel, and its peripheral isolates. Mol. Biol. Evol., 10:590–604.

Nevo E., S. Simson, G. Heth and A. Beiles. 1992. Adaptive pacifistic behaviour in subterranean mole rats in the Sahara Desert, contrasting and originating from polymorphic aggression in Israeli species. Behaviour, 123:70–76.

Nevo E., S. Simson, G. Heth, C. A. Redi, and M. G. Filippucci. 1991. Recent speciation of subterranean mole rats of the *Spalax ehrenbergi* superspecies in the El-Hammam isolate, northern Egypt. Page 43 *in 6th International Colloquium on the Ecology and Taxonomy of small African Mammals, 11–16 August 1991, Mitzpe Ramon, Israel.*

Reyment, R. A. 1991. Multidimensional palaeobiology. Pergamon Press: Oxford.

Richtsmeier, J. T., B. D Corner, H. M. Grausz, J. M. Cheverud, and S. E. Danahey. 1993. The role of postnatal growth pattern in the production of facial morphology. Systematic Biology, 42:307–330.

Rohlf, F. J. 1993. NTSYS-pc. Numerical taxonomy and multivariate analysis system, version 1.8. Exeter: Setauket, New York.

Rohlf, F. J and F. L. Bookstein (eds.). 1991. Proceedings of the Michigan morphometrics workshop. University of Michigan Museum of Zoology Special Publication 2.

Rohlf F. J and D. Slice. 1990. Extensions of the Procrustes method for the optimal superimposition of landmarks. Systematic Zoology, 39:40–59.

SAS for Personal Computers. 1993. SAS Institute, version 6.08 (Windows). SAS Institute: Cary, North Carolina

Slice, D. 1993. GRF-ND Generalised rotational fitting of n-dimensional landmark data. Department of Ecology and Evolution, State University of New York. Stony Brook, New York 11794.

Streissguth, A. P., F. L. Bookstein, P. D. Sampson, and H. M. Barr. 1993. The enduring effects of prenatal alcohol exposure on child development. University of Michigan Press: Ann Arbor.

Survey of Israel 1970. Atlas of Israel, Ministry of Labor, Jerusalem, Elsevier: Amsterdam

Tchernov E. 1968. Successions of Rodent faunas during the Upper Pleistocene of Israel. Paul Parey.

PRINCIPAL COMPONENTS OF BODY SHAPE VARIATION WITHIN AN ENDEMIC RADIATION OF THREESPINE STICKLEBACK

Jeff A. Walker*

Department of Zoology
Field Museum of Natural History
Roosevelt Road at Lakeshore Drive
Chicago, Illinois 60605

ABSTRACT

Principal component analysis of superimposition coordinates and partial warp scores (i.e., RWs) are used to explore patterns of body shape variation among 32 samples of lacustrine threespine stickleback *Gasterosteus aculeatus*, from Cook Inlet, Alaska. Principal components from orthogonal least squares, orthogonal resistant fit, affine least squares and affine resistant fit superimpositions are compared with each other and with the RWs. The pattern of body shape variation described by the first component of all five methods is extremely similar. This pattern describes a body shape with caudally positioned dorsal spines, a poorly developed pelvic girdle and short median fins at one extreme of an axis and a body shape with cranially positioned dorsal spines, a robust pelvis and long median fins at the opposite extreme. This pattern suggests a predator effect, which is supported by differences in first component scores between sticklebacks inhabiting lakes with and without native predatory fish. The second component of the orthogonal analyses and the uniform parameters of the affine analyses describe similar patterns of body form variation. This component describes a body shape with a long snout, shallow body and long caudal peduncle at one extreme of an axis and a body shape with a short snout, deep body, and short, deep caudal peduncle at the opposite extreme. This pattern suggests a foraging effect. Interestingly, a foraging effect is supported only in lakes with native predatory fish. Sticklebacks inhabiting lakes without native predatory fish tend to have body shapes representing open-water foragers regardless of habitat type.

* e-mail: walker@fmppr.fmnh.org

Advances in Morphometrics, Edited by L. F. Marcus et al.
Plenum Press, New York, 1996

INTRODUCTION

Threespine stickleback, *Gasterosteus aculeatus*, form a heterogeneous complex of northern temperate marine and coastal freshwater fishes (Wootton, 1976; Bell and Foster, 1994a). Threespine stickleback (hereafter, stickleback) inhabit a remarkably diverse array of environments, including open ocean, estuarine, stream, lake and pond. Not surprisingly, genetic, morphological, behavioral, ecological and life history characters vary greatly among populations (Bell and Foster, 1994a). Following the most recent deglaciation of the northwest coast of North America, a diverse radiation of freshwater stickleback has arisen due to the rapid colonization of emergent lakes and streams by the marine form of stickleback (Bell and Foster, 1994b; McPhail, 1994). Subsequent differentiation of body shape among freshwater populations rapidly followed. This interpopulation variation emerges at a local geographic level: among populations inhabiting adjacent lakes (Moodie and Reimchen, 1976; Lavin and McPhail, 1985; Reimchen et al., 1985; Francis et al., 1986) and even between populations within the same lake (Larson, 1976; Bentzen and McPhail, 1984; McPhail, 1984, 1994; Baumgartner et al., 1988; Schluter and McPhail, 1992), suggesting that local differences in microhabitat are more important causal factors than are regional climatic differences or phylogenetic relationships.

In this paper, I use Principal Component Analysis (PCA) to summarize and describe the major features of body shape variation among the mean adult shapes of 32 lacustrine samples from Cook Inlet, Alaska. Principal Component Analysis is a within-group multivariate analysis that is typically applied to a single sample. Various multiple-group PCAs have been developed (Airoldi and Flury, 1988; Thorpe, 1988), but canonical variates analysis, a PCA of among-group variation weighted by pooled within-group variation (Campbell and Atchley, 1981), is the traditional method for summarizing patterns of variation among groups. I chose to ignore within-group variation and analyze only the group means because I envision each of the samples in this study as estimating the location of the population on a multivariate phenotypic fitness surface. If population means are located near local peaks and ridges, a PCA of the sample means serves as a useful exploratory analysis of the patterns of selection on stickleback body form. Fish body shape is a particularly good system in which to study adaptation because (1) the extensive literature on the theoretical hydromechanics of a swimming fish provides the functional basis for observed patterns of shape variation and (2) the extensive literature on the feeding and antipredator ecology and behavior of fish provides the adaptive basis for observed patterns of shape variation.

METHODS

During the summer of 1990, 1992 and 1993 stickleback were collected with minnow traps from 32 lakes and ponds in the Mat-Su Valley and the Kenai Peninsula, Alaska. Minnow traps were either baited with cheddar cheese (1990) or unbaited (1992, 1993) and placed one to three meters off shore at depths of one-half to two meters. Samples were collected during the breeding season when reproductive males are nesting on the lake bottom near the shoreline. All samples were fixed in 10% buffered formalin, stained in Alizarin Red-S and stored in 50% isopropyl alcohol. To reduce variation due to sex and ontogeny, only the largest males from each sample were measured. Sample sizes for each lake ranged from 10 to 25. The total number of fish measured was 604.

Fifteen landmarks on each specimen were digitized in two dimensions by use of a video camera and MorphoSys software (Meacham and Duncan, 1990). The landmarks digitized are (Fig. 1): (1) anterior tip of upper jaw, (2) posterior border of the supraoccipital

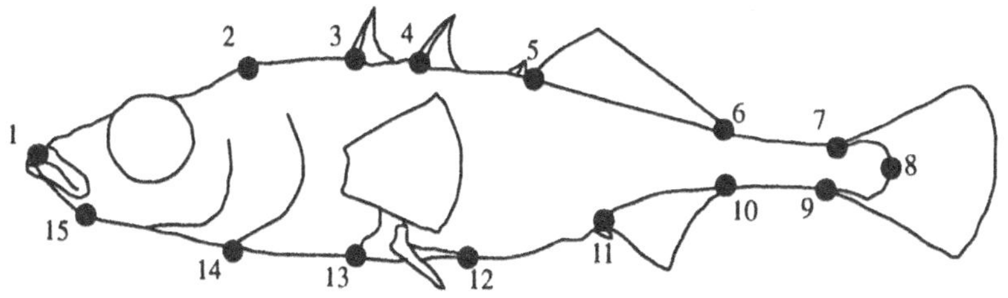

Figure 1. Locations of landmarks measured in this study.

on the dorsal midline, (3) anterior junction of first dorsal spine with the dorsal midline, (4) anterior junction of the second dorsal spine with the dorsal midline, (5) base of the first fin ray of the dorsal fin, (6) insertion of the dorsal fin membrane with dorsal midline, (7) origin of caudal fin membrane on dorsal midline, (8) junction of lateral line with caudal border of hypural plate, (9) origin of caudal fin membrane on ventral midline, (10) insertion of anal fin membrane on ventral midline, (11) base of first fin ray of anal fin, (12) caudal tip of posterior process of pelvic girdle, (13) posterior tip of ectocoracoid, (14) anterior border of ectocoracoid on ventral midline and (15) posterior edge of angular.

I used both superimposition analysis and partial warp analysis to transform the coordinate values into shape variables. All figures were first superimposed by the two-point registration, landmarks 1 and 8 were used as the baseline landmarks, in order to align the length of the fish with the horizontal axis. Principal components of four different superimpositions are compared: the Generalized Orthogonal Least Squares (GOLS) fit, the Generalized Affine Least Squares (GALS) fit, the Generalized Orthogonal Resistant Fit (GORF) and the Generalized Affine Resistant Fit (GARF). Eigenvectors for all analyses were determined from the covariance matrix of the superimposition coordinates.

An ordinary PCA of partial warp scores is also known as a RW analysis with α equal zero (Rohlf, 1993). For heuristic purposes, RWs were calculated and compared from partial warp scores calculated from four different initial alignments. The first three alignments were the Two-Point Registration (TPR, Bookstein, 1991), GOLS and GORF superimpositions. The fourth alignment consisted of the TPR figures recentered at the origin and rotated by a random uniform angle between $\pm \pi/4$ radians. The mean figure of each alignment was used to calculate the principal warp matrix.

Pros and cons of different superimposition methods can be found in Siegel and Benson (1982), Chapman (1990), Rohlf and Slice (1990) and Goodall (1991).

I used a vector correlation, which is the inner product of two normalized vectors, to compare the pattern of coefficients on eigenvectors computed from the different analyses (Krzanowski, 1988; Voss et al., 1990). Geometrically, a vector correlation is equivalent to the cosine of the angle between the two vectors. For the superimposition methods, I compared the normalized eigenvectors of the covariance matrix of the transformed coordinates. In order to compare results from the partial warp analysis with the superimposition analysis, I transformed the normalized eigenvectors of the covariance matrix of partial warp scores into RW loadings (Rohlf, 1993).

Shape variation is a priori partitioned into uniform (or affine) and nonuniform components in both the affine superimposition methods and the RW analysis (Rohlf and Slice, 1990; Bookstein, 1991). To summarize the affine component of shape variation, I calculated the Pearson correlation coefficients between the GALS estimates of the uniform parameters (natural log of anisotropy and θ) and the principal component scores from both the GOLS and GALS analyses.

In traditional morphometrics, the pattern of variation accounted for by a particular component is generally interpreted from the pattern of eigenvector coefficients transformed to loadings (Marcus, 1990). Eigenvector coefficients in superimposition-based morphometrics maintain the geometric information inherent in the coordinates and, therefore, have a strict geometric interpretation. The *j*th eigenvector coefficients for the x and y coordinates of the *i*th landmark are the components of a resultant vector, which is a graphical description of the correlation between the *j*th component and the *i*th landmark (Fig. 2). In order for the length of the resultant vectors to reflect the actual magnitude of covariation, the eigenvector coefficients were rescaled by

$$e^{*}_{i(x,y)j} = \pm \sqrt{|e_{i(x,y)j}|}\, d_j \tag{1}$$

where $e_{i(x,y)j}$ is the normalized coefficient for the *x* or *y* coordinate of the *i*th landmark on the *j*th eigenvector and d_j is the *j*th eigenvalue. I have used these resultant vectors to illustrate patterns of shape variation among the lake stickleback. To facilitate interpretation of the resultant vectors, an outline of a stickleback was deformed using the thin-plate spline (Bookstein, 1989, 1991) to match precisely the landmarks on the outline with the tips of the corresponding resultant vectors. For the RW analysis, the deformation is calculated directly from the eigenvector coefficients (Rohlf, 1993). For the display of the uniform factor, I have plotted the uniform deformation of the thin-plate spline between the pair of figures that lie on opposites ends of the anisotropy axis based on least squares estimates of anisotropy.

Biomechanical and ecomorphological interpretations of the observed patterns of shape variation indicated by the thin-plate spline deformations are based primarily on the review by Webb (1984). I use a scatter plot of the first two principal components of the least squares analysis in order to explore the association between shape variation and ecology. To do this the 32 mean shapes were grouped into one of four classes: (1) shallow lakes with native predatory fish, (2) shallow lakes without native predatory fish, (3) deep lakes with native predatory fish and (4) deep lakes without native predatory fish. Shallow and deep refer not to lake depth but to relative littoral area. Relative littoral area was calculated as the

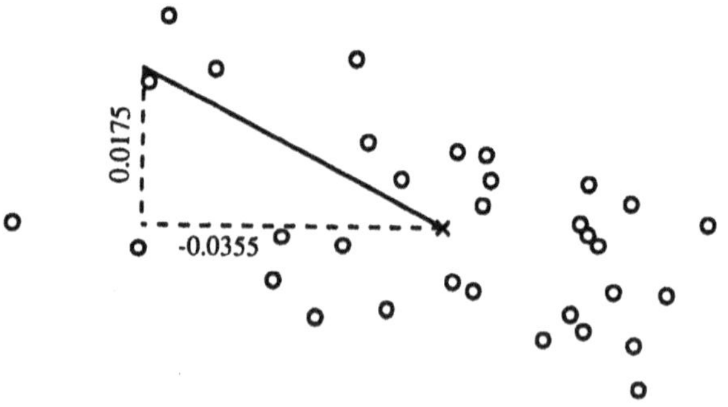

Figure 2. Construction of principal component resultant vectors. In this example, I have zoomed in on the orthogonal least squares scatter at landmark 1, the tip of the premaxilla. The x marks the centroid of this scatter. The eigenvectors of the covariance matrix of all $k * p$ coordinates are computed (not a separate eigenanalysis for each landmark). The first eigenvector coefficient for x1 is -0.0355 and the first eigenvector coefficient for y1 is 0.0175. These coefficients have been rescaled to equal the square of the covariance between the first principal component and the original variables. The resultant vector is simply the resultant of these two coefficients. In traditional PCA, the relative importance of a particular variable is inferred from the relative magnitudes of the coefficients. Patterns of shape variation must be mentally reconstructed from the table of coefficients. In this superimposition based PCA, patterns of shape variation can be physically reconstructed using the resultant of the scaled eigenvector coefficients. The importance of a coordinate on a principal component is inferred by the relative length and direction of the resultant vectors.

Table 1. Correlations between corresponding RW loading vectors calculated from four different initial alignments

	Relative Warp			
Comparison	1	2	3	4
TPR-GOLS	0.99998	0.99998	0.99999	0.99998
TPR-GORF	0.99998	0.99996	0.99997	0.99995
TPR-Rand	0.99627	0.90438	0.76263	0.71269
GOLS-GORF	1.00000	0.99999	0.99999	0.99998
GOLS-Rand	0.99598	0.90446	0.76399	0.71460
GORF-Rand	0.99595	0.90527	0.76478	0.71511

Abbreviations as in text; rand = random

percentage of lake area above the estimated euphotic zone depth. Presence or absence of native predatory fish is based on Alaska Fish and Game records.

All computation and graphics were performed in the software Morphometrika available for Macintosh computers (Walker, 1994).

RESULTS

Comparisons of Analyses

Vector correlations among corresponding RW loading vectors calculated from different alignments are listed in Table 1. A high vector correlation indicates that a pair of RW loading vectors describe similar patterns of shape variation. Correlations between the first four RW loading vectors calculated from the three superimposition alignments are all greater than 0.9999. The correlation between vectors calculated from the random alignment and the corresponding superimposition alignments are extremely high for the first vector ($r > 0.99$) and moderately high for the second vector ($r > 0.90$).

Vector correlations among the first three eigenvectors for the four superimposition analyses and the first three RW loading vectors are reported in Table 2. Highly correlated vectors were not always of the same rank order. In the GALS–GOLS comparison, for example, the second eigenvector of the GALS analysis is highly correlated with the third eigenvector of the GOLS analysis.

The first vectors from all five analyses are highly intercorrelated. First vector correlations are highest between the two orthogonal analyses (GOLS, GORF) and among the three affine analyses (GALS, GARF, RW) but are also relatively high between orthogonal and affine analyses. The pattern of vector correlations for the second and third principal components is more complicated. Within either orthogonal or affine analyses, the correlations are high between the second eigenvectors. In comparisons between orthogonal and affine analyses, however, the second eigenvectors of the orthogonal analyses have only moderate correlations with the third vectors from the three affine analyses. In contrast, second vectors of the affine analyses have high correlations with the third eigenvectors of the orthogonal analyses. Finally, the extremely high vector correlations between corresponding eigenvectors from the GALS and RW analyses should be noted.

Correlations between ln(anisotropy) and θ based on least squares estimation, and the GALS and GOLS scores for the first four principal components are listed in Table 3. The results clearly indicate that a uniform factor is correlated with the third nonuniform principal

Table 2. Vector correlations between eigenvectors of the four superimposition analyses and RW loading vectors of RW analysis. The RW loadings vectors are calculated from the two-point registration alignment. Abbreviations as in text

Analysis	Vector Correlations											
GORF1	0.995	-0.018	-0.004									
GORF2	-0.035	-0.950	-0.024									
GORF3	-0.017	0.063	-0.984									
GALS1	0.876	0.124	0.007	0.869	-0.149	-0.013						
GALS2	0.010	-0.076	-0.994	0.016	0.097	0.973						
GALS3	-0.034	0.466	-0.102	-0.041	-0.438	0.148						
GARF1	-0.931	-0.035	-0.010	-0.941	0.025	-0.002	-0.969	0.000	0.009			
GARF2	-0.072	0.052	-0.964	-0.052	0.032	0.987	0.001	0.956	0.138			
GARF3	-0.108	0.412	0.119	-0.102	-0.357	-0.047	0.026	-0.211	0.921			
RW1	-0.873	-0.133	-0.012	-0.867	0.157	0.017	-1.000	0.005	-0.005	0.968	0.003	-0.033
RW2	0.011	-0.069	-0.995	0.016	0.090	0.974	0.005	1.000	0.012	-0.030	0.957	-0.200
RW3	-0.035	0.461	-0.091	-0.042	-0.434	0.136	-0.005	-0.012	1.000	0.012	0.125	0.921
	GOLS1	GOLS2	GOLS3	GORF1	GORF2	GORF3	GALS1	GALS2	GALS3	GARF1	GARF2	GARF3

component and that the second total shape principal component contains this uniform factor, in addition to the nonuniform shape variation correlated with it.

Patterns of Shape Variation

Because of the extremely high correlations within the orthogonal comparisons and within the affine comparisons, I will illustrate the components by only one of the orthogonal methods and one of the affine methods (Fig. 3 and 4). The patterns of variation described by the first eigenvector of the GOLS analysis ($PC1_{GOLS}$) and the first vector of RW loadings ($PC1_{RW}$) are illustrated in Fig. 3. $PC1_{GOLS}$ and $PC1_{RW}$ have a high (negative) correlation. Both $PC1_{GOLS}$ and $PC1_{RW}$ describe variation in the size of the head, position of the dorsal spines and posterior process of the pelvic girdle and length of the median fins. The position of the posterior process could indicate variation in either length of the posterior process itself or position of the entire girdle. The small variation in the position of the anterior border of the girdle (landmark 13) indicates it is variation in length of the posterior process that accounts for positional differences in landmark 12. The variation in armor phenotype and median fin length is consistent with expected variation given a predation-intensity effect. This observa-

Table 3. Correlations between ln(anisotropy) and θ, both estimated by least squares, and the first four principal components of the orthogonal least squares superimposition (GOLS) and affine least squares superimposition (GALS)

	ln anisotropy	θ
GOLS1	-0.0665	-0.0028
GOLS2	0.9331	-0.5923
GOLS3	0.0055	-0.0739
GOLS4	0.0161	0.0469
GALS1	0.0345	-0.0799
GALS2	-0.1048	0.1328
GALS3	0.6552	-0.3975
GALS4	0.0133	0.0402

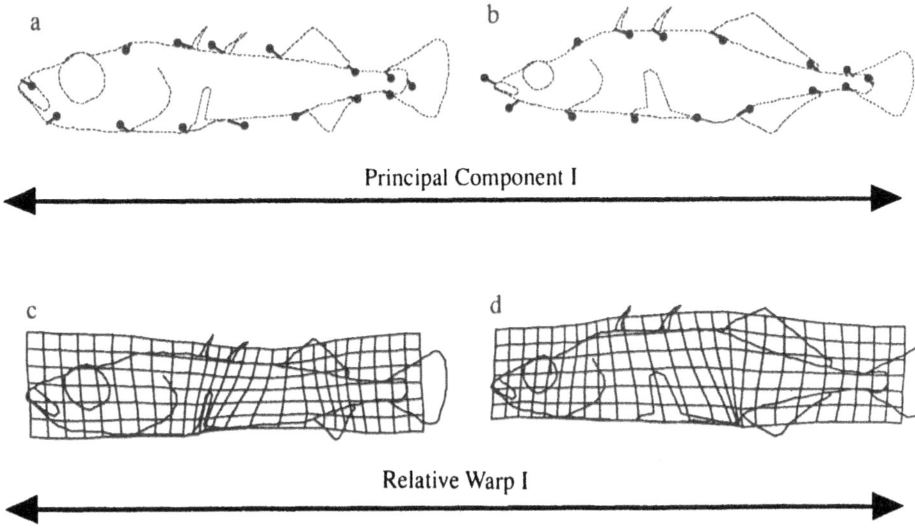

Figure 3. (a, b). Shape variation indicated by first eigenvector of orthogonal least squares analysis. Magnitude of resultant vectors is multiplied by a factor of two. (c, d). First RW of the RW analysis. Magnitude of deformation is multiplied by a factor of two.

tion is supported by the different $PC1_{GOLS}$ scores for sticklebacks inhabiting lakes with and without native predatory fish (Fig. 5).

As mentioned above, the second eigenvector of the GOLS analysis ($PC2_{GOLS}$) has only a moderate correlation with the vector of RW loadings, but the $PC2_{GOLS}$ scores are highly (negatively) correlated with anisotropy. The patterns of variation described by $PC2_{GOLS}$ and the thin-plate spline estimate of the uniform component are illustrated in Fig. 4. $PC2_{GOLS}$

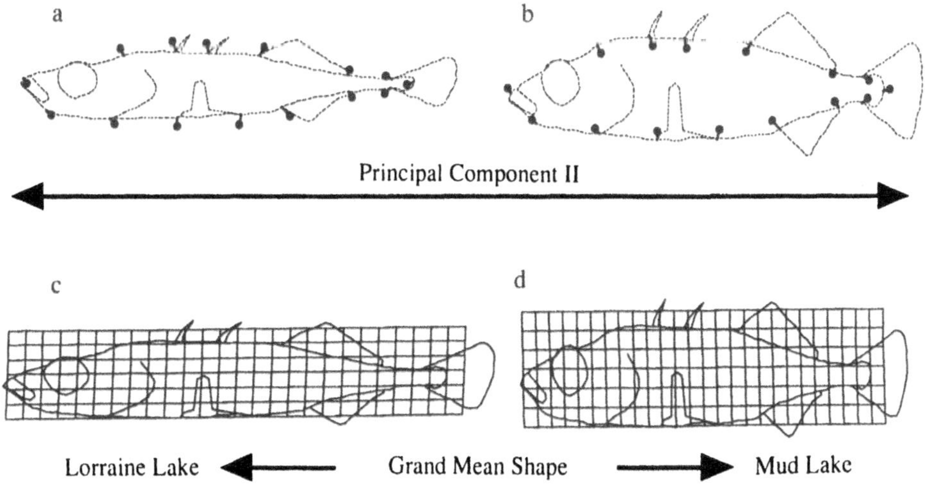

Figure 4. (a, b). Shape variation on the second principal component of the orthogonal least squares analysis. Magnitude of vectors is multiplied by a factor of two. (c). Uniform deformation of the grand mean shape into the mean Lorraine lake shape. Magnitude of deformation is multiplied by a factor of two. (d). Uniform deformation of the grand mean shape into the mean Mud lake shape (see text). Magnitude of deformation is multiplied by a factor of two.

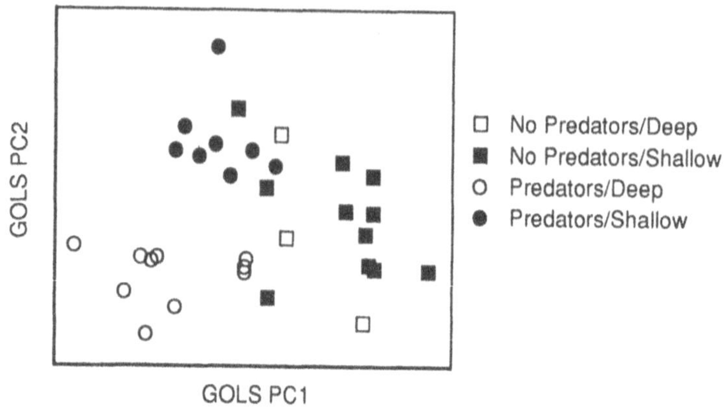

Figure 5. Scatter plot of first and second principal component scores from orthogonal least squares analysis.

describes variation in the length of snout and orientation of the mouth, depth at the posterior edge of the skull, midbody depth and length of the caudal peduncle. In general, the fish with positive $PC2_{GOLS}$ scores are shallow bodied and have long, upturned snouts and long, caudal peduncles. Stickleback with negative $PC2_{GOLS}$ scores are deep bodied and have shorter and more horizontal snouts and short, deep caudal peduncles. The pattern of variation described by $PC2_{GOLS}$ is consistent with expected patterns of variation given a significant foraging effect. Interestingly, differences in PC2 scores differ between sticklebacks inhabiting lakes with high and low littoral areas only if native predatory fish are present (Fig. 5).

DISCUSSION

Comparison of Relative Warps Calculated from Different Initial Alignments

The first p - 3 eigenvectors of the bending energy matrix, known as the principal warps (Bookstein, 1989), estimate a subspace with no affine dimensions. A partial warp analysis (PWA) transforms $k * p$ coordinates into $k * (p - 3)$ nonuniform shape variables, or partial warp scores, by projecting the p coordinates for each dimension onto the p - 3 principal warps (Bookstein, 1991; Rohlf, 1993). This projection effectively removes affine variation by a criterion that minimizes bending energy. As Bookstein (1991: 347) emphasized, "Because rotations are...linear transformations, they too are annihilated by the principal warps to a sufficient degree of precision in the vicinity of any consistent alignment." Thus, PWA is not simply a rearrangement of GOLS or GORF coordinates into k * 3 fewer variables that are ordered by scale. Rather, PWA is an alternative to GALS and GARF for estimating nonuniform shape variation. The results in this comparison indicate that a RWA produces consistent results regardless of the initial superimposition. Surprisingly, even the random rotation of figures by as much as $\pi/4$ radians resulted in similar RW axes.

Comparison of Eigenvectors among Superimposition and Relative Warp Analyses

The high vector correlations among corresponding vectors estimated by the five different analyses in this study indicate that all methods describe similar patterns of shape variation. The first vector correlation between the GOLS and RW analyses, for example, is -0.873. This value is low relative to many of the other comparisons between corresponding components, yet the pattern of shape variation described by $PC1_{GOLS}$ and $PC1_{RW}$ are strikingly similar graphically (Fig. 3).

Vector correlations are higher for the within-orthogonal and within-affine comparisons than for the between orthogonal and affine comparisons. This indicates that, for these data at least, the decision to partition, or not partition, the uniform and nonuniform shape components has more influence than does the particular estimation method on the subsequent PCA. The second vectors of the orthogonal analyses have only moderate correlations with the third vectors of the affine analyses. The high correlation of $PC2_{GOLS}$ with anisotropy, however, indicates that variation in the uniform component of shape differences, in combination with correlated nonuniform shape variation, has emerged as the second principal component of the orthogonal analyses. In the affine analyses, however, this shape component has been partitioned into a uniform factor, represented by the anisotropy and θ scores, and a nonuniform shape factor, represented by the third component. That is, PC2 of the orthogonal analyses is roughly equal to the uniform factor in combination with PC3 of the affine analyses.

Thus we have the crux of the difference between PCA of total shape variation and nonuniform shape variation. In a PCA of orthogonal superimposition coordinates, if a large uniform factor is present in the sample, it will likely emerge as one of the principal components. Non-uniform shape variation correlated with this uniform shape factor will also contribute to the direction of the principal component. In an affine analyses, however, the uniform component is removed a priori and analyzed separately. Although correlations can be computed between the nonuniform (RWs or principal components) and uniform (θ and anisotropy) scores, the two types of information cannot be combined quantitatively, that is, they are incommensurate (Bookstein, 1991).

Functional Basis of Body Shape Variation

Marine and most freshwater stickleback present a robust bony armor complex that functions as a defense against piscine and avian predators. The major components of the complex form a complete bony ring enveloping the midbody of the fish (Reimchen, 1983). Although the ring is composed of different bones, adjacent elements interlock, which gives the ring a rigid structure. The middle of three dorsal spines projects from the dorsal surface of this ring, and two pelvic spines project laterally from the ventral surface of the ring. Of the dorsal spines, the middle spine is the most robust, whereas the third spine is quite reduced, especially in freshwater populations of Cook Inlet. A locking mechanism keeps both the dorsal and pelvic spines erect without active input from the fish. The pelvis, which forms the base of the ring, has a posterior process that extends back toward the vent, thus protecting most of the ventral abdominal wall.

Developmental expression of armor robustness varies tremendously among populations (Bell, 1984, 1987), and there is good field and experimental evidence for selective predation on stickleback armor phenotype (Hoogland et al., 1957; Hagen and Gilbertson, 1972, 1973; Moodie, 1972; Moodie et al., 1973; Moodie and Reimchen, 1976; Gross, 1978; Reimchen, 1988, 1992, 1994; Banbura et al., 1989; Bell et al., 1993). The maintenance of robust armor, however, carries a high metabolic cost. In low calcium lakes, for example, armor robustness may be selected against because of the increased cost of allocating energy to calcium uptake and transport (Giles, 1983; Bell et al., 1993; Bourgeois et al., in press).

Given this trade-off in costs of armor development and function, it is not surprising that we find measures of armor phenotype highly correlated with the major axis of variation among the sticklebacks in this study. Although pelvic robustness was not directly measured in this study, a casual comparison of stickleback pelvic phenotypes indicates that length of the posterior process of the pelvis (i.e., position of landmark x_{12}) is a good indicator of pelvic robustness. The good comparative evidence for an association between predation intensity and armor robustness suggests that the major axis of body shape variation in Cook Inlet stickleback is associated with predation intensity.

Variation in position of the dorsal spines (x_3 and x_4) also has a high correlation with the first principal component. This variation was not expected but can be interpreted in light of functional and behavioral observations. The second spine is the largest of the three dorsal spines. The observed first principal component indicates that stickleback with robust armor structure have relatively anterior dorsal spines, which positions the second spine directly dorsal to the ascending process of the pelvis. This position enables the pterygiophore of the second spine to articulate firmly with the lateral plate that, in turn, firmly articulates with the ascending process of the pelvis (and thus completes the bony ring enveloping the midbody; Reimchen, 1983). In weakly armored stickleback, this ring can be incomplete, and there is no functional reason for the second dorsal spine to maintain its position above the ascending process. Reimchen's (1991) observation that stickleback predators prefer to ingest stickleback headfirst suggests an explanation for the anterior position of the spines in stickleback inhabiting high predation-risk lakes. A more anterior position of the spines may increase the incidence of escape during prey handling by piscine predators. This hypothesis does not explain, however, why stickleback in relatively low predation-risk lakes consistently have more posteriorly positioned spines.

The association between armor phenotype and predation risk indicates only one of several possible antipredator mechanisms in the stickleback: escape due to handling failure during the predation cycle. Stickleback have several antipredator behaviors, but two common behaviors are rapid but short flights, which require strong acceleration with some amount of periodic swimming, and a quick jump, which is a rapid acceleration and powered turn (Huntingford and Giles, 1987; Foster and Ploch, 1990; Bishop and Brown, 1992; Swain, 1992; Huntingford et al., 1994). Although stickleback predominantly use pectoral fin rowing for periodic swimming activities, the median and caudal fins are used for fast starts and turns (Wootton, 1976; Taylor and McPhail, 1986; Whoriskey and Wootton, 1987). Theoretical (Weihs, 1972, 1973), experimental (Webb, 1977) and comparative (Webb, 1978, 1983) studies of fish locomotion indicate that transient locomotor performance increases with median fin area. Theoretical models, in turn, demonstrate that the frequency of successful evasions increases with fast start and turning performance (Howland, 1974; Weihs and Webb, 1984). The high association between length of the median fins and the first principal component, in combination with the armor phenotype contribution to this component, suggests a locomotor mechanism of escape in addition to the defensive armor mechanism.

It is of particular interest that it is median fin size and not caudal body depth that is associated with native predatory fish. Although increased body depth increases fast-start swimming performance, the resulting increase in frictional drag reduces the efficiency of periodic swimming performance (Webb, 1977, 1984). Median fins, however, can be retracted against the body during periodic swimming (Webb, 1977). Indeed, stickleback foraging in the open-water have been observed to collapse both the median fins as well as the caudal fin during steady swimming (Jeff Walker, Bill Cresko, personal observation). Increasing median fin size, as opposed to increasing body depth, should theoretically enhance fast-start swimming performance but have little consequence to steady swimming performance.

The second major axis of shape variation among the 32 samples describes an axis of variation from stickleback with longitudinally shortened bodies, short, horizontal snouts, deep mid- and caudal bodies and short, deep caudal peduncles to stickleback with elongate, shallow bodies, long, upturned snouts and long, shallow caudal peduncles. This pattern of polymorphism is common to many fishes (Ehlinger and Wilson, 1988; Ehlinger, 1990;

Skulason et al. 1989), including stickleback (Lavin and McPhail, 1985, 1986; Schluter and McPhail, 1992; Schluter, 1993), that predominately forage in either the vegetated littoral zone or the open-water of the limnetic zone. Foraging in the littoral zone includes a searching component and an attack component. Aquatic macrophytes, fallen trees and branches and boulders create a complex obstacle course for a fish searching for littoral prey. Increasing maneuverability, therefore, is one strategy for increased foraging performance in the complex structure of the littoral. In lakes with native predatory fish, stickleback inhabiting shallow lakes have shortened and deepened bodies and caudal peduncles. The association of this body shape and habitat is consistent with the expectation of hydromechanical theory (Webb, 1984). A shortened, deep body decreases the radius of curvature for fine turning maneuvers (Weihs, 1972). The major littoral prey type of stickleback in lakes of the Pacific northwest is chironomid larvae, although cyclopoid copepods, ostracods and amphipods may be taken with high frequencies in some populations (Rogers, 1968; Manzer, 1976; Lavin and McPhail, 1986; Schluter, 1993). Optimal shapes for attacking littoral prey include a short and deep caudal peduncle for increased acceleration and powered turning performance (Weihs, 1972; Webb, 1977, 1984) and a short, deep head for increased suction-feeding performance (Norton, 1991).

In sharp contrast to the littoral zone, the limnetic zone presents an open-water environment generally free of any complex structure. Foraging in the limnetic zone requires both cruising the water column in search of patches of zooplankton and attacking individual zooplankters once a patch is found. In lakes with native predatory fish, stickleback inhabiting deep lakes have elongated, shallow bodies and caudal peduncles. Stickleback use the pectoral fins to generate thrust during routine swimming. We can, nevertheless, cautiously interpret the shape of the open-water sticklebacks as consistent with a frictional drag reduction mechanism (Webb, 1984). In the limnetic zone of lakes of the Pacific Northwest, stickleback feed predominately on calanoid copepods, an evasive zooplankter (Rogers, 1968; Manzer, 1976; Lavin and McPhail, 1986; Schluter, 1993). Preliminary observations of stickleback feeding on an evasive copepod indicates that sticklebacks combine a ram-feeding and suction-feeding mechanism for attacking these prey (Jeff Walker, in preparation). Reduced surface area in these stickleback, then, may serve both to increase performance during steady swimming and to reduce frictional drag at the high velocities reached while attacking evasive prey.

ACKNOWLEDGMENTS

I want to thank Susan Foster for providing invaluable field assistance. This research was improved through numerous discussions with M. A. Bell, L. F. Marcus, F. J. Rohlf and F. L. Bookstein. This research was supported by C. Sexton, a Raney Award from the American Society of Ichthyologists and Herpetologists, a Theodore Roosevelt Memorial Fund award from the American Museum of Natural History, National Science Foundation Doctoral Dissertation Improvement Award DEB-9223844 to J. Walker and NSF grant BSR-8905758 to M. A. Bell.

REFERENCES

Airoldi, J.-P. and B. K. Flury. 1988. An application of common principal component analysis to cranial morphometry of *Microtus californicus* and *M. ochrogaster* (Mammalia, Rodentia). Journal of Zoology, London 216:41–43.

Banbura, J., M. Przybylski and P. Frankiewicz. 1989. Selective predation of the pike *Esox lucius*: comparison of lateral plates and some metric features of the three-spined stickleback *Gasterosteus aculeatus*. Zoologica Scripta 18:303–309.

Baumgartner, J. V., M. A. Bell and P. H. Weinberg. 1988. Body form differences between the Enos Lake species pair of threespine sticklebacks. Canadian Journal of Zoology 66:467–474.

Bell, M. A. 1984. Evolutionary phenetics and genetics. The threespine stickleback, *Gasterosteus aculeatus*, and related species. *in* B. J. Turner, (ed.), Evolutionary genetics of fishes. Plenum: New York.

Bell, M. A. 1987. Interacting evolutionary constraints in pelvic reduction of threespine sticklebacks, *Gasterosteus aculeatus* (Pisces, Gasterosteidae). Biological Journal of the Linnean Society 31:347–382.

Bell, M. A. and S. A. Foster (eds.). 1994a. Evolutionary biology of the threespine stickleback. Oxford University Press: Oxford.

Bell, M. A. and S. A. Foster. 1994b. Introduction. *in* M. A. Bell and S. A. Foster, (eds.), Evolutionary biology of the threespine stickleback. Oxford University Press: Oxford.

Bell, M. A., G. Orti, J. A. Walker, and J. P. Koenings. 1993. Evolution of pelvic reduction in threespine stickleback: a test of competing hypotheses. Evolution 47:906–914.

Bentzen, P. and J. D. McPhail. 1984. Ecology and evolution of sympatric sticklebacks (*Gasterosteus*): specializations for alternative trophic niches in the Enos lake species pair. Canadian Journal of Zoology 62:2280–2286.

Bishop, T. D. and J. A. Brown. 1992. Threat-sensitive foraging by larval threespine stickleback (*Gasterosteus aculeatus*). Behavior Ecology and Sociobiology 31:133–138.

Bookstein, F. L. 1989. Principal warps: thin-plate splines and the decomposition of deformations. IEEE Trans. Pat. Anal. Mac. Intel. 11:567–585.

Bookstein, F. L. 1991. Morphometric tools for landmark Data. Geometry and biology. Cambridge University Press: Cambridge.

Bourgeois, J. F., D. M. Blouw and M. A. Bell. In press. Multivariate analysis of geographic covariance between phenotypes and environments in the threespine stickleback, *Gasterosteus aculeatus*, from the Cook Inlet area, Alaska. Canadian Journal of Zoology.

Campbell, N. A. and W. R. Atchley. 1981. The geometry of canonical variate analysis. Systematic Zoology 30:268–280.

Chapman, R. E. 1990. Conventional Procrustes approaches. *in* F. J. Rohlf and F. L. Bookstein, (eds.), Proceedings of the Michigan morphometrics workshop. University of Michigan Museum of Zoology Special Publication 2.

Ehlinger, T. J. 1990. Habitat choice and phenotype-limited feeding efficiency in bluegill: individual differences and trophic polymorphism. Ecology 71:886–896.

Ehlinger, T. J. and D. S. Wilson. 1988. Complex foraging polymorphism in bluegill sunfish. Proceedings of the National Academy of Science 85:1878–1882.

Foster, S. and S. Ploch. 1990. Determinants of variation in antipredator behavior of territorial male threespine stickleback in the wild. Ethology 84:281–294.

Francis, R. C., J. V. Baumgartner, A. C. Havens and M. A. Bell. 1986. Historical and ecological sources of variation among lake populations of threespine sticklebacks, *Gasterosteus aculeatus*, near Cook Inlet, Alaska. Canadian Journal of Zoology 64:2257–2265.

Giles, N. 1983. The possible role of environmental calcium levels during the evolution of phenotypic diversity in Outer Hebridean populations of the three-spined stickleback, *Gasterosteus aculeatus* L. Canadian Journal of Zoology 56: 398–413.

Goodall, C. 1991. Procrustes methods in the statistical analysis of shape. Journal of the Royal Statistical Society, B 53:285–339.

Gross, H. 1978. Natural selection by predators on the defensive apparatus of the three-spined stickleback, *Gasterosteus aculeatus* L. Canadian Journal of Zoology 56:398–413.

Hagen, D. W. and L. G. Gilbertson. 1972. Geographic variation and environmental selection in *Gasterosteus aculeatus* L. in the Pacific Northwest, America. Evolution 26:32–51.

Hagen, D. W. and L. G. Gilbertson. 1973. Selective predation and the intensity of selection acting upon the lateral plates of threespine sticklebacks. Heredity 30:273–287.

Hoogland, R., D. Morris, and N. Tinbergen. 1957. The spines of sticklebacks (*Gasterosteus* and *Pygosteus*) as a means of defense against predators. Behavior 10:205–237.

Howland, H. C. 1974. Optimal strategies for predator avoidance: the relative importance of speed and maneuverability. Journal of Theoretical Biology 47:333–350.

Huntingford, F. and N. Giles. 1987. Individual variation in anti-predator responses in the three-spined stickleback (*Gasterosteus aculeatus* L.). Ethology 74:205–210.

Huntingford, F. A., P. J. Wright and J. F. Tierney. 1994. Adaptive variation in antipredator behavior in threespine stickleback *in* M. A. Bell and S. A. Foster (eds.), Evolutionary biology of the threespine stickleback. Oxford University Press: Oxford.

Krzanowski, W. J. 1988. Principles of multivariate analysis. Clarendon Press: Oxford, U. K.

Larson, G. L. 1976. Social behavior and feeding ability of two phenotypes of *Gasterosteus aculeatus* in relation to their spatial and trophic segregation in a temperate lake. Canadian Journal of Zoology 54:107–121.

Lavin, P. A., and J. D. McPhail. 1985. The evolution of freshwater diversity in the threespine stickleback (*Gasterosteus aculeatus*): site-specific differentiation of trophic morphology. Canadian Journal of Zoology 63:2632–2638.

Lavin, P. A. and J. D. McPhail. 1986. Adaptive divergence of trophic phenotype among freshwater populations of the threespine stickleback (*Gasterosteus aculeatus*). Can. J. Fish. Aquat. Sci. 43:2455–2463.

Marcus, L. F. 1990. Traditional morphometrics. Pages 77–122 *in* F. J. Rohlf and F. L. Bookstein, (eds.), Proceedings of the Michigan Morphometrics Workshop. University of Michigan Museum of Zoology Special Publication 2.

Manzer, J. I. 1976. Distribution, food, and feeding of the threespine stickleback, *Gasterosteus aculeatus*, in Great Central Lake, Vancouver Island, with comments on competition for food with juvenile sockeye salmon, *Oncorhynchus nerka*. Fish. Bull. 74: 647–668.

McPhail, J. D. 1984. Ecology and evolution of sympatric sticklebacks (*Gasterosteus*): morphological and genetic evidence for a species pair in Enos Lake, British Columbia. Canadian Journal of Zoology 62:1402–1408.

McPhail, J. D. 1994. Speciation and the evolution of reproductive isolation in the sticklebacks (*Gasterosteus*) of southwestern British Columbia *in* M. A. Bell and S. A. Foster (eds.), Evolutionary biology of the threespine stickleback. Oxford University Press: Oxford.

Meacham, C. A. and T. Duncan. 1990. MorphoSys. Version 1.29. University Herbarium, University of California: Berkeley.

Moodie, G. E. E. 1972. Predation, natural selection and adaptation in an unusual threespine stickleback. Heredity 28:155–167.

Moodie, G. E. E., J. D. McPhail, and D. W. Hagen. 1973. Experimental demonstration of selective predation on *Gasterosteus aculeatus*. Behavior 47:95-105.

Moodie, G. E. E., and T. E. Reimchen. 1976. Phenetic variation and habitat differences in *Gasterosteus* populations of the Queen Charlotte Islands. Systematic Zoology 25:49–61.

Norton, S. F. 1991. Capture success and diet of cottid fishes: the role of predator morphology and attack kinematics. Ecology 72:1807–1819.

Reimchen, T. E. 1983. Structural relationships between spines and lateral plates in threespine stickleback (*Gasterosteus aculeatus*). Evolution 37:931–946.

Reimchen, T. E. 1988. Inefficient predators and prey injuries in a population of giant stickleback. Canadian Journal of Zoology 66:2036–2044.

Reimchen, T. E. 1991. Trout foraging failures and the evolution of body size in stickleback. Copeia 1991:1098–1104.

Reimchen, T. E. 1992. Injuries on stickleback from attacks by a toothed predator (*Oncorhynchus*) and implications for the evolution of lateral plates. Evolution 46:1124–1230.

Reimchen, T. E. 1994. Predators and morphological evolution in threespine stickleback *in* M. A. Bell and S. A. Foster, (eds.), Evolutionary biology of the threespine stickleback. Oxford University Press: Oxford.

Reimchen, T. E., E. M. Stinson and J. S. Nelson. 1985. Multivariate differentiation of parapatric and allopatric populations of threespine stickleback in the Sangan River watershed, Queen Charlotte Islands. Canadiaon Journal of Zoology 63:2944–2951.

Rogers, D. E. 1968. A comparison of the food of sockeye salmon fry and threespine sticklebacks in the Wood River lakes *in* R. L. Burgner (ed.), Further studies of Alaska sockeye salmon. University of Washington: Seattle.

Rohlf, F. J. 1993. Relative warp analysis and an example of its application to mosquito wings. Pages 131–159 *in* L. F. Marcus, E. Bello, and A. García-Valdecasas, (eds.), Contributions to morphometrics. Monografias del Museo Nacional de Cienceias Naturales 8, Madrid.

Rohlf, F. J. and D. Slice. 1990. Extensions of the Procrustes method for the optimal superimposition of landmarks. Systematic Zoology 39:40–59.

Schluter, D. 1993. Adaptive radiation in sticklebacks: Size, shape and habitat use efficiency. Ecology 74:699–709.

Schluter, D., and J. D. McPhail. 1992. Ecological character displacement and speciation in sticklebacks. American Naturalist 140:85–108.

Siegel A. F. and R. H. Benson, 1982. A robust comparison of biological shapes. Biometrics 38: 341–350.

Skulason, S., D. L. G. Noakes, and S. S. Snorrason. 1989. Ontogeny of trophic morphology in four sympatric morphs of arctic charr *Salvelinus alpinus* in Thingvallavatn, Iceland. Biolological Journal of the Linnean Society 38:281–301.

Swain, D. P. 1992. The functional basis of natural selection for vertebral traits of larvae in the stickleback *Gasterosteus aculeatus*. Evolution 46:987–997.

Taylor, E. B. and J. D. McPhail. 1986. Prolonged and burst swimming in anadromous and freshwater threespine stickleback, *Gasterosteus aculeatus*. Canadian Journal of Zoology 64:416–420.

Thorpe, R. S. 1988. Multiple group principal component analysis and population differentiation. Journal of Zoology, London 216:37–40.

Voss, R. S., L. F. Marcus and P. Escalante. 1990. Morphological evolution in muroid rodents I. Conservative patterns of craniometric covariance and their ontogenetic basis in the neotropical genus *Zygodontomys*. Evolution 44:1568–1587.

Walker, J. A. 1994. Morphometrika. Geometric Morphometrics for the Macintosh. Department of Anatomical Sciences, State University of New York at Stony Brook, New York, USA,

Webb, P. W. 1977. Effects of median-fin amputation on fast-start performance of rainbow trout (*Salmo gairdneri)*. Journal of Experimental Biology 68:123–135.

Webb, P. W. 1978. Fast-start performance and body form in seven species of teleost fish. Journal of Experimental Biology 74:211–226.

Webb, P. W. 1983. Speed, acceleration and maneuverability of two teleost fishes. Journal of Experimental Biology 102:115–122.

Webb, P. W. 1984. Body form, locomotion and foraging in aquatic vertebrates. American Zoologist. 24:107–120.

Weihs, D. 1972. A hydrodynamical analysis of fish turning maneuvers. Proceedings of the Royal Society of London,.B, 182:59–72.

Weihs, D. 1973. The mechanism of rapid starting of slender fish. Biorheology 10:343–350.

Weihs, D. and P. W. Webb. 1984. Optimal avoidance and evasion tactics in predator-prey interactions. Journal of Theoretical Biology 106:189–206.

Whoriskey, F. G. and R. J. Wootton. 1987. The swimming endurance of threespine sticklebacks, *Gasterosteus aculeatus* L., from the Afon Rheidol, Wales. J. Fish Biol. 30:335–339.

Wootton, R. J. 1976. The biology of the sticklebacks. Academic Press: New York.

A LANDMARK-BASED MORPHOMETRICS APPROACH TO THE SYSTEMATICS OF CROCIDURINAE

A Case Study on Endemic Shrews *Crocidura sicula* and *C. canariensis* (Soricidae, Mammalia)

Maurizio Sarà

Institute of Zoology of Palermo University
Via Archirafi
18 - 90123 Palermo, Italy

ABSTRACT

The phenetic relationships between two endemic taxa of Crocidurinae, *Crocidura sicula* and C. *canariensis*, having the same karyotype and very low Nei's genetic distance were studied by geometric morphometrics. Fifteen landmarks were recorded in three populations of *C. sicula* (Sicily and Egadi) and one of *C. canariensis* (Canary) to test the hypothesis, based on distance-based multivariate analyses, of low phenetic difference between the shrews in Western Sicily and surrounding islands (Egadis) and those in the Canary islands. The Generalized Least Squares (GLS) superimposition method was used to align the samples for further use in Thin Plate Spline Relative Warps Analysis (TPSRW, $\alpha = 0$). Canonical variate analysis of the resulting weight matrix and UPGMA clustering of Mahalanobis distances were run to achieve an ordination of the four populations. The analyses showed that shape changes are restricted to some regions of the mandible related to chewing. These differences can be better explained as a result of individual within-population and intraspecific variation. The results of the geometric morphometric analysis substantiated the strict phenetic relationships between the Canary and Sicilian shrews, being congruent with the distance-based analyses. These populations are two peripheral isolates that stem from a Maghrebi ancestor during the Plio-Pleistocene. It is suggested that taxonomically, *C. canariensis* be recognized as a subspecies of *C. sicula*.

INTRODUCTION

Recent studies based on either karyology, allozyme electrophoresis or morphometrics have modified Crocidurinae systematics in the Palearctic region. Three new species of

Advances in Morphometrics, Edited by L. F. Marcus et al.
Plenum Press, New York, 1996

Crocidura were found to be endemic and restricted to islands and archipelagos: in Crete (*C. zimmermanni*, Vogel et al., 1986); in the siculo-maltese archipelago (*C. sicula*, Vogel, 1988; Vogel et al., 1989; Sarà et al., 1990); and in the Canary archipelago (*C. canariensis*, Hutterer et al., 1987). *C. sicula* and *C. canariensis* proved to have exactly the same karyotype and a value of Nei's genetic distance normally considered as an indicator of conspecificity among mammals (see Vogel, 1988; Maddalena, 1990).

The morphometrics of these two taxa has never been studied or reported in published papers, excluding a published assertion found in Molina and Hutterer (1989): "New morphological studies of *C. sicula*, *C. canariensis* and *C. osorio* by one of us (R. H.) show that all three species are clearly separable on qualitative and quantitative characters."

The presence of these two taxa in such distant archipelagos (over some 3,300 kms) is a puzzling zoogeographic and phylogenetic problem. An analysis of the morphometrics and geographic variation among these shrews seemed desirable, if not necessary. It was carried out (Sarà, 1995) by traditional (i.e., distance-based) multivariate and zoogeographical (fossil record) analyses that supported some hypotheses and taxonomic conclusions regarding the close phenetic relationships between these populations. The form of both the calvarium and mandible is very similar among *C. canariensis* and the populations of *C. sicula* in the small islands that were more anciently separated from Sicily (Marettimo in the Egadi and Gozo in the Maltese archipelagos).

In the last few years the development of geometric morphometrics based on landmark coordinates, or the so-called "New Morphometrics" (Rohlf and Marcus, 1993), has seemed to provide a more effective tool for extracting information on the shape of individuals under study and to offer more robust statistical procedures for testing differences in shape. This paper is an exploratory study of the shape relationships of *C. sicula* and *C. canariensis* by means of geometric morphometric methods.

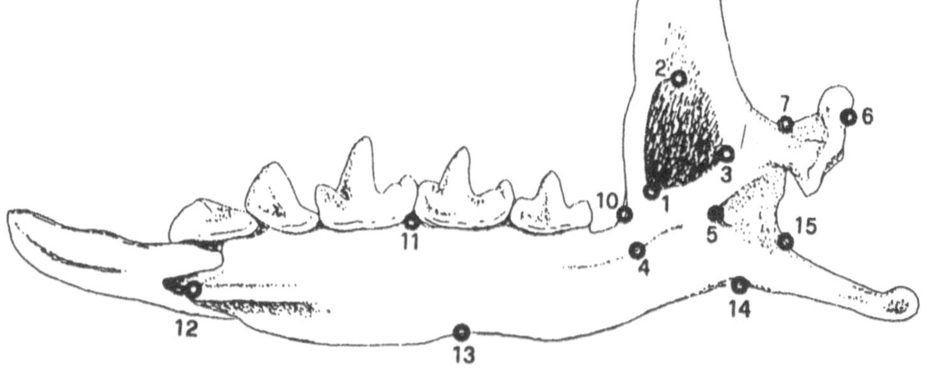

Figure 1. Location of the 15 landmarks on the right shrew mandible: landmarks 1-3 refer to the masseter fossa; 4-5 to the masseter crest; 6-7 to the angular process; 8-9 to the coronoid process; 10 to the insertion of the mandibular branch on the mandibular plate; 11 to the upper mandibular branch ; 12 to the incisive epiphysis; 13 to the lower mandibular branch; 14-15 to the condylar process. Landmarks 11-13 are also referred to in the text as the incisive or mandibular branch region, whereas landmarks 1-10 are referred to as the mandibular plate.

MATERIAL AND METHODS

During the Advanced Study Institute I was able to collect the coordinates of 15 landmarks on 60 mandibles of *C. sicula* and *C. canariensis* (Fig. 1): 16 specimens of *C. sicula* from northwestern Sicily (Salemi, Province of Trapani); 18 specimens from midwestern Sicily (Ficuzza Wood, Province of Palermo); 18 from Marettimo (Egadi Archipelago, Province of Trapani); and 8 from the Canary Archipelago. All the material is stored in the theriological collection at the Museum of the Zoological Institute of Palermo University (MZPA).

Coordinate data were acquired using a video camera, and Morphosys software (Meacham and Duncan, 1990). Four replicates of the data acquisition session on the overall sample were done both for training and reducing digitizing errors. The number of shrew mandibles is small compared with the samples processed in traditional multivariate analysis (480 mandibles and 229 calvaria for 15 OTUs) and thus permits only preliminary analysis of the shape of *C. sicula* and *C. canariensis*.

The hypothesis tested in this paper concerns the general resemblance between the two taxa and, in particular, the closer phenetic proximity between Canary and Marettimo populations as well as Canary and Western Sicily populations (i.e., the founder populations of Egadis should have come from this area).

The raw coordinates were analyzed (coordinate set rotated, translated and scaled) using the Generalized Least Squares (GLS) Procrustes superposition method in the GRF-ND software (Slice, 1992). The output obtained by GLS was then used as input for the Thin-Plate Spline Relative Warp Analysis (TPSRW) software (version October 1993; Rohlf, 1993a); the mean configuration was considered the reference configuration to obtain relative warps as defined by Bookstein (1991). The following options were selected: an α of 0; initial alignment by GLS as a superimposition criterion; weights based on raw data as deviation from the mean reference configuration; and N, extra columns not appended to the weight matrix (Rohlf, 1993b, 1993c). The resulting weight matrix was employed as input for a Canonical Variates Analysis to obtain both the ordination in the three-dimensional-canonical space and the matrix of generalized Mahalanobis distances (NTSYS-pc software, version 1.80; Rohlf, 1993b).

A second TPSRW was also run with the same options as above except for the last; U instead of N was chosen to estimate the uniform components *U*1 and *U*2.

RESULTS

Procrustes distances, obtained by GLS, and giving indication of the degree of fit among the four superimposed populations, are shown in Table 1. They proved to be

Table 1. The Procrustes distances (Rohlf, 1990) give some indication of the degree of fit among the four superimposed populations. The p values for the F-test are according to the sequential Bonferroni adjustment

	Salemi	Ficuzza	Marettimo
Ficuzza	0.00045**		
Marettimo	0.00051*	0.00041*	
Canary	0.00106**	0.00139***	0.00139**

* $p = 0.05$; ** $0.01 < p < .05$; *** $p < .01$

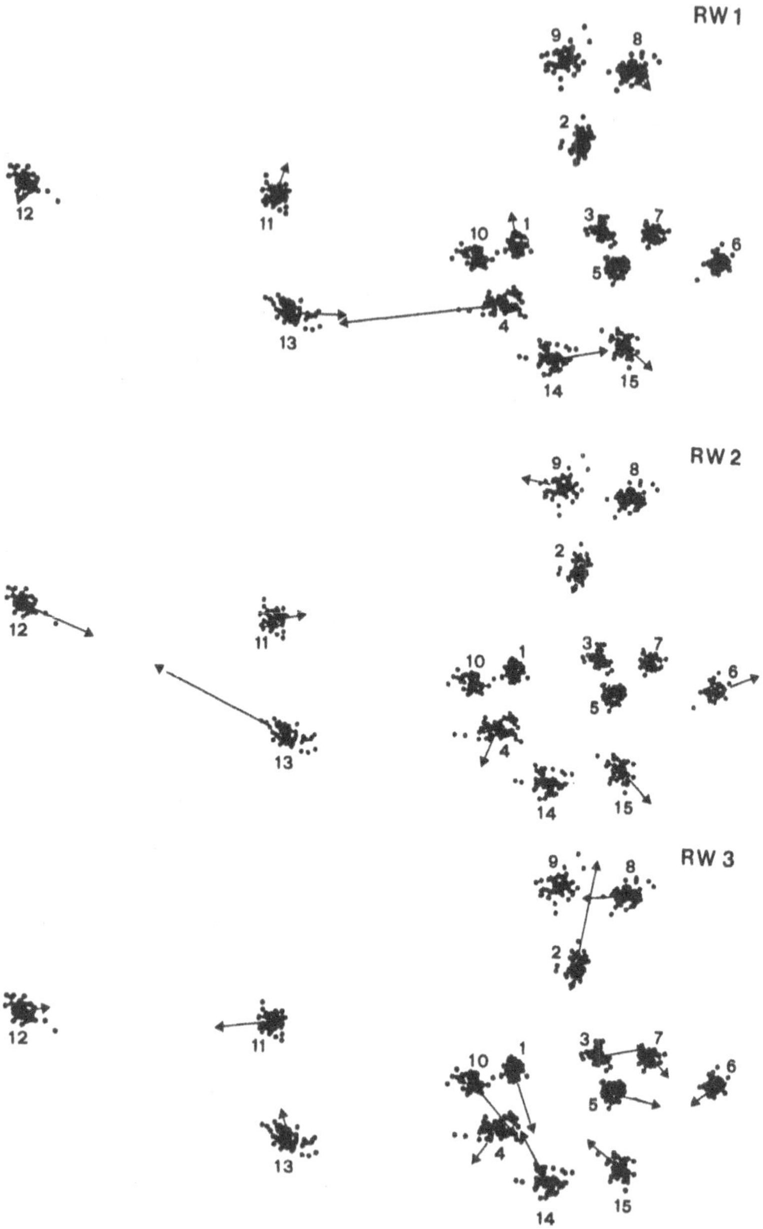

Figure 2. Plots of the first three relative warps (RW1-3), $\alpha = 0$, on the mean reference configuration. Vectors for each landmark display the amount and direction of displacement at the same magnitude, scaled arbitrarily to 0.25.

statistically significant. The highest level of significance resulted in the pair-wise comparison between the Ficuzza and Canary and the Marettimo and Canary populations.

The highest relative warp loadings (without any uniform part) are reported as vectors at each landmark in Fig. 2 and Table 2. These vectors display the directions of maximum variance of the whole sample of mandibles. The longest vectors for relative warps 1 and 2 (32.3% of total explained variance) refer to the masseter crest (landmark 4), whose median bent point varies greatly among individuals. (The long vector on relative warp 1 can be

Table 2. The first 9 relative warps (out of 24) express more than the 75% of the total variance. For a given relative warp the main modifications are described by the reported horizontal and vertical shifting of landmarks with the x and y components of principal warps. Greatest shape changes occur for the displacements of landmarks 8, 14, 6, 13, 11, 10, and 12 and smallest for 1, 2, 4, and 9. Landmarks and principal warps arranged in order of importance

Relative warp	Eigenvalue	Percent of variance	Cumulative percent	Principal warp	Landmarks
1	0.000375	16.62	16.62	3*x* - 11*x*	4 -11 -13
2	0.000353	15.65	32.27	9*x* - 12*x* - 11*x*	11 - 13 - 6 - 12
3	0.000242	10.73	43.00	6*y* - 8*y* - 10*x*	2 - 6 - 10 - 14 - 8
4	0.000182	8.07	51.07	2*x* - 12*x*	1 - 10 - 6 - 12
5	0.000157	6.96	58.03	10*y* - 6*y*	14 - 8 - 2
6	0.000119	5.27	63.30	11*y* - 7*y* - 5*y*	11 - 13 - 8 - 9 - 15 -14
7	0.000105	4.65	67.95	7*x* - 5*y* - 12*y*	8 - 9 - 15 - 14 - 6 -12
8	0.000099	4.39	72.34	10*x* - 8*x* - 11*x*	14 - 8 - 6 - 10 - 11 - 13
9	0.000083	3.68	76.02	2*y* - 10*x*	1 - 10 - 14 - 8

biased, to some extent, by digitizing errors or by different levels of mandible wear.) Other long vectors account for variation in the incisive-mandibular region (11,12,13), in the condylar process insertion (14,15) and in the coronoid process (9). The movements, often opposite each other, of these landmarks indicate that shape variation results on the whole in a more or less squat and short mandible (11,12,13,9) with a different degree of insertion of

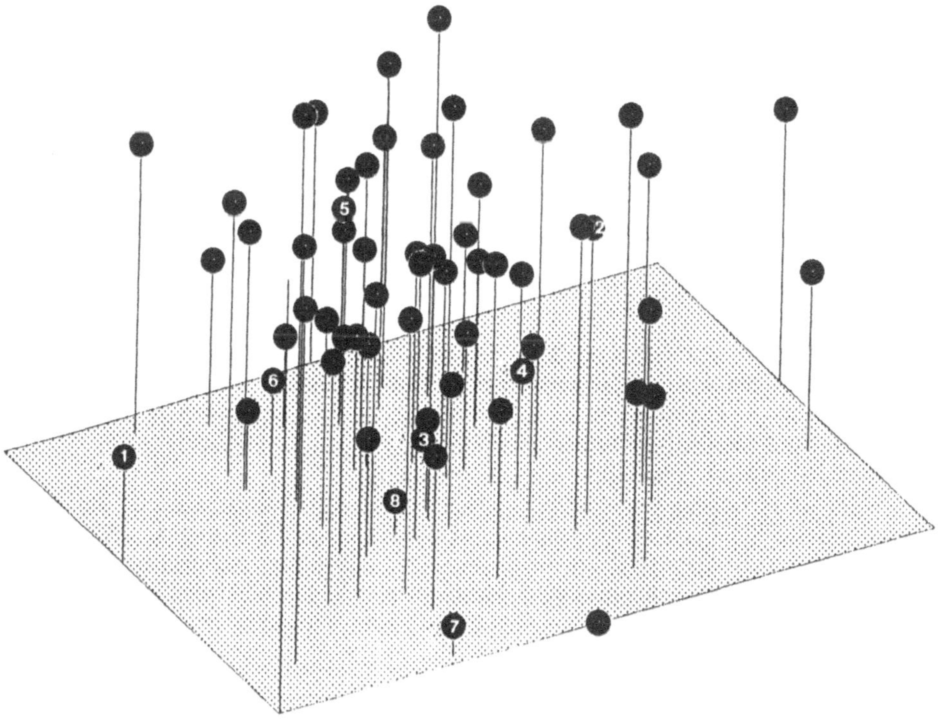

Figure 3. The plot of the scores on the first three relative warps (43 % of total variance explained) give an ordination of the 60 specimens with respect to the nonlinear components of shape change. Only the eight *C. canariensis* specimens (1-8) were marked to show their dispersion within the global sample.

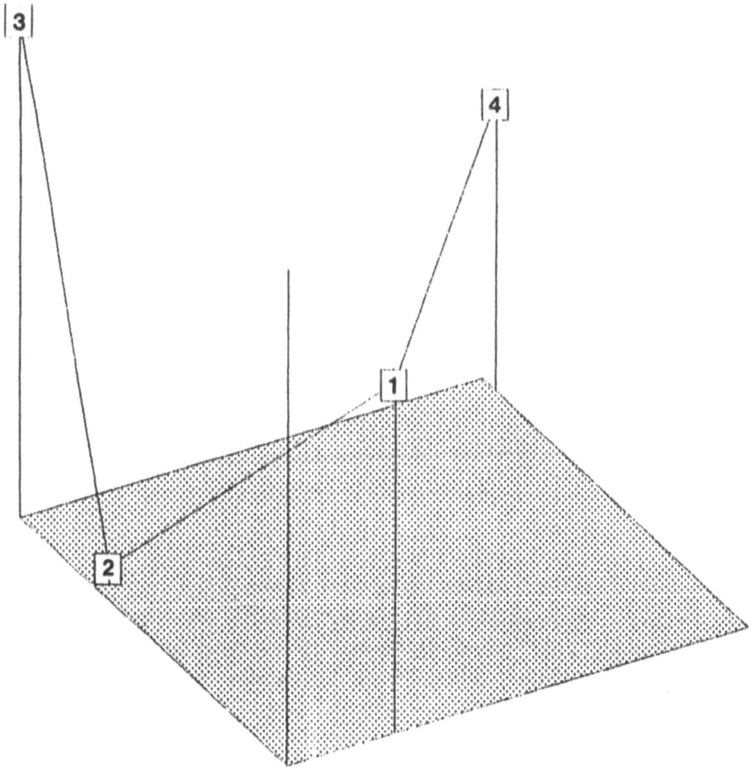

Figure 4. Minimum Spanning Tree of population centroids superimposed on the plot of the first three canonical variates. Abbreviations are 1 = Salemi, 2 = Ficuzza, 3 = Marettimo and 4 = Canary. The first canonical variate explains 63.75 % of total variance, the second expresses 24.53% and the third 11.72%. Differences among group means are statistically significant: Wilks' $\lambda = 0.037$; $p < 0.001$ ($F = 2.79$, df = 72, 99.5).

the condylar process (14, 15) on the mandibular plate and a somewhat larger masseter crest (4). Relative warp 3 shows the displacement of the mandibular plate landmarks. The pattern displayed by relative warp 3 for the most part implies differences determined by the enlargement and modification of the masseter area (1,2,3,4,5) and by the rotation of the angular process (6,7).

Fig. 3 shows the scatter plot of the relative warp scores for the first three relative warps. In the three-dimensional space of relative warps, the 60 specimens are not ordered in relation to their geographic location; the *C. canariensis* are scattered among the *C. sicula* specimens. The shape changes displayed by the TPRSW analysis thus seem to be accounted for more by individual rather than geographic variation.

The weight matrix was used as input for a Canonical Variates Analysis, which is a stronger method for checking the relationships among groups relative to the within-group variation. The four population centroids are well separated in the canonical space. A Minimum Spanning Tree displays their similarity in Fig. 4. The UPGMA clustering of Mahalanobis distances results in the pattern shown in Fig. 5. Only the β-flexible, among the available combinatorial algorithm options, resulted in a different output—clustering together Marettimo and Canary apart from the two Sicilian samples. "Cophenetic correlation" (Rohlf, 1993b) was used as measure of goodness of fit of the two cluster outputs. The UPGMA proved to have a very good fit ($r = 0.99$), whereas the β-flexible had a null fit ($r = 0.01$). The UPGMA among Procrustes distances (Table 1) showed an analogous clustering of the four populations.

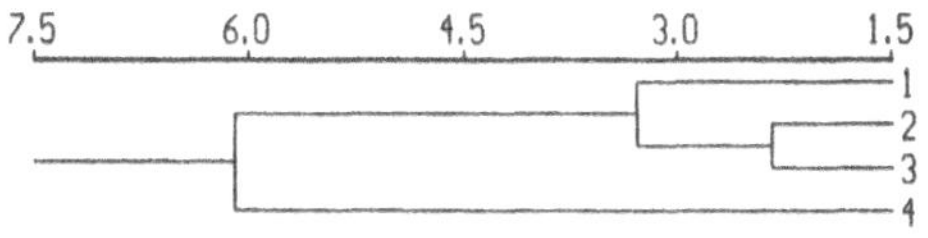

Figure 5. The UPGMA computed from Mahalanobis distances; 1 = Salemi, 2 = Ficuzza, 3 = Marettimo, and 4 = Canary.

The analysis of the uniform part for the global sample shows that its components along the *U1* and *U2* axes are scarcely correlated (Fig. 6), with an $r^2 = 0.033$ ($p = 0.165$). The Canary sample is, however, slightly separated for the affine components, with much more overlap among the other populations.

DISCUSSION

Geometric morphometrics seems promising in the study of geographic variation and systematics of the Crocidurinae. Nonaffine shape changes, so far shown in the mandible, are related to differences: in the mandibular branch; in those areas functionally tied to the insertion and development of the masseter the main chewing muscle; and in the insertion of the condylar apophysis in the glenoid cavity of the braincase, which determines the anterior-posterior gliding of the mandible during chewing.

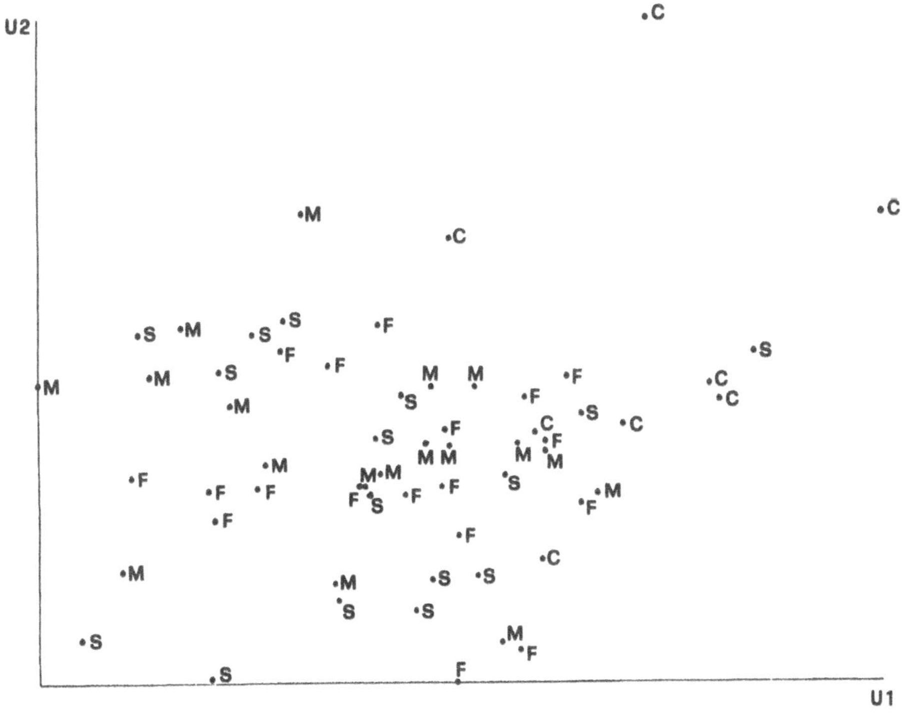

Figure 6. Scatter plot of uniform components *U1* and *U2* for all 60 specimens. Abbreviations are S = Salemi, F= Ficuzza, M = Marettimo and C = Canary.

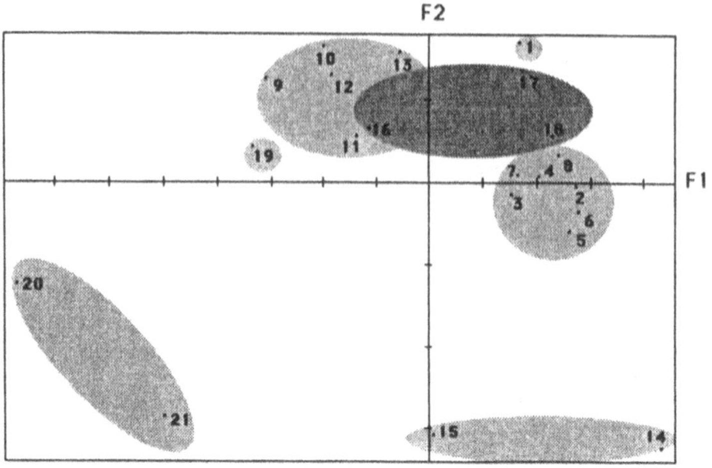

Figure 7. Principal components analysis of 14 metric variables of the skull are reported. The taxa with different karyotypes (shaded areas) are morphologically distinct. The populations (2-8) living in Sicily are also distinct from the pool of the small islands (9-13), in whose variability lies *C. canariensis*. The Pleistocene *C. esuae* is separate from its descendent *C. sicula*. The convergence between the populations of *C. sicula* and *C. russula*, revealed by mandible analysis, should be noted. 1 = *C. esuae*, Spinagallo; 2 - 8 = *C. sicula*, Sicily; 9 = *C. sicula*, Favignana; 10 = *C. sicula*, Gozo; 12 = *C. sicula*, Ustica; 13 = *C. canariensis*, Canaries; 14 = *C. leucodon*, Italy; 15 = *C. leucodon*, Poland; 16 = *C. russula*, Morocco; 17 = *C. russula*, Sardinia; 18 = *C. russula*, Tunisia; 19 = *C. suaveolens*, Italy; 20 = *C. whitakeri*, Morocco; and 21 = *C. whitakeri*, Tunisia; F1 = 65.4% and F2 = 15.8% of total explained variance.

The shrews studied can differ in having a more stout and powerful or a thinner and weaker mandible. These two extreme morphotypes, with a range of intermediate stages, can be recognized in different percentages among all four populations. These differences in shape can perhaps be better explained as the result of individual within-population and intraspecific variations, i.e., local adaptations to different trophic ecologies in environments such as sclerophyll woodland (Ficuzza), open grassland (Salemi) and islands (Marettimo and Canary).

The hypotheses tested and the results obtained from the landmark-based approach, even though done on only a small sample of 36-chromosome shrews and on an incomplete frame — i.e., without an outgroup comparison to scale differences — are congruent with the traditional morphometrics result (Sarà, 1995; Fig. 7). The results corroborate once more the strict phenetic relationships between the Sicilian and Canary shrews.

C. sicula and *C. canariensis* proved to be phylogenetically related, and their present disjunct distribution would arise from an ancestor distributed throughout the Maghreb during the Plio-Pleistocene. The subsequent speciation in modern forms (such as the both yet unkaryotyped *C. tarfayensis* or *whitakeri*) and extinction of this ancestor should have left two peripheral isolates (Mayr, 1963). The centrifugal speciation model (Frey, 1993) predicts, alternatively to others modes of speciation in which peripheral isolates are involved, that peripheral isolates remain relatively plesiomorphic while the large central population becomes highly derived and this is the principal source of evolutionary change.

The ecological adaptations to life on islands similar in size and climate (see Kunkel, 1976)—an "island effect "—should have resulted in steady selection on the gene pools and accounted for the small phenetic differences and Nei's genetic distances among these phylogenetically related isolates, even though the populations are separated in space and probably in time of colonization. Sondaar (1977), Case (1978) and Davis (1983) have made some studies in this regard.

The body of knowledge available so far (karyology, allozyme electrophoresis, genetic distance and geometric morphometrics) is considered sufficient to guarantee conspecificity between *C. sicula* and *C. canariensis*. The Canary shrew is thus taxonomically a subspecies of *C. sicula* (Sarà, 1995) according to the law of priority (ICZN).

ACKNOWLEDGMENTS

Special thanks to Marco Corti for useful comments on the manuscript and technical support in the analyses.

REFERENCES

Bookstein, F. L. 1991. Morphometric tools for landmark data, geometry and biology. Cambridge University Press: Cambridge.

Case, T. J. 1978. A general explanation for insular body size trends in terrestrial vertebrates. Ecology 59:1–18.

Davis, S. J. M. 1983. Morphometric variation of House mice *Mus domesticus* in Britain and Faeroe. Journal of Zoology 199:521–534.

Frey, J. K. 1993. Modes of peripheral isolate formation and speciation. Systematic Biology 42:373–381.

Hutterer, R., L. F. Lopez-Jurado, and P. Vogel 1987. The shrews of the Eastern Canary islands: a new species (Mammalia: Soricidae). Journal of Natural History 21:1347–1357.

Kunkel, G. 1976. Biogeography and ecology in the Canary islands. W. Junk: The Hague, The Netherlands.

Maddalena, T. 1990. Systematics and biogeography of Afrotropical and Palearctic shrews of the genus *Crocidura* (Insectivora: Soricidae): An electrophoretical approach. Pages 297–308 *in* G. Peters and R. Hutterer, (eds.), Vertebrates in the tropics. Museum A. Koenig: Bonn.

Mayr, E. 1963. Animal species and evolution. Belknap Press of Harvard University Press: Cambridge, Massachusetts.

Meacham, C. A. and T. Duncan 1990. MorphoSys. Version 1.29. University Herbarium, University of California: Berkeley.

Molina, O. M., and R. Hutterer. 1989. A cryptic new species of *Crocidura* from Gran Canaria and Tenerife, Canary islands (Mammalia: Soricidae). Bonn. zool. Beitr. 40:85–97.

Rohlf, F. J. 1990. Rotational fit (Procrustes) methods. Pages 227–236 *in* F. J. Rohlf and F. L. Bookstein, (eds.), Proceedings of the Michigan morphometric workshop. University of Michigan Museum of Zoology Publication 2.

Rohlf , F. J. 1993a. TPSRW—Thin-plate spline relative warp. Department of Ecology and Evolution, State University of New York, Stony Brook, NY.

Rohlf, F. J. 1993b. Relative Warp Analysis and an example of its application to Mosquito wings. Pages 131–159 *in* L. F. Marcus, E. Bello and A. García-Valdecasas (eds.), Contribution to Morphometrics. Monografias del Museo Nacional Ciencias Naturales 8, Madrid.

Rohlf, F. J. 1993c. NTSYS-pc - Numerical taxonomy and multivariate analysis system, version 1.80. Exeter: Setauket, New York.

Rohlf, F. J., and L. F. Marcus 1993. A revolution in morphometrics. Trends in Ecology and Evolution 8:129–132.

Sarà, M.1995. The Sicilian (*C. sicula*) and Canary shrew (*C. canariensis*), Mammalia: Soricidae. Peripheral isolate formation and geographic variation. Boll. Zool. 62: 173-182.

Sarà, M., M. Lo Valvo, and L. Zanca. 1990. Insular variation in central Mediterranean *Crocidura* Wagler, 1832 (Mammalia, Soricidae). Boll. Zool. 57:283–293.

Sondaar, P. J. 1977. Insularity and its effect on mammal evolution. Pages 671–707 *in* M. K. Hecht , P. C. Goody and B. M. Hecht, (eds.), Major pattern in vertebrate evolution. Plenum: New York.

Slice, D. E. 1992. GRF-ND Generalized rotational fitting of n-dimensional landmark data. Department of Ecology and Evolution, State University of New York, Stony Brook, New York.

Vogel, P. 1988. Taxonomical and biogeographical problems in Mediterranean shrews of the genus *Crocidura* (Mammalia, Insectivora) with reference to a new karyotype from Sicily (Italy). Bull. Soc. Vaud. Sc. nat. 79:37–48.

Vogel, P., R. Hutterer, and M. Sarà 1989. The correct name, species diagnosis and distribution of the Sicilian shrew. Bonn. zool. Beitr. 40:243–248.

Vogel, P. T., Maddalena, and F. Catzeflis 1986. A contribution to the taxonomy and ecology of shrews (*Crocidura zimmermanni* and *C. suaveolens*) from Crete and Turkey. Acta Theriologica 31:537–545.

THREE-DIMENSIONAL LANDMARK TECHNIQUES FOR THE RECOGNITION OF REEF CORAL SPECIES

Kenneth G. Johnson[1] and Ann F. Budd[2]

[1] Department of Geology and Applied Geology
University of Glasgow
Glasgow, G12 8QQ, United Kingdom
[2] Geology Department
University of Iowa
Iowa City, Iowa 52242

ABSTRACT

In the past, the taxonomy of reef corals has suffered from a lack of reliable quantitative techniques for defining and classifying species. In an attempt to develop a useful protocol for recognizing morphospecies, we used a set of 20 three-dimensional landmarks to test for morphological differences among populations of the common Caribbean reef coral *Porites* from Belize, Miami and St. Croix (U.S. Virgin Islands). Based on molecular techniques, tissue samples from the same collections have been shown to be distinct within collection sites. Generalized Least-Squares (GLS) analysis was used to align the sets of landmarks, and the principal coordinates of GLS residuals were examined to determine modes of variation among the landmarks. Group distinctiveness was examined using principal components analysis of mean residuals. We were able to distinguish four morphological groups that suggest a taxonomy corresponding to local field names. Collections of *P. porites* from Belize, Miami and St. Croix show considerable overlap, but collections of *P. divaricata* from Miami are morphologically distinct from corals assigned to *P. divaricata* in St. Croix. A collection *P. ?furcata* from St. Croix could be reliably distinguished. Analysis of the distribution of corallite size supports these results. Although both the morphological and molecular data suggest similar species groupings, neither yields unambiguous classification results at this point, and new measurement and analytical techniques are required to incorporate colony-level information into the quantitative definition of reef coral species.

Advances in Morphometrics, Edited by L. F. Marcus et al.
Plenum Press, New York, 1996

INTRODUCTION

The recognition of species is an important first step in answering many biological questions. During two centuries of debate, theoreticians have considered the utility and implications of a diversity of species concepts, but practicing biologists continue to work almost exclusively with species defined based on morphological characteristics (Stuessy, 1990). As discussed by Cheetham (1987) and Jackson and Cheetham (1990), discrimination of morphospecies via a strict sequence of multivariate statistical procedures can provide the nonarbitrary taxonomic framework essential to address ecological, evolutionary and management questions. These authors clearly show that analyses of isolated linear measurements or analyses confounding species and higher categories often yield misleading results.

Reef corals are exclusively marine invertebrates that are common in shallow tropical marine ecosystems. Most tropical corals are colonial and live as a series of more or less independent polyps that secrete a calcareous exoskeleton. This modular growth allows relatively small polyps to construct immense wave-resistant structures. Two processes are significant during the development of reef coral colonies: (1) the production of new modules through polyp budding leading to an increase in colony surface area, and (2) the accretion of new carbonate skeleton as polyps grow outward which increases the elevation of the colony above the sea floor. In most reef corals, individual modules (termed "corallites") show little intracolony morphological variability, but local environmental variation may lead to considerable ecophenotypic plasticity among conspecific colonies (Foster, 1984; 1985). Therefore, morphological variation in reef corals occurs on a series of hierarchical levels: within corallite, within colony, within local environment (population), and within species. In general, most species of colonial corals are distinguished by the skeletal architecture of the individual modules; genera tend to be distinguished by the arrangement and degree of integration of corallites within colonies. However, recent work on the common Caribbean genera *Montastraea* and *Siderastrea* has shown colony-level variation to be useful in species discrimination and classification (Budd, in press; Knowlton et al., 1992). Therefore, variation at all levels must be considered in species definition.

Porites is an important reef builder in both the Caribbean and Indo-Pacific regions and has been a major component of reef systems since it first appeared in the middle Eocene (Budd, Stemann, and Johnson, 1994). Approximately 25–35 *Porites* species are generally accepted in the Indo-Pacific and 4–8 in the Caribbean (Veron and Pichon, 1982; Potts et al., 1993), but here we consider only a subset of this diversity. Discrimination of *Porites* species based on conventional taxonomic methods has proven difficult in the past (e.g., Brakel, 1977), and recent work has focused on the development of biochemical techniques for recognizing taxa within the genus (Garthwaite et al., 1994; Weil, 1993). *Porites* is characterized by vertical and horizontal skeletal elements arranged in a bilaterally symmetric pattern (Bernard, 1905). During life, a thin film of living tissue lies across the surface of the skeleton, and these skeletal elements provide points for muscle attachment and support for digestive and reproductive tissue. An outer ring of 12 septa consists of a dorsal directive, 4 lateral fused pairs of septa and a ventral triplet (Fig. 1). Each septum is composed of 3–4 simple vertical rods (trabeculae) that are joined by horizontal elements (synapticulae), and the vertical rods usually project above the upper septal margins. These vertical projections are arranged in circles around a central columella; the inner circle is termed pali and the middle and outer circles are identified as septal and wall denticles. The upper surface of these vertical elements were used as landmarks in this study.

The approach applied here to the discrimination of coral morphospecies is to align groups of corallites by means of rotational fit techniques (Rohlf, 1990; Rohlf and Slice, 1990) and analyze residuals for taxonomic patterns.

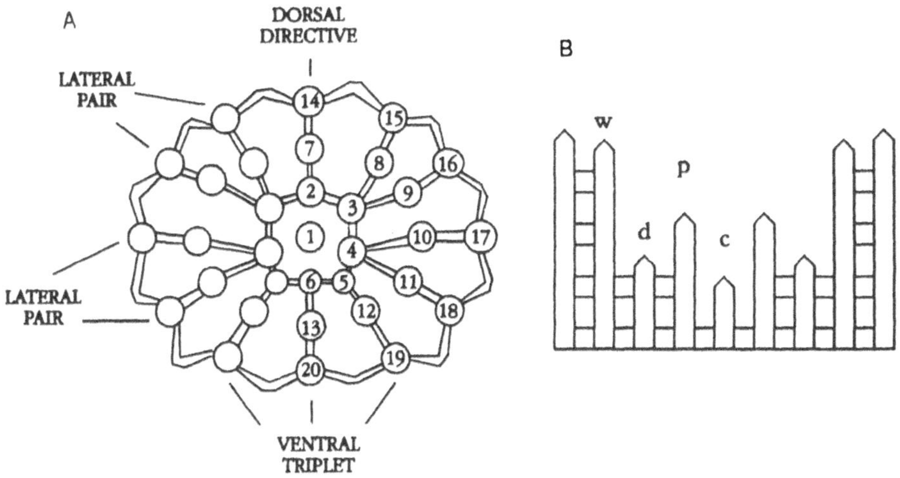

Figure 1. Arrangement of skeletal elements in a *Porites* corallite, showing distinctive bilateral symmetry. (A) transverse view with identification of the dorsal and ventral directions and the 20 landmarks were used in this study. (B) Longitudinal view illustrating how differences in elevation of vertical trabeculae results in surface structures; the columella (c) is in the center and is surrounded by rings of pali (p), septal denticles (d) and wall denticles (w).

MATERIALS AND METHODS

Coral skeletons were collected from shallow reef communities in Miami, St. Croix (U.S. Virgin Islands) and Belize as part of a larger study of *Porites* taxonomy based on both molecular and morphological data (Budd, Johnson, and Potts, 1994; Garthwaite et al., 1994). Colonies were assigned to collection groups on the basis of qualitative differences in colony form, color, polyp behavior and tentacle form, and subsequent analysis of allozymes from tissue samples has shown that, within each of the three geographic locations, the collection groups are electrophoretically distinct (Potts et al., 1993). In the present study, only populations of colonies with branching colony morphology were analyzed (Table 1). A Reflex microscope (Reflex Measurement Ltd., Somerset, England) was used to digitize the upper surfaces of 33 vertical elements in three dimensions on six widely separated corallites from 150 colonies.

Porites corallites are characterized by a distinctive bilateral symmetry imprinted upon the radial symmetry typical of cnidarians. This symmetry was used to split landmark data from each corallite into two halves (with the seven medial landmarks included in both halves), which

Table 1. Localities and sample size for the six collection groups used in this study. Local names assigned during collection

Region	Group	Site (depth)	Local name	Colonies (n)
Belize (Carrie Bow Cay) 1991	1	Reef Flat (< 3 m)	*porites*	29
Miami 1989	2	Soldier Key (LWM[a])	*divaricata*[b]	31
		Sands Key (< 25 cm)		
	3	Bache Shoal (1–3 m)	*?porites*	22
St. Croix 1989	6	Salt River Bay (< 50 cm)	*divaricata*	21
	7	Buck Is. forereef (< 6 m)	*?furcata*	20
	8	Buck Is. forereef (< 6 m)	*?porites*	27

a Low water mark
b Collections merged

resulted in 20 three-dimensional landmarks for each corallite half (Fig. 1). Consensus configurations for the "left" and "right" halves from the six measured corallites were calculated using the nonaffine Generalized Least-Squares rotational fit (GLS) algorithm (GRF-ND, version 11-01-94; Slice, 1993). Residuals from these within-colony analyses were examined to detect outlying and potentially influential observations, and then an overall consensus was derived from GLS analysis on the 300 within-colony consensus configurations. Following the guidelines of Bookstein ("Combining" article, this volume), we tested group differences by means of a multivariate analysis of variance on scores of all but the last seven principal components extracted from the GLS residual matrix. Centroid size (Bookstein, 1991) was also calculated for each corallite half and averaged over all corallite halves for each colony. The distribution of centroid size within and between groups was also examined to determine the utility of corallite size in distinguishing among the collection groups.

The GLS residuals were rotated into a shape space by calculating a set of 53 ($3n$-7) principal components using a singular value decomposition of unstandardized residuals (Marcus, 1993). This analysis is analogous to performing a relative warps analysis with alpha equal zero and testing group differences in this shape space by means of multivariate analysis of variance (Bookstein "Combining" article, this volume). To assess the distinctiveness of the collection groups, we performed a second principal components analysis on mean GLS residuals calculated for each group. All colony half residuals were then rotated into the space of the first two mean principal components and plots were examined for overlap.

RESULTS

Residuals from the least-squares within-colony fit (Fig. 2A) do not vary in magnitude among the landmarks. This is typical for the least-squares rotational fit procedure in which

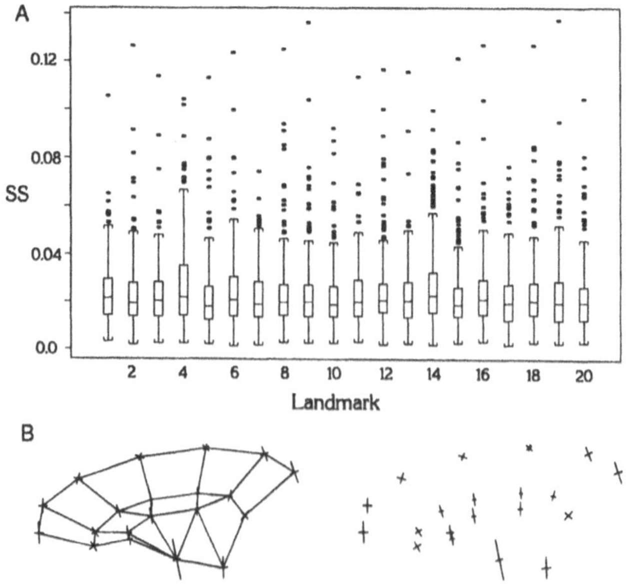

Figure 2. Results of a nonaffine Generalized Least-Squares analysis within and among colony halves. (A) Sum of squared residuals at each landmark for within-colony halves and (B) overall consensus configuration from the between colony-analysis with intralandmark principal components analysis axes calculated using GRF-ND (Slice, 1993).

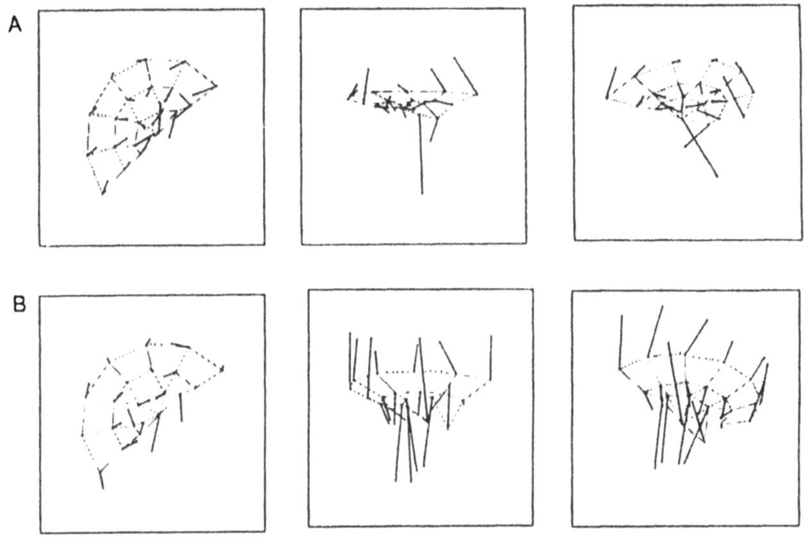

Figure 3. Views of the first two eigenvectors from a principal components analysis of GLS residuals from 300 colony halves. Three views of each eigenvector are shown, with loadings plotted as vectors superimposed on landmarks in the overall mean configuration. (A) Variation along the first axis is concentrated in changes in elevation of the columella and medial wall landmarks and radial variation of the palar ring and denticles. (B) Changes in the relative elevations of wall, denticles and pali are suggested along the second principal component. Note: the scales are adjusted to show all effects—the longest vector in A is in fact more than twice as long as the longest vector in part B.

residual variation will be spread evenly among all landmarks. Four corallites were identified with high residual sum of squares, and examination of the original data suggested error in digitizing, most likely due to movement of the specimen during use of the Reflex microscope. Intralandmark principal components analysis calculated on residuals from the all-colony analysis indicate high residual variability in a direction perpendicular to the living surface of the corallite (Fig. 2B), suggesting significant intergroup variation in that direction. The columella (landmark 1) was especially highly variable. Variation in landmarks associated with the corallite wall (landmarks 14–20) is inclined towards the corallite center, but variation in pali and septal denticles (landmarks 2–13) is inclined away from the center axis of the corallite, suggesting that change in radial spacing and the relative height of the wall are also important modes of variation among these corals.

Results of a principal components analysis of GLS residuals are useful for exploring modes of variation within the entire sample. Much of the shape variation is captured in the first few principal components, with over 50 % of total variation accounted for in the first three components. Examination of the first two eigenvectors suggests two major modes of landmark variation (Fig. 3). The orientation of these plots is not biologically determined, so interpretations of the *x*, *y* and *z* components of eigenvectors must be made with reference to an interpreted life position of the organism. In this case, *x* and *y* components indicate variation in the oral plane of the coral polyp, whereas variation in the *z* coordinate indicates differences in the relative elevation of the skeletal elements. Shape variation along the first axis is concentrated in the relative elevation of the columella (landmark 1), dorsal palus (landmark 2) and medial wall (landmarks 14 and 20). Variation in the remaining palar (landmarks 3 to 6) and denticle elements (landmarks 7 to 13) is concentrated in the oral plane of the corallite and corresponds to constriction and relaxation of these radial elements. Shape variation along

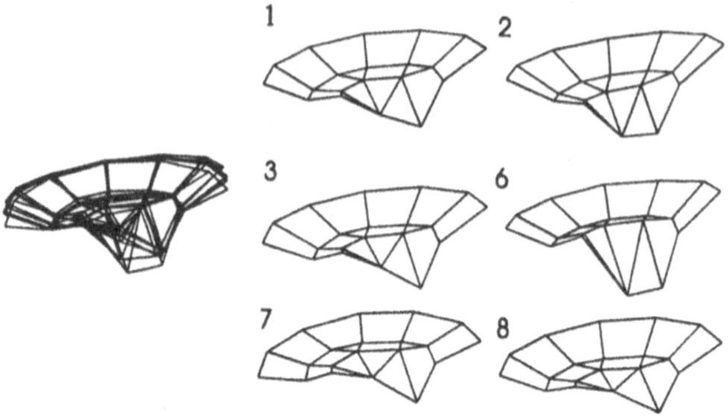

Figure 4. Mean configurations for each collection group for corallite halves aligned in an all-colony least-squares rotational fit.

the second axis mainly involves change in the relative elevation of the skeletal elements, with the elevations of the columella (landmark 1), dorsal palus (landmark 2) and wall elements (landmarks 14–20) inversely related to elevation of the pali and denticles. Multivariate analysis of variance of scores on the principal components axes suggests significant inter-group variation (Wilks' Lambda = 0.011; $F = 6.64$; $p < 0.001$; df 265, 1211).

Mean configurations for each group were calculated from the GLS results and show clear differences among the collection groups (Fig. 4). Little variation occurs in the horizontal plane and much of the variation is concentrated in the columella and palar landmarks. Groups 2 and 6 (both *P. divaricata*) are clearly characterized by a low columella (landmark 1) relative to the dorsal palus (landmark 2), with corallite depth distinguishing between groups 2 and 6. The other groups have tall columellae and low dorsal septa. Group 7 (*P. ?furcata* from St. Croix) has a relatively wide palar ring, with septal denticles developing closer to the corallite wall. Groups 1 and 8 (*P. porites* morphospecies from Belize and St. Croix) have similar palar configurations, but the relative heights of ventral pali, columella and dorsal palus vary to some extent.

To examine overlap among the collection groups, mean GLS residuals for each group were ordinated using principal components analysis. Two axes accounted for 94 % of the variation among the six mean configurations. Significantly, eigenvectors for the first two principal components are very similar to those calculated in the all-colony analysis (Fig. 5), suggesting that group differences are causing much of the overall shape variation in the collections. All colony half residuals were subsequently rotated into the plane defined by the

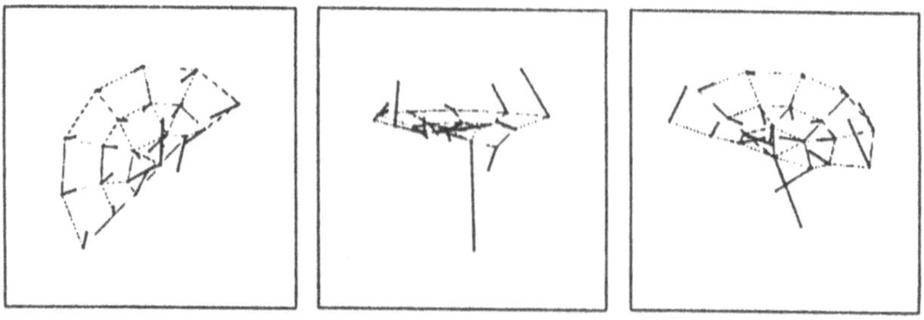

Figure 5. Views of the first three-dimensional eigenvectors for each landmark from a principal components analysis of group mean GLS residuals. The mode of variation is similar to the all-colony analysis (compare with Fig. 3).

first two axes of the group-mean principal components analysis (Fig. 6). Examination of the scatter of colony half scores in this plot suggests that three groups are distinguished along the first axis, with *Porites ?furcata* from St. Croix (group 7) different from *P. divaricata* from Miami (group 2) and *P. divaricata* collected in St. Croix (group 6). However, *P. divaricata* collected in St. Croix is morphologically distinct from *P. divaricata* collected in Miami along the second axis. In contrast, the three other collection groups could not be distinguished. Colonies of P. *porites* from Belize (group 1), *P. ?porites* from Miami (group 3) and *P. ?porites* from St. Croix (group 8) show nearly complete overlap in this space. There is also some overlap between colonies of *P. divaricata* from Miami and the P. *porites* group, possibly due to misidentification of some colonies of *P. divaricata*. These results suggest that four morphospecies are represented in the material: (1) P. *porites* from Belize, Miami and St. Croix, (2) *P. divaricata* from Miami, (3) *P. divaricata* from Miami and (4) *P. ?furcata* from St. Croix.

Analysis of variance of centroid size by group membership suggests significant variation among the groups ($F = 72.85$; df 5, 145). Unplanned comparisons of means based on Tukey's method (Sokal and Rohlf, 1981) suggests that groups 1 and 8 (*P. porites* from Belize and St. Croix) have significantly higher centroid size than the other groups, whereas groups 2 and 6 (*P. divaricata* from Miami and St. Croix) are significantly smaller (Fig. 7). Colonies from collection group 3 (*P. ?porites* from Miami) and group 7 (*P. ?furcata* from St. Croix) cannot be distinguished from one another on the basis of corallite size but are intermediate between groups with larger and smaller corallites.

DISCUSSION

Based on allozyme data, previous work has shown the collection groups to be distinct within each geographic location (Potts et al., 1993; Garthwaite and Potts, unpublished data), but collection group 3 (*P. ?porites* in Miami) could not be distinguished from group 7 (*P.*

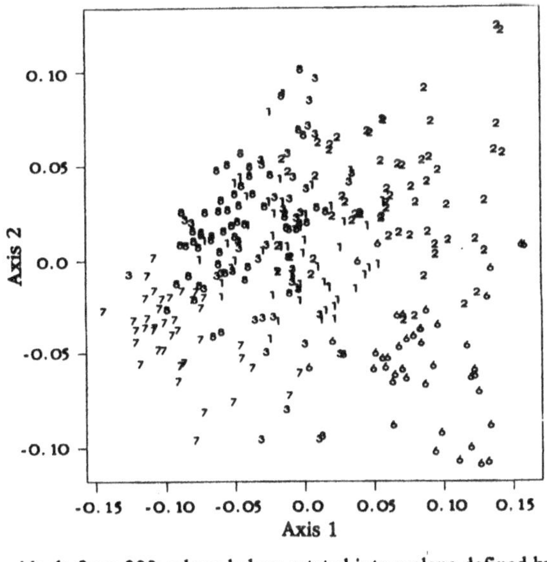

Figure 6. Plot of GLS residuals from 300 colony halves rotated into a plane defined by the first two axes of principal components analysis of group means. Groups 2, 6 and 7 are reasonably discrete, but groups 1, 3 and 8 show considerable overlap.

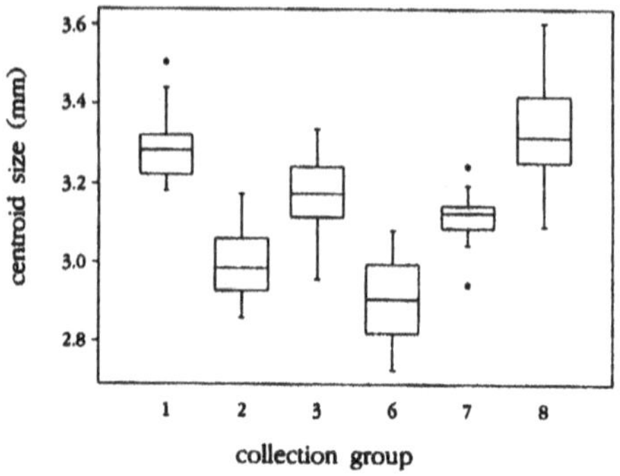

Figure 7. Distribution of root centroid size by collection group indicates that groups 1 and 8 have large corallites, whereas groups 2 and 6 have small corallites and groups 3 and 7 are intermediate in size.

?furcata in St. Croix). For the material considered here, the enzyme data demonstrate the existence of two valid species-level groups (*P. divaricata* and *P. porites–P. furcata*) separated by 5–55% of loci with fixed allelic differences. Collections within species groups differ by significantly different allele frequencies at only 5–11% of loci, but existing data are insufficient to determine whether these differences represent young, recently diverged species or merely variation among local populations within three species.

The landmark-based techniques are successful at distinguishing four morphs among the six collection groups. The proposed taxonomy corresponds to the informal field names currently in use but indicates significant geographic variation in *P. divaricata*. Such previously unrecognized variation has been documented in other Caribbean corals (Weil and Knowlton, 1994) and suggests that the biogeography of the Caribbean coral fauna might be more complex than is currently thought. These results also suggest that traditional field characters may not be reliable when used to identify species from different sites. It is reassuring to note that there is substantial agreement between molecular and morphological data in highlighting differences among coral species within regions. However, molecular and morphological data support conflicting interpretations when comparing collections from across the Caribbean. Although the allozyme and morphometric data are still not in full agreement, the high degree of correspondence between approaches is promising, especially because no information about overall colony morphology or the spatial arrangement of corallites within colonies has yet been included in the morphometric protocol. New measurement and analytic techniques are needed so that this information can be included in future quantitative species classification schemes.

ACKNOWLEDGMENTS

Early drafts were greatly improved with the assistance of L. Marcus, J. Rohlf, and other reviewers. D. C. Potts and R. L. Garthwaite provided collections and molecular analysis. This project was funded in part by a U.K. Natural Environment Research Council Advanced Postdoctoral Fellowship to K. G. Johnson and a U.S. National Science Foundation grant (BSR 9008225) to D. C. Potts.

REFERENCES

Bernard, H. M. 1905. The family Poritidae. II. The genus *Porites*. Part 1. Catalogue of the Madreporarian Corals of the British Museum (Natural History) 5: 1–303.

Bookstein, F. L. 1991. Morphometric tools for landmark data: Geometry and biology. Cambridge University Press: Cambridge.

Brakel, W. H. 1977. Corallite variation in *Porites* and the species problem in corals. Pages 457–462 *in* Proceedings of the Third International Coral Reef Symposium.

Budd, A. F. In press. Variation within and among morphospecies of *Montastraea*. Courier Forschungs-Institut Senckenberg.

Budd, A. F., K. G. Johnson, and D. C. Potts. 1994. Recognizing morphospecies in colonial reef corals: I. Landmark-based methods. Paleobiology 20:484–505.

Budd, A. F., T. A. Stemann, and K. G. Johnson. 1994. Stratigraphic distributions of Neogene to Recent Caribbean reef corals: A new compilation. Journal of Paleontology 68:951–977.

Cheetham, A. H. 1987. Tempo of evolution in a Neogene bryozoan: are trends in single morphologic characters misleading? Paleobiology 13:286–296.

Foster, A. B. 1984. The species concept in fossil hermatypic corals: a statistical approach. Palaeontographica Americana 54:58–69.

Foster, A. B. 1985. Variation within coral colonies and its importance for interpreting fossil species. Journal of Paleontology 59:1359–1383.

Garthwaite, R. L., D. C. Potts, J. E. N. Veron, and T. J. Done 1994. Electrophoretic identification of poritid species (Anthozoa: Scleractinia). Coral Reefs 13:49–56.

Jackson, J. B. C., and A. H. Cheetham. 1990. Evolutionary significance of morphospecies: A test with cheilostome Bryozoa. Science 248:579–583.

Knowlton, N., E. Weil, L. A. Weigt, and H. M. Guzman. 1992. Sibling species in *Montastraea annularis*, coral bleaching, and the coral climate record. Science. 255:330–333.

Marcus, L. F. 1993. Some aspects of multivariate statistics for morphometrics. Pages 95–130 *in* L. F. Marcus, E. Bello, and A. García-Valdecasas (eds.), Contributions to morphometrics. Monografías del Museo Nacional de Ciencias Naturales 8, Madrid.

Potts, D. C., A. F. Budd, and R. L. Garthwaite. 1993. Soft tissue vs. skeletal approaches to species recognition and phylogeny reconstruction in corals. Courier Forschungs-Institut Senckenberg 164:221–231.

Rohlf, F. J. 1990. Rotational fit (Procrustes) methods. Pages 227–236 *in* F. J. Rohlf and F. L. Bookstein (eds.), Proceedings of the Michigan morphometrics workshop. University of Michigan Museum of Zoology Special Publication 2.

Rohlf, F. J., and D. E. Slice. 1990. Extensions of the Procrustes method for the optimal superposition of landmarks. Systematic Zoology 39:40–59.

Slice, D. E. 1993b. GRF-ND: Generalized rotational fitting of N-dimensional data. Department of Ecology and Evolution. State University of New York at Stony Brook, Stony Brook, New York 11794.

Sokal, R. R., and F. J. Rohlf. 1981. Biometry: The principles and practice of statistics in biological research, 2nd. edition. W. H. Freeman: San Francisco.

Stuessy, T. F. 1990. Plant Taxonomy. Columbia University Press: New York.

Veron, J. E. N., and M. Pichon. 1982. Scleractinia of eastern Australia. Part IV. Family Poritidae. Australian Institute of Marine Science Monograph Series 5:1–159.

Weil, E. 1993. Genetic and morphological variation in Caribbean and eastern Pacific *Porites* (Anthozoa, Scleractinia). Preliminary results. Pages 643–656 *in* Proceedings of the Seventh International Coral Reef Symposium.

Weil, E. and N. Knowlton. 1994. A multi-character analysis of the Caribbean coral *Montastraea annularis* (Ellis and Solander, 1786) and its two sibling species, *M. faveolata* (Ellis and Solander, 1786) and *M. franksi* (Gregory, 1895). Bulletin of Marine Science 55:151–175.

PART II

Section 2. Low Taxonomic Level and Intraspecific Comparisons

INTRODUCTION TO COMPARISON OF SIMILAR OBJECTS WHOSE GROUP IDENTITY IS KNOWN

Marco Corti

Comparisons between well-known, unambiguously classified organisms is the subject of all but one article in this section of the volume. The topics discussed are human sexual dimorphism, in one article by Ahlstrom and another by Wood and Lynch; ontogeny of fish by Loy et al.; species differences between closely related mice by Demeter et al. and between closely related weasels by Reig; and some fossil comparisons by Reyment. These all represent known cases of morphological differences whose cause or correlates are well identified. The shape differences here are frequently less subtle then those discussed in the section on geographic and intraspecific systematics of the previous section. Also included in this section is an analysis of hand axes by Brande and Saragusti. Although hand axes are inorganic, they are the product of ancient human hands and show a considerable evolution.

There are many important points and question raised in these articles. One that has been around since the origin and early spread of geometric methods concerns the comparison of traditional morphometric characters collected as linear distances with two- or three-dimensional landmarks in the context of discrimination between groups of closely related organisms and the identification of unknowns The papers by Reig, Demeter et al., and Reyment suggest quite definitive answers. though with the caveat that ambiguity cannot be excluded a priori in the design of an experiment.

To start with, distances computed between two landmarks (either in two or three dimensions) do not necessarily correspond to the distance between the same points recorded with a caliper; there are errors in the instruments, whether caliper or digitizer, and there errors made by the person collecting data, both of which cannot be avoided. There is, however, better correspondence between distances collected from landmarks in three dimensions and caliper measurements than between caliper measurements and distances from landmarks in two dimensions. The authors all agree that there is nothing wrong with distance data, but it is impractical to construct simple models for a comparison of the results from one kind of data set to the other.

Nevertheless, there might be cases where distance data—especially certain kinds of distance data, such as those proposed by Festing (1972) and later used by others (for example, Corti and Thorpe, 1989)—are better for estimating differences than are landmarks. This problem is relevant when it is impossible to locate a landmark on a curved outline whereas distances are able to capture aspects of these particular features. Edgels are expected to provide a good addition to landmarks, but there has been practically no

Advances in Morphometrics, Edited by L. F. Marcus et al.
Plenum Press, New York, 1996

experimentation with these new constructs by morphometricians. These uses of measurement data do not imply that the major criticism of traditional morphometrics (sensu Marcus, 1990) is wrong. Implicit in distance data is an inadequate description of the direction of variation; landmarks may show high variability, but this is not recorded by the distance simply because both landmarks involved move in the same direction. This is elegantly discussed by Reig in this section.

Landmark-based morphometrics is accepted as the best tool to report differences in shape among organisms; differences are displayed graphically and reported statistically. Visualization is practical and readily accessible for two-dimensional landmark data. In their respective articles Reyment discusses shape and size variation in fossil and living species of Ostracoda and the evolution of asymmetry in the right and left shells and Loy et al. show how differences in the shape of the sea bass, *Dicentrarchus labrax*, occur during postlarval ontogenetic growth if one major ecological parameter, salinity, changes in the environment where these fish are raised. Major significant trends in shape changes are visualized as splines by Reyment, who uses relative warps (principal components of the weight matrix; Bookstein, 1992; Rohlf, 1993). Loy et al. regressed growth stages, as represented by centroid size, on partial warp scores.

Both the ostracod shells and teleost fish analyzed in these studies are sufficiently flat that the third dimension can be ignored a priori. The two-dimensional landmarks are not expected to induce an important distortion when shape changes are represented using two-dimensional graphics. Also see Reig for a discussion of this matter.

Three data sets discussed in this section pertain to objects that are exquisitely three dimensional. Two data sets involve human skulls and the third hand axes, and different solutions are adopted. Three-dimensional digitizing is still not available and affordable for many researchers, although a new variety of devices will be available by the time this book is printed (see Dean this volume). Different solutions have been adopted. Wood and Lynch used lateral midsagittal radiographs of skulls of African and Romano–British modern humans and found that shape coordinates are able to identify a consistent sexual dimorphism in the cranio-facial region. However, the discriminant function derived is not adequate to sex unknown individuals. Similarly, Brande and Saragusti found for hand axes that two-dimensional landmarks and their derived shape coordinates are more satisfactory than are the linear measurements traditionally used by archaeologists and anthropologists in their study. However, Brande and Saragusti invoke the need to add a third dimension to the analysis. These tools were used by human hands, and the third dimension should provide additional information. Ahlstrom presents the only paper in this section with an analysis of three-dimensional landmarks. He uses the thin-plate spline to investigates sexual dimorphism in the skull of a Swedish population from the Middle Ages. Statistical analysis is straightforward, as the thin-plate spline algorithms extend easily to three dimensions. However, the visualization of shape differences as deformations is difficult, especially in text figures. Ahlstrom uses displacement vectors for visualization of the total warp in his figures. Practically, he restricted his presentation to the sagittal plane. This turns out to be an acceptable solution because most of variation and differences occur in that plane. There will be much experimented work in the near future with acquisition and analysis of three-dimensional data.

One final point, I would suggest that the reader take heed of Reyment's examination of the most invariant principal component, which, in a geometric morphometric perspective, is the last (the tenth, in his case) relative warp. As a matter of fact, this quasi-invariant relative warp proved to be one of the most interesting shape indicators. There are too few morphometric studies that have reserved the same kind of attention to data.

REFERENCES

Bookstein, F. L. 1991. Morphometric tools for landmark data: Geometry and biology. Cambridge University Press: Cambridge.

Corti, M. and R. S. Thorpe. 1989. Morphological clines across a karyotypic zone of house mice in central Italy. Journal of Evolutionary Biology 2:253–264.

Festing, M. 1972. Mouse strain identification. Nature 238: 351–352.

Marcus, L. F. 1990. Traditional morphometrics. Pages 77–122 *in* F. J. Rohlf and F. L. Bookstein, (eds.), Proceedings of the Michigan morphometrics workshop. University of Michigan Museum of Zoology Special Publication 2.

Rohlf, F. J. 1993. Relative warp analysis and an example of its application to mosquito wings. Pages 131–159 *in* L. F. Marcus, E. Bello and A. García-Valdecasas, (eds.), Contribution to Morphometrics. Monografias del Museo Nacional de Ciencias Naturales 8, Madrid.

IDENTIFICATION OF HOUSE MICE, *Mus musculus*, AND MOUND-BUILDING MICE, *M. spicilegus*, BASED ON DISTANCE AND LANDMARK DATA

András Demeter,[1,2] Gábor Rácz,[2,3] and Gábor Csorba[2]

[1] Secretariat of the Hungarian Academy of Sciences
Budapest, Nádor u. 7
H-1051 Hungary
[2] Zoological Department
Hungarian Natural History Museum
Budapest, Baross u. 13
H-1088 Hungary
[3] Department of Genetics
Eötvös University, Budapest
Muzeum krt. 4/A
H-1088 Hungary

ABSTRACT

Festing's (1972) distance measurements and Bookstein's (1991) shape variables, based on 15 landmarks digitized from the left mandible were used as input for discriminant function analyses of house mice, *Mus musculus*, and mound-building mice, *M. spicilegus*, in Hungary. The distance variable resulted in better separation of the training sets than did the shape variables, but both methods performed equally poorly when allocating unknowns from owl pellets. The lack of discriminatory power is probably due to heterogeneity in size (and probably age) distribution among the unknown samples.

INTRODUCTION

There is disagreement as to whether four (Boursot et al., 1993; Marshall, in prep.) or five (Sage et al., 1993) species of house mice and their wild-living relatives (Muridae: *Mus* spp.) occur in Europe. The known ranges of distribution of these taxa have been repeatedly mapped (e.g., Bonhomme et al., 1984; Auffray et al., 1990). Hungary is depicted to be

Advances in Morphometrics, Edited by L. F. Marcus et al.
Plenum Press, New York, 1996

inhabited by the mound-building mouse, *Mus spicilegus*, and the eastern form of the house mouse, *M. musculus musculus.*

At least for the Carpathian Basin, this map is of a rough scale and was based on scanty records and anecdotal evidence. Therefore we decided to use the large collection of mammalian remains from owl pellets deposited in the Hungarian Natural History Museum in Budapest and the computerized database associated with this collection (Demeter et al., 1989) to elaborate in detail the ranges of the two species. The distribution of the two species had not been documented in Hungary because previously, the distinctness of the two had not been realized. Schmidt (1976) found "house mice" to be distributed all over the country as evidenced by owl-pellet analyses.

Orsini et al. (1983) reported a diagnostic character, the so-called zygomatic index (a ratio of two measurements on the zygomatic arch) to be useful for separating *M. spicilegus* from *M. musculus*. Rácz et al. (unpublished) found this index to be not as reliable as purported by its proponents, and in the owl-pellet collection only about half as many crania are preserved as are mandibles. Therefore we decided to proceed with a large-scale identification process based on shape differences in the mandible.

There are additional reasons for focusing our analyses on the mandible:

1. The rodent mandible provides an excellent object for studies of the evolution of morphological form because it is a complex morphological structure that has been studied extensively from embryological, developmental, morphometric and quantitative genetic points of view (Atchley et al. 1992). The masticatory apparatus is very important in mammals, and many evolutionary processes are clearly manifested in dental, mandibular and cranial characters.
2. Shape and size variation in the mandible has been successfully used to differentiate between different laboratory strains (Fleming, 1972), transgenic and control mice (Vogl et al., 1993), feral island populations (Davis, 1983; Schriven and Bauchau, 1993) and chromosomal races (Corti and Thorpe, 1989) of house mice, *M. m. domesticus*. Therefore we were hopeful that the mandible would prove to carry a promising subset of characters for identifying mice different at the species level.
3. The mouse mandible is a more or less two-dimensional (2D) object with little information in the third dimension, so the lateral view (2D simplification) of this bone accurately reflects the three-dimensional object.

In this paper we report a pilot study exploring the utility of traditional distance (cf. Marcus, 1990) versus landmark-based data (cf. Bookstein, 1991) for identification by discriminant function analysis of mandibles of the two species of mice in question. Previously, the senior author successfully used a simple system consisting of a digitizing tablet attached to a personal computer for computer-aided identification of three species of wood mice, *Apodemus spp.* (briefly described in Demeter and Topál, 1987) based on three mandible measurements found to be discriminating (Demeter and Lázár, 1984). Therefore it was a natural option to try a similar approach with house and mound-building mice.

MATERIALS

There were two groups of "knowns." *M. spicilegus* were collected by digging out mice from their characteristic mounds ($n = 38$). *M. musculus* were collected indoors ($n = 40$). The zygomatic index values of the training sets were nonoverlapping and thus confirmed the species identifications (Rácz et al., unpublished). The "unknowns" ($n = 137$) were mouse mandibles derived from owl pellets. One pooled sample ($n = 36$) was formed from five owl-pellet samples (numbers 489, 416, 613, 706 and 710 in the database) for which the presence of only *M. musculus* was detected using the zygomatic index (Rácz et al., unpublished). Another pooled sample ($n = 29$) of four lots (numbers 78, 93, 540 and 595) had only *M. spicilegus* according to the zygomatic index. Finally the rest of the unknowns ($n = 62$) were indicated to contain both species (numbers 395, 689 and 741). (Localities and other collection data are available from the authors upon request. All pellets are from the central part of the Carpathian Basin, an area of about 300 by 200 km.) The pellets were from both the barn owl, *Tyto alba,* and the long-eared owl, *Asio otus*, and are part of the collection on which Schmidt (1976) based his report of house mice distribution in Hungary.

METHODS

Images of left disjointed mandibles viewed from the inner, lingual side were captured and digitized using VDAS and software IMAGOES described by Demeter et al. (this volume). Auffray et al. (this volume) found fluctuating asymmetry to be negligible in *M. m. domesticus*, therefore using only the left mandible in this study was probably not confounded by the effect of asymmetry. Fifteen landmarks were collected from each mandible (Fig. 1): most of these are equivalent to those used by Corti (1992), Atchley et al. (1992), and Vogl et al. (1993). The 13 distance measurements of Festing (1972) were calculated from the captured landmark coordinates by rigid rotation of the coordinates to new axes with origin at landmark 13 and direction of the x-axis towards landmark 15. There are two differences from Festing's scheme: since the incisor is often missing in owl-pellet remains, character 3 was calculated as the distance from landmark 2 rather than from the tip of the incisor; and because the molars are often missing or they rock loose in the alveoli in owl-pellet remains, character 4 was read from landmark 4 at the alveolus rather than from the corner of the crown of M_1.

ANALYSES

Coordinate data were transformed to shape coordinates (Bookstein, 1991) relative to the baseline formed by landmarks 1 and 12 by means of the program XY written by

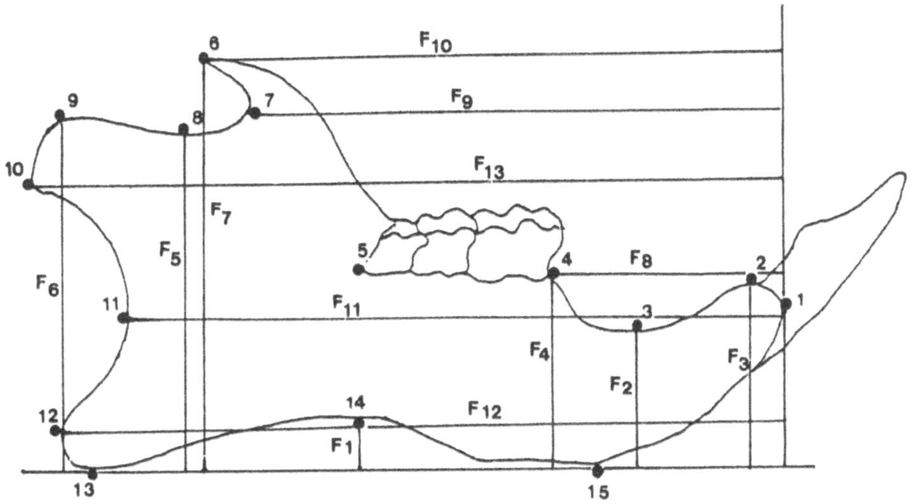

Figure 1. Landmarks and distance measurements collected from the left mandibles of the mice.

M. Baylac; estimates of centroid size and the affine components of shape differences were generated by the same program.

Some of the mandibles had one or two processes broken; missing values were estimated by regression. Statistical analyses were performed using SPSS/PC+ (Norusis, 1990) and SYSTAT (Wilkinson, 1990). During the discriminant function analyses, in addition to the resubstitution method for estimating the error rate of the allocation rules, leave-one-out cross-validation (Krzanowski, 1988) was also carried out. Unbiased estimates of Mahalanobis D^2, and F-tests were computed using the formulae given in Marcus (1993).

Digitizing error was assessed in the following manner. The employed software, IMAGOES, works in such a way that the landmark coordinates collected from an object are aligned to two internal system coordinates of the object (Demeter et al., this volume). In this case, the endpoints of the baseline along the longitudinal axis of the mandible also served as internal coordinates. Landmark 12 was the origin of the new coordinate system (0,0) and landmark 1 was at x,0 (x = distance from origin in microns). Redigitization at these two landmarks during the data collection phase gives us an estimate of the error of capturing the coordinates of these landmarks (by calculating the differences of the x and y coordinates of landmark 12 and the y coordinate of landmark 1 from zero).

RESULTS

Digitizing Error

We found low error in the data capture of the coordinates (all data in microns; landmark 12x: mean difference -5.62, standard deviation (sd) 23.93, range -60 to 64; landmark 12y: mean 9.59, sd 26.3, range -63 to 152; and landmark 1y: mean 0.53, sd 22.34, range -63 to 63). Given the size of the mouse mandibles studied (ca 10 mm in length), this error rate is negligible.

Specific Differences

Figure 2 shows the mean digitized coordinates of the landmarks of the two training sets aligned with an origin at landmark 12 and orientation towards landmark 1. It can be seen that although there are general differences, mainly the region of the coronoid process seems to be different in the two species of mice. This is indicated by the differences in the positions of landmarks 5–8.

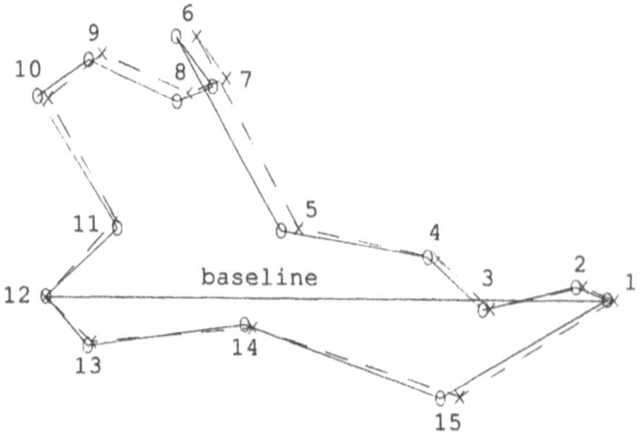

Figure 2. Comparison of mean landmark coordinates of the training sets (*Mus spicilegus* dashed line, *Mus musculus* continuous line) in an overlay plot. Landmarks are numbered 1–15 and are aligned in such a way that 12 is the origin and the horizontal axis of the coordinate system of the mandible lies towards 1.

Table 1. Basic statistics and univariate tests of Festing's distance variables

	Mus spicilegus		*Mus musculus*		
Variable	Mean Minimum	Standard Deviation Maximum	Mean Minimum	Standard Deviation Maximum	*t*
F1	0.76 0.545	0.08 0.957	0.65 0.479	0.10 0.892	5.51***
F2	1.64 1.432	0.12 2.012	1.59 1.396	0.10 1.780	1.72*
F3	2.27 2.039	0.13 2.611	2.21 1.971	0.15 2.540	1.83*
F4	2.40 2.076	0.17 2.738	2.32 1.765	0.19 2.778	1.82*
F5	4.40 3.894	0.27 5.100	4.50 3.797	0.25 4.945	-1.73*
F6	4.88 4.368	0.28 5.581	4.95 4.171	0.28 5.464	-1.04
F7	5.50 5.062	0.32 6.238	5.48 4.959	0.25 5.965	0.32
F8	2.84 2.508	0.16 3.219	2.77 2.485	0.15 3.091	-2.05**
F9	7.23 6.692	0.29 7.915	7.12 6.375	0.29 7.632	-1.75*
F10	8.03 7.271	0.37 8.766	7.78 6.967	0.39 8.746	-2.91***
F11	8.72 8.139	0.29 9.540	8.91 8.022	0.33 9.359	2.63***
F12	9.93 9.120	0.37 10.983	10.07 8.877	0.42 10.848	1.53
F13	10.51 9.738	0.38 11.437	10.44 9.555	0.36 11.012	-0.83

* $p < 0.1$; ** $p < 0.05$; *** $p < 0.01$

Table 2. Mahalanobis D^2, unbiased D_u^2 and F-tests of shape variables of the training sets

Landmark	D^2	D_u^2	F
2	0.0262	0.0000	0.357
3	0.1475	0.0390	2.711*
4	0.9805	0.8392	9.233***
5	1.9743	1.7937	19.966***
6	2.0164	1.8342	17.214***
7	1.8654	1.6891	17.202***
8	1.2477	1.0958	10.418***
9	1.2753	1.1223	11.425***
10	1.5708	1.4062	16.084***
11	0.4952	0.3730	4.331**
13	1.3412	1.1856	11.554***
14	0.2939	0.1797	2.739*
15	1.4135	1.2551	11.890***

* $p < 0.1$; ** $p < 0.05$; *** $p < 0.01$

Table 3. Basic statistics of specific differences in centroid size and affine components of shape difference

Variable	*Mus spicilegus* Mean (St. Dev.)	*Mus musculus* Mean (St. Dev.)	*t*
Centroid size	15595.67 (636.78)	15704.63 (634.07)	-0.76
x affine	-0.0252 (0.055)	0.0352 (0.049)	-5.12***
y affine	-0.0208 (0.056)	-0.0014 (0.0061)	-1.46

*** $p < 0.01$

Table 1 contains the basic statistics of the distance variables for the two training sets. Only three variables are highly significantly different between the training set (F1, F10 and F11) and two additional ones at the $p < 0.05$ level.

Table 2 is a summary of the specific differences in shape variables. To condense information, Mahalanobis D^2 and related statistics of the two species are tabulated instead of mean x and y shape variables. Mahalanobis D^2 between the means of the shape variables of the two training sets are significantly different in 9 out of 13 cases.

Table 3 contains the group statistics for centroid size and the affine components of variation. There is no significant group difference in centroid size or the y component, but the two training sets are significantly different as regards the x affine component. This group difference may be confirmed visually by inspecting Fig. 2. Mean positions of the landmarks on the coronoid and condylar processes are especially different along the x-axis.

Analyses of variance were carried out to test for differences in distribution of centroid size (a proxy for age) in the training sets and the samples of unknowns. It appears that the sample containing both species of mice in the owl pellets is different from the training sets and the other unknown samples (Table 4).

Discrimination and Allocation of Unknowns

The univariate analyses detailed above indicate important group differences. Based on the training sets, we should be able to allocate unknown individuals to either of the two species. A series of discriminant function analyses were carried out towards this goal. Because of the small sample sizes relative to the number of variables, the unbiased estimates of D^2 are important for evaluating the results (Table 5). With the given sample sizes of the training sets, it was not possible to have all landmark coordinates for the shape variables (26)

Table 4. Analyses of variance of centroid size among the training sets and unknown samples. 1 = *Mus spicilegus*, 2 = *M. musculus*, 3 = owl pellet (*M. spicilegus*), 4 = owl pellet (*M. musculus*), 5 = owl pellets (mixed species)

Sample	2	3	4	5
1	0.573	2.086	1.511	2.019*
2		0.180	0.823	2.379*
3			0.134	3.294**
4				3.398*

* $p < 0.1$, ** $p < 0.05$

Table 5. Mahalanobis D^2 and related statistics for discriminant function analyses of the training sets

Analysis	Number of variables	D^2	F	D_u^2	F
Distance measurements	13	10.679	13.481	8.045	10.155
Shape variables	12	4.449	6.180	3.073	4.268

included in the analysis. Therefore the number of shape variables was reduced to a subset found to be highly significantly different by univariate tests (a total of 12 x and y coordinates).

Based on Festing's distance measurements, the discriminant function analysis produced significant separation of the two known groups. The classification matrix (Table 6) shows a low apparent error rate of the allocation rule. Cross-validation produced a slightly greater rate of misclassification. Figure 3a shows the distributions of discriminant function scores of the training sets and the unknowns. The known groups are fairly well separated, but the unknowns are not. The lack of separation was not due to error introduced by the procedure of estimation of the missing values because the distribution of the scores was not improved when the specimens with missing but estimated data for the coronoid process were excluded.

Figure 3b shows the results of the discriminant function analysis of the Bookstein shape variables. The distribution of the discriminant function scores of the unknowns is nowhere near bimodality, both as regards the total sample and the subsample with the coronoid process intact. The classification matrix (Table 6) shows that the probability of misclassification is underestimated by the resubstitution method in comparison with cross-validation.

Both the distance and the landmark-based analyses detected the presence of the other mouse species in similar proportions in samples considered to contain only one of the species, as inferred from the zygomatic index (*Mus spicilegus* sample: 24–26 mound-building mice to 2–3 house mice; *Mus musculus* sample: 23–27 house mice to 3–10 mound-building mice). There was greater disparity in allocating individuals of the mixed sample, which is not

Table 6. Classification tables for the discriminant function analyses

	Mus spicilegus		*Mus musculus*	
Analysis	*n*	(%)	*n*	(%)
Distance measurements				
Resubstitution method				
Mus spicilegus	37	(97.4)	1	(2.6)
Mus musculus	4	(10.0)	36	(90.0)
Cross-validation				
Mus spicilegus	36	(94.7)	2	(5.3)
Mus musculus	5	(12.5)	35	(87.5)
Shape variables				
Resubstitution method				
Mus spicilegus	30	(78.9)	2	(21.1)
Mus musculus	5	(12.5)	35	(87.5)
Cross-validation				
Mus spicilegus	29	(76.3)	9	(23.7)
Mus musculus	8	(20.0)	32	(80.0)

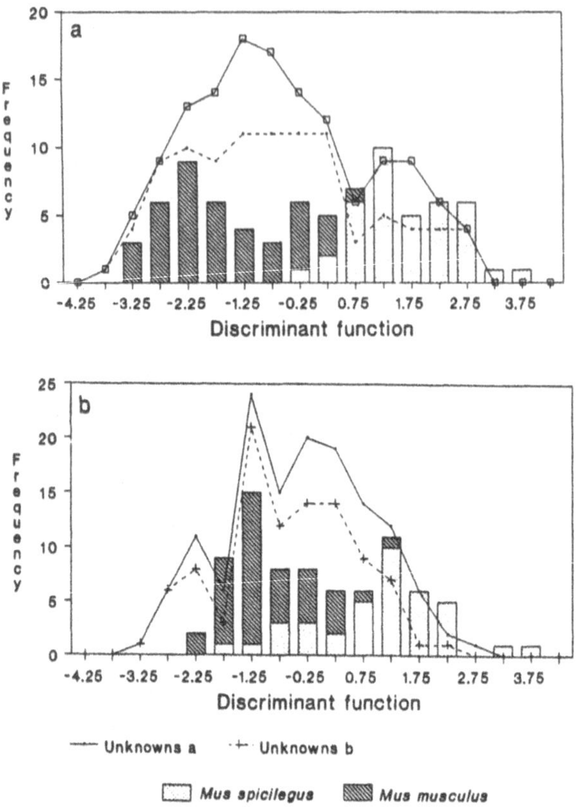

Figure 3. Histograms of the discriminant scores of the training sets and the unknowns of mice from owl pellets based on Festing's distance variables (a) and (b) Bookstein's shape variables. Unknowns *a* = all unknowns; Unknowns *b* = specimens with missing values for landmarks 5-8 excluded.

surprising given the outlying nature of this sample as regards size distribution (see Table 4). The distance measurements produced a ratio of one mound-building mouse to five house mice, whereas the proportion was approximately one to three with the shape variables.

DISCUSSION

Why did discriminant function analysis of the distance variables separate the training sets more successfully than did analysis of shape variables? Should we have experimented with other variants of discriminant function analysis used in classification and image analysis (Krzanowski, 1988, McLachlan, 1992) or mixture analysis (Flury et al. 1992)?

Bookstein (1991 and references therein) argues at length against morphometric inference based on distance data. Rohlf and Marcus (1993) concede that there is nothing "wrong" with traditional multivariate analysis, but the conceptual framework, the theoretical foundations and the graphic capabilities of geometric morphometric techniques certainly put the latter at an advantage.

Atchley et al. (1992) and Vogl et al. (1993) found that traditional multivariate techniques and finite element scaling analysis (FESA), a variant of geometric morphometrics criticized by Bookstein (1991) and Rohlf and Marcus (1993), yielded comparable results in their studies of rodent mandibles. In this study, we used morphometrics to differentiate morphologically very similar sibling species. After having digitized many mandibles, we had a feeling for the slight differences in curvatures of the outlines that might not have been effectively sampled by the set of landmarks employed.

Bookstein (1991) gives an example of the characterization of mandibular shape of the normal condition in children and in those with Apert syndrome or Crouzon syndrome. Distance measures of the medial axis resulted in a much better separation of the two groups than did landmark-based shape variables. Although the distance variables used in this study were derived from the same coordinates as the landmarks, Festing's (1972) scheme is somewhat different from the strictly speaking inter-landmark distances (see Fig. 1). A possible explanation for the superior performance of these point-to-point distances is that they may have sampled better the information in the curves of the outlines of the mouse mandibles.

Landmark-based analyses are more prone to problems of degrees of freedom than are the distance-based analyses. Bookstein (1991) points out this important problem and the need for large sample sizes. Our study is an example of the conjunction of the "biometric domain" and the "reality of the biological world." Not only was sampling constrained (essentially by our inability to locate and dig out more mound-building mice), but some of the unknowns behaved unexpectedly.

James and McCullogh (1990) cite the unsubstantiated rule of thumb of having three times as large sample sizes per group as the number of variables. Bookstein ("Combining" article, this volume) recommends sample sizes four times the number of landmarks. With the 15 landmarks sampled in our case, the number of variables in the discriminant function analyses would have been prohibitively large for the given sample sizes. Therefore we had to reduce them. We resorted to selecting a subset based on univariate tests because stepwise selection procedures have been condemned with strong words by James and McCullogh (1990) and Wilkinson (1990). Nevertheless, with the given samples sizes of the training sets, the number of selected shape variables and the number of distance variables hovered at the lower bound set by the empirical rule of thumb.

Analyses of both the distance variables and the shape variables have pointed out differences between the two mouse species in the shape of the coronoid process. Unfortunately, this process is the most vulnerable part of the mandible, and it is often broken. Several studies have shown that the coronoid process is a very important part of the mandible in rodents. Muscle attachments are found especially on the coronoid and the angular processes. The size and the shape of the coronoid process is dependent on the size and extent of the temporal muscle (Vogl et al. 1993). Atchley et al. (1992) tested four different hypotheses in an attempt to answer the question whether differences in the adult rodent mandible (of mice and rats) can be explained by changes in specific morphogenetic components. Three of the four hypotheses could not be ruled out, suggesting a complex and multidimensional differentiation in mandibular form involving isometric size scaling, muscle hypertrophy and intrinsic control of the developmental precursors (mesenchymal condensation).

Why were the unknowns classified equally poorly by the distance and the landmark data? House mice and mound-building mice have been shown to interbreed to produce fertile F_1 females in captivity and hybridization may play an important role in the evolution of mice (Boursot et al., 1993; Sage et al., 1993). Sexual differences are negligible, but age differences are appreciable in mandible size and shape in house mice (Davis, 1983). Owls are documented to predate differentially on young and subadult rodents (Mikkola, 1983), age classes that may not have been well represented in the training samples. It appears that the mixed sample is different from the training sets and the other unknown samples as regards distribution of centroid size. This heterogeneity may be responsible for the poor performance of both the traditional and the landmark-based methods for allocating the unknowns. It appears that the traditional and the geometric battery of morphometric techniques equally lose discriminating power when confounding effects such as heterogeneity in size (viz age) distribution is present.

ACKNOWLEDGMENTS

Funds for this project were provided in part by OTKA grant number 3177/91 to A. Demeter. We thank the participants of the NATO Advanced Study Institute for useful suggestions for improving the work presented here. Part of the writing of this contribution was carried out while A. D. was recipient of a Fulbright fellowship in the National Museum of Natural History, Smithsonian Institution, Washington, D.C., under the mentorship of Robert S. Hoffmann. Ralph Chapman of the Applied Morphometrics Laboratory of this museum is thanked for welcoming A. D. to facilities under his supervision. Leslie F. Marcus, Ralph Chapman and members of the editorial committee are thanked for making useful comments on earlier versions of the manuscript.

REFERENCES

Atchley, W. R., D. E. Cowley, C. Vogl, and T. McLellan. 1992. Evolutionary divergence, shape change, and genetic correlation structure in the rodent mandible. Systematic Biology 41:196–221.

Auffray, J.-C., F. Vanlerberghe and J. Britton-Davidian. 1990. The house mouse progression in Eurasia: A palaeontological and archaeozoological approach. Biological Journal of the Linnean Society 41:13–25.

Bonhomme, F., J. Catalan, J. Britton-Davidian, V. M. Chapman, K. Moriwaki, E. Nevo and L. Thaler. 1984. Biochemical diversity and evolution in the genus *Mus*. Biochemical Genetics 22:275–303.

Bookstein, F. L. 1991. Morphometric tools for landmark data: Geometry and biology. Cambridge University Press: Cambridge.

Boursot, P., J.-C. Auffray, J. Britton-Davidian and F. Bonhomme. 1993. The evolution of house mice. Annual Review of Ecology and Systematics 24:119–152.

Corti, M. 1992. Data analysis in systematics: A workshop and a manual to introduce geometric morphometrics. ICOLATER, Caracas, Venezuela (Document accessible through anonymous from life.bio.sunysb.edu—see Appendix III for instructions)

Corti, M. and R. S. Thorpe. 1989. Morphological clines across a karyotypic zone of house mice in central Italy. Journal of Evolutionary Biology 2:253–264.

Davis, S. J. M. 1983. Morphometric variation of populations of house mice *Mus domesticus* in Britain and Faroe. Journal of Zoology (London) 199:521–534.

Demeter, A. and P. Lázár. 1984. Morphometric analysis of field mice *Apodemus*: character selection for routine identification (Mammalia). Annales Historico-Naturales Musei Nationalis Hungarici 76:297–322.

Demeter, A., G. Lőrincz, Zs. Krajcsik and Cs. Moskát. 1989. Mapping mammal distribution with the Hungarian Natural History Museum's computerized database. Abstract. Fifth International Theriological Congress, Rome:900.

Demeter, A. and G. Topál. 1987. Mammals of the Kiskunság. Pages 446–452 *in* S. Mahunka, (ed.), The fauna of the Kiskunság National Park, Volume II. Akadémiai Kiadó: Budapest.

Festing, M. 1972. Mouse strain identification. Nature 238:351–352.

Flury, B. D., J.-P. Airoldi, and J.-P. Biber. 1992. Gender identification of water pipits (*Anthus spinoletta*) using mixtures of distribution. Journal of Theoretical Biology 158:465–480.

James, F. C. and C. E. McCullogh. 1990. Multivariate analysis in ecology and systematics: Panacea or Pandora's box? Annual Review of Ecology and Systematics 21:129–166.

Krzanowski, W. J. 1988. Principles of multivariate analysis. A user's perspective. Oxford University Press: Oxford.

Marcus, L. F. 1990. Traditional morphometrics. Pages 77–122 *in* F. J. Rohlf and F. L. Bookstein, (eds.),. Proceedings of the Michigan morphometrics workshop. University of Michigan Museum of Zoology Special Publication 2.

Marcus L. F. 1993. Some aspects of multivariate statistics for morphometrics. Pages 95–130 *in* L. F. Marcus, E. Bello and A. Garcia-Valdecasas, (eds.), Contributions to morphometrics. Monografias del Museo Nacional de Ciencias Naturales 8, Madrid.

Marshall, J. T. (in prep.) Taxonomy of European *Mus* (Mammalia: Muridae).

McLachlan, G. J. 1992. Discriminant analysis and statistical pattern recognition. Wiley: New York.

Mikkola, H. 1983. Owls of Europe. Buteo Books: Vermillion.

Norusis, M. J. 1990. SPSS/PC+ Advanced Statistics TM 4.0 for the IBM PC/XT/AT and PS/2. SPSS: Chicago.

Orsini, F., J. Bonhomme, J. Britton-Davidian, H. Croset, S. Gerasimov and L. Thaler. 1983. Le complexe d'espéces du genre *Mus* en Europe Centrale et Orientale. II. Critéres d'identification, répartition et caractéristiques écologiques. Zeitschrift für Säugetierkunde 48:86–95.

Rohlf, F. J., and L. F. Marcus. 1993. A revolution in morphometrics. Trends in Ecology and Evolution 8:129–132.

Sage, R. D., W. R. Atchley and E. Capanna. 1993. House mice as models in systematic biology. Systematic Biology 42:523–561.

Schmidt, E. 1976. Kleinsäugefaunistische Daten as Eulengewöllen in Ungarn. Aquila 82:119–144.

Schriven, P. N. and V. Bauchau. 1992. The effect of hybridization on mandible morphology in an island population of the house mouse. Journal of Zoology (London) 226:573–583.

Vogl, C., W. R. Atchley, D. E. Cowley, P. Crenshaw, J. D. Murray, and D. Pomp. 1993. The epigenetic influence of growth hormone on skeletal development. Growth, Development and Aging 57:163–182.

Wilkinson, L. 1990. SYSTAT: The system for statistics. Systat: Evanston, Illinois.

CORRESPONDENCE BETWEEN INTERLANDMARK DISTANCES AND CALIPER MEASUREMENTS

Santiago Reig*

Museo Nacional de Ciencias Naturales, CSIC
José Gutiérrez Abascal 2
28006 Madrid, Spain

ABSTRACT

Differences between distances measured with calipers and corresponding distances calculated from two-dimensional (2D) or three-dimensional (3D) coordinates were studied in a sample of 39 skulls of *Mustela erminea* and *M. frenata*. The caliper data set consisted of 12 length and width measurements. Coordinate data were obtained with a Polhemus 3D system and included 12 landmarks on the dorsal side of the skull and 10 on the ventral side. Distances between pairs of coordinate points were always shorter than corresponding distances taken with calipers; differences between data sets were highly significant in most cases. As expected, distances calculated from 3D landmarks were always closer to caliper values than were distances obtained from 2D points. Average root mean square error obtained from a regression of coordinate-based distances on caliper measurements was 0.422 mm (3.2% incongruence) for the whole set of 12 variables. The coefficient of variation was much larger for interlandmark distances than it was for corresponding caliper data. When differences between species were tested, analysis of variance (ANOVA) and multivariate analysis of variance (MANOVA) F values were larger for caliper than coordinate-based distances. Mahalanobis D^2 distances between the two species were smaller for coordinate data sets, and the coefficients of the canonical variate separating skulls of the two species were also different in caliper compared with 2D and 3D data sets. Analysis of displacement along the x-, y- and z- coordinates of each landmark confirm the existence of variation overlooked by caliper measurements. Thus, although traditional measurements can be used to describe differences between these two closely related species, the caliper traits provide a deficient, sometimes misleading, description of the variability. Finally, differences between coordinate and caliper data sets, and the precision of caliper measurement estimation from coordinates, varies from one character to another. This compromises the utility of simple models that are designed to exchange results from one data set to another.

* e-mail: MUSTELA@PINAR1.CSIC.ES

INTRODUCTION

Increasingly, two-dimensional (2D) or three-dimensional (3D) coordinates of landmarks are displacing caliper measurements as input data of choice for morphometric studies. However, caliper distances are still widely used in comparative morphology and systematics. Although using coordinates solely to compute interlandmark distances may not appear to represent the best use of data (given that more descriptive and powerful geometric methods are available for coordinate data), there remains an interest in relating caliper measurements to linear distances computed from coordinate data.

This paper explores the relationship between linear distances measured with calipers and corresponding distances calculated from 2D or 3D coordinates. Knowledge of this relationship would be necessary should one want to convert coordinate data into interlandmark distances so that the data could be contrasted, in an unbiased way, with results from studies based on caliper measurements. I also evaluate the limitations of linear measurements and assess the extent to which they are incomplete descriptions of the variability occurring at the landmarks associated with them. It is anticipated that this evaluation, in highlighting potential problems with caliper data, will help to improve the future design of studies employing traditional methods. This review on the use of landmark coordinates versus traditional measurements is extended briefly to triangulation schemes, another method based on distances between pairs of points. Triangulation schemes are suggested to be superior to standard caliper traits because a more even and geometrically reasonable coverage of the form is attained (Strauss and Bookstein, 1982; Zelditch et al., 1989).

The biological problem investigated was the difference between skulls of two closely related weasel species (Mammalia: Carnivora, Mustelidae) from North America, *Mustela frenata* and *M. erminea*. These two species are phenotypically very similar, and correct identification based only on skull morphology can be very difficult. The morphometric challenge is therefore to describe patterns of variation and find morphological differences that can be used to distinguish between the two species.

MATERIAL AND METHODS

This study is based on a sample of 19 *M. erminea* and 20 *M. frenata* skulls obtained from the U.S. National Museum of Natural History (USNMNH), Smithsonian Institution. Skulls were chosen to represent as homogeneous a sample as possible, excluding variability due to factors other than individual variation and differences between species. The skulls for each species were from male specimens belonging to a single subspecies, *M. erminea bangsi* and *M. frenata noveboracensis*, from the north-central United States. Geographic variation within the range of these subspecies should be insignificant (Hall, 1951). Only skulls from adults, more than one year old were selected. Age was determined using relative age criteria based on suture obliteration, date of capture, tooth wear and bone texture (Hall, 1951). Over one year of age, growth of weasel skulls can be ignored (King, 1991).

Caliper Data

Twelve caliper measurements were made on each skull (Fig. 1): bimolar width, condylobasal length, condylomolar length, ectoorbital width, interorbital width, mastoid width, maxillary toothrow length, postglenoid length, rostrum width, tympanic bulla length, tympanic bulla width and zygomatic width. This list includes most standard mammal skull traits and was chosen to achieve a complete coverage of the different regions of the skull

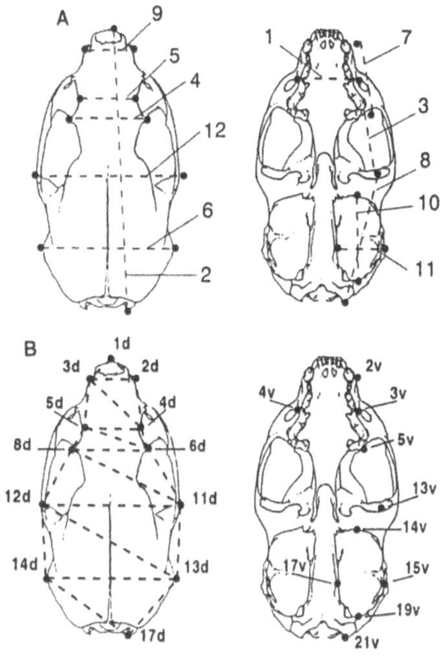

Figure 1. Dorsal and ventral view of a weasel skull showing the two sets of data used. (A) Set of 12 caliper measurements: (1) bimolar width; (2) condylobasal length; (3) condylomolar length; (4) ectoorbital width; (5) interorbital width; (6) mastoid width; (7) maxillary toothrow length; (8) postglenoid length; (9) rostrum width; (10) tympanic bulla length; (11) tympanic bulla width; and (12) zygomatic width. (B) Location of the 12 dorsal and 10 ventral landmarks digitized. This set of landmarks corresponds to the endpoints of the set of caliper measurements. Dashed line shows the triangulation scheme of 21 distances estimated from *x*, *y* and *z* dorsal coordinate points.

along its length and width and to record the variability of skull features known to be important for classification of skulls into species (Hall, 1951). Measurements were recorded to the nearest 0.01 mm by means of a digital caliper connected to a portable computer that used the automatic data capturing software, EASYCAL (Marcus and Reig, 1991).

An estimate of the precision of caliper measurements was made by taking two repeated measurements on a series of 16 *M. nivalis* specimens. Because skulls of *M. nivalis* are smaller than those of *M. erminea* and *M. frenata*, root mean square errors (RMS) were adjusted for size difference by multiplying each RMS by the ratio between means of *M. nivalis* and those of the other two species for each character. This represents a conservative transformation because measurement error should be smaller for larger skulls. Average RMS between replicates of the 12 caliper measurements was 0.072 mm. Two characters showed large RMS: tympanic bulla width (0.228 mm) and upper tooth row length (0.148 mm). The rest of the characters showed small values, ranging from 0.119 mm (bimolar width) to 0.012 mm (interorbital width) (Fig. 3). Other morphometric studies using calipers on mammals of this size report RMS values ranging from 0.034 to 0.011 mm (Voss et al., 1990).

Coordinate Data

The 3D coordinate data set consisted of 12 landmarks digitized on the dorsal side of the skull and 10 on the ventral side (Fig. 1) of the same skulls. Landmarks were selected to reproduce the set of points used to define caliper measurements. Most of the landmarks selected represent extrema that maximize lengths or widths and correspond to type 3 landmarks in Bookstein's (1991) classification. As such, their homology across specimens may be uncertain. However, the problem of landmark homology is less critical within a series of specimens of the same species and between closely related species because of the close morphological resemblance. Landmarks were digitized by means of a 3Space digitizer (Polhemus, Inc.) available at the USNMNH. This system has a sensor in a stylus point that is placed directly on the landmark. The system records *x*-, *y*- and *z*- coordinates and stores them in a file. Two sets of interlandmark distances, corresponding to the caliper measurements were computed from the 3D coordinates, the first based on only *x*- and *y*- coordinates, and the second based on *x*-, *y*-, and *z*- coordinates.

Digitizing precision was estimated by treating the same distances calculated from both ventral and dorsal views as replicates. Precision was worse than that recorded for calipers. Based on the whole set of 39 specimens, RMS between dorsal and ventral condylobasal length was 0.114 mm and was 0.121 mm in zygomatic width.

Triangulation Scheme

A scheme of triangles was constructed to contrast species discrimination based on caliper measurements with methods that use distances calculated from coordinate data (Zelditch et al., 1989; Dos Reis, 1990). Triangulation shares most of the attributes of the box truss for morphometric analysis of size and shape change (Strauss and Bookstein, 1982).

The triangulation scheme constituted 21 distances for the dorsal view of the skull, based on the same 12 landmarks that define caliper measurements taken on the dorsal side (Fig. 1). The triangulation scheme has the advantage of achieving a more comprehensive coverage of the skull morphology.

Direction of Change and Species Differentiation

A main shortcoming of caliper measurements compared with coordinate data is the incomplete description of the direction of variation (Bookstein et al., 1985). Calipers can only register variation along an axis from one extreme point to the other. The degree to which each caliper measurement provides an incomplete description of the variability occurring in that region of the skull was evaluated by an analysis of the displacements at each landmark that defined caliper characters. A multivariate test of simultaneous landmark variation along the *x*-, *y*- and *z*- axes based on Hotelling's T^2 was combined with separate analyses of variance (ANOVAs) of each coordinate for each landmark to test for differences along the *x*-, *y*- and *z*- axis independently. Significance of these tests was Bonferroni adjusted. Minimum significant probability was adjusted to a p of 0.002 (0.05 / 22).

Analysis of absolute landmark differences in the series of skulls required an optimal superimposition of the coordinates of each specimen in 3D space. Alignment was achieved using the GRF-ND program (Slice, 1993) by a least-squares fit of all configurations to an arbitrary specimen used as reference (Rohlf and Slice, 1990). Dorsal and ventral sets of landmarks were aligned independently of one another because they were recorded separately. Because the goal was to analyze absolute landmark displacement, the size of each specimen's configuration was restored after alignment. This alignment, translation and rotation was the only transformation applied to coordinate data.

Means and coefficients of variation were used to summarize variability patterns and ANOVA, multivariate analysis of variance (MANOVA) and Hotelling's T^2 to test the significance of this variability associated with species differentiation. The efficiency of using coordinate data to estimate caliper distances was assessed by regressing the coordinate data on the caliper mean, with the intercept set to zero.

In the coordinate data set a few specimens had some missing values; these were estimated by stepwise regression of the remaining variables available in each case. Significance for the regression procedures was always very high. Computations were made using the BMDP-7 (Dixon, 1992) and SAS 6.03 (SAS, 1989) packages.

RESULTS

Comparison of Coordinate and Caliper Distance Means

Distances between pairs of coordinate points were always shorter than equivalent distances taken with calipers. Differences between means of the same skulls were highly significant in most cases. Distances calculated from 3D landmarks were always closer to caliper values than distances obtained from 2D points (Fig. 2). This is to be expected because some information is lost by projection of 3D to 2D. Width measurements defined as maximum distances at certain regions of the skull also showed larger differences than did

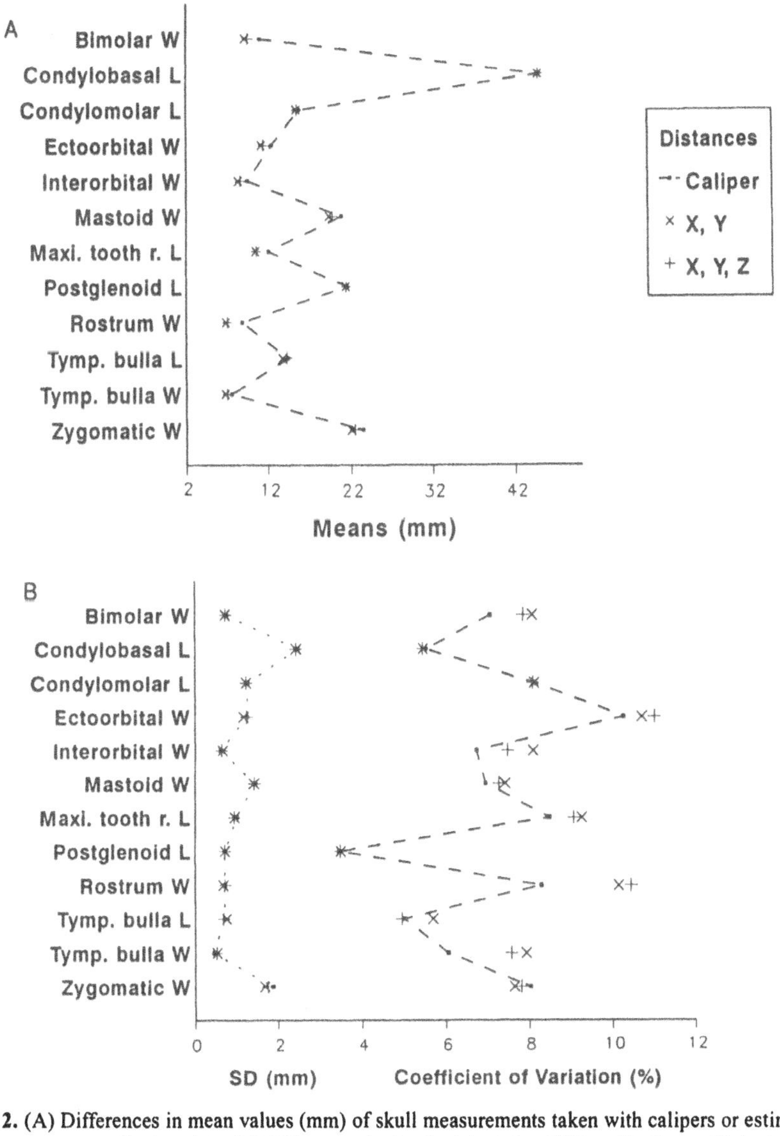

Figure 2. (A) Differences in mean values (mm) of skull measurements taken with calipers or estimated from *x-*, *y-*, *z-* or *x-*, *y-* coordinate points; (B) differences in standard deviation (SD) and in the coefficient of variation between the same data sets.

Table 1. Mean and coefficient of variation (CV) of 12 caliper measurements for *M. erminea* and *M. frenata*. Slope and error terms (in mm) of a regression through zero model to estimate caliper measurements from their corresponding interandmark distances are also given

	Caliper measurements				Estimation from coordinates			
	Mean		CV		X-Y		X-Y-Z	
	M. e.	*M. f.*	*M. e.*	*M. f.*	Slope	Error	Slope	Error
Bimolar W	10.41	11.57	5.64	3.82	1.189	0.472	1.146	0.356
Condylobasal L	42.81	46.61	4.17	2.90	1.002	0.249	1.001	0.251
Condylomolar L	14.73	16.77	5.61	4.30	1.022	0.226	1.016	0.226
Ectoorbital W	11.35	13.46	6.30	5.32	1.110	0.371	1.085	0.279
Interorbital W	9.25	9.84	6.20	5.90	1.133	0.348	1.100	0.243
Mastoid W	19.62	22.04	4.49	3.16	1.078	0.477	1.055	0.205
Maxi. tooth r. L	11.18	12.93	5.13	3.71	1.144	0.348	1.137	0.332
Postglenoid L	21.18	21.91	3.10	3.01	1.009	0.313	1.008	0.311
Rostrum W	8.34	9.43	7.22	3.89	1.282	0.396	1.244	0.391
Tymp. bulla L	14.05	14.84	4.49	4.01	1.048	0.295	1.010	0.292
Tymp. bulla W	7.46	7.73	6.35	5.43	1.111	0.502	1.072	0.496
Zygomatic W	21.92	24.86	5.75	4.28	1.063	0.550	1.044	0.281

length measurements. There was no apparent association between scale and size of the difference (Fig. 2).

Measurements defined as maximum or minimum distances between points showed the largest differences (Fig. 2). The most divergent variables, rostrum width and bimolar width, showed means 12 to 19% shorter for coordinate data than caliper data. Length measurements, which are not strictly defined as maximum distances between points but rather as well-delineated distances between precise points, showed less differences between coordinate and caliper data. Maxillary tooth row length, one of the well-defined group of skull traits, shows values 12% shorter for coordinate than for caliper data. It was later discovered that this difference arises from landmark 2v, which represents a different point than that used with the caliper measurement of tooth row length.

Means of the tympanic bulla width for each species show a distinct size pattern in caliper and coordinate data sets. Based on coordinate data, mean width of the tympanic bulla was larger in *M. erminea* than in *M. frenata*, whereas the opposite pattern holds for the caliper data (Table 1), suggesting that the points used were different between the caliper and the digitizer. The variability shown by this character also differs among the two data sets (Fig. 2).

Comparison of Coefficient of Variation Values

Differences in variability among data sets were larger than the differences among means. Variability of interlandmark distances was much larger than caliper data in terms of the coefficient of variation (CV). Only zygomatic width showed more variability in caliper than in coordinate data (Fig. 2). The same distinction between well-defined length measurements versus characters associated with maxima, such as most widths, was seen for differences in variability. Length measurements show similar variability in coordinate and caliper values. Largest differences are shown by rostrum width and tympanic bulla width, which had CVs for coordinate data over 25% larger than for caliper values. As with mean values, size of the trait or of CV seems to be independent of the magnitude of the differences between data sets (Fig. 2).

Estimation of Caliper Measurements from Coordinate Data

Slopes of regression through zero models were larger than one in all variables (Table 1), supporting the observation that caliper measurements have larger values (Fig. 2). Largest slopes of this regression were those for rostrum width and bimolar width, whereas condylobasal length and postglenoid length showed slopes close to one in the 2D and 3D data set models (Table 1). Errors in the estimation of caliper distances from coordinate data were independent of mean size and of absolute differences between data sets. The largest error term in the regression model was for width of the tympanic bulla and zygomatic and rostrum widths (Table 1). Rostrum width shows both a large slope and a large estimation error, suggesting little correspondence between data sets. As expected, using the 3D data set improves the estimation: slopes approach one and error terms were smaller using 3D data (Fig. 3; Table 1). This suggests that caliper measurements do not follow the main axis of the

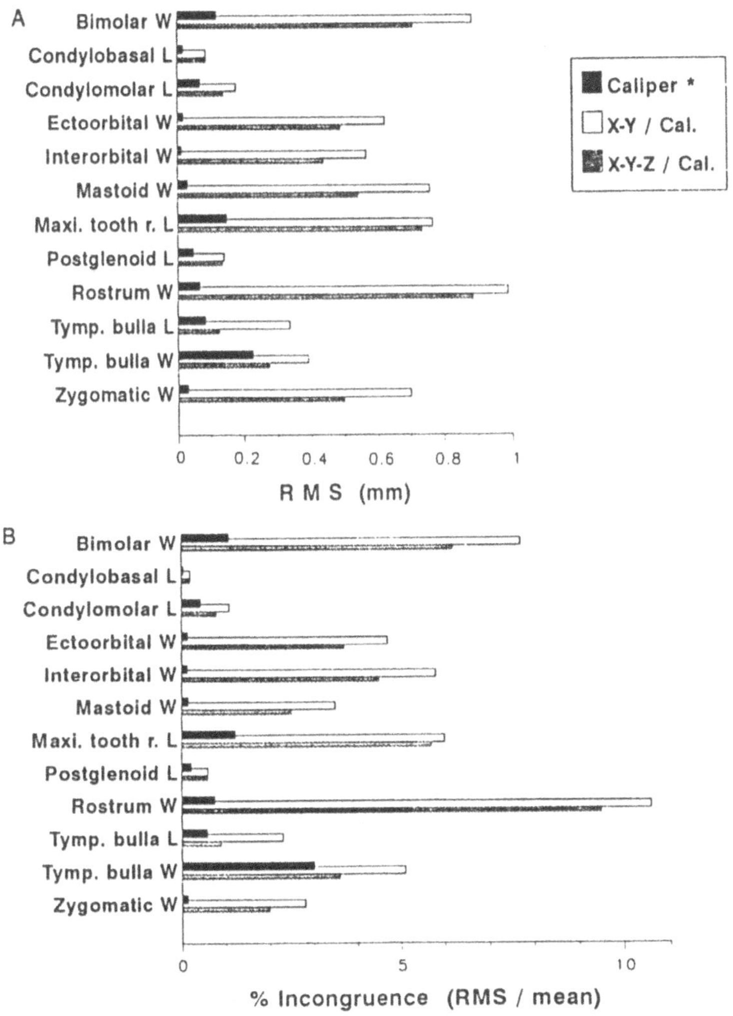

Figure 3. (A) Root mean square error (RMS) differences between skull measurements taken with calipers versus distances estimated from *x*-, *y*-, *z*- or *x*-, *y*- coordinate points. Caliper RMS indicates measurement error using calipers (see text for details); (B) percent incongruence of the RMS relative to the means of each variable.

skull (dorsoventral, sagittal) but rather run in diagonal directions to this axis. This also explains why the 2D projection represents a deficient approximation. The most extreme example of this observation is shown by zygomatic and mastoid widths, for which the error term is reduced by half when going from 2D to 3D (Fig. 3). By contrast, condylobasal length and postglenoid length show little difference between 2D and 3D data sets (Fig. 3).

The relative significance of RMS values was estimated by computing percent incongruence (100 % RMS / mean; Hildebolt and Vannier, 1988). Incongruence values ranged from 0 to 10.6% (Fig. 3). In terms of percent incongruence, the error of zygomatic and mastoid widths becomes less important. The minimum incongruence is shown by condylobasal and postglenoid lengths (Fig. 3). Compared with caliper RMS, error of the estimation was always large, except for tympanic bulla width (Fig. 3).

Morphometric Differentiation Between Species

On average, *M. frenata* is about 10% larger than *M. erminea*. Largest differences in caliper measurements of single traits were shown by ectoorbital width (17%) and length of the maxillary tooth row (14.5%) (Table 1). A similar pattern of size differences appears in 2D and 3D data sets except for the width of the tympanic bulla which showed no significant difference between species. Based on ANOVA most caliper measurements show highly significant differences between species; only upper tooth row length shows no overlap between the ranges of *M. erminea* and *M. frenata*. Multivariate analysis of variance confirms these differences (Table 2).

Statistical significance of species separation was lowest based on coordinate data. The ANOVAs performed on each data set showed larger F values for caliper than for interlandmark distances, and the same result extends to MANOVA (Table 2). Species discrimination based on Mahalanobis D^2 showed lower values for coordinate data sets: D^2 based on caliper data was 56.3, whereas in 2D and 3D data sets, it was 48.9 and 48.5,

Table 2. Results of an ANOVA test of differences between species for each measurement based on the three data sets. All F-values were highly significant ($p < 0.001$) unless otherwise stated. Multivariate ANOVA and standardized coefficients of the canonical variate (CV I) describing interspecies differences and Mahalanobis D^2 distances are also given

	F df (1,37)			CV I		
	Caliper	*x-y*	*x-y-z*	Caliper	*x-y*	*x-y-z*
Bimolar W	48.8	22.4	39.8	-0.790	-0.211	-0.111
Condylobasal L	56.4	64.8	64.6	0.462	3.293	3.610
Condylomolar L	67.0	59.4	64.5	-1.519	-1.511	-1.809
Ectoorbital W	84.1	73.4	69.8	1.416	1.683	1.806
Interorbital W	10.1**	4.5*	5.0*	-1.150	-1.396	-1.258
Mastoid W	90.7	59.7	70.0	2.856	0.936	1.266
Maxi. tooth r. L	106.3	95.5	87.5	2.082	1.219	0.763
Postglenoid L	11.9**	9.6**	9.6**	-0.912	-0.720	-0.880
Rostrum W	47.0	21.5	21.6	-1.400	-1.055	-1.145
Tymp. bulla L	16.0	19.1	10.3**	1.647	-0.145	-0.234
Tymp. bulla W	ns	ns	ns	-1.888	-0.850	-0.884
Zygomatic W	62.4	48.4	51.5	1.957	0.727	0.762
MANOVA df(12,26)	32.1	27.9	27.7			
Mahalanobis D^2	56.3	48.9	48.5			

(* $p < 0.5$; ** $p < 0.01$; ns, not significant).

respectively. The canonical variate extracted from the between-group covariance matrix was quite different for each data set. Mastoid width and upper tooth row length had the largest standardized coefficients for the caliper data set and also showed the most significant differences between species in the ANOVA (Table 2). By contrast, condylobasal length, ectoorbital width and condylomolar length showed the largest standardized coefficients when the 2D and 3D coordinate data sets were used (Table 2).

Triangulation Scheme

Most of the 21 distances included in the triangulation scheme showed highly significant differences between species. Mahalanobis D^2 obtained from the triangulation scheme and D^2 obtained from a set of the six dorsal caliper measurements defined for the same set of 12 landmarks were contrasted to assess the relative performance of the triangulation scheme versus standard caliper measurements. This set of six measurements shares five distances with the triangulation scheme, bimolar width, ectoorbital width, interorbital width, mastoid width, rostrum width and zygomatic width, and also one length variable expressing total skull length, condylobasal length. Coordinate data (x, y, z) were used to estimate these 6 caliper distances. Mahalanobis D^2 between species based on the entire triangulation scheme was 45.1, whereas based on the six dorsal caliper measurements, D^2 was 20.0.

Table 3. Results of univariate (ANOVA) and multivariate (Hotelling's T^2) tests of differences between species at each landmark. The T^2 values indicate the significance of landmark displacement simultaneously along the x-, y- and z- axis, whereas F-values indicate the significance of landmark displacement separately along each x-, y- and z- axis. Minimum significant probability was Bonferroni adjusted to p of 0.002 (0.05 / 22). All F (df 1, 37) and T^2 (df 3, 35) values were significant ($p < 0.002$) unless otherwise stated (ns, not significant). See Fig. 4 for location of landmarks

		F		
Landmark	T^2	x	y	z
1d	22.3	67.6	ns	31.7
2d	20.2	61.5	23.4	41.4
3d	29.7	77.8	12.5	37.7
4d	27.9	ns	ns	88.1
5d	17.1	20.8	ns	44.0
6d	17.4	ns	54.2	ns
8d	22.7	ns	70.1	ns
11d	18.6	ns	49.3	ns
12d	15.4	ns	47.5	ns
13d	23.0	25.8	70.9	ns
14d	20.7	22.4	59.5	ns
17d	20.8	49.3	18.4	28.6
2v	24.8	63.8	18.6	11.7
3v	15.0	26.9	24.3	ns
4v	18.8	33.8	37.5	25.2
5v	18.5	11.1	52.2	ns
13v	30.6	83.4	ns	14.7
14v	20.9	ns	30.0	35.9
15v	7.7	10.4	ns	ns
17v	10.8	ns	15.1	13.6
19v	18.3	22.8	18.4	ns
21v	16.3	51.1	ns	ns

Direction of Change and Species Differentiation

Hotelling's T^2 test revealed significant displacements between species in all the landmarks when the *x*-, *y*- and *z*- coordinates were considered simultaneously (Table 3). Results of separate ANOVAs for the *x*-, *y*- and *z*- coordinates of each landmark confirm the existence of variation overlooked by caliper measurements. For instance, variation along the dorso ventral axis cannot be detected from the caliper data because there was no measurement following that axis. However, significant displacements of the *z*-coordinate between species were recorded in most landmarks (Fig. 4; Table 3). This variation along the *z*-axis explains the better fit to caliper measurements of the 3D data over the 2D data, and indicates a loss of information in 2D data that should be considered. Note that these analyses concern only the variability of landmark displacement associated with species differences. This does not necessarily reflect variability for that landmark within species.

There are some clear examples of deficient caliper measurements. Analysis of 2D and 3D landmarks suggests that the variation observed in width of the rostrum is in fact produced by an anteroposterior displacement of the canines rather than from the broadening of the rostrum (Fig. 4; Table 3). The low significance of differences in interorbital width are explained by the lack of displacement along the *y*-axis at landmarks 4d and 5d (Fig. 4). However, there is significant variability of these two landmarks along the *z*-axis and along the *x*-axis in landmark 5d. Thus, a measurement recording height of the skull in the interorbital region should be more informative for species recognition than one recording its width. Maximum width at the mastoid apophyses also seems to yield an incomplete description of the variation in this region of the skull. Although landmarks 13d and 14d show highly significant displacements along the *y*-axis suggesting width variation, they also show displacement along the anteroposterior axis (Fig. 4; Table 3). Condylobasal length also provides a deficient description of variation. Landmarks 17d and 21v both show a major displacement along the *x*- axis and also differ in their *y*- and *z*- axis locations, despite being

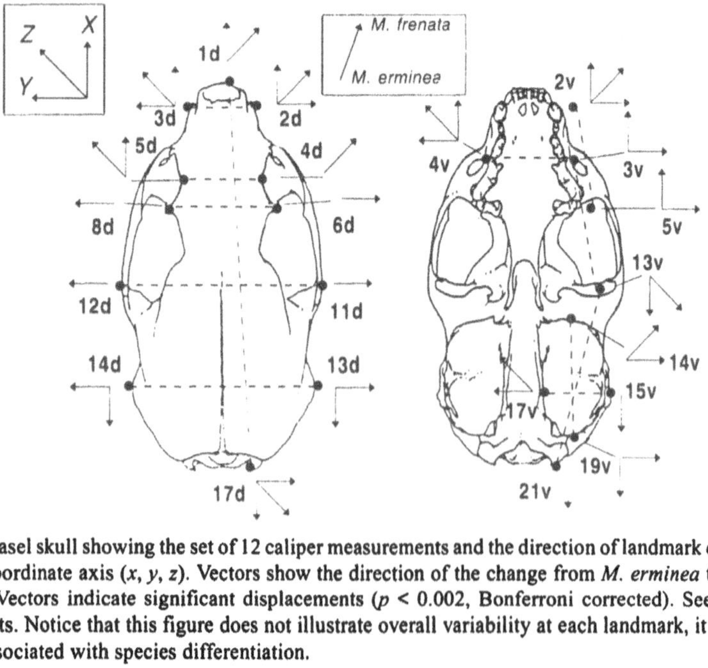

Figure 4. Weasel skull showing the set of 12 caliper measurements and the direction of landmark displacement along each coordinate axis (*x, y, z*). Vectors show the direction of the change from *M. erminea* to *M. frenata* coordinates. Vectors indicate significant displacements ($p < 0.002$, Bonferroni corrected). See Table 3 for detailed results. Notice that this figure does not illustrate overall variability at each landmark, it only reflects variability associated with species differentiation.

theoretically the same landmark viewed from the ventral and dorsal side (Table 3). This could be interpreted either as a mistake in locating the landmark on either side of the skull or as a reflection of real morphological differences between the condyles of both species, which can only be seen from the dorsal view. There seems to be some support for this latter inference. Condylobasal length showed little RMS errors, suggesting that variation was not an artifact. In addition, this measurement was the only one showing a higher F-value for the coordinate data set in the ANOVA. Its standardized coefficients on the canonical variate were very different between caliper and coordinate data (Table 2).

There are also examples of caliper traits that seem to be effective morphometric descriptors of variation. Ectoorbital width reflects accurately the variability at this region of the skull because there is only significant displacement in the y direction at landmarks 6d and 8d (Fig. 4; Table 3). Zygomatic width also seems to be a reliable caliper trait. Maximum width of the skull at the zygomatic arch does not seem to move along the x- axis at landmarks 11d and 12d but does in the y direction, suggesting that only changes in width take place at those landmarks (Fig. 4; Table 3).

On the ventral side, landmarks 2v, 5v, 14v and 19v show displacements along the y-axis and also along the anteroposterior axis. The small displacements at landmark 15v explains the lack of differences between species found in the width of the tympanic bulla (Tables 2, 3). However, displacement of landmark 17v on the y-axis suggests a widening of the bulla that would be missed if measured with a caliper (Fig. 4; Table 3).

Another shortcoming of caliper traits compared with coordinate displacements concerns symmetry of the form and its variation. Measurements estimate morphological variation solely upon changes in the size of the distance between the two points bounded by the measurement itself. Therefore, variation has to be uniformly allocated to the two extreme points or to the whole anatomical structure that lies in between when that structure can be described (Zelditch et al., 1989). This assumption of simultaneous and symmetric variation—displacement of the two extreme landmarks of each measurement—is not always likely to hold.

There were some clear examples of these problems in the set of skull characters. For instance, the F-value for the displacement in the x direction of landmark 2v was 64.7, whereas for landmark 5v F was 15.6 (Table 3). This suggests that variation observed in length of the maxillary tooth row is primarily caused by the anteroposterior displacement of the canine rather than by an even enlargement of the whole tooth row. Displacement of the molar (5v) was less important than that of the canine (2v) (Fig. 4; Table 3).

Another example of misinterpretation of caliper data can be found in the two caliper measurements ending at landmark 13v. Postglenoid length shows little species difference, but, at the same time, condylomolar length shows a highly significant difference (Table 2). From this, one might deduce that there had been a forward displacement of landmark 3v producing the variation in condylomolar length, whereas 13v had not changed (Fig. 4). But this is not the case: landmark 13v shows a highly significant displacement along the x-axis (Table 3). Thus, the low variability between species in postglenoid length comes from the concerted displacements of landmarks 13v and 21v. Both landmarks 13v and 21v are displaced from *M. erminea* to *M. frenata* in the same direction, toward the back of the skull, reducing the apparent variability exhibited by this measurement (Fig. 4).

DISCUSSION

For the same sample of specimens means obtained from caliper data may be significantly different from those derived from corresponding interlandmark distances. This fact must be considered before any attempt is made to compare values from caliper and coordinate data sets. Moreover, the precision of caliper measurement estimation from coordinates varies from one character to another (Fig. 2, 3). These issues pose practical problems for transforming results from one data type to another.

As expected, interlandmark distances of 3D coordinates provided better estimates of caliper measurements than did 2D coordinates, though some characters showed similar results for both 2D and 3D data. Although one should expect a random error in the estimation of caliper data from coordinates (Bookstein et al., 1991), caliper distances were consistently underestimated by coordinate pairs, regression slopes being greater than one in all variables. This result is primarily explained by the nature of the set of measurements used, which mainly consisted of distance maxima. Finding the two points that define the maximum distance of a given feature of the skull is easily achieved by expanding the gape of the caliper jaws until the largest distance is found. However, precisely and independently locating the two points that define the maximum distance with a digitizer stylus is more challenging. Thus, discrepancies between calipers and interlandmark distances arise from difficulties in locating the same landmarks that were used with the calipers. Essentially this reflects poorer landmark definition for traits comprising distance maxima and minima than for those measured between certain anatomical features.

Caliper traits can be divided into two groups according to the difficulties in locating their landmark endpoints. Measurements defined upon well-delimited points show much better agreement with coordinate distances than characters that measure maximum distances between regions of the skull. In the weasel skull, the best examples of accurate estimation were condylobasal, condylomolar and postglenoid lengths. Among the less reliable estimates were rostrum and bimolar widths (Fig. 3).

An example of poor correspondence between coordinate and caliper traits was the width of the tympanic bulla, explained in part by the large measurement error associated with this character. Variation in size and shape of the tympanic bulla is a relevant source of information in taxonomic and evolutionary studies of mammals (Pocock, 1921; Burda et al., 1990). However, the irregular shape and position of the bulla in the skull makes it difficult to measure with calipers. There is evidence suggesting that the discrepancy between caliper and coordinate distance arises from the use of different endpoints to measure it with the caliper, not the landmarks 15v and 17v. Caliper measurement error of this variable was large, suggesting that despite its potential taxonomic relevance, results should be interpreted cautiously. The practice of taking repeated measurements on each specimen is recommended to reduce measurement error in this character (Yezerinac et al., 1992).

Average RMS error of the estimation of caliper measurements from 3D interlandmark distances was 0.422 mm. This value falls within the range obtained in other studies. Hildebolt and Vannier (1988) report a RMS of 0.494 mm based on a series of five human skulls and a similar digitizing system.. Bookstein et al. (1991) reported a RMS of 0.377 mm for estimated caliper distances on human skulls, based on stereolocation in 3D cephalograms. These values are close to the theoretical minimum RMS of 0.323 mm, which corresponds to a series of 40 repeated digitizations of the same skull of *Macaca fascicularis* made by two observers (Corner et al., 1992). However, the small size of weasels introduces a larger average incongruence (3.4%) compared with other studies, 1.8% (Hildebolt and Vannier, 1988), and 1.3% (Corner et al., 1992), although it lies between the limits of 3.0 and 5.0% recommended by Corner et al. (1992). According to this criterion, five measurements should not be estimated form coordinate data: bimolar width, interorbital width, maxillary tooth row length, rostrum width and tympanic bulla width (Fig. 3).

Caliper measurements are more limited than are landmark data for describing variability in the form of an organism (Strauss and Bookstein, 1982; Bookstein et al., 1985). This is mainly because caliper measurements are uni-dimensional representations of the displacement seen in two landmarks that actually exist in three dimensions. For this reason, only when landmark displacement occurs (1) exclusively in one direction, (2) simultaneously for the two landmarks and (3) in a direction that corresponds to the axis defined by the measurement, do caliper distances characterize the variation that takes place at those landmarks. In this study, only 2 of the 12 caliper traits, zygomatic and ectoorbital width, fully conform to these requirements enough to be considered good descriptors of variability, at least at the level of species differences. Most of the caliper distances showing significant differences between species conform to the actual direction of landmark displacement (Fig. 4). Only two caliper measurements, rostrum and interorbital width, miss landmark variation that occurs in a different direction from the one being measured (Fig. 4).

The problem of distortion in the description of variability caused by asynchronized displacement of landmarks is illustrated here by postglenoid length and is discussed in the previous section. Although landmarks defining this distance showed highly significant displacements, the distance itself showed little variability simply because both landmarks were displaced in the same direction, canceling out the absolute change in the net distance (Fig. 4). This finding has special consequences for weasel taxonomy because the ratio of postglenoid length to condylobasal length has been used as a taxonomic character to distinguish between the two species studied here (Hall, 1951). Variation of this ratio was understood as a relative reduction of the brain case (Hall, 1951). However, after examination of coordinate data it becomes clear that instead of a change associated with the brain case, the glenoid fossa has been displaced toward the back of the skull. This statement has very different implications than those suggested by a reduction in the brain case, and therefore the anatomical interpretation is quite different (Fig. 4). This displacement of the glenoid fossa suggests, for instance, a morphological rearrangement to achieve greater mandible force that enables the larger species, *M. frenata*, to feed upon larger prey items than are taken by *M. erminea* (Hall, 1951).

Quantitative analysis of landmark displacements also provides evidence of overlooked variation in the symmetry of the form and its variability, which may be highly relevant biologically. For instance, caliper measurements assume that an increase in width of the skull reflects a symmetric displacement of both right and left endpoints. However, landmarks at the orbital and rostral region of the skull (5d–4d, 3d–2d) show an important asymmetry compared with the rest of the skull (raw data of landmark mean locations not included here) and also show differences in this asymmetry between species (Fig. 4).

Statistical analyses of species differences also show interesting differences between data sets. Univariate and multivariate ANOVA showed less resolution in coordinate than in caliper data sets (Table 2). Results of canonical variate analysis also show D^2 between skulls of *M. erminea* and *M. frenata* to be larger for the caliper data set than for the corresponding distances estimated from 2D or 3D data (Table 2). Contribution of each skull distance to the canonical variate (CV I) also reflected differences between the caliper and coordinate data sets (Table 2). Two factors likely account for these differences: the larger variability shown by distances estimated from landmarks (Fig. 2) and the different discrepancy between data sets shown by each morphometric character. A combination of these two factors is probably responsible for the slightly better results, in terms of multivariate F-values and D^2, shown by 2D distances despite the closer correspondence to caliper measurements shown by 3D distances.

Despite their obvious limitations, caliper measurements are a valid morphometric option for the description of variation and species differentiation in the weasel skull. Within the framework of inter-landmark distances, triangulation schemes do not seem superior to caliper traits. The D^2 between species based on the triangulation scheme of 21 distances was much larger than the D^2 obtained based on a small set of the six caliper traits that were based upon the same set of 12 dorsal landmarks used for the triangulation scheme. However, this reduced set of six caliper measurements was sufficient to separate skulls of each species.

This result is particularly meaningful because this set of six caliper distances illustrate the major biases and weaknesses of traditional characters pointed out by Strauss and Bookstein (1982). Five of these characters depict width of the skull and are aligned with one axis only (cross-sagittal), resulting in repetitious information. The sixth character, condylobasal length, represents the whole length of the skull and ignores the variation occurring within the regions spanned. The largest D^2 (55.3) was obtained when the whole set of 12 caliper traits was used. Thus, discrimination between the two species based on the caliper data set was better than that seen with the triangulation scheme. This is particularly noteworthy given the large number of variables included in the triangulation scheme. Other studies in small mammals have also shown that discriminating power of truss networks is not necessarily better than that obtained by traditional measurements (Dos Reis, 1990). This finding persists in spite of the fact that the homology of many landmarks used in traditional measurements are poorly defined (i.e., are maximum distances).

On the other hand, considering the advances that have been made in geometric morphometrics since the time that the box truss was first proposed as a progressive method for morphometric studies (Strauss and Bookstein, 1982), the use of 3D coordinate data simply to estimate linear distances between points represents a waste of information about the variability of the form. Other newer methods, such as shape coordinates, finite element analysis and thin-plate spline mapping, which use the relational information embedded in whole configurations of landmarks, are better suited for data analysis (Bookstein, 1991; Richtsmeier et al., 1992; Rohlf and Marcus, 1993).

In summary, though traditional measurements may provide adequate evidence to solve biological problems related to morphological variation, one should be aware that caliper traits provide an incomplete, sometimes misleading, description of the variability. Complementary analysis of coordinate data exploring the spatial variability that occurs at each landmark can improve considerably the interpretations made from distance data about their functional or evolutionary implications.

ACKNOWLEDGMENT

I am very grateful to Les Marcus for his patience and support and to all members of the editorial board for their conclusive criticisms to previous versions of the manuscript. Collection of data was possible thanks to the kind and helpful hospitality of Ralph Chapman and of the curatorial staff of the Mammal Division of the National Museum of Natural History, Smithsonian Institution. The study was supported by FPI research fellowships from the Spanish Ministry of Education and initiated during a post-doctoral stay at the Carnegie Museum of Natural History. I thank also Carmen Sainz for improving the English.

REFERENCES

Bookstein, F. L. 1991. Morphometric tools for landmark data: Geometry and biology. Cambridge University Press: Cambridge.

Bookstein, F. L., B. Chernoff, J. M. Humphries Jr., G. R. Smith, and R. E. Strauss. 1985. Morphometrics in evolutionary biology: The geometry of size and shape change, with examples from fishes. Academy of Natural Sciences of Philadelphia Special Publication 15.

Bookstein, F. L., B. Grayson, C. B. Cutting, H. C. Kim, and J. G. McCarthy. 1991. Landmarks in three dimensions: Reconstruction from cephalograms versus direct observation. American Journal of Orthodontics and Dentofacial Orthopedics 100:133–140.

Burda, H., E. Nevo, and V. Bruns. 1990. Adaptive differentiation of ear structures in subterranean Mole-rats of the *Spalax ehrenbergi* superspecies in Israel. Zool. Jb. Syst. 117:369–382.

Corner, B. D., S. Lele, and J. T. Richtsmeier. 1992. Measuring precision of three-dimensional landmark data. Quantitative Anthropology 3:347–359.

Dixon, W. J. (ed.). 1992. BMDP Statistical software. University of California Press: Berkeley, California.

Dos Reis, S. F. 1990. Mammalian cranial morphometrics: A comparative study of traditional and truss networks. Zoologisches Anzeiger 224:225–235.

Hall, E. R. 1951. American weasels. University of Kansas Publication 4.

Hildebolt, C.F., and M. W. Vannier. 1988. Three-dimensional measurement accuracy of skull surface landmarks. American Journal of Physical Anthropology 76:497–503.

King, C. M. 1991. A review of age determination methods for the stoat *Mustela erminea* . Mammal Reviews 21:31–49.

Marcus, L. F. and S. Reig. 1991: Data capturing software for digital calipers. Abstracts. Seventy-first Annual Meeting of the American Society of Mammalogists: Manhattan, Kansas.

Pocock, R. I. 1921. The auditory bulla and other cranial characters in the Mustelidae. Proceedings of the Zoological Society of London 1921:473–486.

Richtsmeier, J. T., J. M. Cheverud, and S. Lele. 1992. Advances in anthropological morphometrics. Annual Review of Anthropology 21:283–305.

Rohlf, F. J., and L. F. Marcus. 1993. A revolution in morphometrics. Trends in Ecology and Evolution 8:129–132.

Rohlf, F. J., and D. Slice. 1990. Methods for comparison of sets of landmarks. Systematic Zoology 39:40–59.

SAS. 1989. SAS user's guide statistics, version 6.03. SAS Institute: Cary, North Carolina.

Slice, D. E. 1993. GRF-ND: Generalized rotational fitting of N-dimensional data. Department of Ecology and Evolution. State University of New York at Stony Brook, Stony Brook, New York 11794.

Strauss, R. E., and F. L. Bookstein. 1982. The truss: Body form reconstruction in morphometrics. Systematic Zoology 31:113–135.

Voss, R. S., L. F. Marcus, and P. Escalante. 1990. Morphological evolution in muroid rodents I. Conservative patterns of craniometric covariance and their ontogenetic basis in the Neotropical genus *Zygodontomys. Evolution 44:1568–1587.*

Yezerinac S. M., S. C. Lougheed, and P. Handford.1992. Measurement error and morphometric studies: Statistical power and observer experience. Systematic Biology 41:471–482.

Zelditch, M. L., R. W. Debry, and D. Straney. 1989. Triangulation-measurement schemes in the multivariate analysis of size and shape. Journal of Mammalogy 70:571–579.

SOME APPLICATIONS OF GEOMETRIC MORPHOMETRICS TO OSTRACODA

Richard A. Reyment

Institute of Earth Sciences
Department of Historical Geology and Palaeontology
University of Uppsala
Uppsala, Sweden

ABSTRACT

Case histories of the application of geometric and algebraic morphometrics to the study of shape variation in fossil and living Ostracoda are reviewed. Evidence for polymorphism in shape is presented for the luminescent species *Vargula hilgendorfii* from Japan. Divergence in valve shape as a function of time for the fossil (Oligocene) species *Neobuntonia airella* from southeastern Australia is demonstrated. Many of the geometrically obtained results for *Vargula hilgendorfii* could be duplicated by the multivariate analysis of distance measures.

INTRODUCTION

The accepted standard approach to morphometric studies is to base the analysis on measures of dimensions, distances between selected points, made in the usual manner in taxonomic work. That these traits encompass information on **size** and **shape** is not in dispute, but these components are inextricably entangled. Hence, any study based on distance-traits and concerned with establishing significant change in a morphological structure, is going to face the difficulty posed by partitioning changes resulting from size differentiation and those due to shape. We should prefer to concentrate attention on innovations solely ascribable to differentiation in shape. A sound reason for wishing to realize this ambition is that shape mutability can be expected to be more relevant for assessing significant evolution in a lineage.

The partitioning of size and shape by mathematical means was first taken up by Wright (1932) and followed by Penrose (1954). The most interesting attempt, in my opinion, is that of Burnaby (1966), supported by Rao (1966), who separated variability in standard distance traits into a subspace containing differences due to size (e.g., different growth stages and sexual dimorphism) and a complementary subspace containing the taxonomically relevant differentiation. Burnaby's interests lay mainly with organisms, such as mollusks and

Advances in Morphometrics, Edited by L. F. Marcus et al.
Plenum Press, New York, 1996

foraminifers, that lack a terminal growth stage and, consequently, pose a special problem in multivariate statistical applications in taxonomic work (cf. Reyment et al., 1984).

The above and other attempts at separating size from shape, such as the principal component solution of Teissier (1938) and Jolicoeur (1963), have not taken issue with the fundamental question of how to describe pure shape and its variability. This situation changed when Bookstein (1989, 1991) developed a powerful technique for the analysis of within-population shape based on landmark data. Landmarks are defined as observations made at diagnostically important sites and identified by their coordinate pairs, in a Cartesian system. Such geometric variables contain all essential information on size and shape and should, therefore, be valuable for registering variations in the crustacean carapace.

The Ostracoda constitute a class of shell-secreting bivalve crustaceans that are of wide distribution today. The two examples reviewed below expose many of the variational characteristics of the group.

POLYMORPHISM IN A LUMINESCENT OSTRACOD SPECIES

Vargula hilgendorfii is a luminescent ostracod species that is of relatively common occurrence in the Bay of Tokyo and other parts of Honshu, the largest of the Japanese islands. It is a fairly "big" species (around 3 mm in length). The problem is (1) to examine the area in the vicinity of the rostrum, which seems to be more variable than is the rest of the organism and (2) to ascertain whether the individuals of the two demes sampled differ in size, on the average (Reyment and Abe, 1995). The specimens analyzed were collected all in one night at Tateyama in February 1993 and at Misaki in October 1990. These localities are close geographically, being located on opposite sides of the mouth of Tokyo Bay. All individuals analyzed here are adult females. At the intrapopulational level, indications of shape polymorphism occur. A consistent, and pronounced, difference in centroid size and a less pronounced, though significant, difference in shape between populations could be demonstrated. Granted that Misaki and Tateyama are not geographically distant from each other, and that the analysis was confined to females, this marked divergence may have its origin in some kind of persistent ecological factor.

Studying Shape Variation in *Vargula hilgendorfii*

The initial problem faced here concerns the quantification of a "smooth" ostracod in a shape-analytical context. At first sight, the carapace seems to have about as many landmarks as an egg, but closer inspection discloses that there are several: four good landmarks located around the rostrum and another four relatively reliable ones at points around the circumference of the carapace. Visual inspection of the specimens from Misaki and Tateyama did not

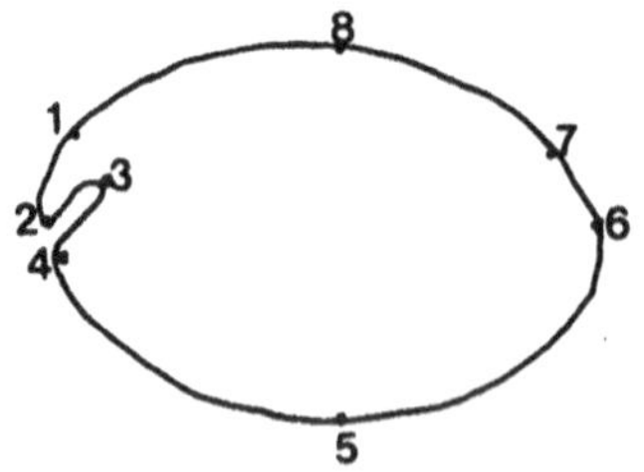

Figure 1. The eight landmarks determined on *V. hilgendorfii*. The glands that secrete luciferine and luciferase are located in the immediate vicinity of landmarks 2, 3 and 4.

suggest any self-evident variations in shape to occur; on just plain observational grounds, there would have been slight justification for assuming any form of morphological polymorphism or regional morphological differentiation in the configuration of the carapace to exist. Obviously, careful attention to measuring detail is an absolute necessity in studying such material.

The Landmarks

A set of standard measures on the carapace would, no doubt, settle the issue as far as size differences are concerned but not the issue of shape variation. Fig. 1 shows the sites chosen as landmarks. They are (1) the anterodorsal angularity, (2) the upper tip of the rostral incision, (3) the inner extremity of the rostral incision, (4) the lower lip of the rostral incision, (5) the site of maximum ventral inflation, (6) the posterior tip, (7) the posterodorsal angularity and (8) the site of the maximum dorsal inflation. The sites of maximum dorsal and ventral inflation are more subjective but, with a little care, could be measured with satisfactory statistical accuracy.

Within-Sample Variability

The plot of *RW1* against *RW2* (*RW* signifies relative warp) for the sample of left valves from Tateyama shows several subdivisions to occur. These are outlined on the graph shown in Fig. 2.

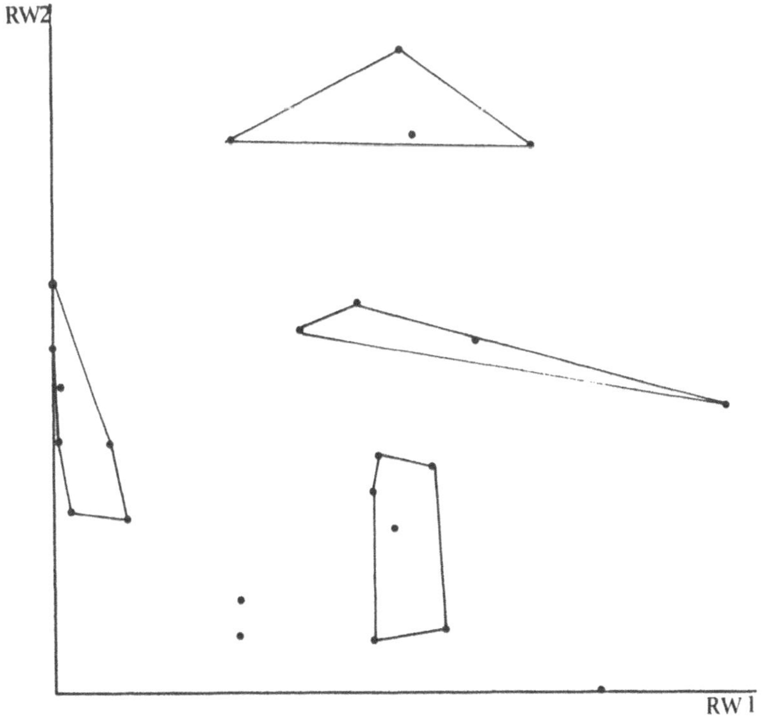

Figure 2. Plot of *RW1* against *RW2* for the Tateyama sample of *V. hilgendorfii*. Suggested groupings indicated.

Table 1. Generalized distances of weight matrices for the Tateyama sample morphs of *V. hilgendorfii*

Weight matrix	D^2	F	p
Nonaffine+affine	40.83	$8.88_{12,10}$	>0.001
Nonaffine	23.27	$7.25_{10,12}$	0.001

One group comprises seven quite clearly segregated specimens. Other indications are persuasive, but the sample is not complete enough to permit definite conclusions. Relative warp 2 can be seen to be the axis along which most of the separation occurs. Likewise, the plot of *RW1* against *RW2* for the left valve of the Misaki sample, although smaller than the Tateyama sample, indicates similar groupings to exist in the material. The same plot of *RW1* against *RW2* for right valves of the two samples also indicates differentiation in shape to occur. The most interesting shape indicator for right valves is, however, the quasi-invariant component, *RW10*. The two major groupings in the Tateyama collection obtained from that warp were tested for significant Mahalanobis statistical distance by means of the weight matrices. The results are listed in Table 1. These values indicate that much of the differences in shape within the Tateyama sample derive from the nonaffine component.

The 10th Relative Warp

The deformation expressed by the 10th relative warp, which was found to be expressive for illustrating invariant shape relationships for the present material, the pooled sample of right valves of specimens from both sites, is shown in Fig. 3. The analogous figure for left valves is almost identical, but the rostrum in that case is slightly more compressed than that of right valves.

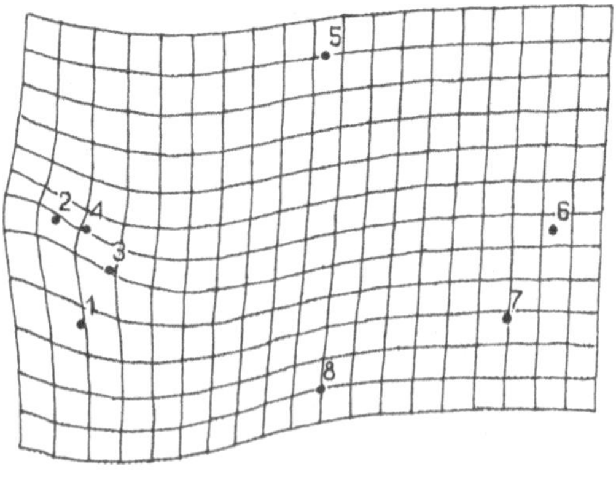

Figure 3. Tenth relative warp for all right valves (Tateyama and Misaki) of *V. hilgendorfii*. The actual distances between landmarks can be a constructive step to contemplate. Some of these are listed in Table 2. Only the distance between landmarks 2 and 3 differs much relatively. The wisdom of using a principal component type of analysis with such a slight variance as that represented by the 10th relative warp might well be questioned. However, assuming that the logic of the method of principal components carries over to the principles of geometric morphometrics, it is useful to examine not only those components that have large variances but also those that have variances which do not greatly differ from naught. These are the invariant principal components of some statisticians (Gower, 1967; Reyment and Jöreskog, 1993).

Table 2. Relative distances between landmark deformations (left and right valve of *V. hilgendorfii*) for the 10th relative warp (unit free)

Between landmarks	Left valves	Right valves
6 and 7	2.55	2.61
5 and 6	7.50	7.65
7 and 8	6.50	5.40
1 and 8	7.40	6.75
1 and 3	1.80	1.65
2 and 3	1.30	1.95

Relative Warps and Centroid Size

The next step in the analysis is to inquire whether there are significant size differences between the two sites. This is not something one would expect to occur, given that both sampling localities are located near each other, on opposite sides of Tokyo Bay, only females were included in the study, and *V. hilgendorfii* is a fairly mobile benthic species that has a highly developed optical-recognition mechanism connected with the luminescence function. In this section, plots between selected relative warps and centroid size are examined.

First Warp. The convex hulls for the plot of left valves are illustrated in Fig. 4A. The points for the two localities are almost completely separated with only slight overlap along the size axis; one specimen from Tateyama groups with large Misaki individuals. The axis

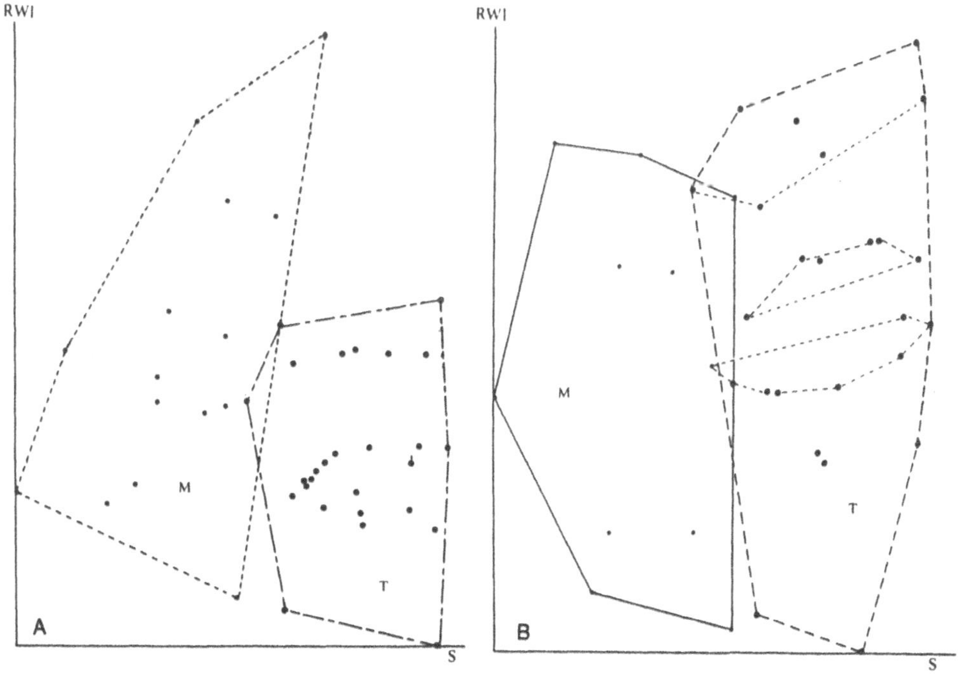

Figure 4. (A) Relative warp 1 plotted against centroid size for all left valves. Observe separation on the size (S) axis. (B) Relative warp 1 against centroid size for all right valves. In both cases, the Tateyama data display clusters (outlined). M = Misaki; T = Tateyama.

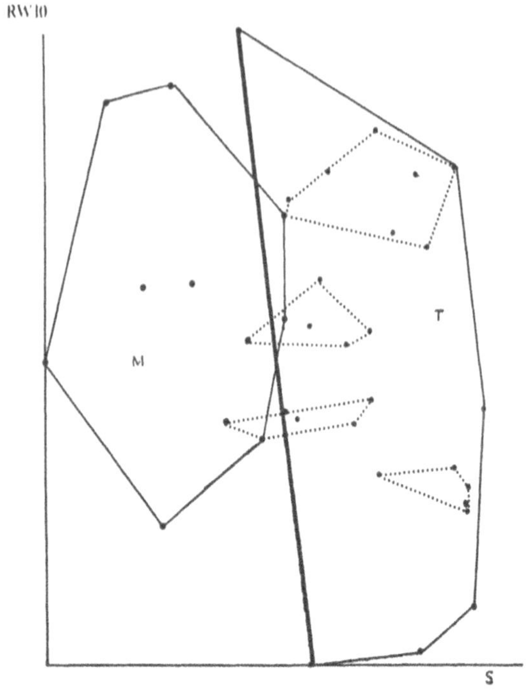

Figure 5. Relative warp 10 plotted against centroid size (*S*) for all right valves of *V. hilgendorfii*.

for *RW1* distinguishes two large Tateyama groups. Right valves (Fig. 4B) display likewise slight overlap (the same specimen as above). In this projection, three Tateyama groups are separated out by *RW1*.

Tenth Warp. The relevant graph is shown in Fig. 5. For left valves, three Tateyama specimens lie inside the convex hull for Misaki. There are, as before, discrete Tateyama shape clusters—four in all and two of which encompass both sites. In the case of the right valves, there is no overlap between sites, and there are four shape clusters that encompass both of the samples.

Discussion

The graphical analyses indicate that there is a pronounced difference in size between the females from the two localities. Additionally, there are hierarchically expressed differences in shape (hierarchical in the sense of the principal component decomposition of the warps). The reason for the strong difference in size, with practically no overlap between samples, awaits elucidation. The same method of multivariate analysis as used before (Table 1) was applied to the weight matrices of left and right valves for both of the samples. The results for the comparisons are displayed in Table 3. In all cases, the generalized statistical distances are significantly different for affine plus nonaffine shape and for affine shape on its own. This somewhat unexpected result fuels further the suspicion that there are notable dissimilarities between factors of the marine environments at Tateyama and Misaki.

Table 3. Generalized distances for comparisons of weight matrices for Misaki and Tateyama samples of *V. hilgendorfii*

Valve and weight matrix	D^2	F	p
Left			
Nonaffine+affine	29.79	$10.22_{16,22}$	> 0.01
Nonaffine	4.74	$3.31_{10,28}$	0.05
Right			
Nonaffine+affine	17.69	$6.80_{16,19}$	> 0.01
Non-affine	6.19	$3.47_{10,25}$	0.05

Principal Component Analysis of Distance Measures

Four distances were computed from the landmarks. These are *D1*: the distance between landmarks 1 and 5; *D2*: the distance between landmarks 2 and 6; *D3*: the distance between landmarks 5 and 7; and *D4*: the distance between landmarks 5 and 8.

A standard multivariate analysis of distance characters can be expected to yield about the same results as the landmark-based procedures as regards differences in size (cf. Reyment et al., 1984, for a summary of the application of these methods to morphometrics). Four of the many possible distances available for the eight landmarks selected above, were analyzed by principal components. Table 4 lists the latent roots and vectors for the pooled covariance matrices of right and left valves of Misaki and Tateyama specimens. As to be expected, the elements of the latent vectors of the two sets correspond closely, as do also the magnitudes of matching latent roots. The smallest latent vector expresses invariant relationships, which are almost identical for the two valves. The angles between the pairs of latent roots are listed in Table 4; it can be seen that first and fourth latent roots are separated by small angles, thus showing good agreement in variation along those axes. The second and third latent vectors diverge a little more.

The results shown in Fig. 6 for distance measures of left valves almost exactly match those obtained by geometric methods. Misaki and Tateyama are well separated in all diagrams. The first of each pair shows the projection in the plane of the first two latent vectors and the second of the pair illustrates the scatter of points in the plane of the first and third latent vectors.

The "algebraic model" for the partition of size and shape (cf. Reyment, 1991, chapter 4 for a review and Reyment and Jöreskog, 1993) assigns shape-ordinating prop-

Table 4. Latent roots and vectors for four distance measures on left and right valves of the Misaki and Tateyama specimens of *V. hilgendorfii*

	Right valves I	Right valves II	Right valves III	Right valves IV	Left valves I	Left valves II	Left valves III	Left valves IV
Variable	Latent Vectors							
D1	0.358	0.830	-0.028	0.426	0.394	0.717	0.225	0.529
D2	0.562	-0.109	-0.759	-0.311	0.685	0.097	-0.611	-0.383
D3	0.500	-0.543	0.182	0.649	0.413	-0.671	-0.030	0.615
D4	0.553	0.064	0.624	-0.548	0.452	-0.160	0.758	-0.442
	Latent roots							
	7.404	0.967	0.521	0.074	5.880	0.896	0.489	0.061
% Trace	82.58	10.79	5.81	0.83	80.26	12.24	6.67	0.83

Angles between pairs of latent vectors for right and left valves (in degrees)

I	II	III	IV
10.74	20.17	22.33	8.08

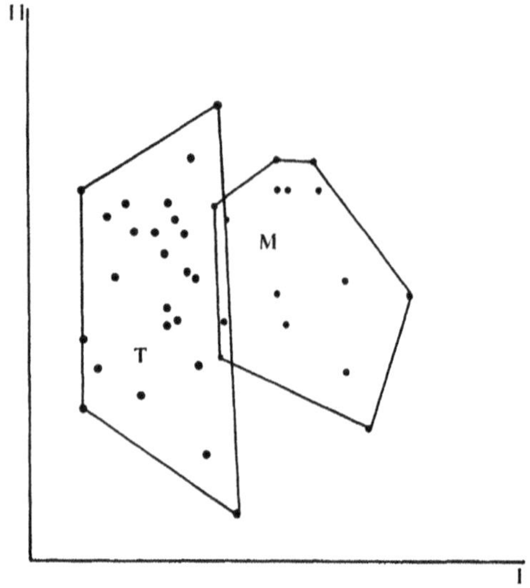

Figure 6. Plots of the scores for the principal component analysis of distance measures obtained from selected landmarks on the left valves of the Misaki and Tateyama samples of *V. hilgendorfii*. The appropriate convex hulls are indicated and projection on the plane of the first and second latent vectors is shown. There is slight overlap along the first principal axis. The first principal axis is denoted by **I** and the second principal axis by **II**.

erties to the second and subsequent latent vectors. It is of interest to inquire how the standard principal component decomposition of the four distance measures succeeds in reproducing the shape relationships indicated by the relative warps. There are several points of agreement with the ordination yielded by the plot of *RW1* against *RW2*, not least being that the main clusters are reproduced and notwithstanding the reduced body of information used for the analysis.

The significance of the results summarized briefly in this section is not that the algebraic solution of the size–shape dichotomy is in any way better than the geometric one, but rather that it does accomplish part of the task successfully and to a greater extent than

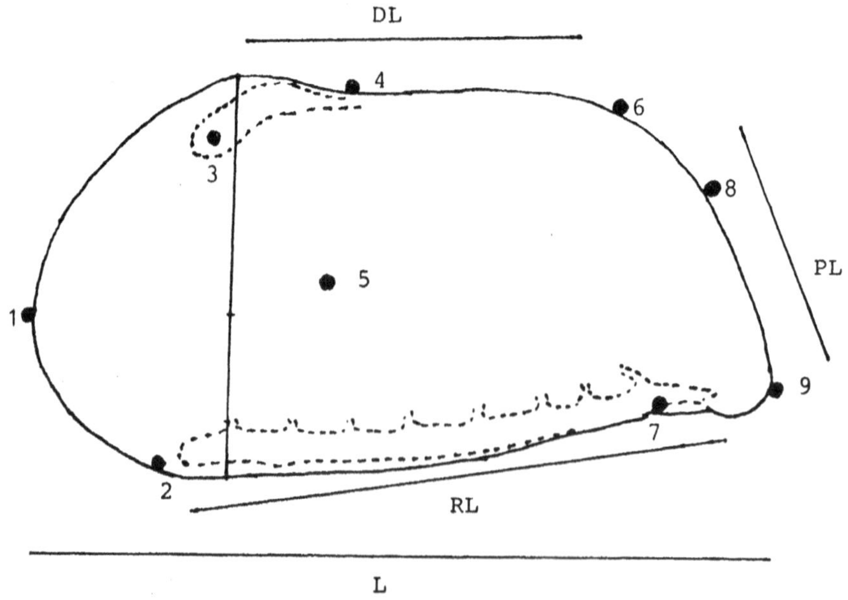

Figure 7. The landmarks on *Neobuntonia airella*.

Table 5. Multivariate statistics for total shape weight matrix for *Neobuntonia airella*. L. Olig = Lower Oligocene; U. Olig. = Upper Oligocene; inv = mirror image of left valve; LV = left valve; and RV = right valve

Samples compared	D^2	T^2	F	n_1	n_2
LV (L. Olig. w. U. Olig.)	38.00	253.30	4.50	18	8
RV (L. Olig. w. U. Olig.)	29.56	225.70	5.66	18	14
Inv LV w. RV, L. Olig.	111.40	668.39	8.44	18	5
Inv .LV w. RV, U. Olig.	79.61	696.56	19.35	18	17
LV L. Olig. w. RV U. Olig.	44.12	336.92	8.45	18	14
RV L. Olig. w. LV U. Olig.	20.50	136.68	2.43	18	8

one might suppose. The analysis presented here achieves several purposes. First, it is successful in unveiling shape variability within samples of a moderately mobile species. Second, it proves the existence of geographically linked size differences, a result that is all the more remarkable when the short distance between sampling sites is weighed in the balance. The observed differences are neither sexually dimorphic nor ontogenetic in nature, given that all material measured consisted of adult females. The variability identified in the rostral area could reflect the morphological expression of three factors, to wit, (1) the swimming function in that the second antennae are protruded there when in use, (2) the secretion of luminescent substances, and (3) the light-sensor ability involved in infraspecific signaling (Vannier and Abe, 1993).

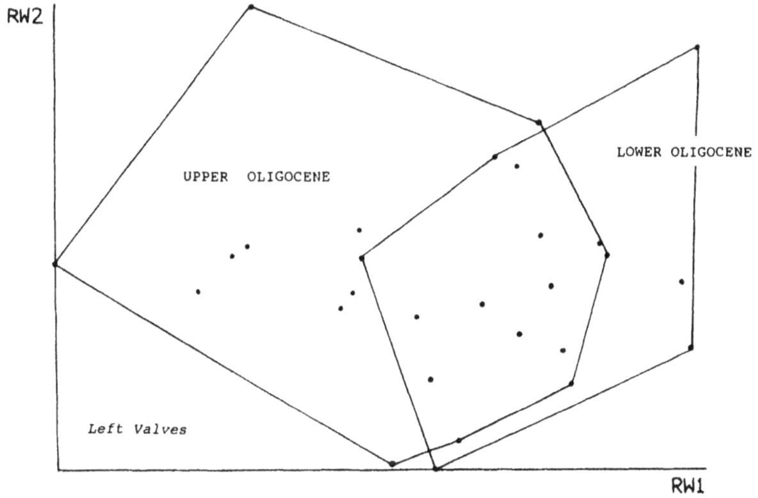

Figure 8. Asymmetric shape example of *N. airella*. Plot of *RW1* against *RW2* for left valves.

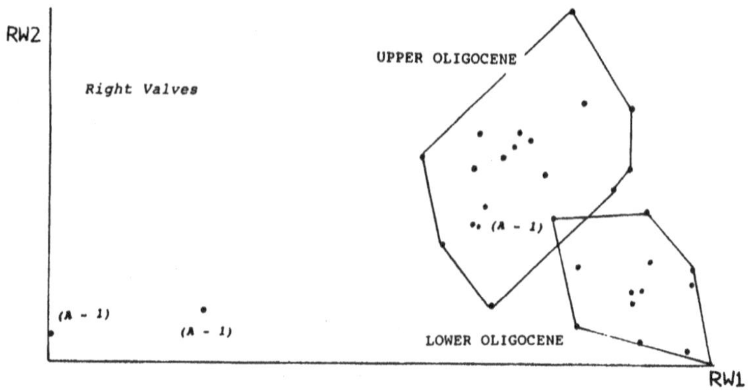

Figure 9. Asymmetric shape of *N. airella*. Plot of *RW1* against *RW2* for left valves. The pre-adult molt-stage is denoted by A-1.

EVOLUTION IN OSTRACOD VALVE ASYMMETRY

Ostracod shells are always asymmetric, with the left valve being slightly larger and different in shape from the right valve in the overwhelming majority of genera. It is of interest to examine how the geometry of evolutionary differentiation in valve features takes place. Specific transitions are accompanied by shape change in many lineages, but it is not so clear how much of this divergence is the outcome of accentuated (or, conversely, relaxed) asymmetry. It should be understood that the asymmetry discussed here has nothing whatsoever in common with fluctuating asymmetry such as can be found in rodent skulls. It is a fundamental structural property of the class.

The data considered here consist of left and right valves of *Neobuntonia airella* from the Lower and Upper Oligocene of southeastern Australia. The measurements made on the shells are shown in Fig. 7.

Asymmetry in the present example encompasses developmental differentiation in shape and hence the degree of overlap of the right valve by the left. In order to compare left valves with right valves, it is necessary to use the mirror image of measurements of one of

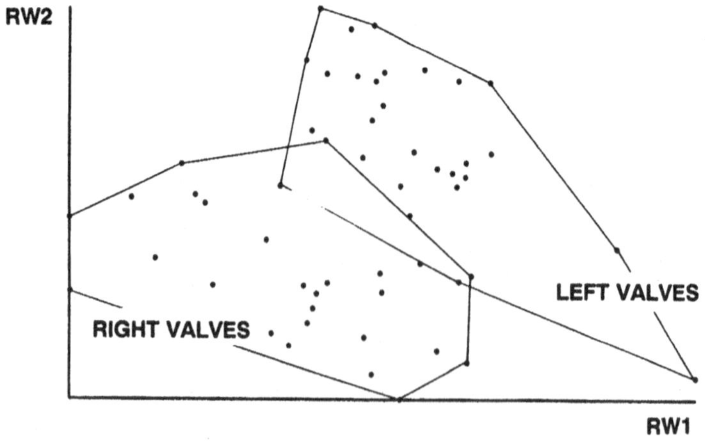

Figure 10. Asymmetric shape of *N. airella*. Left and right valves of Upper Oligocene; contrasted.

the classes. In the present analysis, I have used inverted left valves, which have been compared directly with the landmarks observed on right valves.

The generalized distances for the weight matrices for total shape variation (affine and nonaffine) are as given in Table 5. These differences are highly significant. There is shape differentiation between both valve kinds at the two stratigraphic levels. The impression yielded by the generalized distance comparisons for the affine and non-affine weight matrices is that the difference in shape decreases over time. That is, right valves tend to become more like left valves.

Relative Warp Analysis

The ordinations on the projections of the first and second relative warps show the following salient features:

1. Right valves (Fig. 9) are completely separated into discrete groups on warps 1 and 2 (*RW1* plotted against *RW2*). (For the plot of *RW1* against barycentric size, there is complete separation on the former axis but complete overlap on the latter.)
2. Rather unexpectedly, perhaps, the same plot for left valves displays overlap (Fig. 8). The points for the Upper Oligocene specimens are many times more widely dispersed than for the Lower Oligocene due to the presence of precocious final male instars in the latter collection (such instars tend to simulate the morphology of adult females).

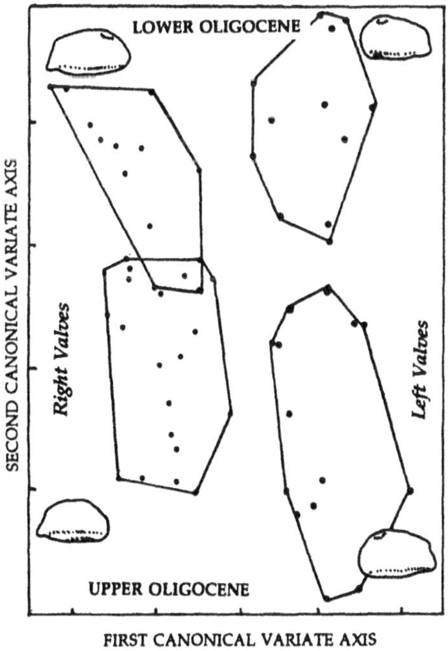

Figure 11. Asymmetric shape of *N. airella*. Canonical variate ordination for left and right valves; landmark coordinates were used as input. Note that the left valves were "inverted" in order to make the coordinates compatible.

3. The *RW1–RW2* ordination for left valves contrasted with right valves (Fig. 10) supports the results of the generalized distances. Figure 10 shows slight overlap for the Upper Oligocene specimens. The plot for Lower Oligocene left and right valves yields, however, complete separation.
4. A canonical variate analysis for the four samples of valves, to wit, two samples of right valves and two samples of inverted left valves, displays the relationships outlined in (3) more succinctly. The ordination in Fig. 11 shows the two right-valve samples to overlap slightly, whereas the two left-valve ones are separated. The "traditional" ordination on these observations gives the same result as the affine+non-affine weight matrix submitted to a canonical variate analysis.

REFERENCES

Bookstein, F. L. 1989. Principal warps: thin-plate splines and the decomposition of deformations. IEEE Transactions on Pattern Analysis and Machine Intelligence 11:567–585.

Bookstein, F. L. 1991. Morphometric tools for landmark data: Geometry and biology. Cambridge University Press: Cambridge.

Burnaby, T. P. 1966. Growth invariant discriminant functions and generalized distances. Biometrics 22:96–110.

Gower, J. C. 1967. Multivariate analysis and multidimensional geometry. The Statistician 17:13–28.

Jolicoeur, P. 1963. The multivariate generalization of the allometry equation. Biometrics 19:497-499.

Penrose, L. S. 1954. Distance, size and shape. Annals of Eugenics 18:337–343.

Rao, C. R. 1966. Discriminant functions between composite hypotheses and related problems. Biometrika 53:339–345.

Reyment, R. A. 1991. Multidimensional palaeobiology. Pergamon Press: Oxford.

Reyment, R. A. and K. Abe. 1995. Morphometrics of *Vargula hilgendorfii (Müller)* (Ostr., Crust.). Mitt. Hamb. Zool. Mus. Inst., 92: 325-336.

Reyment, R. A., R. E. Blackith, and N. A. Campbell. 1984. Multivariate morphometrics. 2nd edition, Academic Press: London.

Reyment, R. A. and K. G. Jöreskog 1993. Applied factor analysis in the natural sciences. Cambridge University Press: New York.

Vannier, J. and K. Abe. 1993. Functional morphology and behavior of *Vargula hilgendorfii* (Ostracoda: Myodocopida) from Japan, and discussion of its crustacean ectoparasites: Preliminary results from video recordings. Journal of Crustacean Biology 13:51–76.

Wright, S. 1932. General, group and special size factors. Genetics 17:603–619.

SHAPE CHANGES DURING THE GROWTH OF THE SEA BASS, *Dicentrarchus labrax* (TELEOSTEA: PERCIFORMES), IN RELATION TO DIFFERENT REARING CONDITIONS

An Application of Thin-Plate Spline Regression Analysis

Angelo Loy,[1] Stefano Cataudella,[1] and Marco Corti[2]

[1] Stazione di Acquacoltura ed Ecologia Sperimentale
II Università di Roma 'Tor Vergata'
via di Passolombardo 430
00133 Rome, Italy
[2] Dipartimento di Biologia Animale e dell'Uomo
Università di Roma 'La Sapienza'
via Borelli 50
00161 Rome, Italy

ABSTRACT

Thin-plate spline regression analysis is applied to sample of sea bass *Dicentrarchus labrax,* reared at two different salinities, i.e., marine and freshwater, in order to show shape changes and to test statistically morphological differences. All specimens were derived from the same breeding stock and were sampled at five different ages. Centroid size is used as the independent variable in the thin-plate spline regression analysis, and splines at extreme values of centroid size are computed and plotted. Differences in centroid size, for Bookstein's uniform components (*U1* and *U2*) as well as for the pure nonuniform components of shape change are tested for significance. These analyses allow a visualization of allometry and description and testing of significance of the morphological plasticity of the sea bass. In this sense they can be valuable tools in the study of shape change during ontogeny.

INTRODUCTION

The sea bass, *Dicentrarchus labrax*, is an euryhaline and eurytherm teleost widely distributed throughout the Mediterranean and eastern Atlantic coasts from 30 to 55° of northern latitude (Barnabè, 1980). At present, the sea bass represents one of the prime species in Mediterranean aquaculture. Intensive rearing techniques and extensive management of coastal lagoons are the most important ways to breed this fish. The rearing of this species in freshwater is considered here.

In this study we consider different effects of rearing sea bass in waste marine and fresh water. The shortage of rearing sites on the coasts and the possibility of exploiting the fresh cooling waters of electrical power plants for intensive breeding of the species, both provided a strong stimulus to analyze the effect of freshwater on shape change during fish growth. It is well known that form in bony fishes may be strongly altered during growth by changes in a few environmental conditions, such as temperature and salinity (Barlow, 1961; Holliday, 1988). This is generally known as ecophenotypism (Mayr, 1963). Therefore of great interest for morphometricians is the planning of experiments in which one can relate changes in form of the organism—the size and shape of bony fishes—to a few external conditions kept under strict control. We used landmark-based geometric morphometrics to analyze size and shape changes during growth of samples of fishes reared at different salinities. All fishes descended from the same breeding stock and were reared to commercial size. The model presumes that all differences found are consequent to individual adaptation to freshwater, given a common genetic background of the entire sample. Centroid size (the square root of the sum of squares of the landmarks distances from the centroid, Bookstein, 1991) was used to estimate the increase in size during growth and to represent graphically growth allometry through the new methods of geometric morphometrics (Rohlf and Marcus, 1993).

MATERIALS AND METHODS

In total, 466 individuals were collected at different ages (Table 1). A subsample of 9-month old fishes was transferred from the original stock and gradually acclimated to freshwater according to the explorative trials of Yashuov (1969) and Chervinsky (1974, 1979). Acclimation to freshwater lasted 17 days, and the highest mortality occurred during the first 24 hours. After this period the rate of mortality remained within the usual range for intensive breeding. All other rearing conditions were held constant. Following this procedure it was possible to study size and shape differences between the marine and freshwater samples.

Samples were collected at different periods representing different age classes, i.e., 7, 12, 15, 19 and 24 months (Table 1). Age class 19 is represented by only marine specimens.

Pictures were taken of all specimens on 35mm film, and a scale was present in each photograph. The image filled the frame, and the 70mm lens was centered as nearly

Table 1. Number of individuals measured in each age class (7, 12, 15, 19, 24 months) for both rearing environments

Age Class	7	12	15	19	24
Marine	46	44	53	47	33
Freshwater		58	88		49

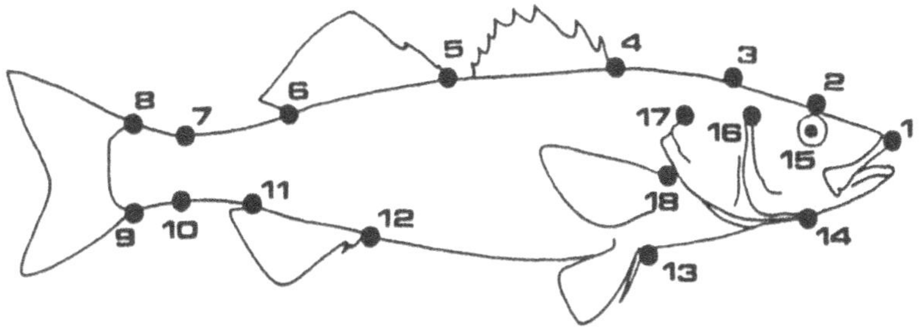

Figure 1. The 18 landmarks collected on the sea bass, *Dicentrarchus labrax*.

as possible to the center of the image. Negatives were projected in a darkroom and images were enlarged to their natural magnitude. The 18 landmarks were marked on paper and were collected using a Calcomp 2200 digitizing tablet and the software DS-Digit (Slice, 1991). The landmarks were collected on the right side of the fish to avoid the effect of lateral asymmetry (Fig. 1).

For a detailed description of each landmark see Corti et al. (1996). These landmarks can be classified according to Bookstein (1991): landmarks 4, 5, 6, 8, 9, 11, 12, 13, 16, 17 and 18 are of type one; 7, 10 and 14 are of type two; and 1 and 15 are of type three (Fig. 1). Centroid size was computed for each individual from the original raw coordinates by means of the program UNICOORD written in SAS-IML (by L. F. Marcus, see appendix of Reyment, 1991), as modified by Corti (1992). Mean and standard deviations of centroid size were computed and differences between rearing groups at each age class were tested with the TTEST procedure of SAS.

Landmarks were then translated, rotated and scaled by means of the Generalized Least Square (GLS) method (Rohlf and Slice, 1990) and a consensus configuration was computed for all individuals (GRF-ND; Slice, 1993). Uniform components of shape change were computed as *U1* and *U2* based on Bookstein's method (Bookstein, "Standard formula" article, this volume) as an option of the updated version of the TPSRW analysis program (Rohlf, 1993a). Differences in *U1* and *U2* between rearing groups were tested with a multivariate analysis of variance (MANOVA).

The weight matrix (Rohlf, 1993b) containing the partial warp scores was computed for the combined trials with the TPSRW program by use of the overall consensus as the reference and the residuals of the GRF-ND GLS as input. Because no particular assumption about large-scale shape differences was made, an α of 0 was used as the exponential weight of the bending energy matrix.

Scores of individuals in the weight matrix of the marine and the freshwater fishes were compared in a MANOVA to test for significance the effect of the different salinity conditions (procedure GLM, MANOVA option in SAS) within each age class.

As the thin-plate spline related techniques allows us to display shape difference through a series of transformations of grids, the two new sets of variables were analyzed with the thin-plate spline regression analysis (TPSREGR, Rohlf, 1993c). Centroid size was the independent variable, α was equal to 0, and the overall consensus was the reference configuration. In this way, the TPSREGR program provides output showing how shape would change from the youngest to the oldest class. Plots of both uniform and nonuniform deformations from age class 7 to 24 months were obtained and compared among rearing samples.

Table 2. Analysis of variance of centroid size: *t* values and significance of mean differences among rearing samples of the same age class

t value, 12 months df 100	*t* value, 15 months df 139	*t* value, 24 months df 80
1.48 ns	26.36 ***	18.05 ***

$p < .0001$(***); non significant at $p = 0.05$(ns)

RESULTS

As shown in Table 2, differences in centroid size are not significant at age class 12 months, whereas they are significant both at age classes 15 and 24 months. Values in freshwater (FW) are slightly smaller at age class 12 (FW centroid size mean = 19.99 cm, marine (M) mean = 20.26 cm). After age class 12, the freshwater trial always has higher centroid size values (age class 15, FW centroid size mean = 24.62 cm, M mean = 22.38 cm; age class 24, FW mean = 41.56 cm, M mean = 38.76 cm).

The results of the analysis of thin-plate spline regression on centroid size are represented as splines in Fig. 2 and 3. Values of centroid size were extrapolated to highlight differences between the early and the late growth stages in the two rearing trials.

There is a common pattern relative to the two trials depicted by the purely nonaffine component that concerns the head, the operculum and the anterior insertion of the dorsal and pelvic fins (landmarks 1, 2, 3, 4, 13, 14, 15, 16, 17 and 18; Fig. 2).

All of these landmarks tend to move forward while the fishes are growing. There is, as well, a nonuniform backward movement of the peduncular region (landmarks 6, 7, 8, 9, 10 and 11). This results in a non-uniform overall elongation of the body leading to the fusiform shape of the late stages. Although nonuniform shape changes depicted by the splines

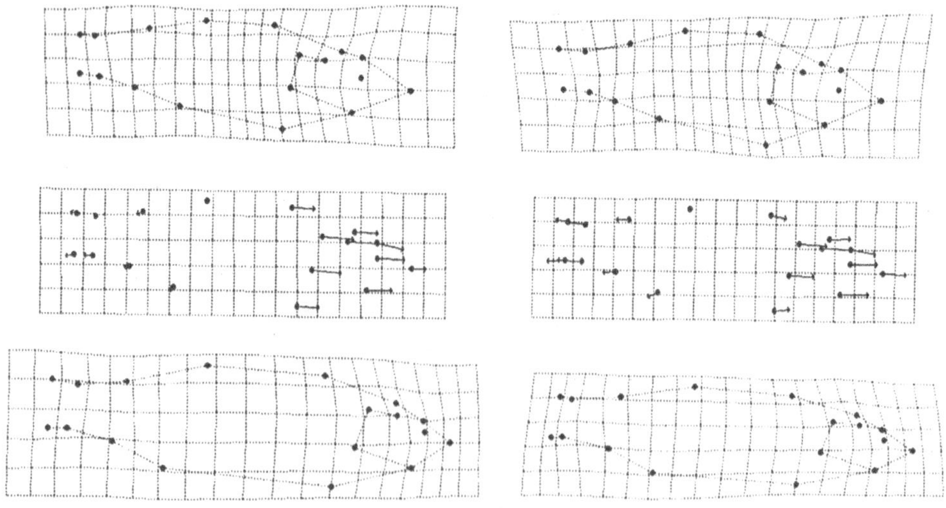

Figure 2. Nonuniform part of shape change during growth of the marine sample (left in side view) and freshwater sample (right in side view). Splines extrapolated from extreme values of centroid size. On top, spline relative to small values of centroid size; in the middle, vectors of landmark displacements; and on bottom left, spline relative to high values of centroid size.

Table 3. Multivariate analysis of variance (*F* approximation values based on Wilks' Lambda statistics) of the weight matrix and Bookstein's uniform components (*U1* and *U2*) of the two rearing samples at different age classes

	F approx., 12 months	*F* approx., 15 months	*F* approx., 24 months
Non-uniform	12.07 ***	9.08 ***	7.04 ***
Uniform	10.57 ***	10.09 ***	4.63 *

$p < 0.0001$(***), $p < 0.05$(*)

seem to be similar in both samples, MANOVA of the weight matrix gives significant differences between rearing samples at each age class (Table 3).

Interesting and contrasting patterns are depicted by the splines relative to the uniform shape changes (Fig. 3). Uniform shape changes are mainly perpendicular to the anteroposterior axis in both samples, but they are of opposite signs; in the marine sample uniform growth result in an overall slendering of the body (Fig. 3, left in side view) while in freshwater the uniform growth pattern is depicted as the acquisition of a rounded shape (Fig. 3, right in side view). There is also an opposite sign shearing in an oblique direction in both samples.

Uniform shape changes are significantly different ($\alpha = .05$) at age classes 12, 15 and 24 months (Table 3).

DISCUSSION

It is noteworthy that differences in size are established soon after acclimation to freshwater and increase through time. Here we are not concern about the explanation of size differences per se but for the purpose of analyzing shape changes in relation to size, i.e., allometry.

Traditional approaches would have given us a list of allometric coefficients making it hard to reconstruct the growth pattern and the nature of growth allometry. The splines

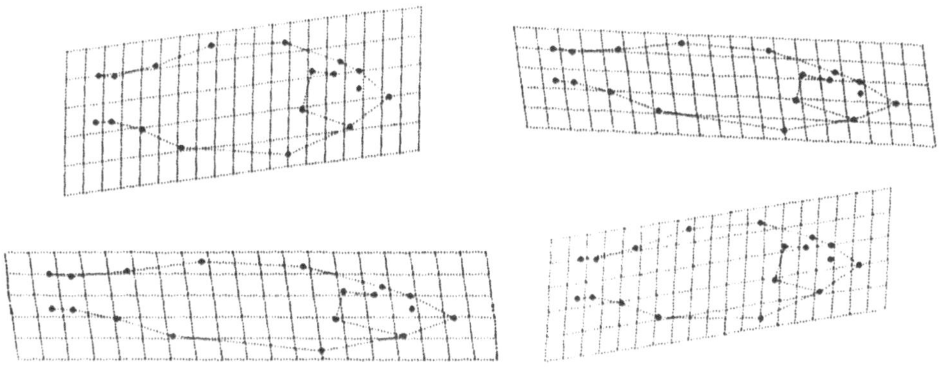

Figure 3. Uniform part of shape change of the marine sample (left in side view); and freshwater sample (right in side view). On top and bottom, splines of respectively small and high values of centroid size.

obtained with TPSREGR give a visual representation of allometry simultaneously for all landmarks. However, it still difficult to explain functionally the uniform and nonuniform differences in shape. Do these modifications reflect metabolic constraints, different feeding strategies or different swimming ability required in freshwater ?

There are two major conclusions of interest for evolutionary and fishery biology. If our assumption is correct, shape changes are a pure ecophenotypic effect related to salinity. This opens a question of paramount importance for taxonomy and systematics. Can there can be a confusing effect of ecophenotypism over morphometric characters normally used in fish taxonomy, such as, for instance, length of the head and height of the body, which needs to be explained before drawing any inference?

Furthermore, besides the possible economic implications of such shape plasticity in the marketing of freshwater-reared sea bass, the ability to extrapolate shape changes relative to centroid size may offer the opportunity to choose among rearing and acclimation practices.

REFERENCES

Barlow, G. W. 1961. Causes and significance of morphological variation in fishes. Systematic Zoology 10:105–117.

Barnabè, G. 1980. Expose synoptique des donnees biologiques sur le loup ou bar *Dicentrarcus labrax* (Linne, 1758). FAO Fisheries. Synopsis. 126.

Bookstein, F. L., 1989. Principal warps: Thin-plate splines and the decomposition of deformations. IEEE Transactions on Pattern Analysis and Machine Intelligence 11:567–585.

Bookstein, F. L. 1991. Morphometric tools for landmark data: Geometry and biology. Cambridge University Press: Cambridge.

Chervinsky, J. 1974. Sea bass (*Dicentrarchus labrax* L., Pisces, Serranidae) a "police fish" in freshwater ponds and its adaptability to various saline conditions. Bamigdeh 26:110–113.

Chervinsky, J. 1979. Preliminary experiments on the adaptability of juvenile European sea bass (*Dicentrarchus labrax*) and gilthead sea bream (*Sparus aurata*) to brackish water. Bamigdeh 31:14–17.

Corti, M. 1992. Data analysis in systematics: A workshop and a manual to introduce geometric morphometrics. ICOLATER., Caracas, Venezuela (Document accessible through anonymous FTP from life.bio.sunysb.edu—see Appendix 3 for instructions)

Corti, M. A. Loy, and S. Cataudella. 1996. Ecophenotypism in the sea bass, *Dicentrarchus labrax* (Moronidae: Teleostea), after acclimation to freshwater. A geometric morphometric analysis using Shape Coordinates. Envir. Biol. Fish.

Holliday, F. G. T. 1988. The effects of salinity on the eggs and larvae of teleosts. *in* W. S. Hoar and D. J. Randall (eds.), The physiology of developing fish. Volume I. Academic Press: New York.

Mayr, E. 1963. Animal species and evolution. Bellknap Press of Harvard University Press: Cambridge, Massachusetts.

Reyment, R. A. 1991. Multidimensional paleobiology. Pergamon Press: Oxford.

Rohlf, F. J. 1993a. TPSRW—Thin-plate spline relative warp. Department of Ecology and Evolution, State University of New York, Stony Brook, New York 11794

Rohlf, F. J. 1993b. Relative warp analysis and an example of its application to mosquito wings. Pages 131–159 *in* L. F. Marcus, E. Bello and A. García-Valdecasas, (eds.), Contribution to morphometrics., Monografias del Museo Nacional de Ciencias Naturales 8, Madrid.

Rohlf, F. J. 1993c. TPSREGR: A program for regression of partial warp scores. Department of Ecology and Evolution, State University of New York, Stony Brook, New York 11794.

Rohlf, F. J. and L. F. Marcus. 1993. A revolution in morphometrics. Trends in Ecology and Evolution 8:129–132.

Rohlf, F. J. and D. Slice, 1990. Extension of the Procrustes method for the optimal superimposition of landmarks. Systematic Zoology 39:40–59.

Slice, D. 1991. DS-DIGIT: Basic digitizing software. Department of Ecology and Evolution, State University of New York, Stony Brook, New York 11794.

Slice D. 1993. GRF-ND. Generalized rotational fitting on n-dimensional landmark data. Department of Ecology and Evolution, State University of New York, Stony Brook, New York 11794.

Sokal, R. R., and F. J. Rohlf, 1981. Biometry: The principles and practice of statistics in biological research. 2nd edition. W. H. Freeman: San Francisco.

Yashuov, A. 1969. Preliminary report on induced spawning of *M. cephalus* L. reared in captivity in fresh water ponds. Bamigdeh 21:19–24.

SEXUAL DIMORPHISM IN THE CRANIOFACIAL SKELETON OF MODERN HUMANS

Chris G. Wood[1] and John M. Lynch[2]

[1] Hominid Palaeontology Research Group
Department of Human Anatomy and Cell Biology
The University of Liverpool
PO Box 147
Liverpool, L69 3BX, England
[2] Interdisciplinary Humanities Program
Arizona State University
Tempe, Arizona 85287-1501

ABSTRACT

Sexual dimorphism is a major component of infraspecific variation in nonhuman primates. Among humans, differences in the pattern and degree of sexual dimorphism may provide a means for discriminating between populations. In this paper we explore the variance dimorphism, sex-specific size and shape differences, and between-population differences in patterns of sexual dimorphism in a sample of Romano–British (n = 47) and modern African (n = 53) crania. Shape coordinates, uniform components of variation and centroid size were derived for nine midsagittal landmarks obtained from lateral radiographs of the skull. No evidence was found for size-variance or shape-variance dimorphism in either population. Significant sexual size dimorphism was detected in the Romano–British sample alone. No significant sexual dimorphism in uniform components of variation was found in either sample. Multivariate analyses of variance for the effect of sex on shape coordinate pairs detected significant variation in the African sample, particularly in those landmarks that measure the amount of prognathism. Although the discriminant functions derived from the shape coordinate pairs correctly reallocated 81% of specimens to their sex, this value was not found to differ significantly from that of randomized data. The results of this study suggest that the consistency in patterns of between-population sexual dimorphism warrant caution when extrapolating from one sample to others for the purposes of sex discrimination.

Advances in Morphometrics, Edited by L. F. Marcus et al.
Plenum Press, New York, 1996

INTRODUCTION

Sexual dimorphism is, perhaps, one of the most dramatic illustrations of intraspecific variation in nature (Hanken and Hall, 1993). Its pattern and degree have been examined in many primate taxa, often with a view to extrapolating the findings to fossil hominids (e.g., Lieberman et al., 1988; Wood et al., 1991). Such studies have primarily used "traditional" morphometric techniques (e.g., O'Higgins et al., 1990), although of late, landmark-based studies have begun to appear. Many of these more recent analyses have used finite element scaling (FES) methods (e.g., Cheverud and Richtsmeier, 1986; Corner and Richtsmeier, 1991; Leigh and Cheverud, 1991) despite objections to the use of this technique for morphological investigations (Bookstein, 1987; 1991). Recently, the technique of Euclidean distance matrix analysis (EDMA) has been applied to dimorphism and growth patterns in *Macaca fascicularis* (Richtsmeier and Lele, 1993). We regard both FES and EDMA as relatively complicated compared with shape coordinate analysis and thus favor the latter for our current studies. We are aware of only one published work that uses shape coordinate analysis to examine dimorphism in primates (O'Higgins and Dryden, 1993).

The expression of sexual dimorphism in present and archaeological populations of modern humans has been the focus of considerable interest among anthropologists (see Calgano, 1981, and Uytterschaut, 1986, for reviews of the literature). Much of this research has been concerned with providing accurate and reliable means for the sexing of individuals from their skeletal remains—a prerequisite for paleodemographic studies (see Meindl et al., 1985). Methods include both the use of nonmetric traits (e.g., White, 1991) and the statistical analysis of linear measurements (see Hamilton, 1982); both approaches agree on the paramount importance of within-population size differences for assessing sex (Hall, 1982). The use of discriminant functions appears to be one of the more popular of the multivariate sexing techniques. Although there has been some debate as to applicability of discriminant functions to populations beyond those from which they were derived (e.g., Calgano, 1981) —ironically because of the influence of between-population size differences—it has nonetheless proved possible to derive "race-independent" discriminators (e.g., Uytterschaut, 1986) by removing such "extraneous variation" beforehand (Burnaby, 1966; Rohlf and Bookstein, 1987; Bookstein, personal communication).

In this paper we examine sexual dimorphism within African and Romano-British specimens of *Homo sapiens*. We examine a number of questions. (1) Is there any evidence for variance dimorphism (Leutenegger and Cheverud, 1985) in the human craniofacial region? (2) Can the sexes be separated based on cranial size, shape or both? (3) What is the possible utility of any such separation for sexing material of unknown sex? (4) How constant are the patterns of dimorphism across two modern populations?

MATERIAL AND METHODS

Lateral (mid-sagittal) radiographs were available for 100 crania of modern adult *Homo sapiens*. The African specimens (26 male, 27 female) date from recent times (post-1947) and the Romano–British sample (28 male, 19 female) from between the first and third centuries AD (Luboga, 1986). The African material is drawn from the Galloway Museum collection, housed at Makerere Medical School, Kampala (Uganda). All skeletons in this collection come from the unclaimed bodies of persons who died while undergoing treatment at Mulago Hospital. The sex of this material is known from the hospital records. The Romano-British material originates from two cemeteries at Poundbury (Dorchester, Dorset) and are housed at the Natural History Museum in London. These specimens were sexed by the curatorial staff at the museum from

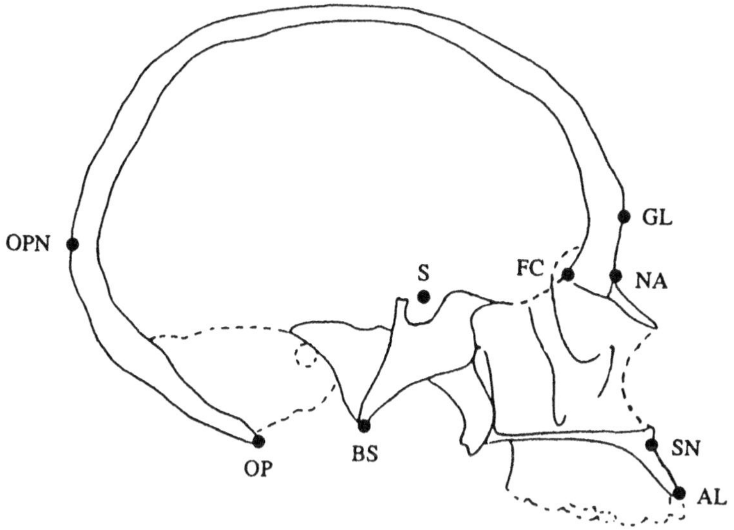

Figure 1. Nine craniofacial landmarks used in the study (adapted from Luboga and Wood, 1990).

osteological and dental characteristics, that is, using variables other than those examined in the present study. Details of the sample can be found in Farwell and Molleson (1993). Although ages were unknown, all specimens were judged to be adult on the basis of alveolar eruption of the third molar and fusion of the spheno-occipital synchondrosis (Luboga and Wood, 1990). The Cartesian coordinates of the nine landmarks (see Fig. 1 and Table 1) were obtained from the radiographs by use of a GRAF/BAR™ Mark II sonic digitizer and the DS-DIGIT program,

Table 1. Definitions of nine craniofacial landmarks used in the study (from Luboga, 1986). Landmark types follow Bookstein (1991)

Landmark and Abbreviation	Landmark type	Definition
Glabella (GL)	3	The most prominent point between the superciliary arches in the sagittal plane.
Nasion (NA)	1	The junction of the internasal suture with the nasofrontal suture.
Foramen caecum (FC)	3	The anterior limit of the cribriform plate of the ethmoid bone, immediately in front of the crista galli.
Sella (S)	1/2	The geometric centre of the sella turcica: the estimated intersection of the anteroposterior and superoinferior diameters.
Subnasospinale (SN)	3	The point at which the tangent to the lower margins of the nasal apertures intersects with the midline (i.e. the deepest point in the fossa immediately below the anterior nasal spine).
Alveolare (AL)	1	The lowest point on the alveolar process between the sockets of the two upper central incisors.
Opisthocranion (OPN)	3	The most prominent point on the posterior aspect of the cranial vault in the sagittal plane.
Opisthion (OP)	2	The point at which the internal and external surfaces of the occipital bone meet, on the posterior margin of the foramen magnum, in the sagittal plane.
Basion (BS)	2	The lowest point at which the internal and external surfaces of the occipital bone meet, on the anterior margin of the foramen magnum, in the sagittal plane.

version 1.0 (Slice, 1992). These points represent landmarks from all three landmark classes (Bookstein, 1991; Table 1).

Shape coordinates were calculated for all specimens based on the baseline OPN–GL. Although these landmark points are highly variable in life, an analysis of the data set based on alternative baselines indicated that the results were virtually identical for all baselines (unpublished data). Because the OPN–GL baseline allowed a more "natural" visualization of craniofacial variation, it was used for all results reported here. Shape coordinates, uniform components of variation and the squared form of centroid size (Bookstein, 1991) were obtained using the XY: Coordinates Utilities program, version 21/07/93 (Baylac, 1993). Mean forms were obtained by direct averaging of the data sets.

The two samples were tested for variance dimorphism by means of Levene's test (for size; Nie et al., 1975) and Box's M (for shape; Nie et al., 1975) and for magnitude dimorphism (by means of analysis of variance [ANOVA] for size and multivariate analysis of variance [MANOVA] for shape). In addition, discriminant functions (Marcus, 1990) were calculated for the separation between the sexes for each sample, based on the seven non-baseline shape coordinate pairs. All standard statistical procedures were carried out using version 5.02 of SPSS for Windows™.

RESULTS

Variance Dimorphism

There was no evidence for size-variance dimorphism in either sample (Levene's test: $F = 1.283$, df = 1, 51 for the African material; $F = 1.259$, df = 1, 45 for the Romano–British material; $p > 0.05$). Similarly, there was no shape-variance dimorphism in either the Romano–British (Box's $M = 186.65$; $F = 1.14$; df = 105, 4691; $p > 0.05$) or African (Box's $M = 170.51$; $F = 1.14$; df = 105, 8078; $p > 0.05$) samples.

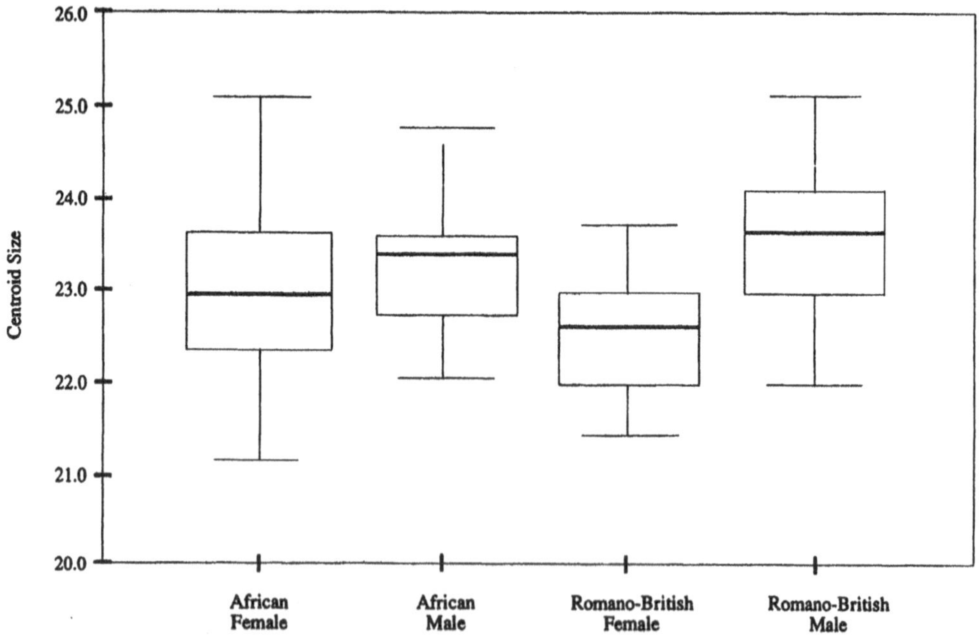

Figure 2. Box plot for centroid size across the four subsamples.

Table 2. Shape dimorphism in the African and Romano–British samples of *Homo sapiens*. Degrees of freedom are 2 and 44 for the African sample and 2 and 50 for the Romano–British sample

	Hotelling's T^2	
Shape coordinate	African	Romano-British
Opisthocranion	0.07	0.10
Basion	0.15*	0.09
Sella	0.09	0.12
Foramen caecum	0.06	0.04
Nasion	0.31***	0.07
Subnasospinale	0.21**	0.04
Alveolare	0.29***	0.02

* $p < 0.05$; ** $p < 0.01$; *** $p < 0.005$

Size and Shape Dimorphism

Significant size dimorphism (as measured by centroid size) was detected between the sexes in the Romano–British sample ($t = 3.70$; $df = 1$ and 45; $p < 0.001$) but not in the African material ($t = 1.57$; df = 1, 51; $p > 0.05$). Figure 2 presents box plots for centroid size across the four subsamples. Analysis of variance detected a significant effect of sex ($F = 14.65$; df = 1; $p < 0.0001$) but not sample ($F = 0.69$; df = 1; $p > 0.05$) on centroid size. The significant size difference in the Romano–British sample was not sufficient to produce a significant interaction between sex and sample ($F = 3.01$; df = 1; $p = 0.09$).

There was no significant difference between the sexes in the X or Y uniform components for either sample (African: $T^2 = 0.29$, df = 2, 44; Romano-British: $T^2 = 0.02$, df = 2, 50; $p > 0.05$ for both). These uniform components were insignificantly correlated with centroid size for all four subsamples (r varied between 0.16 and 0.45, F between 0.34 and 2.13; all $p > 0.05$).

Table 2 gives the results of MANOVAs for the effect of sex on shape coordinate pairs for each sample, that is, tests for shape dimorphism. No shape dimorphism occurred in the Romano–British sample, whereas significant shape variation occurred between the African males and females. A MANOVA detected significant shape variation between the sexes (Wilks' $\Lambda = 0.764$; df = 14, 83; $p < 0.05$) and samples (Wilks' $\Lambda = 0.197$; df = 14, 83; $p < 0.0001$), although there was no interaction between these two factors (Wilks' $\Lambda = 0.823$; df = 14, 83; $p = 0.44$).

Four shape coordinate pairs (BS, NA, SN and AL) exhibited significant sexual dimorphism within the African sample. Male specimens were significantly more prognathous than were their female counterparts (Fig. 3). Female specimens have a basion that is shifted back and up, while the nasion is shifted down and back (Fig. 3).

The seven shape coordinate pairs were entered into a discriminant function analysis, which resulted in a function that correctly reallocated 81.13% of specimens to their sex. This value is not, however, significantly different from the result if functions are calculated from randomized data (100 randomized replicates; $p = 0.08$). In addition, this discriminant function failed to sex the Romano–British material correctly (with 22 of the 47 specimens being misidentified). Thus, the observed shape differences, though significant, are insufficient to allow correct classification of unknown specimens.

Given the insignificance of the interaction terms in the above analyses, it would appear that the apparent difference in size and shape dimorphism between the two samples are relatively insignificant. Thus, we believe that the pattern of sexual dimorphism in the

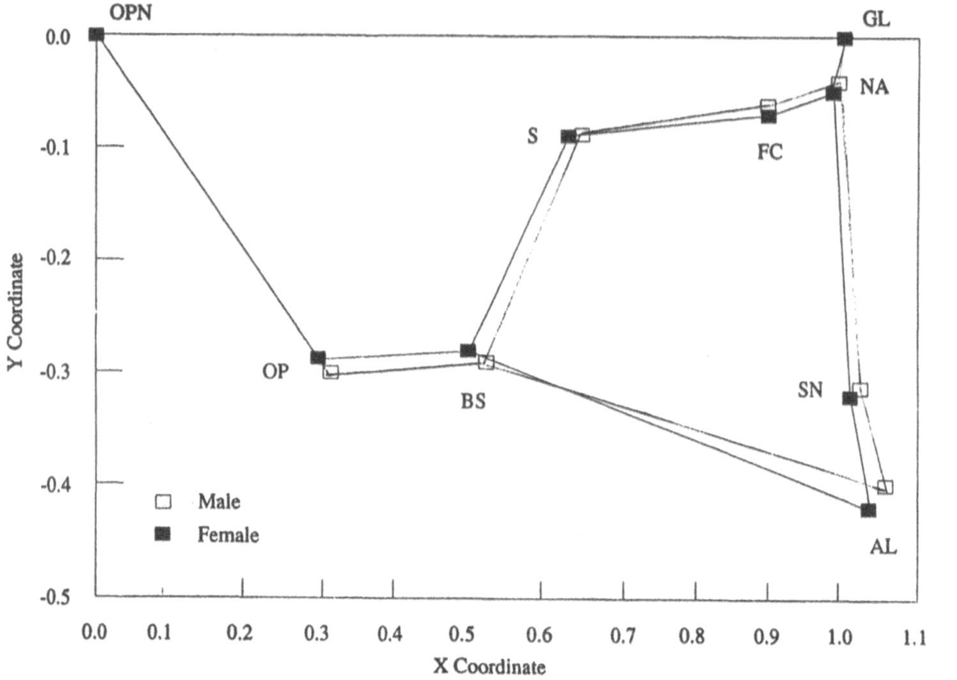

Figure 3. Mean shape coordinate forms for crania of male and female African *Homo sapiens*, generated using OPN and GL as a baseline. Sample sizes are 26 males and 27 females. Differences at SN, AL, BS and NA are statistically significant (see text for details).

landmarks used, though significant and relatively constant across samples, is insufficient to allow sexing of unknown material.

DISCUSSION

This study considered two types of sexual dimorphism, namely variance and magnitude dimorphism (Leutenegger and Cheverud, 1985). The results indicate that there is no evidence for variance dimorphism in the craniofacial region of male and female *Homo sapiens* in the relatively small samples considered here. There was some inconsistency in the observed patterns of magnitude dimorphism, with the African material being dimorphic in shape but not size and the Romano–British material being dimorphic in size alone. However, an examination of interaction terms in MANOVA detected this inconsistency to be insignificant. The shape variation, though statistically significant in the African material, was insufficient to allow sexing of unknown material.

Studies of a number of species of monkey and ape have shown a similar lack of craniofacial shape variation between the sexes in some taxa (e.g., *Pan* - O'Higgins and Dryden, 1993; *M. fascicularis*—Dryden, 1993; and *Papio cynocephalus*—Leigh and Cheverud, 1991), whereas shape dimorphism has been found in *Gorilla* and *Pongo* (O'Higgins and Dryden, 1993). It should be noted that the studies on *Papio* and *Macaca* used Procrustes analysis and finite element scaling analysis (FESA), respectively, and may therefore not be directly comparable with studies that use shape coordinates. For example, studies of *M. fascicularis* have shown both shape variation (based on EDMA; Richtsmeier and Lele, 1993) and a lack of such variation (based on shape coordinate analysis; Dryden, 1993), although few landmarks were shared between the two studies. There is an obvious need to compare the various methods available for simplicity and biological tractability.

Using Bookstein's shape coordinate analysis, O'Higgins and Dryden (1993) found size variance dimorphism in *Gorilla* but not in *Pan* or *Pongo*, and there was no evidence for shape variance dimorphism in any taxa. Greater character variance has been noted in males of a number of anthropoids (for review, see O'Higgins and Dryden, 1993). Leutenegger and Cheverud (1985) have suggested that variance dimorphism combined with selection dimorphism could prove a potent source of sexual dimorphism. The presence of variance or magnitude dimorphism in the craniofacial region is probably partly a function of the importance of sexual selection (via intermale competition or combat) within a species or population. This is generally negligible in modern humans, although the occurrence of shape dimorphism in the African sample may be of some adaptive significance.

From the above it is obvious that the pattern of sexual dimorphism in the craniofacial region of modern *Homo* is relatively consistent across the two populations analyzed in this study. Analysis of secular trends within populations would help to isolate any size and shape factors that are a function of a sample's temporal span: this is particularly important when considering variation in fossil hominid taxa (Wood et al., 1994). Other factors that are likely to play a role in producing size and shape differences both within- and between-populations include biological age of the individuals constituting the sample, climatic and cultural influences, nutritional status and the incidence of disease (see Hall, 1982, and papers therein), although the latter can usually be discounted by removing clear pathological specimens from the study.

The discriminant function derived for the African material failed to prove useful in sexing either its own specimens or those from the Romano–British sample. It may be that populations not examined in this study would prove highly dimorphic at the seven landmarks used here. In such a case, it may be likely that discriminant functions are specific for populations as, for example, in a number of mustelid carnivores (Lynch, 1993). While we agree with Uytterschaut's (1986) belief that sample-independent discriminants may exist, the results of the present study would suggest that a degree of caution must be exercised in extrapolating from patterns of sexual dimorphism in single samples to those in other populations.

ACKNOWLEDGMENTS

We would like to thank Sam Luboga for kindly making the radiographs available for study, and Alan Turner, Gabriele Macho, Malcolm Hall, Bernard Wood and the editorial committee for critical comments and discussion.

REFERENCES

Baylac, M. 1993, XY: Coordinates utilities. Laboratoire d'Entomologie. Museum National d'Histoire Naturelle, 45 Rue Buffon, F-75005 Paris.

Bookstein, F. L. 1987. Describing a craniofacial anomaly: Finite elements and the biometrics of landmark locations. American Journal of Physical Anthropology 74: 495-509.

Bookstein, F. L. 1991. Morphometric tools for landmark data: Geometry and biology. Cambridge University Press: Cambridge.

Burnaby, T. P. 1966. Growth-invariant discriminant functions and generalized distances. Biometrics 22: 91–110.

Calgano, J. M. 1981. On the applicability of sexing human skeletal material by discriminant function analysis. Journal of Human Evolution 10:189–198.
Cheverud, J. M. and J. T. Richtsmeier. 1986. Finite-element scaling applied to sexual dimorphism in rhesus macaque (*Macaca mulatta*) facial growth. Systematic Zoology 35:381–399.
Corner, B. D. and J. T. Richtsmeier. 1991. Morphometric analysis of craniofacial growth in *Cebus apella*. American Journal of Physical Anthropology 84:323–342.
Dryden, I. L. 1993. Statistical shape analysis of landmark data. Paper presented at the European Young Statisticians Meeting (EYSM-8), Palanga, Lithuania.
Farwell, D. E., and T. I. Molleson. 1993. Poundbury Volume 2. The Dorset Natural History and Archaeological Society, Dorset County Museum, Dorchester, U. K.
Hall, R. L. (ed.) 1982. Sexual dimorphism in *Homo sapiens*. A question of size. Praeger: New York.
Hamilton, M. E. 1982. Sexual dimorphism in skeletal samples. Pages 107–163 *in* R. L. Hall, (ed.), Sexual dimorphism in *Homo sapiens*. A question of size. Praeger: New York.
Hanken, J. and B. K. Hall. 1993. Mechanisms of skull diversity and evolution. Pages 1–36 *in* J. Hanken and B. K. Hall, (eds.), The skull. Volume 3: Functional and evolutionary mechanisms. University of Chicago Press: Chicago.
Leigh, S. R., and J. M. 1991. Sexual dimorphism in the baboon facial skeleton. American Journal of Physical Anthropology 84:193–208.
Leutenegger, W., and J. Cheverud. 1985. Sexual dimorphism in primates: the effects of size. Pages 33-50 *in* W. L. Jungers (ed.), Size and scaling in primate biology. Plenum Press: New York.
Lieberman, D. E., D. R. Pilbeam and B. A. Wood. 1988. A probabilistic approach to the problem of sexual dimorphism in *Homo habilis*: A comparison of KNM-ER 1470 and KNM-ER 1813. Journal of Human Evolution 17:503–511.
Luboga, S. A. 1986. Morphometric variation in the cranial base and facial skeleton of higher primates with special reference to modern humans. Ph.D. dissertation, University of Makerere.
Luboga, S. A., and B. A. Wood. 1990. Position and orientation of the foramen magnum in higher primates. American Journal of Physical Anthropology 81:67–76.
Lynch, J. M. 1993. Craniometric variation in selected mustelidae. Ph.D. dissertation, National University of Ireland.
Marcus, L. F. 1990. Traditional morphometrics. Pages 77–122 *in* F. J. Rohlf and F. L. Bookstein, (eds.), Proceedings of the Michigan morphometrics workshop. University of Michigan Museum of Zoology Special Publication 2.
Meindl, R. S., C.O. Lovejoy, R. P. Mensforth, and L. D. Carlos. 1985. Accuracy and direction of error in sexing of the skeleton: Implications for paleodemography. American Journal of Physical Anthropology 68:79–85.
Nie, N. H., C. H. Hull, J. G. Jenkins, K. Steinbrenner, and D. H. Bennet. 1975. Statistical package for the social sciences. McGraw-Hill: New York.
O'Higgins, P., and I. L. Dryden. 1993. Sexual dimorphism in hominoids: Further studies of craniofacial shape differences in *Pan*, *Gorilla*, and *Pongo*. Journal of Human Evolution 24:183–205.
O'Higgins, P., W. J. Moore, D. R. Johnson, T. J. McAndrew and R. M. Flinn. 1990. Patterns of cranial sexual dimorphism in certain groups of extant hominoids. Journal of Zoology (London) 222:399–420.
Richtsmeier, J. T. and S. Lele. 1993. A coordinate-free approach to the analysis of growth patterns: models and theoretical considerations. Biological Reviews 68:381–411.
Rohlf, F. J. and F. L. Bookstein. 1987. A comment on shearing as a method of "size correction." Systematic Zoology 36:356–367.
Slice, D. 1990. DS-DIGIT: Basic digitizing software. Department of Ecology and Evolution, State University of New York, Stony Brook, New York 11778.
Uytterschaut, H. T. 1986. Sexual dimorphism in human skulls. A comparison of sexual dimorphism in different populations. Human Evolution 3:243-250.
White, T. D. 1991. Human osteology. Academic Press: San Diego.
Wood, B. A., Yu Li, and C. Willoughby. 1991. Intraspecific variation and sexual dimorphism in cranial and dental variables among higher primates and their bearing on the hominid fossil record. Journal of Anatomy 174:185–205.
Wood, B. A., C. G. Wood, and L. W. Konigsberg. 1994. *Paranthropus boisei*: An example of evolutionary stasis? American Journal of Physical Anthropology 95:117–136.

SEXUAL DIMORPHISM IN MEDIEVAL HUMAN CRANIA STUDIED BY THREE-DIMENSIONAL THIN-PLATE SPLINE ANALYSIS

Torbjörn Ahlström

Osteological Research Laboratory
Stockholm University
Royal Castle Ulriksdal
S-170 71 Solna, Sweden

ABSTRACT

Sexual dimorphism with regard to size and nonuniform shape of the human skull was studied by means of a three-dimensional TPS analysis based on 42 landmarks of the skull. Significant differences with regard to both size and nonuniform shape could be substantiated. These results, although preliminary, suggests that the muscles of mastication and neck constrain the possible sexual shape variation encompassing anteroposterior as well as superoinferior dimensions of the skull, but not the lateral dimensions.

INTRODUCTION

This paper reports the findings of an exploratory study into the nature of sexual dimorphism in crania from the medieval cemetery at Västerhus, province of Jämtland, central Sweden, based on a three-dimensional (3D) extension of thin-plate spline (TPS) analysis. This cemetery was used during a time span of approximately 300 years, from the introduction of Christianity in the region at circa 1050 A. D. to 1350 A. D. when the pandemic known as the Black death struck the region (Gejvall, 1960). What will be specifically addressed is the question of sexual dimorphism with regard to both size and shape. Some aspects of the application of the TPS algorithm to 3D data will also be discussed.

Advances in Morphometrics, Edited by L. F. Marcus et al.
Plenum Press, New York, 1996

Table 1. List of landmarks used

1. Inion	22. Frontomalare orbitale dexter
2. Opisthion	23. Frontomalare temporale sinister
3. Basion	24. Frontomalare temporale dexter
4. Sphenobasion	25. Maxillofrontale sinister
5. Hormion	26. Maxillofrontale dexter
6. Stenion sinister	27. Zygoorbitale sinister
7. Stenion dexter	28. Zygoorbitale dexter
8. Asterion sinister	29. Zygomaxillare inferior sinister
9. Asterion dexter	30. Zygomaxillare inferior dexter
10. Auriculare sinister	31. Jugale sinister
11. Auriculare dexter	32. Jugale dexter
12. Mastoidale sinister	33. Sphenion sinister
13. Mastoidale dexter	34. Sphenion dexter
14. Lambda	35. Krotaphion sinister
15. Prosthion	36. Krotaphion dexter
16. Subspinale	37. Zygotemporale superior sinister
17. Nasion	38. Zygotemporale superior dexter
18. Bregma	39. Zygotemporale inferior sinister
19. Stephanion sinister	40. Zygotemporale inferior dexter
20. Stephanion dexter	41. Entomion sinister
21. Frontomalare orbitale sinister	42. Entomion dexter

MATERIAL AND METHODS

Twenty adult specimens were selected, 10 female and 10 male, primarily on the basis of preservation. These individuals were sexed using postcranial evidence and following standard osteological techniques (Gejvall, 1960; Krogman and Iscan, 1986). From these 20 specimens, coordinate data for 42 landmarks were collected (Table 1) by means of a diagraph. Landmarks are represented in Fig. 1.

Most landmarks are standard craniometric landmarks that have been in use since the beginning of this century, and their description can be found in Woo (1937), Howells (1973) and Bräuer (1988). The size of the individual specimens was measured by the centroid size of Bookstein (1989a). Centroid size is measured as the square root of the sum of all the $k(k-1)/2$ squared interlandmark distances of the individual specimens; k is the number of landmarks. Subsequently, for each sex separately, all 10 specimens were scaled, translated, and rotated into a common coordinate system in which a consensus, or mean, specimen was defined. For this procedure I used the generalized Procrustes fit supplied in the software GRF-ND (Slice, 1993). As the TPS algorithm is invariant to translations and rotations, the subsequent TPS analysis will not be affected by the fact that the male and female consensus forms were defined in two separate coordinate systems. After finding the two consensus forms, I rotated the female specimens into the male consensus form by use of a Procrustes fit. Thus, all residuals that are referred to below are Procrustes residuals with respect to the male consensus.

Bookstein (1989b) introduced an algorithm based upon the TPS interpolant for the mapping of a target form onto a reference form. Its extension to 3D data is straightforward (Bookstein, 1989b). It involves the kernel function $U_{(r)} = |r|$, the absolute value of the distance between landmarks i and j. Other than this, the extension to 3D data involves a redimensioning of the relevant matrices to accommodate a third coordinate. In this analysis, the male consensus form served as the reference into which the female consensus form was mapped. Although the principal and partial warps allow a decomposition of the nonuniform deformation into successively lower levels of scale, I restrict the discussion in this paper to the total nonuniform warp. For this total warp I computed scores by multiplying the coefficients of the total warp by the Procrustes residuals of all specimens with respect to the male consensus. This results in three scalars (*x*, *y*, *z*) for every specimen that may be plotted and serve as a foundation for the statistical tests discussed below.

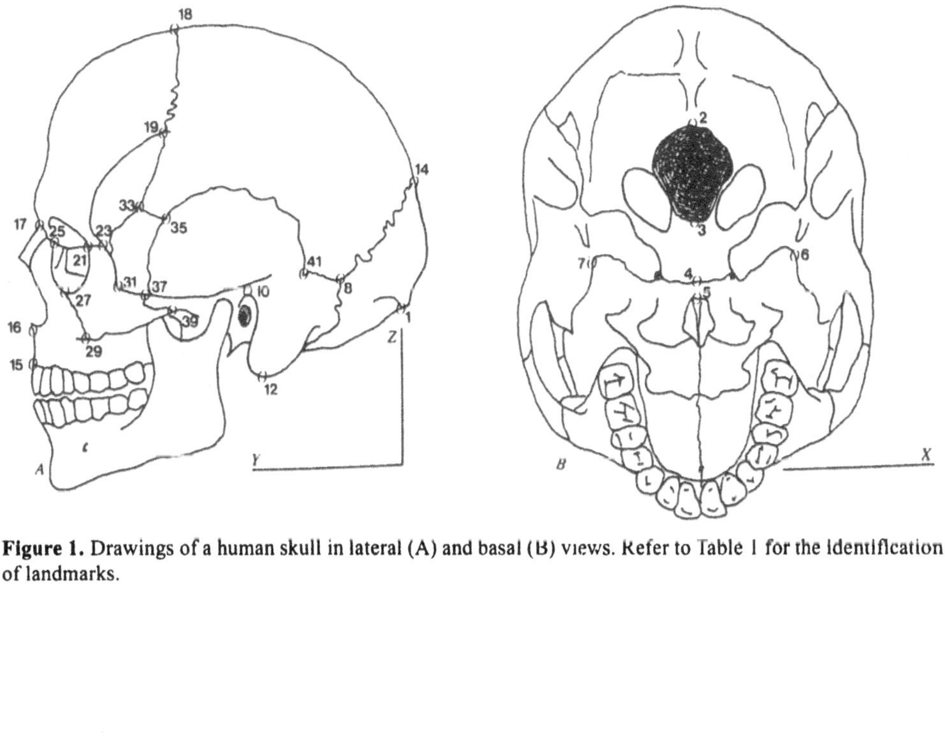

Figure 1. Drawings of a human skull in lateral (A) and basal (B) views. Refer to Table 1 for the identification of landmarks.

For a two-dimensional analysis, it is sufficient to use one grid defined in the plane of the data for the visualization of the deformation. Form changes are thereby inferred by the behavior of the deformed grid in the vicinity of landmarks. The 3D analog to this grid would be curtains of grid points passing through the reference archive. However, curtains of grid points might not be the most effective way to visualize 3D form changes because the view will be obstructed by the curtains. In this study, the total warp was visualized as vectors, analogous to Fig. 7.5.14 of Bookstein (1991), rather than by driving the function that deforms curtains of grid points.

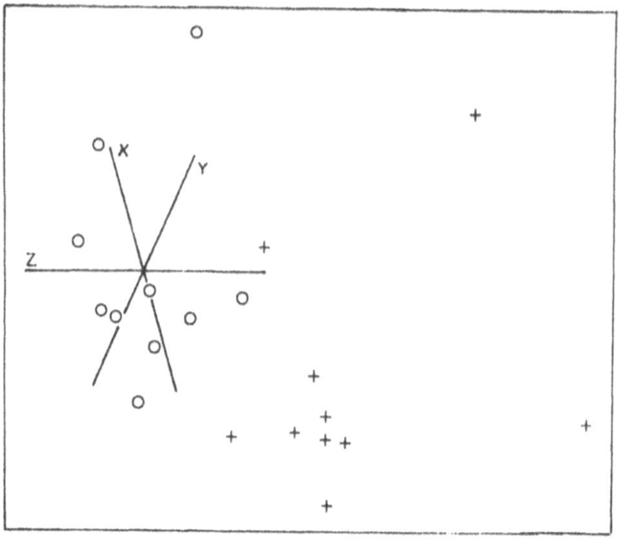

Figure 2. Scores for the individual specimens with respect to the total warp (+ = females, o = males). The origin of the plot represents the male consensus form.

RESULTS AND DISCUSSION

With regard to centroid size, the male mean was 46.701 with a standard deviation of 1.178 and the female mean was 45.130 with a standard deviation of 0.899. The coefficients of variation are 2.52 and 1.99, respectively; the males show more variation in size than do the females. These results were subjected to a one-sided t-test ($t = 3.353$, $df = 18$, $p < 0.05$) which revealed significant differences between males and females with respect to centroid size. Centroid size is presented in Fig. 3.

The principal strains corresponding to the affine part of the transformation are 1.0237, 0.9887 and 0.9808 respectively. The dilatation, the product of all principal strains, is 0.9926, signifying a volume decrease of 1%. This is not surprising as all specimens were scaled in the initial Procrustes fit. Anisotropy, the ratio of the largest to the smallest principal strain, for the most anisotropic plane section, is 1.0437. This corresponds to a shape change of 4%. For this comparison, based on the mapping of consensus forms only, I conclude that the affine component is negligible. Judging by the principal strains, we are left with a spheroid rather than an ellipsoid. There are 38 principal warps corresponding to deviations from the affine part of the mapping. The eigenvalues associated with these principal warps range from 90.6563 to 2.5271. The first principal warp involves features at a small scale; in this case it involves the landmarks sphenobasion and hormion. The last principal warp, corresponding to large-scale features, involves mainly the following bilateral landmarks; asterion, auriculare, mastoidale, frontomalare orbitale and temporale as well as entomion. The bending energies associated with the partial warps range from 0.0031 to 0.0001, summing up to a total bending energy of 0.0178. All partial warps add up to a significant total warp as judged by the distribution of the scores (Fig. 2). In the following exposition I shall restrict the discussion to the total warp.

The scores of the total warp are provided in Fig. 2, and, he descriptive statistics for the scores of the total warp are printed in Table 2. The male reference constitutes the origin of the plot. Females are drawn as pluses and males as circles. The total warp scores for the two sexes were subsequently tested by a two group Hotelling's T^2, and the group differences

Table 2. Descriptive statistics for the total warp scores

Statistic	*x*-coordinate	*y*-coordinate	*z*-coordinate
Males			
Mean	0.000	0.000	0.000
SD	0.001	0.006	0.002
Females			
Mean	0.003	0.009	0.005
SD	0.001	0.010	0.006

were found to be significant ($F = 17.589$, df = 3, 16, $p < 0.05$). It can be seen on the basis of the standard deviations that the *y*- and *z*-coordinates are more involved than is the *x*-coordinate. The practical importance of this is that the pictorial representation of the total warp can be restricted to the *y*- and *z*-coordinates. In biological terms this translates into shape variation along the length and height dimensions of the cranium rather than the breadth dimension.

To assess the relationship between the variation with regard to centroid size and the nonuniform shape variation further I plotted centroid size against the sum of scores for each specimen (Fig. 3).

The results of the total deformation as the female consensus is deformed into the male consensus is drawn for the sagittal plane in Fig. 4. Only the left bilateral landmarks are shown along with the unilateral landmarks. The major features of the nonuniform deformation are the posterosuperior deformation at inion, an anterior deformation at the bilateral landmark asterion, a posteroinferior deformation at the bilateral landmark entomion, a posteroinferior deformation at the bilateral landmark stephanion and an anterosuperior deformation at nasion. Some landmarks are not involved at all in the deformation, i.e. lambda, both auriculare and the bilateral landmarks zygomaxillare superior and inferior.

For these 20 human crania, sexual dimorphism is a question of size and shape. Most of the nonuniform shape changes are located to the posterior part of the neurocranium. These differences suggest a relaxation of the lower parts of the occipital region in the females

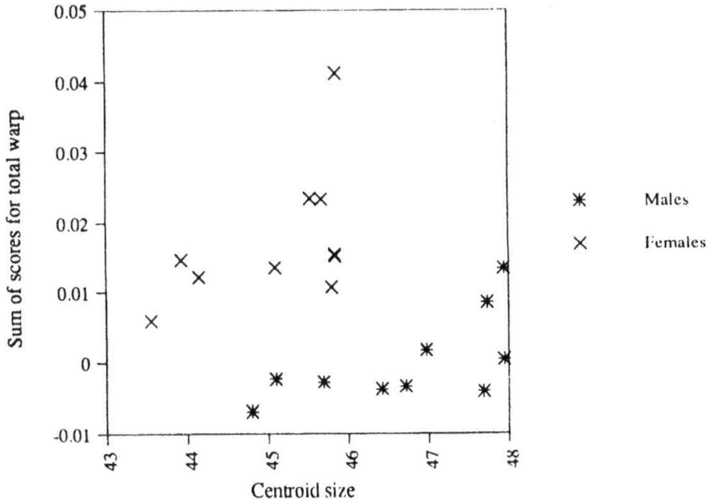

Figure 3. Sum of scores for all specimens with regard to the total warp plotted against centroid size.

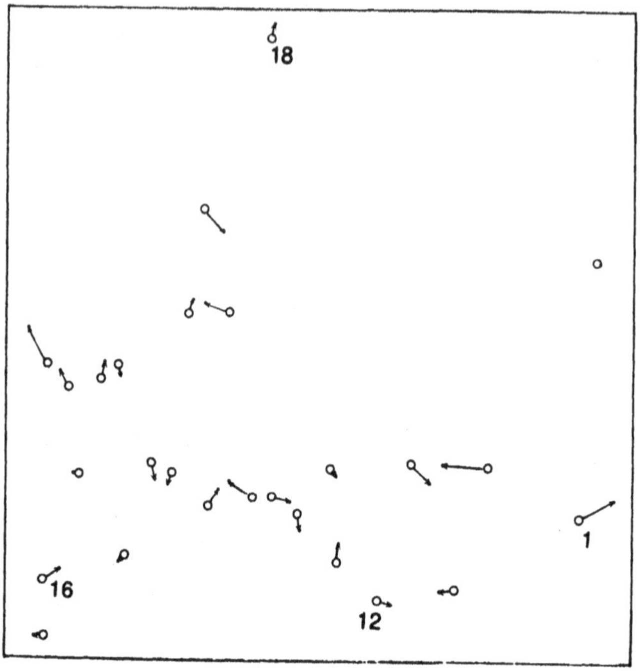

Figure 4. Total warp drawn as vectors. Refer to Fig. 1 and Table 1 for the identification of landmarks.

compared with the males. This can be given a functional interpretation because the nuchal muscles arise in the region. The bilateral landmark stephanion, which is connected to the arising of the temporalis muscle, is deformed in a posteroinferior direction. This could suggest that the temporalis muscle is displaced in a posteroinferior direction in the female skull. Because there are no more anatomical landmarks over the relevant region of the parietal bone to calibrate this feature, interpretations should be cautious. One effect of these implied changes could be a relative lengthening of the female skull compared with the male, as judged collectively by the landmarks nasion and inion. This study has demonstrated that the shape variation due to sexual dimorphism mainly involves the *y*- and *z*-coordinates, but not the *x*-coordinate. These results, although preliminary, suggest that the muscles of mastication and neck constrain the possible sexual shape variation encompassing anteroposterior as well as superoinferior dimensions of the skull but not the lateral dimensions.

ACKNOWLEDGMENTS

I thank Fred Bookstein, David Dean, Anna Loy, Leslie F. Marcus, F. James Rohlf, Torstein Sjøvold and Dennis E. Slice for constructive criticism and software. Travel costs in association with this work was defrayed by a generous grant from the Wallenberg's Jubileums-fund, Stockholm.

REFERENCES

Bookstein, F. L. 1989a. "Size and shape": A comment on semantics. Systematic Zoology 38:173–180.

Bookstein, F. L. 1989b. Principal warps: Thin-plate Splines and the decomposition of deformations. IEEE Transactions on Pattern Analysis and Machine Intelligence. 11:567–585.

Bookstein, F. L. 1991. Morphometric tools for landmark data: Geometry and biology. Cambridge University Press: Cambridge.

Bräuer, G. 1988. Osteometrie. Pages 160–232 *in* R. Knussmann, (ed.), Anthropologie. Handbuch der vergleichenden biologie des menschen. Band I, Teil 1., Gustav Fischer Verlag: Stuttgart.

Gejvall, N.-G. 1960. Westerhus. Medieval population and church in the light of skeletal remains. Håkan Ohlssons Boktryckeri: Lund.

Howells, W. W. 1973. Cranial variation in man. A study by multivariate analysis of patterns of difference among recent human populations. Peabody Museum Papers. Harvard University.

Krogman, W. M., and M. Y. Iscan. 1988. The human skeleton in forensic medicine. 2nd edition Charles C Thomas: Springfield, Illinois.

Slice, D. E. 1993. GRF-ND: Generalized rotational fitting of N-dimensional data. Department of Ecology and Evolution, State University of New York, Stony Brook, New York 11794.

Woo, T. I. 1937. A biometric study of the human malar bone. Biometrika 29:113–123.

A MORPHOMETRIC MODEL AND LANDMARK ANALYSIS OF ACHEULIAN HAND AXES FROM NORTHERN ISRAEL

Scott Brande[1] and Idit Saragusti[2]

[1] Department of Geology
University of Alabama at Birmingham
Birmingham, Alabama 35294
[2] Institute of Archaeology
Hebrew University - Mt. Scopus
Jerusalem 91905, Israel

ABSTRACT

The numerical abundance of artifacts in the archaeological record is in some ways analogous to the occurrence of individuals in many species populations. Methods devised by biometricians for the geometric description of organism form may be applied to artifacts, provided that coordinates of landmarks on individual specimens are available. We have conceptualized two- and three-dimensional geometric models of a bifacial tool, the hand axe, defined landmark locations and from them transformed traditional linear measures into geometric coordinates, illustrated individual variations in hand-axe shape, and found statistically significant differences in mean shape between two stratigraphic units in an excavation in northern Israel. The concept of a geometric model for a hand axe may be easily applied to other types of artifacts for which geometric coordinates may be calculated and may prove a useful tool for the archaeologist who desires to illustrate and analyze artifact form.

INTRODUCTION

More than 1.5 million years ago, a new type of tool appeared in the archaeological record—the hand axe. This tool type is bifacially retouched (Fig. 1) and may have been used for a wide variety of functions, including among others, chopping, hide scraping or skinning and meat cutting (Schick and Toth, 1993, chapter 8). Bifacially retouched tools in general, and hand axes in particular, are the most characteristic tools of the Acheulian industries of the Lower Paleolithic period. Assemblages belonging to Acheulian hand-axe industries are generally dated to the Lower and Middle Pleistocene, between about 1.5 million years ago, when they first appeared in East Africa, and about 200,000 years ago. Assemblages assigned

Advances in Morphometrics, Edited by L. F. Marcus et al.
Plenum Press, New York, 1996

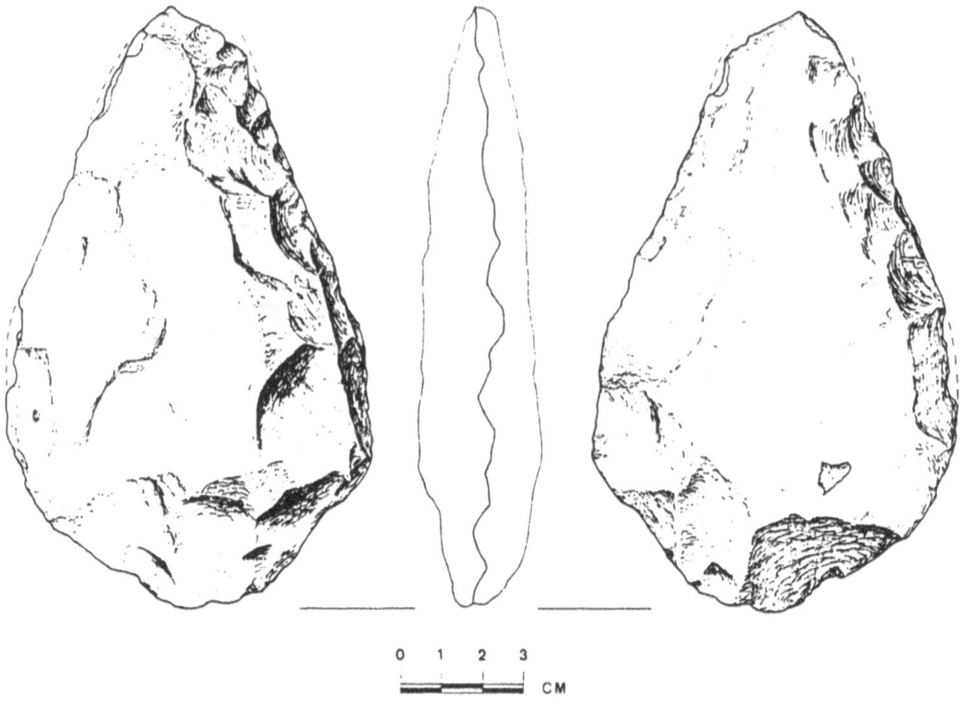

Figure 1. A drawing of a hand axe made of basalt from unit II-6-L4 at the site of Gesher Benot Ya'aqov, northern Israel.

to these industries are known from England to South Africa, from Portugal to India and beyond in the Orient. Acheulian hand-axe industries are thus broadly dispersed both in time and space. Hand axes may be recovered in abundance at many archaeological sites (Potts, 1993, Table 3.1) and are found in museum collections in large numbers.

Hand-axe morphology must certainly provide abundant, indirect evidence for hominid behavioral choices—from the selection of raw material to manufacturing methods to temporal morphometric changes. The systematic biologist may collect and store many thousands of specimens of a species of insect, but not each of them is usually figured in a systematic publication. Rather, the systematist today relies on a wide variety of morphometric techniques that statistically or graphically summarize a sample of individual specimens in ways which result in a compact description of the species. We believe the study of hand-axe morphology may benefit from methods developed for the quantitative summary and analysis of biological variation.

Previous approaches to the morphological study of hand-axe populations range from a typological classification based on quantitative measures of shape (computed from such simple length and position measures as maximum length, maximum width, location of maximum width, width at half the length and maximum thickness) of many individual specimens (Bordes, 1961) to more extensive statistical studies supplemented by graphical analysis of bivariate scattergrams (Roe, 1964, 1968; Isaac, 1977) that would permit quantitative comparisons of population samples of hand axes within and among different sites. Crompton and Gowlett (1993) briefly reviewed previous approaches, and applied a number of multivariate statistical techniques to quantitative studies of bifacial tools.

With advances in computer technology, there have been many developments in quantitative methods used by biometricians, resulting in the recognition of a field now termed geometric morphometrics (reviewed in Rohlf and Marcus, 1993). We have begun studies to apply some of these methods to the morphometrical analysis of hand axes. We begin with a formulation of a generalized morphometric model of a hand axe based in part upon variables selected by Bordes (1961) and Roe (1964, 1968) and then convert various length and position measurements to two-dimensional (2D) coordinates. Following Bookstein (1991: page 2), we recognize the (x,y) location of the maximum width on a length-normalized baseline as a pair of Galton shape coordinates. For two samples of hand axes from an Acheulian site of northern Israel, we compute shape coordinates and test for a statistically significant difference in mean landmark position based on Hotelling's T^2 statistic in a general multivariate analysis of variance. We illustrate hand-axe shape variation in graphical displays. Finally, we conceptually extend our 2D, landmark-based model to a three-dimensional (3D) one.

TWO-DIMENSIONAL GEOMETRIC MODEL OF A HAND AXE

We conceived of modeling the geometry of hand axes by imposing the various measured lengths upon a 2D, Cartesian coordinate system (Fig. 2).

Following Bookstein (1991: page 6), we locate the origin ($x = 0, y = 0$) at the hand-axe base and orient the stone with its longest axis (base to tip) parallel to the y-axis, thus defining a geometric baseline.

Computed shape coordinates are located at what we refer to as hand-axe landmarks (type 3, sensu Bookstein, 1991: page 63ff, as we apply the concept to nonbiologic objects) at the ends of the measured intervals of length. We sequentially number landmarks from the base (x_1=0, y_1=0) in a counterclockwise fashion (Fig. 3). Roe (1964, 1968) made linear measurements on the hand-axe length (maximum length, the distance from base to tip), the vertical location along the length (height) where the largest width occurred and the width at three specific locations: at 20 and 80% of the height from base to tip and the largest width. Bordes (1961) measured an additional width at half (50%) of the length. Using Bordes' and Roe's measures and our model, we can compute x,y coordinates (symmetrical on x) around the periphery of each stone. Although actual specimens may be asymmetrical, whereby widths are unequal from an edge to the centerline at some specific height above the base,

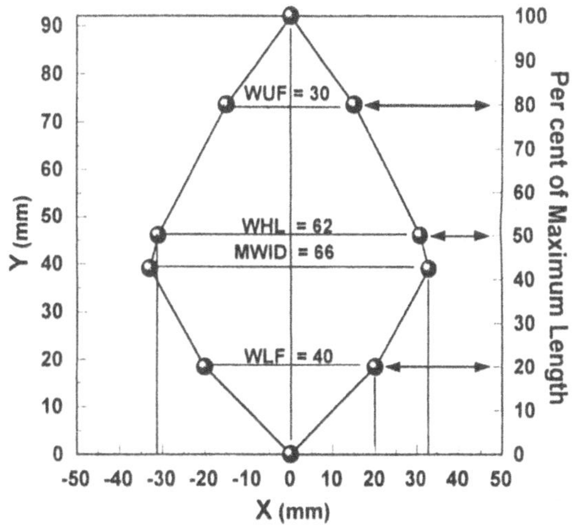

Figure 2. Conversion of length and position measurements of the plan view of a hypothetical hand axe to x, y coordinates. Hand axe is oriented with base at (0, 0), and base-to-tip line on y-axis. Width measurements are divided by two to compute positive and negative x-coordinate values. All units are in millimeters (mm). Maximum length (MLEN) = 92, maximum width (MWID) = 66, width at lower fifth of maximum length (WLF) = 40, width at half the maximum length (WHL) = 62, width at upper fifth of maximum length (WUF) = 30 and location on maximum length of maximum width (LMW) = 39.

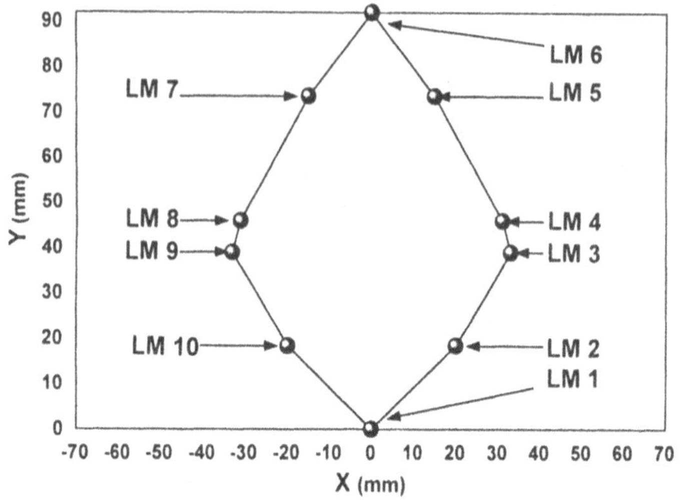

Figure 3. Sequential locations of landmarks (LM) on a hypothetical hand axe, beginning with hand axe base and continuing in a counterclockwise fashion.

such asymmetries are not detected by Bordes' and Roe's measurements. (Hand axe asymmetry may, of course, be studied by taking separate measurements from the centerline to each edge of the stone.) We therefore reflect the x-coordinates across the y-axis, thus generating equal but negative x-coordinates on the opposite side of the stone, as in Fig. 3 for landmarks 2 and 10. For example, if the width of a hand axe is 40 mm at the lower fifth (20%) of maximum length ($y = 18$ mm $= 0.2 \bullet 92$ mm), then the positive x-coordinate is 20 mm and the negative x-coordinate is -20 mm (Fig. 2, WLF).

x-coordinates are similarly computed from other linear measurements. The y-coordinates are derived from specified locations of Bordes' and Roe's measurements or from direct measurement on the stone. For example, if the maximum length of the stone is 92 mm, then the vertical location (y-coordinate) at 80% of the height (to accompany the x-coordinate based on the width of the upper fifth of maximum length) is 74 mm. The maximum width (twice the x-coordinate) and its location along the y-axis (y-coordinate location of the maximum width) is measured, not derived.

MORPHOMETRIC AND STATISTICAL ANALYSES OF HAND AXES

A geometrical analysis of hand axe morphology proceeds from our model formulation through two additional steps: (1) baseline length normalization and (2) statistical analysis of shape coordinates. Our study of hand-axe shape begins with the scaling of hand-axe coordinates to a baseline of unit length (simply divide all coordinate values by the baseline length). Shape differences among individual hand axes, and between groups, are then reflected in the variation of landmark (shape) coordinates relative to the unit length baseline.

Figure 4a illustrates a hypothetical sample of three hand axes. Each is a different size as measured by the baseline length, and each is shaped differently in that the location of the maximum width (landmark 3) varies both in x and y. In Fig. 4b, each of the three hand axes is scaled to a baseline of unit length by dividing all measures (and therefore coordinates) by each stone's maximum length. Raw (x, y) coordinates have now been transformed to Galton

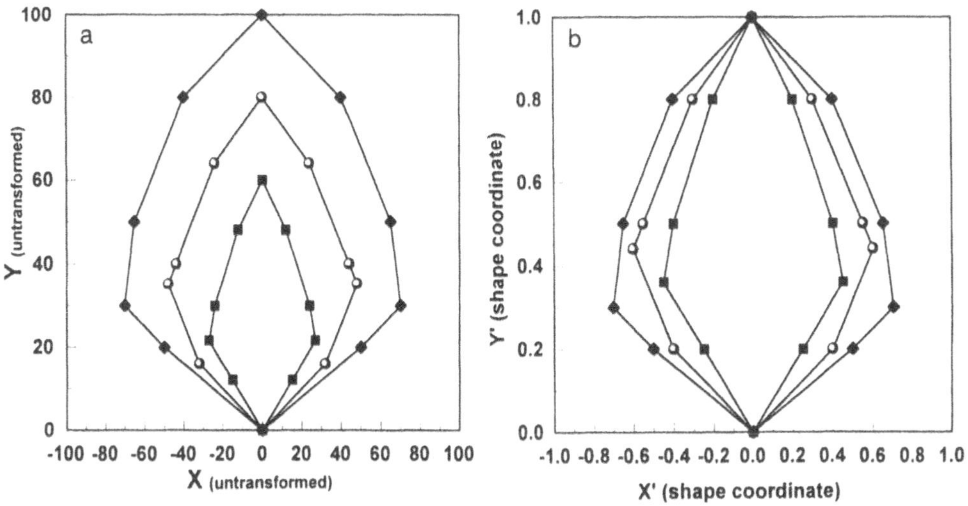

Figure 4. Sample of three hypothetical hand axes, with landmarks indicated by symbols located on outline corners. Scale units (*x*- and *y*-axis) are arbitrary. (a) Unscaled (raw data) coordinates, (b) coordinates scaled to baseline of unit length.

shape coordinates (Bookstein, personal communication). Variation in positions of the landmarks relative to the baseline is now size-independent. For example, the narrowness of the smallest hand axe compared with the largest remains even after scaling to unit baseline length.

If all three hand axes are assumed random samples from a single population, it would be of interest to illustrate and to analyze statistically geometric variation of landmark position. In Fig. 5, only three landmarks (1, 3 and 6) for each of the three hand axes are illustrated. The three landmarks, two of which terminate the hand-axe's major axis, form for

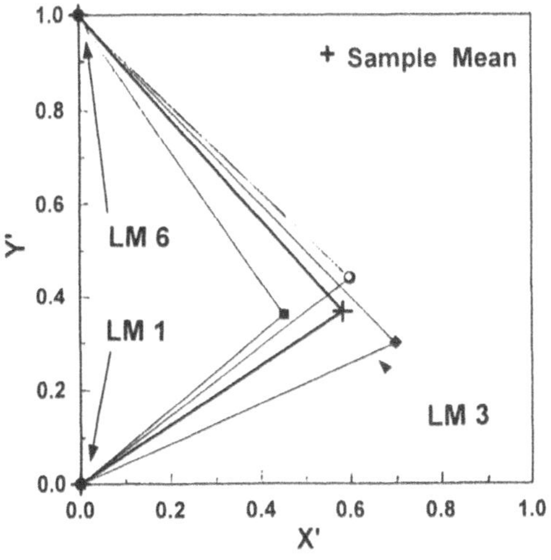

Figure 5. Shape triangles, based on landmarks 1, 3 and 6, for a sample of three hand axes and sample mean shape triangle.

each item a shape triangle. Landmark 3 is the location of maximum width, and we can see a scatter of points that reflects the sample geometric variation of its position. A mean position for this landmark may be calculated as the arithmetic mean of the x and y shape coordinates for the three items in the sample. If the scatter of individual hand axes (in landmark 3) about this bivariate mean is observed to be approximately equal in both Cartesian coordinates, we may infer circular normal scatter (Bookstein, 1991: section 5.3), analogous to the univariate standard deviation, a measure of variation of sample items about a mean.

If we had two random samples of hand axes, say from two layers, and observed in each a circular scatter of items about a mean landmark of interest, we may wish to assess statistically the null hypothesis that the population landmark means from which the samples are drawn are the same. The alternative hypothesis is that the populations from which the samples are drawn are different in means of landmark positions. Because each landmark is composed of two variables (shape coordinates x and y), a statistical comparison of two groups is necessarily a multivariate one. An analogue of the univariate, two-group Student's t-test for the difference between two means is the two-group, multivariate analysis of variance (MANOVA) for the difference between two mean vectors (more than one independent variable). In our hand-axe example, each landmark adds two independent variables, x and y, to the analysis. The dependent variable is the group, for which each sample will have a different value. The null hypothesis is assessed by the probability of the value of a test statistic, in the two-group case, Hotelling's T^2. The null hypothesis is rejected if the probability of Hotelling's T^2 falls below the conventional limit of 0.05.

A two-group, multivariate statistical test is subject, of course, to a variety of assumptions, including multivariate normality, and equality of variance–covariance matrices. Following Bookstein (1991: page 159), however, we proceed with the MANOVA as long as the scatter of shape coordinates appears approximately circular about the landmark and if there are no striking outliers.

MATERIAL AND METHODS

Two assemblages of hand axes (designated L1 and L4) were recovered from different stratigraphic units during renewed excavation at the Acheulian site of Gesher Benot Ya'aqov, Hula Valley, northern Israel (Fig. 6; Goren-Inbar and Belitzky, 1989; Belitzky et al. 1991; Goren-Inbar et al., 1992). As discussed above, various length and position measurements

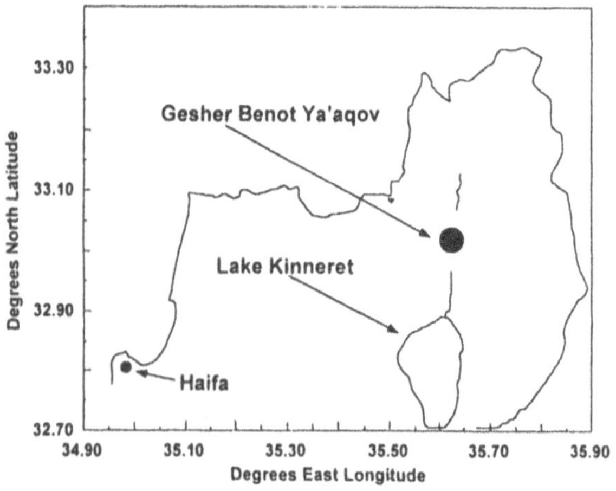

Figure 6. Location of Gesher Benot Ya'aqov excatation in northern Israel. For detailed description of site location see Goren-Inbar et al. (1992).

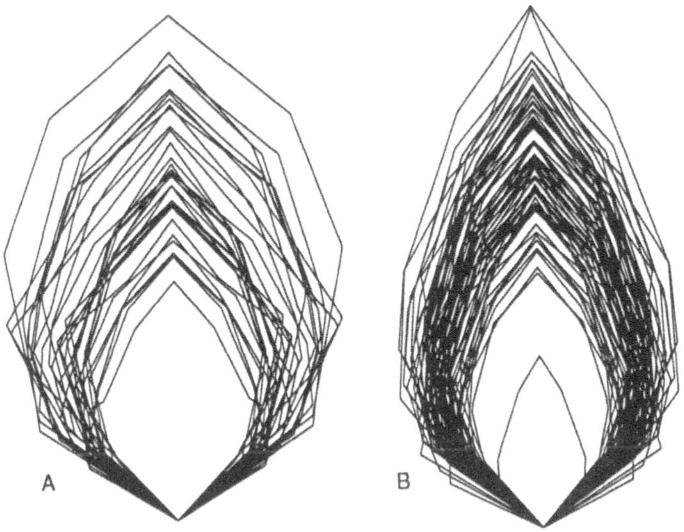

Figure 7. Outline maps of raw peripheral coordinate data derived from length and position measurements for hand axes excavated from two stratigraphic units at the Gesher Benot Ya'aqov site, northern Israel. (A) Layer II-6-L1, $n = 33$ and (B) layer II-6-L4, $n = 73$.

defined by Bordes (1961) and Roe (1964, 1968) and a few additional ones were made to the nearest millimeter (mm) on each specimen (Fig. 2).

We wrote a computer program (in QBASIC) to calculate landmark coordinates according to our 2D model for each specimen from the length measures. Several data files of different formats were prepared for use in other programs. One file of (x, y) coordinates was prepared in a format for input to GRF-ND, a program written for the graphical display and geometric analysis of landmark data (Slice, 1993).

In this study, we restrict our analysis to landmarks 1, 3 and 6, the baseline and location of maximum width of each hand axe, on the plan (2D) view of each tool. These three landmarks thus constitute a shape triangle for each hand axe. Raw coordinates were then divided by maximum length for baseline normalization and calculation of shape coordinates. Descriptive statistics for hand-axe measures and calculated shape coordinates and MANOVA computations were performed using Minitab for Windows (Release 9.2).

RESULTS

Descriptive statistics of hand axes excavated from two stratigraphic units of the Gesher Benot Ya'aqov are presented in Table 1, and outlines based on raw coordinates are illustrated in Fig. 7. Our qualitative impression is of a general uniformity in "geometric shape" within, and of a difference in shape between, these units.

Mean shape triangles for the two groups and individual item scatters for shape coordinates of landmark 3 are illustrated in Fig. 8. Results of a two-group multivariate analysis of variance are presented in Table 2. The samples of hand axes from the two stratigraphic units, L1 and L4, are drawn from significantly different populations of mean shape, as defined by the position of landmark 3, the location of maximum width relative to a baseline of unit length. Univariate analyses of variance (ANOVAs) for both x and y shape

Table 1. Descriptive statistics for two groups of Acheulian hand axes from the Gesher Benot Ya'aqov of northern Israel. L1 is layer II-6-L1; L4 is layer II-6-L4

Variable and sample statistic	Value
L1 ($n = 33$)	
Maximum length (MLEN)	
Minimum	77
Mean	114.94
Maximum	163
Standard deviation	21.49
Maximum width (MWID)	
Minimum	57
Mean	80.58
Maximum	115
Standard deviation	17.39
Location of maximum width (LMW)	
Minimum	20
Mean	50.91
Maximum	87
Standard deviation	13.37
Landmark 3 shape coordinate X'	
Minimum	0.287
Mean	0.350
Maximum	0.440
Standard deviation	0.0361
Landmark 3 shape coordinate Y'	
Minimum	0.253
Mean	0.444
Maximum	0.639
Standard deviation	0.0923
L4 (n = 73)	
Maximum length (MLEN)	
Minimum	63
Mean	137.82
Maximum	193
Standard deviation	23.45
Maximum width (MWID)	
Minimum	35
Mean	82.68
Maximum	111
Standard deviation	13.67
Location of maximum width (LMW)	
Minimum	17
Mean	51.75
Maximum	94
Standard deviation	14.03
Landmark 3 shape coordinate X'	
Minimum	0.225
Mean	0.302
Maximum	0.440
Standard deviation	0.0315
Landmark 3 shape coordinate Y'	
Minimum	0.105
Mean	0.379
Maximum	0.588
Standard deviation	0.0894

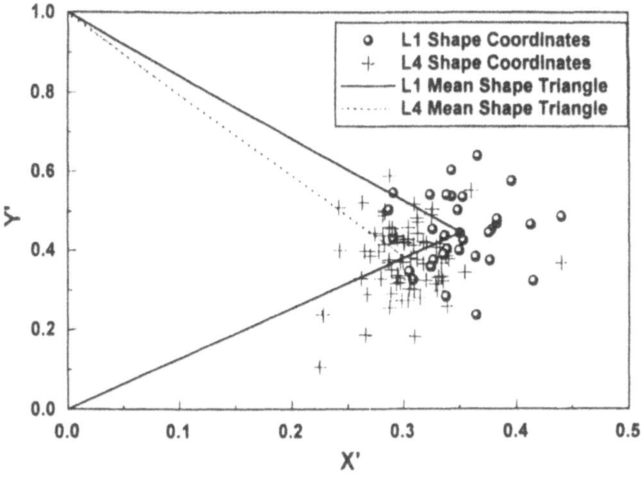

Figure 8. Shape coordinates and mean shape triangles based on landmarks 1, 3 and 6 for two samples of Acheulian hand axes.

coordinates are highly significant ($p < 0.001$ and $p \sim 0.001$, respectively). Hotelling's T^2 statistic for the hypothesis of equality of multivariate means is also highly significant at $p <$ 0 .000.

Hand axes from L1 are significantly wider than are the hand axes from L4, and the position of maximum width is significantly higher in hand axes from L1 than in hand axes from L4.

DISCUSSION

Some previous approaches to quantitative analyses of hand-axe morphometry (e.g. Bordes, 1961; Roe, 1964, 1968) focused on the measurement of simple lengths, calculation and graphing of various ratios to express what was thought to reflect essential components of hand-axe shape, and computation of two-sample *t*-tests (or nonparametric Mann–Whitney

Table 2. Two-group univariate and multivariate analysis of variance of Gesher Benot Ya'aqov hand axes (L1, $n = 33$; L4, $n = 73$)

Source of variation	df	Statistic	*p*-value
ANOVA on *x* shape coordinate			
Between groups	1	F = 50.00	<0.000
Error	104		
ANOVA on *y* shape coordinate			
Between groups	1	$F_s = 12.03$	~0.001
Error	104		
MANOVA on *x*, *y* shape coordinates			
Between groups	2	$T^2 = 58.66$ $F_s = 29.05$	<0.000
Error	103		

tests) for assessment of intersample variation. Similar studies, somewhat elaborated, were conducted by Isaac (1977). In a recent study of the variation of size and shape of African bifacial tools, Crompton and Gowlett (1993) found several multivariate techniques, such as multivariate allometry, principal components and discriminant function analyses, helpful in discerning the presence of allometric variations within and among sites.

Our formulation of a model of hand-axe geometry, and utilization of the growing body of theory surrounding the morphometric analysis of organisms (e.g., Bookstein, 1991) permits a different approach that may offer substantial benefits to the analysis of artifacts, such as the hand axes in this study, as well as alternative methods to those used in previous studies.

Our geometric model permits conversion of classical length measures to geometric coordinates, thus making possible a graphical display of the geometry of the hand axe, individually and in groups.

As hand axes in any given sample tend to vary in size (measured by maximum length, for example), one would like to assess geometric differences in shape in a way that is independent of overall size. Baseline normalization and construction of shape coordinates satisfies that objective. Whereas previous workers have sought quantification of hand-axe shape in such derived variables as ratios based on maximum length, some pairs of ratios are together equivalent to a shape coordinate pair, such as for the geometric position of the maximum width: x = maximum width divided by maximum length and y = the location of maximum width along the baseline divided by maximum length. Shape coordinates may then be used to illustrate directly the geometry of the objects.

Extension of the 2D to a 3D Hand Axe Model

We have focused on the geometric analysis of hand axes in two dimensions by considering only landmarks derived from maximum length and various widths around the periphery of a specimen. An extension of our 2D model to three dimensions is easily made by incorporating measurements of thickness into the model.

Consider a thickness value measured perpendicular to the (x, y) plane of the hand axe periphery (Fig. 9) along what we define as the z-axis. As the stone's peripheral coordinates all lie in the (x, y) plane, their third-dimensional coordinates in z are all zero, and may be specified as $(x, y, 0)$. Let us suppose we have measured thickness along the hand-axe baseline at three different locations: near the tip end (landmark 5 at 80% of the height) at half the

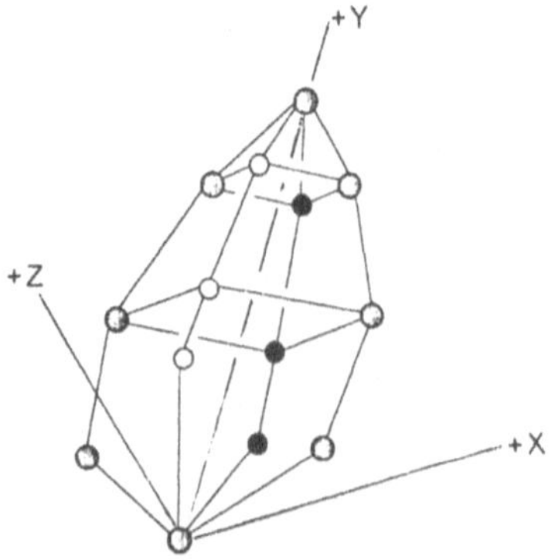

Figure 9. Geometric model for 3D hand axe, showing landmarks (corners marked by circles) derived from length and thickness measurements. Thickness measures taken along z axis, perpendicular to x-y plane.

height (landmark 4), and at the maximum thickness (and its location), wherever it occurs along the baseline (y) axis. Because the x-coordinate of all positions on the baseline is 0, coordinates of thickness will be specified as (0, y, z). If one wanted to study asymmetry in hand axe-thickness, one could measure the individual thickness of each side away from the (x, y) plane. If only total thickness is measured, as in the classical studies, no study of thickness asymmetry is possible. For example, if the thickness of a stone 100 mm in height is 35 mm, measured at half its height (landmark 4), then the 3D coordinates (x, y, z) would be (0, 50, 17.5) and (0, 50, -17.5).

Two-dimensional shape coordinates (x', y') may be easily extended to three (x', y', z'). As with 2D data, 3D shape coordinates of a single group of specimens may be averaged to a "mean" specimen and illustrated along with within-group scatter. For two or more groups of specimens, shape coordinates may be analyzed within a traditional multivariate statistical analysis program for assessing the distinctiveness of individual samples or specimens.

Landmark Density

We measured hand axes with some traditional, and a few other, measurements (e.g., maximum length and maximum width). As a result, the number of landmarks around the periphery of our specimens is small ($n = 6$ independent, 2D landmarks with some y-coordinates fixed by the nature of traditional measures, e.g., 80% of maximum height). As a consequence, any shape differences that may exist between or among population samples may remain undetected if shape variation occurs between the sparse landmarks we measured. It is obvious that with a greater number of landmarks about the periphery of a specimen, one would have a greater chance to detect significant differences in shape, should they exist.

Our model could easily accommodate an increase in the number of landmarks, as long as some quantitative basis is defined for each landmark position. For example, using the hand axe base as the coordinate system origin, one might video-digitize coordinates around a hand-axe periphery, and collect coordinates at the intersection of a line from the base to the periphery at some constant angle of rotation, say every 1 degree. A quantitative measure of the peripheral shape could thus be made as accurate as desired, depending of course, upon the resolution of the hardware.

Application of Landmark-Based Models to Other Artifacts

Although the archaeological objects we measured are relatively simple in shape, our conceptual model could be easily applied to other classes of objects of more complicated form. The Bookstein notions of a biological landmark (defined by one of the following—intersections of two structures, maximal curvature or extremal points) may perhaps be more readily applied to such artifacts as pots and jars than to stone hand axes.

We are looking forward to a more widespread application of geometrical morphometric methods to the quantitative analysis of shape of archaeological artifacts, for we see great promise in the removal of personal bias and qualitative description from dense population samples (not unlike those studied in recent decades by biometricians). Following Richtsmeier et al. (1992: page 302), we believe the incorporation of biometric methods into archaeological research programs may result in at least four advances.

1. With quantification will come an increase in accuracy and precision and, consequently, the ability to discriminate more similar classes of artifacts.
2. Multivariate statistical significance of intergroup comparisons will replace qualitative visual assessments.

3. Our simplified model may be used to calculate geometric coordinates from which graphical displays may be made.
4. With the advent of less expensive computer hardware for video-digitizing, the collection of shape data will exponentially increase. Such advances may parallel those in, for example, geochemistry, where the use of programmable analytical devices (such as the electron microprobe) make possible the collection and analysis of immense data sets, within and among sites and regions.

CONCLUSIONS

We have conceptualized a geometric model of a simple hand axe. We have written a computer program to transform traditional linear measurements to 2D and 3D coordinates based on our model. We have used our model, computer program and other available software to display outlines of hand axes and have applied traditional multivariate methods to test for statistically significant, multivariate differences between two groups. We believe extensions of these studies will prove a valuable addition to the archaeologist's tool chest of quantitative techniques and may help to improve description of artifact form.

ACKNOWLEDGMENTS

We thank Na'ama Goren-Inbar for inviting us to join the excavation at the Gesher Benot Ya'aqov and for providing access to hand axes for measurement, Eitan Tchernov and Na'ama Goren-Inbar for providing S. Brande space and support during his sabbatical at Hebrew University, inmates of the Zooarchaeology Lab (Hebrew University); and F. James Rohlf and Dennis Slice for helpful discussions; Fred Bookstein for his clarification of Galton shape coordinates and advice on multivariate analysis; Steve Weiner for his hospitality to the senior author while at the Weizmann Institute and Derek Roe for his inspirational papers written in the precomputer era. The senior author is grateful for encouragement and support from his wife, Barbara, with whom he spent an extended stay at University Hospital for the birth of their daughter, Sara Michal, during which a major portion of this paper was written. And thanks to Rita, a true friend.

REFERENCES

Belitzky, S., N. Goren-Inbar, and E. Werker. 1991. A Middle Pleistocene wooden plank with man-made polish. Journal of Human Evolution 20:349–353.

Bookstein, F. L. 1991. Morphometric tools for landmark data: geometry and biology. Cambridge University Press: Cambridge.

Bordes, F. 1961. Typologie du Paleolithique Ancien et Moyen. Delmas: Bordeaux, France.

Crompton, R. H., and J. A. J. Gowlett. 1993. Allometry and multidimensional form in Acheulean bifaces from Kilombe, Kenya. Journal of Human Evolution 25:175–199.

Goren-Inbar, N., and S. Belitzky. 1989. Structural position of the Pleistocene Gesher B'not Ya'acov site in the Dead Sea Rift zone. Quaternary Research 31:371–376.

Goren-Inbar, N., S. Belitzky, K. Verosub, E. Werker, M. Kislev, A. Heimann, I. Carmi, and A. Rosenfeld. 1992. New discoveries at the Middle Pleistocene Acheulian site of Gesher Benot Ya'aqov, Israel. Quaternary Research 38:117–128.

Isaac, G. 1977. Olorgesailie: Archaeological studies of a Middle Pleistocene lake basin. University of Chicago: Chicago.

Potts, R. 1993. Archeological interpretations of early hominid behavior and ecology. Pages 49–74 *in* D. Tab Rasmussen, (ed.), The origin and evolution of humans and humanness. Jones and Bartlett: Boston.

Richtsmeier, J. T., J. M. Cheverud, and S. Lele. 1992. Advances in anthropological morphometrics. Annual Reviews of Anthropology 21:283–305.

Roe, D. A. 1964. The British Lower and Middle Palaeolithic: Some problems, methods of study and preliminary results. Proceedings of the Prehistoric Society (New Series) 30(13):245–267.

Roe, D. A. 1968. British Lower and Middle Palaeolithic hand axe groups. Proceedings of the Prehistoric Society (New Series) 34(1):1–82.

Rohlf, F. J., and L. F. Marcus. 1993. A revolution in morphometrics. Trends in Ecology and Evolution. 8:129–132.

Schick, K. D. and N. Toth. 1993. Making silent stones speak: human evolution and the dawn of technology. Simon and Schuster: New York.

Slice, D. E. 1993. GRF-ND, Generalized rotational fitting of N-dimensional landmark data. Department of Ecology and Evolution, State University of New York, Stony Brook, New York 11794.

PART II

Section 3. Higher Taxonomic Level Comparisons

INTRODUCTION TO COMPARISONS AMONG TAXA AT THE LEVEL OF SPECIES, GENUS AND FAMILY

Gavin J. P. Naylor

The power of landmark-based morphometrics lies in their potential for comparative description. Description has been the cornerstone of comparative biology for the past 200 years, so it might seem at least superficially, that the new landmark-based methods don't provide biologists with anything they didn't have already. Ironically, nothing could be further from the truth. Geometric morphometrics allow us to recognize features of organisms that we may previously not have noticed by providing analytical and graphical means to decompose variation into components that reflect differences at different spatial scales. At one end of the scale, these components reflect shape differences that are restricted to highly localized subregions, at the other, they reflect differences that are globally and diffusely spread over the entire form. In essence the new methods provide a set of filters that can be used, individually or in combination, to sequester particular aspects of the ensemble morphological variation so as to leave a clearer, less confounded view of the aspects that remain.

The contributions in this section all center around studies at taxonomic levels that are generally higher than those described in earlier sections. Higher level taxonomic study involves comparing organisms that differ more widely in form and therefore brings with it a concern that landmark homology is less reliable. For example, although it might be reasonable to assume that a pair of "type 3" landmarks, such as the two points defining the maximum width of the skull, are homologous for a study of within-species variability, it is probably not reasonable to do the same for a study of morphological variation at the family level. Studies at higher levels must therefore be restricted to a subset of landmarks whose correspondences are less equivocal across divergent forms. The use of a subset rather than a full suite of landmarks generally means that less of the variability among forms is captured. This can be a problem when divergent taxa show considerable variation but have too few "type 1" landmarks to capture the latent shape differences among them. When presented with the choice of either (1) documenting uninteresting variation in a subset of morphological features that are characterized by reliable landmarks or (2) exploring more interesting variation dispersed over the entire form, but which requires the inclusion of landmarks whose homology is questionable, most of us err toward the curious. As a result it is especially important in higher-level studies to temper conclusions about differences among taxa with a careful inspection of landmarks used.

In most of the studies presented in this section traits known to be important in the classification of the various organisms studied are rediscovered though landmark-based

Advances in Morphometrics, Edited by L. F. Marcus et al.
Plenum Press, New York, 1996

methods; in several cases new aspects of morphological variation are discovered that corroborate established taxonomic arrangements; and in a few cases new traits are found that are able to provide support for controversial taxonomic hypotheses. Berge uses thin-plate spline analysis to show how the ilium has undergone a predominantly non-linear shape change in its evolutionary transformation from *Australopithecus* to *Homo*. She shows that these differences in shape are not as striking when the *Australopithecus* ilium is compared with that of a human neonate, supporting the notion that heterochrony has been important in the evolution of the human pelvis. Van Dam uses landmark-based morphometrics to describe shape change in a geological sequence of fossil murid teeth. He shows how allometry accounts for a major part of the observed shape change through the sequence and finds that the ordinations derived from geometric methods are largely consistent with classical taxonomy. He finds a trend toward increasing alignment of "valley landmarks" on the teeth throughout the sequence and hypothesizes that this may reflect a functional adaptation for a more efficient masticatory "power stroke" that better handles grasses. This hypothesis is consistent with the concomitant increase in hypsodonty seen in the teeth and with the contemporaneous expansion of steppe dwellers among large mammals. Hugot and Baylac show how landmark-based methods permit recognition of previously described parasitic pinworm taxa based on the morphology of the caudal bursa in males. The landmark-based phenetic groupings support a hypothesis of coevolution between these parasites and their primate hosts. These authors show how aspects of shape change implied by thin-plate spline analyses are concordant with existing hypotheses of evolutionary transformation and further show how previously undescribed evolutionary shape transformations are also implied. As a result of their findings, these authors conclude that geometric methods may provide a way to uncover subtle morphological traits for subsequent use in phylogenetic analysis. Carpenter finds a correlation between centroid size and the uniform component of shape differences in a group of emperor fishes suggesting, like Van Dam, that allometry may account for much of the shape variation seen. He also finds a correlation between body depth and tooth type in these fishes, suggesting a biomechanical relationship between body shape and feeding mode. Bogdanowicz and Owen, studying morphological divergence in plecotine bats based on shape coordinates and relative warp analysis, find an unweighted pair-group method using arithmetic averages (UPGMA) dendrodgram based on Mahalanobis D^2 of shape coordinates to be entirely consistent with a phylogenetic hypothesis of relationships based on a parsimony analysis of skin and skull characters. They find that Mahalanobis D^2 of shape coordinate and relative warp analyses yields the same phenetic arrangement of species within their respective generic groups, but only when uniform components of shape differences were excluded from the relative warp analysis. When the uniform component was included, the concordance between the phenograms diminished. This result is unexpected because shape coordinates include the uniform component and should therefore be consistent with a relative warp analysis with the uniform component included. It is possible that the inconsistency is due to the particular arbitrary method of including uniform components in relative warp analyses (Rohlf., personal communication). Gubanyi describes a method of automatic feature extraction in which outlines and landmarks are derived from scanned images by means of pattern recognition software. He uses these methods to collect data for a study of morphological variation in the attachment hooks of taeniid tapeworms. He subjects the outline data to elliptical Fourier analysis and the landmark data to shape coordinate analysis and shows that a canonical variates analysis of the two data types yield different results. Like Van Dam and Carpenter, he finds a linear allometric association between size and shape for the hooks of these parasites. Becerra et al. contrast use of classical morphological distance measurements with landmark-based methods for alpha taxonomy of the water mite genus *Torrenticola*. They find that landmarks fail to show the differences that are summarized in distances but conclude that the failing is due to single specimens being used to represent taxa. Finally, Naylor considers Zelditch et al.'s (1995) claim that partial warps can serve as cladistic characters for phylogenetic reconstruction. He evaluates the claim by generating a known, albeit contrived, phylogeny of fishes. His phylogeny incorporates a perfect phylogenetic signal of nested shape differences among the various fishes. Following Zelditch et al. (1995) he derives partial warp characters for the taxa and subjects these characters to cladistic analysis. Two equally most parsimonious trees result, one of which matches the true tree. However, considerable phylogenetic noise in shape characters is implied by the parsimony analysis, contrasting starkly with the perfect phylogenetic signal used to generate the true phylogeny.

Clearly, there are pitfalls and limitations associated with the unbridled use of any method. In general, landmark-based methods provide extremely powerful and appealing analytical tools for description. They should be regarded exactly as such—"tools to assist with description"—not "protocols to replace description." Nevertheless, this caution notwithstanding, it is clear that these new methods are providing fresh insights into the nature of form differences among organisms. Many of the contributors in this last section discover, through careful use of landmark-based methods, aspects of shape variation that they had not previously noticed, but that they were subsequently able to confirm upon reinspecting specimens. These studies exemplify how landmark-based methods can enhance our powers of description. Because description is the fundamental basis for all of comparative biology, landmark-based morphometrics promise considerable enrichment for the field.

REFERENCE

Zelditch, M. L., W. L. Fink, and D. L. Swiderski. 1995. Morphometrics, homology and phylogenetics: Quantified characters as synapomorphies. Systematic Biology 44(2):179–189.

THE EVOLUTION AND GROWTH OF THE HOMINID PELVIS

A Preliminary Thin-Plate Spline Study of Ilium Shape

Christine Berge

URA 1137, Centre National de la Recherche Scientifique
Laboratoire d'Anatomie Comparée
Muséum National d'Histoire Naturelle
55 rue Buffon
Paris, 75005, France

ABSTRACT

Previous studies have shown that the pelvis of an adult *Australopithecus* resembles that of a human neonate both in morphometric features and qualitative anatomical characters (Berge, 1993). In this study, the thin-plate spline was used to compare the shape of the ilium of *Australopithecus* (adult) and *Homo* (adult and neonate). The spline highlights three aspects of the ilium (anterior, medial and posterior), that feature significantly in the comparisons. The adult australopithecine and the human neonate are similar in all three of these aspects in the broad upper portion of the ilium. However, differences in shape are more apparent when the australopithecine is compared with the human adult. This result is consistent with the hypothesis that hominid evolution has been strongly influenced by heterochrony.

INTRODUCTION

It is well known that the pelvic morphology of hominids reflects common adaptive factors that result from stress exerted on the pelvic girdle due to vertical bipedalism and obstetric influences that are different from other primates (McHenry, 1975; Jungers and Stern, 1983; Berge, 1984, 1991; Berge et al., 1984; Tague and Lovejoy, 1986). *Australopithecus* differs markedly from *Homo* in the shape and proportions of its ilium. This difference is particularly noteworthy given that the entire hominid lineage is characterized by a common pelvic bauplan that persists for more than 3 million years. From a functional viewpoint, the pelvic morphology of *Australopithecus* reflects different static and dynamic adaptations to bipedalism, probably related to a partly arboreal behavior (Stern and Susman, 1983; Berge,

Advances in Morphometrics, Edited by L. F. Marcus et al.
Plenum Press, New York, 1996

1991, 1994). From an evolutionary viewpoint, the pelvic morphology of *Australopithecus* can be understood in relation to changes in human ontogeny.

In a previous paper in which I used multivariate analyses to compare the ontogenetic and phylogenetic changes in the pelvic proportions of hominids and pongids, I demonstrated that the pelvic proportions of adult *Australopithecus* are closer to those of human fetuses and neonates than they are to those of adult *Homo sapiens* (Berge, 1993). Principal component analyses of form and shape—after elimination of isometric size (Mosimannn and James, 1979; Berge and Kazmierczak, 1986)—suggest that growth processes (which are close to pongids for *Australopithecus* if we compare the juvenile MLD 7 to adult fossils) can account for the increased pelvic differences between adult *Australopithecus* and adult *Homo*.

The aim of this study is to investigate, using Thin-Plate spline methods, overall shape differences in the ilium between representatives of two different stages of the hominid evolutionary lineage, *Australopithecus* and *Homo sapiens*. The overall form of the ilium structure is strongly related to the expansion of muscle attachments on its internal and external faces and on the iliac crest. An analysis of the overall form of the structure, based on geometric methods, should help to delineate those portions of the ilium which are most different in the comparison. The thin-plate spline method (Bookstein, 1991) is known to be an effective way to depict deformations that are less readily captured by classical multivariate approaches (Bookstein, 1991, 1993). Because only three specimens are analyzed, the study should be regarded as preliminary.

MATERIAL AND METHODS

The material comprises the pelvis of *Australopithecus afarensis* (cast of "Lucy" reconstructed by Schmid, 1983) and two human pelves (of European origin), one of an adult female and one of a neonate. The choice of the three specimens was based on previous multivariate results and made in order to avoid any possible sampling bias. The cast had been loaned by the Anthropologishes Institut der Universität, Zurich; the adult pelvis belongs to the collection of the Laboratoire d'Anatomie Comparée, Paris (MNHN); and the neonate pelvis belongs to the collection of Institut d'Anatomie, Centre Universitaire des Saints-Pères, Université Paris 5 (Musé Orfila, Paris). The

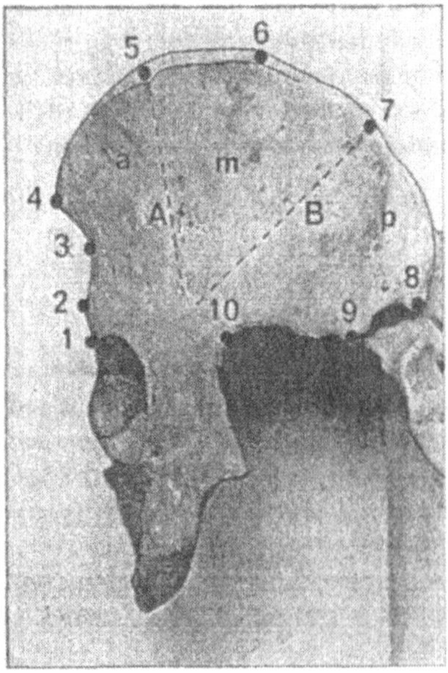

Figure 1. The 10 landmarks recorded on the left ilium (external surface) of *Homo sapiens*: (1) upper lip of the acetabulum; (2) anterior inferior iliac spine; (3) maximum of curvature of the notch of the anterior border of the ilium; (4) anterior superior iliac spine; (5) iliac tubercle; (6) maximum of curvature of the iliac crest between landmarks 5 and 7; (7) limitant spine; (8) posterior superior iliac spine; (9) posterior inferior iliac spine; and (10) maximum of curvature of the greater sciatic notch constructed between landmark 9 and the ischial spine. Dotted line A is the acetabulo-cristal buttress; dotted line B is the linea limitans (on the internal surface); a is the anterior part of the ilium; m the medial part; and p the posterior part.

iliac blade is a relatively flat bone, allowing a two-dimensional analysis of the intrinsic ilium shape. The ilia were photographed from the external surface (Fig. 1).

Ten landmarks have been defined on the external and internal lips of the iliac crest to draw the outline of the ilium. Seven of them are anatomically homologous (landmarks 1, 2, 4, 5, 7, 8 and 9; type 1, Bookstein, 1991), whereas three of them are geometrically constructed to complete the outline (landmarks 3, 6 and 10; type 2). The broad upper portion of the ilium is the iliac blade, which is bordered above by the iliac crest which terminates at either end in a spinous process, the anterior superior and posterior superior iliac spine. Two dotted lines, A and B, have been added to Fig. 1 to show more clearly the three different parts on the iliac blade. Dotted line A is situated on the external surface of the ilium. It begins at a prominence on the external lip of the iliac crest (iliac tubercle: landmark 5), follows a thickness of the iliac wall (acetabulo-cristal buttress) and ends close to the acetabulum. Dotted line B is situated on the internal surface of the ilium and is projected here onto the external surface. It delimits the anterior portion of the iliac blade (fossa iliaca) and the posterior portion (sacral tuberosity and auricular surface). Line B begins at a small prominence on the internal lip of the iliac crest (spina limitans, landmark 7), follows a rough line (linea limitans) and ends in the arcuate line. These two lines help to describe the changes in the ilium shape. The portion of the ilium lying to the left of dotted line A on Fig. 1 is named the anterior part of the ilium (a), the portion located between the two dotted lines is named the medial part (m), and the portion lying to the right of dotted line B is named the posterior part (p). The reference terms and orientations are those of human anatomy. For example, the affine component on Fig. 4 is described as an oblique compression directed anterosuperiorly, i. e., from the upper limit of the anterior border of the iliac blade (anterior superior iliac spine) to the lower limit of the posterior border (posterior inferior iliac spine).

Landmarks were digitized using the TPSPLINE program (version 07/06/93, Rohlf, 1990). Splines were computed using the same program for the following three paired contrasts: (1) adult *Australopithecus* and adult *Homo sapiens*; (2) neonate and adult *Homo sapiens* and (3) adult *Australopithecus* and neonate *Homo sapiens*.

RESULTS

Comparison of *Australopithecus* and an Adult *Homo sapiens*

The australopithecine specimen ("ancestor"; Fig. 2) was chosen as the reference configuration. The total deformation (Fig. 3) involves a minor affine component (Fig. 4), that corresponds to a slightly oblique compression directed antero-superiorly.

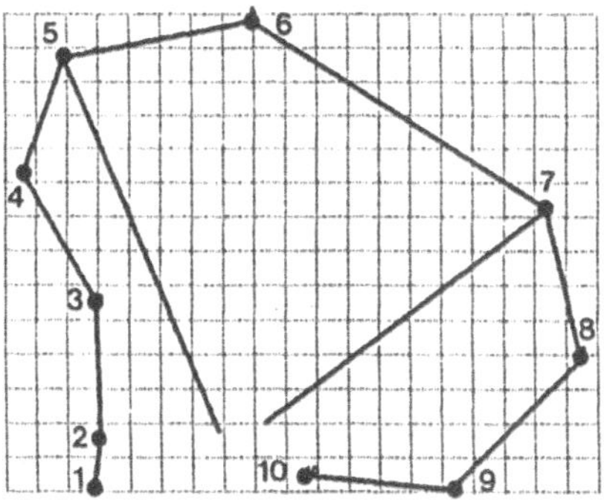

Figure 2. Display of *Australopithecus* in the original untransformed space.

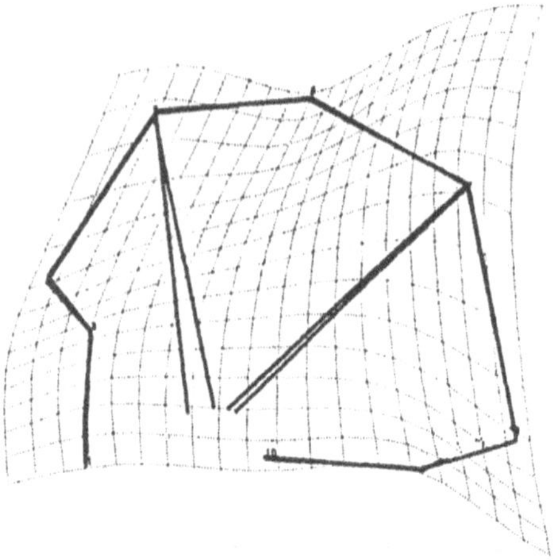

Figure 3. Comparison of *Australopithecus* and an adult *Homo sapiens*. Grid is of total deformations.

The human ilium appears, on the whole, narrower than the australopithecine one between the anterior superior and the posterior inferior iliac spines. The main part of the deformation corresponds to the nonaffine component (Fig. 5) which has a "butterfly" shape, a constricted medial region and dilated lateral regions, that allows one to recognize three main parts: anterior, medial and posterior. However, the overall deformation is not symmetrical. (1) The anterior region of the ilium is constricted in its inferior part and dilated in its superior part: the distance between landmarks 3 and 4 on the anterior border of the ilium is reduced, while the distance from landmarks 4 and 5 is substantially increased. (2) The medial region of the ilium is relatively constricted in its upper portion (landmarks 5 to 6 and 7) and has a slight posterior orientation of landmarks 5 to 7 relative to the center of the form. This portion of the human ilium becomes relatively lower and narrower in contrast with the anterior and posterior portions, which are more dilated. (3) The posterior region is strongly dilated in three directions, and the distances between landmarks 7, 8 and 9 are increased considerably: landmark 7 is displaced superiorly, landmark 8 posteroinferiorly; and landmark 9 posteriorly. The posterior part of the ilium is connected with the sacrum by its internal

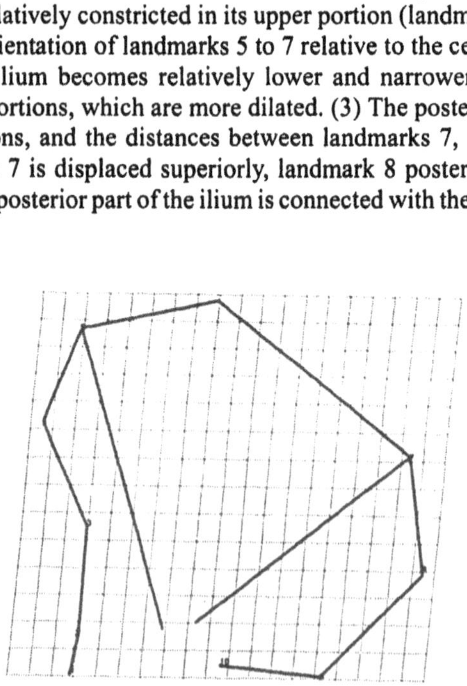

Figure 4. Comparison of *Australopithecus* and an adult *Homo sapiens*, affine component.

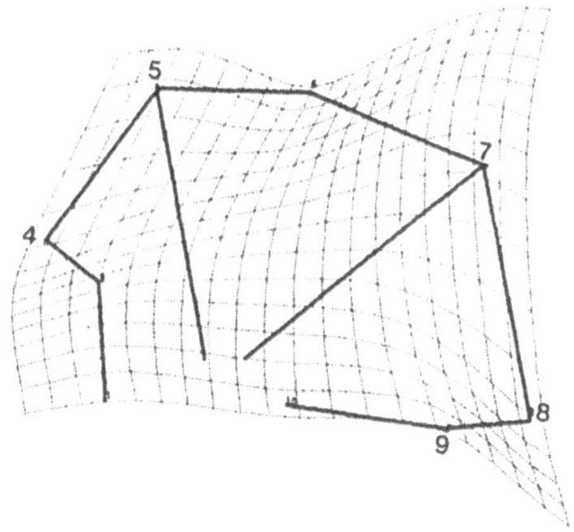

Figure 5. Comparison of *Australopithecus* and an adult *Homo sapiens*, total nonaffine component.

face. This iliac portion is broader on the human pelvis and directed downward and backward relative to the vertebral axis.

Comparison of a Neonate and Adult *Homo sapiens*

The neonate ilium is used as the reference configuration. The affine component (not shown, but identical to Fig. 4) is minor and corresponds to a slightly oblique compression directed anterosuperiorly as in the previous analysis. By contrast, the displacement of landmarks in the nonaffine component is quite dramatic (Fig. 6).

The shape of the deformation grid is also comparable to a butterfly shape. However, when compared with Fig. 5, the shape is more symmetrical. (1) The whole anterior region of the ilium is substantially dilated, being characterized by the opposite movements of landmark 4 (anteroinferiorly) and landmark 5 (posterosuperiorly), which increases the distance between landmarks 4 and 5. The distance between landmark 5 and 6 seems to be unmodified. (2) The medial region is constricted only in its upper portion corresponding to landmark 6, which moves closer to the center of the shape. (3) The posterior region is less

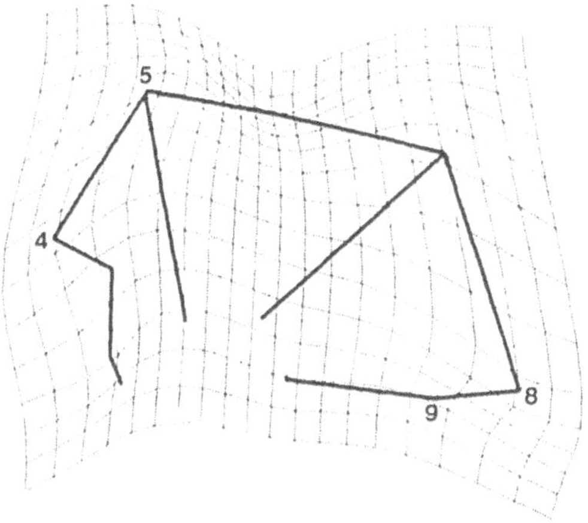

Figure 6. Comparison of a neonate and an adult *Homo sapiens*, total nonaffine component.

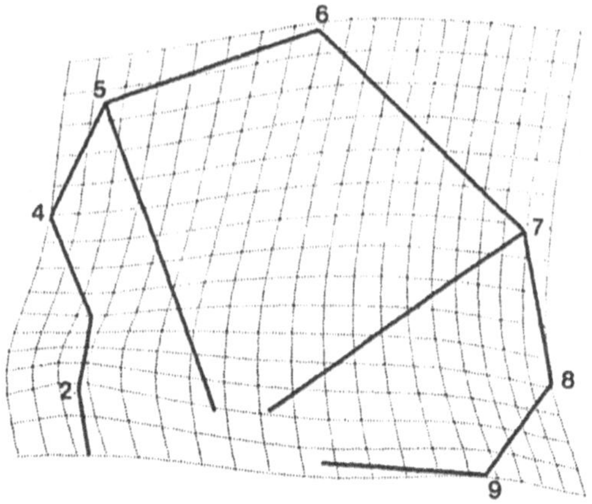

Figure 7. Comparison of *Australopithecus* and a neonate *Homo sapiens*, total nonaffine component.

dilated than in the previous analysis. It is slightly displaced superiorly (landmark 7) and more markedly postero-inferiorly (landmarks 8 and 9).

Comparison of *Australopithecus* and a Neonate *Homo sapiens*

The reference organism here is the adult *Australopithecus*. The affine change is almost zero. The nonaffine component (Fig. 7) indicates that the deformation is much smaller than those involving the adult *Homo sapiens* and does not exhibit the same butterfly figure.

The anterior and posterior regions are not dilated and the medial region not constricted (landmark 6 is the highest landmark on the iliac crest). The changes are slight and overall. The adult *Australopithecus* seems much more similar to the human neonate. For example, landmark 5, which was seen to move markedly posterosuperiorly in the previous analysis contrasting *Australopithecus* with the adult *Homo* is unchanged here. Moreover, we observe that the whole upper portion of the iliac blade between the anterior superior and posterior superior iliac spines shows few differences in shape. The narrow lower portion of the ilium is more dilated on the neonate ilium, with a slight constriction between landmarks 2 and 3 and a slight increase of the distance between landmarks 9 to 10.

DISCUSSION

Although limited to a simple comparison between pairs of individuals, the present study demonstrates the usefulness of the geometric approach in the study of ilium morphology. The thin-plate splines highlight the location of deformations and allow a visualization of their extensions on the ilium surface.

The deformation of the anterior portion of the human ilium in the comparison between *Australopithecus* and adult *Homo* had already been noted by anthropologists when they described the "beak" form of the anterior superior iliac spine of the australopithecines relative to the condition in humans (Zihlman, 1971; Sigmon, 1986; Berge, 1993). The differences in the pelvic proportions of *Australopithecus* and adult *Homo* that result from the shape change of the anterior portion of the ilium can be seen in Fig. 8 (A, C). The opposite movements of the anterior superior iliac spine (s) in a more anteroinferior position and of the iliac tubercle (t) in a more posterosuperior position lead to an increase in the curvature

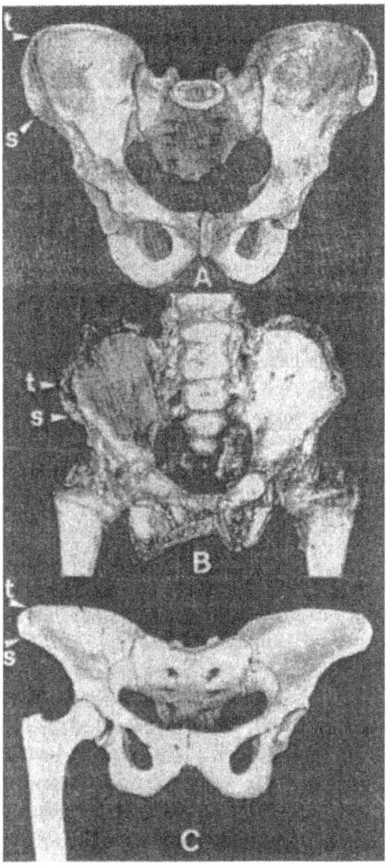

Figure 8. The hominid pelves (anterior superior view). (A) Adult *Homo sapiens* (pelvic height 20.5 cm); (B) neonate *Homo sapiens* (pelvic height 4 cm); (C) *Australopithecus* ("Lucy"; pelvic height 17 cm). Shown are (s) the anterior superior iliac spine (landmark 4) and (t) the iliac tubercle (landmark 5).

of the ilium with the formation of the internal iliac fossa. The human ilia are consequently more sagittally positioned.

The results here confirm the shape change of the anterior part of the human ilium and, more precisely, the relative movements of the anterior superior iliac spine and the iliac tubercle. Moreover, the results show that the shape change concerns the whole ilium and is described as a butterfly-shape distortion. The results of the study reveal that there are fewer differences in the comparison of the ilium shape of *Australopithecus* and *Homo* if the second individual is a neonate. This indicates that some features which have been previously described by authors as australopithecine ones when *Australopithecus* was compared with an adult *Homo* are actually present in humans at the beginning of life; for example, the condition in which the iliac tubercle (landmark 5) lies close to the anterior superior iliac spine (landmark 4). The overall change of shape in humans (between the neonate and the adult) may also be described as a butterfly-shape distortion but with marked differences in the general shape. We principally observe that the shape obtained is more symmetrical than in the comparison of adult *Homo* with *Australopithecus*. This indicates that the posterior part of the human ilium is more dilated in reference to the neonate than in reference to *Australopithecus*. The similarity between the fossil *Australopithecus* and the neonate *Homo* is clearly visible on the upper portion of the ilium corresponding to the ilium blade but less visible in the lower portion which is broader in neonate humans than in australopithecines.

Evolutionary processes including heterochrony have been previously proposed to explain the likeness between the ilium of the human fetus and neonate and that of *Australopi-*

thecus. The results presented here support this hypothesis, and demonstrate that the similarity between the ilium morphology of *Australopithecus* and the neonate *Homo* involves the entire upper portion of the ilium.

ACKNOWLEDGMENTS

My thanks go to P. Schmid for the loan of the fossil and to J. P. Lassau for the neonate pelvis. I wish to thank M. Baylac, F. Bookstein, D. Dean, and X. Penin for their helpful comments on and corrections to the manuscript. Thanks also go to J. W. Underhill for stylistic corrections to the English. I am indebted to the organizers and the Groupe d'Etude Morphométrique et Analyse de Forme (MNHN, Paris) for allowing me to participate in the activities of the NATO Advanced Study Institute in Il Ciocco. This research was supported by the CNRS (URA 11 37, Equipe "Locomotion Animale").

REFERENCES

Berge, C. 1984. Multivariate analysis of the pelvis for hominids and other extant primates: Implications for the locomotion and systematics of the different species of Australopithecines. Journal of Human Evolution 13:555–562.

Berge, C. 1991. Size and locomotion-related aspects of hominid and anthropoid pelves: An osteometrical multivariate approach. Human Evolution 6:365–376.

Berge, C. 1993. L'Evolution de la hanche et du pelvis des Hominidés. Bipédie. Parturition. Croissance, Allométrie. Cahiers de Paléoanthropologie. Centre National de la Recherche Scientifique Paris.

Berge, C. 1994. How did the australopithecine walk? A biomechanical study of the hip and thigh of *Australopithecus afarensis*. Journal of Human Evolution 26:259–273.

Berge, C., and J. K. Kazmierczak. 1986. Effects of size and locomotor adaptations on the hominid pelvis: Evaluation of australopithecine bipedality with a new multivariate method. Folia Primatologica 46:185–204.

Berge, C., R. Orban-Segebarth, and P. Schmid. 1984. Obstetrical interpretation of the australopithecine pelvic cavity. Journal of Human Evolution 13:573–587.

Bookstein, F. L. 1991. Morphometric tool for landmarks data: Geometry and Biology. Cambridge University Press: Cambridge

Bookstein, F. L. 1993. A brief history of the morphometric synthesis. Pages 15–40 *in* L. F. Marcus, E. Bello, and A. Garcia-Valdecasas, (eds.), Contributions to morphometrics. Monografias del Museo Nacional de Ciencias Naturales 8, Madrid.

Jungers, W. L., and J. T. Stern, Jr. 1983. Body proportions, skeletal allometry and locomotion in the Hadar hominids. A reply to Wolpoff. Journal of Human Evolution 12:673–684.

McHenry, H. M. 1975. Biomechanical interpretations of the early hominid hip. Journal of Human Evolution 4:343–355.

Mosimann, J. E., and F. C. James. 1979. New statistical methods for allometry with application to Florida red-winged blackbirds. Evolution 33:444–459.

Rohlf, F. J. 1990. TPSPLINE: a program to compare two shapes using a thin-plate spline. Department of Ecology and Evolution, State University of New York, Stony Brook, New York 11794.

Schmid, P. 1983 Eine reconstruktion der skelettes von A.L. 288-1 (Hadar) und deren Konsequensen. Folia Primatologica 40: 283–306.

Stern, Jr., J. T., and R. L. Susman. 1983. The locomotor anatomy of *Australopithecus afarensis*. American Journal of Physical Anthropology 60:279–317.

Sigmon, B. A. 1986. Evolution in the hominid pelvis. Paleont. Afr. pages 25–32.

Tague, R. G., and C. O. Lovejoy. 1986. The obstetric pelvis of AL 288-1 (Lucy). Journal of Human Evolution 15:237–255.

Zihlman, A. L. 1971. The question of locomotor differences in *Australopithecus*. Proceedings of the 3rd International Congress on Primatology, Zürich, 1970: 1:54–66.

STEPHANODONTY IN FOSSIL MURIDS

A Landmark-Based Morphometric Approach

Jan van Dam

Department of Geology
Institute of Earth Sciences
Utrecht University
Post Office Box 80.021
3508 TA Utrecht, The Netherlands

ABSTRACT

Landmark based morphometric methods were used to study shape variation in fossil murid teeth. Nineteen landmarks were defined and recorded on the occlusal surfaces of first upper molars belonging to eight populations from Late Miocene and Pliocene localities in Spain. The study focused on an evolutionary sequence characterized by development towards a specialized crown structure known as stephanodonty. Population differences were investigated using Bookstein shape coordinates and analyzed using analysis of variance and Mahalanobis distances. Relative warp analysis was used to study the nature of the shape variation among the landmarks. The results are consistent with the accepted taxonomic classification, although the taxonomic position of one of the populations is problematic. The analyses show that the development towards stephanodonty (a feature defined by the presence of ridges) is correlated with changes in shape, which are themselves correlated with size in the evolutionary sequence studied. It is apparent that landmark-based approaches show promise for functional morphological and paleoenvironmental interpretations of fossil murid teeth.

INTRODUCTION

The number of applications of landmark-based morphometrics is growing, and more biological shapes are being described and analyzed with these new geometric methods (Bookstein, 1991). This study explores the potential of these methods for interpreting shape differences among lineages of fossil teeth from murid rodents during their Old World radiation evidenced in Late Miocene and Pliocene sediments of Spain (Misonne, 1969; Michaux, 1971; Van de Weerd, 1976). The study focuses on an extinct group of south-western European murids that show an evolutionary development towards stephanodonty (Schaub,

Advances in Morphometrics, Edited by L. F. Marcus et al.
Plenum Press, New York, 1996

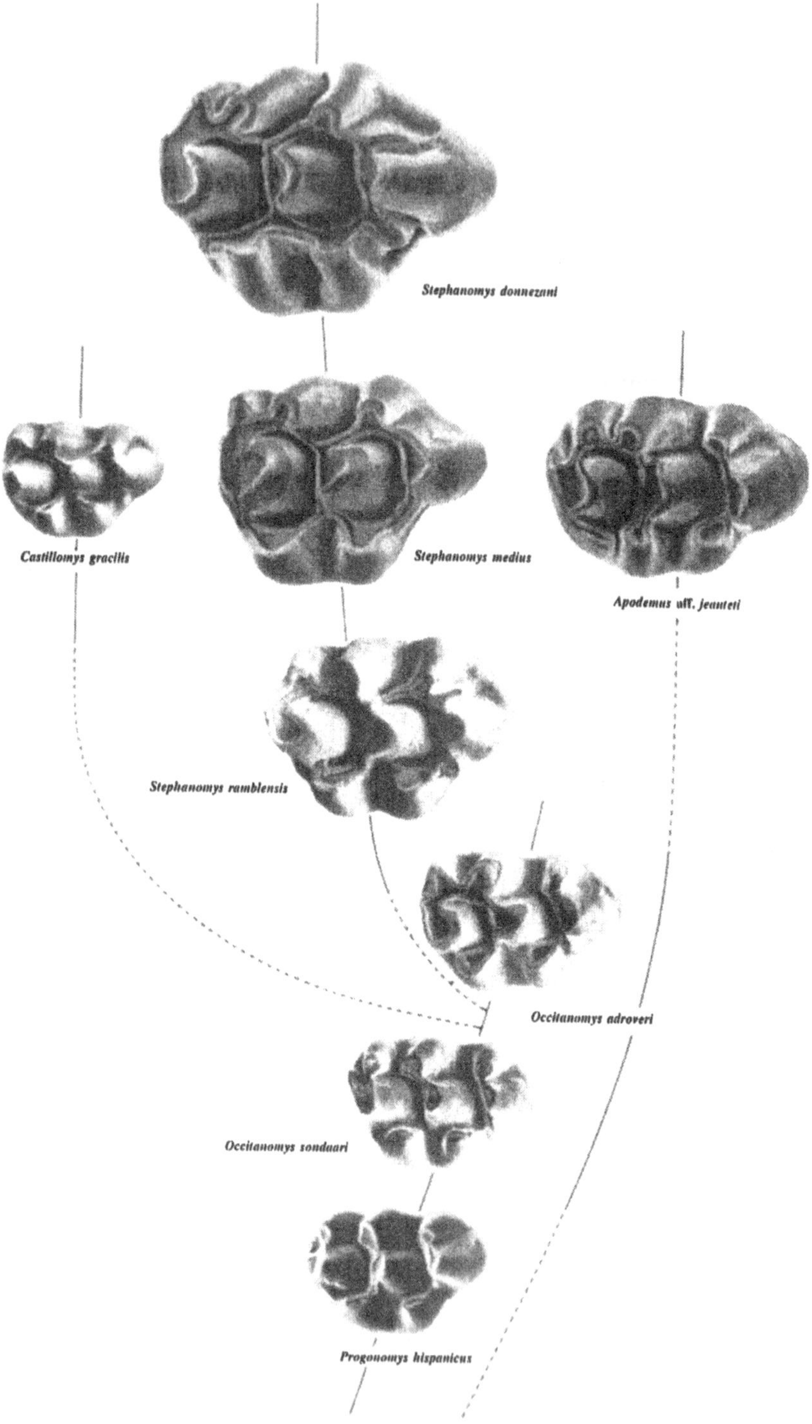

Figure 1. Evolutionary relationships among murid species used in this study, represented by their right upper molars. Dashed lines represent likely lineages. Enlargement approximately × 20.

1938; Cordy, 1976). In complete stephanodonty (Sensu Cordy, 1976) all cusps of the upper molars are connected by ridges, resulting in a garland-type dental pattern. Figure 1 shows this evolutionary trend in the first right upper molars of the sequence *Progonomys hispanicus-Occitanomys sondaari-Occitanomys adroveri-Stephanomys ramblensis-Stephanomys medius-Stephanomys donnezani.* This trend is accompanied by a size increase and the development of a special type of hypsodonty (elongation of the cusps in an oblique, posterior direction.)

The *P. hispanicus-O. sondaari-O. adroveri* lineage was described by Van de Weerd (1976) and represents an excellent example of anagenetic evolution (Chaline and Mein, 1979). The sequence *S. ramblensis-S. medius-S. donnezani* was interpreted as an evolutionary lineage by Cordy (1976, 1978) and Adrover (1986). The suggested relationship of these two lineages is shown in Fig. 1, with *S. ramblensis* descending from *Occitanomys* (Van de Weerd, 1976; Cordy, 1976).

A *Castillomys* species (*C. gracilis*) is included in this study because the genus shows stephanodonty, but the molars are small and brachiodont (Van de Weerd, 1976; Martín-Suárez and Mein, 1991). An *Apodemus* species is included as an outgroup. *Castillomys gracilis* and *A.* aff. *jeanteti* are shown in their respective phylogenetic positions in Fig. 1.

MATERIAL AND METHODS

In total, 250 first upper molars from eight populations were studied. Each population consisted of a different species. These eight species belong to five genera (Fig. 1; Table 1). The first upper molar was chosen because it represents the most diagnostic element of the murid dentition.

The oldest four populations come from the basin of Teruel in northeast central Spain (Van de Weerd, 1976). The most recent population comes from the Layna locality in central Spain (Cordy, 1976). The Caravaca locality of southwestern Spain is represented by three species: *Stephanomys medius* (Adrover, 1976), *Castillomys gracilis* (Van de Weerd, 1976) and *Apodemus* aff. *jeanteti* (de Bruijn et al, 1975). I included one locality for each MN biochronologic zone (Mein, 1990; de Bruijn et al., 1992) from MN 10 to MN 15. Absolute ages range from circa 9.5 to 3 million years before present (MYBP). All material examined is stored in the collection of the Institute of Earth Sciences of the Utrecht University. Nineteen landmarks were defined on the occlusal surface of the first upper molar (Fig. 2), 13 along

Table 1. Species used in this study and their collection data

Species	Locality	MN biozone[a]	Approx. age[b] (MA)	Sample size
Apodemus aff. *jeanteti*	Caravaca	14	5	57
Castillomys gracilis	Caravaca	14	5	9
Stephanomys donnezani	Layna	15	3	12
Stephanomys medius	Caravaca	14	5	32
Stephanomys ramblensis	Villalba Baja 1	13	5.5	3
Occitanomys adroveri	Masada de Valle 2	12	7.5	66
Occitanomys sondaari	Tortajada A	11	8.5	37
Progonomys hispanicus	Masia del Barbo 2B	10	9.5	34
Total				250

[a] MN biozones according to de Bruijn et al. (1992).
[b] Approximate ages estimated by the author on the basis of Pevzner and Vangenheim (1993) (magnetostratigraphy), Steininger et al. (1990) (correlation table), Van de Weerd (1976), and De Bruijn et al. (1992).

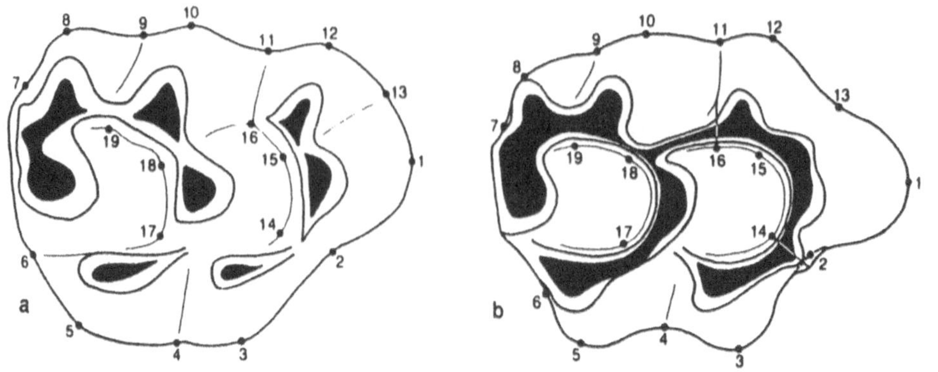

Figure 2. Location of landmarks on the occlusal surface of the right upper molar of (a) an *Occitanomys*-type molar and (b) an advanced *Stephanomys*-type molar. Orientation: right is anterior and bottom is lingual.

the outline and 6 on top, all in occlusal view. The 6 landmarks on top of the occlusal surface were defined in valleys between the cusps, rendering their location virtually independent of wear.

Landmarks 3, 5, 8, 10 and 12 are type 2 landmarks in the classification of Bookstein (1991): they are points of maximum curvature of the outline and are associated with cusps. Similarly, 4, 6, 7, 9, 11 and 13 are type 2 landmarks: they are points of maximum curvature and are associated with valleys. (Note that these last 6 landmarks could also be defined as intersections of a valley with the outline. Interpreted this way they become type 1 landmarks. However, the bottoms of these valleys usually do not form distinct lines but rather represent lines of maximum curvature between adjacent cusps). Landmarks 14, 15, 17, 18 and 19 come close to type 1 landmarks: they are defined as the 'intersections' of the central valleys (which do approach distinct lines) with lines perpendicular to these valleys running through the lowest part of the connections between the two adjacent cusps. In a number of specimens one or two additional cuspules have developed next to the central anterior cusp (i.e., *S. ramblensis* in Fig. 1): a cuspule called t1-bis at landmark 2 and a cuspule called t3-bis between landmarks 13 and 15. In these cases, landmarks 14 and 15 are defined by the intersections of the central valleys with lines perpendicular to these valleys running through the points of maximum curvature of the cuspules. Figure 2b shows the construction in the case of t1-bis. Landmark 16 is best classified as type 2. It is defined as the meeting point of three valleys in the *Progonomys*, *Occytanomys* and *Apodemus* species. For the *Stephanomys* and *Castillomys* specimens a construction had to be made for this landmark (Fig. 2b). Landmark 2 is a true type 2 landmark, being a point of maximum curvature of the outline. However, in some specimens the t1-bis was so large that it filled up the inner curve of the outline where landmark 2 was defined. When this was the case the position of this landmark was estimated as indicated in Fig. 2b. Landmark 1 is the anterior extreme of the molar in the orientation of Fig. 2. It is a type 3 landmark.

The *S. medius* and *S. donnezani* samples consist of specimens characterized by a moderate to high state of wear, because landmarks 14, 15, 17 and 18 would not have been visible in specimens with little wear.

Specimens were placed horizontally, and *x* and *y* coordinates of the landmarks were recorded using a Reflex three dimensional measuring microscope. Coordinates of left-side specimens were re-calculated as though they came from the right side.

Landmarks 4 and 11 were chosen as endpoints for the baseline (Bookstein, 1991). This baseline was chosen instead of one in the direction of the long axis of the tooth because landmarks such as 1, 13, 6 or 7 could be measured with relatively less precision.

Coordinates were rotated and scaled to the baseline with end points (0,0) and (1,0) to produce "size-free" Bookstein shape coordinates. Calculations were carried out using the program XY written by Baylac (1993), which was also used to calculate centroid sizes.

Uni-and multivariate analysis of variance were used to test for population differences. This was done with the SAS procedure CANDISC, which also produced Mahalanobis distances, *D*, between the populations. These distances were transformed into *F* values and were adjusted to compensate for bias according to Marcus (1993). An UPGMA cluster analysis was performed on the adjusted *D*.

Generalized Least Square (GLS) fitting was performed to find consensus specimens for each of the eight populations. Consensus specimens were fitted by Least Square fitting to the consensus specimen of the most primitive species, *Progonomys hispanicus*. The fits were calculated using the program GRF-ND written by Slice (1993).

The fitted consensus configurations were compared using relative warps (Bookstein, 1991; Rohlf, 1993a) with the TPSRW program of Rohlf (1993b).

Different options within the program were explored, such as the use of the parameter, α, whose value influences the scale of the nonlinear deformations, and the option to include the uniform component (the linear part of shape change) as estimated by the method of Bookstein ("Standard formula" article, this volume). When the latter option is used, the relative warps summarize non-linear as well as linear shape variation between the consensus specimens. The program TPSPLINE was used to visualize deformations of grids between different consensus specimens.

Partial warp scores (Bookstein, 1991, Rohlf, 1993a) and uniform components of the consensus specimen shapes were regressed on centroid size and geological time for the *P. hispanicus-S. donnezani* evolutionary sequence by means of TPSREGR (Rohlf, 1993 c). The consensus configuration of *P. hispanicus* was used as the reference. Deformation grids were produced for values of the independent variable. In each case, the initial condition was the average landmark pattern.

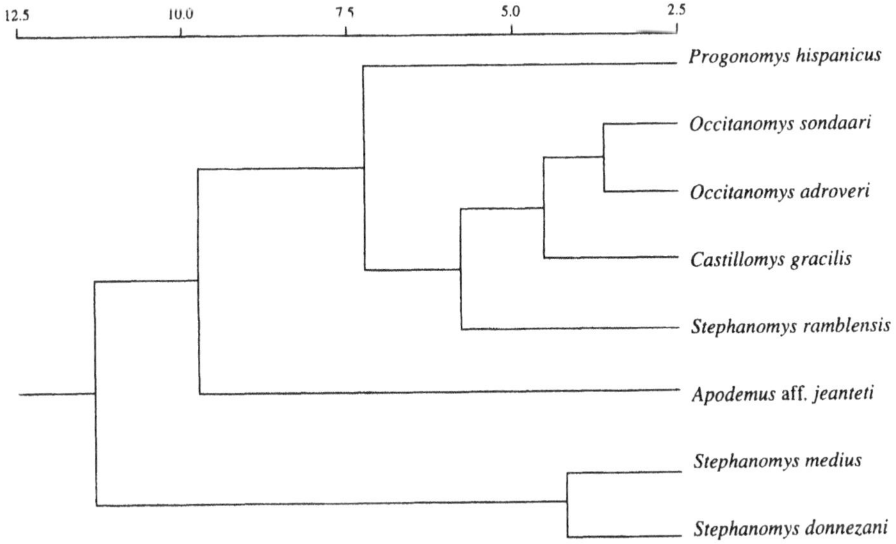

Figure 3. The UPGMA phenogram of populations based on adjusted Mahalanobis distances (see text) computed on 34 Bookstein shape coordinates.

Table 2. The F statistics and probability values (df 7, 242) from one-way ANOVA (SAS procedure CANDISC) testing population variation for the Bookstein shape coordinates

Variable	*F*	*p>F*	Variable	*F*	*p>F*
X1	36.37	0.0001	Y1	19.01	0.0001
X2	16.83	0.0001	Y2	7.62	0.0001
X3	23.09	0.0001	Y3	9.32	0.0001
X5	5.83	0.0001	Y5	22.07	0.0001
X6	112.34	0.0001	Y6	39.97	0.0001
X7	20.60	0.0001	X7	121.46	0.0001
X8	4.07	0.0003	Y8	60.62	0.0001
X9	3.45	0.0015	Y9	64.05	0.0001
X10	4.17	0.0002	Y10	54.81	0.0001
X12	6.52	0.0001	Y12	32.26	0.0001
X13	30.56	0.0001	Y13	13.31	0.0001
X14	7.68	0.0001	Y14	34.52	0.0001
X15	1.24	0.2802	Y15	32.30	0.0001
X16	15.37	0.0001	Y16	9.75	0.0001
X17	4.51	0.0001	Y17	120.16	0.0001
X18	25.86	0.0001	Y18	103.19	0.0001
X19	14.57	0.0001	Y19	48.78	0.0001

RESULTS

Statistical Differences between the Populations

Table 2 shows the univariate F ratios and probability values for the separate x and y Bookstein shape coordinates (except those of the baseline, which have fixed values of (0,0) and (1,0)) that resulted from a one way analysis of variance (ANOVA). Population differences were clearly significant for all the coordinates except for x_{15}, and for all coordinates together (Table 3). Table 4 shows the pairwise Mahalanobis D between the populations. The D are all significant with probability values less than 0.0001. Figure 3 shows the phenogram resulting from an UPGMA cluster analysis based on the adjusted Mahalanobis distances. It is interesting that *S. ramblensis* does not cluster with the other *Stephanomys* populations but with *Occitanomys* and *Castillomys*. That the *Apodemus* species does not form a tight cluster with the other species is consistent with its relatively more distant taxonomic position.

Landmark Shape Differences between the Populations

Figure 4a shows the GLS fitted consensus specimen for the *Progonomys hispanicus* population along with all the fitted specimens of this population. Figures 4b-h show the

Table 3. Multivariate statistics, F approximations and probability values (df 7,242) from multivariate analysis of variance (SAS procedure CANDISC) testing population variation for the Bookstein shape coordinates

Statistic	Value	*F*	Numerator df	Denominator df	*p*
Wilks' Lambda	0.00	15.00	238	1448.41	0.0001
Pillai's Trace	3.91	8.04	238	1505.00	0.0001
Hotelling-Lawley's Trace	35.96	31.32	238	1451.00	0.0001
Roy's Greatest Root	19.84	125.46	34	215.00	0.0001

Table 4. Mahalanobis distances between populations computed using Bookstein shape coordinates. Values below the diagonal are adjusted for bias

	*A.*aff *j.*	*C.g.*	*S.d.*	*S.m.*	*S.r.*	*O.a.*	*O.s.*	*P.h.*
Apodemus aff. *jeanteti*	0.0	10.0	14.2	13.0	12.5	9.6	9.5	7.5
Castillomys gracilis	9.8	0.0	10.8	8.8	6.2	4.3	4.7	7.0
Stephanomys donnezani	13.9	10.5	0.0	4.3	10.9	11.1	11.6	14.0
Stephanomys medius	12.8	8.6	4.1	0.0	9.5	9.5	10.0	12.3
Stephanomys ramblensis	12.2	5.9	10.6	9.2	0.0	5.0	6.9	10.3
Occitanomys adroveri	9.5	4.2	10.9	9.3	4.6	0.0	3.6	6.9
Occitanomys sondaari	9.4	4.5	11.4	9.8	6.6	3.5	0.0	5.8
Progonomys hispanicus	7.4	6.8	13.7	12.1	10.0	6.8	5.6	0.0

landmark shape differences of the consensus specimens of the other species relative to the *P. hispanicus* consensus specimen. The trends shown by Fig. 4a-f are the general broadening of the molars, the straightening of the posterior border, the relative broadening and the anterior shift of the central region and the sharpening of the anterior part in *Stephanomys medius* and *S. donnezani* by means of posterior-buccal movement of landmarks 10, 11 and,

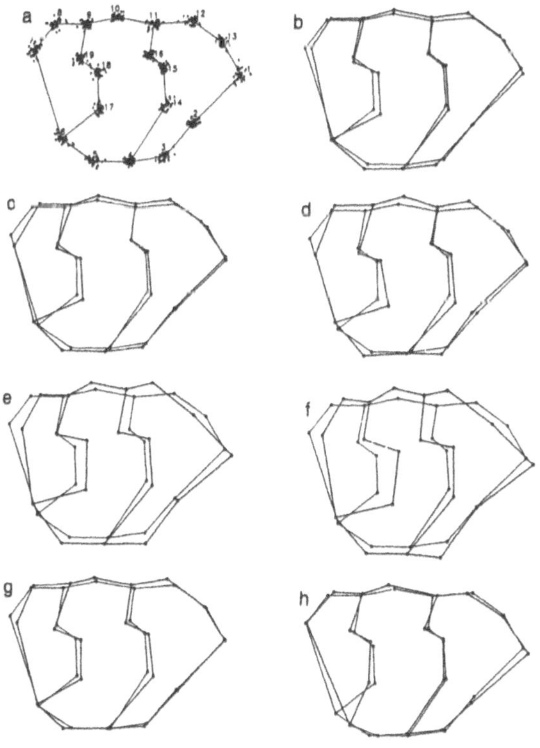

Figure 4. (a) Generalized Least Square fitted consensus specimen for the *Progonomys hispanicus* population together with the landmark points of all the fitted specimens. The landmark shape differences relative to the *P. hispanicus* consensus specimen (thin line) of the consensus specimens (thick line) of the populations of (b) *Occitanomys sondaari*, (c) *O. adroveri*, (d) *Stephanomys ramblensis*, (e) *S. medius*, (f) *S. donnezani*, (g) *Castillomys gracilis* and (h) *Apodemus* aff. *jeanteti*.

in particular 12. The shape of *Castillomys gracilis* (Fig. 4g) is very similar to that of the *Occitanomys* species. The largest difference between *A.* aff. *jeanteti* and the reference specimen is the position of landmark 6, which has moved in the antero-lingual direction. This shift is associated with the formation of a new cusp, the so-called t7 (see Fig. 1).

The results of relative warp analysis are presented in Fig. 5 and Table 5. Table 5 shows that in all three cases a large part of the variation is explained by the first relative warp (80.8, 77.6 and 78.9 %, respectively). Inspection of the relative warp loadings (not shown) indicate that the value of α, had little effect on the first relative warp. However, inclusion of the uniform component estimate resulted in significant differences between relative warp 2 and those higher, and yielded a meaningful second relative warp that did not reflect statistical error. The largest differences between the loadings of the first relative warp from analyses with and without the uniform component were localized in the posterior-lingual area of the teeth. Figure 5a shows the loadings as vectors for the first relative warp with estimation of the uniform component included. In the analysis excluding this uniform component, the vectors of landmarks 4, 5 and 6 point more to the posterior side of the teeth (not shown here). Figures 5a and b show that transverse deformation is mainly represented by relative warp 1 whereas longitudinal deformation is represented in relative warp 2. The unique morphology of *Apodemus* is reflected in relative warp 2 by the anteriorly directed vectors from landmarks 5 and especially 6 (compare with Fig. 4h). Figure 5c shows the first relative warp as a deformed grid (all eight populations included, α = 1). The inflation and relative anterior movement of the central region and the generally symmetric shape (caused by the posterior movement of landmark 12 and the straightening of the posterior border) compared with the reference specimen were already observed in Fig. 4e and f. Figures 5d and e show a similar clustering of populations, indicating that the x and the y part of the uniform component contribute strongly to relative warp 2 and 1 respectively. Both plots are concordant with the phenogram in Fig. 3. It can be seen from the positions of the points in Fig. 5d that the order of the evolutionary sequence *P. hispanicus-S. donnezani* is reflected in the scores on relative warp 1, except for the relatively "advanced" position of *O. sondaari*. Relative warp separates the oldest four populations of the sequence (*Progonomys hispanicus*, the two *Occitanomys* species and *Stephanomys ramblensis*) from the two most recent populations (*S. medius* and *S. donnezani*), which like *Apodemus* aff. *jeanteti*, are characterized by stronger longitudinal shifts of landmarks. Both *A.* aff. *jeanteti* and *C. gracilis* remain relatively close to the primitive form, *P. hispanicus*.

Figures 6 a, b and c show grid deformations from the primitive form to *O. adroveri*, *S. donnezani* and *A.* aff. *jeanteti*, respectively. The grids in Fig. 6a and b reflect the first

Table 5. Results of relative warp analysis on consensus specimens, based on the consensus configuration of the *Progonomys hispanicus* population as the reference. Reported for the first three relative warps are singular values (SV), proportion of explained variance (prop. var.) and probability values for equality of the remaining singular values (prob.). Results are given for three different combinations of program options. The , = 0 and , = 1 options give more weight to small-scale and large-scale variations, respectively. In the case of , = 1 singular values are adjusted (see TPSRW documentation; Rohlf 1993b)

	Relative warp analysis								
	α = 0, uniform component not included			α = 0, uniform component included			α = 1, uniform component not included		
	Singular value	Proportion variance	*p*	Singular value	Proportion variance	*p*	Singular value	Proportion variance	*p*
1	0.158	0.808	0.000	0.078	0.776	0.000	0.269	0.789	0.000
2	0.066	0.141	0.001	0.029	0.104	0.384	0.101	0.110	0.164
3	0.030	0.030	0.234	0.023	0.067	0.528	0.067	0.049	0.411

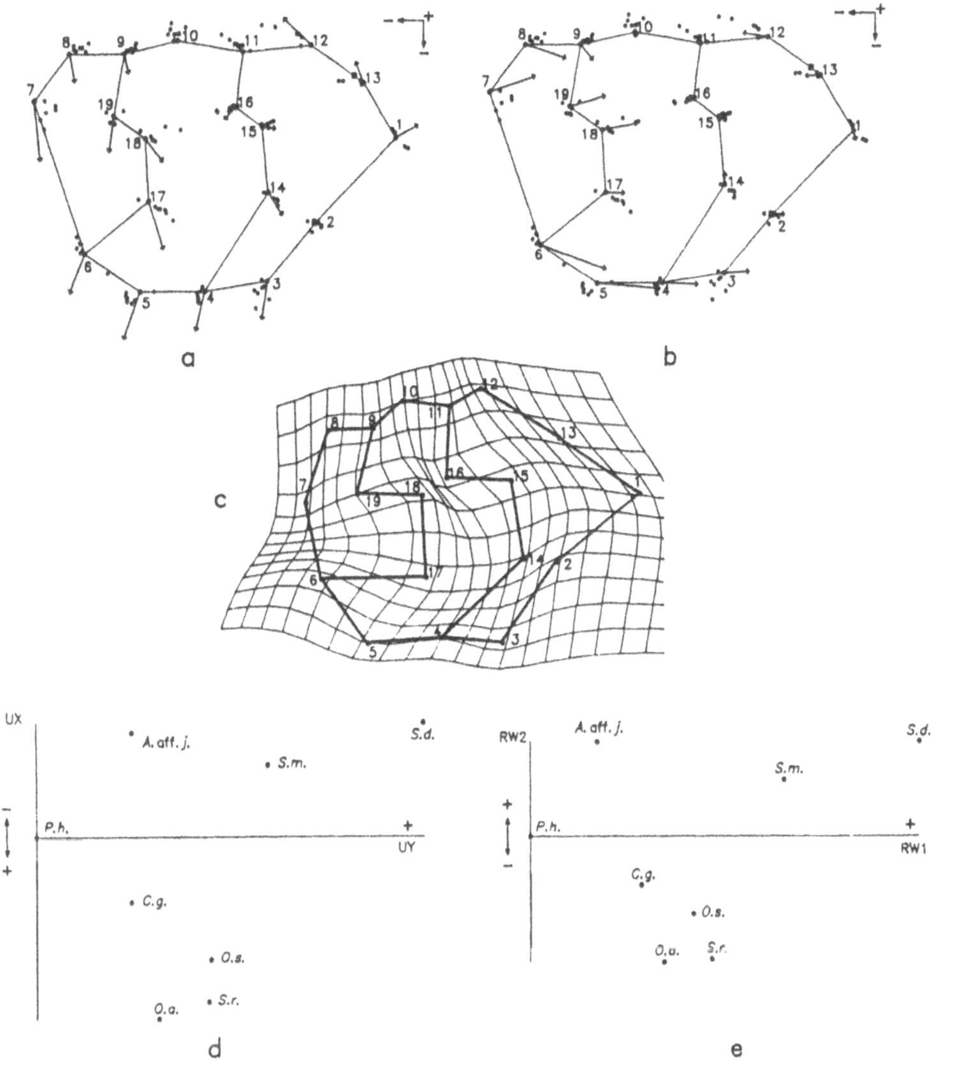

Figure 5. (a) Vectors of landmark changes for the first relative warp. Estimation of uniform component included, $\alpha = 0$, and all populations used. (b) Vectors of landmark changes for the second relative warp. Options as in (a). (c) Deformation grid for the first relative warp. Estimation of uniform component not included, $\alpha = 1$, and all populations used. (d) Plot of uniform x (Ux) and uniform Y (Uy). Options as in (a). (e) Plot of the first two relative warps. Options as in (a).

relative warp (compare Fig. 5c). The deformation seen in Fig. 6b appears essentially to be a magnification of the deformation seen in Fig. 6a, and corresponds to a more advanced evolutionary state. The deformation in Fig. 6c is of a different type, reflecting relative warp 2.

Regressions of Landmark Shape on Size and Geological Age

Least-square-regressions of the partial warp scores and uniform components of the consensus specimens for the populations of the *P. hispanicus-S. donnezani* sequence were

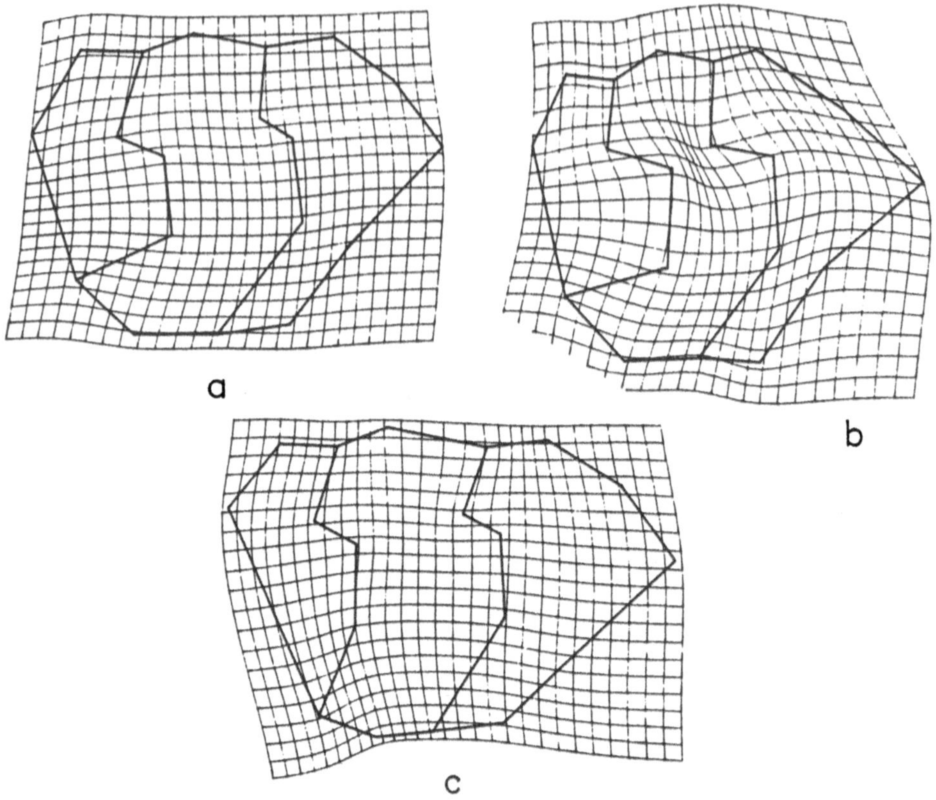

Figure 6. Thin-plate spline grid deformations from the *P. hispanicus* consensus specimen into consensus specimens for (a) *Occitanomys adroveri*, (b) *Stephanomys donnezani*, (c) *Apodemus* aff. *jeanteti*.

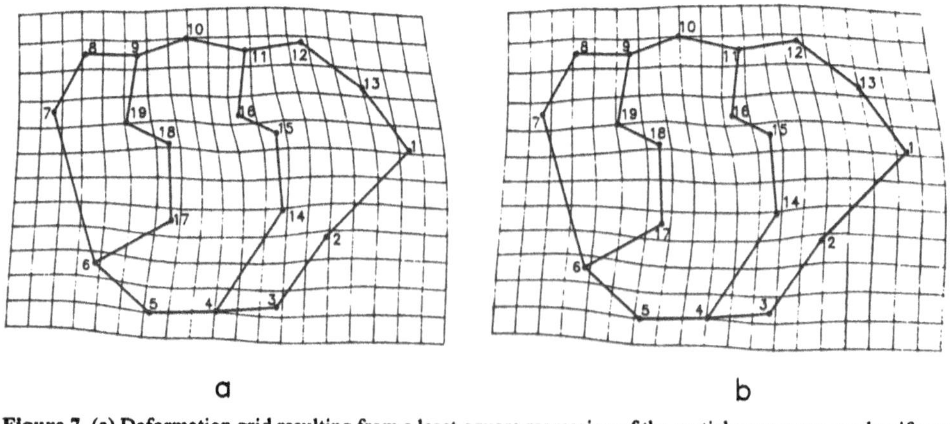

Figure 7. (a) Deformation grid resulting from a least-square regression of the partial warp scores and uniform component for consensus specimens of the *P. hispanicus-S. donnezani* sequence on centroid size (centroid size = 4.483). The *P. hispanicus* consensus configuration was used as the reference. (b) Deformation grid resulting from a regression as in (a) but with absolute age as independent variable (absolute age = 3.25 MYBP).

performed on centroid size and on geological age by means of the TPSREGR program (Rohlf, 1993c). (See Table 1 for ages. Average centroid sizes are 2.62 (*P. hispanicus*), 2.75 (*O. sondaari*), 2.99 (*O. adroveri*), 3.59 (*S. ramblensis*), 3.79 (*S. medius*) and 4.75 (*S. donnezani*)). The initial average conditions of the deformations in both Fig. 7a and b correspond to stages between *O. adroveri* and *S. ramblensis*, whereas in both cases the resulting conditions were chosen so that they were close to those of *S. donnezani*. Both grids cannot be distinguished from each other by eye. This is not at all surprising because these two variables, centroid size and geological age, are highly correlated for this sequence (r = 0.98).

Figures 7a and b strongly resemble Fig. 6a, which represents the older part of the studied sequence. It is evident that the results of this regression analysis express, in another way, the effects of the first relative warp described above. However, as the shape changes are now explicitly related to size and time, we are justified in saying that allometry has played a major role in this evolutionary sequence.

DISCUSSION

Taxonomy

The taxonomy of the murid species studied is largely confirmed, except for *S. ramblensis* Van de Weerd, 1976, which clusters not so much with the other *S.* species as with the *Occitanomys* species. In this respect it is interesting that this author considers this species to be intermediate between *Occitanomys* and *Stephanomys* on the basis of size and stage of stephanodonty. He classifies it as *Stephanomys* because it fits in with the supposed evolutionary lineage toward the type species of this genus, *S. donnezani*. The sample size of the *S. ramblensis* population examined was very small ($n = 3$), but this does not account for the observed clustering. New material of this species will be available soon for further work. The results presented here suggest a shape of *S. ramblensis* that is much closer to *Occitanomys* than it is to *Stephanomys*. The shape of *C. gracilis* as defined by the landmarks is also similar to that of *Occitanomys*. This is in agreement with Van de Weerd's (1976) observation that the molars of the two genera closely resemble one another in shape but not in size.

Evolutionary Trends

The data presented above show that the degree of stephanodonty is correlated with changes in the pattern of landmarks as defined along the outline and on the surface of the molars. For instance, the anterior shift of the central landmarks allow an increase of oblique hypsodonty. Similarly, the lingual-buccal symmetry of the garland-like dental pattern in the true stephanodont forms corresponds to the lingual-buccal symmetry in the landmark patterns. Trends towards broadening of the molars and straightening of the posterior border, associated with increasing stephanodonty, have been mentioned in the literature. Results presented here show that the evolution towards a stephanodont pattern, is adequately described by the first relative warp.

Functional Morphology

Increase in stephanodonty has been interpreted as adaptation to grass eating (Michaux, 1982). This interpretation is based on comparison of the dental morphology and the diet of certain living African murids showing similar specializations (Misonne, 1969) and on the observation that the *Occitanomys-Stephanomys* cladogenesis occurs concurrently with the expansion of the steppe dwellers among the large mammals. In addition to their increase in stephanodonty, the molars of the *P. hispanicus-S. donnezani* evolutionary sequence also show a general increase in hypsodonty. The association of hypsodonty in herbivores with the bulk consumption of calorie-poor and relatively abrasive food (such as grasses) is well established for rodents (Rensberger, 1975) and mammals in general (Janis and Fortelius, 1988).

An interesting finding revealed by this study is the trend towards alignment of the "valley" landmarks 7, 19, 18, 16 and 15 on the one hand and 6, 17, and 14 on the other (Fig. 5c). Presumably, such a configuration allows a longer and more efficient power stroke, with increased guidance of movement by longitudinal ridges (see also Weijs and Dantuma, 1975 on mastication in *Rattus norvegicus*). During the power stroke the cusp rows of the lower molars move anteriorly and longitudinally across the upper molars. The two anterior cuspules, the t1-bis and the t3-bis of stephanodont forms, allow an even longer power stroke. In order to test such correlations and to make inferences about diet and paleoenvironment, more information concerning the function of the masticatory apparatus and wear patterns in recent murids is needed.

GENERAL CONCLUSIONS

This study shows that complex three-dimensional shapes like those of teeth can be described by landmark patterns. The example of stephanodonty makes it clear that landmark-based morphometrics are a very useful tool to describe and analyze shape in fossil murid molars. Population differences, evolutionary trends and allometry can be adequately described. The results are promising for taxonomic, evolutionary and functional morphological studies. This publication is no. 960402 of the Netherlands Research School of Sedimentary Geology (NSG).

ACKNOWLEDGMENTS

I would like to thank the organizers and staff of the NATO Advanced Study Institute for the opportunity they gave in Il Ciocco for applying the morphometric tools to my data and the assistance they offered with my trials (and errors). I thank H. de Bruijn, A. J. van der Meulen and P. Pelaez-Campomanes for reading and discussing parts of the manuscript, W. den Hartog for making electron microscope photographs, J. Luteyn for preparing Fig. 1 and 2 and T. van Hinte for his AUTOCAD modifications of graphical output.

REFERENCES

Adrover, R. 1986. Nuevas faunas de Roedores en al Mio-Pliocene continental de la región Teruel (España). Interés bioestratigráfico y paleoecológico. Instituto de Estudios Turolenses: Teruel.

Bookstein, F. L. 1991. Morphometric tools for landmark data: Geometry and Biology. Cambridge University Press: Cambridge.

de Bruijn, H., C. Montenat and A. van de Weerd. 1975. Correlations entre les gisements des Rongeurs et les formations marines du Miocene terminal d'Espagne meridional (Prov. d'Alicante et de Murcia). Kon. Ned. Akad. v. Wetensch., Proc., Series B 78: 282–313.

de Bruijn, H., R. Daams, G. Daxner-Höck, V. Fahlbusch, L. Ginsburg, P. Mein, and J. Morales. 1992. Report of the RCMNS working group on fossil mammals, Reisensburg, 1990. Newsl. Stratigr. 26:65–118.

Chaline, J., and P. Mein. 1979. Les rongeurs et l'évolution. Doin Éditeurs, Paris.

Cordy, J. M. 1976. Essai sur la micoévolution du genre *Stephanomys* (Rodentia, Muridae). Thesis, Liège.

Cordy, J. M., 1978. Caractéristiques générales de la microévolution du genre *Stephanomys* (Rodentia, Muridae). Bull. Soc. géol. France 20(6):815–819.

Janis, C. M., and M. Fortelius. 1988. On the means whereby mammals achieve increased functional durability of their dentitions, with special reference to limiting factors. Biological Reviews 63:197–230.

Marcus, L. F., 1993. Some aspects of multivariate statistics for morphometrics. Pages 95–130 *in* L. F. Marcus, E. Bello and A. Garcia-Valdecasas, (eds.), Contribution to morphometrics. Monografias del Museo Nacional de Ciencias Naturales 8, Madrid.

Martín-Suárez, E., and P. Mein. 1991. Revision of the genus *Castillomys* (Muridae, Rodentia). Scripta Geologica 96:47–81.

Mein, P. 1990. Updating of MN zones. Pages 73-90 *in* E. H. Lindsay, F. Fahlbusch and P. Mein, (eds.), European neogene mammal chronology. Plenum Press: New York.

Michaux, J. 1971. Muridae (Rodentia) néogènes d'Europe sud-occidentale. Évolution et rapports avec les formes actuelles. Paleobiol. Contin. 2 I:1–67.

Michaux, J. 1982. Aspects de l'évolution des Muridés (Rodentia, Mammalia) en Europe sud-occidentale. Pages 195–199 *in* J. Chaline, (ed.), Coll. intern. C.N.R.S. Dijon, 330. Centre National Recherche Scientifique. Dijon.

Misonne, X. 1969. African and Indo-Australian Muridae, Evolutionary trends. Annales Mus. Roy. Afr. Cent. Belgique. Ser. Sc. zool. 172.

Pevzner, M. A., and E. A. Vangenheim. 1993. Magnetochronological age assignments of Middle and Late Sarmatian mammalian localities of the eastern Paratethys. Newl. Stratigr. 29:63–75.

Rensberger, J. M. 1975. Function in the cheek tooth evolution of some hypsodont geomyoid rodents. Journal of Paleontology 49:10–22.

Rohlf, F. J. 1993a. Relative warp analysis and an example of its application to mosquito wings. Pages 131–160 *in* L. F. Marcus, E. Bello and A. Garcia-Valdecasas, (eds.), Contributions to morphometrics. Monografias del Museo Nacional de Ciencias Naturales 8, Madrid.

Rohlf, F. J. 1993b. TPSRW–Thin-plate spline relative warp. Department of Ecology and Evolution, State University of New York, Stony Brook, New York 11794.

Rohlf, F. J. 1993c. TPSREGR: A program for regression of partial warp scores. Department of Ecology and Evolution, State University of New York, Stony Brook, New York 11794.

Schaub, S. 1938. Tertiäre und Quartäre Murinae. Abh. Schweiz. Pal. Ges. 61:1-39.

Slice, D. 1993. GRF-ND: General rotational fitting of N-dimensional landmark data. Department of Ecology and Evolution, State University of New York, Stony Brook, New York 11794.

Steininger, F., F. Bernor and V. Fahlbusch. 1990. European Neogene marine/continental chronologic correlations. Pages 15–46 *in* E. H. Lindsay, F. Fahlbusch and P. Mein (eds.), European neogene mammal chronology. Plenum Press: New York.

Van de Weerd, A. 1976. Rodent faunas of the Mio Pliocene continental deposits of the Teruel-Alfambra region, Spain. Utrecht Micropaleontological Bulletins, special publication 2.

Weijs, W. A., and R. Dantuma. 1975. Electromyography and mechanics of mastication in the albino rat. Journal of Morphology 146:1–34.

COMPARATIVE LANDMARK ANALYSIS OF VARIOUS OXYURIDAE PARASITES OF PRIMATES AND RODENTS

Patterns of Variation Seen in the Caudal Bursa of Males

J. P. Hugot*[1] and M. Baylac[2]

[1] Muséum National d'Histoire Naturelle
Laboratoire de Biologie Parasitaire, Protistologie et Helminthologie
Associé au Centre National Recherche Scientifique (URA 114)
61 rue Buffon
75231 Paris cedex 05, France
[2] Muséum National d'Histoire Naturelle
Laboratoire d'Entomologie
45 rue Buffon
75231 Paris cedex 05, France

ABSTRACT

The patterns of shape variation in the caudal bursa of males of various species of pinworms were analyzed using geometric morphometric methods. Thirty-four of a group of 45 oxyurid species that are primarily parasitic on primates, and are considered to constitute a monophyletic group, were studied. This study was designed to evaluate (1) the application of geometric morphometric techniques to cuticular characters in soft-bodied invertebrates; (2) the current classification for the group; (3) Cameron's (1929) hypothesis of close coevolution between Primates and their oxyurid parasites; and (4) Chabaud and Petter's (1961) hypothesis that male genital papillae of phasmidian nematodes provide useful characters for systematic and phylogenetic studies. Analyses identified the same groups and subgroups that had been described previously using different morphological characters. The host specificity of different parasites was found to be consistent with Cameron's hypothesis. The thin-plate spline analyses revealed shape changes of the genital papillae that are consistent with the evolutionary transformations proposed by Chabaud and Petter, suggesting that geometric methods might be useful for identifying characters for phylogenetic reconstruction.

* e-mail: hugot@mnhn.fr

Table 1. List of parasites and host taxa. Source denotes the author(s) who first described the species. When other descriptions were used, corresponding authorship is given. Honacki et al. (1982) was used as the reference for identification of the hosts

Parasites taxa	Number	Abbreviation	Source	Host taxa
Colobenterobius		COLO	Quentin et al., 1979	
C. colobis	1	COLCOLO	Vuylstéke, 1964 - Hugot, 1987a	*Poliocolobus penantii*
C. entellus	2	COLENT1	Hugot, 1987b	*Semnopithecus entellus*
C. guerezae	3	COLGUER	Hugot, 1987a	*Colobus guereza*
C. longispiculum	4	COLLONG	Quentin et al., 1979	*Tachypithecus obscura*
C. paraguerezae	5	COLPARA	Hugot, 1987a	*Colobus guereza*
C. presbytis	6	COLPRES3	Yin, 1973 - Hugot 1987b	*Tachypithecus phayrei, T. cristata*
C. zakiri	7	COLZAKI	Siddiqi & Mirza, 1954 - Hugot, 1987b	*Semnopithecus entellus*
Enterobius		ENTE	Leach, 1853	
E. anthropopitheci	8	ENTANTH	(Gedoelst, 1916) - Hugot, 1993	*Pan troglodytes, Pan paniscus*
E. bipapillatus	9	ENTBIPA	(Gedoelst, 1916)	*Cercopithecus sabeus, C. aethiops*
E. brevicauda	10	ENTBREV	(L., 1758)	*Papio porcarius*
E. gregorii	11	ENTEGREG	Hugot & T.-Schaeffer, 1983	*Homo sapiens*
E. macaci	12	ENTEMACA	Yin, 1973	*Macaca mulatta*
E. vermicularis	13	ENTEVERM	(L., 1758) - Hugot & T.-Schaeffer, 1983	*Homo sapiens*
Lemuricola		LEM	Chabaud & Petter, 1959	
L. baltazardi	14	LEMBALT	Chabaud et al., 1965	*Lemur fulvus*
L. bauchoti	15	LEMBAUC	Chabaud et al., 1965	*Hapalemur sp.*
L. contagiosus	16	LEMCONT	Chabaud & Petter, 1959	*Cheirogaleus major*
L. lemuris	17	LEMLEM	(Baer, 1935)	*Lemur macaco*
L. nana	18	LEMNANA	Hugot (1996)	*Lemur macaco*
L. nycticebi	19	LEMNYCT	(Baylis, 1928)	*Nycticebus coucang*
L. vauceli	20	LEMVAUC	Chabaud et al., 1965	*Lemur fulvus*
L. daubentoniae	21	LEMDAUB	Petter et al., 1972	*Daubentonia madagascariensis*
Hapaloxyuris		HAP	Inglis & Cosgrove, 1965	
H. goeldii	22	HAPGOEL	Inglis & Cosgrove, 1965 - Hugot, 1984b	*Callimico goeldii*
H. oedipi	23	HAPOEDI	Inglis & Cosgrove, 1965 - Hugot, 1984b	*Oedipomidas oedipi*
H. tamarini	24	HAPTAMA	(Inglis & Dunn, 1964) - Hugot, 1984b	*Tamarinus nigricollis*
Trypanoxyuris		TRY	Vevers, 1923	
T. atelis	25	TRYATEL	(Cameron, 1929) - Hugot, 1985	*Ateles ater*
T. croizati	26	TRYCROI	Hugot, 1994	*Callicebus torquatus*
T. microon	27	TRYMICR	(Linstow, 1907) - Hugot, 1985	*Aotes trivirgatus*
T. minutus	28	TRYMINU	(Schneider, 1866) - Hugot, 1985	*Alouatta seniculus*
T. satanas	29	TRYSATA	Hugot, 1985	*Chiropotes satanas, C. chiropotes*
T. sceleratus	30	TRYSCEL	(Travassos, 1925) - Hugot, 1985	*Saimiri orstedii, S. sciureus*
T. trypanuris	31	TRYTRYP	Vevers, 1923 - Hugot, 1985	*Pithecia monachus, P. pithecia*

Rodentoxyuris		ROD	Quentin, 1974	
R. sciuri	33	RODSCIU	(Cameron, 1932) - Hugot, 1984a	*Sciurus vulgaris, S. carolinensis*
R. bicristata	32	RODBICR	Hugot, 1984a	*Glaucomys sabrinus, G. volans, S. niger*
Xeroxyuris		XER	Hugot (1995)	
Xeroxyuris parallela	34	XERPARA	(Linstow, 1907)	*Xerus inauris*

INTRODUCTION

The patterns among the caudal bursa of the males of various species of pinworms were analyzed using geometric morphometric methods. Thirty-four of 45 species that constitute a monophyletic group were studied. The group's monophyly is supported by the unique and derived morphology of the genital tract of females (Hugot and Tourte-Schaeffer, 1985). Most of these species (42 out of 45) are parasites on Primates, each one being specific to a different host genus. We surveyed parasites of most of the primate families and genera, namely Colobidae, Cercopithecidae, Pongidae, Hominidae, Callithrichidae, Atelidae, Cebidae, Lemuridae, Daubentoniidae and Lorisidae (Table 1). The aims and objectives of the study were to evaluate (1) the application of geometric morphometric techniques to cuticular characters in soft invertebrates (particularly the relationship, if any, between the taxonomic level and the type of shape differences observed both within and among taxa); (2) the current classification for the group; (3) Cameron's (1929) hypothesis of close co-evolution between Primates and their oxyurid parasites; and (4) Chabaud and Petter's (1961) hypothesis that male genital papillae of phasmidian nematodes can be used for systematic and phylogenetic studies.

The characters explored in this study are all associated with the caudal bursa of the males: namely, caudal papillae, phasmids and opening of the cloaca. Osche (1958) and Chabaud and Petter (1961) have shown these characters to be useful for systematic and phylogenetic studies of Nematoda. They have suggested that the male genital papillae of phasmidian nematode parasites are homologous to the cloacal papillae of free-living Rhabditida and that the number and arrangement of caudal papillae in free-living Rhabditida represent the plesiomorphic condition for all phasmidian phyla. They forward hypotheses describing putative sequences of evolutionary change in these structures and propose that the number of genital papillae found in free-living phasmidians has become secondarily reduced in some parasitic lineages. The disappearance of these papillae has been described by Chabaud (personal communication) "as if the posterior part of the cuticle peripheral to the cloaca opening progressively entered into the aperture." The movement of the cuticle, considered a pliable surface, causes the genital papillae to move toward the cloaca following different tracks that are based on their respective primitive locations The caudal (postcloacal) part tends to shorten, along its median axis, leading to a migration of the ventral postcloacal papillae toward the anterior extremity, accompanied by a displacement of the lateral caudal papillae toward the caudal point.

MATERIAL AND METHODS

Table 1 lists the taxa included in this study. Honacki et al. (1982) was used for identification of the hosts. A large number of these species were recently redescribed (see Hugot 1984a, b, 1985, 1987a,b, 1993; Hugot and Tourte-Schaeffer, 1985; Hugot and Vaucher,

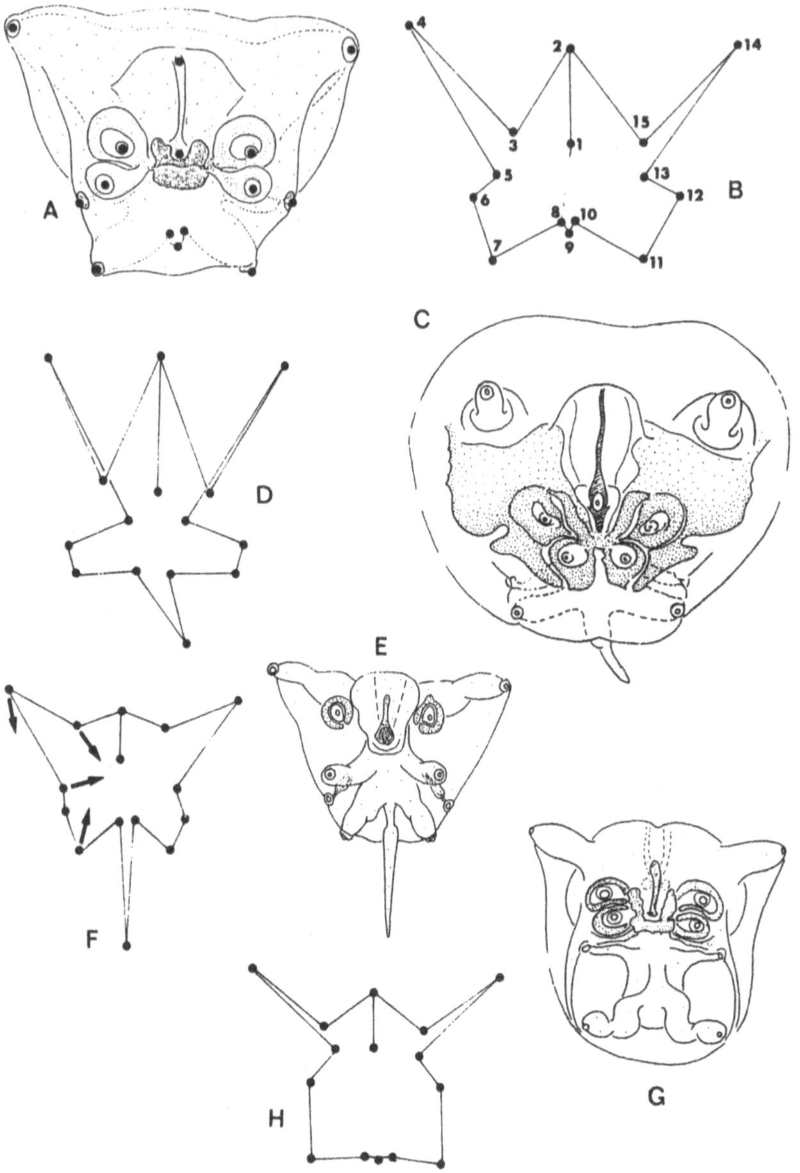

Figure 1. Bursal cauda in ventral view and corresponding landmarks and outline of (A) and (B) *Colobenterobius colobis;* (C) and (D) *Lemuricola nycticebi;* (E) and (F) *Trypanoxyuris microon;* (G) and (H) *Enterobius gregorii.* Landmarks 1–15 are shown in (B).

1985; Hugot et al., 1994). Eleven species were excluded because either the male was unknown or the description of the caudal bursa was incomplete. The preliminary analysis included, in a few cases, three individuals per species. No significant within-species variability was found in these cases so within-species replicates were not included in the study. Fifteen landmarks (Fig. 1 B) were recorded on a digitizing tablet by means of the DS-DIGIT program developed by Slice (1991). The landmarks were (1) the posterior lip of the cloaca; (2) the anterior lip of the cloaca; (3) and (15) the anterior juxta-cloacal papillae; (4) and (14) the precloacal papillae; (5) and (13) the posterior juxta-cloacal papillae; (6) and (12) the openings of the phasmids; (7) and (11) the postcloacal papillae; (8) and (10) the connections of the caudal appendix with the pedoncules of the postcloacal papillae; and (9) the tip of the caudal appendix. The length of this caudal appendix varies among the species: it may be almost absent, reduced or five times longer than the other part of the tail. Landmark 9 exhibits the most variation in the data set—so much so that its influence masks the variation seen at other landmarks. Landmarks 9 and 10 were therefore removed from the data set after preliminary analysis. Landmarks 8 and 1 were conserved among taxa and were used as a baseline for comparison. Consensus coordinates were produced for landmarks 3, 4, 5, 6 and 7 by averaging the coordinates of their respective mirror-image counterpart landmarks (15, 14, 13, 12 and 11) after reflection and superimposition. The data set used for analysis was thus restricted to eight landmarks: 1, 2, 3, 4, 5, 6, 7 and 8.

Relative warps (Bookstein, 1991) and uniform parameters were calculated from shape coordinates by use of the TPSRW package (version 27/09/93; Rohlf, 1993). Because of the exploratory nature of this study, the value of α (Bookstein, 1991; Rohlf, 1993) was set to 0. The mean configuration of the shape coordinate matrix was used as the reference configuration. The differences between genera were analyzed by multivariate analysis of variance (MANOVA), discriminant analyses and stepwise discriminant analyses, on both the affine and nonaffine components. The splines illustrating deformations along the canonical axes were calculated using the TPSREGR package (version 12/11/92; Rohlf, 1992). Contrasts with centroid size (Bookstein, 1991) were used to test the canonical axes and relative warps for allometry. Statistical analyses were performed using SAS 6.04 and NTSYS 1.7. There was no attempt to study the allometric patterns in detail because there were no within-species replicates in the data set.

RESULTS

Centroid Sizes

Centroid size was determined for each taxonomic group (Table 2). The differences among groups were not statistically significant (analysis of variance of the log of the centroid size, $F = 0.59$, df 6, 27, $p = 0.732$).

Analyses of the Affine Component

Neither *U*1 nor *U*2 affine parameters show significant differences among groups when taken separately. However, a MANOVA indicates that they are highly significant when taken together (Table 3). The plot of *U*1 versus *U*2 (Fig. 2) shows a canalization of the

Table 2. Centroid size for each taxonomic group

Groups	Min	Max	Mean	Coefficient of variation
COLO	2.03	3.85	2.82	22.90
ENTE	1.98	3.07	2.51	17.06
LEM	1.85	4.17	2.79	25.32
HAP–TRY–ROD	2.12	3.07	2.60	18.25
HAP	2.38	4.89	3.48	30.10
TRY	2.30	3.32	2.77	12.16
ROD	2.47	2.78	2.63	5.89
XER	2.94	2.94	2.94	0.00

Table 3. Percentage of variability for $U1$ and $U2$

Parameter	F	df	p
$U1$	1.71	6.27	0.157
$U2$	2.03	6.27	0.096
$U1 + U2$	0.33	12.52	0.002*

* indicates significance.

uniform dispersion both within and between the seven genera, which are incompletely separated. The main direction of affine transformation results from the contrast between narrow elongated profiles as seen in LEMDAUB (21), and short dilated ones (with a relatively long 3-4-5 part; see Fig. 1B) as seen in HAPTAMA (24). No secondary trends are readily apparent from the plot. Regression on the log of centroid size shows that the direction of elongation ($U1$) is correlated with size ($r = 0.56$, $p = 0.006$) and $U2$ ($r = 0.35$ $p = 0.045$). Nevertheless, the percentage of variance explained remains low (31 % and 12 % respectively). Most of the correlation is due to within-group allometric patterns exhibited by the LEM and HAP groups. The affine part separates the genera *Enterobius* and *Colobenterobius*, and divides the genus *Lemuricola* into two distinct subgroups.

Relative Warp Analysis

Most of the taxonomic information at the generic level is seen in the four first relative warps which account for nearly 82 % of the total variation. Relative warps 1, 2, 3 an 4,

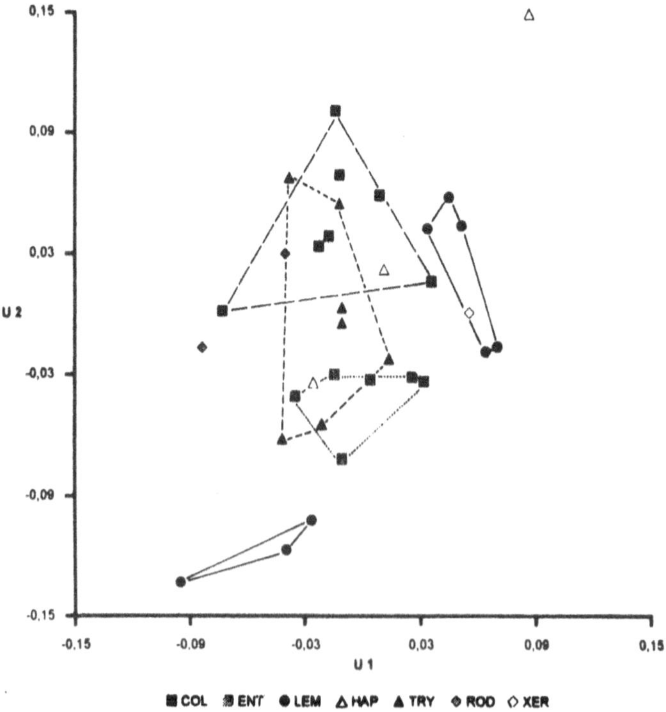

Figure 2. Bivariate plot showing affine components $U1$ and $U2$. The human parasites (genus *Enterobius*) are represented by gray squares with a black dot.

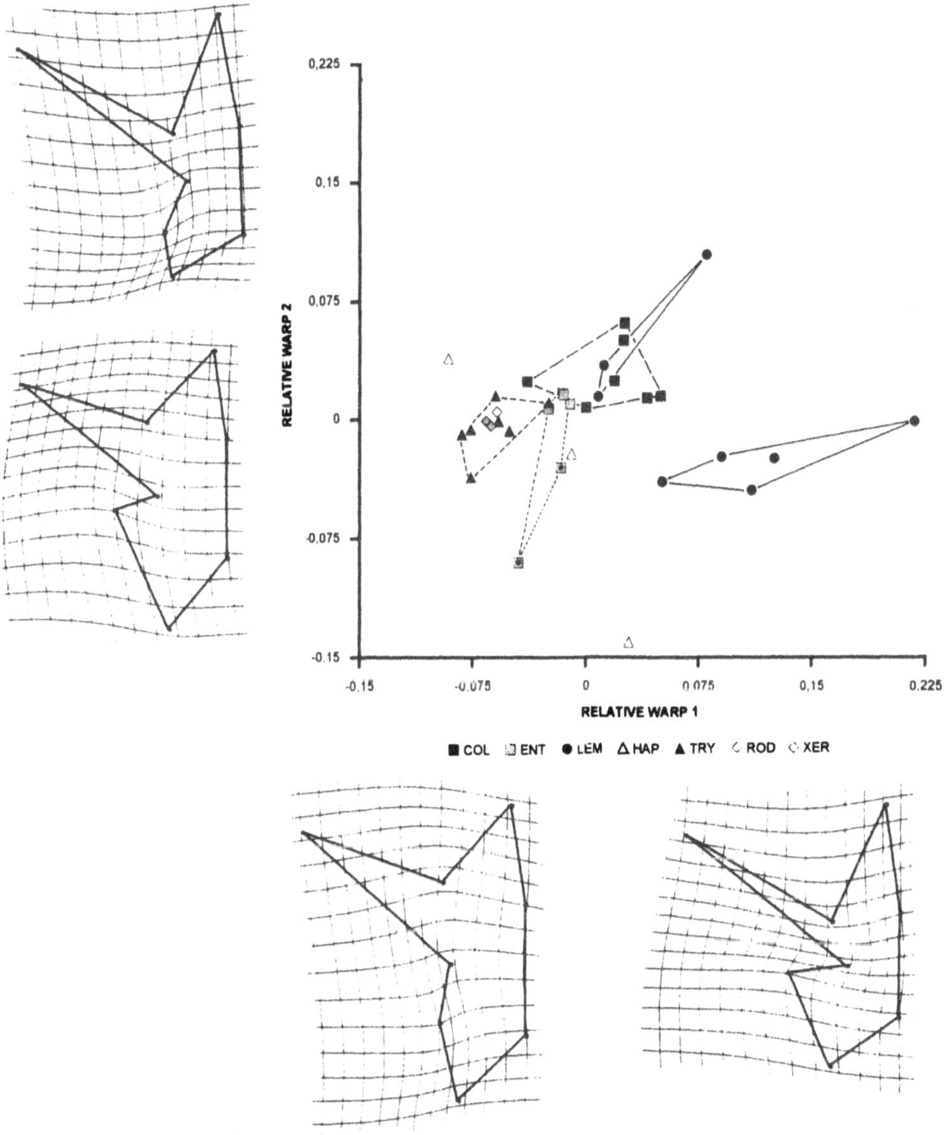

Figure 3. Bivariate plot showing Relative warps 1 and 2. The human parasites (genus *Enterobius*) are represented by gray squares with a black dot.

respectively account for 43, 16, 14 and 8% of the total variation. The 90 % level is reached only when the first six relative warps are considered.

Whole Sample Consideration. Relative warps 1 and 2 (Fig. 3) reflect the following shape differences as one proceeds positively along the axes: landmark 5 moves posteriorly relative to landmark 1 and closer to axis 1–8 than it is to landmark 3; part 1–2–3–4–5 becomes longitudinally stretched; and angle 3–4–5 becomes progressively more acute and landmarks 3 and 5 assume a more posterior position.

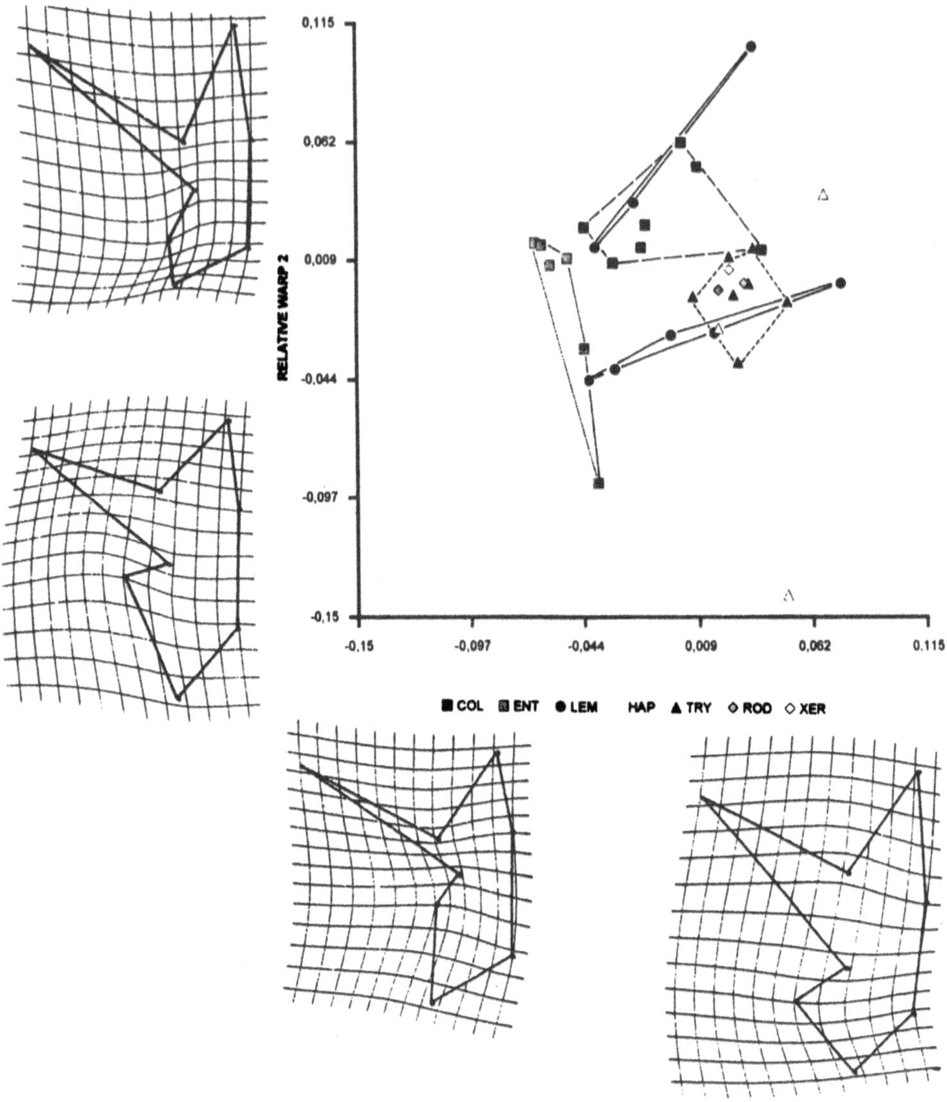

Figure 4. Bivariate plot showing relative warps 2 and 3. The human parasites (genus *Enterobius*) are represented by gray squares with a black dot.

Relative warps 2 and 3 (Fig. 4) mainly reflect differences in the relative positions of landmarks 6 and 7. In the positive direction, landmark 7 is progressively shifted toward axis 1–8 while landmark 6 moves farther away. These changes are accompanied by a reduction in part 5–6–7–8–1. The combination of relative warps 2 and 3 allows a better separation of the taxonomic groups: The two genera *Enterobius* and *Colobenterobius* are completely separated. The plot depicts a distribution of *Enterobius* along relative warp 2 and of *Colobenterobius* along relative warp 3. *Rodentoxyuris, Xeroxyuris* and *Trypanoxyuris* overlap in this plot, and *Hapaloxyuris* generally falls well outside the group.

Patterns Within Enterobius and Colobenterobius. Whatever the combination of axes, these two subgroups always appear as distinct clusters but also as close neighbors. A plot of relative warps 2 and 3 shows the best separation between *Colobenterobius* and *Enterobius* (Fig. 4). *Enterobius* is characterized by a longitudinally stretched region 5–6–7–8–1 and a backward extension, whereas *Colobenterobius* exhibits a longitudinal compression of the same region. In *Colobenterobius* landmark 6 is generally placed laterally to landmarks 3 and 5. As one progresses from the top to the bottom of Fig. 4, landmark 6 (the phasmid) becomes more lateral and landmarks 3 and 5 move closer to the cloacal aperture and nearer to the median axis.

The species are distributed along relative warp 3 as follows: COLGUER–COLPRES3–COLZAKI–(COLONG–COLENT1–COLPARA) –COLCOLO. Different species within the genus *Enterobius* are dispersed as two subgroups along relative warp 2, the human parasites being separate from the other four species. The shape differences along this axis involve contrasts in segments 2–1 and 1–8 parallel to the sagittal plane. In all species, with the exception of the two human parasites, landmark 3 is level with, or posterior to landmark 2, and landmark 5 is posterior to landmark 1. In the human parasites landmark 3 is anterior to landmark 2 and landmark 5 is anterior to landmark 1.

Patterns Within Lemuricola. The entire group is characterised by a landmark configuration in which landmark 5 is posterior to landmark 1 and closer to the median axis than to landmark 3, the region 1–2–3–4–5 is longitudinally stretched, the angle 3–4–5 is reduced, landmarks 3 and 5 are posteriorly placed and landmark 4 is at the same level or weakly anterior to landmark 2. Plotting relative warps 1 and 2 allows separation of most species of the genus, except LEMCONT (16) and LEMDAUB (21). Two subgroups can be distinguished. (1) The first one includes LEMCONT (16), LEMNYCT (19) and LEMDAUB (21). In the positive part of relative warp 2 these three species are characterized by the lateral position of landmarks 7 and 6, which results in a lateral dilation of the region 5–6–7–8. This stretching is more accentuated in LEMNYCT (19) because of the more anterior position of landmark 7. (2) The second subgroup includes the five remaining species in the negative part of relative warp 2, LEMBALT (14), LEMBAUC (15), LEMLEM (17), LEMNANA (18), LEMVAUC (20). These species are characterized by landmark 6 moving forward up to the level of landmark 5, a shortening of segment 1–8 and a slight movement of landmark 7 closer to the median axis.

Patterns Within Hapaloxyuris - Trypanoxyuris - Rodentoxyuris. This is the most homogeneous group. All the species are closely grouped, particularly those in *Trypanoxyuris* and *Rodentoxyuris*. The *Hapaloxyuris* species are often peripheral to the group, especially HAPGOEL (22) and HAPOEDI (23). The variability of the ratio (5–7)/(2–6) in this group is very small, between 0.50 and 0.52. An analysis of the ratio ((5–6 + 6–7))/(5–7) shows that *Hapaloxyuris* accounts for much of the heterogeneity within the group due to the relative placement of landmark 6 which is either in line with, or anterior to, segment 2–5. This point is in line with segment 5–7 in *Trypanoxyuris* and *Rodentoxyuris*.

Canonical Discriminant Analysis

Owing to the low number of species in some groups, conclusions based on these analyses should be regarded as tentative.

Affine Component Excluded. Only canonical axes 1 and 2 are statistically significant. The first axis represents 70.1% of the variance and the second 16.6% (86.7% cumulative). The corresponding plot reveals two discrete clusters, the first comprising *Enterobius*,

Colobenterobius and *Lemuricola*, the second *Hapaloxyuris, Trypanoxyuris* and *Rodentoxyuris* (Fig. 5). Within each of these two groups components are recognizable but incompletely separated. *Xeroxyuris* falls distantly from all other groups. No significant correlation was observed between any of the axes and the log of centroid size ($p = 0.15$), suggesting that the discrimination does not include a statistically significant allometric component.

The first group, comprising taxa *Enterobius, Colobenterobius* and *Lemuricola*, is characterized by a compression of the region delimited by landmarks 3–4–5 as a result of landmarks 3 and 5 drawing closer together and a relative expansion of segment 1–2 and, to a lesser extent, segment 1–8. In the group *Hapaloxyuris–Trypanoxyuris–Rodentoxyuris* the reverse is observed: an expansion of region 3–5 and a relative compression of segments 1–2 and 1–8, together with a marked lateral shift in landmark 6. The anterior and posterior regions of the genital plate appear to evolve somewhat independently of one another. *Xeroxyuris* appears completely different from the rest of the taxa, with a broad genital plate without

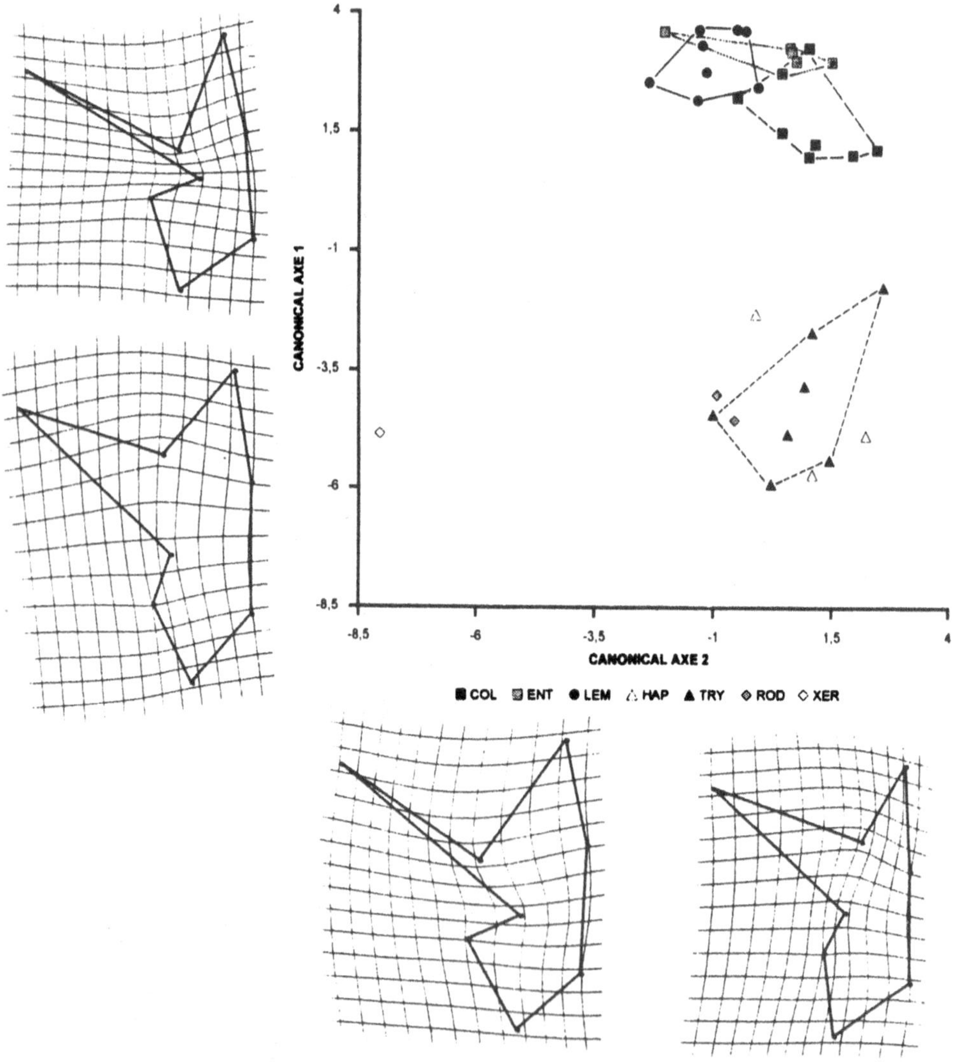

Figure 5. First two canonical variates from a canonical discriminant analysis with affine part excluded. The human parasites (genus *Enterobius*) are represented as gray squares with a black dot.

anterior or posterior narrowing shifts. The region delimited by the landmarks 3–4–5 is very narrow as observed in the *Enterobius–Colobenterobius–Lemuricola* group. Nevertheless this pattern differs in *Xeroxyuris* where landmarks 3 and 5 are very distant.

Affine Components Included. Axes 1, 2 and 3 are statistically significant. The first axis represents 74.4% of variance, the second 12.8% and the third 6.1% (93.3% cumulative). The plot of the canonical axes 1 and 2 (Fig. 6) allows recognition of the same two groups, but in this case *Colobenterobius* is distinct from *Lemuricola* and *Enterobius*. In the group *Hapaloxyuris–Trypanoxyuris–Rodentoxyuris* the subgroups are better isolated, but the discrimination is not complete. *Xeroxyuris* is distant from both groups. The plane defined by canonical axes 1 and 3 (Fig. 7) completely separates *Enterobius*, *Colobenterobius* and *Lemuricola*. In the *Hapaloxyuris-Trypanoxyuris–Rodentoxyuris* group, *Hapaloxyuris* and

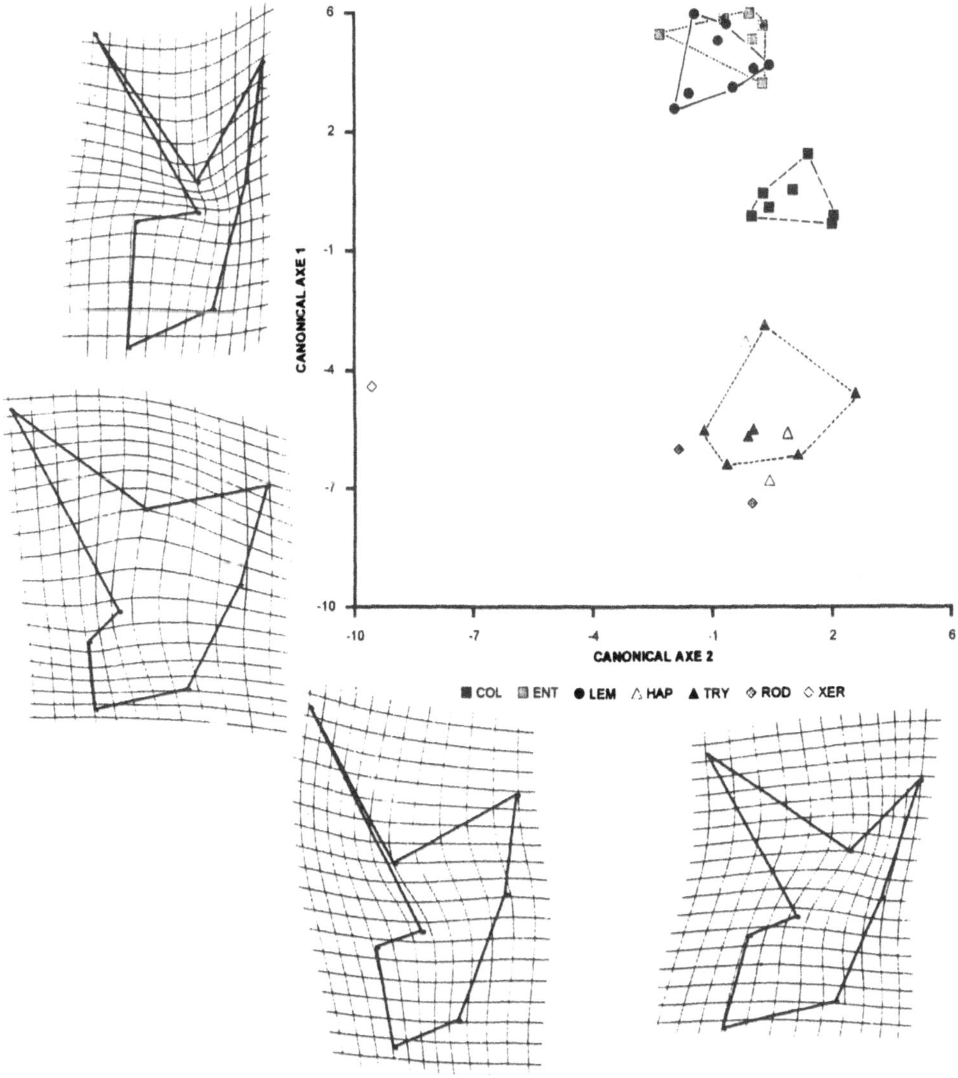

Figure 6. First two canonical variates from a canonical discriminant analysis with affine part included. The human parasites (genus *Enterobius*) are represented by gray squares with a black dot.

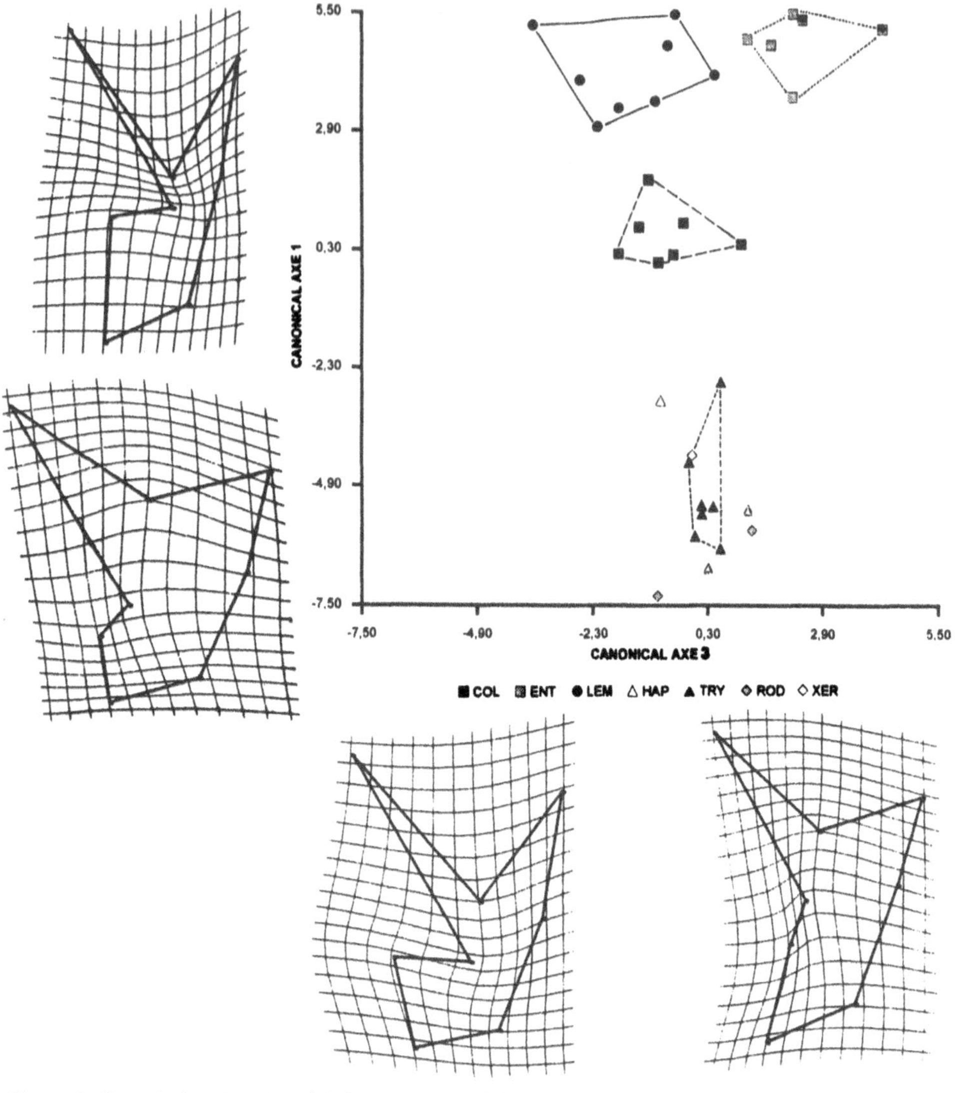

Figure 7. Canonical variates 1 and 3 from a canonical discriminant analysis with affine part included. The human parasites (genus *Enterobius*) are represented as a gray squares with a black dot.

Rodentoxyuris are restricted to an area surrounding a more compact group formed by *Trypanoxyuris*. Hence this plot can be considered the most informative for discrimination of the different groups. None of the axes is correlated with the log of centroid size ($p = 0.41$).

In contrast to the preceding results, the inclusion of the affine component indicates that the group *Enterobius–Colobenterobius–Lemuricola* is characterized by considerable lateral compression of the anterior region (landmarks 1–5) and an anteroposterior dilatation of the posterior region (landmarks 6–1). Conversely, the group *Hapaloxyuris–Trypanoxyuris–Rodentoxyuris* exhibits a strong lateral dilatation of the anterior region and no anteroposterior dilatation. The characters of *Xeroxyuris* are the same as those reported above but are more accentuated: particularly the narrowing of the region 3–4–5 and the large anteroposterior gap between landmarks 3 and 5. The separation of *Enterobius* from *Lemuri-*

cola appears to be due mainly to a lateral affine component, which is reflected in a strong anteroposterior shortening of the posterior part of the genital plate (region 6–1) in *Lemuricola* and the anterior region (landmarks 1–5) being shifted forward in *Enterobius*.

CHARACTERIZATION OF THE TAXA

Enterobius and *Colobenterobius*

The genus *Enterobius* is anatomically characterized by an elongated postcloacal part and a short cloacal aperture (segment 1–8 being longer than segment 1–2 and segment 5–6 being shorter than segment 6–7); unlateralized phasmids (landmark 6); and the third pair of papillae (point 5) being vertically in line with the second pair (point 3) or slightly more medially placed. The two human parasites can be distinguished from the four other species by their very short cloacal apertures and their more lateralized phasmids. The parasite of the chimpanzees, ENTANTH (8) is intermediate for these characters.

The genus *Colobenterobius* is anatomically characterized by a short postcloacal region; lateralized phasmids (point 6), except in COLGUER (3); the third pair of papillae (point 5) being vertically in line with, or weakly more lateral to, the second pair (point 3) except in COLZAKI (7). COLCOLO (1), COLENT1 (2), COLLONG (4) and COLPARA (5) have a very short segment: (1–8)<(1–2). Regardless of the axes considered, no differences between the species can be identified to reflect a possible separation between the Ethiopian and the Asiatic species.

The ratio (5–7)/(2–6) permits complete separation of the two genera. The values for this ratio range between 0.40 and 0.64 in the *Colobenterobius* species and between 0.82 and 1.19 in the *Enterobius* species. The same holds for the ratio ((5–6) +(6–7))/(5–7); values between 1.13 and 1.29 are seen for *Colobenterobius* and values between 0.91 and 1.08 for the *Enterobius*. The ratio (3–5)/(5–7) falls between 0.36 and 0.69 for *Colobenterobius* and between 0.19 and 0.30 for the *Enterobius* species.

The ratio (5–7)/(2–6) completely separates *Enterobius* into two groups, the first comprising ENTANTH (8), ENTBIPAP (9), ENTBREV (10) and ENTMACA (12) (ratio < 0.88) and the second comprising the human parasites ENTGREG (11) and ENTVERM (13) (ratio >1).

Angle 5–6–7 allows separation of the *Colobenterobius* and the human parasites; it is reflex in COLGUER (3), ENTANTH (8), ENTBIPAP (9), ENTBREV (10) and ENTMACA (12). In the *Colobenterobius*, angle 2–3–4 is progressively wider from 90° up to greater than 120°.

Lemuricola

The genus *Lemuricola* is characterized by (1) a longitudinally stretched cloacal aperture; (2) a markedly anterior placement of the first pair of papillae; and (3) the placement of the second and third pairs of papillae close to the posterior lip of the cloaca, both close together and close to the median axis. In LEMCONT (16), LEMDAUB (21) and LEMNYCT (19) the medial position of these papillae is so accentuated that landmark 3 becomes posterior to segment 4–5. This closes the angle subtended by landmarks 3–4–5. In these species the posterior part of the caudal bursa is laterally developed and the fourth cloacal papillae is vertically in line with the first pair. LEMVAUC (20), LEMBALT (14) and LEMLEM (12) have a phasmidian aperture anterior to the third pair of papillae and a shortened postcloacal part of the tail; the tail appendix is absent. LEMNANA (18) shares the characters of this subgroup with the exception of a relatively short cloacal aperture and a third pair of genital papillae that are more laterally situated. LEMBAUC (15) differs from the four other species in the position of landmark 6 (phasmid). The group LEMCONT (16), LEMDAUB (21) and LEMNYCT (19) can be distinguished from the group LEMNANA (18), LEMVAUC (20), LEMBALT (14) and LEMLEM (12) according to the angle 4–5–6 if it is greater than 90° for the former and is less than 45° for the latter. LEMBAUC (15) is intermediate.

Hapaloxyuris–Trypanoxyuris–Rodentoxyuris

This group is anatomically characterized by a marked lateral placement of the first pair of papillae (landmark 4); the placement of the second pair (landmark 3) anterior to the posterior lip of the cloaca (landmark 1), resulting in the angle 2–3–4 always greater than 90°; and the third pair of papillae (landmark 5) more lateral than the second pair. Compared with *Hapaloxyuris*, *Trypanoxyuris* and *Rodentoxyuris* are characterized by the relatively more posterior position of the phasmid. Ratio (3–5)/(5–7) completely separates *Trypanoxyuris* and *Rodentoxyuris* from parasites of humans, apes and other monkeys: This ratio has values between 0.60 and 1.22 in *Trypanoxyuris* and *Rodentoxyuris*; between 0.36 and 0.69 in *Colobenterobius*; and between 0.19 and 0.30 in *Enterobius*. The ratio (3–5)/(1–2) completely separates *Trypanoxyuris* and *Rodentoxyuris* from the *Colobenterobius* and *Enterobius* but does not distinguish *Colobenterobius* from *Enterobius* : the ratio falls between 0.70 and 1.50 in *Trypanoxyuris* and *Rodentoxyuris*; between 0.39 and 0.65 in *Colobenterobius* and between 0.39 and 0.54 in *Enterobius*.

Xeroxyuris

This genus in monotypic. The single species, XERPARA (34) was previously classified in the genus *Enterobius*. Our analysis indicates that this species is more often associated with the group *Hapaloxyuris–Trypanoxyuris–Rodentoxyuris*, with which it shares the position of the second and third pairs of papillae and general form of the region 5–6–7–8. It differs from these species by the general lateral stretching of the outline and the very anterior position of the first pair of papillae (point 4).

DISCUSSION

Testing the Classification: Relationships between Taxonomic Level and Deformation

The techniques used in this study allowed classification of species into three groups, each one characterized by a particular spatial distribution of the phasmids and genital papillae: (1) *Colobenterobius–Enterobius*; (2) *Lemuricola;* and (3) *Hapaloxyuris–Trypanoxyuris–Rodentoxyuris*.

Colobenterobius and *Enterobius*, which were previously classified in the same genus, are clearly differentiated. This confirms the proposal of Quentin et al. (1979) to isolate the parasites of the Colobinae into a separate taxon. In *Enterobius* the human parasites are separated from the four other species, specific to *Pan, Cercopithecus, Papio* and *Macaca*. The genus *Lemuricola* could be separated into two subgroups. These sub-groups do not fit with the three subgenera previously proposed by Chabaud and Petter (1959). However, they separate pinworms with a caudal appendix from all others, even though the landmarks describing this appendix were eliminated from the data set. This finding reveals that the presence or absence of an appendix is linked with other aspects of the morphology of the caudal bursa. It is noteworthy that the species whose males possess an appendix parasitize lemurs, which are considered to have branched off from the main primate lineage at an early stage. This character distribution consistent with Chabaud and Petter's (1959) interpretation that the presence of this appendix represents the retention of a plesiomorphic character.

In the *Hapaloxyuris–Trypanoxyuris–Rodentoxyuris* group the previously proposed subdivisions are not apparent. However, it should be noted that RODBICR (32), RODSCIU (33) and XERPARA (34), which are all parasites of squirrels, are often associated. The parasites of marmosets, HAPGOEL (22), HAPOEDI (23) and HAPTAMA (24), are relatively peripheral, although they are not closely grouped. XERPARA (34), the parasite of the South African squirrel, appears very distinct from the *Colobenterobius–Enterobius* group.

Testing Cameron's Hypothesis

Cameron's (1929) hypothesis supposes close coevolution between primates and their oxyurid parasites. This hypothesis was based on the distribution of the parasites into different taxa, each one specific for a particular host group. Our study identifies the same groups and thus supports this hypothesis.

Testing Chabaud and Petter's Hypothesis

The evolutionary transformations hypothesized by Chabaud and Petter (1961) are shown on Fig. 1F: papillae anterior and (or) lateral to the cloaca would first move backwards and then converge toward the posterior lip of the cloacal aperture, and papillae posterior to the cloaca and close to the ventromedian axis would move directly to the posterior lip as represented. The deformations reflected in the splines (Fig. 3 and 4) are entirely consistent with these hypothesized transformations, particularly for the juxtacloacal papillae, which follow a curve around the cloacal aperture from an anterior and lateral position to a more posterior and medial one. The apparent movement of the lateral precloacal and caudal papillae, which is generally parallel to the median axis, is also consistent with the hypothesized evolutionary transformations forwarded by Chabaud and Petter. Moreover, this study also shows an enlargement or a shortening of the cloacal aperture, which was not previously reported.

CONCLUSION

The geometric morphometric techniques used in this study permitted recognition of taxa that had previously been described based on different morphological characters. This suggests that geometric methods can be usefully applied to studies of morphological variation in soft-bodied animals. Moreover, the thin-plate spline analysis is concordant with some of the evolutionary shape changes of the genital papillae proposed by Chabaud and Petter (1961) to have taken place, and further suggests some additional shape changes. This implies that geometric methods may help to discover characters that might be useful for phylogenetic reconstruction.

REFERENCES

Bookstein, F. L. 1991. Morphometric tools for landmark data: Geometry and biology. Cambridge University Press: Cambridge.

Cameron, T. W. 1929. The species of *Enterobius* Leach, in Primates. Journal of Helminthology 7:161–182.

Cameron, T. W. 1932. On a new species of oxyurid from the grey squirrel in Scotland. Journal of Helminthology 10:29-32.

Chabaud, A. G. and Petter, A. J. 1961. Evolution et valeur systématique des papilles cloacales chez les nématodes phasmidiens parasites de vertébrés. Comptes Rendus de l'Académie des Sciences 252:1684–1686.

Hugot, J. P. 1984a. Sur le genre *Trypanoxyuris* (Oxyuridae, Nematoda). I. Parasites de Sciuridae : sous-genre *Rodentoxyuris*. Bulletin du Muséum National d'Histoire Naturelle, Paris, 4e série, 6:711–720.

Hugot, J. P. 1984b. Sur le genre *Trypanoxyuris* (Oxyuridae, Nematoda). II. Sous-genre *Hapaloxyuris* parasite de Primates Callitrichidae. Bulletin du Muséum National d'Histoire Naturelle, Série A, Zoologie 6:1007–1019.

Hugot, J. P. 1985. Sur le genre *Trypanoxyuris* (Oxyuridae, Nematoda). III. Sous-genre *Trypanoxyuris* parasite de Primates Cebidae et Atelidae. Bulletin du Muséum National d'Histoire Naturelle, Série A, Zoologie 7:131–155.

Hugot, J. P. 1987a. Sur le genre *Enterobius* (Oxyuridae, Nematoda): sous-genre *Colobenterobius*. I. Oxyures parasites de Primates Colobinae en région éthiopienne. Bulletin du Muséum national d'Histoire naturelle, Paris, 4e série, 9:341–352.

Hugot, J. P. 1987b. Sur le genre *Enterobius*: sous genre *Colobenterobius*. II. Oxyures parasites de Singes Colobinae en région orientale. Bulletin du Muséum National d'Histoire Naturelle, Paris, 4e série, 9:799–813.

Hugot, J. P. 1993. Redescription of *Enterobius anthropopitheci* (Nematoda, Oxyurida), parasite of the chimpanzees. Systematic Parasitology 26:201–207.

Hugot, J. P. 1995. Redescription of *Xeroxyuris parallela* (Linstow, 1907) n.gen., n.cb., parasite of *Xerus inauris*. Parasite 2:323-329.

Hugot, J. P. 1996. Description of *Lemuricola nana* n. sp. (Nematoda, Oxyurida), parasite of *Lemuris macao*. Research and Review in Parasitology (in press).

Hugot, J. P., S. Morand and R. Guerrero. 1994.*Trypanoxyuris croizati* n.sp. and *T. callicebi*, two vicariant forms parasite of *Callicebus*_spp. (Primate, Cebidae). Systematic Parasitology 27(1):35–43

Hugot J. P. and C. Tourte-Schaefer. 1985. Etude morphologique des deux oxyures parasites de l'Homme: *Enterobius vermicularis* et *E. gregorii*. Annales de Parasitologie Humaine et Comparée 60:57–64.

Hugot, J. P. and C. Vaucher, C. 1985. Sur le genre *Trypanoxyuris* (Oxyuridae, Nematoda). IV. Sous-genre *Trypanoxyuris* parasite de Primates Cebidae et Atelidae (*suite*). Etude morphologique de *Trypanoxyuris callicebi* n. sp. Bulletin du Muséum National d'Histoire Naturelle, Série A, Zoologie 7:633–636.

Osche, G. 1958. Die bursa und Schwanzstrukturen und ihre Aberrationen bei den Strongylina (Nematoda). Morphologische studien zum Problem der pluri und Paripotenzerscheinungen. Zoologische Morph. und Okol. 46:571–635.

Quentin, J. C., C. Betterton and M. Krishnasamy 1979. Oxyures nouveaux ou peu connus, parasites de Primates, de Rongeurs et de Dermoptères en Malaisie. Création du sous-genre *Colobenterobius* n. subgen. Bulletin du Muséum National d'Histoire Naturelle, Série A, Zoologie 1:1031–1050.

Rohlf, F. J. 1992. TPSREGR - Thin-plate splines regression analysis. Department of Ecology and Evolution, State University of New York, Stony Brook, New York 11794.

Rohlf, F. J. 1993. TPSRW - Thin-plate splines relative warp analysis. Department of Ecology and Evolution. State Univ. New York, Stony Brook, New York 11794.

Slice, D. E. 1991. DS-DIGIT: Basic digitizing software. Department of Ecology and Evolution. State University of New York, Stony Brook, New York 11794.

Vevers, C. M. 1923. Some new and little known helminths from British Guiana. Journal of Helminthology 1:35–45.

MORPHOMETRIC PATTERN AND FEEDING MODE IN EMPEROR FISHES (Lethrinidae, Perciformes)

Kent E. Carpenter

Department of Biological Sciences
Old Dominion University
Norfolk, Virginia 23529-0058

ABSTRACT

The relationship between body shape and tooth type is examined for the 28 species of emperor fishes in the genus *Lethrinus* by means of shape coordinates and thin-plate splines. Both methods delineate two major groups within the genus. One group consists of low-bodied species with conically shaped lateral teeth. The other group consists of high-bodied species with both molariform, conical and intermediate lateral tooth types. Uniform shape components appear to account for much of the inferred evolutionary shape changes in these fishes. The primary nonuniform component of shape difference among fishes describes an arching of the body upwards and downwards, which may reflect a preservation artifact.

INTRODUCTION

General trends describing the evolution of body shape in teleost fishes in relation to their swimming behavior, habitat and feeding type are well known (e.g., Alexander, 1974). However, the relationship between body form and feeding mode within a closely related group of marine fishes has yet to be examined. The purpose of this paper is to investigate the variation of body shape in relation to feeding type, assessed through differences in dentition, among the 28 known species of emperor fishes, genus *Lethrinus*, by means of geometric morphometric methods (Rohlf and Bookstein, 1990; Bookstein, 1991.)

Emperor fishes are tropical and subtropical marine fishes that occur mostly in the Indo-West Pacific (Carpenter and Allen, 1989). They are benthic, carnivorous, coastal fishes ranging primarily on or near reefs. They generally possess large, strong jaws, and food preference is correlated with the type of teeth in the lateral jaw. Species with well-developed molars are high bodied and usually consume hard- or soft-shelled sessile or slow-moving invertebrates, such as mollusks, sea urchins and certain crustaceans. Those that have exclusively conical lateral jaw teeth feed mostly on fishes and cephalopods. They tend to

Advances in Morphometrics, Edited by L. F. Marcus et al.
Plenum Press, New York, 1996

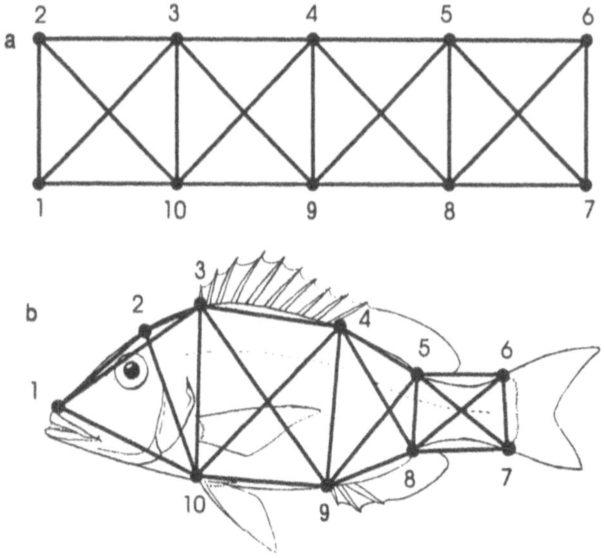

Figure 1. The truss network and landmarks measured on species of *Lethrinus*.

chase or ambush their mobile prey and have more fusiform bodies than do the species with molars. There are also species with intermediate tooth types and body shapes that exhibit considerable variation in their diets.

The Lethrinidae, together with the Sparidae, and Nemipteridae are a closely related perciform group (Akazaki, 1962) and have been placed with the Centracanthidae in the superfamily Sparoidea (Johnson, 1981). Among these sparoid fishes, lethrinids are unique in their shared possession of 11 epipleural ribs. Although the phylogenetic position of lethrinids among sparoids is at least partially hypothesized, relationships within lethrinids are unknown. Furthermore, nothing is known about phylogenetic relationships within the genus *Lethrinus* because very few characters are available for phylogenetic analysis (Carpenter and Allen, 1989). I have therefore restricted this paper to a phenetic investigation of morphometric variation in the hope of defining characters that may subsequently be used to infer evolutionary relationships.

METHODS

Interlandmark caliper distances forming a 10-landmark truss network (Strauss and Bookstein, 1982; Fig. 1) were measured for 413 specimens representing the 28 species of emperor fishes. These interlandmark distances were converted to Cartesian coordinates by means of a simplified multidimensional scaling method (Carpenter et al., this volume). The landmarks are (1) tip of the snout, (2) posterior margin of the skull on the midline, (3) origin of the dorsal fin, (4) origin of the soft dorsal fin, (5) terminus of the dorsal fin, (6) intersection of upper margin of caudal fin on a vertical at the level of the end of hypural plates, (7) intersection of lower margin of caudal fin on a vertical at the level of the end of hypural plates, (8) terminus of anal fin, (9) origin of anal fin and (10) origin of pelvic fin (Fig. 1). Table 1 lists the species abbreviations used for subsequent tables and figures, the general lateral tooth type for each species and the number of specimens examined per species.

Table 1. The species of *Lethrinus* examined, abbreviations (Abbr.) used in all subsequent tables and figures, number of specimens examined, and primary lateral tooth type possessed by each species

Species	Abbrev.	n	Tooth type
Lethrinus amboinensis	lambo	4	conical
L. atkinsoni	latki	18	molar
L. atlanticus	latla	16	conical
L. borbonicus	lborb	25	molar
L. concyliatus	lconc	2	conical
L. crocineus	lcroc	5	molar
L. enigmaticus	lenig	9	intermediate
L. erythropterus	leryo	25	molar
L. erythracanthus	leryt	15	conical
L. genivittatus	lgeni	22	conical
L. haematopterus	lhaem	14	conical
L. harak	lhara	16	molar
L. laticaudis	llati	9	intermediate
L. lentjan	llent	25	molar
L. mahsena	lmahs	18	molar
L. microdon	lmicr	14	conical
L. miniatus	lmini	8	conical
L. nebulosus	lnebu	28	intermediate
L. obsoletus	lobso	18	intermediate
L. olivaceus	loliv	18	conical
L. ornatus	lorna	20	molar
L. reticulatus	lreti	13	conical
L. rubrioperculatus	lrubr	13	conical
L. semicinctus	lsemi	17	conical
L. sp1	lsp1	6	conical
L. sp2	lsp2	5	conical
L. variegatus	lvari	17	conical
L. xanthochilus	lxant	13	conical

Carpenter and Allen (1989) described and illustrated the species studied here and list the museum sources for the material examined.

Bookstein shape coordinates were computed with landmarks 1 and 7 as end points for the baseline (Fig. 1) by means of the SAS program UNICOORD written by L. F. Marcus, listed in Corti (1992) and modified here for use with the *Lethrinus* data. This program was also used to compute centroid size and the uniform x and y factors (Bookstein, 1991) of the truss. Pearsonian correlations were computed among baseline size, centroid size and both x and y uniform factors for each species. A nested analysis of variance (ANOVA) was used to test for effects of species differences and tooth type on centroid size and shape coordinates.

Canonical variate analysis was used to explore the combined effects of species differences and tooth type on shape coordinates by use of PROC CANDISC in SAS (1988). The resulting Mahalanobis squared distances were clustered using UPGMA in NTSYS-pc (Rohlf, 1993a) to summarize shape differences among species. Consensus landmark configurations were computed from raw coordinates for each species, by use of the Generalized Least Squares Procrustes method in GRF-ND (Slice, 1993). The consensus shapes for each species were input to TPSRW (Rohlf, 1993 version 2 October 1993) for relative warp analysis, and the plots provided in the program were used to visualize the direction and magnitude of shape differences (warps were equally weighted [α, = 0] following Rohlf's (1993b) guidelines for exploratory systematic studies). Generalized Least Squared Procrustes residuals of the raw coordinates for all 413 specimens were also subjected to two separate relative warp analyses, one including and one excluding the affine components. Both univariate and multivariate differences among species and tooth types were analyzed using canonical variate analysis of the uniform and partial warps components by means of SAS (1988). Finally, canonical scores on axes 1 and 2 from these analyses were regressed against the residuals from the Procrustes analysis by means of Rohlf's TPSREG program.

RESULTS

Centroid size was highly correlated with baseline size ($p < 0.001$). The choice of landmarks 1 and 7 for the baseline was thus considered acceptable as a measure of overall size. The uniform x component derived from the shape coordinates (Bookstein, 1991) was significantly correlated ($p < 0.05$) with centroid size for *Lethrinus atlanticus*, *L. borbonicus*, *L. erythropterus*, *L. harak* and *L. miniatus*. The uniform y component was significantly correlated with centroid size for *L. erythracanthus*, *L. miniatus*, *L. ornatissimus* and *L. xanthochilus*. Centroid size was significantly correlated with the uniform x and y components for all specimens pooled ($r = -0.19048$, $p < 0.001$ for x; $r = -0.10844$, $p = 0.0275$ for y).

Univariate tests of shape coordinates showed significant differences ($p < 0.001$) among species for all coordinates and significant differences ($p < 0.05$) for tooth types at 13 of the 16 coordinates (all coordinates except x_6, y_8, and x_{10}). Multivariate tests of these shape coordinates were also highly significant ($p < 0.001$) for species (df = 432) and tooth types (df = 32). Clustering of Mahalanobis distance showed two distinct clusters, one consisting of all low-bodied species with conically shaped lateral teeth in their jaws, the other with high-bodied species having molariform, conical and intermediate lateral tooth types (Fig. 2). Results from both univariate and multivariate tests for differences among principal warps and uniform components were similar to those for the shape coordinates; all multivariate differences were significant ($p < 0.001$) all univariate differences among species were

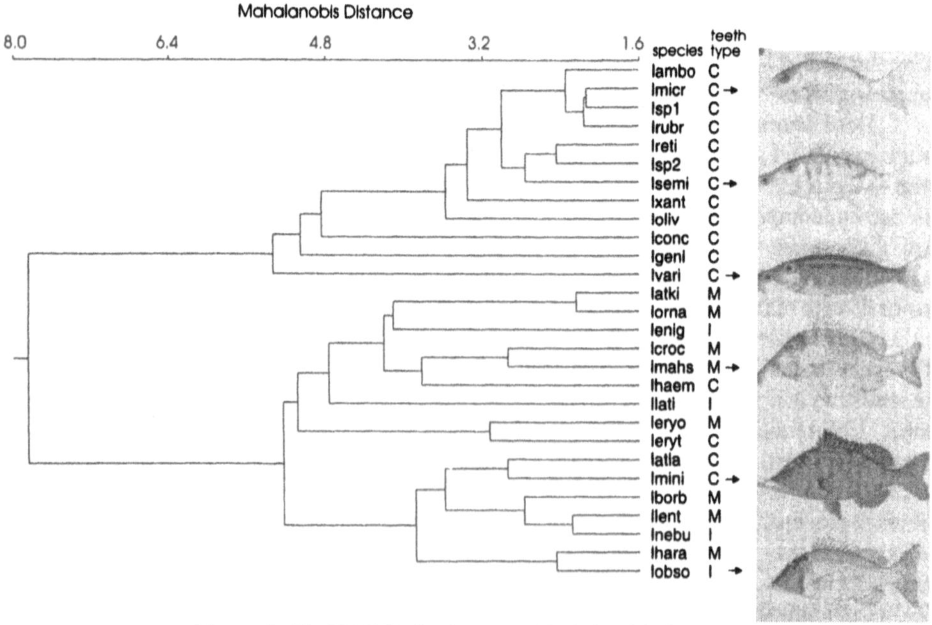

Figure 2. The UPGMA dendrogram of Mahalanobis distances.

Table 2. Results from two TPSRW analyses of the 413 specimens, including Bookstein's uniform component and excluding the uniform component. The singular standard deviations (D) among specimens for each relative warp vector (k) are given along with degrees of freedom (df); * $p < 0.05$, *** $p < 0.001$

	With Uniform		Without uniform	
k	*D*	df	*D*	df
1+	0.70851	135***	0.02312	104***
2+	0.20900	119***	0.01645	90***
3+	0.02098	104***	0.01428	77***
4+	0.01536	90***	0.01271	65***
5+	0.01285	77***	0.01002	54***
6+	0.01134	65***	0.00891	44***
7+	0.00932	54***	0.00729	35***
8+	0.00766	44***	0.00567	27***
9+	0.00630	35***	0.00537	20***
10+	0.00563	27***	0.00481	14***
11+	0.00486	20***	0.00462	9***
12+	0.00462	14***	0.00370	5
13+	0.00413	9***	0.00347	2
14+	0.00362	5*	0.00315	
15+	0.00333	2		
16+	0.00298			

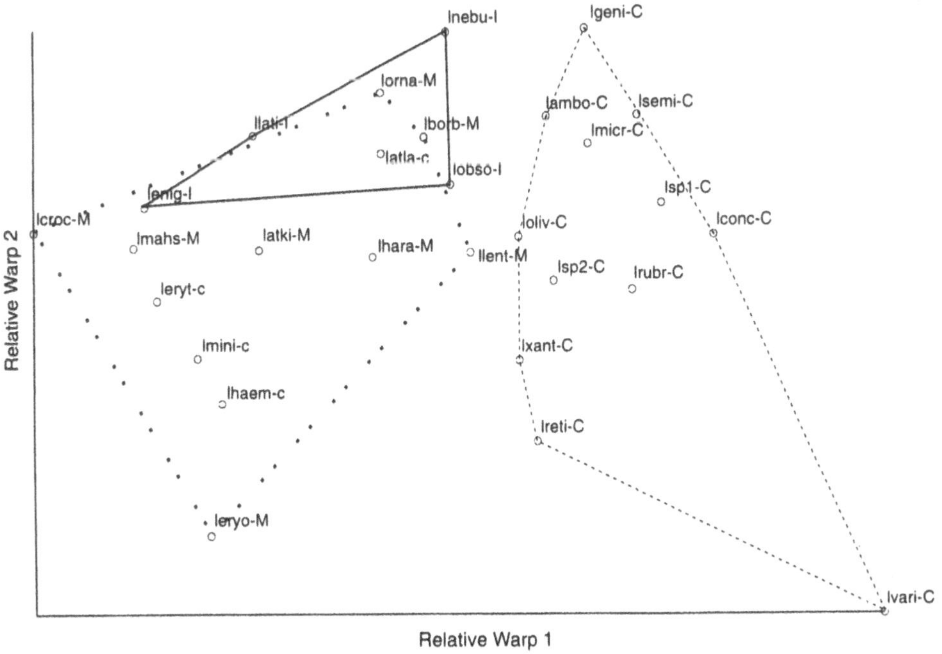

Figure 3. Relative warp 1 versus relative warp 2 for consensus shapes; the affine component is included. Dashed lines delineate low-bodied species with conical lateral teeth, dotted lines delineate high-bodied species with molariform lateral teeth and solid lines delineate species with intermediate tooth types.

significant ($p < 0.001$) and 13 of the 16 principal warps and uniform components were significantly different in univariate tests of tooth type ($p < 0.01$).

The TPSRW analyses including and excluding the uniform components clearly show that most of the morphometric variation is due to the two uniform components, which account for over 90% of the observed variation (Table 2). A plot of relative warp 1 against relative warp 2 (or an approximation of U_I versus U_2) when the uniform component is included based on the consensus shape configuration for each species, shows a pattern (Fig. 3) similar to the UPGMA of Mahalanobis distances derived from shape coordinates. A group consisting of the same 12 low-bodied species with conical lateral jaw teeth are separated from all other species.

The pattern of shape difference on the first three warps, with affine components included, are shown in Fig. 4 (Fig. 4a and b approximately correspond to the two uniform components). The largest amount of variation in shape (about 70%) is explained by differences in overall body depth (Fig. 4a). The second uniform component is characterized by a posterior stretching of caudal and ventral landmarks and a concomitant anterior stretching of the three forward dorsal landmarks (Fig. 4b). The first nonaffine component (Fig. 4c) in

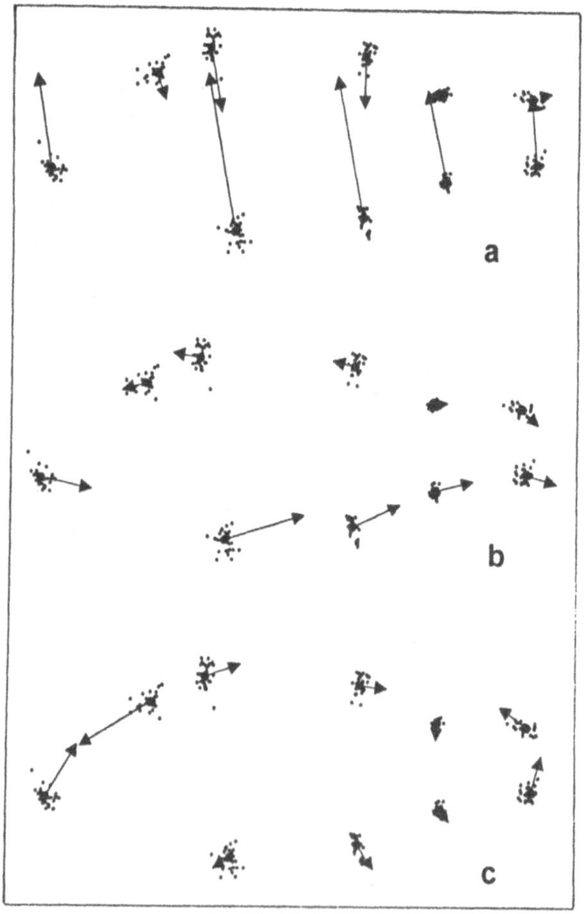

Figure 4. Vectors from a TPSRW analysis with the affine component included for (a) relative warp 1 (approximately equal to U_I; magnification = 0.5), (b) relative warp 2 (approximately equal to U_2; magnification = 0.25) and (c) relative warp 3 (first nonaffine component; magnification = 0.25).

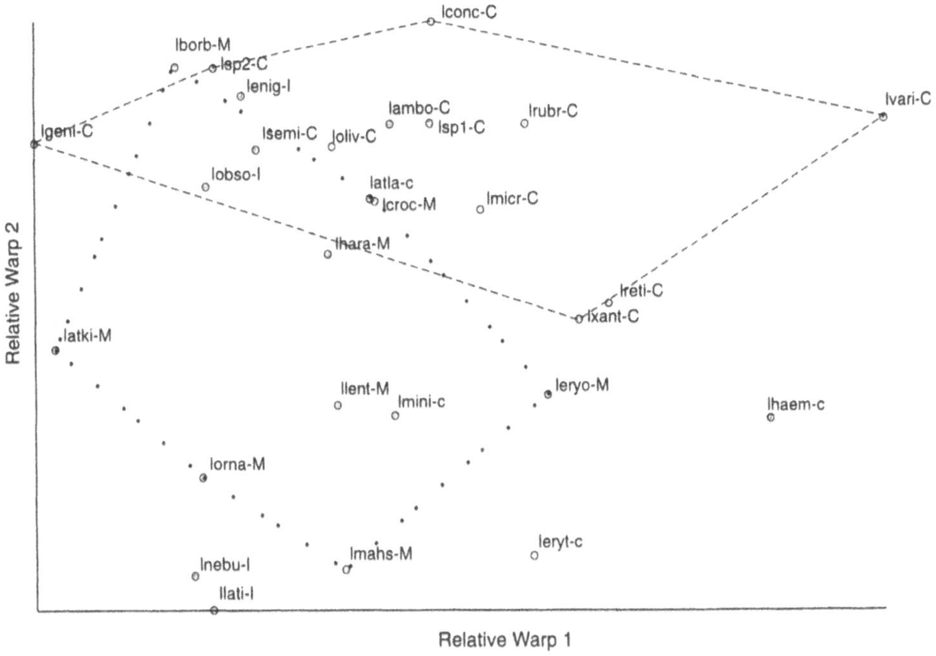

Figure 5. Relative warp 1 versus relative warp 2 for consensus shapes with the affine component excluded. Dashed lines delineate low-bodied species with conical lateral teeth and dotted lines delineate high-bodied species with molariform lateral teeth (species with intermediate tooth types are not delineated).

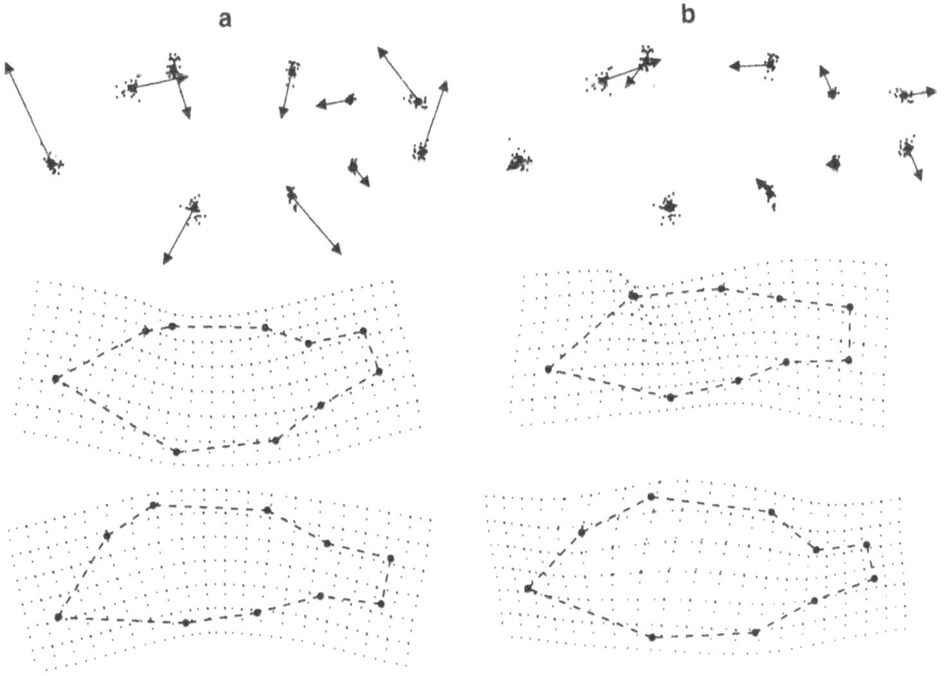

Figure 6. Vectors and thin-plate splines from a TPSRW analysis with the affine component excluded for (a) relative warp 1 (magnification = 0.5) and (b) relative warp 2 (magnification = 0.25).

this analysis is characterized by a vertical shift of the snout and tail and a change in the length of the snout.

When affine components are excluded there is no clear separation of species and tooth types (Fig. 5). The shape change vectors and thin-plate splines from relative warp 1 show an upward and downward arching of the body (Fig. 6a). The shape change on relative warp 2 (Fig. 6b) is dominated by a change in distance between landmarks 2 and 3 (posterior margin of skull to dorsal-fin origin).

The results from a TPSREG analysis of the canonical scores on the Procrustes coordinates are essentially the same as those shown in Figures 3 through 6. When the affine component is included, canonical axis 1 separates the low-bodied species with conical lateral teeth from all other species and describes a stretching of the body primarily along a vertical axis. When the affine component is excluded, canonical axis 1 does not give a clear separation of tooth types among species and primarily depicts an upward and downward arching of the body.

DISCUSSION

Centroid size was correlated with certain uniform components, suggesting that allometric growth should be considered in the morphometry of the species studied. The significant differences in affine components among species suggest that this component of shape change may be expressed differently among species. However, significant differences in centroid sizes among samples from different species may have confounded the effect. There were only a few individuals available for some species, and therefore a complete investigation into allometric pattern was not possible. The results presented here should be regarded as tentative until the patterns of allometry can be more fully investigated within individual species.

The significant differences in body shape among species is most evident in differences in body depth. This is described by the first uniform component (Fig. 4a), which accounts for about 70% of the observed variation. Separation of groups for this component of shape difference is shown in both the UPGMA dendrogram (Fig. 2). Note that the low-bodied group of *Lethrinus* all have conical lateral teeth whereas the higher-bodied group have molariform, conical or intermediate lateral teeth. This indicates that there may be a constraint in the biomechanical relationship between shape and feeding type in these fishes. Low-bodied species are restricted to feeding on soft-bodied prey whereas high-bodied species can accommodate a wider range of prey types.

The upward and downward arching of the body depicted in the first relative warp excluding the affine components (Fig. 6a) could either be an artifact of data collection or could reflect a functional difference. Although every effort was made to standardize measurement procedures it is possible that certain species arch upward when preserved whereas others arch downward. This can be tested by reexamining the specimens from species that are widely separated on this relative warp (for example *L. variegatus* versus *L. genivittatus* as shown in Fig. 6) and scoring them for a tendency to arch upward or downward. If the arching truly reflects a functional difference among species then it would be possible to account for this arching by hypothesizing that fishes feeding on bottom-living organisms have a caudal peduncle and body shape which facilitates downward propulsion, whereas those that feed higher in the water column have a caudal peduncle and body shape which facilitates upward propulsion. However, this functional hypothesis should be considered only if the possibility of a measurement artifact is eliminated.

Shape differences among species of *Lethrinus* are explained mostly by uniform components indicating that heterochrony may have played an important role in the evolution of shape in this group. A test of an hypothesis of paedomorphic influence would be possible if a complete series of developmental stages were available. This would provide an independent test from outgroup analysis of the basal origins of this group. Assuming that the group *Lethrinus* is holophyletic, there are two general hypotheses that would account for character transformations in tooth type and body shape. In the first hypothesis the ancestral form is a low-bodied species with conical lateral teeth and gives rise to forms with other tooth types and body shapes. In the alternate hypothesis the ancestral form is a high-bodied species with some other lateral dentition type and subsequently gives rise to all other combinations of body shapes and tooth types. In the latter case, the low-bodied species with conical lateral teeth would constitute a specialized group within *Lethrinus*. An examination of a series of larval and juvenile stages of individual species may provide clues to the phylogenetic transformation of the body shapes delineated in this geometric morphometric study.

REFERENCES

Akazaki, M. 1962. Studies on the spariform fishes—anatomy, phylogeny, ecology, and taxonomy. Misaki Marine Biology Institute, Kyoto University. Special Report 1: 1–368 (In Japanese with English summary).

Alexander, R. McN. 1974. Functional design in fishes, 3rd edition. Hutchinson: London.

Bookstein, F. L. 1991. Morphometric tools for landmark data: Geometry and biology. Cambridge University Press: Cambridge.

Carpenter, K. E. and G. R. Allen. 1989. FAO species catalogue volume 9. Emperor fishes and large-eye breams of the world. An annotated and illustrated catalogue of lethrinid species known to date. FAO: Rome.

Corti, M. 1992. Data analysis in systematics: A workshop and a manual to introduce geometric morphometrics. University of Rome. Privately Published.

Johnson, G. D. 1981. The limits and relationships of the Lutjanidae and associated families. Bulletin of the Scripps Institute of Oceanography 24: 1–114.

Rohlf, F. J. and F. L. Bookstein. (eds.) 1990. Proceedings of the Michigan morphometric workshop. University of Michigan Museum of Zoology Special Publication 2.

Rohlf, F. J. 1993a. NTSYS-pc: Numerical taxonomy and multivariate analysis system, version 1.8. Exeter Software: Setauket, New York.

Rohlf, F. J. 1993b. Relative warp analysis and an example of its application to mosquito wings. Pages 131-159 *in* L. F. Marcus, E. Bello, and A. García-Valdecasas, (eds.), Contributions to morphometrics. Monografías del Museo Nacional de Ciencias Naturales 8, Madrid.

SAS (Statistical Analysis Systems). 1988. Statistical Analysis System for Personal Computers. version 6.03. SAS Institute: Cary, North Carolina.

Slice, D. E. 1993. GRF-ND Generalized rotational fitting of N-dimensional landmark data. Department of Ecology and Evolution,. State University of New York, Stony Brook, New York 11794.

Strauss, R. E., and F. L. Bookstein. 1982. The truss: Body form reconstruction in morphometrics. Systematic Zoology 31(2):113–135.

LANDMARK-BASED SIZE AND SHAPE ANALYSIS IN SYSTEMATICS OF THE PLECOTINE BATS

Wiesław Bogdanowicz[1] and Robert D. Owen[2]

[1] Mammal Research Institute
Polish Academy of Sciences
17-230 Białowieża, Poland
[2] Department of Biological Sciences
Texas Tech University
Lubbock, Texas 79409

ABSTRACT

Morphometric divergence among the skulls of 10 species of plecotine bats (n = 105) was studied using x, y-coordinates of 11 homologous landmarks recorded from the left half dorsal view of each skull. Univariate and multivariate analyses of shape coordinates provided estimates of differences among taxa and between sexes. The differences in size among species were correlated with uniform shape differences. The centroid size to uniform factor allometry was more pronounced longitudinally (i.e., along the midline of the skull) than it was laterally. Significant shape differences among species were also detected in both uniform and non-uniform components. Non-uniform shape variation involved lateral rather than longitudinal displacement of landmarks on the skull. Sexual dimorphism was reflected by centroid size and was seen only in *Otonycteris hemprichi*, in which females were about 4.4% larger than males. The UPGMA phenograms of Mahalanobis D^2 of shape coordinates and of relative warp scores (in which the uniform shape component had been removed) revealed two consistent phenetic clusters. One was formed by *O. hemprichi*, with the remaining genera grouping in a second cluster. The relationships among genera in this second cluster varied depending on the phenogram generated. Nevertheless, the UPGMA phenogram derived from Mahalanobis D^2 computed on Bookstein shape coordinates (sexes combined) was entirely congruent with the current systematic hierarchy and phylogenetic hypothesis of the Plecotini sensu stricto recently put forward based on a parsimony analysis of 32 skin and skull characters. In our analysis, the most divergent species group was the monotypic *O. hemprichi*. A second group was formed by the genus *Barbastella*. The remaining clusters were composed of species of *Corynorhinus; Plecotus;* and a cluster containing *Idionycteris* and *Euderma*.

Advances in Morphometrics, Edited by L. F. Marcus et al.
Plenum Press, New York, 1996

INTRODUCTION

The tribe Plecotini (Chiroptera: Vespertilionidae) consists of 10 or 11 species of mainly "long-eared" bats, that are widely distributed throughout the Holarctic (e.g., Yoshiyuki, 1991; Frost and Timm, 1992; Tumlison and Douglas, 1992; Zima et al., 1992; Qumsiyeh and Bickham, 1993). Handley (1959) first examined the relationships among the tribe in detail and proposed the phylogeny (*Barbastella* (*Euderma Plecotus*)). His taxonomic arrangement included the genera *Barbastella*, *Euderma* and *Plecotus*, with the last containing three subgenera: *Plecotus* of the Old World, and *Corynorhinus* and *Idionycteris* of the New World. Subsequent works suggested that *Idionycteris* should be given full generic status (Williams et al., 1970; Nader and Hoffmeister, 1983) on the basis of a closer relationship between *Idionycteris* and *Euderma* (Stock, 1983). Yet, Handley's (1959) arrangement is still generally accepted (reviewed by Frost and Timm, 1992; Tumlison and Douglas, 1992; see also Menu, 1987; Topál, 1989). Recently, though, three different classifications of the Plecotini have been proposed based on cladistic treatments of morphological and chromosomal traits. Tumlison and Douglas (1992) examined 32 cranial and external characters. These data yielded the most parsimonious tree (*Barbastella* (*Corynorhinus* (*Plecotus* (*Idionycteris Euderma*)))). Frost and Timm (1992) used transformation series of 25 morphological and 11 karyological characters to propose the phylogeny (*Euderma* [including *Idionycteris*] (*Barbastella* (*Plecotus Corynorhinus*))). Studies of G-banded chromosomes (Zima et al., 1992; Qumsiyeh and Bickham, 1993; Volleth and Heller, 1994) indicated the phylogeny (*Euderma Idionycteris* (*Otonycteris Barbastella Plecotus Corynorhinus*)). Recent advances in geometric morphometrics provide systematists with potentially powerful tools for testing evolutionary patterns and processes (e.g., Tabachnick and Bookstein, 1990; Rohlf, 1993a; Rohlf and Marcus, 1993; Swiderski, 1993). These new techniques offer useful means to describe evolutionary shifts in the morphology of organisms and can provide new sets of characters with which to hypothesize systematic relationships. They also can show how anatomy, as depicted by the geometric relations between homologous landmarks, differs among populations and taxa in a way that may be related to their history (phylogeny) and ecology (e.g., Reyment, 1991; Loy et al., 1993). The aim of this study is twofold. First, we want to assess the amount of morphological variation and resolution seen among species within plecotine bats. This will be investigated by decomposing morphological variation into centroid size, uniform and nonuniform components (Bookstein, 1990, 1991), and distinguishing inter-taxon differences among forms. Second, we will evaluate some of the recently developed geometric methods as tools for evolutionary and phylogenetic inference by comparing our results with previously published arrangements proposed for the Plecotini.

MATERIAL AND METHODS

In total 105 specimens (see Appendix) representing 10 species of the plecotine bats were examined. To avoid the effect of bilateral asymmetry, 11 landmarks were collected from the left half dorsal view of the skull of each specimen (Fig. 1). These landmarks are (1) center of posterior curvature of internasal opening; (2) proximal extremity of premaxilla; (3) point of maximum lateral curvature of maxilla at center of upper canine; (4) center of infraorbital foramen; (5) point of the maximum curvature (width) of the infraorbital plate; (6) point of postorbital constriction inside the orbit; (7) point where internal portion of the posterior zygoma contacts the cranium; (8) most posterior internal curvature of zygomatic arch; (9) most lateral point of the mastoid process; (10) most posterior point of interparietal; and (11) center of coronal suture.

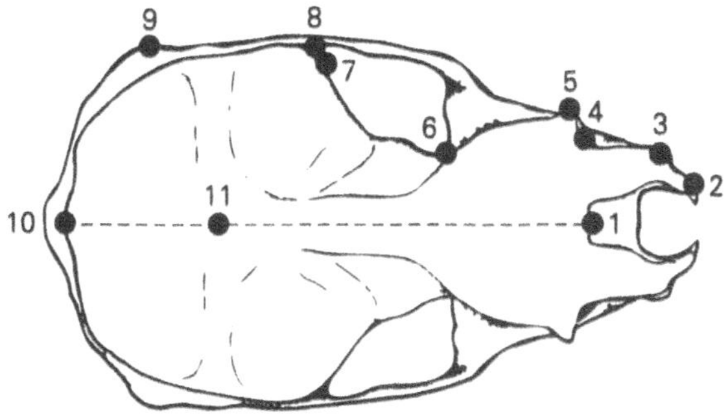

Figure 1. Anatomical positions of the 11 landmarks used in this study. Landmarks 1 and 10 were used as baseline end points.

Data were gathered using a COHU 4815 Series monochrome video camera equipped with a 60-300 mm macro zoom lens and using MorphoSys software (Meacham and Duncan, 1990). Format conversion of files from MorphoSys to other software packages was carried out using the FORCON program designed and written by D. E. Slice. Coordinates of the few missing landmarks were calculated based on the coordinates of existing landmarks (Richtsmeier et al., 1992). All specimens were translated and rotated using the SHAPETU transformation of the NTSYS-pc package (Rohlf, 1993b). This process places one designated landmark at the origin (coordinates 0,0) and a second along the x axis (coordinates x,0; $x > 0$). To predict the coordinates of the missing landmark for a particular specimen, multiple regression was performed on the translated and rotated coordinates for each species separately.

A baseline was fixed to standardize both position and scale by selecting landmarks 1 and 10 (center of curvature of internasal opening and posterior point of the interparietal, respectively) situated along the midline of the skull (Fig. 1). All other landmarks were rotated and scaled to this baseline to produce Bookstein shape coordinates by means of the SAS IML program UNIGRAPH, written by L. F. Marcus (see Appendix III of volume for instructions for obtaining this and other programs from Stony Brook). A size variable expressed as centroid size was generated from the same landmarks before standardization in order to investigate size and shape covariance (Bookstein, 1986, 1991). Uniform x and y components were extracted from the shape coordinates based on Bookstein (1990) and by means of UNIGRAPH. Centroid size, baseline size, uniform x, and uniform y components were analyzed using a two-way analysis of variance (ANOVA unbalanced design) to test for effects of species, sex and species-sex interaction. When no differences between sexes or an interaction were detected, a one-way ANOVA was applied to test for among-species differences. The 95% probability ellipses for bivariate scattergrams were drawn using the SAS IML program ELLIPSE, written by L. F. Marcus. Sexual dimorphism as measured by centroid size was calculated using Storer's (1966) index: the difference between the female average and male average divided by the overall average and then multiplied by 100.

Differences in shape coordinates among species were tested using both univariate and multivariate methods. Multivariate analysis of variance (MANOVA) and canonical variate analysis were used to examine differences among taxa as described by the entire set of coordinates. Differences among species were summarized by unbiased Mahalanobis D^2 values by means of L. F. Marcus' SAS IML program DSQ. These values were then clustered by UPGMA by means of the NTSYS-pc program (Rohlf, 1993b). Significant differences among species means were tested using D^2 statistics and corresponding F-tests. Adjustments in the level of significance for multiple comparisons were made using the Bonferroni inequality (Marcus, 1993). In order to find consensus configurations for each species, digitized coordinates were superimposed by using the Generalized Least Squares (GLS) option in the Procrustes superimposition software written by Slice (1992; see also Rohlf and Slice, 1990). The consensus configurations for each of the 10 species were then compared using relative warps (Thin-Plate Spline Relative Warp Analysis-TPSRW; Rohlf, 1993b, version October 1993). Relative warps were determined with respect to the mean configuration for all Plecotini based on an α, of 0 (Bookstein, 1991). Weights were based on deviations from the sample means. Relative warp scores were computed both retaining and removing the affine portion of the variation. Matrices of the squared average taxonomic distances were then produced and analyzed using the Mantel test in NTSYS-pc (Rohlf, 1993b). For each analysis, taxa were phenetically clustered using UPGMA. No significant differences between sexes were seen for shape coordinates or for the components of size and shape, in any of the species except for *Otonycteris hemprichi*, which showed slight sexual dimorphism in centroid size. Generally, this justified combining the data over sexes for each species and ignoring sex in further analyses. The IML analyses were done with the SAS version 5 (SAS, 1985) running under MS-DOS version 6.0.

RESULTS

Sexual Dimorphism

Results of the two-way ANOVA indicated that none of the Bookstein shape coordinates exhibit statistically significant sexual variation and only two (x_2 and x_6) show significant

Table 1. Two-way ANOVA testing sexual dimorphism against taxon variation for 18 shape coordinates and one-way ANOVA with sexes combined. Mean squares (x 10^4) from SAS Type III GLM procedure for unbalanced designs

	Two-way ANOVA				One-way ANOVA	
Coordinates	Species	Sex	Interaction	Error	Species	Error
x_2	43.93***	3.20	4.22**	1.38	58.99***	1.69
y_2	4.23***	0.23	0.81	1.13	5.01***	1.09
x_3	63.43***	2.53	2.39	1.25	79.51***	1.38
y_3	8.08***	0.02	0.34	0.58	10.22***	0.55
x_4	41.80***	0.31	2.25	1.30	56.67***	1.38
y_4	26.66***	0.03	0.70	0.72	28.88***	0.71
x_5	19.54***	0.50	1.11	1.29	24.91***	1.27
y_5	12.52***	0.00	0.79	0.90	14.33***	0.88
x_6	28.23***	0.97	2.32*	1.12	35.15***	1.24
y_6	14.78***	0.17	0.75	0.57	18.51***	0.58
x_7	73.42***	0.70	2.34	1.86	82.23***	1.89
y_7	29.22***	0.02	2.35	2.38	36.69***	2.36
x_8	53.84***	1.44	2.18	1.65	65.31***	1.70
y_8	10.98***	0.57	0.81	1.96	13.66***	1.83
x_9	71.88***	0.87	3.29	3.84	74.24***	3.75
y_9	44.47***	0.54	1.49	1.70	51.26***	1.67
x_{11}	13.78***	4.07	2.45	3.27	14.28***	3.18
y_{11}	0.76	0.02	0.81	0.63	0.72	0.64
df	9	1	9	85	9	95

* = $p < 0.05$; ** = $p < 0.01$; *** = $p < 0.001$

Table 2. Summary of two-way ANOVA testing sexual dimorphism against species variation both with and without *O. hemprichi* ($n = 5$) for centroid size, baseline size, uniform x, and uniform y components and one-way ANOVA combining sexes. The F-values from SAS Type III GLM procedure for unbalanced designs

	Two-way ANOVA			One-way ANOVA
Variable	Species	Sex	Interaction	Species
O. hemprichi included				
Centroid size	546.50***	5.61*	1.95	516.65***
Baseline size	410.04***	3.20	1.81	411.07***
Uniform x	36.53***	0.10	1.54	42.41***
Uniform y	15.12***	0.07	0.74	19.35***
O. hemprichi removed				
Centroid size	219.65***	0.86	0.90	247.59***
Baseline size	198.45***	0.13	0.73	236.69***
Uniform x	33.05***	0.00	1.63	38.32***
Uniform y	12.97***	0.06	0.81	16.86***

* = $p < 0.05$; ** = $p < 0.01$; *** = $p < 0.001$

interaction (Table 1). Neither significant sexual differences (Hotelling–Lawley Trace [HLT] = 0.197; df 18, 68; $p = 0.75$) nor interaction between sexes and taxa (HLT = 2.343; df 162, 596; $p = 0.62$) were found by MANOVA. This suggests that sexual and species variation are, in general, independent of each other in the morphospace studied.

Sexual dimorphism is exhibited only for centroid size (Table 2). Females are usually bigger than males but the differences are small and do not exceed 2.3%. The one exception

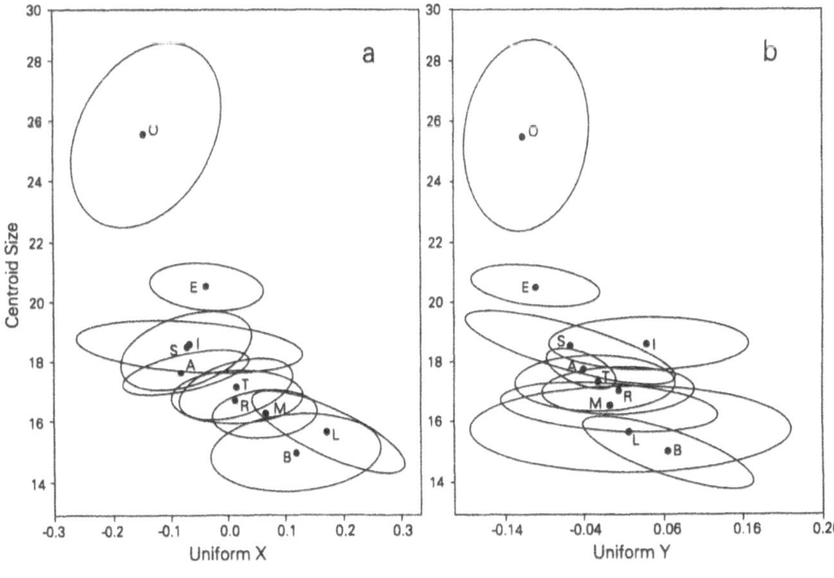

Figure 2. Relationships shown by projections of centroids and 95% probability ellipses of bivariate means between centroid size and (a) uniform x component and (b) uniform y component for 10 species of plecotine bats. Abbreviations used: A–*Plecotus auritus*; B–*Barbastella barbastellus*; E–*Euderma maculatum*; I–*Idionycteris phyllotis*; L–*Barbastella leucomelas*; M–*Corynorhinus mexicanus*; O–*Otonycteris hemprichi*; R–*Corynorhinus rafinesquii*; S–*Plecotus austriacus*; and T–*Corynorhinus townsendii*.

is for females of *O. hemprichi* which are approximately 4.4% larger than males. When this species was removed from the analysis, no significant sex effect was detected. No interaction of centroid size between species and sexes was detected by the ANOVA. Baseline size and the x and y components of the uniform factor are similar in females and males (Table 2).

Interspecific Relationships

Size and Uniform Shape Variation. Comparison of size and uniform shape components reveals a pattern of differences in which *O. hemprichi* is characterized by much larger centroid size as compared with the other taxa of Plecotini (Fig. 2). Next in size is *Euderma maculatum*. The 95% probability ellipses of these two species do not overlap and they form well-differentiated groups. The remaining bats represent a more homogeneous assemblage, with the genera *Plecotus* and *Corynorhinus* showing marked similarity.

Differences in uniform shape are highly correlated with variation of centroid size. Correlations indicate that 48.0 and 39.1% of the total variation of the centroid size is explained by either uniform x ($F = 95.01$; df 1, 103; $p < 0.001$; Fig. 2a) or uniform y ($F = 66.09$; df 1, 103; $p < 0.001$; Fig. 2b) variable, respectively. The pooled effect of these components accounts for as much as 61.8% of the total variance of the centroid size ($F = 82.42$; df 2, 102; $p < 0.001$). The uniform shape components along the x and y axes are significantly intercorrelated, with an $r = 0.415$ and $p < 0.001$. The 95%-probability ellipses around centroids of these components (not shown) indicate largely overlapping bivariate distributions among species. The differences, however, are still highly significant and might at least define groups of similar taxa.

Bookstein Shape Coordinates Space. Highly significant differences among species were found for 17 of the 18 (except y_{11}) standardized shape coordinates (Table 1). These differences are also highly significant when all coordinates are examined simultaneously (HLT = 48.71; df 162, 596; $p < 0.001$). Mahalanobis D^2 is always large and the F-values are significant ($p < 0.001$) for each pairwise comparison (Table 3). The Bonferroni multiple range test adjusted for the number of groups shows no overlap between taxa for an overall 0.05 significance level (see Loy et al., 1993) and with 100% correct assignment for all samples.

Table 3. Unbiased Mahalanobis D^2 (above diagonal) and their corresponding F-values (below) computed among 10 species of Plecotini based on Bookstein shape coordinates. Females and males combined. All values significant at 0.001 level. Abbreviations used: A—*Plecotus auritus*; B—*Barbastella barbastellus*; E—*Euderma maculatum*; I—*Idionycteris phyllotis*; L—*Barbastella leucomelas*; M—*Corynorhinus mexicanus*; O—*Otonycteris hemprichi*; R—*Corynorhinus rafinesquii*; S—*Plecotus austriacus*; and T—*Corynorhinus townsendii*

Species	A	B	E	I	L	M	O	R	S	T
A		120	58	60	146	55	315	66	29	54
B	66.8		128	106	14	54	256	79	131	99
E	29.2	76.1		31	143	73	232	76	42	54
I	32.7	70.4	18.5		135	70	251	61	37	51
L	58.0	7.2	59.8	60.4		66	282	89	169	109
M	35.9	45.3	50.6	55.7	34.6		283	20	66	13
O	111.0	104.0	85.7	99.0	88.6	127.0		282	247	285
R	41.2	62.6	50.5	46.1	45.2	20.3	123.0		55	20
S	15.7	82.1	23.6	22.9	73.1	48.9	94.5	39.3		47
T	28.1	62.2	30.0	31.1	47.6	10.8	109.0	14.8	27.5	

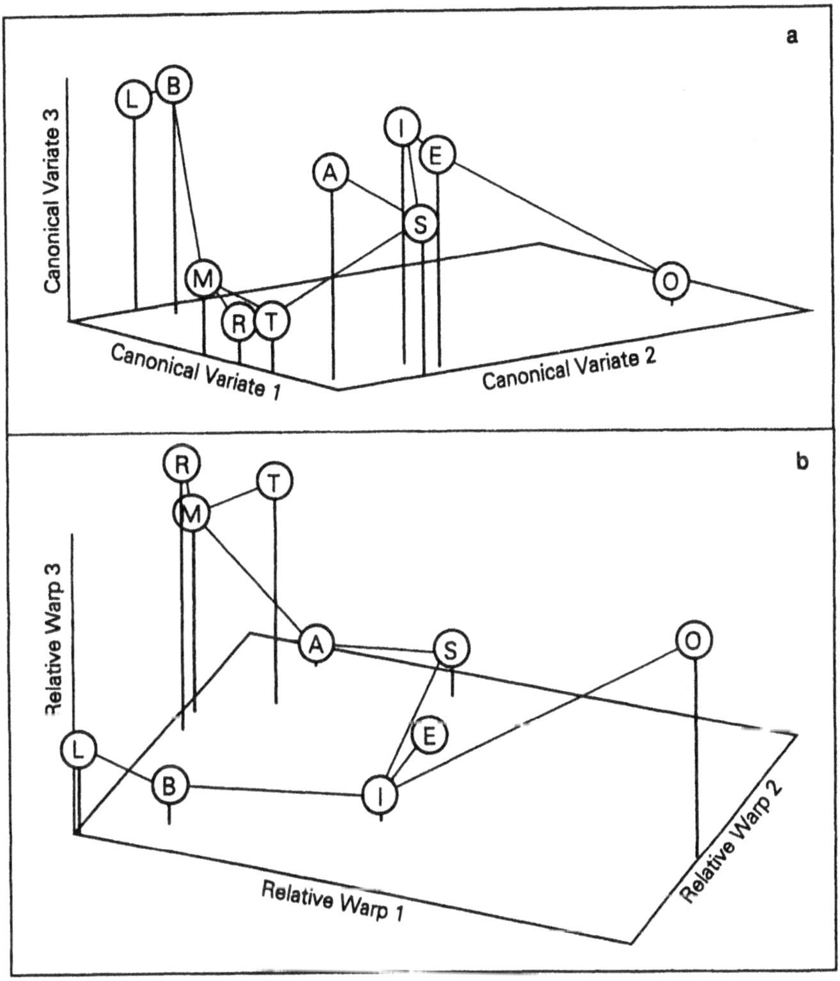

Figure 3. (a) Distribution of canonical–variate centroids of 10 species of plecotine bats in three-dimensional space. The first three canonical–variate axes account for 35.3, 32.0 and 12.8% of the total sample variation, respectively. (b) Relative warp analysis (uniform factor removed) of 11 pairs of *x*, *y*-coordinates from the dorsal view of the skull showing projections of 10 species of Plecotini onto the first three relative warps. The first, second, and third axes account for 42.5, 25.7 and 14.1% of the total variance. Both plots are shown with minimum spanning trees superimposed. Species abbreviations as in Fig. 2.

A canonical variate analysis based on the Bookstein shape coordinates reveals five possible groupings that consist of: (1) *Otonycteris,* (2) *Euderma, Idionycteris,* (3) *Plecotus,* (4) *Corynorhinus,* and (5) *Barbastella* (Fig. 3a). Similar patterns are seen both when a minimum spanning tree is superimposed onto the plot for the first three canonical variates and when a UPGMA phenogram is computed from canonical variate scores (Fig. 4a).

Otonycteris hemprichi is the most distant from all other taxa. *Euderma maculatum* has as its nearest neighbor I. *phyllotis*, and they are connected to the group *P. austriacus* and *P. auritus*. *Corynorhinus townsendii*, *C. rafinesquii* and *C. mexicanus* are intermediate between the genera *Plecotus* and *Barbastella*.

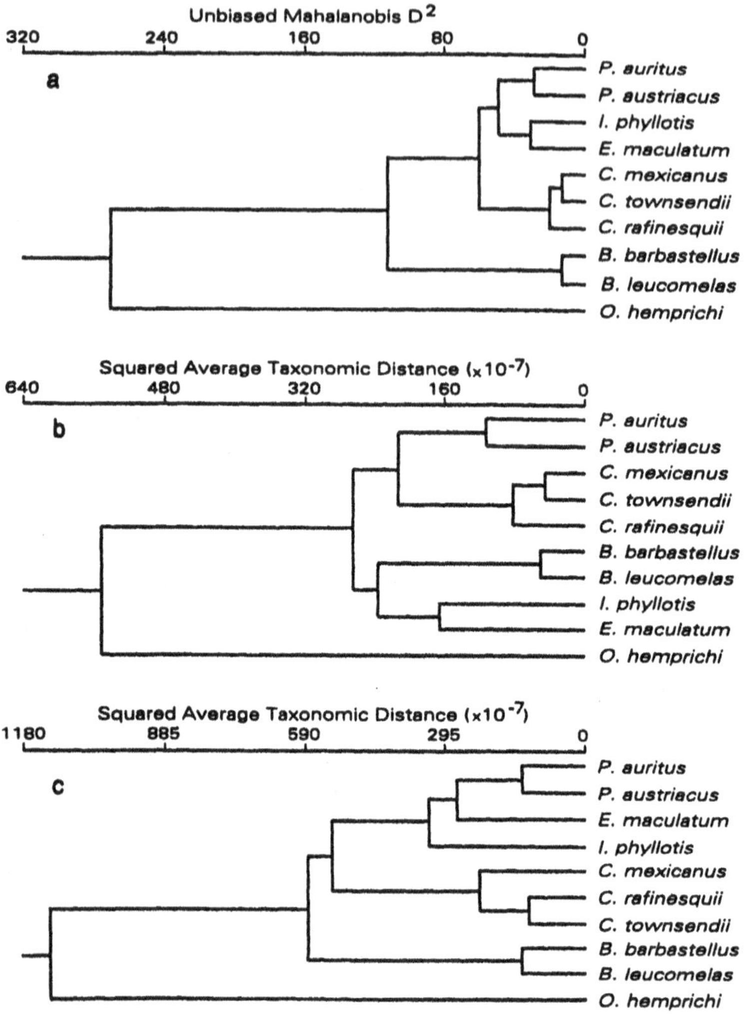

Figure 4. UPGMA phenograms of 10 taxa of Plecotini (a) derived from unbiased Mahalanobis D^2 computed from all 18 Bookstein coordinates; (b) computed on squared average taxonomic distances in relative warp space with the uniform factor removed; and (c) based on the matrix of squared average taxonomic distances computed from relative warp scores with the uniform factor retained. Their cophenetic values are 0.970, 0.859 and 0.834, respectively.

Relative Warps. Most of the variation among species is distributed along the first relative warp axis which accounts for 42.5% of the total variance; *B. leucomelas* is at the extreme left and *O. hemprichi* at the extreme right (Fig. 3b). Displacement vectors for this warp (Fig. 5) suggest that the most obvious difference between these two taxa is a displacement of landmarks 2, 3, 4, 5 and 9 in the direction of the decreased rostrum and mastoid region of *O. hemprichi*. The displacements in relative warp 2 (25.7% of total variance) imply that landmarks 1, 10 and 11 along the midline of the skull are more rostrally positioned (making the distance between internasal opening and supraoccipital larger) in *Plecotus* spp. than they are in *Barbastella* spp. The third relative warp axis (14.1% of total variance) emphasizes the differences between *I. phyllotis* and *P. auritus*, and *Corynorhinus* spp. and *Otonycteris hemprichi* relative to the lengthening and widening rostrum and narrowing of the braincase for the genera *Corynorhinus* and *Otonycteris*.

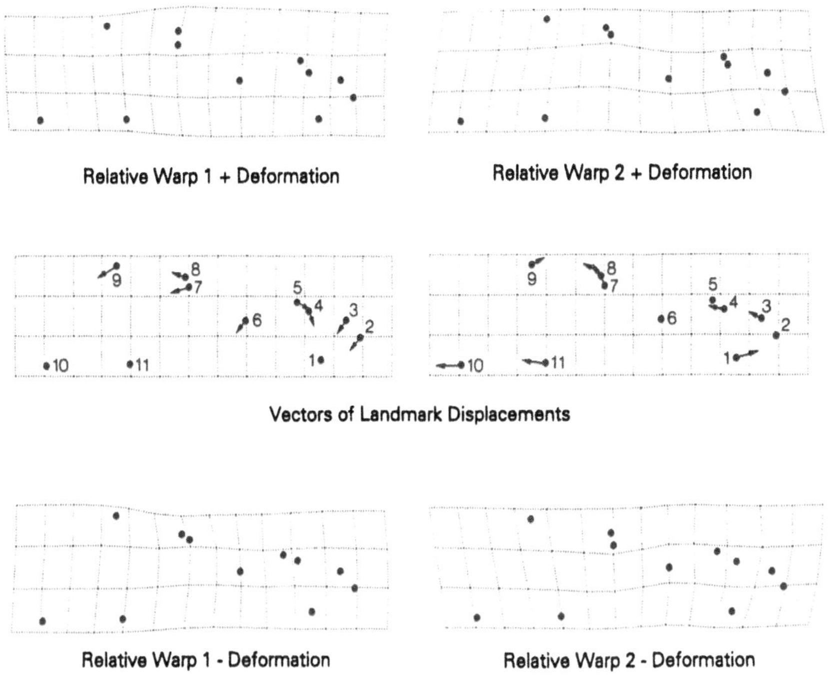

Figure 5. Displacement vectors and deformation grids for the first and second relative warps computed for 10 species of Plecotini based on the mean configuration as the reference and $\alpha = 0$. Landmarks numbered as in Fig. 1.

The plot of the first three relative warps with the minimum spanning tree superimposed produces five different groupings of species (Fig. 3b), the most divergent of which is the monotypic *O. hemprichi*. The second grouping includes *I. phyllotis* and *E. maculatum*. *Plecotus austriacus* and *P. auritus* form another cluster, which is intermediate between *I. phyllotis* and *Corynorhinus* spp. *Idionycteris phyllotis* is connected to *Barbastella barbastellus* which clusters together with *B. leucomelas*. The relative warp loadings suggest complex and localized patterns of deformations. Relative warp 1 indicates displacement of landmarks 2 and 9 towards the center of the neurocranium relative to the displacement of landmarks 4, 5 and 8 in the opposite direction. Relative warp 2 indicates an expansion of the region between internasal opening and the coronal suture (landmarks 1 and 11). The longest vectors for relative warp 3 imply the major modifications in the zygomatic and rostral regions of the skull (landmarks 4, 7 and 8). None of the relative warps examined is significantly intercorrelated with centroid size ($r = 0.261–0.498$, df 8, $p > 0.05$). The relationships obtained from cluster analysis of relative warp scores were similar to those seen for the phenogram of canonical variates scores, with one exception (Fig. 4b). *Idionycteris* and *Euderma* are associated with *Barbastella*, whereas in the previous analysis the first two genera had their closest affinities with the genus *Plecotus*. In both cases, however, the distances between species means are congruent with the current systematic hierarchy; i.e., interspecific distances are smaller than intergeneric ones. However, inclusion of the affine eigenvectors in the computation of relative warp scores affects these relationships. Although the matrices of nonaffine and affine relative warps scores are significantly correlated (Mantel test, $r = 0.89$, $p < 0.001$), the inclusion of affine eigenvectors in computations results in a partial contradiction of the previously generated "natural" generic groupings. This is especially visible in the UPGMA phenogram (Fig. 4c).

DISCUSSION

In this study, interpretation of morphological variation within the plecotine bats is evaluated by decomposing landmark variation into centroid size and uniform and nonuniform shape components. All these variables are significantly different among taxa and express well-defined morphometric divergence within the tribe. The differences in size among species are closely linked to only linear (uniform) shape modifications. More linear deformations were detected longitudinally than laterally. Nonlinear shape variation was independent of centroid size and mainly involved lateral displacement of landmarks associated with the widening of different regions of the skull. Significant sexual dimorphism was indicated by centroid size in only *O. hemprichi*.

One of the key questions in the taxonomy of Plecotini concerns the systematic position of *O. hemprichi*. Handley (1959) and Tumlison and Douglas (1992) did not consider this species in their analyses, probably because Miller (1907) had suggested that *Otonycteris* is only superficially similar to the plecotine bats. The bacular classification of Plecotini, including *Rhogeessa* (along with *Baeodon*), *Nycticeius* and *Otonycteris* (Hill and Harrison, 1987), was rejected due to its subjectivity (Frost and Timm, 1992) and karyological incongruence (reviewed by Horácek, 1991). In the G-band structure of chromosomes, only *Otonycteris* exhibits the highly specific banding pattern typical of *Plecotus*, *Corynorhinus*, and *Barbastella* (Zima et al. 1992; Qumsiyeh and Bickham, 1993). In our analyses, the highly specialized *O. hemprichi* (see Horácek, 1991) is the most divergent taxon of the plecotines. This implies, in contrast to the chromosomal findings, that it may belong to a different ancestral stock.

One of the central aims of this study was to evaluate landmark-based methods as practical tools for systematic inference. The UPGMA phenograms for Bookstein shape coordinates and relative warp scores (computed excluding uniform shape variation) depict an arrangement of species within their respective generic groups that is consistent with cut-off levels for genus recognition. This result is very clear in spite of the low number of specimens examined (Figs. 4a and 4b). This indicates that landmark-based morphometrics effectively capture the information necessary to distinguish genera. Retention of the affine eigenvectors in the computation of relative warp scores increased intertaxon distances and affected these relationships. This was especially evident in UPGMA phenograms (Fig. 4c), which are subject to distortion of distances between major clusters (Sneath and Sokal, 1973).

Phenetic distances computed on Bookstein shape coordinates and relative warp scores (uniform shape change removed) showed two similarities. Each confirmed that *O. hemprichi* is morphologically the most divergent of all Plecotini and *Barbastella* is separated from *Plecotus* and *Corynorhinus*. The main differences in phenograms concerned the relative position of the genera *Idionycteris* and *Euderma*, which either grouped with *Plecotus* or with *Barbastella*. Comparison of these phenograms with recently published systematic arrangements of Plecotini sensu stricto (lacking *Otonycteris*) indicates that the phenogram derived from the analysis of Bookstein coordinates matches perfectly with the cladistic classification proposed by Tumlison and Douglas (1992). On the other hand, the phenogram computed from relative warp scores partly corresponds to two of the four morphology-only most parsimonious trees illustrated by Frost and Timm (1992: Figs. 2.3 and 2.4).

There are some intrinsic and extrinsic reasons that suggest that the phenogram from Bookstein shape coordinates (Fig. 4a) should be treated as a working hypothesis of systematic relationships among Plecotini. First, its goodness of fit to the original data set, expressed as a cophenetic value, is higher (0.970) than that seen for the phenogram of relative warp scores (0.859) and the obtained relationships duplicate those suggested by a three-dimensional plot with the minimum spanning tree superimposed (see Fig. 3). Second, ancestral character states in Frost and Timm's (1992) analyses were inferred from examination of an outgroup represented by several species of only one genus, *Myotis*. This approach excludes the possibility of detecting characters that are derived in all *Myotis* taxa (cf. Tumlison and Douglas, 1992). Third, Frost and Timm's (1992) morphological analyses resulted in several trees of similar length (and therefore with comparable amounts of implied homoplasy) but with considerable variation in topology. This suggests that their data show only marginally preferential support for their chosen tree. These authors opted for the tree (*Euderma* [including *Idionycteris*] (*Barbastella* (*Plecotus Corynorhinus*))) because it was supported by both morphological and karyological data.

Both phenetic and cladistic findings based on morphology are in disagreement with certain aspects of the chromosomal data. The karyological evidence implies that the genera *Otonycteris*, *Plecotus*, *Corynorhinus* and *Barbastella* form one lineage of Plecotini and *Euderma* and *Idionycteris* form the other (Zima et al. 1992; Qumsiyeh and Bickham, 1993; see also Volleth and Heller, 1994). Both morphological (e.g., Tumlison and Douglas, 1992; this study) and chromosomal (Qumsiyeh and Bickham, 1993; Volleth and Heller, 1994) analyses support a close relationship between *Idionycteris* and *Euderma*, but they differ in their support for other relationships. This may be because the karyological tree includes only centric fusion events as synapomorphies. Such conditions may produce misleading phylogenies if the possibility of fissions are not also considered (Qumsiyeh, 1989).

Several problems are also associated with the use of morphometric methods in estimating systematic relationships (e.g., Wiley, 1981; Bookstein et al., 1985; Felsenstein, 1988) and in equating the relative degree similarity with recency of common ancestry (MacLeod and Kitchell, 1990; Bookstein et al., 1985). The distribution of taxa in morphospace can, however, corroborate a phylogenetic hypothesis if monophyletic groups are localized in the morphospace (Bookstein et al., 1985). Further analyses of additional populations of Plecotini will show the extent to which the generic groups identified here remain distinct and whether this phylogenetic hypothesis is being confounded by morphological convergence or our inability to detect accurately the synapomorphic patterns within morphological data.

ACKNOWLEDGMENTS

Leslie F. Marcus, Fred L. Bookstein, and F. James Rohlf provided prompt and thoughtful replies to our questions concerning the analytic methods. This work was conducted while the first author was a Fulbright Fellow at Texas Tech University. Fred Bryant of the Office of Research Services, Texas Tech University, provided funds to extend the first author's residence so that this study could be completed. Mazin B. Qumsiyeh made his *Otonycteris* specimens available to us before they were deposited at the Carnegie Museum of Natural History. We also thank Stephen Kasper for reading a draft of the manuscript. Appreciation is extended to the following institutions and their curators and staffs for the use of specimens and facilities: American Museum of Natural History (AMNH); Field Museum of Natural History (FMNH); Harrison Zoological Museum, Sevenoaks, Great Britain (HZM); University of Kansas Museum of Natural History (KU); Louisiana State University Museum of Zoology (LSUMZ); Carnegie Museum of Natural History (specimens listed as MBQ in Appendix); Moscow State University, The Museum, Russia (MGU); University of New Mexico Museum of Southwestern Biology (MSB); Harvard University Museum of Comparative Zoology (MCZ); Naturhistorisches Museum Wien, 1 Zoologische Abteilung, Austria (NMW); Texas Tech University, The Museum (TTU); United States National Museum of Natural History (USNM).

REFERENCES

Bookstein, F. L. 1986. Size and shape spaces for landmark data in two dimensions (with discussion and rejoinder). Statistical Science 1:181–242.

Bookstein, F. L. 1990. Higher-order features of shape change for landmark data. Pages 237-250 *in* F. J. Rohlf and F. L Bookstein, (eds.), Proceedings of the Michigan morphometrics workshop. University of Michigan Museum of Zoology Special Publication 2.

Bookstein, F. L. 1991. Morphometric tools for landmark data: geometry and biology. Cambridge University Press: Cambridge.

Bookstein, F. L., B. Chernoff, R. L. Elder, J. M. Humphries, Jr., G. R. Smith, and R. E. Strauss. 1985. Morphometrics in evolutionary biology: The geometry of size and shape change, with examples from fish. Academy of Natural Sciences of Philadelphia Special Publication 15.

Felsenstein, J. 1988. Phylogenies and quantitative characters. Annual Review of Ecology and Systematics 19:445–471.

Frost, D. R., and R. M. Timm. 1992. Phylogeny of plecotine bats (Chiroptera: "Vespertilionidae"): Summary of the evidence and proposal of a logically consistent taxonomy. American Museum Novitates 3034:1–16.

Handley, Jr., C. O. 1959. A revision of American bats of the genera *Euderma* and *Plecotus*. Proceedings of the United States National Museum 110:95–246.

Hill, J. E., and D. L. Harrison. 1987. The baculum in the Vespertilioninae (Chiroptera: Vespertilionidae) with a systematic review, a synopsis of *Pipistrellus* and *Eptesicus*, and the description of a new genus and subgenus. Bulletin of the British Museum (Natural History), Zoology 52:225–305.

Horácek, I. 1991. Enigma of *Otonycteris*: Ecology, relationship, classification. Myotis 29:17–30.

Loy, A., M. Corti, and L. F. Marcus. 1993. Landmark data: Size and shape analysis in systematics. A case study on Old World Talpidae (Mammalia, Insectivora). Pages 215–240 *in* L. F. Marcus, E. Bello, and A. Garcia-Valdecasas, (eds.), Contributions to morphometrics. Monografias del Museo Nacional de Ciencias Naturales 8, Madrid.

MacLeod, N., and J. A. Kitchell. 1990. Morphometrics and evolutionary inference: a case study involving ontogenetic and developmental aspects of foraminiferal evolution. Pages 283–299 *in* F. J. Rohlf and F. L Bookstein, (eds.), Proceedings of the Michigan morphometrics workshop. University of Michigan Museum of Zoology Special Publication 2.

Marcus, L. F. 1993. Some aspects of multivariate statistics for morphometrics. Pages 95–130 *in* L. F. Marcus, E. Bello, and A. Garcia-Valdecasas, (eds.), Contributions to morphometrics. Monografias del Museo Nacional de Ciencias Naturales 8, Madrid.

Meacham C. A., and T. Duncan. 1990. MorphoSys. Version 1.26. University Herbarium, University of California: Berkeley.

Menu, H. 1987. Morphotypes dentairs actuels et fossiles des chiroptères vespertilioninés. Deuxième partie: Implications systématiques et phylogéniques. Palaeovertebrata 17(3):77–150.

Miller, Jr., G. S. 1907. The families and genera of bats. Bulletin of the United States National Museum 57:1–281.

Nader, I. A., and D. F. Hoffmeister. 1983. Bacula of big-eared bats *Plecotus*, *Corynorhinus*, and *Idionycteris*. Journal of Mammalogy 64:528–529.

Qumsiyeh, M. B. 1989. Chromosomal fissions and phylogenetic hypotheses: Cytogenetic and allozymic variation between species of *Meriones* (Rodentia, Gerbillidae). Occasional Papers of The Museum of Texas Tech University 132:1–16.

Qumsiyeh, M. B., and J. W. Bickham. 1993. Chromosomes and relationships of long-eared bats of the genera *Plecotus* and *Otonycteris*. Journal of Mammalogy 74:376–382.

Reyment, R. A. 1991. Multidimensional palaeobiology. Pergamon Press: Oxford.

Richtsmeier, J. T., J. M. Cheverud, and S. Lele. 1992. Advances in anthropological morphometrics. Annual Review of Anthropology 21:283–305.

Rohlf, F. J. 1993a. Relative warp analysis and an example of its application to mosquito wings. Pages 131-159 *in* L. F. Marcus, E. Bello, and A. Garcia-Valdecasas, (eds.), Contributions to morphometrics. Monografias del Museo Nacional de Ciencias Naturales 8, Madrid.

Rohlf, F. J. 1993b. NTSYS-pc numerical taxonomy and multivariate analysis system, version 1.80. Exeter Software: Setauket, New York.

Rohlf, F. J., and L. F. Marcus. 1993. A revolution in morphometrics. Trends in Ecology and Evolution 8:129–132.

Rohlf, F. J., and D. E. Slice. 1990. Extensions of the Procrustes method for the optimal superimposition of landmarks. Systematic Zoology 39:40–59.

SAS (Statistical Analysis System). 1985. SAS user's guide: Statistics, version 5 edition. SAS Institute: Cary, North Carolina.

Slice, D. E. 1992. GRF-ND: Generalized rotational fitting of N-dimensional landmark data. Department of Ecology and Evolution, State University of New York, Stony Brook, New York 11794.

Sneath, P. H. A., and R. R. Sokal. 1973. Numerical taxonomy: The principles and practice of numerical classification. Freeman and Company: San Francisco.

Stock, A. D. 1983. Chromosomal homologies and phylogenetic relationships of the vespertilionid bat genera *Euderma, Idionycteris* and *Plecotus*. Cytogenetics and Cell Genetics 35:136–140.

Storer, R. W. 1966. Sexual dimorphism and food habits in three North American accipiters. Auk 83:423–436.

Swiderski, D. L. 1993. Morphological evolution of the scapula in tree squirrels, chipmunks, and ground squirrels (Sciuridae): An analysis using thin-plate splines. Evolution 47:1854–1873.

Tabachnick, R. E., and F. L. Bookstein. 1990. The structure of individual variation in Miocene *Globorotalia*. Evolution 44:416–434.

Topál, G. 1989. Tertiary and Early Quaternary remains of *Corynorhinus* and *Plecotus* from Hungary (Mammalia, Chiroptera). Vertebrata Hungarica 23:33–55.

Tumlison, R., and M. E. Douglas. 1992. Parsimony analysis and the phylogeny of the plecotine bats (Chiroptera: Vespertilionidae). Journal of Mammalogy 73:276–285.

Volleth, M., and K.-G. Heller. 1994. Karyosystematics of plecotine bats: A reevaluation of chromosomal data. Journal of Mammalogy 75:416–419.

Wiley, E. O. 1981. Phylogenetics: The theory and practice of phylogenetic systematics. Wiley: New York.

Williams, D. F., J. D. Druecker, and H. L. Black. 1970. The karyotype of *Euderma maculatum* and comments on the evolution of the plecotine bats. Journal of Mammalogy 51:602–606.

Yoshiyuki, M. 1991. A new species of *Plecotus* (Chiroptera, Vespertilionidae) from Taiwan. Bulletin of the National Science Museum, Series A 17:189–195.

Zima, J., M. Volleth, I. Horácek, J. Cerveny, and M. Macholán. 1992. Karyotypes of two species of bats, *Otonycteris hemprichi* and *Pipistrellus tramatus* (Chiroptera, Vespertilionidae). Pages 237–242 *in* I. Horácek and V. Vohralík, (eds.), Prague studies in mammalogy. Charles University Press: Prague.

APPENDIX

Specimens Examined

Museum acronyms for specimens used in this study are defined in the acknowledgments section. Numbers in parentheses after species name indicate the number of examined males and females, respectively. *Plecotus auritus* (3, 5)—MGU1281; MGU2351; MGU4194; MGU29103; MGU104736; MGU105481–3; *P. austriacus*—USNM476627; USNM476629; USNM476632-4; USNM470707-11; *Barbastella barbastellus* (2, 10)—MNW11516; NMW11586; NMW12028-9; AMNH212187-9; USNM142583; USNM347421; HZM7.606; MCZ37000-1; *B. leucomelas* (3, 3)—AMNH16307; AMNH44562; FMNH34768; FMNH82737-8; HZM2.4415; *Euderma maculatum* (5, 4)—MSB6235; MSB9610; MSB17285; MSB23376-7; MSB24999-25000; MSB27715; MSB37724; *Idionycteris phyllotis* (4, 7)—MSB9474; MSB9518-9; MSB9578; MSB9612-3; MSB11084; MSB11635; MSB13013; MSB13847; MSB14833; *Corynorhinus mexicanus* (7, 11)—KU29848-60; KU29888; KU29890; KU29915; KU29918; KU29923; *C. rafinesquii* (8, 8)—LSUMZ54; LSUMZ6258; LSUMZ6263; LSUMZ8733; LSUMZ9305; LSUMZ1395-6; LSUMZ19793; LSUMZ20390-1; LSUMZ20393; LSUMZ20396; LSUMZ21927; LSUMZ23793; LSUMZ25225; LSUMZ28462; *C. townsendii* (5, 5)—TTU6149; TTU6464; TTU9161; TTU17232; TTU19960-1; TTU23277; TTU34549; TTU43684-5; *Otonycteris hemprichi* (2, 3)—MBQ1154; MBQ1170; MBQ1190; MBQ1201; MBQ1226.

MORPHOMETRIC ANALYSIS OF MICROSCOPIC HOOKS OF TAENIID TAPEWORMS (CESTODA, TAENIIDAE)

Application of Graphics Software for Automated Computation of Landmarks and Outlines

András Gubányi

Zoological Department
Hungarian Natural History Museum, Baross u. 13
H-1088 Budapest, Hungary

ABSTRACT

Hooks of five species of taeniid tapeworms were scanned at high resolution. After tracing the objects, automated extraction of outline and landmark data was carried out using a computer program written in AutoLISP. Morphometric variation was studied by elliptical Fourier analysis, canonical variate analysis and minimum spanning trees for both landmark and outline data. A comparison of centroid size and the affine component of shape differences among specimens was carried out for the landmark data. Canonical variate analysis scores seemed to show different patterns for landmark and outline data. There was a linear association between size and shape for the species studied, which was more pronounced for larger hooks.

INTRODUCTION

Various aspects of image data acquisition for systematic biology have been recently reviewed by Fink (1990), MacLeod (1990), Rohlf (1990a) and Becerra et al. (1993). In general, macroscopic objects are digitized using digitizing tablets or video cameras. To date, scanners have not been widely used, even though they offer a number of advantages and are widely available as integral components of desktop publishing systems. Scanners provide much higher resolution than do video cameras. Scanning also takes less time than does digitizing with tablets. Indeed, it is the time required for digitizing with tablets that has precluded their widespread usage for generating outlines. By contrast, high-resolution bitmap images (e.g., those captured by scanners) can be readily manipulated. Outlines of objects can be traced and point coordinates along contours can be computed.

In this paper I report on the use of commercially available graphics and computer-aided design software (CAD) for reading scanned images, generating contours and digitizing coordinates of both outlines and landmarks. I have used such a system to study microscopic preparations of hooks of parasitic worms (Cestoda: Taeniidae). Specifically, I investigated the size and shape differences in these structures among five European taxa. Rohlf (1993) identified automatic feature extraction as an area of image analysis in need of development. In this paper I demonstrate how landmarks can be automatically extracted from contours by means of a program written in a graphics package language.

The specific and generic status of a number of different parasitic tapeworms (family Taeniidae) have yet to be reliably established. The morphological uniformity of the strobilar stage (except for the scolex) and the great diversity of the larval forms recovered from different host species make it difficult to recognize different generic and subgeneric groupings (cf. Wardle and McLeod, 1952; Abuladze, 1964; Schmidt, 1986; Movsessian, 1989). This has resulted in misidentification and a number of synonymous descriptions (see Verster, 1969; Rausch, 1985; Movsessian, 1989, for reviews).

MATERIAL AND METHODS

The following adult *Taenia* species parasitizing carnivorous mammals have been studied (numbers refer to the inventory numbers of slides deposited in the Parasitological Collections of the Hungarian Natural History Museum, Budapest): *Taenia crassiceps* (Numbers 7797, 2565, 7844, 7849, 6956; $n = 40$ small hooks, $n = 45$ large hooks), *T. hydatigena* (Number 13202; $n = 7$ small hooks, $n = 11$ large hooks), *T. ovis* (Numbers 6940, 7214; $n = 22$ small hooks, $n = 17$ large hooks), *T. pisiformis* (Numbers 6433, 7040 7508, 7610, 7834, 7835 $n = 26$ small hooks, $n = 14$ large hooks), and *T. polyacantha* (Number 647; $n = 29$ small hooks, $n = 32$ large hooks).

Hooks of the tapeworms were dissected and separated from each other in Berlese medium (Baker and Wharton, 1952). After they were mounted, they were photographed in profile by means of an Olympus BH-2 microscope connected to an Olympus OM-4 TI camera equipped with a 25-mm adapter lens. In order to sharpen the outlines of the hooks, a low-speed film (i.e. Agfa copex) was used, and prints were enlarged 100 times on line-art unglazed paper. Photographs were scanned at 300 dots per inch bitmap resolution in black-and-white mode by use of a Microtec Scanmaker II scanner. Hooks were rotated upside down to make further computation simpler. Bitmap files were vectorized using the centerline tracing method of the CorelTRACE program of the commercially available graphics package CorelDRAW (1992). Outlines were calculated with Bézier curve transformation by tracing the drawings by three pixel spaces and saving the outlines into a dxf file format. Corner and straight line thresholds were set at three and two pixels, respectively. Coordinates of landmarks and outlines were generated automatically by a program written in AutoLISP under the CAD package AutoCAD Release 12 (1992). In essence, the program selects points from a complex entity (polyline or outline) by means of a dxf subroutine and four AutoLISP built-in functions: entsel and entget, to obtain the polyline and its head, entmod to close the opened polyline, and entnext and entget to loop through each vertex and to record the coordinates data. Before data acquisition the scanned objects were rotated to a horizontal plane by means of two AutoCAD built-in commands, rotate and move. Outline data were analyzed by means of relational functions and the two AutoCAD system variables extmax and extmin, which represented the maximum and minimum coordinates of the drawing. Five homologous landmarks on the hooks that are taxonomically important were calculated

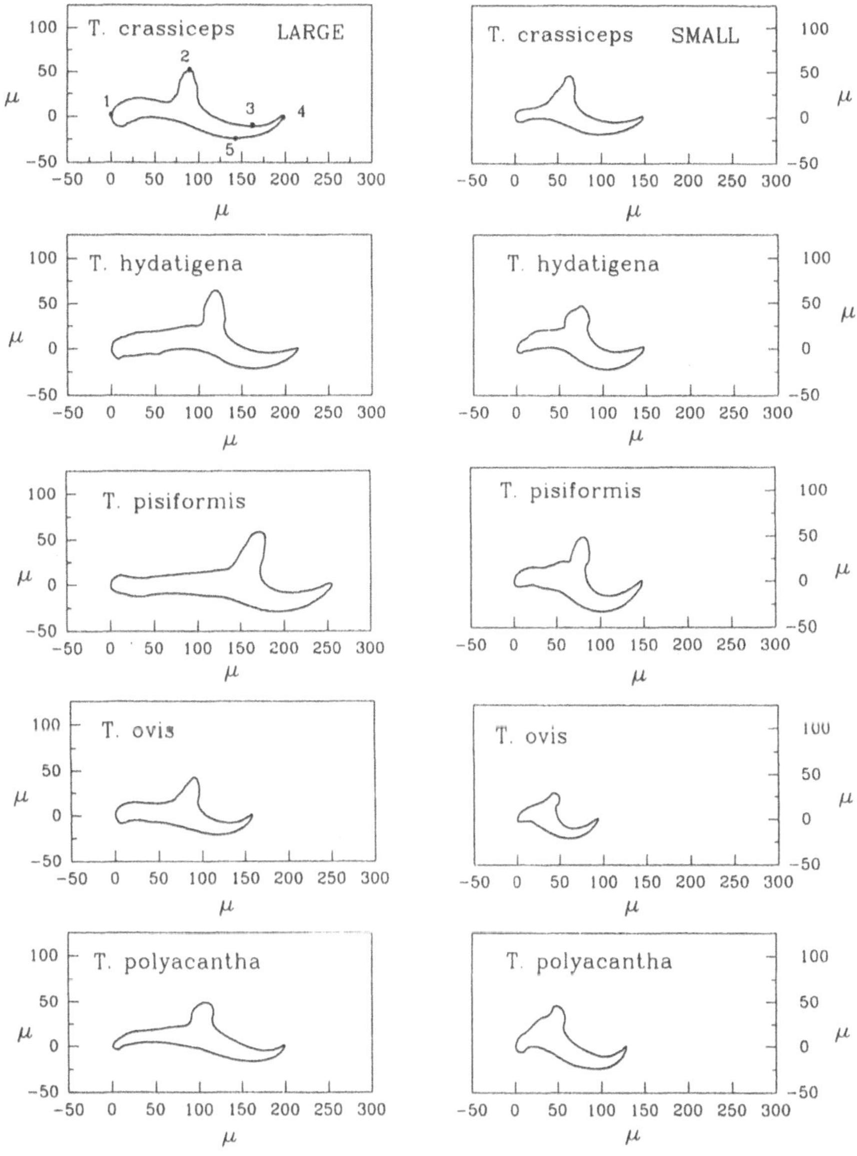

Figure 1. Range of variation in hook shape and size in five species of tapeworms. Landmarks locations are shown for the large hook of *T. crassiceps*.

automatically. They are (1) tip of the proximal root of the hook (handle); (2) tip of ventral root of the hook (guard) ; (3) minimum point of the upper chord of the blade; (4) tip of the hook's blade; and (5) minimum point of the lower chord of the blade (see Fig. 1). All vertex data, comprising more than 200 coordinate points per object, and the landmark coordinates were saved in different data files for further computation. (The program is available from the author upon request).

Shape coordinates (Bookstein, 1991) were computed by a routine in the SPSS/PC+ program package (Norusis, 1990); landmarks 1 and 4 were used as baseline end points. Canonical variates for each group were computed using the same package. Centroid size

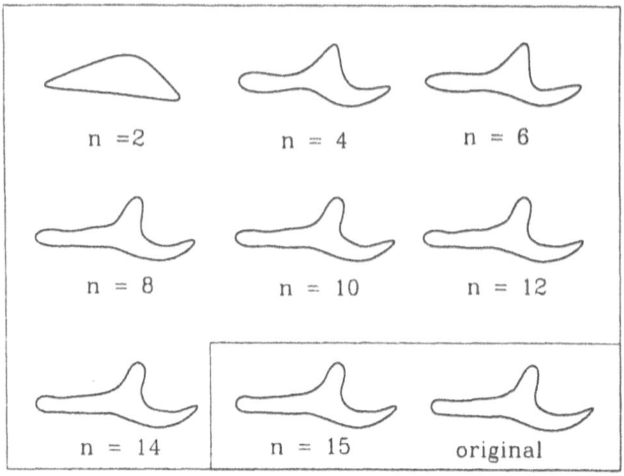

Figure 2. Reconstruction of the hook outline based on different number of harmonics (*n*) of elliptic Fourier analysis.

(Bookstein, 1989) was computed as the square root of the mean squared interlandmark distances by a routine in SPSS/PC +. Minimum spanning trees of the canonical variate centroids were constructed using average taxonomic distance in the NTSYS-pc program (Rohlf 1990b).

Uniform *x* and *y* factors were calculated from the shape coordinates (Bookstein, 1990). Outline data were analyzed by elliptic Fourier analysis (EFA) using the EFA program of Rohlf (1990c). Canonical variates were computed from 15 harmonic amplitudes of the Fourier coefficients (Fig. 2) and minimum spanning trees were constructed as they were for the landmark data described above.

RESULTS

Figure 1 displays the range of variation in mean shape and size in the small and large hooks of the five tapeworm species studied.

As Fig. 3 shows, significant variation in the size of the large hooks is found among species in both baseline length and centroid size, *Taenia ovis* and *T. pisiformis* representing

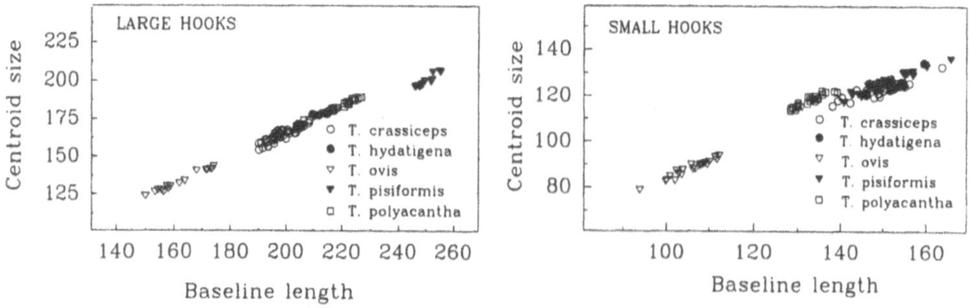

Figure 3. Bivariate scatter plots of centroid size versus baseline length for large and small hooks in five taeniid species.

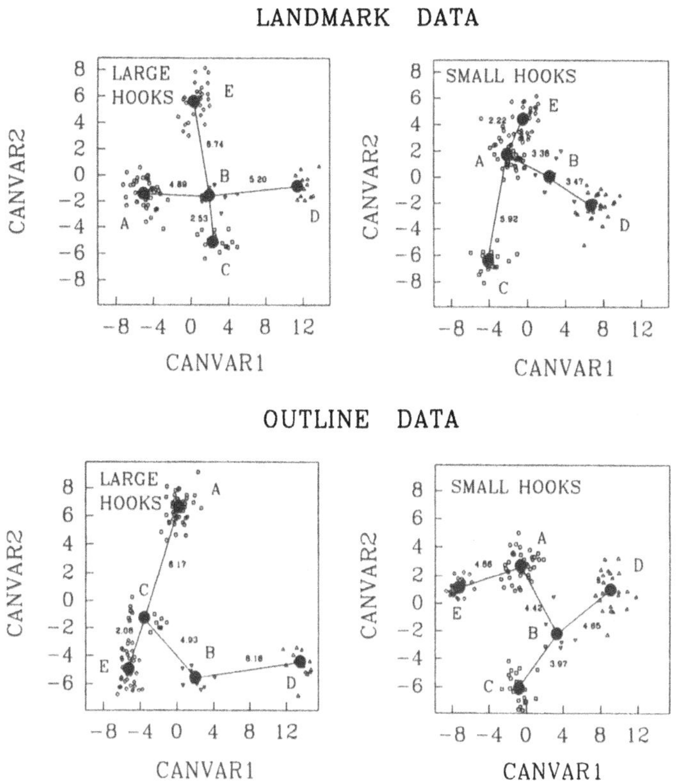

Figure 4. First two canonical variates and minimum spanning trees among centroids for large and small hooks (A = *T. crassiceps*, B = *T. hydatigena*, C = *T. ovis*, D = *T. pisiformis*, and E = *T. polyacantha*).

the two extremes. In the case of small hooks not all species follow the same trend. Three species, *T. crassiceps*, *T. hydatigena* and *T. pisiformis* overlap in the bivariate scatter plot of centroid size versus baseline length. The cloud of points for *T. polyacantha* is somewhat distinct from the other four species.

The first canonical variate analysis (CVA) scores based on landmark data of large hooks clearly separate the species. There is an overlap between CVA scores of *T. crassiceps* and *T. polyacantha* for small hooks. The minimum spanning trees are different for the large and small hooks (Fig. 4). However, the ordination pattern for *T. polyacantha* is similar in both cases. *Taenia hydatigena* occupies a central position with respect to the other four taeniid species if the tree is based on large hooks. The pattern is quite different for the small hooks. Distances among the canonical variate centroids proved to be considerable for both types of hooks.

In the EFA 15 harmonics were sufficient to reproduce the outline accurately (Fig. 2). The CVA of these 15 harmonics resulted in clear discrimination with no overlap among the species. The canonical variates for large hooks of *T. ovis* and *T. polyacantha* are relatively close to one another, in contrast to the pattern observed for the small hooks (Fig. 4).

Figure 5 shows the scatter plots of the uniform x and y components (large hooks: $r = 0.162$, $p = 0.04$; small hooks: $r = 0.165$, $p = 0.03$). As can be seen the species are well separated and form tight clusters in the bivariate scatter plots of the affine factors versus centroid size. The correlations between affine factors and centroid sizes in all possible combinations were at moderate levels ($r = 0.315$-0.776, $p < 0.05$).

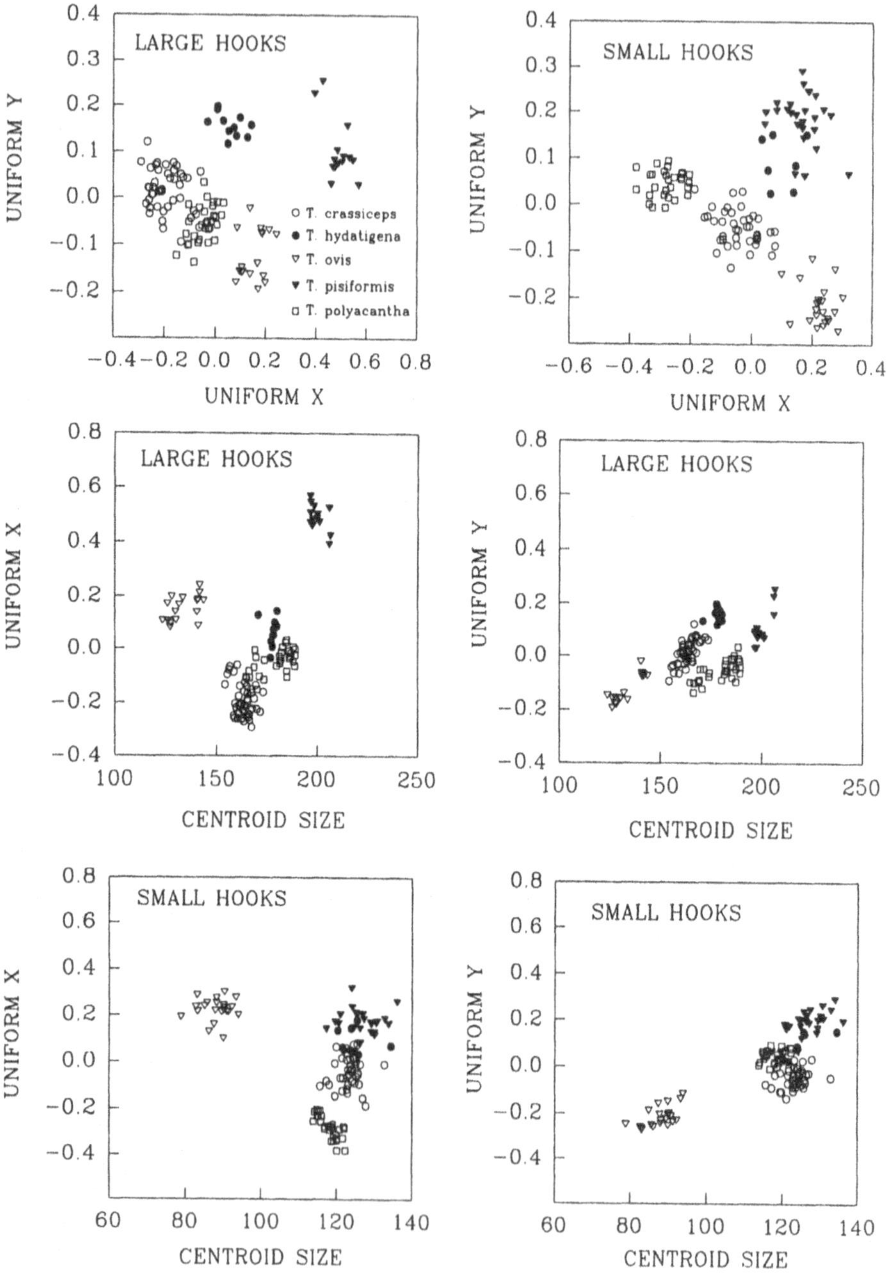

Figure 5. Bivariate scatter plots of affine components and centroid size.

DISCUSSION

The two size measures, baseline length and centroid size, yield comparable bivariate scatter plots, justifying the choice of baseline.

In the plot of baseline length and centroid size the separation of species into distinct species clusters was more apparent for the large hooks than it was for the small ones (Fig. 3).

Affine components of shape variation in both small and large hooks are correlated with size. Thus there is a linear allometric association between size and shape in the species studied. This association is more pronounced for large hooks.

The taxonomic distances calculated from shape coordinates and outline data indicate that the five taeniids are distantly related species. The five species are traditionally considered to belong to distinct groups within the genus *Taenia*. The larval stages (asexual reproduction) of *T. crassiceps* and *T. polyacantha* differ markedly from the other three species (Rausch, 1985). The placement of *T. hydatigena*, *T. ovis* and *T. pisiformis* is also affected by the intermediate host species (Abuladze, 1964). The analyses of landmark and outline data confirmed this notion. The structure of the scolex, which accommodates the hooks, shows no evolutionary adaptations that are clearly related to parasitism (Baer, 1952). However, the variation in shape and size of the two different hook types across taxa is likely the result of adaptive evolution.

The reproductive anatomy of tapeworms is also used extensively by taxonomists to make assignments at the familiar, generic and subgeneric level. Such assignments are usually based upon differences in the shape of the uterus, seminal receptacle, eggs, among other structures. Unfortunately the descriptions of these structures are often very subjective. Automated feature extraction techniques, such as those described in this paper, could be usefully applied to these structures to derive outlines and landmarks for more objective analyses. It seems, then, that these methods might be usefully applied to solving taxonomic problems in tapeworms.

ACKNOWLEDGMENTS

I am grateful to Leslie F. Marcus for making it possible for me to attend the NATO Advanced Study Institute on morphometrics. Part of the research presented here received funding from OTKA (Hungarian National Research Fund) grant 3177/1991 to A. Demeter.

REFERENCES

Abuladze, K. I. 1964. Teniaty-lentochnye gel'minty zhivotnykh i cheloveka i vyzyvaemye imi zabolevaniia. Osnovy Tsestodologii 4:1–530.

AutoCAD release 12. 1992. Autodesk, Inc.

Baker, E. W. and G. W. Wharton. 1952. An introduction to acarology. MacMillan: New York.

Baer, J. G. 1952. Ecology of animal parasites. University of Illinois Press: Urbana.

Beccerra, J. M., E. Bello and A. Garcia-Valdecasas. 1993. Building your own machine image system for morphometric analysis: A user point of view. Pages 65–92 *in* L. F. Marcus, E. Bello and A. Garcia-Valdecasas, (eds.), Contributions to morphometrics. Monografias del Museo Nacional de Ciencias Naturales 8, Madrid.

Bookstein, F. L. 1989. "Size and Shape": A comment on semantics. Systematic Zoology, 38:173–180.

Bookstein, F. L. 1990. Higher order features of shape change for landmark data. Pages 237–250 *in* F. J. Rohlf and F. L. Bookstein, (eds.), Proceedings of the Michigan morphometrics workshop. University of Michigan Museum of Zoology Special Publication 2.

Bookstein, F. L. 1991. Morphometric tools for landmark data: Geometry and biology. Cambridge University Press: Cambridge.

CorelDRAW. 1992. Coreldraw user's manual 3.0. Corel Corporation: Ottawa.

Fink, W. L. 1990. Data acquisition for morphometric analysis in systematic biology. Pages 9–19 *in* F. J. Rohlf and F. L. Bookstein, (eds.), Proceedings of the Michigan morphometrics workshop. University of Michigan Museum of Zoology Special Publication 2.

MacLeod, N. 1990. Digital images and automated image analysis systems. Pages 21–35 *in* F. J. Rohlf and F. L. Bookstein, (eds.), Proceedings of the Michigan morphometrics workshop. University of Michigan Museum of Zoology Special Publication 2.

Movsessian, S. O. 1989. Present views on the systematics of cestodes of the suborder Taeniata. Helminthologia 26:3–14.

Norusis, M. J. 1990. SPSS/PC+ Advanced Statistics TM 4.0 for the IBM PC/XT/AT and PS/2. SPSS: Chicago.

Rausch, R. L. 1985. Presidental address. Journal of Parasitology 71 (2): 139–151.

Rohlf, F. J. 1990a. An overview of image processing and analysis techniques for morphometrics. Pages 37–60 *in* F. J. Rohlf and F. L. Bookstein, (eds.), Proceedings of the Michigan morphometrics workshop. University of Michigan Museum of Zoology Special Publication 2.

Rohlf, F. J. 1990b. NTSYS-pc. Numerical taxonomy and multivariate analysis system, version 1.6 Exeter Software: Setauket, New York.

Rohlf, F. J. 1990c. Fitting curves to outlines. Pages 167–177 *in* F. J. Rohlf and F. L. Bookstein, (eds.), Proceedings of the Michigan morphometrics workshop. University of Michigan Museum of Zoology Special Publication 2.

Rohlf, F. J. 1993: Computer vision needs in systematic biology. (With discussion.) Pages 365–373 *in* R. Fortuner, (ed.), Advances in computer methods for systematic biology, artificial intelligence, databases, computer vision. John Hopkins University Press: Baltimore, Maryland.

Rohlf, F. J. and D. E. Slice. 1990. Extension of the Procrustes method for the optimal superposition of landmarks. Systematic Zoology, 39:40–59.

Schmidt, G. D. 1986. CRC handbook of tapeworm identification. CRC Press: Boca Raton, Florida.

Verster, A. 1969. A taxonomic revision of the genus *Taenia* Linnaeus, 1758 s. str. Onderstepoort. Journal of Veterinary Research 36(1):3–58.

Wardle, R. A. and J. A. McLeod. 1952. The zoology of tapeworms. University of Minnesota Press: Minneapolis.

A TAXONOMIC COMPARISON OF MORPHOMETRIC METHODS

Distances and Shape Coordinates

José M. Becerra, Elisa Bello, and Antonio García-Valdecasas*

Museo Nacional de Ciencias Naturales (C.S.I.C.)
Departamento de Biodiversidad
c/o Jose Gutierrez Abascal, 2.
28006-Madrid, Spain

ABSTRACT

Landmark characters were defined and recorded for 19 species and subspecies of the water mite genus *Torrenticola* (Acari, Hydrachnellae). This set of taxa has been previously studied by Lundblad (1956), who gives detailed morphological descriptions, standardized drawings of different body parts and metric measurements (distances). The distance data and shape coordinates derived from landmarks are compared in terms of their relative usefulness for discerning alpha taxonomic relationships.

INTRODUCTION

Water mite species of the genus *Torrenticola* (Acari, Hydrachnellae) have been previously assigned to a nominal subgenus or to the subgenus *Monatractides* on the basis of qualitative characters (see Cook, 1974 for a worldwide revision). In his 1956 paper, Lundblad describes and gives distance data for 19 species found in central and southern Europe of the subgenera *Torrenticola* and *Monatractides*. In this paper we identify landmark variables for that set of species and use those landmarks to evaluate the assignment of *Torrenticola* species to one of the two subgenera.

* e-mail: MCNJM14@CC.CSIC.ES; MCNEB18@CC.CSIC.ES; MCNHV00@CC.CSIC.ES

Advances in Morphometrics, Edited by L. F. Marcus et al.
Plenum Press, New York, 1996

Table 1. Specimens by subgenus and sex

Subgenus and species	Male	Female
Torrenticola		
T. anomala	+	+
T. elliptica	+	+
T. microphallus	+	+
T. brevirostris	+	+
T. hispanica	+	+
T. guadarramensis	+	+
T. barsica	+	+
T. andrei	+	
T. ungeri	+	+
T. lunbladi	+	
T. amplexa	+	
T. amplexa var. *tenuipalpis*		+
T. amplexa var. *minutivalvata*	+	+
Monatractides		
T. stadleri	+	+
T. stadleri var. stenostoma	+	
T. madritensis	+	+
T. aberrata	+	+
T. lusitanica	+	+

MATERIAL AND METHODS

All too often traditional invertebrate alpha taxonomists overlook individual variation and erect new species based on single representative specimens. In this study we investigate the consequences of this practice by mimicking the procedure for morphometric comparisons among different species of water mites within the genus *Torrenticola*. Lundblad (1956) gives descriptions of type material and redescriptions of some poorly known species of *Torrenticola* (Acari, Parasitengona): 14 belong to the subgenus *Torrenticola* and 5 belong to the subgenus *Monatractides* (Table 1). We examined the microscope slides (from the Swedish Museum of Natural History) upon which these species descriptions were based. Measurements in Lundblad (1956) were used as traditional variables and landmarks were taken from the microscope slides. For every specimen, only distances for the dorsal and ventral shield were given by Lundblad.

The variables measured are shown in Fig. 1 and consist of for the dorsal shield, length, width, length of the anterior plate and length of posterior plate; and, for the ventral shield, depth of the mandibular bay, distance between posterior end of mandibular bay and posterior end of genital field, length between epimera II and III and length of the genital field. These distances were log transformed to improve the linearity of distribution within variables. We chose eight landmarks on the dorsal shield and five on the ventral shield (Fig. 1). The chosen landmarks were unrelated to the distances shown in Fig. 1. They correspond instead to homologous structures that we could readily identify. The landmarks are as follows: on the dorsal shield (landmarks 1 to 8): pore in the right anterior plate, pore in the right posterior plate, four pores in the dorsal shield, pore in the left anterior plate and pore in the left posterior plate; and on ventral shield (landmarks 1 to 5), insertion of the right fourth leg, gonopore, insertion of the left fourth leg, insertion of the right third leg and insertion of the left third leg.

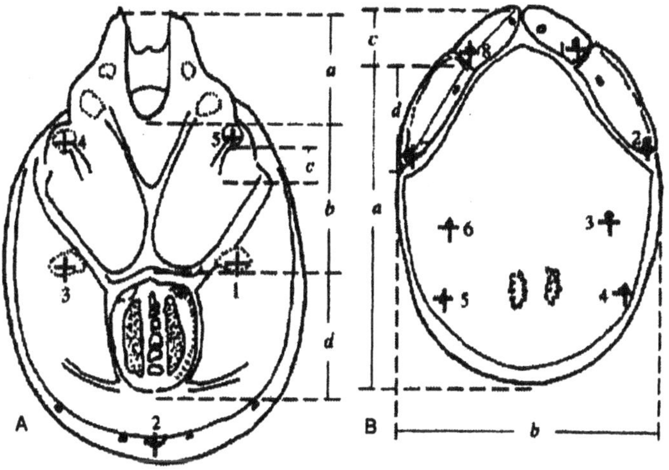

Figure 1. Schematic view of *Torrenticola guadarramensis* female. (**A**) Ventral shield. Numbers specify landmarks 1 to 5 and letters are Lundblad's measurements: (a) mandible sinus depth; (b) distance between mandible sinus and anterior part of the genital field; (c) epimere II to III length; and (d) genital field length. (**B**) Dorsal shield. Landmarks 1 to 8 and Lundblad's measurements: (a) dorsal shield length; (b) dorsal shield width; c) anterior plate length; and (d) posterior plate length.

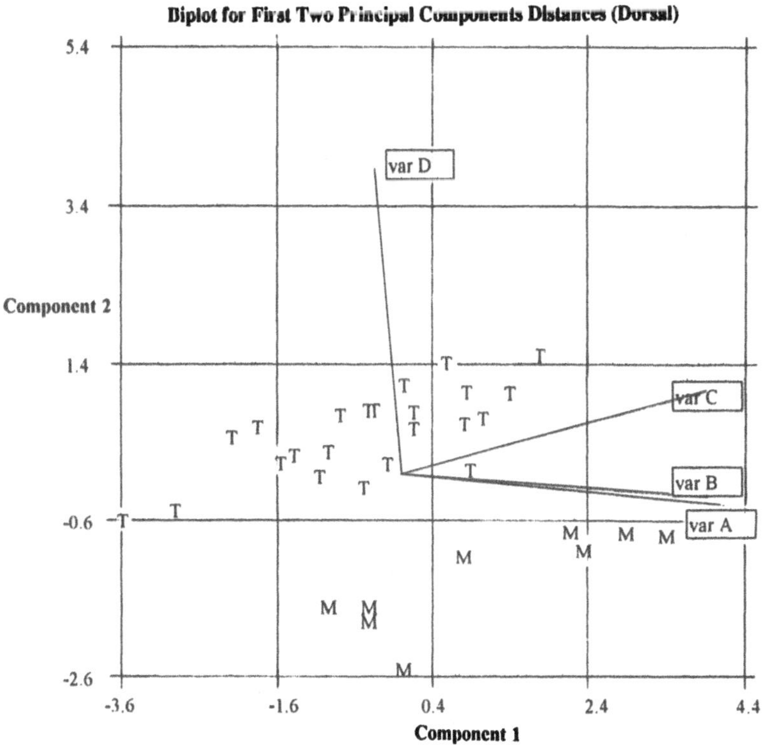

Figure 2. Biplot for the log of distances on the dorsal shield. Abbreviations are M = *Monatractides* species; T = *Torrenticola* species; var A = length of dorsal shield; var B = width of dorsal shield; var C = length of anterior plate; and var D = length of posterior plate.

Landmark coordinates were collected using a three-dimensional Reflex microscope (MacLarnon, 1989, Dean this volume). Three coordinates (*x, y*, and *z*) were captured for each landmark, but only the *x*- and *y*- coordinates were retained to ensure that the data were comparable to the measurements given by Lundblad. The landmark data were transformed to shape coordinates (Bookstein, 1991) by means of the program UNIGRAPH, version March 6, 1993 (Marcus, 1993). For each specimen and for each set of dorsal and ventral landmarks, we obtained all possible sets of shape coordinates, one for each different baseline choice. We compared baseline length with centroid size to determine which baseline was most appropriate for subsequent comparisons. We applied a principal component analysis to both the distance data and to the shape coordinate data. Biplots for dorsal and ventral variables (for distances and shape coordinates) where plotted. STATGRAPHICS version 5.0 was used for all these analyses.

RESULTS

A baseline between landmarks 3 and 6 was chosen for the dorsal shield based on the high correlation between centroid size and its length. Landmarks 4 and 5, representing the insertion points for the third pair of legs, were selected as baseline end points for the ventral shield (also based on a high correlation with centroid size). The ordination of species based

Table 2. Percent of variance and scores obtained in the principal components analyses. Note: Not all the principal components are shown in C and D

(A) Dorsal shield distances

	Percent of variance	Cumulative. percentage
PC1.	62.70.	62.7
PC2.	26.64.	89.3
PC3.	7.72.	97.0
PC4.	2.92.	100.0

	PC1	PC2	PC3	PC4
varA	0.573	-0.045	0.731	0.365
varB.	0.597	-0.113	-0.081	0.789
varC.	0.557	0.259	0.646	0.451
varD.	-0.052	0.958	-0.200	-0.198.

(B) Ventral shield distances

	Percent of variance	Cumulative. percentage
PC1.	60.05	.60.0
PC2.	30.23	90.2
PC3.	.6.55	.96.8
PC4.	3.16	100.0

	PC1	PC2	PC3	PC4
var A	0.448	.0.572	-0.672	0.139
var B	-0.503	.0.522	-0.033	-0.686
var C	-0.547	.0.416	.0.137	0.712
var D	0.494	.0.474	.0.726	-0.037

(C) Shape coordinates dorsal shield (baseline 3-6)

	Percent of variance	Cumulative. percentage
PC1.	61.6	61.6
PC2.	20.9	82.5
PC3.	8.2	90.6
PC4.	4.0	94.6
PC5.	1.9	96.6
PC6.	1.4	98.0
PC7.	0.9	.98.9
PC8.	0.5	.99.3
PC9.	0.3	.99.6
PC10.	0.2	99.8
PC11.	0.1	99.9
PC12.	0.1	100.0

	PC1	PC2	PC3	PC4
var X1	0.356	0.056	0.103	-0.124
var Y1	-0.314	0.224	0.022	0.156
var X2	0.205	0.394	0.3120	-0.596
var Y2	-0.318	0.248	.001	0.3440
var X4	-0.280	0.350	0.2140	-0.194
var Y4	0.301	0.161	0.270	0.4770
var X5	-0.244	-0.324	0.542	-0.0452
var Y5	-0.083	-0.517	0.526	-0.001
var X7	-0.313	0.281	0.170	0.155
var Y7	-0.2390	-0.325	-0,380	-0.435
var X8	-0.338	0.129	0.083	0.041
var Y8	-0.355	-0.084	-0.146	0.061

(D) Shape coordinates ventral shield (baseline 4–5)

	Percent of variance	Cumulative percentage
PC1.	74.50	74.5
PC2.	19.01	93.5
PC3.	3.76	97.3
PC4.	1.74	99.0
PC5.	.68	99.7
PC6.	.32	100.0

	PC1	PC2	PC3	PC4
var X1	0.413	0.283	.0.754	.0.384
var Y1	-0.457	.0.052	0.380	-0.288
var X2	-0.138	.0.889	-0.057	-0.323
var Y2	-0.466	-0.061	.0.135	-0.086
var X3	-0.435	0.249	-0.261	.0.808
var Y3	-0.442	-0.247	0.443	.0.054

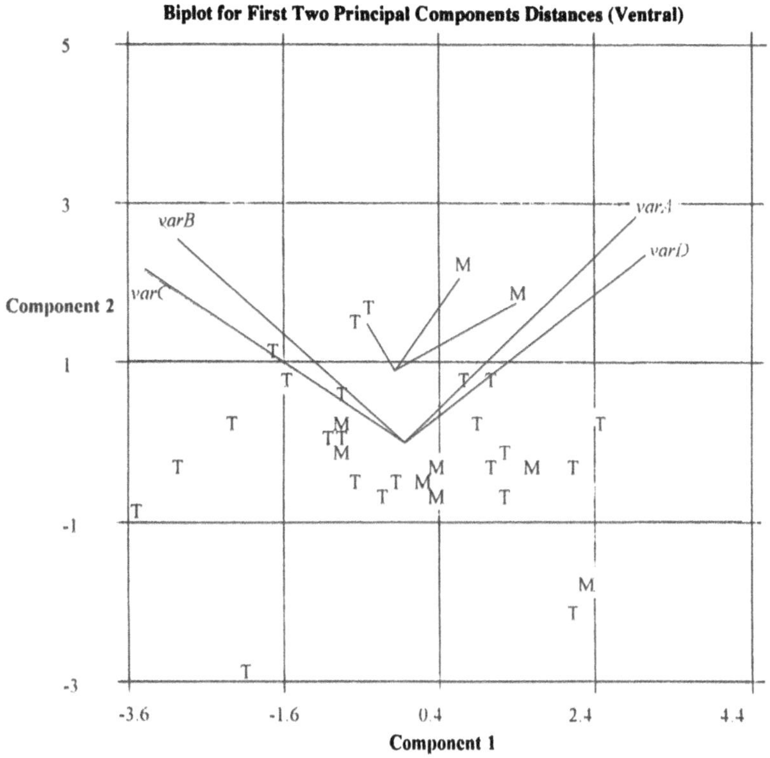

Figure 3. Biplot for the log of distances on the ventral shield. Abbreviations are M = *Monatractides* species; T = *Torrenticola* species; var A = depth of mandibular bay; var B = distance between posterior end of mandibular bay and posterior end of genital field; var C = length between epimera II and III; and var D = length of the genital field.

on dorsal shield distances is shown in the biplot in Fig. 2, in which the two subgenera are clearly separated. This separation is evident along the second principal component, which accounts for 26.6% of variance (Table 2) and predominantly reflects variation in the length of the small posterior plate. *Torrenticola* species have longer posterior plates than do *Monatractides* species.

Figure 3 shows the comparable plot for distances on the ventral shield. No clear separation of the species among subgenera is seen in this plot. Landmark analysis for dorsal and ventral shields are shown in Figs. 4 and 5, respectively. Only the landmarks of the ventral shield for baseline 4-5 give a clear grouping of species. This separation is mostly due to the first principal component, which accounts for 74.5 % of the total variance. The effect is due primarily to coordinate variation in the *y*-axis among specimens: y_1 (-0.46), y_2 (-0.47) and y_3 (-0.44). This variation is highly correlated reflecting the fact that displacements are coordinated. *Torrenticola* specimens have these landmarks displaced farther from the baseline than do *Monatractides* specimens.

DISCUSSION

Taxonomic identification in the genus *Torrenticola* is traditionally based on qualitative characters: presence versus absence or the different shapes or locations of particular

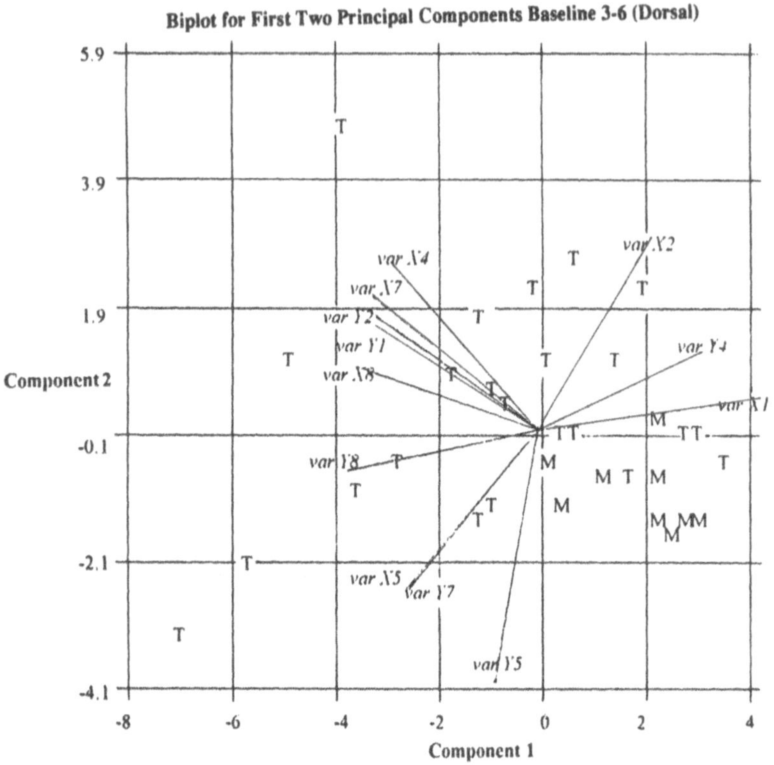

Figure 4. Biplot for the shape coordinates based on the baseline 3–6 on the dorsal shield. Abbreviations are M = *Monatractides* species; T = *Torrenticola* species; var *x*number = shape coordinate *x* for landmark number; and var *y*number = shape coordinate *y* for landmark number.

structures. Distance data have also been used by some classical taxonomists. Good classifications are sometimes obtained based solely on these types of data. We have discovered a hitherto undescribed difference in the length of the small posterior plate that clearly differentiates the two subgenera of *Torrenticola*. The subgenus *Torrenticola* has a longer small posterior plate than does the subgenus *Monatractides*. It remains to be seen if this difference applies to all the species in the subgenera. The landmarks chosen for analysis in this paper do not appear to be especially powerful for taxonomic purposes. Only the ventral landmarks classified the species of both subgenera, although with some overlap. We feel that the lack of differentiation among subgenera based on landmark descriptions does not reflect a failing of the landmark methods but rather is a consequence of using single specimens to represent taxa. We specifically set out to mimic the way classical taxonomists have worked: using single specimens for species nomination. When such an approach is adopted analyses cannot effectively address differences in landmark variation (or variation in distances, for that matter) among taxa. Only when intraspecific variation is simultaneously considered with differences among taxa can meaningful conclusions be drawn. The fact that dorsal distance data and ventral landmark data roughly separate the subgenera, while ventral distances and dorsal landmarks do not, warrants further, more detailed study.

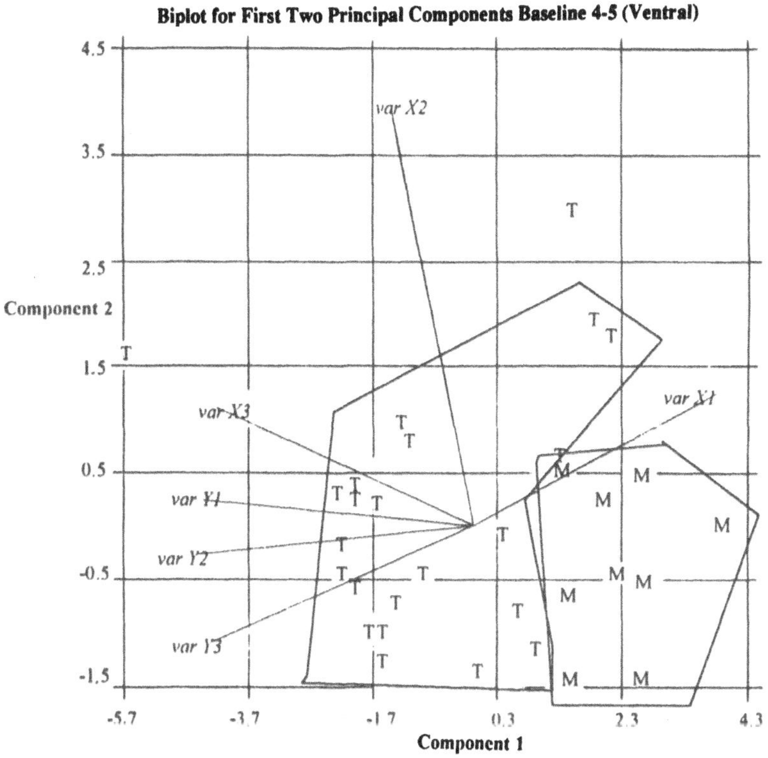

Figure 5. Biplot for the shape coordinates based on the baseline 4–5 on the ventral shield. Abbreviations are M = *Monatractides* species; T = *Torrenticola* species; var *x*number = shape coordinate *x* for landmark number; and var *y*number = shape coordinate *y* for landmark number. Note almost complete separation of the two subgenera.

ACKNOWLEDGMENTS

Many people helped to complete this paper. Les Marcus invited us and gave us financial support to attend the NATO Advanced Study Institute "Advances in Morphometrics". He also provided many suggestions for the improvement of this paper. We are in debt to him for this (and for many more things). The comments of F. J. Rohlf helped improve the manuscript. F. Bookstein, D. Slice, M. Corti and L. Barrios have also helped in one way or another, and Santiago Reig has been especially helpful throughout. T. Kronestedt (Swedish Museum of Natural History) loaned the *Torrenticola* microscope slides. This work is supported in part by a grant from the Consejo Superior de Investigaciones Cientificas and by the project PB92-0082 (DGYCIT), a CAM grant to E. Bello and J. M. Becerra.

REFERENCES

Bookstein, F. 1991. Morphometric tools for landmark data: Geometry and biology. Cambridge University Press: Cambridge.

Cook, D. R. 1974. Water mite genera and subgenera. Memoirs of the American Entomological Institute 21.

Lunblad, O. 1956. Zur kenntnis sud - und mitteleuropischer Hydrachnellen. Arkiv for Zool. 10:1–306.

MacLarnon, A. M. 1989. Applications of the Reflex instrument in quantitative morphology. Folia Primatologica 53:33 Special Publication No. 2,49.

Marcus, L. F. 1993. UNIGRAPH, a program to compute shape or Bookstein coordinates from 2 dimensional coordinates from a digitizer. Version 6 March 1993.

CAN PARTIAL WARP SCORES BE USED AS CLADISTIC CHARACTERS?

Gavin J. P. Naylor

Osborn Memorial Laboratory
Department of Biology
Yale University, P.O. Box 208104
New Haven, Connecticut 06520-8104

ABSTRACT

Zelditch et al. (1995) have recently proposed that partial warp scores derived from shape differences among organisms can be used as cladistic characters for phylogenetic inference. The current study represents a test of Zelditch et al.'s proposal. A group of different–shaped imaginary fishes, interrelated by a known phylogeny, was artificially generated by subjecting an ancestral fish to evolutionary change along a predetermined tree. Care was taken to ensure that invoked shape changes occurred only once on the tree and that shape changes occurred independently of each other. These conditions were set in order to maximally predispose any emergent "shape characters," which might be derived from shape differences among terminals, to cladistic analysis. Thirty-eight landmarks were recorded for each of the eight terminals resulting from the contrived evolutionary tree. These landmarks were transformed into shape coordinates. The shape coordinates for each of the terminals were then individually contrasted with those of the ancestor by means of the computer program TPSPLINE (Rohlf, 1994). The resulting partial warp scores were recoded as cladistic characters following the protocol of Fink et al.(1994) and subjected to cladistic analysis. Two most parsimonious trees (MPTs) resulted, one of which matched the "true" tree. Although this tree matched the true tree in topology, the reconstructed ancestral shape states implied by parsimony for this tree were quite different from those of the true tree. Whereas the true tree contained no homoplasy whatsoever, the two inferred MPTs implied considerable shape homoplasy (Retention Index = 0.48). An UPGMA of raw partial warp scores also retrieved the correct tree topology.

INTRODUCTION

Given that geometric morphometrics allows a quantitative and fully efficient description of shape differences between organisms, the question arises—can geometric methods be used to identify homologous quantitative features for phylogenetic analysis? Zelditch,

Advances in Morphometrics, Edited by L. F. Marcus et al.
Plenum Press, New York, 1996

Fink and Swiderski have recently proposed that they can be (Fink and Zelditch, 1994; Zelditch and Fink, 1994; Zelditch and Fink, 1995; Zelditch et al., 1995) and have outlined a protocol in which partial warp scores (Bookstein, 1991) are used as character data for cladistic analysis (Fink et al., 1994). Their (paraphrased) protocol is as follows.

1. Select landmarks that can be identified on all of the specimens to be used in the study. The ideal distribution of landmarks should capture as much of the shape variation in the data set as is possible.
2. Select one individual (an outgroup) to serve as the reference specimen against which all specimens are to be contrasted. The location of the landmarks on this outgroup specimen determines the principal warps (Bookstein, 1991). The fact that all shape contrasts are made against this one specimen ensures that the shape space, and thus the partial warp scores, of different pairwise contrasts will be comparable (and thus implicitly homologous).
3. Compute shape coordinates for all of the specimens by use of a selected baseline (Bookstein, 1991).
4. Compare each of the ingroup taxa to the selected outgroup and compute their respective partial warp scores.
5. For each partial warp, plot the scores and examine the scatter of all specimens for all of the taxa. Any partial warp that shows separation between the outgroup and ingroup taxa is potentially phylogenetically informative. When two or more of the ingroup taxa have similar scatters and those scatters are distinct from the outgroup, the taxa can be putatively regarded as sharing the same derived character state (synapomorphy).
6. Construct a character matrix in which each partial warp corresponds to a separate character. Code all of the taxa that have similar scatters for any given partial warp as having the same character state for that character.
7. Subject the resultant character matrix to a cladistic parsimony analysis, treating states within characters as unordered.
8. Examine the partial warp character distributions and recode as a single character any partial warps that show the same distribution on the cladogram—unless evidence is available to suggest that they are historically independent.

Under this protocol any putative synapomorphy seen to be congruent with the inferred cladogram is regarded as homologous (i.e., corroborated by the inferred tree).

If this protocol does indeed result in the discovery of homologous shape characters that can be used to infer the pattern of evolutionary diversification among a group of organisms, it will represent a major contribution toward the integration of morphometric methods into phylogenetic analysis. The achievement would be all the more remarkable given the many attempts at integration that seem to have met with little success in the past (Smith, 1990).

Bookstein (1994) has argued that geometric morphometrics (indeed all biometric procedures) are inherently incompatible with biological notions of homology and thus cannot yield homologous characters for phylogenetic inference. The core of Bookstein's argument is that biological homology is a nonmetric concept and that any metric approach couched in terms of covariance, estimation and approximation will fail to capture the essence of what it is to be "homologous". He is quite clear about this. He states, "The notion of homology allows no approximations. Structures either are or are not 'the same'-no notion of 'near miss' can apply" (Bookstein, 1994:221-222). He explains that the biological concept of homology spans a range of shape variation which exceeds the effectiveness of morphometric methods to capture it. Although he allows that studies at the subspecific level where modifications are "smooth" and minor, can be effectively treated using "geometrizable factors," he asserts that "no important evolutionary change can be captured persuasively in the language of biometrics" (Bookstein, 1994:220).

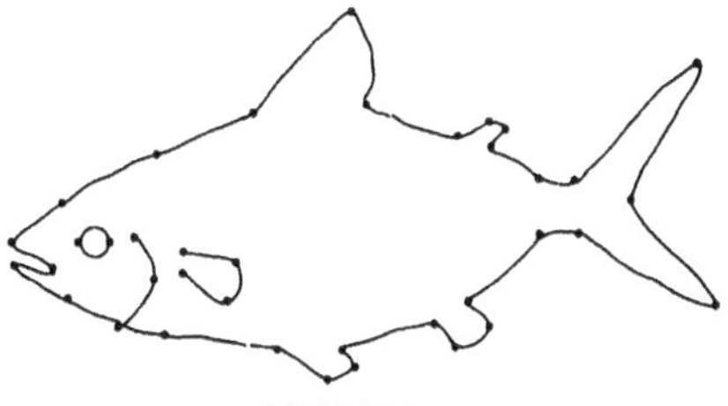

Figure 1. The ancestral-shaped fish used as the starting point for the evolutionary changes which resulted in the eight ingroup taxa. Shape changes were invoked by expanding different regions of the form. The 38 landmarks used to capture the form are shown as points along the outline.

Bookstein's arguments are philosophically important and compelling. However, they do not leave us with any clear sense of the breakpoint at which shape differences among organisms become too large to approximate homology effectively based on geometric morphometrics. It is, after all, conceivable that the sort of shape variation typically encountered in systematic studies will be "within bounds" for the sorts of analyses that Zelditch et al. (1995) advocate. In an effort to shed light on this issue I have attempted to evaluate empirically the efficacy of the proposed protocol. I have applied Zelditch et al.'s method to organisms with a known (simulated) phylogeny and whose shape variation is akin to that which might be encountered in a typical systematic survey.

METHODS

I simulated morphological evolution for a group of imaginary fishes by taking an ancestral fish form (Fig. 1) and subjecting it to shape changes at different points along a

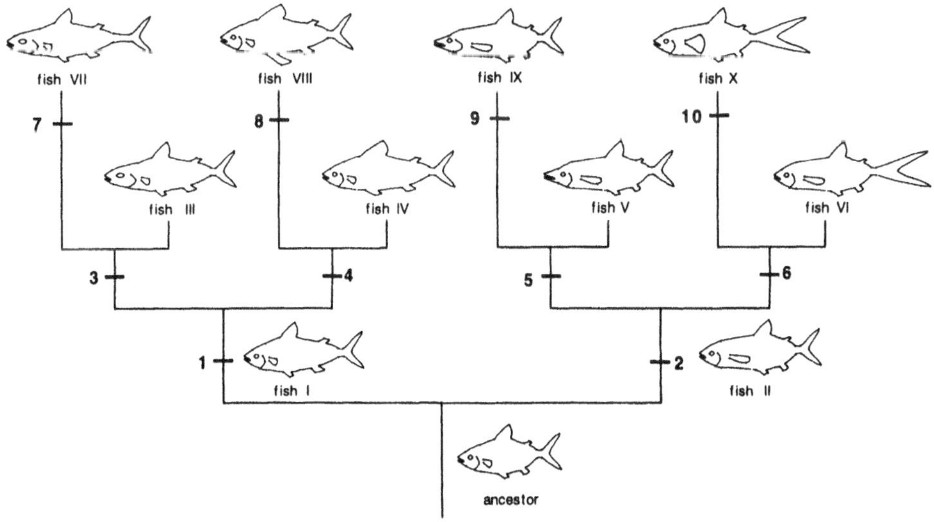

Figure 2. Evolutionary tree used as a basis for inherited shape change. Each numbered tick mark represents a unique shape change that is inherited by descendant lineages. Shape changes consisted of a dilatation (or a compression) of a localized region of the form. Shape changes consist of (1) lengthening of the posterior body region immediately in front of peduncle, (2) lengthening of the body region immediately behind the gills, (3) lengthening of the region containing gill and eye, (4) lengthening of the midregion containing the first dorsal fin, (5) lengthening of the snout, (6) lengthening of the tail, (7) lengthening of the caudal peduncle, (8) expansion of the pelvic fin, (9) movement of gill and eye anteriorly and (10) "fanning out" of pectoral fin.

predetermined evolutionary trajectory such that shape changes were "inherited" by descendant lineages. This resulted in eight taxa within the ingroup, each of which had a different shape (Fig. 2). Each of the shape changes used in this contrived evolutionary tree was brought about by expanding one part or another of the form. Steps were taken to ensure that each of the invoked shape changes occurred in discrete nonoverlapping regions and that each kind of shape change occurred only once on the tree. These conditions—character independence and lack of homoplasy—are known to be optimal for the retrieval of phylogenetic signal by means of cladistic parsimony (Farris, 1983). Indeed, they were specifically chosen to maximally predispose the collected data to cladistic interpretation.

All eight terminal taxa and the ancestor were digitized for 38 homologous landmarks (Fig. 1) by means of a video-digitizing system. The standard length of the fish (a line between the tip of the snout and the fork of the tail) was used as a baseline to compute shape coordinates for all of the taxa. Shape coordinates from each of the eight ingroup taxa were then individually contrasted with those of the ancestor by means of the computer program TPSPLINE (Rohlf, 1994). The partial warp scores were recorded for each of these eight contrasts (Table 1).

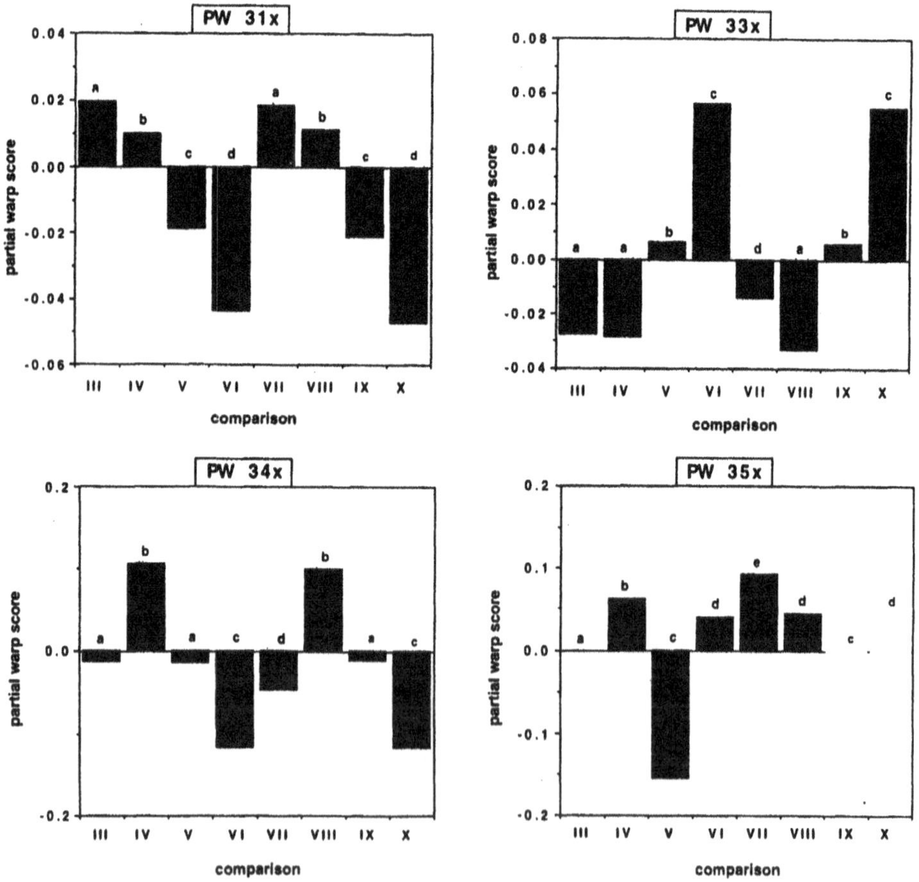

Figure 3. Partial warp scores for four different partial warps. Partial warp scores were obtained by comparing the shape coordinates of each of the terminal taxa (fishIII–fishX) to those of the ancestor. The four plots show how taxa with similar partial warp scores were recoded as having the same alphameric character state.

Table 1. Partial warp scores for *x* and *y* components of 35 partial warps. Partial warp scores were obtained by contrasting shape coordinates of each of the terminal taxa (fishIII–fishX) to those of the ancestor

PW	IIIx	IIIy	IVx	IVy	Vx	Vy	VIx	VIy	VIIx	VIIy	VIIIx	VIIIy	IXx	IXy	Xx	Xy
1	0.0027	0.0036	0.0019	0.0036	0.0042	0.0023	0.0050	0.0023	0.0039	0.0026	0.0025	0.0022	0.0001	-0.0005	0.0095	-0.0053
2	-0.0027	0.0007	-0.0013	-0.0003	0.0028	-0.0025	0.0032	-0.0002	-0.0013	0.0027	0.0018	-0.0023	0.0068	-0.0003	0.0052	-0.0039
3	-0.0042	-0.0023	-0.0069	-0.0036	0.0020	-0.0023	0.0018	0.0001	-0.0052	-0.0017	-0.0042	-0.0040	0.0025	-0.0036	0.0051	-0.0023
4	0.0034	-0.0019	0.0054	0.0003	0.0023	0.0005	0.0031	-0.0001	0.0001	-0.0010	0.0403	0.0087	0.0056	-0.0002	0.0011	-0.0037
5	-0.0023	-0.0005	-0.0007	-0.0007	-0.0031	-0.0016	-0.0007	-0.0005	-0.0018	0.0012	-0.0042	-0.0032	-0.0029	0	0.0031	-0.0008
6	0.0001	-0.0017	0.0007	-0.0002	0.0034	0.0012	0.0057	0.0001	-0.0009	-0.0004	0.0142	-0.0003	0.0042	-0.0006	0.0025	-0.0020
7	-0.0048	0.0017	-0.0054	0.0008	0.0001	0.0017	-0.0075	-0.0008	0.0112	0.0010	-0.0026	0.0019	-0.0041	-0.0006	-0.0078	-0.0019
8	-0.0028	-0.0003	-0.0005	-0.0004	0.0012	-0.0012	-0.0022	-0.0004	-0.0025	-0.0011	-0.0011	-0.0020	0.0016	0.0011	-0.0008	-0.0045
9	-0.0046	-0.0014	0.0009	0.0014	-0.0058	0.0030	-0.0066	0.0021	-0.0026	0.0001	-0.0029	0.0007	0.0071	0.0073	-0.0128	-0.0132
10	0.0032	-0.0018	0.0027	-0.0020	0.0036	-0.0001	0.0086	0.0004	-0.0094	-0.0012	0.0033	-0.0002	-0.0003	0.0037	0.0034	0.0001
11	0.0099	-0.0020	0.0023	0.0003	0.0042	-0.0009	0.0038	0.0008	0.0072	-0.0003	0.0028	0.0038	-0.0039	-0.0063	-0.0007	0.0025
12	-0.0087	-0.0014	-0.0032	-0.0003	0.0032	-0.0016	0.0007	-0.0013	-0.0009	0.0005	0.0080	0.0108	0.0051	0.0015	0.0087	0.0304
13	-0.0013	-0.0004	-0.0028	-0.0010	-0.0002	-0.0029	-0.0207	-0.0027	0.0382	-0.0033	-0.0032	-0.0001	-0.0008	-0.0009	-0.0202	-0.0127
14	0.0013	-0.0034	0.0021	-0.0023	-0.0075	-0.0022	-0.0099	-0.0013	0.0072	-0.0002	-0.0210	-0.0223	-0.0045	-0.0023	-0.0073	0.0110
15	-0.0006	-0.0022	0.0048	-0.0003	0.0176	0.0026	0.0155	0.0017	0.0025	-0.0030	-0.0049	-0.0034	0.0081	-0.0044	0.0123	-0.0164
16	0.0042	0.0033	0.0051	0.0030	0.0064	0.0012	0.0073	-0.0010	0.0058	0.0043	0.0107	0.0117	-0.0033	0.0025	0.0058	-0.0019
17	0.0011	0	-0.0072	-0.0006	-0.0029	0.0003	0.0030	-0.0006	-0.0016	-0.0008	-0.0016	-0.0003	-0.0211	-0.0061	-0.0058	-0.0060
18	0.0059	-0.0017	0.0001	-0.0027	-0.0014	-0.0003	-0.0103	-0.0026	0.0111	-0.0021	-0.0060	-0.0083	-0.0067	-0.0032	-0.0125	0.0072
19	-0.0086	0.0016	-0.0019	-0.0007	0.0142	0.0025	-0.0055	0.0004	-0.0036	0.0009	0.0130	0.0101	0.0200	0.0028	-0.0095	-0.0165
20	-0.0089	-0.0014	-0.0030	-0.0014	0.0070	-0.0008	0.0093	0	-0.0014	-0.0009	-0.0016	-0.0023	0.0123	-0.0018	0.0072	-0.0084
21	-0.0068	0.0011	0.0021	-0.0007	-0.0080	-0.0004	-0.0073	-0.0006	-0.0088	0.0012	-0.0036	-0.0050	-0.0070	-0.0018	-0.0123	-0.0076
22	0.0019	-0.0001	-0.0015	-0.0002	0.0121	-0.0010	0.0103	0.0024	-0.0009	-0.0013	0.0011	0.0019	0.0050	-0.0014	0.0162	0.0076
23	0.0250	0.0041	0.0087	0.0048	-0.0605	-0.0049	-0.0476	0.0009	0.0184	0.0031	0.0190	0.0093	-0.0519	-0.0022	-0.0366	0.0068
24	-0.0008	-0.0018	0.0065	0.0009	0.0185	0.0001	0.0124	0.0014	-0.0238	0.0008	0.0152	0.0036	0.0174	-0.0002	0.0127	-0.0030
25	0.0050	-0.0003	-0.0034	0.0019	-0.0148	0.0003	-0.0068	0.0002	0.0124	0.0002	0.0104	0.0080	-0.0123	0.0041	-0.0072	-0.0108
26	-0.0393	0.0025	-0.0060	0.0047	0.0151	0.0017	0.0137	-0.0017	-0.0167	0.0016	-0.0007	0.0052	0.0023	-0.0026	0.0123	-0.0067
27	-0.0049	0	0.0061	0.0014	0.0025	-0.0007	0.0031	0.0031	-0.0037	-0.0022	-0.0004	-0.0039	0.0085	0.0015	0.0023	-0.0119
28	-0.0137	-0.0020	0.0092	0.0015	-0.0338	0.0004	-0.076	-0.0042	-0.0102	0	0.0321	0.0177	-0.0558	-0.0079	-0.0686	0.0183
29	0.0217	0.0056	0.0050	0.0034	0.0160	-0.0021	-0.0345	-0.0018	0.0567	0.0058	0.0053	0.0023	0.0211	0.0011	-0.0394	0.0008
30	0.0015	0.0010	0.0057	0.0039	-0.0305	0.0017	-0.0295	0.0022	-0.0224	0.0013	-0.0215	-0.0159	-0.0394	-0.0042	-0.0311	0.0028
31	0.0196	0.0013	0.0101	0.0001	-0.0185	-0.0012	-0.0435	0.0015	0.0186	-0.0014	0.0113	-0.0008	-0.0210	-0.0033	-0.0470	-0.0072
32	-0.0722	-0.0031	-0.0194	-0.0008	0.0349	0.0008	0.1129	0.0035	-0.0672	-0.0018	0.0078	0.0176	0.0419	0.0069	0.1168	0.0166
33	-0.0273	0.0038	-0.0284	0.0031	0.0061	0.0016	0.0563	0.0013	-0.0140	0.0043	-0.0332	0	0.0059	0.0061	0.0550	0.0010
34	-0.0132	-0.0027	0.1055	0.0011	-0.0144	-0.0021	-0.1156	-0.0079	-0.0468	-0.0015	0.0989	-0.0014	-0.0111	-0.0008	-0.1163	0.0119
35	-0.0008	0.0031	0.0623	0.0063	-0.1538	-0.0104	0.0407	-0.0005	0.0938	0.0020	0.0456	-0.0051	-0.1574	-0.0120	0.0449	0.0068

Table 2. Character matrix of 35 partial warp scores recoded as discrete characters for phylogenetic analysis

Characters and their states based on X partial warp scores

	1	2	3	4	5	6	7	8	9	10	11	12	13	14	15	16	17	18	19	20	21	22	23	24	25	26	27	28	29	30	31	32	33	34	35
FISH III	A	B	A	A	B	A	B	B	C	A	C	B	A	A	C	A	C	C	A	A	A	A	A	D	D	E	A	A	A	A	A	A	A	A	A
FISH IV	B	A	B	A	A	A	B	C	D	A	B	C	A	A	D	A	A	D	D	B	B	A	C	E	E	C	C	C	B	A	B	D	A	B	B
FISH V	C	C	C	A	C	B	C	A	A	A	A	D	A	B	A	A	D	D	B	C	A	B	B	A	A	A	B	D	A	B	C	B	B	A	C
FISH VI	D	C	C	A	A	C	A	B	A	B	A	E	B	B	A	A	E	A	C	D	A	B	B	B	B	A	B	B	C	B	D	C	C	C	D
FISH VII	E	A	A	B	B	A	D	B	B	C	D	F	C	C	E	A	B	C	C	B	A	A	A	F	C	D	A	A	D	C	A	A	D	D	E
FISH VIII	A	D	A	C	D	D	E	D	B	A	B	A	A	D	F	B	B	B	B	B	C	A	A	C	C	B	D	E	B	C	B	E	A	B	D
FISH IX	F	E	C	A	C	B	F	A	E	D	E	G	A	B	B	C	F	B	E	E	A	C	B	A	A	B	E	F	A	D	C	B	B	A	C
FISH X	G	F	D	A	E	E	A	E	F	A	F	A	B	B	B	A	A	A	A	C	D	D	B	B	B	A	B	B	C	B	D	C	C	C	D
ANCESTOR	F	G	E	B	F	A	C	F	D	D	F	H	A	E	C	D	G	D	E	F	E	A	D	D	F	B	D	G	B	A	E	E	B	A	A

Characters and their states based on Y partial warp scores

	1	2	3	4	5	6	7	8	9	10	11	12	13	14	15	16	17	18	19	20	21	22	23	24	25	26	27	28	29	30	31	32	33	34	35
FISH III	A	B	B	B	A	C	A	A	C	B	B	A	A	A	B	B	A	B	C	B	A	A	A	C	A	C	A	C	A	A	B	C	A	B	C
FISH IV	A	A	C	A	A	B	B	A	D	B	A	A	A	A	A	B	A	B	B	B	B	A	A	B	B	B	A	D	B	C	C	A	A	A	B
FISH V	B	C	B	A	C	D	A	B	B	A	C	A	B	A	C	A	A	A	A	C	C	B	B	A	A	A	A	A	D	B	A	A	B	A	D
FISH VI	B	A	A	A	A	A	C	A	B	C	D	A	B	A	C	A	A	B	B	A	B	C	C	D	A	D	B	E	D	B	B	D	B	C	A
FISH VII	B	D	D	C	D	B	B	B	A	D	A	A	B	A	B	B	A	B	B	C	A	B	D	B	A	A	C	A	A	A	A	E	A	A	C
FISH VIII	B	C	C	D	E	B	A	C	A	A	E	B	A	B	B	C	A	C	D	D	D	C	E	E	C	B	D	B	B	D	D	B	C	A	E
FISH IX	C	A	C	A	B	E	C	D	E	E	F	A	A	A	D	B	B	B	A	E	E	B	F	A	D	E	A	F	C	E	E	F	D	A	D
FISH X	D	E	B	E	A	C	D	E	F	A	G	C	C	C	E	D	B	D	E	F	F	D	G	F	E	F	E	B	C	B	F	B	B	D	B
ANCESTOR	C	A	A	A	B	A	E	A	A	A	A	A	A	A	A	A	A	A	B	A	G	A	H	A	A	G	A	A	E	F	C	A	C	A	A

Because each taxon was represented by a single specimen it was not possible to examine the variation among and between taxa by means of scatter plots, as suggested by Zelditch et al. (1995). Thus, I have assumed that each fish represents an average form and that variation in partial warp scores for each taxon is evenly and identically distributed about this average. When two or more taxa had similar scores for a particular partial warp (that is, within 0.5 standard errors of each other based on the assumed distribution), they were scored as having the same character state for that partial warp. Examples of this type of recoding are shown in Fig. 3.

The character matrix that resulted from the recoding of partial warp scores as character states (Table 2) was subjected to phylogenetic analysis by means of PAUP version 3.1.1 (Swofford, 1990). The *x* and *y* components of each partial warp were scored separately. The partial warps inferred to change were plotted on the most parsimonious trees that resulted from the analysis.

RESULTS AND DISCUSSION

Two equally most parsimonious trees (MPTs) resulted from the cladistic analysis (Fig. 4). One of these was identical to the true tree, whereas the other differed from the true tree in the placement of one branch. The retention index (Farris, 1989) for the two inferred MPTs was 0.48, reflecting considerable inferred homoplasy.

One of the key advantages of cladistic parsimony over other methods of phylogenetic analysis is that it allows for the inferential reconstruction of ancestral character states at

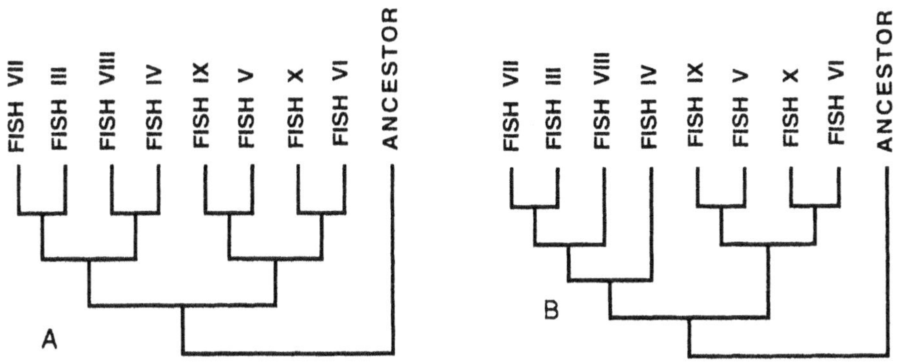

Figure 4. The two equally most parsimonious trees that resulted from a cladistic analysis of the partial warp characters shown in Table 2.

internal nodes (Donoghue, 1989; Maddison and Maddison, 1992). In an effort to evaluate this aspect of cladistic analysis for the partial warp character data, using the computer program MacClade (Maddison and Maddison, 1992), I contrasted the shape changes implied by parsimony (Fig. 5) with those that had actually occurred on the true tree.

Little interpretable correspondence was seen when individual partial warps were contrasted in this way. It is important to realize that such a correspondence should not necessarily be expected. Each partial warp represents only an aspect of the overall shape difference. The different aspects cannot always be readily equated with observed shape differences among forms. This is best appreciated with an example. The plot for partial warp

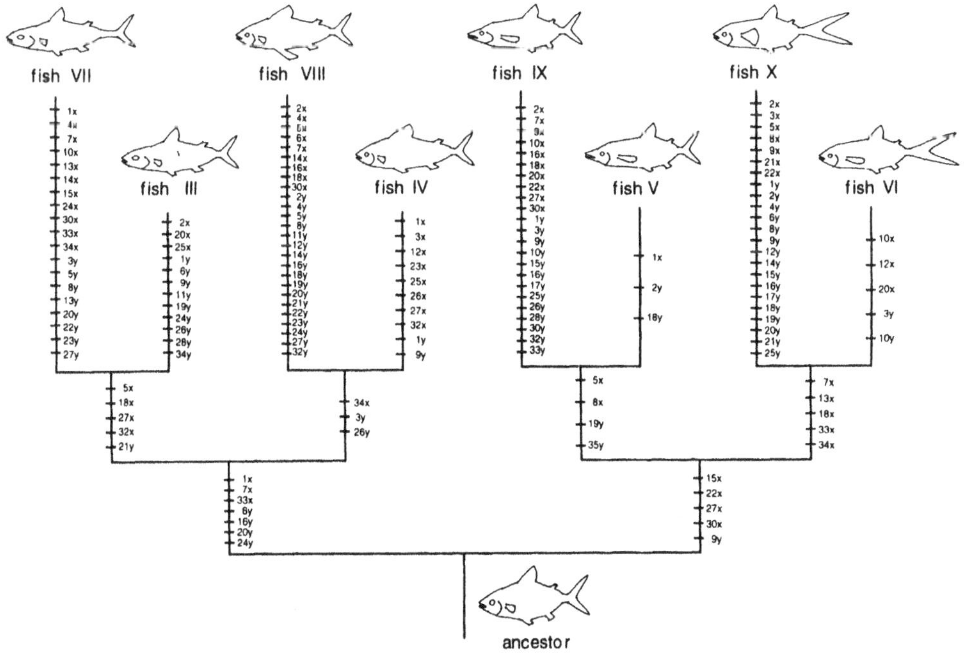

Figure 5. Partial warps that were inferred to have changed at different internodes of the correct most parsimonious tree.

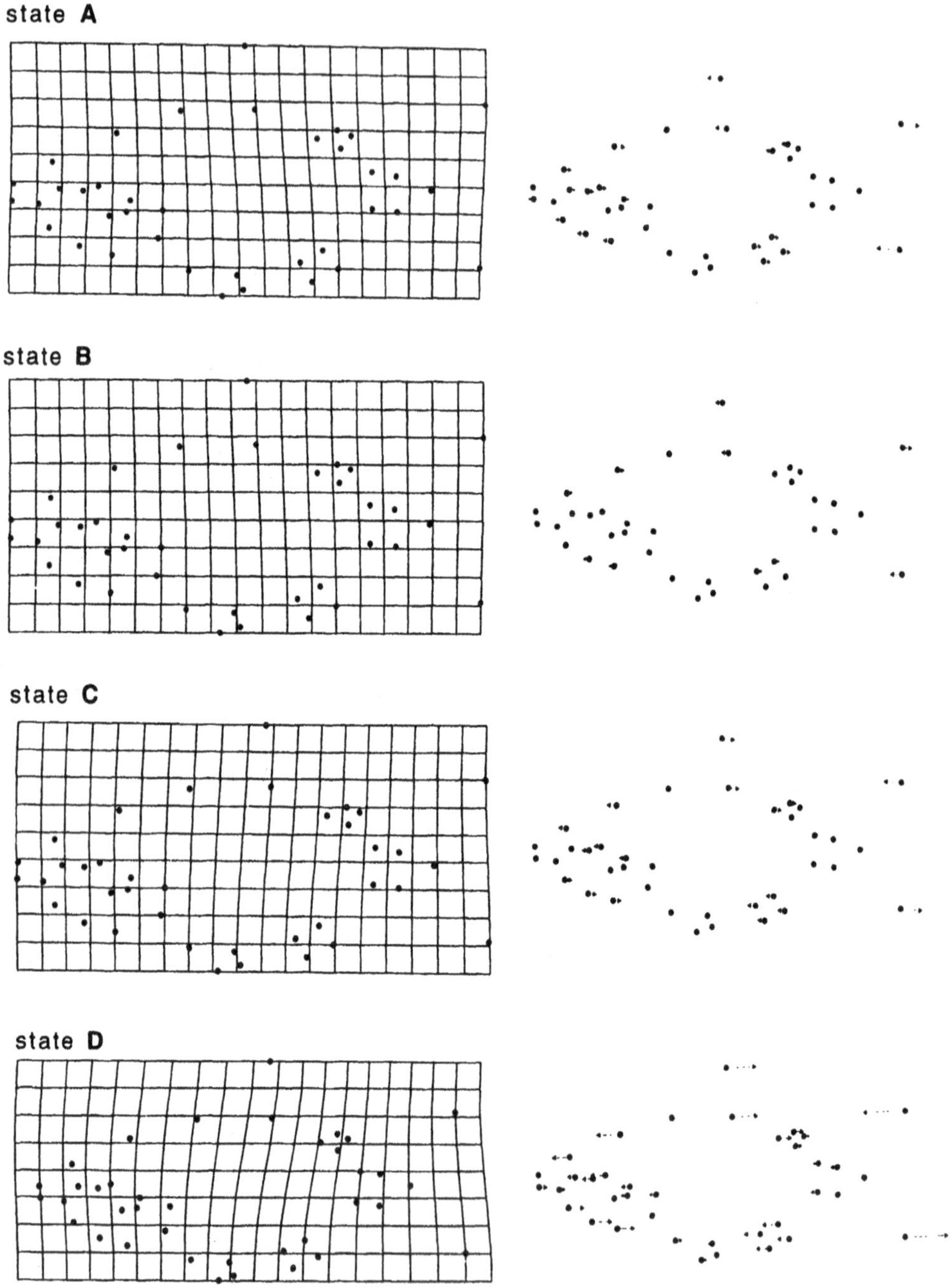

Figure 6. Graphical representation of the four different states for partial warp 31. Each of the four character states is represented by a deformation grid on the left and as (exaggerated) landmark displacements relative to the ancestral form on the right. Note there is little or no discernible correspondence between these representations and the true shape differences shown among the four pairs of taxa in Fig. 2.

31 (Fig. 3) shows a distribution of scores that clearly defines four pairs of taxa (III, VII), (IV, VIII), (V, IX) and (VI, X). Figure 2 shows that each of these pairs are indeed closely related taxa with similar body forms. However, when the shape differences corresponding to the four different states for partial warp 31 are plotted graphically (Fig. 6), there is little or no discernible correspondence between grid deformation or landmark displacements and the true shape differences among the four pairs of taxa. The reason for the disparity between shape differences implied by the partial warp scores and shape differences that are visibly apparent is that the different individual partial warps are no more than mathematical decompositions that describe shape differences at different spatial scales, akin to the different harmonics of a Fourier analysis (F. J. Rohlf, personal communication). It should be clear from this that partial warp scores are not at all comparable to the biological characters that might be used in more traditional phylogenetic analysis.

The perceived correspondence between the shape changes implied by partial warp scores and the shape changes that truly occurred on the tree improved when the partial warps inferred to change at each internode were pooled using the Vector Specter program of Humphries (1994). The reason for this is simply that as more aspects of the original shape differences were reconstituted, so the overall shape differences became more completely described and, therefore, more recognizable.

Although pooling of partial warp scores improved the interpretability, it also revealed a tendency of the method to clump together shape changes that had actually occurred at successive internodes. This is seen in a contrast between the ancestor and fishVI (Fig. 7). In the true tree, the evolutionary shape changes between these forms occurred in two steps at successive internodes: the first step involved a local stretching of the region immediately behind the gill region and the second step involved a lengthening of the tail (see Fig. 2). The

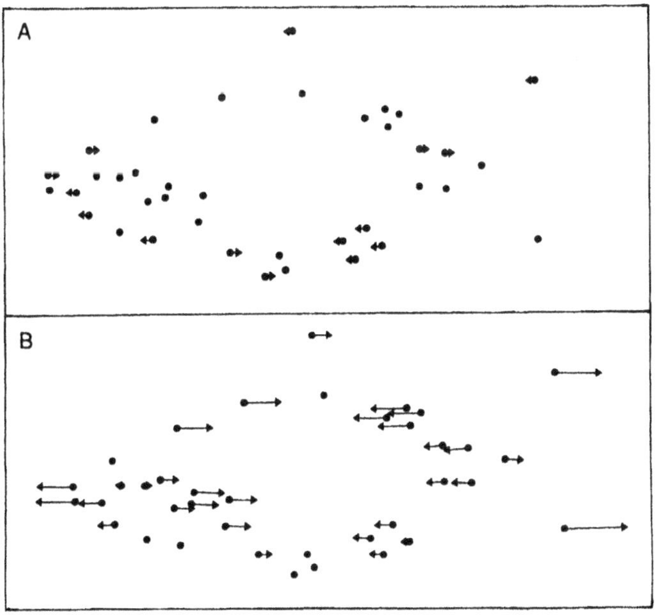

Figure 7. Inferred shape change implied by partial warps for the contrast between the ancestor and fishVI. (A) The implied shape changes inferred for the lower internode (partial warps 15*x*, 22*x*, 27*x*, 30*x*, and 9*y*), which indicate a slight compression of the anterior nasal region. (B) The implied shape changes inferred for the upper internode (partial warps 7*x*, 13*x*, 18*x*, 33*x*, and 34*x*), which indicate an expansion of the region behind the gill and a lengthening of the tail.

inferred shape changes, by contrast, implied a minor compression of the anterior nasal region in the basal internode (partial warps 15*x*, 22*x*, 27*x*, 30*x*, and 9*y*; Fig. 5) and an expansion of the region behind the gill together with a lengthening of the tail at the second internode (partial warps 7*x*, 13*x*, 18*x*, 33*x* and 34*x*). The character state reconstructions thus correctly identified both of the shape changes that occurred (expansion of the region behind the gill and the lengthening of the tail) but incorrectly implied that both changes occurred at the second internode.

On a more optimistic note, the fact that one of the two MPTS was identical to the true tree is promising. It seems that at least some of the shape change implemented in the true tree was retrieved using the method of Zelditch et al. (1995). However an UPGMA phenogram of all of the original partial warp scores also yielded the true tree, suggesting that the signal in this particular data set is such that most methods would likely retrieve the correct tree.

More troubling is the observation that so much homoplasy was inferred to occur (retention index = 0.48). This is particularly curious given that the original shape changes were arranged in such a way as to predispose the collected shape data to cladistic interpretation. Indeed, the original shape changes were distributed with no homoplasy whatsoever. Clearly there is an aspect of the protocol, at least as implemented herein, that leads to the inference that there is more phylogenetic noise in the data than one would expect.

I have been able to identify seven potential causes for this inferred homoplasy and have suggested ways in which they might be ameliorated, where possible. It is likely that more influences are involved than those listed.

1. It is important that there be an even distribution of landmarks over the entire form. Landmarks that are close together could conceivably have undue influence on the bending energy matrix. For example, it is possible that biologically unimportant perturbations between closely spaced landmarks (such as digitizing error) might influence inferred homoplasy.
2. The interpolant nature of the thin-plate spline is such that localized shape changes can often exert an analytical influence over a wider area of the form than is biologically appropriate if insufficient landmarks are available to "contain" the localized change within bounds. (This shortcoming is related to Bookstein's criticism that homology is essentially a nonmetric concept that cannot be approximated—or, in this case, interpolated—with morphometric methods.) It may be possible to minimize this effect by having a denser sampling of landmarks or by employing edgels (Bookstein and Green, 1993).
3. The recoding of the original partial warp scores as discrete character states likely results in the loss of information about shape differences.
4. Decomposing the partial warp scores into their *x* and *y* components may have contributed toward the inferred homoplasy because character states that appear to be shared (potentially synapomorphic) on one axis could conceivably appear distinct and unique (autapomorphic) if both *x*- and *y*-axes were considered together. However, preliminary tests carried out on *x*- and *y*- axes together did not improve the high levels of inferred homoplasy but yielded many more most parsimonious trees.
5. Cladistic parsimony makes use of Manhattan space which is not invariant under rotation. The partial warp methods are based on a rotation of the data. Thus there is an intrinsic degree of incommensurability between the data (the partial warp scores) and the inference model used to transform these data into a hypothesis of evolutionary relationships. Presumably this problem could be resolved by using

a phylogenetic inference method that is commensurate with the original shape space.

6. The use of shape coordinates to align specimens in order to render their shapes comparable may have contributed to inferred homoplasy. Baseline choice is known to exert an influence on partial warp scores when shapes vary widely among forms. Generalized Procrustes procedures are currently preferred for specimen alignment because they are more robust where shape variation among forms varies considerably. (F. J. Rohlf., personal communication).
7. The landmark-based methods have their foundations in Kendall's shape space. As pointed out previously by Slice ("Three dimensional" article, this volume), this space is curved, noneuclidean and not amenable to the same intuitions that characterize traditional morphometrics. Most problems of biological interest range over a small region of this shape space, and their examination can be adequately conducted in a linear tangent space that touches the shape space at the point of the mean form (Bookstein 1991). However, as contrasts are made over an increasingly wide range of shapes, the adequacy of the tangent space to approximate Kendall's shape space is compromised, and the statistical, numerical and graphical summaries that result can no longer be regarded as equivalent to those contrasts based closer to the point where tangent space touches shape space. The use of an ancestor or outgroup (rather than a consensus or mean form) as the reference specimen expands the extent of shape space collectively embraced by the various shape contrasts. Although seemingly appropriate from a cladistic standpoint, this approach may necessitate using regions of the tangent space that are less meaningfully interpreted biologically. (This shortcoming is related to Bookstein's criticism that the biological concept of homology spans a range of shape variation that exceeds the effectiveness of morphometric methods to capture it [Bookstein, 1994]).

It could be argued that because one of the two MPTs yielded the correct tree, the method is resilient to the above listed problems. Although this might be the case, such optimism should be tempered by the knowledge that real data sets will likely have both more homoplasy and more interdependence among shape characters than did the example used herein.

SUGGESTIONS FOR FURTHER WORK

There are many ways in which this study could be usefully furthered. First, the example presented herein needs to be replicated for a variety of differently shaped organisms, with different landmark distributions, with different evolutionary tree topologies and with different numbers of taxa, in order to assess the generality of the presented findings. In the current study, shape changes were brought about simply by stretching or expanding by hand different regions of the forms. It is possible that these various shape changes contain embedded, although inadvertent, correlations (F. J. Rohlf, personal communication). An exploration of a more diverse range of shapes, landmark configurations and tree topologies would go some way toward diminishing such data-set-specific influences. Second, the performance of the method will need to be evaluated in the face of real-life situations that include homoplasy. (A reasonable approach might be to test the method on a group of real organisms whose phylogeny is well established on the basis of independent data). Finally, it will be important to evaluate the extent to which other phylogenetic methods are effective at retrieving phylogenetic signal from partial warp data.

ACKNOWLEDGMENTS

Fink et. al.'s (1994) protocol as presented herein was taken from a relatively informal "MORPHMET" communication, not a published work. Any errors or oversights intrinsic to the protocol should therefore not be attributed to Fink et al. unless the same errors or oversights are formally published. I would like to thank Miriam Zelditch for her enthusiasm and willingness to explain the intricacies of the protocol, Jim Rohlf for patiently and constructively pointing out flaws in this study, Julian Humphries for promptly customizing his Vector Specter program to my capricious requirements (on several occasions), John Lynch for editorial improvement of the manuscript, and finally Fred Bookstein for being the inadvertent impetus for this contribution.

REFERENCES

Bookstein, F. L. 1991. Morphometric tools for landmark data: Geometry and biology. Cambridge University Press: Cambridge.

Bookstein, F. L. 1994. Can biometrical shape be a homologous character? *In*: B.K. Hall (ed.), Homology: the hierarchical basis of comparative biology. Academic Press: London.

Bookstein, F. L. and Green W. D. K.1993. A feature space for edgels in images with landmarks. Journal of Mathematical Imaging and Vision. 3:231–261.

Donoghue, M. J. 1989. Phylogenies and the analysis of evolutionary sequences, with examples from seed plants. Evolution 43:1137–1156

Humphries, J. M.1994. Jspline and Vector. Thin-plate spline programs, version 1.3. Section of Ecology and Systematics, Corson Hall, Cornell University, Ithaca, New York 14840.

Farris, J. S. 1983. The logical basis of phylogenetic analysis. *In* N. I. Platnick and V. A. Funk, (eds.), Advances in cladistics. Volume II. Columbia University Press: New York.

Farris, J. S. 1989. The retention index and the rescaled consistency index. Cladistics 5:417–419.

Fink W. L., M. L. Zelditch, and D. L. Swiderski. 1994. Archive file LOG9404 MORPHMET, at CUNYVM.CUNY.EDU, April 21, 1994. [See Appendix III for details]

Fink W. L. and M. L. Zelditch. 1994. Phylogenetic analysis of ontogenetic shape change in the piranha genus *Pygocentrus*. American Society of Ichthyologists and Herpetologists, Annual Meeting Abstracts..

Maddison, W. P. and D. R. Maddison. 1992. MacClade: Analysis of phylogeny and character evolution, version 3.0. Sinauer: Sunderland, Massachusetts.

Rohlf, F. J. 1994. TPSPLINE computer program. Department of Ecology and Evolution, State University of New York, Stony Brook New York 11794.

Smith, G. R. 1990. Homology in morphometrics and phylogenetics. *In* F. J. Rohlf and F. L. Bookstein, (eds.), Proceedings of the Michigan morphometrics workshop. University of Michigan Museum of Zoology Special Publication 2.

Swofford, D. L.1990. PAUP computer program, version 3.11. Laboratory of molecular systematics, MRC 534, MSC, Smithsonian Institution, Washington, D. C. 20560.

Zelditch M. L. and W. L. Fink. 1994. Heterochrony and heterotropy in the evolution of piranha body form. American Society of Ichthyologists and Herpetologists, Annual Meeting Abstracts.

Zelditch M. L. and W. L. Fink. 1995. Allometry and developmental integration of body growth in a piranha. *Pygocentrus nattereri* (Teleostei: Ostariophysi). Journal of Morphology 223:341–355.

Zelditch, M. L., W. L. Fink, and D. L. Swiderski. 1995. Morphometrics, homology and phylogenetics: Quantified characters as synapomorphies. Systematic Biology 44(2):179–189.

APPENDIX I

A Glossary for Geometric Morphometrics

Dennis E. Slice,*[1] Fred L. Bookstein,†[2] Leslie F. Marcus,‡[3] and F. James Rohlf****[1]

[1] Department of Ecology and Evolution
State University of New York at Stony Brook
Stony Brook, NY 11794-5245
[2] Institute of Gerontology
University of Michigan
Ann Arbor, Michigan 48109-2007
[3] Department of Biology
Queens College of the City University of New York
Flushing, New York 11367

INTRODUCTION

This glossary provides definitions for terms, concepts, and methods frequently encountered in morphometric literature and discussions. It includes entries for technical terms with more-or-less special meaning in shape analysis and biological morphometrics (e.g., preshape, warps, anisotropy) and some of the casual jargon that may be completely foreign to newcomers to the field (e.g., books of various color – Red, Blue, Orange, and Black). Many definitions provide the general idea behind each entry instead of a technically or mathematically rigorous treatment. As such, they are intended to give readers an intuitive understanding of a particular entry that will allow them to follow the main ideas in the literature without becoming unduly distracted, at first, with technical details. Unless otherwise indicated, the following general notation has been used: *n*—number of specimens; *p*–number of points/landmarks; *k*–number of dimensions; and $\mathbf{A}^t$ will refer to the transpose of the matrix **A**. Members of the morphometrics community, especially the subscribers to the MORPHMET electronic mailing list, have helped greatly in the selection of terms to be included in the glossary.

* e-mail: Dennis.Slice@sunysb.edu

† e-mail: fred@brainmap.med.umich.edu

‡ email: lamqc@cunyvm.cuny.edu; marcus@amnh.org

**** e-mail: Jim.Rohlf@sunysb.edu

The latest revision of this glossary is available from the Stony Brook morphometric archives (see Appendix III).

Glossary

α – In a relative warps analysis, α is the exponent used to rescale partial warps before computing their principal components, the relative warps (see Rohlf's chapter in the Black Book). Scale-invariant multivariate analyses using rescaled principal warp scores, such as canonical variates analysis, are not affected by the choice of α (see Rohlf, this volume).

$\otimes$ – The Kronecker tensor product or direct product. The Kronecker tensor product of matrices **X** and **Y**, written as $\mathbf{X} \otimes \mathbf{Y}$, results in a large matrix formed by taking all possible products of the elements of **X** and those of **Y**. For example, if **X** and **Y** are 2×2 then $\mathbf{X} \otimes \mathbf{Y}$ results in a 4×4 matrix:

$$\mathbf{X} \otimes \mathbf{Y} = \begin{bmatrix} x_{11} & x_{12} \\ x_{21} & x_{22} \end{bmatrix} \otimes \begin{bmatrix} y_{11} & y_{12} \\ y_{21} & y_{22} \end{bmatrix}$$
$$= \left[\begin{array}{c|c} x_{11}\mathbf{Y} & x_{12}\mathbf{Y} \\ \hline x_{21}\mathbf{Y} & x_{22}\mathbf{Y} \end{array}\right]$$
$$= \left[\begin{array}{cc|cc} x_{11}y_{11} & x_{11}y_{12} & x_{12}y_{11} & x_{12}y_{12} \\ x_{11}y_{21} & x_{11}y_{22} & x_{12}y_{21} & x_{12}y_{22} \\ \hline x_{21}y_{11} & x_{21}y_{12} & x_{22}y_{11} & x_{22}y_{12} \\ x_{21}y_{21} & x_{21}y_{22} & x_{22}y_{21} & x_{22}y_{22} \end{array}\right].$$

accuracy – The closeness of a measurement or estimate to its true value. See precision.

affine superimposition – A superimposition for which the associated transformations are all affine. See affine transformation.

affine transformation - A transformation for which parallel lines remain parallel. Affine transformations of the plane take squares into parallelograms and take circles into ellipses of the same shape and orientation. Affine transformations of a three-dimensional space take cubes into parallelepipeds (sheared bricks) and spheres into ellipsoids all of the same shape and orientation. Similar results are produced in higher dimensional spaces. Equivalent to "uniform transformation" or "shear."

allometry – Any change of shape with size. It describes any deviation of the bivariate relation from the simple functional form $y/x = c$, where c is a constant and x and y are size measures in units of the same dimension. See Klingenberg, this volume.

anisotropy – An Anisotropy is a descriptor of one aspect of an affine transformation. In two dimensions, it is the ratio of the axes of the ellipse into which a circle is transformed by an affine transformation. In general, it is the maximum ratio of extension of length in one direction to extension in a perpendicular direction.

asymptotically unbiased estimator – An estimator, $\hat{\theta}$, with an expected value that converges in probability on the parametric value it is estimating, θ, as sample size goes to infinity: $E(\hat{\theta}_n) \rightarrow \theta$ as $n \rightarrow \infty$. See unbiased estimator and consistent estimator.

baseline – For a system of two-point shape coordinates (see below) for landmarks in a plane, the baseline is the line connecting the pair of landmarks that are assigned to fixed locations (0,0) and (1,0) in the construction. In general, baselines work better if they are closely aligned with the long axis of the mean landmark shape and pass near the centroid of that mean shape (see the Orange Book).

bending energy – Bending energy is a metaphor borrowed for use in morphometrics from the mechanics of thin metal plates. Imagine a configuration of landmarks that has been printed on an infinite, infinitely thin, flat metal plate, and suppose that the differences in coordinates of these same landmarks in another picture are taken as vertical displacements of this plate perpendicular to itself, one Cartesian coordinate at a time. The bending energy of one of these out-of-plane "shape changes" is the (idealized) energy that would be required to bend the metal plate so that the landmarks were lifted or lowered appropriately.

Whereas in physics bending energy is a real quantity, measured in appropriate units (g cm^2 sec^{-2}), there is an alternate formula that remains meaningful in morphometrics. Bending energy is proportional to the integral of the summed squared second derivatives of the "vertical" displacement—the extent to which the vertical displacement varies from a uniform tilt. The bending energy of a shape change is the sum of the bending energies that apply to any two perpendicular coordinates in which the metaphor is evaluated. The bending energy of an affine transformation is zero because it corresponds to a tilting of the plate without any bending. The value obtained for the bending energy corresponding to a given displacement is inversely proportional to scale. Such quantities should not be interpreted as measures of dissimilarity (e.g., taxonomic or evolutionary distance) between two forms.

bending energy matrix – The formula for bending energy (see above)—the formula whose value is proportional to that integral of those summed squared second derivatives—is a quadratic form (usually written $\mathbf{L}_k^{-1}$) determined by the coordinates of the landmarks of the reference form. That is, if **h** is a vector describing the heights of a plate above a set of landmarks, then bending energy is $\mathbf{h}^t \mathbf{L}_{k_{-1}} \mathbf{h}$. In morphometrics, the bending energy of a general transformation is the sum $\mathbf{x}^t \mathbf{L}_k^{-1} \mathbf{x} + \mathbf{y}^t \mathbf{L}_k^{-1} \mathbf{y}$ of the bending energy of its horizontal *x*-component, modeled as a "vertical" plate, plus the bending energy of its vertical *y*-component, modeled similarly as a "vertical" plate. See **L**.

biplot – A single diagram that represents two separate scatterplots on the same pair of axes. One scatter is of some pair of columns of the matrix **U** of the singular value decomposition of a matrix **S**, and the other scatter is of the matching pair of columns of **V**. When **S** is a centered data matrix, the effect is to plot principal component loadings and scores on the same diagram. See Marcus (Black Book) for an in-depth discussion.

Black Book – Marcus, Bello, García-Valdecasas, 1993. See also Blue Book, Orange Book, Red Book, and Reyment's Black Book.

Blue Book – Rohlf and Bookstein, 1990. See also Black Book, Orange Book, Red Book, and Reyment's Black Book.

Bookstein coordinates – See two-point shape coordinates.

canonical – A canonical description of any statistical situation is a description in terms of extracted vectors that have especially simple ordered relationships. For instance, a canonical correlations analysis describes the relation between two lists of variables in terms of two lists of linear combinations that show a remarkable pattern of zero correlations. Each score (linear combination) from either list is correlated with no other combination from its list and with only one score from the other list.

canonical correlation analysis – A multivariate method for assessing the associations between two sets of variables within a data set. The analysis focuses on pairs of linear combinations of variables (one for each set) ordered by the magnitude of their correlations with each other. The first such pair is determined so as to have the maximal correlation of any such linear combinations. Subsequent pairs have maximal correlation subject to the constraint of being orthogonal to those previously determined.

canonical variates analysis – A method of multivariate analysis in which the variation among groups is expressed relative to the pooled within-group covariance matrix. Canonical variates analysis finds linear transformations of the data that maximize the among-group variation relative to the pooled within-group variation. The canonical variates then may be displayed as an ordination to show the group centroids and scatter within groups. Canonical variates analysis may be thought of as a "data reduction" method in the sense that one wants to describe among-group differences in few dimensions. The canonical variates are uncorrelated, however the vectors of coefficients are not orthogonal as in principal component analysis. The method is closely related to multivariate analysis of variance (MANOVA), multiple discriminant analysis, and canonical correlation analysis. A critical assumption is that the within-group variance–covariance structure is similar, otherwise the pooling of the data over groups is not very sensible.

centroid size – Centroid size is the square root of the sum of squared distances of a set of landmarks from their centroid, or, equivalently, the square root of the sum of the variances of the landmarks about that centroid in x- and y-directions. Centroid size is used in geometric morphometrics because it is approximately uncorrelated with every shape variable when landmarks are distributed around mean positions by independent noise of the same small variance at every landmark and in every direction. Centroid size is the size measure used to scale a configuration of landmarks so they can be plotted as a point in Kendall's shape space. The denominator of the formula for the Procrustes distance between two sets of landmark configurations is the product of their centroid sizes.

cluster analysis – A method of analysis that represents multivariate variation in data as a series of sets. In biology, the sets are often constructed in a hierarchical manner and shown in the form of a treelike diagram called a dendrogram.

coefficient – A coefficient, in general, is a number multiplying a function. In multivariate data analysis, usually the "function" is a variable measured over the cases of the analysis, and the coefficients multiply these variable values before we add them up to form a score. A coefficient is not the same as a loading.

complex numbers – Complex numbers are an algebraic way of coding points in the ordinary Euclidean plane so that translation (shift of position) corresponds to the addition of complex numbers and both rescaling (enlargement or shrinking) and rotation correspond to multiplication of complex numbers. In this system of notation, invented by Gauss, the x-axis is identified with the "real numbers" (ordinary decimal numbers) and the y-axis is identified with "imaginary numbers" (the square roots of negative numbers). When you multiply points on this latter axis by themselves according to the rules, you get negative points on the "real" axis just defined. Many operations on data in two dimensions can be proved valid more directly if they are written out as operations on complex numbers.

consensus configuration – A single set of landmarks intended to represent the central tendency of an observed sample for the production of superimpositions, a weight matrix, or some other morphometric purpose. Often a consensus configuration is computed to optimize some measure of fit to the full sample: in particular, the Procrustes mean shape is computed to minimize the sum of squared Procrustes distances from the consensus landmarks to those of the sample.

consistent estimator – An estimator, $\hat{\theta}_n$, that converges in probability on the parametric value it is estimating, θ, as sample size goes to infinity:

$$\lim_{n \to \infty} P\left(|\theta_n - \theta| < \varepsilon \right) = 1$$

for any positive ε. Asymptotically unbiased estimators are consistent estimators if their variance goes to zero as sample size goes to infinity. See unbiased estimator.

coordinates – A set of parameters that locate a point in some geometrical space. Cartesian coordinates, for instance, locate a point on a plane or in physical space by projection onto perpendicular lines through one single point, the origin. The elements of any vector may be thought of as coordinates in a geometric sense.

correlation – Relation between two or more variables. Frequently the word is used for Pearson's product-moment correlation, which is the covariance divided by the product of the standard deviations,

$$r_{XY} = \frac{s_{XY}}{s_X s_Y}.$$

This correlation coefficient is ± 1 when all values fall on a straight line, not parallel to either axis. However, there are also for example Kendall, Spearman, and tetrachoric correlations that measure other aspects of the relation between two variables.

covariance – A measure of the degree to which two variables vary together. Computed as

$$s_{XY} = \frac{1}{n} \sum_{i=1}^{n} \left(X_i - \bar{X} \right)\left(Y_i - \bar{Y} \right)$$

for two variables X and Y in a sample of size n. See correlation.

covariant – A covariant of a particular shape change is a shape variable whose gradient vector as a function of changes in any complete set of shape coordinates lies precisely along the change in question. For transformations of triangles, the relation between invariants and covariants is a rotation by 90 degrees in the shape-coordinate plane. For more than three landmarks, a given transformation has only one direction of covariants; but has a full plane (four landmarks) or hyperplane (five or more landmarks) of invariants (see the Orange Book). See invariant.

curved space – A space with coordinates and a distance function such that the area of circles, volume of spheres, etc., are not proportional to the appropriate power of the radius, e.g., Kendall's shape space. In curved spaces, the usual intuitions about what "straight lines" can be expected to do will be faulty. For instance, corresponding to every triangular shape in Kendall's shape space, there is another that is "as far from it as possible," just as there is a point on the surface of the earth as far as possible from where you now sit.

D – See (1) generalized distance or (2) fractal dimension.

***D*²** – Squared Mahalanobis, or generalized, distance.

deficient coordinate – In addition to landmark locations, a digitizer can be used to supply information of other sorts. For example, a point can be used to encode part of the information about a curving arc by identifying the spot at which the arc lies farthest from some other image structure (perhaps another such curving arc). The null model of independent Gaussian noise does not apply to position along the tangent direction of the curve that is digitized in this way, and so that Cartesian coordinate is "deficient." The usual model of independent Gaussian noise is inapplicable in principle for such points. See Type III landmark.

degrees of freedom – Given a set of parameters estimated from the data, the "degrees of freedom" of some statistic is the number of independent observations required to compute the statistic. For example, the variance has $n-1$ degrees of freedom because only $n-1$ of the observations are needed for its computation given the sample mean. The missing observation can be computed as

$$X_n = n\overline{Y} - \sum_{i=1}^{n-1} X_i \quad .$$

dilation – Increase of length in a particular direction or along a particular interlandmark segment.

discriminant analysis – A broad class of methods concerned with the development of rules for assigning unclassified objects or specimens to previously defined groups. See discriminant function.

discriminant function – A discriminant function is used to assign an observation to one of a set of groups. Linear discriminant functions take a vector of observations from a specimen and multiply it by a vector of coefficients to produce a score that can be used to classify the specimen as belonging to one or another predefined group. See discriminant analysis.

distance – This term has several meanings in morphometrics; it should never be used without a prefixed adjective to qualify it, e.g., Euclidean distance, Mahalanobis distance, Procrustes distance, or taxonomic distance.

edgel – An extension of the notion of landmark to include partial information about a curve through the landmark. An edgel specifies rotation of a direction through a landmark, extension along a direction through a landmark, or both. The formula for thin-plate splines on landmarks can be extended to encompass data about edgels as well. They are intended eventually to circumvent any need for deficient coordinates in multivariate morphometric analysis. See Little (this volume) and Bookstein and Green (1993).

EDMA – See Euclidean distance matrix analysis.

eigenshapes – Principal components for outline data. An eigenshape analysis begins with the selection of a distance function between pairs of outlines. At the end one gets "eigenshapes," which have the properties of principal component vectors (uncorrelated, describing the sample in decreasing order of variance) and also are outline shapes themselves, so that the scores for each specimen of the sample can be combined to produce a new outline shape that approximates it in some possibly useful way. Eigenshapes apply to curves as relative warps apply to landmark shape. See the chapter by Lohmann and Schweitzer in the Blue Book and that by Sampson, this volume.

eigenvalues – Eigenvalues, λ_i, are the diagonal elements of the diagonal matrix Λ in the equation: $\mathbf{SE} = \mathbf{E}\Lambda$. In the common data analysis case, $\mathbf{S}$ is a symmetrical variance–covariance matrix, $\mathbf{E}$ is a matrix of eigenvectors, $\lambda_i \geq 0$, and

$$\sum \lambda_i = \sum s_i^2.$$

The order of the columns of $\mathbf{E}$ and Λ is arbitrary, but by convention they are usually sorted from largest to smallest eigenvalue. See eigenvectors and singular value decomposition.

eigenvectors – In the equation given to define eigenvalues, $\mathbf{E}$ contains the eigenvectors. In the common data analysis case, $\mathbf{E}$ is an orthonormal matrix (i.e., $\mathbf{E}^t\mathbf{E} = \mathbf{I}$ and $\mathbf{E}\mathbf{E}^t = \mathbf{I}$). When sorted by descending eigenvalues, the first eigenvector is that linear combination of variables that has the greatest variance. The second eigenvector is the linear combination of variables that has the greatest variance of such combinations orthogonal to the first, and so on. See eigenvalues and singular value decomposition.

elliptic Fourier analysis – A type of outline analysis in which differences in x and y (and possibly z) coordinates of an outline are fit separately as a function of arc length by Fourier analysis. The chapter by Rohlf in the Blue Book provides an overview of various methods of fitting curves to outline data.

Euclidean distance – Euclidean distance is defined as:

$$d_{lm} = \left(\sum_{i=1}^{k} \left(x_{li} - x_{mi} \right)^2 \right)^{\frac{1}{2}}$$

for coordinates of points x_l and x_m on the axes of a k-dimensional space. This can be expressed in matrix notation as

$$d_{lm} = \left((\mathbf{x}_l - \mathbf{x}_m)(\mathbf{x}_l - \mathbf{x}_m)' \right)^{\frac{1}{2}},$$

where $\mathbf{x}_l$ and $\mathbf{x}_m$ are the $1 \times k$ row vectors of the coordinates of points l and m in some coordinate system.

Euclidean distance matrix analysis (EDMA) – A method for the statistical analysis of full matrices of all interlandmark distances, by averaging elementwise within samples, and then comparing those averages between samples by computing the ratios of corresponding mean distances. See Lele and Richtsmeier (1991).

Euclidean space – A space in which distances between two points are defined as Euclidean distances in some system of coordinates.

factor analysis – Factor analysis is a multivariate technique for describing a set of measured variables in terms of a set of causal or underlying variables. A factor model can be characterized in terms of path diagrams to show relations between measured variables and factors. See the chapter by Marcus in the Blue Book and Reyment and Jöreskog (1993).

FESA – see finite element scaling analysis.

fiber – In geometric morphometrics, the set of preshapes (configurations that have been centered at the origin and scaled to unit centroid size) that differ only by a rotation. It is the path, through preshape space, followed by a centered and scaled configuration under all possible rotations.

figure – A representation of an object by the coordinates of a specified set of points, for instance, its landmarks.

figure space – For landmark data, the $2p$- or $3p$-space of figures, i.e., the original coordinate data vectors.

finite element scaling analysis (FESA) – Without the word "scaling," finite element analysis is a computational system for continuum mechanics that estimates the deformation (fully detailed changes of position of all component particles) which are expected to result from a specified pattern of stresses (forces) upon a mechanical system. As applied in morphometrics, FESA solves the inverse problem of estimating the strains representing the hypothetical forces that deformed one specimen into another. These results are a function of the "finite elements" into which the space between the landmarks is subdivided. Finite element scaling analysis can be compared with the thin-plate spline, which interpolates a set of landmark coordinates under an entirely different set of assumptions.

form – In morphometrics, we represent the form of an object by a point in a space of form variables, which are measurements of a geometric object that are unchanged by translations and rotations. If you allow for reflections, forms stand for all the figures that have all the same interlandmark distances. A form is usually represented by one of its figures at some specified location and in some specified orientation. When represented in this way, location and orientation are said to have been "removed."

form space – The space of figures with differences due to location and orientation removed. It is of $2p-3$ dimensions for two-dimensional coordinate data and $3p-6$ dimensions for three-dimensional coordinate data.

Fourier analysis – In morphometrics, the decomposition of an outline into a weighted sum of sine and cosine functions. The chapter by Rohlf in the Blue Book provides an overview of this and other methods of analyzing outline data.

fractal dimension (D) – A measure of the complexity of a structure, assuming a consistent pattern of self-similarity (structural complexity at smaller scales is mathematically indistinguishable from that at larger-scales) over all scales considered. See the chapter by Slice in the Black Book.

generalized distance (D) – A synonym for Mahalanobis distance. Defined by the equation for two row vectors $\mathbf{x}_i$ and $\mathbf{x}_j$ for two individuals, and p variables as

$$\left((\mathbf{x}_i - \mathbf{x}_j)\, \mathbf{S}^{-1}\, (\mathbf{x}_i - \mathbf{x}_j)'\right)^{\frac{1}{2}},$$

where **S** is the $p \times p$ within group, variance-covariance–matrix. Generalized distance takes into consideration the variance and correlation of the variables in measuring distances between points, i.e., differences in directions in which there is less variation within groups are given greater weight than are differences in directions in which there is more variation.

generalized superimposition – The superimposition of a set of configurations onto their consensus configuration.

geodesic distance – The length of the shortest path between two points in a suitable geometric space (one for which curving paths have lengths). On a sphere, it is the distance between two points as measured along a great circle. In a Euclidean space, geodesic distance is ordinary (straight-line) Euclidean distance.

geometric morphometrics – Geometric morphometrics is a collection of approaches for the multivariate statistical analysis of Cartesian coordinate data, usually (but not always) limited to landmark point locations. The "geometry" referred to by the word "geometric" is the geometry of Kendall's shape space: the estimation of mean shapes and the description of sample variation of shape based on the geometry of Procrustes distance. The multivariate part of geometric morphometrics is usually carried out in a linear tangent space to the non-Euclidean shape space in the vicinity of the mean shape.

More generally, it is the class of morphometric methods that preserve complete information about the relative spatial arrangements of the data throughout an analysis. As such, these methods allow for the visualization of group and individual differences, sample variation, and other results in the space of the original specimens.

great circle – A circle on a sphere with a diameter equal to that of the sphere. The shortest path connecting two points on the surface of a sphere lies along the great circle passing through the points. See geodesic distance.

homology – The notion of homology bridges the language of geometric morphometrics and the language of its biological or biomathematical applications. In theoretical biology, only the explicit entities of evolution or development, such as molecules, organs or tissues, can be "homologous." Following D'Arcy Thompson, morphometricians often apply the concept instead to discrete geometric structures, such as points or curves, and, by a further extension, to the multivariate descriptors (e.g., partial warp scores) that arise as part of most multivariate analyses. In this context, the term "homologous" has no meaning other than that the same name is used for corresponding parts in different species or developmental stages. To declare something "homologous" is simply to assert that we want to talk about processes affecting such structures as if they had a consistent biological or biomechanical meaning. Similarly, to declare an interpolation (such as a thin-plate spline) a "homology map" means that one intends to refer to its features as if they had something to do with valid biological explanations pertaining to the regions between the landmarks, about which we have no data.

Hotelling's T^2 – See T^2 statistic.

hyperplane – A k–1-dimensional subspace of a k-dimensional space. A hyperplane is typically characterized by the vector to which it is orthogonal.

hyperspace – A space of more than three dimensions.

hypersphere – In any space, the set of all points at a fixed distance (the radius) from some fixed point (the center).

hypervolume – A generalization of the idea of volume to a space of more than three dimensions.

invariant – An invariant, generally speaking, is a quantity that is unchanged (even though its formula may have changed) when one changes some inessential aspect of a measurement. For instance, Euclidean distance is an invariant under translation or rotation of one's coordinate system, and ratio of distances in the same direction is an invariant under affine transformations. In the morphometrics of triangles, the invariants of a particular transformation are the shape variables that do not change under that transformation (see the Orange Book). See covariant.

isometry – An isometry is a transformation of a geometric space that leaves distances between points unchanged. If the space is the Euclidean space of a picture or an organism, and the distances are distances between landmarks, the isometries are the Euclidean translations, rotations, and reflections. If the distances are Procrustes distances between shapes, the isometries (for the simplest case, landmarks in two dimensions) are the rotations of Kendall's shape space. For triangles, these can be visualized as ordinary rotations of Kendall's "spherical blackboard."

isotropic – Invariant with respect to direction. Isotropic errors have the same statistical distribution in all directions, implying equal variance and zero correlation between the original variables (e.g., axis coordinates).

Kendall's shape space – The fundamental geometric construction, developed by David Kendall, underlying geometric morphometrics. Kendall's shape space provides a complete geometric setting for analyses of Procrustes distances among arbitrary sets of landmarks. Each point in this shape space represents the shape of a configuration of points in some Euclidean space, irrespective of size, position, and orientation. In shape space, scatters of points correspond to scatters of entire landmark configurations, not merely scatters of single landmarks. Most multivariate methods of geometric morphometrics are linearizations of statistical analyses of distances and directions in this underlying space.

Kronecker product – See ⊗.

L – The crucial matrix for computing the thin-plate spline interpolant between two landmark configurations. In this entry, k stands for the number of landmarks, for historical reasons.

The equation of the thin-plate spline has coefficients $\mathbf{L}^{-1}\mathbf{h}$, where $\mathbf{h}$ is a vector of the x- or y-coordinates of the landmarks in a target form, followed by three 0's (for two dimensional data, four 0's for three-dimensional data). The entries in the matrix **L** are wholly functions of the starting or reference form for the spline.

Bending energy is the upper left $k \times k$ square of $\mathbf{L}^{-1}$. For the complete formula for **L**, see the Orange Book or Rohlf (Black Book).

landmark – A specific point on a biological form or image of a form located according to some rule. Landmarks with the same name, homologues in the purely semantic sense, are presumed to correspond in some sensible way over the forms of a data set. See Type I, Type II, and Type III landmarks.

least-squares estimates – Parameter estimates that minimize the sum of squared differences between observed and predicted sample values.

likelihood ratio test – A test based on the ratio of the likelihood (the probability of the data given the parameters) under a general model to the likelihood when another, specified hypothesis is true. Many of the commonly used statistical tests are likelihood ratio tests, e.g., the *t*-test for comparisons of means, Hotelling's T^2, and the analysis of variance *F*-test.

linear combination – A sum of values each multiplied by some coefficient. A linear combination can be expressed as the inner product of two vectors, one representing the data and the other a vector of coefficients.

linear transformation – In multivariate statistics, a linear transformation is the construction of a new set of variables that are all linear combinations of the original set. In geometric morphometrics, one linear transformation takes Procrustes-fit coordinates to partial warp scores; another takes them to relative warp scores. A linear transformation of a matrix **A** can be written in the form $\mathbf{y} = \mathbf{Ax}$, where **y** is the resulting linear combination of **x**, a column vector, with the rows of **A**.

linear vector space – In morphometrics, the most common *k*-dimensional linear vector space is the set of all real *k* dimensional vectors, including all sums of these vectors and their scalar multiples. More generally, but informally, a linear vector space is a set of elements, usually bits of geometry or whole functions, that can be added together and can be multiplied by real numbers in an intuitive way. The points of a plane don't form a linear vector space (what is "five times a point"?), but lines (vectors) connecting all the points to the origin do form such a space.

loading – The correlation or covariance of a measured variable with a linear combination of variables. A loading is not the same as a coefficient. In general, coefficients supply formulas for the computation of scores, whereas loadings are used for the biological interpretation of the linear combination.

Mahalanobis distance – Also *D* or Mahalanobis *D*. See generalized distance.

MANOVA - See multivariate analysis of variance.

maximum likelihood estimate - A likelihood function is a probability or density function for a set of data and given estimates of its parameters. A maximum likelihood estimate is the set of parameter values that maximize this function. In some cases, such as the arithmetic mean of a sample used as an estimate of the parameter μ for a normally distributed population, the maximum likelihood estimate may be identical to the least-squares estimate.

median size – A size measure based on the repeated median of interlandmark distances. Used in resistant-fit methods.

metric – A nonnegative function, d_{ij}, of two points, *i* and *j*, satisfying the following conditions: $d_{ii} = d_{jj} = 0$, if $i \neq j$ then $d_{ij} > 0$, $d_{ij} = d_{ji}$, and $d_{ik} \leq d_{ij} + d_{jk}$. The last condition is known as the "triangle inequality." The value of the function is referred to as the "distance" between the two points. A space is called "semimetric" if d_{ij} can be equal to zero when $i \neq j$.

metric space – A space and a distance function defined on every pair of points that meets the requirements of the definition of "metric" above.

morphometrics – From the Greek: "morphē, " meaning "shape," and "metron," meaning "measurement." Schools of morphometrics are characterized by what aspects of biological "form" they are concerned with, what they choose to measure, and what kinds of biostatistical questions they ask of the measurements once those measurements are made. The methods of this book emphasize configurations of landmarks from whole organs or organisms analyzed by appropriately invariant biometric methods (covariances of taxon, size, cause or effect with position in Kendall's shape space) in order to answer biological questions. Another sort of morphometrics studies tissue sections, measures the densities of points and curves, and uses these patterns to answer questions about the random processes that may be controlling the placement of cellular structures. A third, the method of "allometry," measures sizes of separate organs and asks questions about their correlations with each other and with measures of total size. There are many others.

multiple discriminant analysis – Discriminant analysis involving three or more a priori-defined groups. See discriminant analysis.

multiple regression – The prediction of a dependent variable by a linear combination of two or more independent variables by means of least-squares methods for parameter estimation. See multivariate regression and multivariate multiple regression.

multivariate analysis of variance (MANOVA) – An analysis of variance of two or more dependent variables considered simultaneously.

multivariate morphometrics – A term historically used for the application of standard multivariate techniques to measurement data for the purposes of morphometric analysis. Somewhat confusing now as any morphometric technique must be multivariate in nature. See traditional morphometrics.

multivariate multiple regression – The prediction of two or more dependent variables based on two or more independent variables. See multiple regression and multivariate regression.

multivariate regression – The prediction of two or more dependent variables based on one independent variable. See multiple regression and multivariate multiple regression.

normalize – To normalize a geometric object is to transform it so that some function of its coordinates or other parameters has a prespecified value. For example, vectors are often normalized by transformation into unit vectors, which have a length of one.

nuisance parameters – Parameters of a model that must be fit but that are not of interest to the investigator. In morphometrics, the parameters for translation and rotation are usually nuisance parameters.

null model – The simplest model under consideration. The null model for shape is the distribution in Kendall's shape space which arises from landmarks that are distributed by independent circular normal noise of the same variance in the original digitizing plane or space and are drawn from a single, homogeneous population. It is exactly analogous to the usual assumption of "independent identically distributed error terms" in conventional linear models (regression, analysis of variance).

oblique – At an angle that is not a multiple of 90 degrees.

Orange Book – Bookstein, 1991. See also Black Book, Blue Book, Red Book, and Reyment's Black Book.

ordination – A representation of objects with respect to one or more coordinate axes. There are many kinds of ordinations depending upon the goals of the ordination and criteria used. For example, plotting objects according to their scores on the first two principal component axes provides the two-dimensional ordination best summarizing the total variability of the objects in the original sample space. Biplots combine an ordination of specimens and an ordination of variables.

orthogonal – At right angles. In linear algebra, being "at right angles" is defined relative to a symmetric matrix **P**, such as the bending-energy matrix; two vectors **x** and **y** are orthogonal with respect to **P** if $\mathbf{x}^t\mathbf{P}\mathbf{y} = 0$.

Principal warps are orthogonal with respect to bending energy, and relative warps are orthogonal with respect to both bending energy and the sample covariance matrix.

orthogonal superimposition – A superimposition using only transformations that are all Euclidean similarities, i.e., that involve only translation, rotation, scaling, and, possibly, reflection.

orthonormal – A set of vectors is orthonormal if each has length unity and all pairs are orthogonal with respect to some relevant matrix, **P**, such as the identity matrix. A matrix is orthogonal if its rows (columns) are orthonormal as a set of vectors.

outline – A mathematical curve that stands for the two-dimensional image of a physical boundary. Outline data can be archived as a sequence of point coordinates, but such points do not share the notion of homology associated with landmarks (but see Sampson, this volume).

parameter – In general, a parameter is a number (an integer, a decimal) indexing a function. For instance, the F-distribution used to test decompositions of variance has two parameters, both integers: the counts of the degrees of freedom for the two variances whose ratio is being tested. In morphometrics, there are four main kinds of parameters: nuisance parameters, which must be estimated to account for differences not of particular scientific interest; the geometric parameters, such as shape coordinates, in which landmark shape is expressed; statistical parameters, such as mean differences or correlations, by which biological interpretation is confronted with that data; and another set of geometric parameters, such as partial warp scores or Procrustes residuals, in which the findings of the statistical analysis are expressed.

partial least squares – Partial least squares is a multivariate statistical method for assessing relationships among two or more sets of variables measured on the same entities. Partial least squares analyses the covariances between the sets of variables rather than optimizing linear combinations of variables in the various sets. Partial least squares computations usually do not involve the inversion of matrices (see the Orange Book).

partial warp scores – Partial warp scores are the quantities that characterize the location of each specimen in the space of the partial warps. They are a rotation of the Procrustes residuals around the Procrustes mean configuration. For the nonuniform partial warps, the coefficients for the rotation are the principal warps, applied first to the x-coordinates of the Procrustes residuals, then to the y-coordinates and, for three-dimensional data, then the z-coordinates. Coefficients for the uniform partial warps are produced by special formulas (see Bookstein's "Uniform" chapter, this volume).

partial warps – Partial warps are an auxiliary structure for the interpretation of shape changes and shape variation in sets of landmarks. Geometrically, partial warps are an orthonormal basis for a space tangent to Kendall's shape space. Algebraically, the partial warps are eigenvectors of the bending-energy matrix that describes the net local information in a deformation along each coordinate axis. Except for the largest-scale partial warp, the one for uniform shape change, they have an approximate location and an approximate scale.

precision – The closeness of repeated measurements to the same value. See accuracy.

preform space – The space corresponding to centered figures, i.e., differences in location have been removed. It is of $k(p-1)$ dimensions.

preshape space – The space corresponding to figures that have been centered and scaled but not rotated to alignment. It is of $k(p-1)-1$ dimensions.

principal axes and strains – A change of one triangle into another, or of one tetrahedron into another, can be modeled as an affine transformation that can be parameterized by its effect on a circle or sphere. An affine transformation takes circles into ellipses. The principal axes of the shape change are the directions of the diameters of the circle that are mapped into the major and minor axes of the ellipse. The principal strains of the change are the ratios of the lengths of the axes of the ellipse to the diameter of the circle. In the case of the tetrahedron, there are three principal axes, the axes of the ellipsoid into which a sphere is deformed. One has the greatest principal strain (ratio of axis length to diameter of sphere) and one the least, and there is a third perpendicular to both, having an intermediate principal strain.

principal components analysis – The eigenanalysis of the sample covariance matrix. Principal components (PCs) can be defined as the set of vectors that are orthogonal both with respect to the identity matrix and the sample covariance matrix. They can also be defined sequentially: the first is the linear combination with the largest variance of all those PCs with coefficients summing in square to 1; the second has the largest variance (when normalized that way) of all PCs that are uncorrelated with the first one; and so on. One way to compute PCs is to use a singular value decomposition. Relative warps are principal components of partial warp scores. There is a lot to be said about PCs; see any of the colored books.

principal warps – Principal warps are eigenfunctions of the bending-energy matrix interpreted as actual warped surfaces (thin-plate splines) over the picture of the original landmark configuration. Principal warps are like the harmonics in a Fourier analysis (for circular shape) or Legendre polynomials (for linear shape) in that together they decompose the relation of any sample shape to the sample average shape as a unique summation of multiples of eigenfunctions of bending energy. They differ from these more familiar analogues in that there are only k-3 of them for a set of k landmarks—they form a finite series. Together with the uniform terms, the partial warps, which are projections (shadows) of the principal warps, supply an orthonormal basis for a space that is tangent to Kendall's shape space in the vicinity of a mean form.

Procrustes distance – Approximately (see Bookstein's "Combining" chapter, this volume) the square root of the sum of squared differences between the positions of the landmarks in two optimally (by least squares) superimposed configurations at centroid size. This is the distance that defines the metric for Kendall's shape space.

Procrustes mean – The shape that has the least summed squared Procrustes distance to all the configurations of a sample; the best choice of consensus configuration for most subsequent morphometric analyses (see Bookstein's "Combining" chapter, this volume).

Procrustes methods – A term for least-squares methods for estimating nuisance parameters. The adjective "Procrustes" refers to the Greek giant who would stretch or shorten victims to fit a bed and was first used in the context of superimposition methods by Hurley and Cattell (1962). Modern workers have often cited Mosier (1939), a psychometrician, as the earliest known developer of these methods. However, Cole (in press) reports that Franz Boas in 1905 suggested the "method of least differences" (ordinary Procrustes analysis) as a means of comparing homologous points to address obvious problems with the standard point-line registrations (Boas, 1905). Cole further points out that one of Boas' students extended the method to the construction of mean configurations from the superimposition of multiple specimens using either the standard registrations or Boas' method (Phelps, 1932). The latter being essentially a Generalized Procrustes Analysis.

Procrustes residuals – The set of vectors connecting the landmarks of a specimen to corresponding landmarks in the consensus configuration after a Procrustes fit. The sum of squared lengths of these vectors is approximately the squared Procrustes distance between the specimen and the consensus in Kendall's shape space. The partial warp scores are an orthogonal rotation of the full set of these residuals.

Procrustes scatter – A collection of forms all superimposed by ordinary orthogonal Procrustes fit over one single consensus configuration that is their Procrustes mean; a scatter of all the Procrustes residuals each centered at the corresponding landmark of the Procrustes mean shape.

Procrustes superimposition – The construction of a two-form superimposition by least squares based on orthogonal or affine transformations.

Red Book – Bookstein, Chernoff, Elder, Humphries, Smith, and Strauss (1985). See also Black Book, Blue Book, Orange Book, and Reyment's Black Book.

reference configuration – In the context of superimposition methods, this is the configuration to which data are fit. It may be another specimen in the sample, an estimated consensus configuration, or an hypothetical ancestor. The construction of two-point shape coordinates does not involve a reference specimen, though the intelligent choice of baseline for the construction usually does. In shape space itself, the reference specimen is a shape in the vicinity of which the linear tangent space is constructed. When splines or warps are part of the analysis, the reference configuration usually serves as the "starting form."

regression – A model for predicting one variable from another. Coined by Francis Galton, the term derives from the fact that when measurements of offspring, whether peas or people, were plotted against the same measurements for their parents, the offspring measurements "went back," or regressed, towards the mean.

relative warps – Relative warps are principal components of a distribution of shapes in a space tangent to Kendall's shape space. They are the axes of the "ellipsoid" occupied by the sample of shapes in a geometry in which spheres are defined by Procrustes distance. Each relative warp, as a direction of shape change about the mean form, can be interpreted as specifying multiples of one single transformation, a transformation that can often be usefully drawn out as a thin-plate spline. In a relative warps analysis, the parameter α (see above) can be used to weight shape variation by the geometric scale of shape differences. Relative warps can be computed from Procrustes residuals or from partial warps (see Bookstein's "Combining" article, this volume).

repeated median – A median of medians. Repeated medians are used to estimate some superimposition parameters in the resistant-fit methods. For example, the resistant-fit rotation estimate is the median of the estimates obtained for each landmark, which is, in turn, the median of angular differences between the reference configuration and the configuration being fit of the line segments defined by that landmark and the other n-1 landmarks. Repeated medians are insensitive to larger subsets of extremely deviant values than are simple medians.

residual – The deviations of an observed value or vector of values from some expectation, e.g., the differences between a shape and its prediction by an allometric regression expressed in any set of shape coordinates.

resistant-fit superimposition – Superimposition methods that use median- and repeated-median-based estimates of fitting parameters rather than least-squares estimates. Resistant-fit procedures are less sensitive to subsets of extreme values than are those of comparable least-squares methods. As such, their results may provide a simple description of differences in shape that are due to changes in the positions of just a few landmarks. However, resistant-fit methods lack the well-developed distributional theory associated with the least-squares fitting methods. See Slice, this volume.

resolution – The smallest scale distinguishable by a digitizing, imaging, or display device.

Reyment's Black Book - Reyment, 1991. See also Black Book, Blue Book, Orange Book, and Red Book.

ridge curve – Ridge curves are curves on a surface along which the curvature perpendicular to the curve is a local maximum. For instance, on a skull the line of the jaw or the rim of an orbit are ridge curves. See Dean, this volume.

rigid rotation – An orthogonal transformation of a real vector space with respect to the Euclidean distance metric. Such transformations leave distances between points and angles between vectors unchanged. A principal components analysis represents a rigid rotation to new orthogonal axes. A canonical variates analysis does not.

score – A linear combination of an observed set of measured variables. The coefficients for the linear combination are usually determined by some matrix computation. Multivariate statistical findings in the form of coefficient vectors can usually be more easily interpreted if scores are also shown case by case—in the form of scatters of scores, their loadings (correlations of the scores with the original variables), and so on.

shape – The geometric properties of a configuration of points that are invariant to changes in translation, rotation, and scale. In morphometrics, we represent the shape of an object by a point in a space of shape variables, which are measurements of a geometric object that are unchanged under similarity transformations. For data that are configurations of landmarks, there is also a representation of shapes per se, without any nuisance parameters (position, rotation, scale), as single points in a space, Kendall's shape space, with a geometry given by Procrustes distance. Other sorts of shapes (e.g., those of outlines, surfaces, or functions) correspond to quite different statistical spaces.

shape coordinates – In the past, any system of distance ratios and perpendicular projections permitting the exact reconstruction of a system of landmarks by a rigid trusswork. Now, more generally, coordinates with respect to any basis for the tangent space to Kendall's shape space in the vicinity of a mean form. See Procrustes residuals, partial warp scores, and two-point shape coordinates.

shape space – A space in which the shape of a figure is represented by a single point. It is of $2p$-4 dimensions for two-dimensional coordinate data and $3p$–7 dimensions for three-dimensional coordinate data. See Kendall's shape space.

shape variable – Any measure of the geometry of a biological form, or the image of a form, that does not change under similarity transformations: translations, rotations, and changes of geometric scale (enlargements or reductions). Useful shape variables include angles, ratios of distances, and any of the sets of shape coordinates that arise in geometric morphometrics.

shear – In a general sense, an affine transformation. For example (see Bookstein's "Uniform" chapter, this volume), a transformation that leaves one Cartesian coordinate, y, invariant and alters x by a translation that is a multiple of y: for instance, what happens when you slide the top of a square sideways without altering its vertical position or the length of the horizontal edges. The score for such a translation, together with a separate score for dilation of the vertical axis, supplies one orthonormal basis for the subspace of uniform shape changes of two-dimensional data.

similarity transformation – A change in a Cartesian coordinate system that leaves all ratios of distances unchanged. The term proper or special similarity group of similarities is sometimes used when the transformations do not involve reflection. Similarities are arbitrary combinations of translations, rotations, and changes of scale. Compare affine transformation.

singular value decomposition – Any $m \times n$ matrix $\mathbf{X}$ may be decomposed into three matrices, $\mathbf{U}$, $\mathbf{D}$, and $\mathbf{V}$ (with dimensions $m \times m$, $m \times n$, and $n \times n$, respectively), in the form: $\mathbf{X} = \mathbf{UDV}^t$, where the columns of $\mathbf{U}$ are orthogonal, $\mathbf{D}$ is a diagonal matrix of singular values, and the columns of $\mathbf{V}$ are orthogonal. The singular value decomposition of a variance–covariance matrix $\mathbf{S}$ is written as $\mathbf{S} = \mathbf{E}\Lambda\mathbf{E}^t$, where Λ is the diagonal matrix of eigenvalues and $\mathbf{E}$ the matrix of eigenvectors.

size measure – In general, some measure of a form (i.e., an invariant under the group of isometries) that scales as a positive power of the geometric scale of the form. Interlandmark lengths are size measures of dimension one, areas are size measures of dimension two, and so on.

space – In statistics, a collection of objects or measurements of objects treated as if they were points in a plane, a volume, on the surface of a sphere, or on any higher-dimensional generalization of these intuitive structures. Examples are Euclidean spaces, sample spaces, shape spaces, and linear vector spaces.

superimposition – The transformation of one or more figures to achieve some geometric relationship to another figure. The transformations are usually affine transformations or similarities. They can be computed by matching two or three landmarks, by least-squares optimization of squared residuals at all landmarks, or in other ways. Sometimes informally referred to as a "fit" or "fitting," e.g., a resistant fit.

SVD – See singular value decomposition.

T^2 statistic – A multivariate generalization of the univariate t^2 statistic. It is the square of the ratio of the group mean difference to the standard error of that difference. Used in the T^2-test.

T^2-test – A test developed by Hotelling for comparing an observed mean vector with a parametric mean or comparing the difference between two mean vectors to a parametric difference (usually the zero vector). If the observations are independently multivariate normal, then the T^2-test may be used to test null hypotheses using the F-distribution. The T^2 statistic is also closely related to Mahalanobis D^2. See Marcus (Black Book).

tangent space – Informally, if S is a curving space and P a point in it, the tangent space to S at P is a linear space T having points with the same "names" as the points in S and in which the metric on S "in the vicinity of P" is very nearly the ordinary Euclidean metric on T. One can visualize T as the projection of S onto a "tangent plane" "touching" at P just like a map is a projection of the surface of the earth onto flat paper.

In geometric morphometrics, the most relevant tangent space is a linear vector space that is tangent to Kendall's shape space at a point corresponding to the shape of a reference configuration (usually taken as the mean of a sample of shapes). If variation in shape is small then Euclidean distances in the tangent space can be used to approximate Procrustes distances in Kendall's shape space. Because the tangent space is linear, it is possible to apply conventional statistical methods to study variation in shape. See Rohlf (this volume) and Bookstein's "Combining" chapter (this volume).

tensor – An example of a tensor in morphometrics is the representation of a uniform component of shape change as a transformation matrix. The transformation matrix assigns to each vector in a starting (or average) form a vector in a second form. A rigorous, general definition of a tensor would be beyond the scope of this glossary, but a reasonably intuitive characterization comes from Misner, Thorne, and Wheeler (1973): a tensor is a "geometric machine" that is fed one or more vectors in an arbitrary Cartesian coordinate system and that produces scalar values (ordinary decimal numbers) which are independent of that coordinate system. In morphometrics, these "numbers" will be ordinary geometric entities such as lengths, areas, or angles: anything that doesn't change when the coordinate system changes. For the representation of a uniform component as a transformation matrix, the "scalars" of the Misner–Thorne–Wheeler metaphor are the lengths of the resulting vectors and the angles among them.

A different tensor representing the same uniform transformation is the relative metric tensor, which you probably know as the ellipse of principal axes and principal strains. This tensor produces the necessary numerical invariants (distances in the second form as a function of coordinates on the first form) directly. Other tensors include the metric tensor of a curving surface, which expresses distance on the surface as a function of the parameters in which surface points are expressed, and the curvature tensor of the same surface, which expresses the way in which the surface "falls away" from its tangent plane at any point.

thin-plate spline – In continuum mechanics, a thin-plate spline models the form taken by a metal plate that is constrained at some combination of points and lines and is otherwise free to adopt the form that minimizes bending energy. (The extent of bending is taken as so small that elastic energy—stretches and shrinks in the plane of the original plate—can be neglected.) One particular version of this problem—an infinite, uniform plate constrained only by displacements at a set of discrete points—can be solved algebraically by a simple matrix inversion. In that form, the technique is a convenient general approach to the problem of surface interpolation for computer graphics and computer-aided design. In morphometrics, the same interpolation (applied once for each Cartesian coordinate) provides a unique solution to the construction of D'Arcy Thompson-type deformation grids for data in the form of two landmark configurations.

traditional morphometrics – Application of multivariate statistical methods to arbitrary collections of size or shape variables such as distances and angles. "Traditional morphometrics" differs from the geometric morphometrics discussed here in that even though the distances or measurements are defined to record biologically meaningful aspects of the organism, the geometrical relationships between these measurements are not taken into account. Traditional morphometrics makes no reference to Procrustes distance or any other aspect of Kendall's shape space. See multivariate morphometrics.

transformation – In general, a replacement of landmark coordinates by another set purporting to pertain to the same landmarks. For example, a matrix of landmark coordinates might be transformed by multiplication by another matrix to produce a new set of coordinates that have been scaled, rotated, and translated with respect to the original data.

two-point shape coordinates – A convenient system of shape coordinates, originally proposed by Francis Galton and rediscovered by Bookstein, consisting (for two-dimensional data) of the coordinates of landmarks 3, 4, and so on after forms are rescaled and repositioned so that landmark 1 is fixed at (0,0) and landmark 2 is fixed at (1,0) in a Cartesian coordinate system. Also referred to as Bookstein coordinates or Bookstein's shape coordinates.

Type I landmark – A mathematical point whose claimed homology from case to case is supported by the strongest evidence, such as a local pattern of juxtaposition of tissue types or a small patch of some unusual histology.

Type II landmark – A mathematical point whose claimed homology from case to case is supported only by geometric, not histological, evidence: for instance, the sharpest curvature of a tooth.

Type III landmark – A landmark having at least one deficient coordinate, for instance, either end of a longest diameter or the bottom of a concavity. Type III landmarks characterize more than one region of the form. The multivariate machinery of geometric morphometrics permits them to be treated as landmark points in some analyses, but the deficiency they embody must be kept in mind in the course of any geometric or biological interpretation.

unbiased estimator – An estimator, $\hat{\theta}$, that has as its expected value the parametric value, θ, it is intended to estimate: $E(\hat{\theta}) = \theta$. See consistent estimator and asymptotically unbiased estimator.

uniform shape component – That part of the difference in shape between a set of configurations that can be modeled by an affine transformation. Once a metric is supplied for shape space one can ascertain which such transformation takes a reference form closest to a particular target form. For the Procrustes metric (the geometry of Kendall's shape space), that uniform transformation is computed by a formula based in Procrustes residuals or by another based in two-point shape coordinates (see Bookstein's "Uniform" chapter, this volume). Together with the partial warps, the uniform component defined in this way supplies an orthonormal basis for all of shape space in the vicinity of a mean form. In this setting, the uniform shape component may also be interpreted as the projection of a shape difference (between two group means or between a mean and a particular specimen) into the plane (or hyperplane for data of dimension greater than two) through that mean form and all nearby forms related to it by affine transformations. For descriptive purposes, the uniform component is parameterized not by a vector, like the partial warps, but by a representation as a tensor, in terms of sets of shears and dilations with respect to a fixed, orthogonal set of Cartesian axes.

weight matrix, W matrix – The matrix of partial warp scores, together with the uniform component, for a sample of shapes. The weight matrix is computed as a rotation of the Procrustes residual shape coordinates; like them, partial warp scores are a set of shape coordinates for which the sum of squared differences is the squared Procrustes distance between any two specimens.

Wright factor analysis – A version of factor analysis, due to Sewall Wright, in which a path model is used to describe the relation between the measured variables and the factors of interest. It is usually exploratory, in that one fits a simple one-factor model iteratively to explain maximally the correlations among variables and then proceeds to find additional factors to fit to the residuals, and so on until the data is adequately fit. See the Orange Book for examples and discussion of the application of this approach to the analysis of size and group factors for morphometric data.

z – Notation for complex numbers in two-dimensional Procrustes formulas.

ACKNOWLEDGMENTS

Work on this glossary by Slice and Rohlf has been supported by grants BSR–89–18630 and DEB–93-17572 from the Systematic Biology Program of the National Science Foundation.

Bookstein's work in morphometrics is supported by NIH grants DA–09009 and GM–37251. The former of these is jointly supported by the National Institute on Drug Abuse, the National Institute of Mental Health, and the National Institute on Aging as part of the Human Brain Project.

This is publication number 944 from Graduate Studies in Ecology and Evolution, State University of New York at Stony Brook.

REFERENCES

Boas, F. 1905. The horizontal plane of the skull and the general problem of the comparison of variable forms. Science 21:862-863.

Bookstein, F. L. 1991. Morphometric tools for landmark data. Geometry and biology. Cambridge University Press: Cambridge.

Bookstein, F. L., B. Chernoff, R. Elder, J. Humphries, G. Smith, and R. Strauss. 1985. Morphometrics in evolutionary biology: The geometry of size and shape change, with examples from fishes. Academy of Natural Sciences of Philadelphia Special Publication 15.

Bookstein, F. L., and W. D. K. Green. 1993. A feature space for edgels in images with landmarks. Journal of Mathematical Imaging and Vision 3:231–261.

Cole, T. M. in press. Historical note: Early anthropological contributions to "Geometric Morphometrics." American Journal of Physical Anthropology.

Hurley, J. R. and R. B. Cattell. 1962. The Procrustes program: Producing a direct rotation to test an hypothesized factor structure. Behavior Science 7:258–262.

Lele, S., and J. T. Richtsmeier. 1991. Euclidean distance matrix analysis: A coordinate free approach for comparing biological shapes using landmark data, American Journal of Physical Anthropology 86:415–428.

Marcus, L. F., E. Bello, and A. García-Valdecasas (eds.) 1993. Contributions to morphometrics. Monografías del Museo Nacional de Ciencias Naturales 8, Madrid.

Misner, C., K. Thorne, and J. Wheeler. 1973. Gravitation. Freeman: New York.

Mosier, C. I. 1939. Determining a simple structure when loadings for certain tests are known. Psychometrika 4:149–162.

Phelps, E. M. 1932. A critique of the principle of the horizontal plane of the skull. American Journal of Physical Anthropology. 17:71-98.

Reyment, R. A. 1991. Multidimensional palaeobiology. Pergamon Press: Oxford.

Reyment, R. A., and K. G. Jöreskog. 1993. Applied factor analysis in the natural sciences. Cambridge University Press: Cambridge.

Rohlf, F. J., and F. L. Bookstein (eds.) 1990. Proceedings of the Michigan morphometrics workshop. University of Michigan Museum of Zoology Special Publication 2.

APPENDIX II

Morphometric Resources Available on the Internet

F. James Rohlf

Department of Ecology and Evolution
State University of New York
Stony Brook, New York 11794-5245

INTRODUCTION

This appendix is prepared with some trepidation as the descriptions given below are guaranteed to be out of date by the time this volume is published!

Becerra (1993) provides an excellent introductory description and tutorial of listservers (electronic discussion groups), FTP (a protocol for transferring files across the Internet), TELNET (terminal emulation to allow one to log onto another computer across the Internet), and ARCHIE (a program that allows one to search for files across the Internet). The purpose of this paper is to update some of the information given in that paper and to describe some of the newer developments–such as Gopher and World Wide Web (WWW) servers.

THE MORPHMET LISTSERVER

The MORPHMET listserver is a forum for exchange of ideas, references and queries about morphometrics. E-mail messages sent to the server (`morphmet@cunyvm.cuny.edu`) are automatically redistributed to all subscribers to the MORPHMET listserver. See Appendix III for instructions on how to use this facility.

FTP

File transfer protocol (FTP) programs allow you to transfer files (both text files and binary files such as computer software) to or from a remote computer located anywhere on the Internet. Becerra's (1993) detailed tutorial on the use of FTP is now largely out of date because there are now friendly FTP programs that allow you to select files by clicking on their names in a file list, setting options for text or binary, and then to transfer the files to a directory on your computer by clicking on a button. In many cases the owner of the directory provides a "readme" or index file that gives some descriptive information about the files to

Advances in Morphometrics, Edited by L. F. Marcus et al.
Plenum Press, New York, 1996

Morphometrics
at SUNY Stony Brook

This server contains news and files related to the field of geometric morphometrics. Please contact F. James Rohlf with suggestions for improvements and with additional contributions to the morphmet archive.

Notices of meetings and workshops related to morphometrics

- **Morphometrics and Evolution**
  Symposium to be held at the Fifth International Congress of Systematic and Evolutionary Biology, Budapest, Hungary, August 17-24, 1996. More info.
- **Image Fusion and Shape Variability Techniques**
  16th Leeds Annual Statistical Workshop organized by Statistics Department and CoMIR, University of Leeds, UK. July 3-5, 1996. More info.
- **Geometric Morphometrics course**
  The graduate level course on geometric morphometrics is planned to be offered again at the State University of New York at Stony Brook by F. James Rohlf for the Fall 1996 semester. Participants from other institutions are welcome. Course description.

Please send notices of any meetings!

Morphometric listservers

- **MORPHMET**
  This a listserver managed by Les Marcus (lamqc@cunyvm.cuny.edu). If your subscribe to it you will be able to take part (or just listen in) on discussions of various issues related to morphometrics. More information.
- **MORPHOLOGY**
  This is a listserver managed by Henk Heijmans (henkh@cwi.nl). FTP archive.

Morphometrics Archive

This contains both software and data files. "Hotlinks" are provided so that one can view the readme files for many of the programs.

- Software
  Various software packages related to geometric morphometrics. There are programs both for the PC and the Mac (as well some Unix systems).
- Datasets
  A collection of several well-known datasets. Useful for checking prior analyses and for testing new methods.
- FTP
  FTP access to MORPHMET directory for those whose WWW software does not permit downloading.

Morphometrics publications

- Books
  A list of books, workshop proceedings, and similar documents concerned with geometric morphometrics.
- Primer
  A contribution by Richard Reyment containing a "Multivariate Statistical Primer for Geologists."

help you determine which files you need. The speed of transmission is limited by the speed of the slowest link in the network between you and the remote computer.

The principle limitation of FTP is that the interface provides the user with only a simple directory listing of file names, which may be somewhat cryptic. The new Gopher and WWW servers (described below) are designed to provide descriptive information about the files. Even with these new developments, FTP is still used because, if you know the name of the file and its location, FTP provides the fastest and most direct way to transfer a file.

Note: The address of the SUNY Stony Brook morphometrics FTP directory listed in Becerra (1993) has changed. It is now the `morph` directory at `Life.Bio.SUNYSB.edu`. Its contents have greatly expanded over the last two years.

GOPHER SERVERS

Gopher is a campus wide information server developed at the University of Minnesota. With the use of special software on your local computer, Gopher provides a hierarchical system of menus leading to files and documents that can be downloaded. There are also search utilities that help you locate files in "Gopher space." Having descriptive menu entries makes Gopher much easier to use than simple FTP programs. The menus can also provide links to other Gopher directories on the same or other computers.

Even though Gopher was a development that became popular after the Becerra (1993) paper was written, its function has already been largely replaced by the World Wide Web (see below).

THE WORLD WIDE WEB

CERN, the European high-energy physics center, developed a new information system it called the World Wide Web (WWW). They made the WWW and "browser" software to access it, publicly available in January 1992. The WWW is based on the concept of displaying information graphically as documents with hypertext links to related documents (whether they are documents on the same computer or around the world).

The hypertext links to documents are of the following format: "`http://computer/directory/file`," where "`computer`" is the Internet address of the computer containing the document and "`directory/name`" gives the name and the subdirectory containing the desired document. This is called a uniform resource locator, or URL. If the directory and file name are not given, then a default document for that site is displayed. In place of the code "`http`" one sometimes sees other protocol codes, such as "`FTP`" (which will cause a directory to be displayed or a file downloaded using the FTP protocol described above), "`Gopher`" (uses the Gopher protocols to display a hierarchical system of menus), "`telnet`" (which will try to establish a telnet session to a remote computer), "`mailto`" (allows one to

Return to Biological Sciences Home Page

You are the 385th person to access this page since 9/30/95
Revised by F. James Rohlf, Rohlf@Life.Bio.SUNYSB.edu, October 28, 1995. *e-mail comments?*

Figure 1. Opening page of the SUNY at Stony Brook morphometrics WWW page as of the end of October 1995. Note: The underlined words correspond to hyptertext links to other WWW documents, FTP directories or files that can be downloaded.

Morphometric software

The purpose of this directory is to make available programs useful for morphometric analysis. Contributions of additional software are welcome. Contact us with suggestions for improvements and with additional contributions to the morphmet archive.

WWW viewers such as Netscape or Mosaic can download these files. Most of the programs in the archive are for PC compatible computers as indicated by the ".ibmpc" suffix on the names of the files. FTP software for the PC will automatically remove this suffix leaving a valid DOS name. In Mosaic set the "Load to disk" option and then click on the file name shown in "[ ]". Netscape will pop-up a window asking you if you want to save a file to disk.

There are also links to directories on other servers (such as the edgewarp directory on Bookstein's server. These names are shown without the "[ ]" characters. If you select one of these links a directory will be displayed from which you can select files to transfer via FTP. No dates are furnished with these entries since they can be updated independently.

Dates indicate the date of the last upload of the file to this server. For many of the programs there is a link to the *readme* file which may be selected to get additional information about the software.

Software
Click on the categories below to obtain lists of available software.

- **Software based on thin-plate splines**
- **Superimpostion methods**
- **Analysis of outlines**
- **Support routines**
- **Misc. collections of programs**

Notes:

- The files with types of "exe.ibmpc" are self-extracting archive files for the IBM PC. Download them in binary and give them an extension of "EXE" on a PC.
- The files with a type of "zip ibmpc" are IBM PC ZIP files that require the PKUNZIP program to unpack. They are also binary files.
- DPMI = DOS protected mode interface that allows programs to use all of memory and not have the normal 640K limitation. Requires computer using 80386 or better CPU. Not compatible with all clone computers.

Modified: October 25, 1995 by F. James Rohlf. Please send comments, suggestions, and contributions by e-mail.

Figure 2. Listing of the software page from the SUNY at Stony Brook morphometrics WWW Server.

send e-mail to a specified account without having to leave the browser), "news" (allows one to read recent news from an Internet newsgroup) and "WAIS" (allows one to search a WAIS database), as well as variations on the above formats. Most browsers display these URLs somewhere on the screen when you pass the mouse cursor over a hypertext field in a document.

World Wide Web has proven to be a very flexible format for many types of information. The WWW grew even faster after the end of 1993 when the U. S. National Center for Supercomputing Applications (NCSA) released the first browsing software (Mosaic) for the Apple Macintosh, for computers using Microsoft Windows, and for Unix computers supporting

X–Windows. There are now many choices of software for browsing the WWW (Netscape is currently popular, but there are many competitors). Some of these browsers support extensions that allow WWW documents to contain tables, pictures, forms, access to databases, animations, sound and interactions with other software running on the host computer. There is also a text-only browser called Lynx (developed at the University of Kansas). Lynx can be used from text-mode terminals, and it is usually quite fast because it does not have to download images in order to display the information on a WWW page. Many individuals and organizations are putting considerable effort into making information available via the WWW. (I hope comparable effort is spent keeping the information up to date.)

SUNY at Stony Brook Morphometrics WWW Pages

An extensive set of WWW pages is available from the URL address `http://life.bio.sunysb.edu/morph/morph.html`. The opening page is

Software based on thin-plate splines

- [TPS]: Thin-plate spline (Windows)
  Program to compare pairs of specimens by displaying a D'arcy Thompson style transformation grid based on a thin-plate spline. The program has a limit of 50 landmarks. By F. James Rohlf. Version 1.02 2/19/96. This program replaces the [TPSPLINE] program (a DOS program with a limit of 40 landmarks and required the [BGI] Graphics drivers) and its DPMI version [TPSPLINEP] (with a limit of 60 landmarks). Readme. By F. James Rohlf.
- [TPSREGR]: **Regression of partial warp scores** (DOS)
  Performs a regression of partial warp scores onto an independent variable. Can be used to study allometry. 12/2/94. [TPSREGRP] DPMI version of TPSREGR which can be used to process larger datasets. Requires the Graphics drivers. By F. James Rohlf. Readme
- [TPSRW]: **Relative warps** (DOS)
  Performs a relative warp analysis. This corresponds to a principal components analysis of variation in a sample (relative to bending energy if the alpha parameter is not equal to zero). Requires the [BGI] Graphics drivers. 10/17/94. [TPSRWP] DPMI version of the TPSRW program which can process larger datasets. 10/17/94. Readme. By F. James Rohlf.
- Edgewarp: **Thin-plane splines with edgels** (HP, HP901,Solaris, Sun, Stellar,Ultrix)
  This entry points to a FTP directory containing software and documentation for the edgewarp program. Readme. By Bill Green and Fred L. Bookstein.
- JSPLINE & VECTOR: Thin-plate spline programs (DOS & Windows)
  This entry points to a FTP directory containing the file SPLINE.ZIP that contains the programs JSPLINE and VECTOR. These compute thin-plate splines, principal and partial warps, *etc.*, to compare two specimens. The VECTOR program displays principal and partial warps as vector diagrams. 1/20/94. By Julian M. Humphries.

Revised: February 19, 1996 by F. James Rohlf. Please send comments, suggestions, and contributions by e-mail.

Figure 3. Listing of the thin-plate spline software page from the SUNY Stony Brook morphometrics WWW server. Selecting the underlined "Readme" fields brings up WWW pages that give additional information about a program. The fields corresponding to the underlined names of programs are used to download the binary files containing the software or to transfer to another directory in whichwhere one can use FTP to transfer the files.

shown in Fig. 1. The underlined words correspond to links to other WWW documents, FTP directories or files that can be downloaded. If you click on them, your WWW browser will take an appropriate action (display a WWW page, display a directory or popup a window that allows you to download a file). For example, clicking on the word "software" under the Morphometrics Archive heading leads to the display shown in Fig. 2 which lists various categories of software. Figure 3 shows the current list of available software that is based on the use of thin-plate splines. If you click on "[TPSPLINE]", a browser such as Netscape will popup a window to allow you to save the file to disk (you will be presented with another window in which you can enter a file name and select the drive and directory where the file is to be saved on your local computer). The name of this particular file is `tpsz.exe.ibmpc`. For use on a PC it should be saved as `tpsz.exe`. After you click the "OK" button the file will be transferred to your computer.

As explained in Fig. 2, most of the entries correspond to files that have been grouped together (e.g., the program itself plus sample data files and documentation) and then compressed to save both space and the time it takes to download the file. As a convenience to the user, most files have been archived as self-extracting files. To retrieve the original files one executes the file as if it were a program, and the files will be extracted and expanded into their original form.

The entry in Fig. 3 for the edgewarp software is an example of a FTP link to another computer (in this case to the directory `/pub/edgewarp` on the computer `brainmap.med.umich.edu`. Figure 4 shows the contents of this directory using the text-only browser Lynx.

A limitation of the WWW is that there is currently no way for a user to know when the WWW pages have been updated. Dates of last revision are shown at the end of all the WWW pages and a "NEW!" graphic is inserted next to each new item, but users still must access the pages from time to time. For this reason, notices are usually sent to the MORPHMET listserver whenever new or updated programs are added or when other significant changes are made.

```
                              edgewarp directory

                                  EDGEWARP

  Up to pub

Sep  8 02:04  text/plain        arg.1  1Kb
Sep  8 02:04  text/plain        COPYRIGHT  1Kb
Sep  8 02:04  text/plain        data.1  3Kb
Sep  8 02:04  Tar File          demo.oct94.tar  888Kb
Sep  8 02:04  text/plain        display.1  6Kb
Sep  8 02:04  text/plain        edgewarp.1  5Kb
Sep  8 02:04  text/plain        edgewarp@aix  576Kb
Sep  8 02:04  text/plain        edgewarp@hp  616Kb
Sep  8 02:04  text/plain        edgewarp@hp901  344Kb
Sep  8 02:04  text/plain        edgewarp@solaris  576Kb
Sep  8 02:04  text/plain        edgewarp@sun  904Kb
Sep  8 02:04  text/plain        edgewarp@ultrix  713Kb
Sep  8 02:04  text/plain        MANUAL  46Kb
Sep  8 02:04  text/plain        README.EDGEWARP  2Kb
Sep  8 02:04  text/plain        SOURCE  4Kb
Sep  8 02:04  text/plain        xhand.1  2Kb

Commands: Use arrow keys to move, '?' for help, 'q' to quit, '<-' to go back
```

Figure 4. Listing of the pub/edgewarp directory at brainmap.med.umich.edu using the Lynx WWW browser. Versions of the edgewarp software are available for several types of computers that use the Unix operating system.

ACKNOWLEDGMENTS

Work on this appendix was supported by a grant (no. DEB 93-17572) from the Systematic Biology Program of the National Science Foundation. Comments on the manuscript by Dean Adams and Leslie Marcus were greatly appreciated.

This paper is contribution number 952 from the Graduate Studies in Ecology and Evolution, State University of New York at Stony Brook.

REFERENCE

Becerra, J. M. 1993. Knowledge for nothing: Absorbing Internet and bitnet resources. Pages 241-256 *in* Marcus, L. F., E. Bello, E., and A. García-Valdecasas, (eds.), Contributions to morphometrics. Monografios del Museo Nacional de Ciencias Naturales 8, Madrid.

APPENDIX III

Morphmet: The On-Line Bulletin Board on Morphometrics

Leslie F. Marcus

The on-line LISTSERV list MORPHMET provides a forum for exchange of ideas, references and queries about morphometrics, the study of form—shape and size—in organisms. Appropriate contributions are queries and statements about morphometric methods and philosophy; new references or queries about the literature; announcements for relevant meetings and relevant job announcements or needs. Because the list is not monitored, it is up to the users to define the content and frequency of use.

In order to subscribe, send a one-line message to the list without a subject or signature. Send the request to LISTSERV@CUNYVM.CUNY.EDU and include the following single line in the body of the message (leave the subject line blank because LISTSERV will send you an error message and ignore what you say if you write anything at all there):

- subscribe morphmet [your name]

For example, I would subscribe by saying

- subscribe morphmet Leslie F. Marcus

The system is smart enough to know your user ID and node from the received message. Please DO NOT send the request to morphmet@cunyvm.cuny.edu!

When you first sign up, LISTSERV will send you further instructions about signing off, getting an index for the archives (all contributions are archived monthly), retrieving archived monthly files and obtaining further information about LISTSERV lists. The earliest archive is LOG9103 and contains all mail for March of 1991.

In order to retrieve the index of MORPHMET archived files send to LISTSERV@CUNYVM.CUNY.EDU the single line message:

- INDEX MORPHMET

You will receive only a list of monthly log files, and you will have to search a number of them to find out what was said during that period. It is easier and faster than it sounds.

Note that when you answer a query or want to enter a discussion on MORPHMET and you use REPLY on your local mail software, the answer will go back only to the PERSON initiating the mail. If you want to send something to everybody on the MORPHMET list then address it to:

- MORPHMET@CUNYVM.CUNY.EDU

You need not be a member of the list to send a query or comment.
All further queries to:

Leslie F. Marcus

Snail Mail to:

Department of Invertebrates
American Museum of Natural History
CPW at 79th
New York, NY 10024, USA

Telephone: 212-769-5721 (answering service after 4 rings)
FAX: 212-769-5842
E-mail: LAMQC@CUNYVM.CUNY.EDU or MARCUS@AMNH.ORG

SOME OTHER COMMANDS OF POSSIBLE INTEREST TO SEND TO LISTSERV AT CUNYVM

All of these are single-line commands to LISTSERV@CUNYVM.CUNY.EDU in the body of the message—no signature or subject. I have indicated what they do with an asterisk (*) after the command—don't type this part of the line. The commands are not case sensitive.

help	*	sends you a short list of commands
index morphmet	*	already mentioned above
signoff morphmet	*	takes you off the list
subscribe morphmet	*	puts you on morphmet
get log9404 morphmet	*	gets the April 1994 archive that contains discussion of phylogenetics; this was conveniently broken into three files by my mail system—remember everything from that month is archived in this file, so the file will contain all entries for that month, which includes other information than the phylogenetic discussion
review morphmet	*	gets a list of the user ids and nodes of all of the persons currently on the list

You may send more than one command in an e-mail message.

PARTICIPANTS

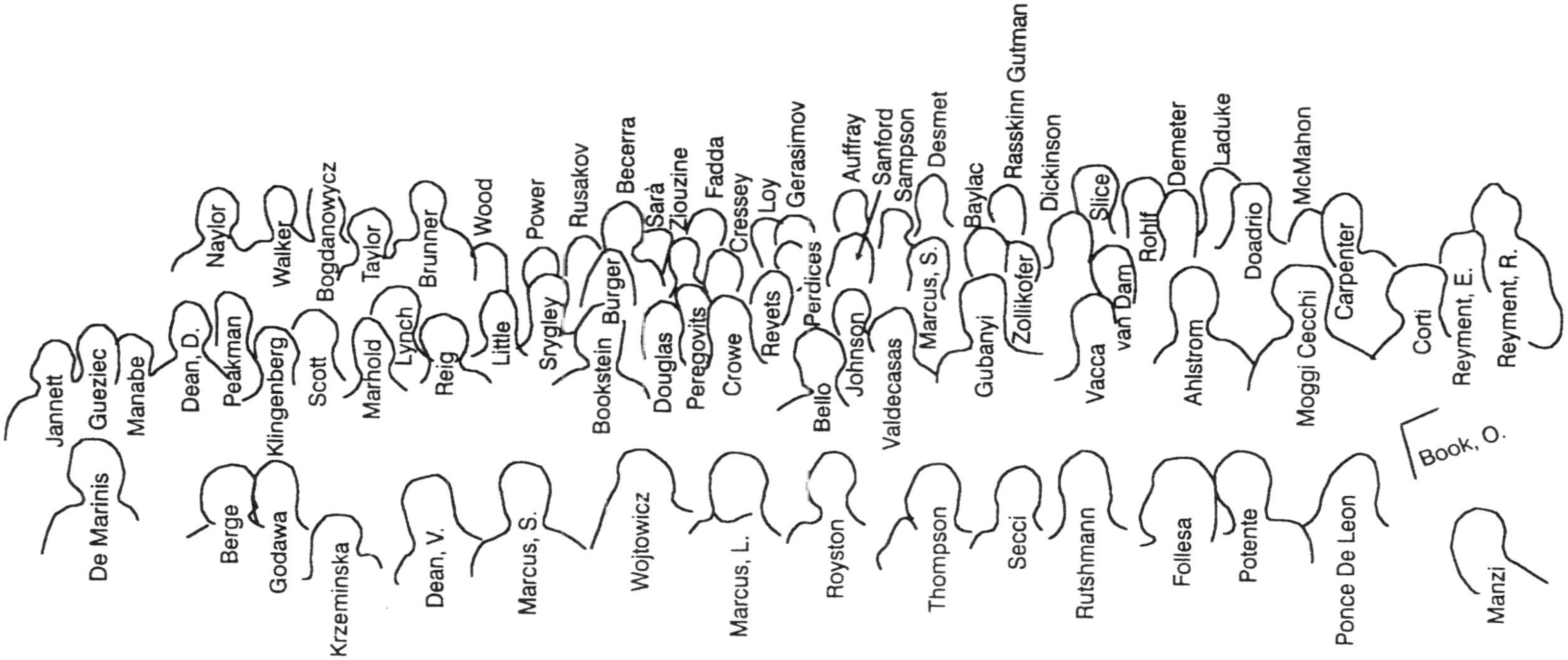

ADVANCES IN MORPHOMETRICS
II Ciocco
July 18 - 30, 1993

PARTICIPANTS

Directors

Marco Corti
Dipartimento di Biologia Animale e dell'Uomo
Universita di Roma "La Sapienza"
via Borelli 50
00161 Rome, ITALY
FAX: 39-6-4457516 e-mail: corti@giannutri.caspur.it

Leslie F. Marcus
Department of Biology
Queens College of CUNY
Mail: Dept. Invertebrates
American Museum of Natural History
CPW at 79th Street
New York, New York 10024, USA
Tel: 212-769-5721 FAX: 212-769-5842
e-mail: lamqc@cunyvm.cuny.edu, marcus@amnh.org

Social Director

Stanley G. Marcus
11192 Ophir Drive
Los Angeles, California 90024, USA
e-mail: sharon@cs.ucla.edu

Faculty

Michel Baylac
Muséum National d'Histoire Naturelle
Laboratoire d'Entomologie
45, rue Buffon
F-75005 Paris, FRANCE
FAX: 1-40-79-36-99 e-mail: baylac@cimrsl.mnhn.fr

Fred Bookstein
Institute of Gerontology
University of Michigan
Ann Arbor, Michigan 48109-2007, USA
FAX: 313-936-9288 e-mail: fred@brainmap.med.umich.edu

David Dean
Department of Anatomy
Case Western Reserve University
10900 Euclid Avenue
Cleveland, Ohio 44106, USA
FAX:216-368-3204
e-mail:dean@lucifer.cwru.edu

Andras Demeter
Hungarian Academy of Sciences
Nador U. 7, H-1088 Budapest, HUNGARY
FAX: 361-1172282 e-mail: h4394dem@ella.hu

Gavin J. P. Naylor
Department of Biology
Osborn Memorial Laboratory
Yale University, P.O. Box 208104
New Haven, Connecticut 06520-8104

Richard Reyment
Institute of Earth Sciences
Department of Historical Geology and Paleontology
Univeristy of Uppsala
Uppsala, SWEDEN
e-mail: Richard.Reyment@pal.uu.se

F. James Rohlf
Department of Ecology and Evolution
State University of New York
Stony Brook, New York 11794, USA
FAX: 516-632-7626 e-mail: rohlf@life.bio.sunysb.edu

Dennis E. Slice
Department of Ecology and Evolution
State University of New York
Stony Brook, New York 11794, USA
e-mail: dslice@life.bio.sunysb.edu

Antonio García-Valdecasas
Departamento de Biodiversidad
Museo Nacional de Ciencias Naturales (C.S.I.C.)
José Guiterrez Abascal 2
Madrid 28006, SPAIN
e-mail: mcnhv00@cc.csic.es

ASI Students

Torbjorn Ahlström
Osteological Ressearch Laboratory
Stockholm University
Ulriksdals kungsgard
S-17071 Solna, SWEDEN
e-mail: ahlstrom@ofl.su-kom.su.se

Jean-Christophe Auffray
Institut des Sciences de l'Evolution, CC064
Université Montpellier II
CC 064 Place Eugene Bataillon
34095 Montpellier Cedex 5, FRANCE
FAX: 33-67-04-20-32 e-mail: auffray@isem.isem.univ-montp2.fr

José Becerra
Departamento de Biodiversidad
Museo Nacional de Ciencias Naturales (C.S.I.C.)
José Gutierrez Abascal, 2.
28006 Madrid, SPAIN
FAX: 34.1.564.50.78 e-mail: MCNJM14@CC.CSIC.ES

Elisa Bello
Departamento de Biodiversidad
Museo Nacional de Ciencias Naturales (C.S.I.C.)
Jose Gutierrez Abascal, 2.
28006 Madrid, SPAIN
FAX: 34.1.564.50.78 e-mail: MCNEB18@CC.CSIC.ES

Christine Berge
Laboratoire d'Anatomie Comparée
URA 1137 CNRS
Muséum National d'Histoire Naturelle
55 rue Buffon
F-75005 Paris, FRANCE
e-mail: berge@cimrs1.mnhn.fr

Patrick Christoph Brunner
Zoological Museum
University of Zurich
Winterthurestrasse 190
8057 Zurich, SWITZERLAND
FAX: 1-364-02-95 e-mail: rutschma@zoolmus.unizh.ch

Wieslaw Bogdanowicz
Mammal Research Institute
Polish Academy of Sciences
17230 Bialowiecz, Poland
FAX (48) 835 12 289 e-mail: wieslaw@bison.zbs.bialowieza.pl

Scott Brande
Department of Geology
University of Alabama
Birmingham, Alabama 35294, USA
FAX: 205-934-2243 e-mail: geof001@uabdpo.dpo.uab.edu

Jill K. Burger
University of Alberta
CW312 Biological Sciences Centre
Edmonton Alberta, T6G 2E9, CANADA
e-mail: userjill@ualtamts

Kent Carpenter
Department of Biological Sciences
Old Dominion University
Norfolk, Virginia 25529, USA
FAX: 804-683-5283
e-mail: kec100f@viper.mgb.odu.edu

Lill Irene Cressey
Department of Anatomy and Cell Biology
University of Bergen
Arstad vn. 19
N-5009 Bergen, NORWAY
FAX: 47-5-20-63-60

Tim Crowe - Deputy Director
FitzPatrick Institute
University of Cape Town
Rondebosch 7700, SOUTH AFRICA
FAX: 27-21-6503295 e-mail: tmcrowe@zoo.uct.ac.za

Valerie Lynn Dean
Dept. of Anthropology
Indiana University
Bloomington, IN 47405, USA
FAX: 812-855-4358 e-mail: vdean@ucs.indiana.edu

Marc Desmet
Institut fur Geologie
1 Rue Blessig
67084 Strasbourg Cedex, FRANCE
FAX: 33-88-36-72-35

Timothy Dickinson
Botany Department, Royal Ontario Museum
100 Queen's Park
Toronto, M5S 2C6, CANADA
FAX: 416-586-5516 e-mail: tim@utstat.toronto.ca

Ignacio Doadrio
Departimento de Biologie
Museo Nacional de Ciencias Naturales (C.S.I.C.)
José Gutierrez Abascal, 2.
28006 Madrid, SPAIN
FAX: 1-5645078 e-mail: mcnid32@cc.csic.es

Michael Douglas
Department of Zoology
Arizona State University
Tempe, Arizona 85287, USA
e-mail: abmed@asuacad.bitnet

Carlo Fadda
Dipartimento di Biologia Animale e dell'Uomo
Universita di Roma "La Sapienza"
Via Borelli 50
00161 Rome, ITALY
FAX: 39-6-4457516

Maria Cristina Follesa
Departimento. Biologia Animale ed Ecologia
Via le Poetto
09126 Cagliari, ITALY
FAX: 33-70-380285

Svetoslav Gerasimov
Institute of Zoology
Academy of Sciences
Boul. Rouski 1, 1000 Sofia, BULGARIA
FAX: 359-2-882897

Joanna Godawa
Institute of Systematics and Evolution of Animals
Slakowska 17
31-016 Cracow, POLAND
e-mail: negodawa@cyfr-kr.edu.pl

Andras Gubanyi
Zoolology Department
Hungarian Natural History Museum
Baross u. 13
H-1088 Budapest, HUNGARY
FAX: 36-1-1138820 e-mail: H9533Gub@ella.hu

André Guéziec
IBM T. J. Watson Research Center
30 Sawmill Road
Hawthorne, New York 10532, USA
FAX: 914-784-6282 e-mail: gueziec@watson.ibm.com

Fred Janett
The Science Museum of Minnesota
30 East 10th Street
Saint Paul, Minnesota 55101, USA
FAX: 221-4777 e-mail:
maft@uminn1.bitnet

Kenneth G. Johnson
Department of Geology and Applied Geology
University of Glasgow
Glasgow, 12 8QQ, UNITED KINGDOM
email: kgj@geology.gla.ac.uk

Christian Klingenberg
Department of Entomology
University of Alberta
Edmonton Alberta T6G 2E3, CANADA
FAX: 403-492-1767 e-mail:
cklingen@gpu.srv.ualberta.ca

Ewa Krzeminska
Institute of Systematics and Evolution of Animals
Polish Academy of Sciences
Piatnizkaya str 1/2
ul. Szwkowska 17 30-016, POLAND

Thomas C. La Duke
Department of Biological Sciences
East Stroudsburg University
East Stroudsburg, Pennsylvania 18301, USA
email: TCLaDuke@AOL.COM

John A. Little
Imaging Processing Group
Division of Radiological Sciences
3rd floor, Guy's Tower
Guy's Hospital
London Bridge, London SE1 9RT, United Kingdom
e-mail: J.Little@lpg.umds.ac.uk

Angelo Loy
Dipartimento di Biologia
II Universita degli Studi di Roma "Tor Vegata"
Via della Ricerca Scientifica, Rome, ITALY
FAX: 39-6-20-26-189 e-mail:
LABAQUA@TOVVX1.INFN.IT

John M. Lynch
Interdisciplinary Humanities Program
Arizona State University
Tempe, Arizona 85287, USA
FAX: 602-965-2519 e-mail:
atjml@asuvm.mrp.asu.edu

Makoto Manabe
National Science Museum
3-23-1 Hyakunin-cho, Shinjuku-ku
Tokyo 169 Japan
FAX: 81-3-3364-7104 e-mail:
manabe@kahaku.go.jp

Giorgio Manzi
D. of B. A. U. (Anthropology)
Universita di Roma "La Sapienza"
P. A. Moro 5, 05185 Rome, ITALY
FAX: 6-499-12273

Karol Marhold
Institute of Botany
Slovak Academy of Science
Dubravszd cesto 14
84223 Bratislava, SLOVAK REPUBLIC
FAX: 42-7371948 e-mail:
botukmar@sauba.sauba.cs

Anna De Marinis
Zoological Section
Museum of Natural History
UIA Romana 17
I-50125 Florence, ITALY
FAX: 39-55-225325

Jocopo Moggi-Cecchi
Istituto di Antropologia.
Universita di Firenze
via del Proconsolo 12
50122 Firenze, ITALY
FAX: 39-55-239-8065

Jim McMahon
Department of Anthropology
Hunter College of CUNY
695 Park Avenue
New York, NY 10021, USA
FAX: 212-772-5797 e-mail:
jmchc@cunyvm.cuny.edu

Anabel Perdices
Departimento de Biologia Evolutiva
Museo Nacional de Ciencias Naturales (C.S.I.C.)
José Gutierrez Abascal, 2.
28006Madrid, SPAIN
FAX: 1-5645078 e-mail: Perdices@PINAR1.csic.es

L. Peregovits
Zoology Department
Hungarian Natural History Museum
Baross u. 13
H-1088 Budapest, HUNGARY
FAX: 36-1-1138820

Marcia Ponce de Leon
Anthropological Institut and Museum
University Zurich-Irchel
WInterthurestr. 190
CH-8057 Zurich, SWITZERLAND

James Power
Coastal Fisheries Institute- CCEER
Louisiana State University
Baton Rouge, LA 70803, USA
e-mail: coepow@sn01.sncc.lsu.edu

Filomena Potente
Consorzio Digamma
C. so Alcide de Gasperi 449/A
70125 Bari, ITALY
FAX: 39-80-5023499

Diego Rasskin-Gutman
Unidad de Paleontología, B-101
Departamento de Biología, Facultad de Ciencias
Universidad Autónoma de Madrid, 28049
Cantoblanco, SPAIN
FAX: 397-8344 e-mail: rasskin@vm1.sdi.uam.es

Santiago Reig
Museo Nacional de Ciencias Naturales (C.S.I.C.)
Jose Gutierrez Abascal 2
28006 Madrid, SPAIN
FAX: 41-1-5645078 e-mail: mcnsr24@cc.csic.es

Stefan A. Revets
Department of Geology
University of Western Australia
Nedlands, Western Australia A6009
FAX: 61-9-380-1037 e-mail: srevets@uniwa.uwa.edu.au

M. Claire Royston
Department of Psychology
Charing Cross Medical School
St. Dunstan's Road
London W6 8RP, UNITED KINGDOM
FAX: 81-846-7025

Dmitri Rusakov
Department of Biology
The Open University
Walton Hall - Milton Keynes, MK7 6AA
UNITED KINGDOM
FAX: 44-908-654-167 e-mail: d.rusakov@open.ac.uk

Marlis Rutschmann
Zoological Museum
University of Zurich
Winterthurestrasse 190
8057 Zurich, SWITZERLAND

Paul Sampson
Department of Statistics, Box 354322
University of Washington
Seattle, Washington 98195-4322, USA
e-mail: pds@stat.washington.edu

Marizio Sarà
Istituto di Zoologia
Via Archirafi 18
90123 Palermo, ITALY
FAX: 39-91-6172009

Elisabetta Secci
Dept. Biologia Animale ed Ecologia
Via le Poetto
09126 Cagliari, ITALY
FAX: 39-70-370263

Robert B. Srygley
Department of Zooology
University of Cambridge
Downing Street
Cambridge, CB2 3EJ, UNITED KINGDOM
FAX: 223-33-66-76 e-mail: rbs1000@phx.cam.ac.uk

Andrea Taylor
School of Physical Therapy
Slippery Rock University
Slippery Rock, PA 16057, USA
e-mail: abt@sru.bitnet

Jennifer L. Thompson
Scarborough Campus University of Toronto
1265 Military Trail
Scarborough, Ontario M1C 1A4, CANADA
FAX: 416-287-7283 e-mail: jl-thompson@macpost.scar.utoronto.ca

Eligio Vacca
Consorzio Digamma
C. so Alcide de Gaspari 449/A
70125 Bari, ITALY

Jan van Dam
Institute of Earth Sciences
Utrecht University, P.O. 80021
3500 TA Urecht, NETHERLANDS
FAX. 31-30-735020 e-mail: jdam@earth.ruu.nl

Jeff A. Walker
Department of Zoology
Division of Fishes
Field Mueum of Natural History
Roosevelt Rd. at Lakeshore Dr.
Chicago, Illinois 60605, USA
e-mail: walker@fmppr.fmnh.org

Chris G. Wood
Dept. of Human Anatomy
Liverpool Univ.
P.O. Box 147
Liverpool L69 3BX, UNITED KINGDOM
FAX: 51-794-5517 e-mail: boisei@uxb.liverpool.ac.uk

Andrei A. Ziouzine
The Institute of Developmental Biology
Russian Academy of Sciences
26 Vavilov St.
Moscow 117808, RUSSIA
FAX: 9523007 e-mail: anzuz@biotest.uucp.free.msk.su

Christoph Zollikofer
Insitut fur Informatik - Multimedia Lab.
Universitat Zurich-Irchel
Winterthresstr. 190
CH-8057 Zurich, SWITZERLAND
FAX: 41-1-3610112 e-mail: zolli@ifi.unizh.ch

AUTHOR INDEX

SUBJECT INDEX

GPSR Compliance

The European Union's (EU) General Product Safety Regulation (GPSR) is a set of rules that requires consumer products to be safe and our obligations to ensure this.

If you have any concerns about our products, you can contact us on ProductSafety@springernature.com

In case Publisher is established outside the EU, the EU authorized representative is:

Springer Nature Customer Service Center GmbH
Europaplatz 3
69115 Heidelberg, Germany

Batch number: 10761599

Printed by Printforce, the Netherlands